CHIMPANZEES AND HUMAN EVOLUTION

Chimpanzees *and* Human Evolution

Edited by

MARTIN N. MULLER, RICHARD W. WRANGHAM,

and DAVID R. PILBEAM

THE BELKNAP PRESS OF HARVARD UNIVERSITY PRESS

Cambridge, Massachusetts

London, England

2017

Copyright © 2017 by the President and Fellows of Harvard College
All rights reserved
Printed in the United States of America

First printing

Library of Congress Cataloging-in-Publication Data
Names: Muller, Martin N., 1971– editor. | Wrangham, Richard W., 1948– editor. |
Pilbeam, David R., editor.
Title: Chimpanzees and human evolution / edited by Martin N. Muller,
Richard W. Wrangham, and David R. Pilbeam.
Description: Cambridge, Massachusetts : The Belknap Press of Harvard University
Press, 2017. | Includes bibliographical references and index.
Identifiers: LCCN 2017017308 | ISBN 9780674967953 (alk. paper)
Subjects: LCSH: Chimpanzees—Behavior. | Human evolution. |
Human behavior. | Social evolution.
Classification: LCC QL737.P94 C47 2017 | DDC 599.93/8—dc23
LC record available at https://lccn.loc.gov/2017017308

*To the trustees and staff of the Leakey Foundation,
for their continuing support of research
on human origins*

Contents

I. Was the Last Common Ancestor of Chimpanzees and Humans Chimpanzee-Like?

1. Introduction: Chimpanzees and Human Evolution   3
   Martin N. Muller

2. Reconstructing the Last Common Ancestor of Chimpanzees and Humans   22
   David R. Pilbeam and Daniel E. Lieberman

3. Equal, Similar, but Different: Convergent Bonobos and Conserved Chimpanzees   142
   Brian Hare and Richard W. Wrangham

II. Chimpanzees and the Evolution of Human Uniqueness

4. Introduction: Chimpanzees and Human Uniqueness   177
   Martin N. Muller

5.	Mortality, Senescence, and Life Span *Michael D. Gurven and Cristina M. Gomes*	181
6.	Fertility and Fecundity *Melissa Emery Thompson and Peter T. Ellison*	217
7.	Locomotor Ecology and Evolution in Chimpanzees and Humans *Herman Pontzer*	259
8.	Evolution of the Human Dietary Niche: Initial Transitions *Sherry V. Nelson and Marian I. Hamilton*	286
9.	Evolution of the Human Dietary Niche: Quest for High Quality *Rachel N. Carmody*	311
10.	From *Pan* to Man the Hunter: Hunting and Meat Sharing by Chimpanzees, Humans, and Our Common Ancestor *Brian M. Wood and Ian C. Gilby*	339
11.	The Evolution of the Human Mating System *Martin N. Muller and David R. Pilbeam*	383
12.	From Chimpanzee Society to Human Society: Bridging the Kinship Gap *Bernard Chapais*	427
13.	Violent Cousins: Chimpanzees, Humans, and the Roots of War *Michael L. Wilson and Luke Glowacki*	464
14.	Cooperative and Competitive Relationships within Sexes *Richard W. Wrangham and Joyce Benenson*	509
15.	Cooperation between the Sexes *Adrian V. Jaeggi, Paul L. Hooper, Ann E. Caldwell, Michael D. Gurven, Jane B. Lancaster, and Hillard S. Kaplan*	548
16.	Sexual Coercion in Chimpanzees and Humans *Martin N. Muller*	572

17. Tool Use and Manufacture in the Last Common Ancestor of
    Pan and *Homo*   602
    Campbell Rolian and Susana Carvalho

18. Cultural Evolution in Chimpanzees and Humans   645
    Joseph Henrich and Claudio Tennie

19. Chimpanzee Cognition and the Roots of the Human Mind   703
    Alexandra G. Rosati

20. Ancestral Precursors, Social Control, and Social Selection
    in the Evolution of Morals   746
    Christopher Boehm

21. Communication and Language   791
    Katie E. Slocombe and Thom Scott-Phillips

    CONTRIBUTORS   825
    INDEX   827

PART I

Was the Last Common Ancestor of Chimpanzees and Humans Chimpanzee-Like?

1

Introduction

Chimpanzees and Human Evolution

MARTIN N. MULLER

> We are also quite ignorant at how rapid a rate organisms ... may under favourable circumstances be modified: we know, however, that some have retained the same form during an enormous lapse of time. From what we see going on under domestication, we learn that within the same period some of the co-descendants of the same species may be not at all changed, some a little, and some greatly changed. Thus it may have been with man, who has undergone a great amount of modification in certain characters in comparison with the higher apes.
> —CHARLES DARWIN, *The Descent of Man* (1871)

Almost sixty years ago, Jane Goodall initiated the first long-term study of wild chimpanzees, in what was then Tanganyika's Gombe Stream Reserve. The work was sponsored by paleoanthropologist Louis Leakey, who believed that naturalistic observations of such a human-like creature would provide insights into the behavior of "early man" (Goodall 1986). Specifically, Leakey hypothesized that behaviors shared between chimpanzees and humans were likely to have been present in the last common ancestor (LCA) of the two. From this perspective, subsequent observations of hunting, meat sharing, tool use, and intergroup killing at Gombe were especially compelling.

In the intervening decades, primatologists have published data on wild chimpanzees from at least 120 sites across Africa, and long-term studies have accumulated hundreds of community-years of observation (Wilson et al. 2015; McGrew 2017). In at least ten sites, chimpanzees can be followed continuously, and at close range, from before they leave their sleeping nests in the morning to the time they fall asleep at night (McGrew 2017). Chimpanzee growth, health, endocrine function, and energetics are regularly monitored using noninvasive methods. Consequently, we now know an astonishing amount about the life history, ecology, genetics, physiology, behavior, and cognition of wild chimpanzees.

There is little agreement, however, about what these chimpanzee data can tell us about human evolution. Some researchers follow Louis Leakey in assuming that "anything that a chimpanzee can do today, the LCA could have done" (McGrew 2010: 3268). Others question whether chimpanzees have any particular relevance in this context, arguing that we can learn just as much about, for example, hominin tool use, from capuchin monkeys (Sayers and Lovejoy 2008). And still others doubt that nonhuman primates generally have any special relevance to human behavioral evolution: "[I]t is highly questionable whether human beings carry a significant primate legacy at all when it comes to behavior.... [W]e are animals and as such, subject to a diverse array of evolutionary constraints, of which phylogeny seems remarkably unimportant" (Barash 2003: 513).

A more extreme version of this final argument is widespread in the social sciences (Pinker 2002). Cultural anthropologists have long taken the lead in arguing that human behavior is so variable that it cannot productively be compared to that of nonhuman primates. "Human Nature is the rawest, most undifferentiated of raw material," said Margaret Mead, implying infinite malleability in the hands of culture (Freeman 1983: 101; see also Mead 1935: 280). "[H]uman nature appears to be everything and everywhere," concluded Jonathan Marks (2002: 178) in a book devoted to disputing the value of chimpanzee-human comparisons.

Even a cursory glance at the behavior and social systems of nonhuman primates, however, reveals vast areas of social space that remain unoccupied by *Homo sapiens* (Rodseth et al. 1991). There is no evidence, for example, that there has ever been a human society in which women universally dominate men (Rosaldo and Lamphere 1974). Yet this is the norm for many lemurs, in which "all adult females are able to consistently elicit submissive behavior from all adult males in dyadic interactions in all behavioral contexts" (Kappeler 2012: 29). There is no human society in which sexual behavior approaches anything like the level of promiscuity typified by chimpanzee society—let alone the more lubricious bonobo (Muller and Pilbeam, this volume). Only systematic comparisons of humans with our primate relatives can delineate such limits to human variation, exposing the central tendencies of human social life (what Chapais refers to in this volume as the *deep structure* of human society). The proper rejoinder to Marks's (2002: 179) glib assessment that "you can't get at human nature from chimpanzees" because "they're not human," might best be Kipling's—"what should they know of England who only England know?"

But how much does phylogeny matter in human social life? In the review quoted above, Barash (2003: 513) asserted that "what is interesting about human behavior is less a consequence of our specifically primate ancestry than of our mammalian ancestry." Why not just compare humans with other mammals? The problem here is that compared to most mammals, primates are more diurnal, more gregarious, bigger brained, larger bodied, longer living, slower growing, and less fertile (van Schaik and Isler 2012). They also show lower total energy expenditure per unit body mass (Pontzer 2015). All of these traits have implications for behavior and sociality. Many primates maintain complex social relationships and exhibit advanced cognitive abilities, precisely the traits that interest us most about our own species. Comparisons with cetaceans, elephants, hyenas, and other highly intelligent, social mammals are informative, but as Wrangham (2012: ix) notes, in the absence of a rich Primate order, "our attempts to understand the biological sources of our social behavior would ... be relatively feeble," because "we could only guess at how components of our exceptional mental ability had been foreshadowed, constrained, or favored in related species."

Humans are not just primates, of course, but hominoids.[1] And hominoids show systematic differences from other primates that would have imposed distinct constraints during early hominin evolution. To apprehend such constraints, one must consider humans in phylogenetic context. Human interbirth intervals, to take one example, don't look unusual from a broad primate perspective. Figure 1.1 shows primate reproductive rates plotted against body size on a log-log scale. Humans fall directly on the best-fit line for non-ape species, implying a typical interbirth interval for a primate of our size. At any given body size, however, apes reproduce more slowly than the average non-ape primate. Consequently, humans maintain strikingly short interbirth

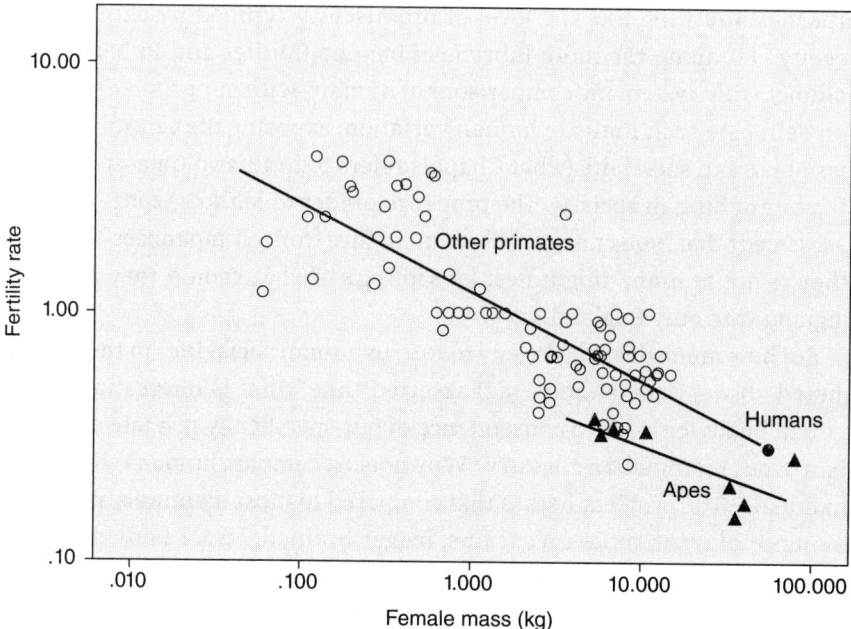

FIGURE 1.1. Birth rates (offspring per year for a mature individual) and body mass (in kilograms) in primates on log-log axes. Birth rates decrease with body size in primates, with humans (shaded circle) showing the predicted relationship for a non-ape (open circles). Apes (triangles) maintain lower fertility rates than other primates of the same body size. Humans are an outlier in the hominoid clade, showing relatively fast reproduction. Figure based on analyses in Walker et al. (2008), using primate data from Lindenfors (2002).

intervals for a hominoid. This major puzzle of human reproduction is obscured if our recent ancestry is ignored.

Consider, too, that hominoid digestive strategies are highly conserved, resulting in shared dietary and energetic constraints (Milton 1999; Wrangham 2006). For reasons discussed by Nelson and Hamilton (this volume), hominoids generally, but great apes especially, require relatively high-quality diets, which are found in forests where the dry season is no longer than four months. Additional evidence suggests that chimpanzees have a lower tolerance for antifeedants (e.g., secondary compounds like tannins and terpenoids) than do cercopithecines living in the same forest, though this result has not yet been tested in other apes (Wrangham et al. 1998). These constraints suggest that, for example, early hominins could not have moved out of rainforests and into drier, savanna-woodland habitats by simply adopting a baboon diet and lifestyle, even if baboons prove useful in thinking about possible primate adaptations to such habitats (e.g., DeVore and Washburn 1963).

Chimpanzees and the Last Common Ancestor

As adumbrated above, observations of nonhuman primates inform studies of human evolution in several ways. Primates provide a useful (though not the sole) standard against which to assess claims of human uniqueness. Are humans really the only species that make and use tools, imitate, infer mental states in others, share intentions at the group level, have a conscience, and so on? Through their diversity, primates give us a sense of what is possible, clarifying human social and behavioral tendencies. And primates provide a rich comparative dataset for testing hypotheses for the evolution of particular human traits. For example, identifying commonalities across primates in species that are sexually dimorphic and in those that are monomorphic can help us to understand why men are larger, on average, than women (Muller and Pilbeam, this volume).

This book makes the case that observations of African apes are critical for reconstructing human social evolution, and that chimpanzees in particular are important for reconstructing the last common ancestor of chimpanzees and humans (hereafter the LCA). Specifically, the opening chapters argue that the LCA closely resembled a modern chimpanzee. It most likely lived in African rainforests around eight million years ago, walking on its knuckles on the ground, and often eating fruits in trees as chimpanzees do, while hanging from one arm. It would have chewed with its small, relatively thin-enameled cheek teeth. If we were to find its fossilized remains tomorrow, we would probably place it in the genus *Pan*. In short, to walk with chimpanzees in an African rainforest today is to open a window on the past (Pilbeam 1996; Wrangham and Pilbeam 2001; Pilbeam and Young 2004).

However, the idea that modern chimpanzees might be similar to the LCA is frequently dismissed out of hand, because of the long time span involved. In a typical comment, Henry Gee insisted that we should not cast chimpanzees in the "role of Our Ancestors," because they "have been evolving away from our common ancestor for precisely as long as we have" (2013: 714). This sentiment, with its underlying assumption that millions of years of evolution must have altered the chimpanzee lineage as much as they have the human one, is now widely shared.[2] Robert Sussman wrote that "chimpanzees have been evolving for as long as humans and gorillas, and there is no reason to believe that ancestral chimps were highly similar to present-day chimps" (2013: 103). Meredith Small's variant (1993: 128) was more measured: "Although we'd like to use chimps as models for our distant forebears, we often forget that they evolved down their own path, shaped physically and behaviorally by pressures slightly different from those our ancient ancestors experienced" (for more or less identical statements, see Ehrlich 2000: 166; Zuk 2013: 41).

The problem with these arguments is that molecular evolution is not the same as phenotypic change. And even in closely related species, rates of morphological or behavioral change can differ substantially over time. Darwin was aware of this latter point, as illustrated by the quotation that opens this chapter. Pilbeam and Lieberman (this volume) observe that multiple primate lineages are morphologically conservative over extended periods of evolutionary time. The gibbon (Hylobatidae) radiation, for example, has been dated to around 7 Ma, putting it close in time to the *Pan/Homo* divergence (Israfil et al. 2011). Few would contend, however, that the common ancestor of the fourteen-plus gibbon species alive today looked very different from a modern gibbon, on the basis that gibbons have been "evolving for as long as humans and gorillas."[3]

The inference that chimpanzees are also a conservative species, little changed since their split from the hominin line, is based on careful analysis and straightforward logic (Pilbeam 1996; Wrangham and Pilbeam 2001; Pilbeam and Young 2004). Chimpanzees and gorillas are extremely similar morphologically, to such an extent that for decades they were considered to be monophyletic, fundamentally size variants of the same animal. Genetic data, however, are unambiguous in showing that chimpanzees and humans are more closely related to each other than either is to the gorilla. This points to one of two conclusions: the extensive similarities between chimpanzees and gorillas represent evolutionary convergence, or the last common ancestor of chimpanzees, gorillas, and humans was very much like a chimpanzee or a gorilla. Because such extensive convergence is unlikely, and because the earliest hominins are all chimpanzee-sized, the LCA is inferred to be chimpanzee-like.

In Chapter 2 of this volume, David Pilbeam and Daniel Lieberman trace the history of ideas about the last common ancestor, presenting detailed evidence for each component of the argument just summarized. In the process, they respond to a range of criticisms that have been directed at the concept of a chimpanzee-like LCA, including the recent pronouncement that such an ancestor is "nullified by the discovery of *Ardipithecus*" (White et al. 2015: 4877). Their primary focus is on morphology, however. Here I briefly address why the nature of the LCA matters, review previous attempts to reconstruct the social system and behavior of the LCA using data from living apes, and look at the relationship between morphology and behavior in the African apes.

Reconstructing Social Organization and Behavior

Reconstructing the last common ancestor of humans and chimpanzees is important, because it provides a starting point for thinking about the evolution of derived or unique human traits, constraining our hypotheses in useful ways. Humans do things that other primates rarely or never do, and such traits obviously cannot be explained in reference to broader primate patterns. Take, for example, habitual bipedal locomotion. Hypotheses for the evolution of bipedality in hominins look different whether one starts with a chimpanzee-like, knuckle-walking ancestor, a gibbon-like upright climber, or a tarsier-like upright leaper (to mention just a few of the models that have been proposed for the LCA; see Pilbeam and Lieberman, this volume). Similar considerations apply to the evolution of social organization and behavior (Foley 1989; Chapais, this volume).

Characterizing the common ancestor of any group of closely related species involves identifying potential homologies (traits that are similar because of common descent). In the hominoid case, any trait or behavior that is found in humans, chimpanzees, bonobos, and gorillas is likely to have been present in their shared common ancestor. The alternative requires that the trait evolved independently at least twice. Such convergence is possible, but for many traits thought to be unlikely, given the ecological differences among these species (Wrangham 1987; Ghiglieri 1987).

Characterizing the common ancestor of the African apes based on presumed behavioral homologies is difficult, however, because the living apes are so variable socially. Gorillas live in cohesive, polygynous groups and form long-term pair-bonds. Males use aggression to monopolize females, but are not territorial (Harcourt and Stewart 2007a, 2007b). Chimpanzees live in fluid, multi-male, multi-female communities, in which males are universally dominant to females, who are less social. Males cooperate with each other to defend territories and kill strangers from neighboring groups (Muller and Mitani 2005). Bonobos live in communities similar to those of chimpanzees, but females form stronger intrasexual bonds than males, and have more social power than chimpanzee females. Nonconceptive sexual behavior is prevalent. Intergroup encounters are hostile, but less so than in chimpanzees (Gruber and Clay 2016). Depending on which of these traits one chooses to emphasize, humans could be characterized as looking similar to any of these species. That is, modern humans do not in totality resemble any one of the African apes more than the others.

It is possible to use only the most phylogenetically conservative traits, that is, those shared by all three of the African apes and humans, to characterize the common ancestor of this group. Wrangham (1987) pioneered this approach, focusing on variables related to social organization and social relationships. He described an ancestor that (1) lived in a closed social network in which (2) males sometimes traveled alone, (3) females tended to emigrate at sexual maturity, (4) males had multiple mates, (5) females did not form alliances with each other, (6) relationships with other groups were hostile, and (7) males took the lead in intergroup encounters, (8) sometimes attacking strangers. At the time bonobos were not well studied, and subsequent observations have shown that females form both close bonds and coalitions with each other. Whether these qualify as alliances, however (point 5), is not clear (see Wrangham and Benenson, this volume). In either case, the overall result is a sketchy description of this ancestor, but one in which some form of male-led hostile intergroup interaction predominates. Wrangham noted that many more variables could have been added to the analysis, but that variation among the apes would continue to exclude most of these from the inferred ancestor.

Subsequent phylogenetic analyses adopted similar methods, but failed to convincingly expand on Wrangham's original characterization. Ghiglieri (1987), for example, provided a slightly longer list of traits for the inferred LCA, including fission-fusion sociality, communal territoriality, moderate sexual dimorphism in body mass, and male alliances. This was the result of excluding gorillas, however, and focusing on the comparison between chimpanzees, bonobos, and humans. Foley (1989) took the opposite approach, emphasizing similarities between male-female relationships in gorillas and humans, and characterized polygynous pair bonds as ancestral. This was justified by considering the basic social organization of gibbons, orangutans, and gorillas to be fundamentally similar: "small social units made up of one male, one or more females, and young" (Foley 1996: 100). This characterization of orangutan sociality has not held up well, however, with additional field data (e.g., Knott and Kahlenberg 2010).

In the most recent iteration of the phylogenetic approach, Duda and Zrzavý (2013) employed formal methods to analyze sixty-five life history and behavioral characters for all of the extant hominids (i.e., members of the family Hominidae: humans, chimpanzees, bonobos, mountain gorillas, lowland gorillas, and both species of orangutan), along with a number of gibbons and cercopithecoids as an outgroup. They produced a detailed list of traits that they categorized as "confident," "likely," "debatable," or "unknown" for every possible common ancestor on their ape tree. Some of the traits that they reconstructed as "confident" for the African ape common ancestor would simply extend Wrangham's original list, as they are present in all of the relevant species—for example, "population specific behavioral traditions" and "positive correlation of male rank and copulation rate." However, they additionally concluded that the common ancestor of the African apes lived in a single-male, multi-female (i.e., gorilla-like) mating system ("confident"), and exhibited strong sexual dimorphism in body mass ("likely"). They noted that in their analyses gorillas appeared to be the least derived of the African apes (in contrast to Groves 1986, who inferred that chimpanzees are the least derived).

Although Duda and Zrzavý's descriptions provided much more detail than earlier efforts, the reliability of all such analyses is low. As Lieberman (1999) noted, valid phylogenetic inference requires discrete, quantifiable, independent characters that actually convey information about ancestry and descent (like genes). Such characters are notoriously difficult to define, even for morphology, and incorrect results are common. For example, most phenotypic analyses concluded that the great apes were monophyletic (placing them in the family Pongidae), until genetic analyses conclusively falsified this hypothesis (Pilbeam and Lieberman, this volume). The problem of defining characters and character states for behavior is more difficult. It is often unclear whether behaviors are comparable across species. Is "nest building" in apes the same thing as "shelter construction" in humans? What about "male rank"?

Lieberman (1999) also noted that phylogenetic methods make the questionable assumption that evolution is as parsimonious as predicted by such analyses. That is, they assume that homology is more likely than homoplasy, and they favor scenarios that minimize the number of evolutionary reversals. Whether such assumptions are true for any particular species or traits is not clear, particularly in small samples where there may be little difference between the most and least parsimonious scenarios. Even inferring parsimony is difficult because of character definition. Duda and Zrzavý's dataset, for example, contained many traits that are potentially correlated consequences of some simpler underlying genetic or developmental change. As discussed by Hare and Wrangham (this volume), many of the numerous traits that distinguish chimpanzees and bonobos are likely the result of a smaller number of changes in the timing of early development. For other traits, such as many of the life history variables, differences may simply reflect variation in body size.

An additional complication is that this kind of analysis is extremely sensitive to the particular characters that are included. In Duda and Zrzavý's dataset, more than a third of the traits were directly related to the mating system (Dixson 1998), whereas diet was represented by two characters, and cognition one. A different set of characters or a different parsing of characters into states could produce different conclusions about which species are more or less derived.

Finally, some of the specific conclusions of this analysis were inconsistent with the known fossil record. The African ape common ancestor, for example, was characterized as "likely" to have exhibited strong sexual dimorphism in body mass. The earliest putative hominin fossils, however, such as *Ardipithecus,* have been argued to show low levels of body mass dimorphism (White et al. 2015), with increases occurring later in the australopithecine lineage (reviewed in Muller and Pilbeam, this volume). Even though three to four million years might separate the LCA from *Ardipithecus,* including such fossil evidence would affect any estimation of parsimony.

Derived Characteristics of Gorillas

A more productive approach to distinguishing derived versus ancestral characters in the living apes may be to rely on our growing knowledge of the behavioral ecology of these species to test hypotheses about their evolution. Pilbeam and Lieberman (this volume), for example, review the evidence that, from a morphological perspective, gorillas are mostly (though not entirely) very large chimpanzees. A robust literature on great ape behavioral ecology similarly suggests that almost all of the differences between chimpanzees and gorillas in diet, social organization, life history, and behavior are the result of large body size in gorillas (Wrangham 1979; Harcourt and Stewart 2007a, 2007b).

The fundamental difference between chimpanzee and gorilla diets is that gorillas, because of their large bodies and guts, can subsist on much lower-quality food. Both apes prefer to eat fruit when it is available. However, comparisons in the same forests show that, as fruit becomes rarer, gorillas fall back on increasing amounts of pith, leaves, and woody stems, while chimpanzees continue to search for higher-quality food (reviewed in Wrangham 2006; Harcourt and Stewart 2007a).

Further differences between gorillas and chimpanzees are the direct result of gorillas having access to superabundant, low-quality vegetation (Harcourt and Stewart 2007a, 2007b). First, gorillas travel shorter distances each day to find food. Second, gorillas are not territorial, because there is little point in defending low-quality, abundant resources. Third, female gorillas, unlike female chimpanzees, can remain in permanent association with other gorillas, because feeding competition is minimal. Fourth, because their food is reliable and easily obtained, gorillas can grow faster and reproduce earlier than chimpanzees and bonobos, despite their larger size (as is true for other mammals that rely on such resources; see Sibly and Brown 2007).

Differences in the mating system between gorillas and chimpanzees follow directly from the differences in grouping. Because female gorillas form cohesive groups, it is possible for males to continuously defend females from rival males. This leads to the evolution of long-term polygynous pair-bonds, intense male-male competition, and sexual selection for increased male body mass and canine height (reviewed in Harcourt and Stewart 2007a, 2007b). Although there is still disagreement over some of the details of this scenario (for example, do females ultimately choose to associate with males to avoid predation or infanticide?), the basic effects of body size and diet on social organization and behavior have been rigorously supported (Harcourt and Stewart 2007a, 2007b).

From this perspective, whether the gorilla mating system and its accompanying dimorphism is ancestral or derived depends entirely on whether large body size in gorillas is derived. No other living primate, and potentially only one fossil ape, is as large as the gorilla (Pilbeam and Lieberman, this volume). Female body size is remarkably consistent across orangutans, chimpanzees, bonobos, and known hominins prior to *Homo erectus* (Grabowski et al. 2015), and it is therefore likely that large body size in gorillas is derived, along with the socioecological traits that arise from it.

Derived Characteristics of Bonobos

Bonobos maintain striking differences from chimpanzees in their social relationships, including stronger female bonds, higher female status, more elaborate sociosexual behavior, and less intense aggression between communities (Gruber and Clay 2016; Hare and Wrangham, this volume). Bonobos share their most recent common ancestor with chimpanzees, however, and as a result both species are equally closely related to humans. Consequently, many researchers have argued that chimpanzees and bonobos make equally good behavioral models for an LCA that was morphologically similar to a chimpanzee (e.g., Parish and de Waal 2000). Others have argued that because these closely related species are ostensibly so different, there is no reason to think that the LCA resembled either (Sayers and Lovejoy 2014).

In Chapter 3 of this volume, however, Hare and Wrangham present detailed evidence that bonobos, like gorillas, are behaviorally and morphologically derived from a more chimpanzee-like ancestor. Chimpanzees and bonobos share important core similarities in social organization, relationships, and behavior, resembling each other far more closely than either does any other ape (Stanford 1998; Gruber and Clay 2016). Where chimpanzees and bonobos differ, bonobos generally exhibit juvenilized versions of chimpanzee traits. This juvenilization, or paedomorphism, is also reflected in aspects of the cranium (Lieberman et al. 2007). Hare et al. (2012) note that the differences in morphology and behavior between bonobos and chimpanzees are similar to those commonly observed between a range of domesticated animals and their wild ancestors. They argue that the juvenilized pattern of development in bonobos was the result of selection against aggression, via a process of "self-domestication" (see also Wilkins et al. 2014).

The lack of evidence for juvenilized skeletal traits in the hominin fossil record prior to *Homo erectus* suggests that the bonobo condition is not ancestral and that any exclusive similarities between bonobos and humans represent evolutionary convergence (Wrangham and Pilbeam 2001). As Hare and Wrangham (this volume) stress, this does not mean that bonobos are unimportant for thinking about human evolution. As de Waal (2005: 16) has insisted, "chimps and bonobos are equally relevant for understanding human evolution." But bonobos are relevant because their unique similarities with humans probably represent homoplasy, meaning that they can potentially shed light on the selection pressures that led to convergence. Chimpanzees are relevant because their similarities with humans are more likely to represent homology (though, as noted below, this possibility must be tested against the alternative of convergence).

Chimpanzees as Conceptual Models

We can confidently assume that an LCA that was morphologically more similar to a modern chimpanzee than to a bonobo or a gorilla would have behaved more like a chimpanzee than a bonobo or a gorilla. Because the earliest hominins are chimpanzee-sized and lack the juvenilized traits of bonobos, chimpanzees are the best available model for thinking about the behavior and life history of the LCA. But is the best model "good enough" (Wrangham 1987)?

In an influential book chapter, Tooby and DeVore (1987) criticized the use of primate models in human evolution, including those based on chimpanzees. They distinguished between (1) referential modeling, in which a modern species (or referent) stands in for a fossil species, and (2) conceptual modeling, in which basic evolutionary principles and concepts that have been tested across multiple taxa are applied to understand an extinct species. They argued that referential models are essentially arbitrary, because there is "no validated principle to govern the selection of an appropriate living species" (Tooby and DeVore 1987: 186), and that a focus on chimpanzee models is ultimately unhelpful for two reasons. First, because it leads to a disproportionate focus on the earlier ape-like phases of human evolution, and second because it neglects the differences, or discontinuities, between apes and humans (see also Van Reybrouck 2012).

In the same volume, Potts (1987) was similarly concerned that focusing on a single referent could lead one to overlook the unique features present in a specific hominin species. He noted that if one tried to reconstruct the socioecology of hamadryas baboons by looking at olive baboons, one "would miss unique aspects of hamadryas socioecology and would preclude an evolutionary study of what made *P. hamadryas* a distinct species" (Potts 1987: 45). Others have echoed these points, maintaining that chimpanzees are of limited use in reconstructing the behavior of hominins that exhibited unique traits such as bipedalism (Sayers and Lovejoy 2014: 11).

As Pilbeam and Lieberman make clear in Chapter 2 of this volume, however, the choice of chimpanzees as a model for the LCA is not referential. Nor is it metaphorical (White et al. 2015) or doctrinal (Sayers et al. 2012). Rather, a chimpanzee-like LCA is *conceptually* inferred from viewing extensive similarities among the extant African apes in phylogenetic context. Such inferential processes may indeed lead to a focus on the early ape-like phase of human evolution, but simply provide a starting point for thinking about adaptation in later hominins. Early hominins obviously differed from chimpanzees in important ways, but understanding how these differences evolved is made easier if one can characterize the ancestral state. Returning to Potts's example, predicting the unique features of the hamadryas social system might not be possible by looking solely at olive baboons, but understanding the evolution of the hamadryas system is made easier by taking the olive baboon social system as a starting point, within the broader context of extant *Papio* species (e.g., Grueter et al. 2012).

What about the risk of neglecting unique traits in early hominins, or in humans? The chapters in the second half of this book demonstrate that this concern has been exaggerated. The authors of those chapters systematically review similarities and differences between chimpanzees and humans across a wide range of traits. Where they find similarities, they do not assume homology, but consider evidence from both the fossil record and a range of living species, to test the alternative of convergence. Where they find differences, they explore hypotheses for the evolution of human traits from a chimpanzee-like ancestor, again informed by broad comparative data. At the same time they evaluate the plausibility of such an ancestor. The detailed data in these chapters would have delighted Louis Leakey, who might well have hoped that such abundance would spring from the initial seeds planted in Gombe.

First, however, we consider, in the next chapter, the history of ideas about the LCA, and the evidence for a chimpanzee-like ancestor.

Endnotes

1. "Hominoid" refers to members of the superfamily Hominoidea, which includes the great apes, the lesser apes, and humans. "Hominin" refers to members of the tribe Hominini, species on the human lineage more closely related to humans than chimpanzees (see Pilbeam and Lieberman, this volume).
2. Although before genetic analyses established that the great apes were not monophyletic, it was generally assumed that they were collectively conservative.
3. This is also true for many other speciose mammalian genera.

Acknowledgments

I thank Keith Hunley, Daniel Lieberman, Sherry Nelson, David Pilbeam, Jessica Rothman, and Richard Wrangham for helpful discussions and comments.

References

Barash, D. P. 2003. Book review: Tree of Origin. *Physiology & Behavior* 78: 513–514.
DeVore, I., and S. L. Washburn. 1963. Baboon ecology and human evolution. In F. C. Howell and F. Bourlière, eds., *African Ecology and Human Evolution*, 335–367. Chicago: Aldine Publishing.
Dixson, A. F. 1998. *Primate Sexuality*. Oxford: Oxford University Press.
Duda, P., and J. Zrzavý. 2013. Evolution of life history and behavior in Hominidae: Towards phylogenetic reconstruction of the chimpanzee-human last common ancestor. *Journal of Human Evolution* 65: 424–426.
Ehrlich, P. 2000. *Human Natures: Genes, Cultures and the Human Prospect*. Washington, DC: Island Press.
Foley, R. A. 1989. The evolution of hominid social behavior. In V. Standen and R. A. Foley, eds., *Comparative Socioecology: The Behavioural Ecology of Humans and Other Animals*, 474–493. Oxford: Blackwell Scientific.
Foley, R. A. 1996. The evolution of hominid social behavior. *Proceedings of the British Academy* 88: 95–117.

Freeman, D. 1983. *Margaret Mead and Samoa*. Cambridge, MA: Harvard University Press.
Gee, H. 2013. *The Accidental Species: Misunderstandings of Human Evolution*. Chicago: University of Chicago Press.
Ghiglieri, M. 1987. Sociobiology of the great apes and the hominid ancestor. *Journal of Human Evolution* 16: 319–357.
Goodall, J. 1986. *The Chimpanzees of Gombe: Patterns of Behavior*. Cambridge, MA: Harvard University Press.
Grabowski, M., K. G. Hatala, W. L. Jungers, and B. G. Richmond. 2015. Body mass estimates of hominin fossils and the evolution of human body size. *Journal of Human Evolution* 85: 75–93.
Groves, C. P. 1986. Systematics of the great apes. In D. Swindler, ed., *Primate Biology*, vol. 1: *Systematics, Evolution, and Anatomy*. New York: Alan R. Liss.
Gruber, T., and Z. Clay. 2016. A comparison between bonobos and chimpanzees: A review and update. *Evolutionary Anthropology* 25: 239–252.
Grueter, C. C., B. Chapais, and D. Zinner. 2012. Evolution of multilevel social systems in nonhuman primates and humans. *International Journal of Primatology* 33: 1002–1037.
Harcourt, A. H., and K. J. Stewart. 2007a. *Gorilla Society: Conflict, Compromise, and Cooperation between the Sexes*. Chicago: University of Chicago Press.
Harcourt, A. H., and K. J. Stewart. 2007b. Gorilla society: What we know and don't know. *Evolutionary Anthropology* 16: 147–158.
Hare, B., V. Wobber, and R. Wrangham. 2012. The self-domestication hypothesis: Evolution of bonobo psychology is due to selection against aggression. *Animal Behaviour* 83: 573–585.
Israfil, H., S. M. Zehr, A. R. Mootnick, M. Ruvolo, and M. E. Steiper. 2011. Unresolved molecular phylogenies of gibbons and siamangs (Family: Hylobatidae) based on mitochondrial, Y-linked, and X-linked loci indicate a rapid Miocene radiation or sudden vicariance event. *Molecular Phylogenetics and Evolution* 58: 447–455.
Kappeler, P. M. 2012. The behavioral ecology of strepsirrhines and tarsiers. In J. C. Mitani, J. Call, P. M. Kappeler, R. A. Palombit, and J. B. Silk, eds., *The Evolution of Primate Societies*, 17–42. Chicago: University of Chicago Press.
Knott, C. D., and S. M. Kahlenberg. 2010. Orangutans: Understanding forced copulations. In C. Campbell, A. Fuentes, K. MacKinnon, S. Bearder, and R. Stumpf, eds., *Primates in Perspective*, 313–326. Oxford: Oxford University Press.
Leigh, S. R. 1994. Ontogenetic correlates of diet in anthropoid primates. *American Journal of Physical Anthropology* 94: 499–522.
Lieberman, D. E. 1999. Homology and hominid phylogeny: Problems and potential solutions. *Evolutionary Anthropology* 7: 142–151.
Lieberman, D. E., J. Carlo, M. P. de Leon, and C. P. E. Zollikofer. 2007. A geometric morphometric analysis of heterochrony in the cranium of chimpanzees and bonobos. *Journal of Human Evolution* 52: 647–662.

Lindenfors, P. 2002. Sexually antagonistic selection on primate size. *Journal of Evolutionary Biology* 15: 595–607.

Marks, J. 2002. *What It Means to be 98% Chimpanzee.* Berkeley: University of California Press.

McGrew, W. C. 2010. In search of the last common ancestor: New findings on wild chimpanzees. *Philosophical Transactions of the Royal Society of London B* 365: 3267–3276.

McGrew, W. C. 2017. Field studies of *Pan troglodytes* reviewed and comprehensively mapped, focussing on Japan's contribution to cultural primatology. *Primates* 58: 237–258.

Mead, M. 1935. *Sex and Temperament in Three Primitive Societies.* New York: William Morrow.

Milton, K. 1999. A hypothesis to explain the role of meat-eating in human evolution. *Evolutionary Anthropology* 8: 11–21.

Muller, M. N., and J. C. Mitani. 2005. Conflict and cooperation in wild chimpanzees. *Advances in the Study of Behavior* 35: 275–331.

Parish, A. R., and F. B. M. de Waal. 2000. The other "closest living relative": How bonobos *(Pan paniscus)* challenge traditional assumptions about females, dominance, intra- and intersexual interactions, and hominid evolution. In D. LeCroy and P. Moller, eds., *Evolutionary Perspectives on Human Reproductive Behavior,* 97–113. New York: New York Academy of Sciences.

Pilbeam, D. R. 1996. Genetic and morphological records of the Hominoidea and hominid origins: A synthesis. *Molecular Phylogenetics and Evolution* 5: 155–168.

Pilbeam, D., and N. Young. 2004. Hominoid evolution: Synthesizing disparate data. *Comptes Rendus Palevol* 3: 305–321.

Pinker, S. 2002. *The Blank Slate: The Modern Denial of Human Nature.* New York: Viking.

Pontzer, H. 2015. Energy expenditure in humans and other primates: A new synthesis. *Annual Review of Anthropology* 44: 169–187.

Potts, R. 1987. Reconstructions of early hominid socioecology: A critique of primate models. In W. G. Kinzey, ed., *The Evolution of Human Behavior: Primate Models,* 28–48. Albany, NY: SUNY Press.

Rodseth, L., R. Wrangham, A. Harrigan, and B. Smuts. 1991. The human community as a primate society. *Current Anthropology* 32: 221–254.

Rosaldo, M. Z., and L. Lamphere, eds. 1974. *Woman, Culture and Society.* Stanford, CA: Stanford University Press.

Sayers, K., and C. O. Lovejoy. 2008. The chimpanzee has no clothes: A critical examination of *Pan troglodytes* in models of human evolution. *Current Anthropology* 49: 87–114.

Sayers, K., and C. O. Lovejoy. 2014. Blood, bulbs, and bunodonts: On evolutionary ecology and the diets of *Ardipithecus, Australopithecus,* and early *Homo. Quarterly Review of Biology* 89: 319–357.

Sayers, K., M. A. Raghanti, and C. O. Lovejoy. 2012. Human evolution and the chimpanzee referential doctrine. *Annual Review of Anthropology* 41: 119–138.

Sibly, R. M., and J. H. Brown. 2007. Effects of body size and lifestyle on evolution of mammal life histories. *Proceedings of the National Academy of Sciences* 104: 17707–17712.

Small, M. F. 1995. *Female Choices: Sexual Behavior of Female Primates.* Ithaca, NY: Cornell University Press.

Stanford, C. B. 1998. The social behavior of chimpanzees and bonobos: Empirical evidence and shifting assumptions. *Current Anthropology* 39: 399–420.

Sussman, R. 2013. Why the legend of the killer ape never dies. In D. Fry, ed., *War, Peace, and Human Nature.* Oxford: Oxford University Press.

Tooby, J., and I. DeVore. 1987. The reconstruction of hominid behavioral evolution through strategic modeling. In W. G. Kinzey, ed., *The Evolution of Human Behavior: Primate Models,* 183–238. Albany, NY: SUNY Press.

Van Reybrouck, D. 2012. *From Primitives to Primates: A History of Ethnographic and Primatological Analogies in the Study of Prehistory.* Leiden, NL: Sidestone Press Dissertations.

van Schaik, C. P., and K. Isler. 2012. Life history evolution in primates. In J. C. Mitani, J. Call, P. M. Kappeler, R. A. Palombit, and J. B. Silk, eds., *The Evolution of Primate Societies,* 220–244. Chicago: University of Chicago Press.

Walker, R. S., M. Gurven, O. Burger, and M. J. Hamilton. 2008. The trade-off between number and size of offspring in humans and other primates. *Proceedings of the Royal Society B* 275: 827–834.

White, T. D., C. O. Lovejoy, B. Asfaw, J. P. Carlson, and G. Suwa. 2015. Neither chimpanzee nor human, Ardipithecus reveals the surprising ancestry of both. *Proceedings of the National Academy of Sciences* 112: 4877–4884.

Wilkins, A. S., R. W. Wrangham, and W. T. Fitch. 2014. The "Domestication Syndrome" in mammals: A unified explanation based on neural crest cell behavior and genetics. *Genetics* 197: 795–808.

Wilson, M. L., C. Boesch, T. Furuichi, J. C. Gilby, C. Hashimoto, C. L. Hobaiter, G. Hohmann, N. Itoh, K. Koops, J. N. Lloyd, T. Matsuzawa, J. C. Mitani, D. C. Mjungu, D. Morgan, R. Mundry, M. N. Muller, M. Nakamura, J. Pruetz, A. E. Pusey, J. Riedel, C. Sanz, A. M. Schel, N. Simmons, M. Waller, D. P. Watts, F. White, R. Wittig, K. Zuberbühler, and R. W. Wrangham. 2014. Lethal aggression in *Pan* is best explained by adaptive strategies, rather than human impacts. *Nature* 513: 414–417.

Wrangham, R. W. 1979. On the evolution of ape social systems. *Social Science Information* 18: 336–368.

Wrangham, R. W. 1987. The significance of African apes for reconstructing human social evolution. In W. G. Kinzey, ed., *The Evolution of Human Behavior: Primate Models,* 51–71. Albany, NY: State University of New York Press.

Wrangham, R. W. 2006. The Delta Hypothesis: Hominoid ecology and hominin origins. In D. E. Lieberman, R. J. Smith, and J. Kelley, eds., *Interpreting the Past: Es-*

says on Human, Primate and Mammalian Evolution in Honour of David Pilbeam, 231–242. Boston: Brill Academic Publishers.

Wrangham, R. W. 2012. Foreword. In J. C. Mitani, J. Call, P. M. Kappeler, R. A. Palombit, and J. B. Silk, eds., *The Evolution of Primate Societies*. Chicago: University of Chicago Press.

Wrangham, R. W., N. L. Conklin-Brittain, and K. D. Hunt. 1998. Dietary response of chimpanzees and cercopithecines to seasonal variation in fruit abundance: I. Antifeedants. *International Journal of Primatology* 19: 949–970.

Wrangham, R. W., and D. R. Pilbeam. 2001. African apes as time machines. In B. M. F. Galdikas, N. E. Briggs, L. K. Sheeran, G. L. Shapiro, and J. Goodall, eds., *All Apes Great and Small*. Berlin: Springer.

Zuk, M. 2013. *Paleofantasy: What Evolution Really Tells Us about Sex, Diet, and How We Live*. New York: W. W. Norton.

2

Reconstructing the Last Common Ancestor of Chimpanzees and Humans

DAVID R. PILBEAM *and* DANIEL E. LIEBERMAN

To make inferences and test hypotheses about why the human lineage diverged from the African great apes (AGAs), we address three fundamental questions concerning the last common ancestors (LCAs) of *Homo* and *Pan* and *Gorilla:* When did these LCAs live, where did they live, and what did they look like? These questions, which have been debated for over a century, remain contentious primarily because we lack an adequate fossil record for the African ape lineages or their obvious ancestors. Instead, the few sources of available data to reconstruct the nature of the LCAs include the comparative genetics of living hominoids, especially *Homo, Pan,* and *Gorilla;* the comparative anatomy of living hominoids; and the fossil record, including Miocene catarrhines, Pliocene hominins from the genus *Australopithecus,* and proposed Late Miocene and Early Pliocene hominins: *Ardipithecus, Orrorin,* and *Sahelanthropus.* Researchers typically use these lines of evidence to address questions about the LCA using two complementary and ideally interwoven approaches. Top-down approaches use living species, especially *Pan* and *Gorilla,* within a cladistic framework to reconstruct likely character-states of ancestors; bottom-up approaches use the fossil record to constrain, test, or modify top-down hypotheses (McNulty 2010); Andrews and Harrison

(2005) proposed a similar dichotomy: "bird's eye perspective" versus "worm's eye view."

Ever since molecular data firmly established that humans and chimpanzees are sister taxa, top-down approaches have generally concluded that their LCA was similar to the African great apes. However, many paleontologists prefer bottom-up approaches that make inferences from fossils. To quote McNulty (2010: 329): "In the absence of a robust fossil record of gorillas and chimpanzees, it is only through a [complementary] bottom-up approach—using the morphological diversity generated through 20 million years of ape evolution—that we can build reasonable hypotheses about the ancestral form which gave rise to the first fossil humans." At first glance, such approaches yield a much greater range of inferences about the LCA. Fossil apes from the Miocene along with fossil hominins from the Late Miocene and Pliocene have been used to reconstruct the LCA as an orthograde climber and terrestrial walker similar to the African apes (discussed as one option in Washburn 1968), an orthograde climber similar to gibbons (e.g., Tuttle 1981), or a pronograde climber similar to many monkeys (e.g., Straus 1949). However, from a logical and methodological perspective, using fossil hominins to reconstruct the LCA is actually a top-down method of inference. Further, the utility of using Miocene ape fossils to make or test predictions about the LCA is constrained by their phylogenetic contexts, which are often ambiguous and distant relative to the African apes and humans, and thus of questionable relevance for reconstructing the LCA.

In this review, we evaluate the many lines of evidence used to reconstruct the LCA using both approaches. Before considering the hypothesis-testing framework necessary to reconstruct the LCA, we begin with a brief review of ideas about the anatomy of the LCA that have been proposed since the mid-nineteenth century. We then consider separately the top-down (molecular and comparative anatomical) evidence and the bottom-up (Miocene fossil) evidence. We conclude that this integrated approach strongly supports the hypothesis that the LCA was probably most similar to the African great apes.

A prefatory note on terminology, which can be a source of confusion and obfuscation: Some researchers continue to use the term "hominid" to refer to species more closely related to humans than chimpanzees, but since molecular evidence unequivocally shows that humans and chimpanzees are monophyletic, and since there is general agreement that taxonomic nomenclature should reflect phylogenetic relationships (Wood 2010), species on the human lineage are members of the Tribe Hominini, whose correct colloquial term is "hominin." Humans, chimps, and gorillas belong in the Subfamily Homininae ("hominines"), which includes the informal *Pan-Gorilla* paraphyletic group, the African apes. The Homininae, in turn, are part of the Family Hominidae ("hominids," which includes humans plus all the great apes) and the Superfamily Hominoidea ("hominoids," which also includes the gibbons). In order to distinguish among various last common ancestors, we generally use the term LCA to refer to the common ancestor of humans and chimps, but when needing to distinguish between this species and the LCA of all the hominines (chimpanzees, humans, and gorillas), we use the more precise terms LCA_{C-H} and LCA_{G-CH}. Finally, we use the colloquial term chimpanzee to refer to *Pan,* chimp to refer to *P. troglodytes,* and bonobo to refer to *P. paniscus.*

Historical Context

Over the last 150 years, reconstructions of the LCA have mostly been addressed within a range of alternative primate phylogenies. Despite long-standing agreement about the monophyly of the major primate clades, proposed relationships among the hominoids have varied widely (Pilbeam and Young 2004), largely reflecting different opinions on the position of the human lineage. Thus, before discussing recent shifts that have stimulated new approaches to reconstructing the LCA, we briefly discuss the first 100 years or so of thinking about hominin origins.

Based on comparative anatomy, Huxley (1863) recognized the African apes, in particular the gorilla, as the closest relatives of humans. Darwin (1871) reflected this relationship in his characteristically qualified inference about the probable origin of the human lineage from the African apes. Half a century later, Keith (1923) continued to favor close relationships between humans and the apes, especially the great apes, and traced the human lineage through several grade-based stages ("grade" referring to equivalence in such features as social organization, diet, positional repertoire, with no necessary link to evolutionary relationships; Wood 2010). These stages progressed from a cercopithecoid-like pronograde quadruped to a small brachiating orthograde gibbon-like "hylobatian," to a large chimp-like "troglodytian":

> No one who accepts evolution as a working hypothesis will have any difficulty in regarding the three great anthropoids—the gorilla, the chimpanzee, and the orang—as divergent branches from the common giant stem. Seeing that man shares so many characters in common with these we are compelled I think, to regard man as an aberrant branch from the troglodytian stock or stem, and it is therefore probable that the plantigrade posture was evolved as soon as the common giant stock began to break up into its various living and fossil forms. (Keith 1923: 453)

Gregory (1927: 555) also favored a close morphological "kinship" with the African apes, particularly the gorilla, yet his attempt at a phylogeny (his figure 10) shows the human lineage as part of a trichotomy with orangutans and the common ancestor of chimps and gorillas (of incidental interest, the figure also shows *Australopithecus,* a mere two years after its description, as linked specifically to the chimpanzee).

Despite initial hypotheses that humans were close to the African apes, other phylogenetic relationships generating alternative reconstructions of the LCA were proposed during the first half of the twentieth century. These reconstructions stemmed largely from the view that human phenotypic differences from the apes and monkeys were too marked to have arisen from a relatively recent LCA. Thus Wood Jones (1923) proposed that modern humans were most closely related to tarsiers, separating from anthropoid primates very early, and affirming his view that humans evolved from small leapers without ever having had a quadrupedal phase during their evolutionary history (Wood Jones 1916). Osborn (1927) also argued that humans had diverged very early from other primates, and were not related to any particular apes. He considered apes too derived to be either ancestors or close relatives of humans. Similarly, Straus (1949) argued that modern humans, cercopithecoids, and hominoids diverged long ago from a common ancestor that was essentially monkey-like rather than ape-like. As late as 1963, Schultz proposed that modern humans and apes diverged about the same time as the great ape and gibbon lineages split (with great apes forming a monophyletic group). The consensus among most but not all paleontologists well into the 1960s was that the great apes were monophyletic (e.g., Simons and Pilbeam 1965), although as early as the 1920s Weinert (1927) proposed a close chimpanzee-human relationship.

In the past half century, two major shifts have stimulated new approaches to the question of hominin ancestry that are still reflected in ongoing debates. The first was the molecular revolution. Starting in the late 1950s, Goodman showed unequivocally that *Homo, Pan,* and *Gorilla* formed a monophyletic group, and that humans were African hominoids rather than being distantly related (Goodman 1963). A few years later, Sarich and Wilson (1967b) demonstrated that genomic change could be quantified and used as a "clock" (stochastic rather than metronomic), and argued (1967a) that the LCA_{G-CH} probably lived no earlier than the Late Pliocene or latest Miocene. In 1989, Caccone and Powell, using DNA-DNA hybridization analyses, resolved the African hominoid trichotomy and showed that chimps were the closest relatives of modern humans. Although some questioned the reliability of DNA-DNA hybridization, the genomic analyses of Ruvolo (1997) unequivocally established that the human and chimp lineages are monophyletic. The second shift was the Hennigian revolution, which catalyzed a transition from grade- to clade-based approaches to inferring and testing hypotheses about ancestors (Delson et al. 1977; Gundling 2005), clade referring to a group of related species descended from a common ancestor. Since the adoption of cladistic logic and methods, it is no longer acceptable to base reconstructions of the LCA on grade-based "phases" of human evolution; instead, it is necessary to propose hypotheses based on analyses of primitive versus derived characters within a well-established and robust phylogenetic framework.

These shifts are first evident in the contributions to an important book, *Classification and Human Evolution* (Washburn 1963), following a seminal 1962 conference, which revealed tensions generated by the first molecular data. Goodman's chapter (1963) presented the comparative immunological analyses demonstrating that humans, chimps, and gorillas formed a monophyletic group. Schultz (1963: 112), however, maintained his morphology-based interpretation of ape phylogeny, and argued that Goodman's evidence, "though of great interest, is more than counterbalanced by the mass of profound differences found in all sorts of other characters of recognized reliability." Simpson (1963) accepted Goodman's phylogenetic framework, yet showed (figure 5) *Homo* diverging from the common ancestor of more closely related chimp and gorilla (classified together in *Pan*). In arguing against including *Pan* and *Homo* in the same (immediately) higher taxon, Simpson (1963: 28) noted that "*Pan* is the terminus of a conservative lineage ... [and] is obviously not ancestral to *Homo* [yet their] common ancestor was almost certainly more *Pan*-like than *Homo*-like."

As noted above, Sarich and Wilson (1967a, 1967b) affirmed Goodman's interpretation concerning the *Pan-Gorilla-Homo* trichotomy, and added a temporal component to the molecular analyses, showing that the LCA of gorillas, chimps, and humans lived more recently than had previously been believed. Using this new molecular-based framework, Washburn (1967: 23) argued that the abundant postcranial similarities between humans and African apes were homologies, which indicated "they shared a common knuckle-walking stage with the ancestors of the chimpanzee and gorilla." Along the same lines, Tuttle (1969: 961) recognized the similarities of chimpanzees and gorillas (which he classified in the same genus, *Pan*), and proposed that gorillas "probably evolved from chimp-like knuckle-walking troglodytians through advanced adaptation for terrestrial foraging and feeding." However, he considered a knuckle-walking stage in human evolution improbable. Later, Tuttle (1981: 89) proposed that bipedalism preceded the emergence of the human lineage and that modern humans were derived from "hylobatians," siamang-sized but less suspensory, and significantly more bipedal.

Over the last few decades, the range of proposed LCAs has remained diverse. Many researchers still argue that the LCA resembled African apes, with many favoring a chimp-like reconstruction (e.g., Groves 1988; Gebo 1996; Richmond and Strait 2000; Wrangham and Pilbeam 2001; Begun 2010; Williams 2010; Lieberman 2011), but some preferring a more gorilla-like LCA (Sarmiento 1994). However, some still support hylobatian models (Filler 2007), and others favor a close orangutan-human phylogenetic relationship (Grehan and Schwartz 2009) or an orangutan-like LCA, at least in terms of locomotion (Thorpe et al. 2014). Somewhat less easy to characterize is the reconstruction proposed in the description and analyses of *Ardipithecus ramidus* by White, Lovejoy, and colleagues (e.g., White et al. 2009, 2015; Lovejoy et al. 2009c, 2009d). According to these analyses, *Ar. ramidus* is characterized by numerous "surprising" primitive features that are not shared with chimps and gorillas but are present in Miocene apes such as *Proconsul* (see below). They reconstruct the LCA as distinctly unlike any living ape: an arboreal palmigrade climber and clamberer, occasionally bipedal as well as quadrupedal, with a touch of pronograde monkey. As discussed below, one implication of this reconstruction is that almost all postcranial similarities among the extant hominoids must be homoplastic, similarities *not* due to inheritance from a common ancestor (White et al. 2009; McCollum et al. 2010).

Hypothesis-Testing Framework

To evaluate alternative hypotheses about the nature of the LCA, it is necessary to agree upon an appropriate hypothesis-testing framework. As noted above, it is convenient to recognize two complementary approaches, top-down and bottom-up, the former based on inferences from extant species, the latter based on the fossil evidence. However, the logical distinction between these two approaches is somewhat artificial, because most of the available hominin fossil record, including those contested as definite hominins, also provides indirect inferences about the LCA when viewed in appropriate phylogenetic context. Both extant and fossil sources of data are necessary, and both are challenged by the problem of distinguishing homologies (similarities due to inheritance from a common ancestor) from homoplasies (independently evolved similarities).

Comparative Data from Extant Species (Top-Down)

The essence of top-down approaches is that they use genetic and phenotypic data from extant species, especially humans and the African apes, to make inferences (in effect hypotheses) about the LCA. Before the advent of molecular data and cladistic methods, this approach consisted primarily of considering particular extant species such as gibbons, gorillas, or even tarsiers as referential models for the LCA (see above). Such approaches are obviously problematic because even a diverse group of modern species may sample only a fraction of its past morphological diversity. Modern referential models can also fail to account for how species have changed over their evolutionary history since divergence from a LCA, and ignore which features of extant species are primitive or derived. Over the last few decades, however, rigorous phylogenetic methods have been developed to generate hypotheses about extinct taxa using comparative data from extant species without reconstructing fossil taxa as simply copies of modern species. All these methods begin with an accepted phylogeny of relevant taxa, then map character states onto the tree, and finally use parsimony, maximum likelihood, or Bayesian methods to reconstruct and estimate the probability of alternative character states at ancestral nodes in the tree, including the LCA (for review, see Nunn 2011).

Before reviewing the predictions and critiques of methods that use comparative data to reconstruct the LCA, it is important to consider three major challenges of this approach: (1) the need to have an accurate a priori tree with the appropriate taxa on which to map character states, (2) difficulties in defining characters to include in the analysis, and (3) interpreting similarities as homologies or homoplasies.

The first of these challenges is, fortuitously, the least contentious because of overwhelming molecular evidence for the cladogram illustrated in Figure 2.1, in which chimpanzees and humans are monophyletic with respect to gorillas. According to cladistic logic, characters (genetic and phenotypic) that are shared in gorillas and chimpanzees are likely also to be present in the LCA_{G-CH} as well as the LCA_{C-H} unless they evolved independently in the *Pan* and *Gorilla* lineages since their last common ancestor.

One potential complexity involves incomplete lineage sorting (ILS), in which a fraction of gene trees (or parts of genes) does not match the species tree (Prüfer et al. 2012; Mailund et al. 2014). Thus, humans may share some genes or partial genes with gorillas rather than with chimpanzees, or chimpanzees may share genetic sequences with gorillas rather than with humans. The phenotypic effects of ILS remain to be explored, but are likely to be both interesting and significant.

One complexity of phenotypic data is that because gorillas are significantly larger than chimps, it is essential to account for the scaling effects of size differences between the African apes when assessing their similarities and differences. Many of these size-related differences are allometric, but some also derive from the complex functional effects of size. For example, since body mass scales to the power of three but muscle forces and bone cross-sectional geometry scale to the power of two, larger animals must adopt different, less flexed postures than smaller animals (Biewener 1989), which in turn can have effects on morphology. Therefore, to make hypotheses about the LCA_{G-CH} and the LCA_{C-H} from human, chimp, and gorilla phenotypes, it is necessary to assess how similarities and differences among these species are size-related before inferring whether they are shared primitive features.

The second challenge affecting top-down inferences is a widespread problem that affects all phylogenetic analyses: homoplasy. As noted above,

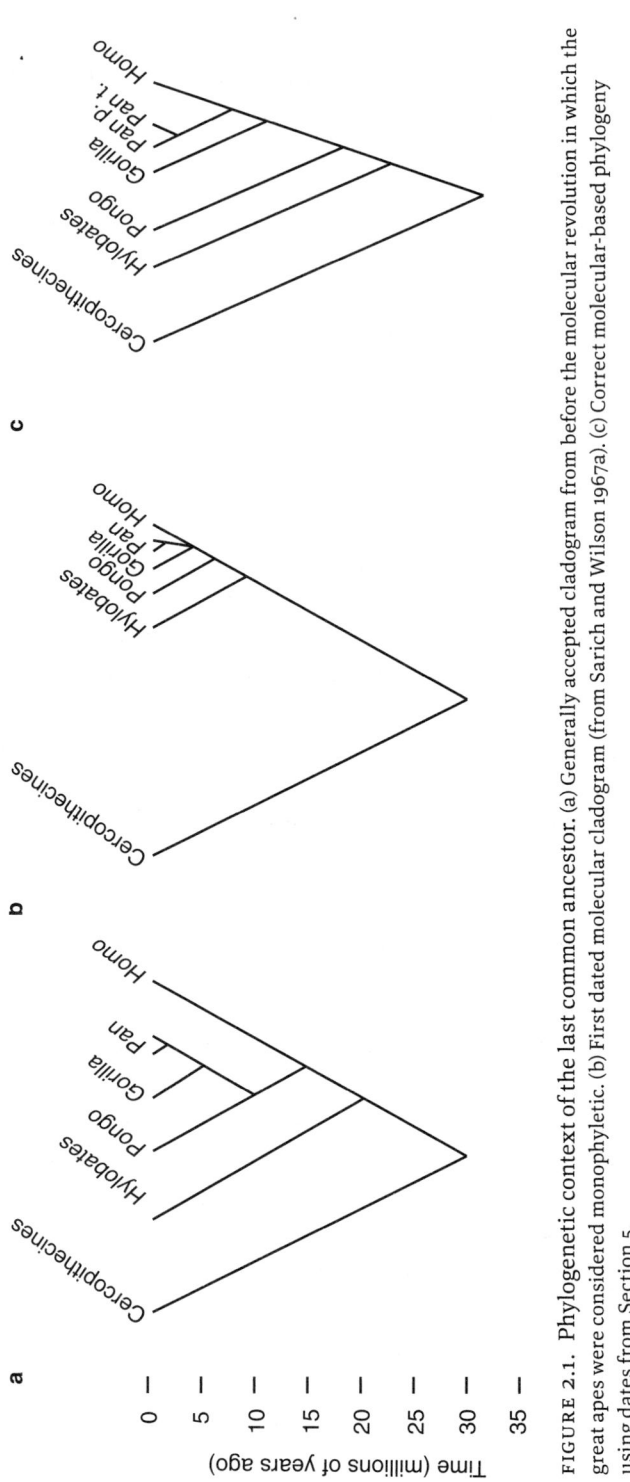

FIGURE 2.1. Phylogenetic context of the last common ancestor. (a) Generally accepted cladogram from before the molecular revolution in which the great apes were considered monophyletic. (b) First dated molecular cladogram (from Sarich and Wilson 1967a). (c) Correct molecular-based phylogeny using dates from Section 5.

any similarities and differences between gorillas and chimpanzees are either shared primitive or homoplastic. If we assume, as most do, that homology is a more common explanation for similarity among closely related species than homoplasy, then it is most reasonable to infer initially that they are mostly shared primitive. However, homoplasy presents a thorny problem for reconstructing the LCA, because there is little agreement on how prevalent homoplasy is and how to recognize its occurrence. Gregory (1927: 555) presciently noted this issue ninety years ago:

> It has been suggested, by myself and others, that some of the characters common to man and one or more of the anthropoids may well be due to "parallelism," that is, to the independent acquisition of similar characters after the divergence from a common stock; but in view of the many positive agreements the burden of proof must rest upon those who would class all of the resemblances as "parallels." In fact, the special resemblances between man and ape, outside of those common to typical mammals, are found in so many different parts of the organism, that the assumption of such extensive parallelism is a mere begging of the question.

Although Gregory was writing before the advent of modern cladistics, his quote is a useful reminder of the ever-present challenges of interpreting similarities as homoplasies, ancestral homologies, or derived homologies. The similarities of a less-preferred hypothesis can be interpreted as homoplasies rather than homologies. In the absence of developmental and functional data necessary to evaluate hypotheses of homology and homoplasy (see below), we must strive to apply consistent, objective criteria to this problem.

This brings up the third and fundamental challenge, the problem of defining and selecting phylogenetically useful characters to consider when reconstructing the LCA. It is widely appreciated that atomizing complex phenotypes into discrete and phylogenetically informative characters is a challenge (e.g., Sarich 1983; Cartmill 1992, 1994; Asfaw et al. 1999; Lieberman 1999; Collard and Wood 2000; Pilbeam 2002; Young 2005). This problem is especially true for skeletal phenotypes, which have been shown to provide less objective and more ambiguous characters and character states than soft tissues (e.g., Groves 1986; Gibbs et al. 2002; Pilbeam and Young 2004; Young 2005; Diogo and Wood 2011; Diogo et al. 2017). Given paleontology's reliance on skeletal characters, some researchers advocate analyzing these characters using developmental and functional data. Lieberman (1999) proposed evaluating characters on a case-by-case basis in terms of key criteria including independence, heritability, plasticity, and the developmental processes that generate them. Lovejoy et al. (1999) went a step further by classifying five kinds of traits, ranging from those that are purely genetic in origin (Type 1), and thus useful for making phylogenetic inferences, to those that result almost entirely from interactions with the environment (Types 4 and 5), and thus useless for making phylogenetic inferences. A problem with these classifications, however, is that the functional, genetic, and developmental data necessary to classify traits according to this scheme are unknown for almost all traits. For example, Lovejoy et al. (1999: 13251) argue that the height of the ilium is a Type 1 trait, ideal for phylogenetic analysis, "that differs in two taxa because its presence and/or expression are downstream consequences of differences in the [positional information] of its cells and its resultant effects on local pattern formation." There are two problems with this classification. First, the definition of a Type 1 trait is essentially just a description of morphogenesis that provides no criteria for recognizing phylogenetically useful aspects of phenotype to define characters as well as character states. Second, the evidence used to test whether iliac height is a Type 1 trait was a virtual experiment in which images of pelves were "morphed" using Photoshop.

The challenges of selecting and evaluating phylogenetically useful characters and of reliably distinguishing homologies from homoplasies are not easy, especially given our limited understanding of the relationship between genotype and phenotype. Note also that defining a character involves implicit assumptions of homology. Nonetheless, a top-down approach is necessary to make predictions about the nature of the LCA in the absence of a fossil record for chimpanzees and gorillas. As noted above, when chimpanzees and gorillas were considered monophyletic, the many similarities between these genera were almost universally considered shared derived (e.g., Simpson 1963; Tuttle et al. 1978a, 1978b, 1992). However, the recognition that chimpanzees and modern humans are monophyletic has required reinterpretation of these similarities as either shared primitive or homoplastic; similarities between chimpanzees and gorillas were either shared by the LCA_{G-CH} or they evolved independently in the two lineages. Since we consider it improbable that so many features would evolve independently, it is most likely that these similarities are shared primitive, which means the LCA_{G-CH} as well as the LCA$_{C-H}$ also had these shared primitive features (Washburn 1968; Groves 1986, 1988; Wrangham and Pilbeam 2001; Richmond et al. 2001; Begun 2010; Williams 2010). Note that better understanding of the phenotypic effects of ILS may have important consequences for phylogenetic analysis.

The top-down inference that the LCA of modern humans and chimpanzees was like the African apes has been controversial based largely on the criticism that it is problematic to assume that an extinct Late Miocene species resembled an extant species such as the chimpanzee (McNulty 2010; White et al. 2009). White et al. (2015) specifically censure chimpanzee-like models of the LCA on the grounds that this inference is based on an "embedded metaphor." However, the hypothesis that the LCA resembled chimpanzees or gorillas is based neither on a metaphor nor on the assumption that the African apes are model species, and working hypotheses tend to be "embedded." Rather, it is based on principles of parsimony following evidence that chimpanzees plus humans are a monophyletic group, the many similarities shared between chimpanzees and gorillas notwithstanding. Put differently, the argument that the LCA is more likely to have resembled chimps or gorillas is an inference, not an assumption. We note that the admonition to not uncritically use particular species as a model for the LCA can also be applied to the fossil record, including the inference that many if not most of the features that *Ardipithecus* does not share with later australopiths (here used in top-down perspective as an undoubted hominin) must be primitive characters also present in the LCA (White et al. 2009, 2015).

A related critique of the top-down inference that the LCA of modern humans and chimps resembled the African apes is discomfort with the implication that these species are "living fossils" or "time machines" that have not changed much since their divergence from the LCA$_{G-CH}$ as well as the LCA$_{C-H}$. To quote McNulty (2010: 329):

> Identifying the earliest humans means, by necessity, that we can differentiate them from ancestral chimpanzees and gorillas. With little or nothing known about the ancestral forms of our closest relatives, we can at best impute their features using the modern species: a top-down approach. This is wrong on logical grounds—gorillas and chimpanzees have evolved over as long a time period as humans have—but is intuitively attractive because it seems obvious that the magnitude of human evolutionary change over this time has been substantially greater. The

implications of this fact are easily misunderstood: while chimpanzees may be more similar to our last common ancestor than are modern humans, they still may not be very similar. Using chimpanzees as a model for the ancestral human may be greatly misleading (Senut 2007; White et al. 2009), and this is precisely why a detailed knowledge of ape evolution is so important to human origins research (Andrews and Harrison 2005).

Although we agree with McNulty and others about the importance of the fossil record, the argument that no living species within a monophyletic group can be identical to the ancestor of that group is true, but only in a trivial sense. The more useful question is: To what extent does any living species in a monophyletic group differ from the LCA of that group?

To appreciate this point, consider the following thought experiment. When a group of species is classified as members of a genus, this inference is based on assumed shared derived similarities of those species. Characters used are necessarily present in their LCA, which means that, by definition, the LCA must also be classified in that genus based on those characters. Equally clearly, these species will differ from each other, each with its own apomorphies, and it is to be expected that the LCA would also differ at the species level from the living members of the genus. Such variation must be characteristic of any speciose genus. Among the primates, *Hylobates, Colobus, Macaca, Eulemur,* and *Microcebus* are all examples of monophyletic generic radiations that, based on genetic analyses, originated in the late Miocene. Although species of those genera would have differentiated before ~5 Ma, the subtle musculoskeletal differences among species in these radiations indicate that few species in these genera have undergone much morphological change since the Miocene. In each of these radiations, the LCA is probably not very different from its extant descendants unless parallelism is extraordinarily rampant. There is no a priori reason why the morphology of the chimpanzee and gorilla lineages has not been as conservative as that of other radiations such as gibbons, macaques, and colobines.

Data from Fossil Taxa (Bottom-Up Approaches)

Making inferences about extinct species such as the LCA solely from living taxa is obviously problematic because inferences that derive from top-down approaches described above are hypotheses. Fossils are necessary not only to test these predictions but also to generate additional hypotheses, since extant taxa sample only those clades that have not gone extinct.

Testing hypotheses about the LCA using the fossil record is far from simple, especially in the near absence of any fossil record for the *Pan* and *Gorilla* lineages and the challenges of placing fossil species in an agreed phylogenetic context. Three general groups of fossils are relevant to generating and testing hypotheses about the LCA. The first is the australopiths, almost universally recognized as hominins, which date to the Pliocene and thus sample hominin diversity several million years after the divergence of the human and ape lineages (see below). The second group includes more primitive putative hominin species currently classified in three genera: *Sahelanthropus, Orrorin,* and *Ardipithecus*. These taxa, which we label *earliest hominins,* date to between 7.2 and 4.3 Ma and are proposed by us and other researchers to be hominins largely based on evidence for bipedalism, reduced canines, and postcanine morphology (e.g., White et al. 1994, 2009; Brunet et al. 2002; Zollikofer et al. 2005; Lovejoy et al. 2009c; Lieberman 2011). Several researchers, however, have questioned whether these taxa are hominins and have instead proposed that at least some similarities to the australopiths are homoplastic (Andrews and Harrison 2005; Wood and Harrison 2011). The final group comprises the diverse and abundant Miocene apes, including the frequently mentioned *Proconsul* (now *Ekembo*), *Kenyapithecus, Nacholapithecus, Sivapithecus, Dryopithecus, Oreopithecus, Pierolapithecus, Chororapithecus, Nakalipithecus* and *Ouranopithecus* (McNulty 2010; Harrison 2010a, 2010b). Except for *Sivapithecus,* which most paleontologists consider a crown pongine, none are Asian, and there is little agreement over their phylogenetic relationships to each other or to extant apes (see below).

Although the fossil record is essential for testing hypotheses about the LCA, it is a necessary condition to interpret fossil taxa within an agreed phylogenetic context. To begin with, unquestioned hominins such as the Pliocene australopiths are useful because they help identify and evaluate derived features of the putative earliest hominins and their timing (hence can be used in a top-down approach). It follows that features present in *Sahelanthropus, Orrorin,* and *Ardipithecus* that are not most parsimoniously interpreted as shared derived with later hominins are hypothetically shared primitive with the LCA. Remember, however, that the earliest hominins would still be derived to varying extents relative to any LCA. For example,

Ar. ramidus is dated to between 4.4 and 4.3 Ma, which means it may postdate the separation of the chimp and human lineages by as much as 3 million years (see below). If the hominin lineage was subject to considerable selection, then *Ar. ramidus* may not be very similar to the LCA (see below).

Using Miocene hominoid taxa to test hypotheses about the LCA is even more difficult, because the phylogenetic position of virtually all these species is debated and unclear. Consider, for example, the argument that many Miocene hominoids are more similar to the LCA than the African apes because the latter are too "specialized" (White et al. 2015; Almécija et al. 2013). The term "specialized" is here a synonym for "derived." Gorillas and chimpanzees are unquestionably derived relative to most Miocene taxa, but hominins evolved within this derived branch of the hominoid radiation. To repeat, if the LCA lacks these derived features (e.g., of postcanine teeth, axial skeleton, hands, and feet), then it would have to be the case that all the putative derived (specialized) features of postcanine teeth, axial skeleton, hands, and feet of chimps and gorillas evolved independently. Thus, while it would be useful to consider relevant Miocene taxa when evaluating the polarity of features in African apes, hominins, and the hypothetical LCA, such evaluations must always be placed in the context of these taxa's unknowable relationships to the LCA and the African apes, and they also depend on hypotheses about relationships among Miocene taxa, about which there is scarce consensus.

A related cautionary issue when using fossil taxa (as well as extant species) to evaluate hypotheses of the LCA based on the top-down approach is the importance of distinguishing primitive versus derived features in multivariate analyses. There should be no debate that primitive features are problematic for inferring phylogenetic relationships, but the growing use of geometric morphometric (GM) methods to quantify multivariate shape has led to a resurgence of phenetic comparisons among species including extant apes, fossil hominins, and Miocene apes (e.g., Richmond and Jungers 2008; Almécija et al. 2013, 2015a). These analyses are useful, but principal components of GM analyses are phenetic measures, and principal component scores are not biological characters that can be assigned polarity (Bookstein 1994; Adams et al. 2011).

Integrating Comparative and Fossil Data

As described above, this chapter integrates both fossil (bottom-up) and extant species comparative (top-down) data and approaches to evaluate hypotheses about the morphology of the LCA. We begin by reviewing the effects of size on comparative anatomical evidence from gorillas and chimpanzees to ask to what extent similarities and differences between these taxa support the inference of a chimp-like LCA. We then consider the genetic evidence for when the human and chimpanzee lineages diverged, and review the phylogenetic context of the Miocene fossil record for testing hypotheses about the LCA. Finally, we review—region by region—key features in gorillas, chimpanzees, early hominins, and australopiths that are usable as characters in phylogenetic analyses. We conclude by mapping character-states for these features onto the known molecular-based phylogeny and assess using parsimony the most likely character state of the LCA.

Size-Independent Similarities and Differences between Gorillas and Chimpanzees

Notwithstanding their size difference, chimpanzees and gorillas are similar in many aspects of morphology, and share numerous other characteristics including the same unique mode of locomotion, knuckle-walking. These many similarities were long assumed to be shared derived features that reflected *Pan* and *Gorilla* monophyly, and were the primary basis for disputing the molecular evidence that *Homo* and *Pan* were monophyletic (e.g., Andrews and Martin 1987; Hartwig-Scherer 1993). Although *Homo-Pan* monophyly is now firmly established, evaluating similarities and differences between chimpanzees and gorillas remains one of the most critical issues for using comparative data (top-down approaches) for reconstructing the LCA. It bears repeating that because *Pan* and *Homo* are monophyletic, with *Gorilla* more distantly related (Figure 2.1c), any similarities between *Pan* and *Gorilla* are either shared primitive or convergent. Which alternative is correct has profound implications for inferring the morphology of the LCA: if the similarities between the two genera are shared primitive, then it is probable that the LCA, if chimp-sized, was also somewhat chimp-like; if the similarities are mostly convergent, then the LCA was likely rather different from either of the AGAs. To test either of these hypotheses, it is therefore first necessary to review and evaluate how morphologically similar *Pan* and *Gorilla* are.

Assessing similarity, however, is hardly trivial. In addition to the thorny problem of homology versus homoplasy (see above), an equally complex problem is size. Male gorillas average approximately 170 kg and females about 75 kg; male chimps average approximately 50 kg and females about 40 kg (Smith and Jungers 1997). Since no other extant apes and perhaps just one extinct ape are as heavy as gorillas, it is likely that gorillas, not chimps, are derived in terms of size. As discussed later, some fossil apes are large-toothed *(Samburupithecus, Chororapithecus, Ouranopithecus, Gigantopithecus)*, but no estimates of their body size are currently available. When assessing the nature of similarities between chimps and gorillas (and other hominoids), it is therefore necessary to correct for any allometric effects of size. Testing the extent to which similarities and differences between *Gorilla* and *Pan* derive from allometric scaling requires studying ontogenetic allometries that compare size-shape relationships among samples of chimpanzees and gorillas from all postnatal ontogenetic stages (for review, see Fleagle 1985).

There are several ways in which a feature in two species such as chimpanzees and gorillas can differ morphologically because of scaling (Figure 2.2). In the simplest case of *common (classic) allometry,* the feature can scale identically with size in similar ways in both species in terms of slope and intercept, and differ only in terms of the observed size range (Figure 2.2a). In such cases, any differences in shape are solely "size-related," and thus probably reflect selection for size only (Gould 1966). If the LCA_{G-CH} was chimp-sized and chimp-like, and if the gorilla lineage became larger than the LCA_{G-CH}, then one would expect many aspects of gorilla morphology to exhibit common allometry with chimpanzees.

Although common allometry provides generally unambiguous evidence for size-independent similarity, many studies have shown that selection can act in multiple ways to increase a species' size as well as to act on size-related functions. It is therefore necessary to consider other reasons why size changes can affect shape. As noted above, bigger animals tend to have more extended limb postures relative to body mass to reduce muscular effort (Biewener 1989), which in turn alters how bones are loaded, thus potentially influencing selection on bone cross-sectional shape. Two patterns of size-required scaling are of particular importance. First, closely related species can share the same

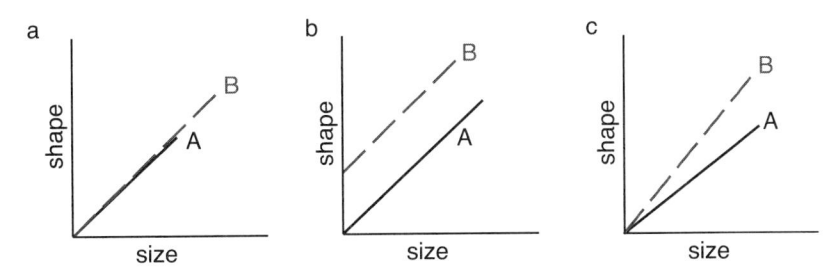

FIGURE 2.2. Alternative ontogenetic allometries between size and shape for two species, A and B. (a) Common allometry, in which B has a different shape than A solely from differences in size. (b) Change in shape, often termed a "grade-shift" because of a change in intercept rather than slope. (c) Change in shape because of a change in slope.

allometry in terms of slope but differ in intercept (Figure 2.2b). Such "grade-shifts" (Pilbeam and Gould 1974) are hypothesized to occur when species share the same basic developmental mechanisms, but differ in the relative amount of tissue initially involved in growth (Atchley and Hall 1991). Second, species can share the same allometry in terms of intercept, but differ in slope (Figure 2.2c). Such shifts may be the result of selection for accelerated growth (Atchley and Hall 1991). Both scaling patterns, which result in a larger-sized descendent taxon having a different shape relative to size than a smaller ancestral taxon, have been observed in comparisons of closely related taxa. For example, many body proportions in *Cebus apella* scale with a different intercept but a slope identical to that seen in *C. albifrons* (Cole 1992; Jungers and Fleagle 1980). Scaling of body proportions between siamangs and gibbons also differs in intercept but not in slope (Jungers and Cole 1992). In addition, ontogenetic scaling patterns that differ in terms of slope but not intercept have been documented among closely related breeds of chickens (Cock 1966), mice (Shea et al. 1990), and domestic dogs (Wayne 1986). For reasons that are still unclear and that need to be explored, scaling relationships in the cranium between closely related species or breeds may be more likely to differ in terms of slope than intercept as compared to relationships in the postcranium (Hartwig-Scherer 1993). However, our review of scaling studies has found little evidence for closely related species having ontogenetic allometries that differ in both slope and intercept.

In short, if the LCA_{G-CH} was chimp-sized and chimp-like in terms of morphology, then we predict that many features in the two species would share common ontogenetic allometries. However, we also predict the ontogenetic allometries of some features to differ in either intercept or slope, the latter especially in the skull. Although ontogenetic allometries with different slopes and intercepts may indicate convergence, they are not necessarily evidence that the two species did not derive from a similar common ancestor. Instead, they are the basis for functional hypotheses underlying "size-required" scaling relationships that require further testing. With this framework in mind, we briefly review the long history of research on scaling differences among the great apes in general, and between gorillas and chimpanzees in particular, to evaluate the extent to which differences between these species are size related, and hence likely to be shared primitive retentions that help reconstruct the $LCA_{G\text{-}CH}$ and the $LCA_{C\text{-}H}$.

The first researcher to address whether allometry accounted for shape differences between gorillas and chimps was Giles (1956), who measured seven cranial distances in 170 chimpanzees, bonobos, and gorillas of varying ages, and found nearly identical coefficients of allometry in these species. He concluded: "The evidence presented suggests that the allometric growth patterns in the chimpanzee would, given an over-all size increase, produce results quite similar to the exaggerated osteological morphology of the gorilla" (Giles 1956: 57). This study inspired Gould (1975) to characterize the gorilla as "an overgrown chimpanzee."

The first rigorous tests of Giles' scaling hypothesis were by Shea (1981, 1983a, 1983b, 1985), who measured an array of cranial and postcranial features in a large ontogenetic sample of chimpanzees, bonobos, and gorillas. Shea scaled craniofacial dimensions against the length of the cranial base (basion-nasion), which he used as a proxy of size, and he also examined scaling relationships among a range of postcranial dimensions. For the skull, he reported that "most of the interspecific variation in craniofacial form among the African apes results from 'ontogenetic scaling,' or the extension of a common growth trajectory of size/shape change into new size ranges" (1983a: 61), and that "overall, gorillas have the facial proportions expected in chimpanzees scaled to their skull sizes" (1985: 186). Among the postcranial scaling relationships he analyzed, Shea found none that differed significantly in slope, but many that differed slightly in intercept (Shea 1981, 1983b). A similar analysis by Jungers and Hartman (1988) that examined five postcranial ratios also found that chimps and gorillas followed ontogenetic scaling for the ratio of femur to tibia length as well as the ratio of ilium to pubis length, but that the species differed in intercept but not slope in humerus to radius length, humerus to femur length, and scapula to clavicle length.

One problem with these analyses is that the scaling relationships were not calculated relative to body mass, which is often unavailable for museum specimens, but is nonetheless the most important size variable to use for most cases of allometry (Jungers 1984). In an effort to redress this problem, Hartwig-Scherer and Martin (1992) devised a series of equations to predict wild body weights of hominoids from skeletal dimensions. Hartwig-Scherer (1993) then used these estimates of body masses to assess differences in scaling for numerous cranial and postcranial dimensions among hominoids (see Figure 2.3a). Overall, she found that for the overwhelming majority of postcranial dimensions, gorillas and chimpanzees shared the same common ontogenetic allometries, with the primary exceptions relating to relatively shorter legs in gorillas. In the skull, however, she found that gorillas tended to be "modified allometric size variants" (Hartwig-Scherer 1993: 129), with gorillas generally having slightly greater slopes than chimpanzees. Gorilla and chimp scaling relationships, however, were much more similar to each other than to either *Pongo* or *Homo*. These results led Hartwig-Scherer to conclude on the basis of scaling that the similarities of chimps and gorillas were likely shared derived and that therefore the two were monophyletic, to the exclusion of *Homo*.

Recently two studies have extended these cranial results using multivariate allometry, in which shapes are quantified using geometric morphometrics and size is assessed using overall cranial size (usually centroid size). Berge and Penin (2004) calculated multivariate allometries from a Procrustes analysis of twenty-nine landmarks in ontogenetic samples of fifty gorillas and fifty chimpanzees. They found that 64 percent of the total shape variation was explained by a single multivariate allometry in which, for a similar stage of growth, smaller nonadult gorillas had the size and shape of larger, older chimpanzees. In addition, as described by Shea (1983a, 1983b, 1985), Berge and Penin (2004) found that the major allometric differences of gorillas are a more posteriorly oriented foramen magnum and a taller, more vertically oriented nuchal plane. The major differences of gorillas unrelated to size are a relatively flatter and longer superior contour of the vault, with a more prominent occipital. In a slightly different analysis using similar methods on ontogenetic samples of chimps ($n = 40$), gorillas ($n = 41$), as well as humans ($n = 24$), Guy et al. (2005) found that the first principal component

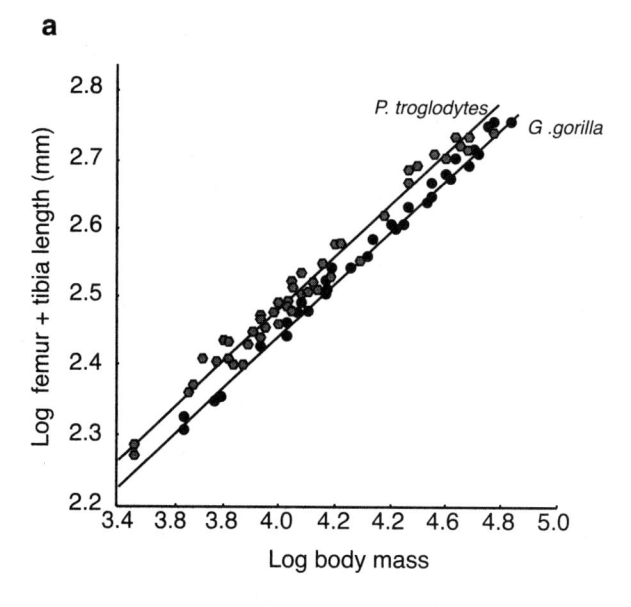

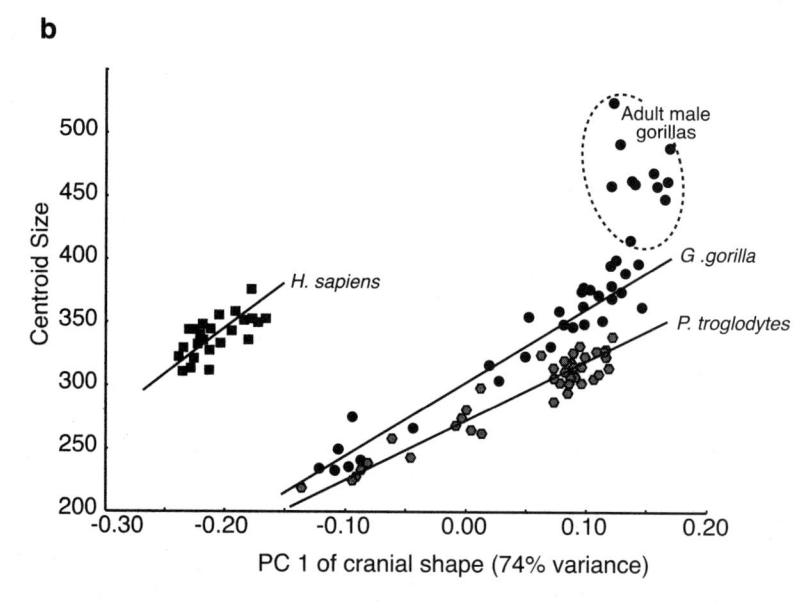

FIGURE 2.3. Examples of ontogenetic allometries for *Pan* and *Gorilla*. (a) Representative example of differences in allometry between a postcranial shape variable for *G. gorilla* and *P. troglodytes*, in this case leg length (femur plus tibia length) versus body mass (modified from Hartwig-Scherer 1992). (b) Cranial allometry based on the first principal component of shape (which explains 74 percent of the sample variance) for *G. gorilla*, *P. troglodytes*, and *H. sapiens* (from Guy et al., 2005). Note that apart from *Gorilla* males, which differ from females because of secondary sexual characteristics that arise in adolescence, *Gorilla* differs from *Pan* primarily because of a grade shift (sharing a similar allometric slope, but differing in intercept).

of variation explains 78 percent of total sample variance, and scales more similarly in chimps and gorillas than in modern humans, but with a slightly greater slope in gorillas (Figure 2.3b). We can therefore conclude that variations in cranial shape between *Pan* and *Gorilla* are mostly but not completely the result of scaling. Multivariate scaling analyses indicate that the same is also true of many differences between *P. troglodytes* and *P. paniscus* (Lieberman et al. 2007).

Finally, a few studies have examined scaling in particular postcranial features. Several of these analyses have focused on the hand and wrist, largely motivated by interest in knuckle-walking. Although chimpanzees and gorillas both knuckle-walk, chimpanzee knuckle-walking tends to be more kinematically variable, involving more weight-bearing on the medial (radial) side of the hand as opposed to the more lateral (ulnar) side in gorillas; chimpanzees also walk with more extended wrist postures (Tuttle 1969; Inouye 1992, 1994; Kivell and Schmitt 2009). We hypothesize that these contrasts reflect differences in body mass and perhaps also locomotor repertoires. As in other large mammals, gorillas probably adopt more columnar, less extended postures in the wrist and other joints because of the scaling of effective mechanical advantage (Biewener 1989). Further, gorillas travel less than chimpanzees, and as they age also spend less time in trees (Doran 1997). Perhaps unsurprisingly, Inouye (1992) found that a majority (but not all) of dimensions in the metacarpal and proximal phalanx of the fifth digit scaled with common allometry in the two species, but that these bones were relatively wider in gorillas. Dainton and Macho (1999) analyzed scaling using twenty-seven measurements of four carpal bones. No coefficients of allometry differed significantly between chimps and gorillas, and intercepts were not significantly different in 56 percent of these analyses, but 44 percent of intercepts were significantly higher.

Another region of focus for scaling studies among apes is the scapula. Alemseged et al. (2006) analyzed ontogenetic scaling in the scapula using glenoid size as a proxy for body mass. Leaving aside the concern that proxies for body mass can lead to incorrect results (Hartwig-Scherer and Martin 1992), Alemseged and colleagues found that glenoid angle scales differently in gorillas and chimps, but that most other scapular variables scaled with common allometry, including supraspinous fossa breadth and infraspinous breadth. A more comprehensive analysis of the scapula by Young et al. (2015) found that shape variation in African apes as well as hominins falls along a single multivariate morphocline that explains nearly half the shape variation among primates, with early hominins intermediate between the AGAs and humans. The most parsimonious model for the evolution of the shoulder in hominins as well as the African apes is from an LCA with an African apelike scapula, although see Almécija (2016) for an alternative interpretation.

Because of the mosaic nature of phenotypes, the numerous scaling studies involving *Pan* and *Gorilla* cannot be synthesized into a single conclusion about scaling, and more research is needed, especially to more comprehensively analyze scaling in the postcranium using multivariate methods. That said, the balance of evidence indicates that gorillas are mostly but not exclusively scaled-up chimpanzees. The majority of postcranial studies, especially those that regress shape against body mass, indicate that *Pan* and *Gorilla* share common ontogenetic allometries for a wide range of limb dimensions. The minority of postcranial variables that have different allometries, either of slope or of intercept, tend to reflect the relative size of the hindlimb as well as "size-related" functional variables that one expects to be targets of selection because of size-related differences in locomotion. Because gorillas are so much larger than chimpanzees, they have predictably more vertical limb postures, engage in less arboreal locomotion, and thus load their skeletons slightly differently than chimpanzees. Many cranial features in the two species also share common allometries, but a number of facial dimensions in gorillas scale more steeply relative to overall cranial size. Such differences may have a functional explanation, because gorillas eat more leaves and other tough, displacement-limited foods that require and generate higher masticatory forces. It is thus reasonable to hypothesize that the *Gorilla* face departs from common allometric scaling of the *Pan* face because of adaptations that reduce craniofacial strain and notably increase prognathism (Preuschoft et al. 1985; Lieberman 2011). The relatively more vertical nuchal plane of gorillas may be another size-required adaptation, since the bigger gorilla face generates higher moments around the atlanto-occipital joint, which thus need to be countered by the nuchal muscles as well as longer cervical spines (Berge and Penin 2004).

To return to the larger question about the nature of the LCA, the top-down prediction based on allometric comparisons of chimpanzees and gorillas is that anatomically both the LCA_{G-CH} and the LCA_{C-H} were probably like chimpanzees or small gorillas (which amount to effectively the same thing), unless all the many similarities between these taxa are the result of homoplasy. Put differently, the major anatomical differences between chimpanzees and gorillas relate to size, which is probably derived in gorillas rather than chimpanzees. One implication of this conclusion is that, as in other Miocene-originating primate radiations such as gibbons, macaques, and colobines, there has not been much morphological evolution in the *Pan* and *Gorilla* lineages over the last 5 to 10 million years. It is also worth emphasizing that the top-down prediction that the LCA was generally chimp-like is not based on any assumption or referential model, but rather on inferences from comparative data given known phylogenetic relationships.

Molecular Clock Estimates for the LCA

Inferences about the LCA from living taxa provide hypotheses that need to be tested with fossil evidence. Before turning to the fossil record, however, we first review the evidence from molecular comparisons of living species for when the LCA might have existed. For over half a century it has been understood that the genome on average accumulates change in an approximately clocklike manner (Zuckerkandl and Pauling 1965; Mailund et al. 2014). According to Morgan (1998), Zuckerkandl had come to the conclusion by the early 1960s that molecular change was regular enough to be considered clocklike. Of equal importance, by the late 1950s and early 1960s Goodman had demonstrated (1) the value of molecular data in determining phylogenetic relationships, (2) that the great apes were not monophyletic, and (3) that modern humans were phylogenetically African apes (Goodman 1963).

Sarich's critical contribution to the molecular clocks debate was the "rate test" (Sarich and Wilson 1967b; Sarich 1983), a way of determining the extent to which the rates of molecular change vary along related lineages. This test involves measuring the genetic difference (distance) between species, earlier determined indirectly through protein or immunological differences (i.e., molecular morphology) and now directly as genomic sequence differences. Any differences in distances between an outgroup and any ingroup lineages reveal the extent to which one ingroup lineage has changed more or less than another (Sarich and Wilson 1967b). The early work using cross-species albumin antigen-antibody interaction levels showed, for example, that relative to an outgroup such as a ceboid, catarrhine species showed similar amounts of molecular difference from the outgroup. The rate test provides the means of determining to what degree change is regular enough to be used as a clock for estimating the dates of lineage divergence times.

Sarich and Wilson (1967a) used the relatively regular rate of change of albumins to date the trichotomous divergence of *Gorilla, Pan,* and *Homo*. They did this by using the fact that the cercopithecoid-hominoid distance was six times that of the distances among *Gorilla, Pan,* and *Homo*. Dating the cercopithecoid-hominoid divergence at 30 Ma (based on vague assessments of the known primate fossil record), they interpolated a date of 5 Ma for the hominine trichotomy.

Sarich and Wilson (1967a, 1967b) and other early studies drew no distinction between the divergence of the ancestral genome and the separation of the descendent lineages (speciation). King and Wilson (1975) was the first paper involving primate evolution of which we are aware that addressed the value of distinguishing the two, doing so in reference to *Pan-Homo* divergence. They followed Nei and Roychoudhury (1974) in recognizing that the total genetic difference ("distance") between chimps and modern humans should be parsed into two parts: that which accumulated in the ancestor before separation, and that which accumulated in the descendant lineages following speciation. Given assumptions about the proportional relationship between genetic difference and time, this is equivalent to parsing total divergence time into that represented during genome segregation within the ancestral population and that represented in the lineages after speciation. King and Wilson (1975) adopted the simplest possible approach and averaged the chimp and modern human within-species genetic variation to estimate that represented in the ancestor. Calculated this way, the contribution of ancestral variation was just 5 percent of the total, and they argued that such a small percentage could be effectively ignored in estimating speciation times from divergence times.

In the decades following King and Wilson (1975), great strides were made in several critical areas. Sequence data accumulated steadily, and population geneticists developed increasingly sophisticated models, particularly important in estimating the time represented in the ancestral population before separation of the descendant lineages (e.g., Edwards and Beerli 2000). For example, Nachman and Crowell (2000: 299) described the two components of genetic difference between species, that which accumulated in the diverging ancestral population, and that after separation of the descendant species as

$$k = 2\mu T + 4N_{e\mu}$$

where k is the total genetic difference between two species, T the separation (speciation) time, N_e the ancestral effective population size, and μ the mutation rate in time units used for T (usually per year). Because mutation rates are now usually measured in units per generation, the other necessary parameter is generation interval. This formula articulates the two contributing genetic components: that which accumulates within the ancestral population, and that which accumulates after the populations separate. (This model makes the assumption of no genetic exchange after speciation. More recent modeling approaches are able to take such contingencies into account; see Mailund et al. 2014; de Manuel et al. 2016.)

Ancestral Population Genetic Diversity

Assessing ancestral diversity has become easier as more genomic data were generated and it became possible to estimate more realistically the fraction of total variation between species present in the ancestral population. A critical parameter in all current discussions of "clocks" is N_e, the effective ancestral population size (sometimes denoted as N_a). Effective population size is the number of individuals in a population contributing offspring to the next generation, averaged over many generations, and is always less than the census population size, N. (A multispecies meta-analysis by Frankham [1995] estimated that the ratio N_e/N averaged around 10 percent.) In real populations, however, generations overlap, change in size, can be subdivided, and are unpredictably more complex than can generally be modeled. As used by geneticists, N_e is an idealized and simplified representation: populations are of constant size, generations nonoverlapping, and mating opportunities and number of offspring per individual are randomly distributed (see Hey and Machado 2003).

N_e is directly related to maintained population genetic diversity (Burgess and Yang 2008), larger populations containing more variation. Wall (2003) presented an early exercise in how to estimate both ancestral population size and separation (divergence) time, N_e and ancestral diversity depending on estimates based on several variables: the extent of incomplete lineage sorting, ILS (the degree to which gene and species trees are mismatched), the genetic variation within the descendant populations, the mutation rate in generation intervals, and generation interval (see Mailund et al. 2014). Several different current modeling approaches specify the ways these variables can be combined to estimate the effective population size of the ancestor (see Prado-Martinez et al. 2013), results depending on the particular model and the algorithms used in calculations.

More detailed discussion is not necessary here, but it is important that a consumer be aware of the assumptions used to estimate N_e and ancestral population genetic diversity. For example, table 2 in Langergraber et al. (2012) lists speciation times for various common ancestors of *Gorilla, Pan troglodytes, Pan paniscus,* and *Homo sapiens,* based on data from nine different studies. The then newest plausible mutation rate was used by Langergraber et al., but because estimates of N_e varied across studies the estimated speciation times also varied. In Prado-Martinez's supplemental table 13.12 (2013), three different mutation rates yield quite different estimates of N_e and speciation times.

Generation Intervals

One important component of estimates of the timing of speciation, generation interval, is defined as the mean breeding age for males and females. Fenner (2005) published values for 360 non-hunter-gatherer societies and 157 hunter-gatherer societies, estimating female generation intervals as the midpoint between ages of first and last reproduction, and male intervals by adding the difference in male/female marriage age to female-estimated intervals (both male and female intervals were reduced slightly using a "mortality" factor). Estimated this way, the mean interval for the non-hunter-gatherer societies was 29.1 years. Five hunter-gather societies, similarly estimated, had intervals averaging 28.6 years. Genealogically determined data from three societies, Iceland (Helgason et al. 2003), Quebec (Tremblay and Vézina 2000), and Polar Eskimo (Matsumara and Forster 2008), had averages of 30.3, 31.7, and 29.5 years, respectively. Generation intervals of 29 or 30 years for humans are now frequently used in recent papers. Using DNA from ancient genomes, Moorjani et al. (2016c) estimated that over the past 45,000 years the human generation interval averaged 26 to 30 years.

Langergraber et al. (2012) reported generation intervals for eight wild populations of chimpanzees and two of mountain gorillas. The eight chimpanzee population intervals average 24.6 years (population means varying between 22 and 29), with ratios of male to female intervals ranging from 1.22 to 0.79 (grand mean of 0.96). Mountain gorillas average 19.3 years. Only scanty data are currently published for lowland gorillas, but both interbirth intervals and female age at first birth are, respectively, longer and older than in mountain gorillas (Breuer et al. 2009; Stoinski et al. 2013), suggesting that lowland gorilla generation intervals may be closer to those of some chimpanzee populations. Generation interval data are available only for female orangutans: 26.7 years (Langergraber et al. 2012). They are unknown for hylobatids, but a long-term study of *Hylobates lar* (Reichard and Barelli 2008) shows that important aspects of life history, such as female age at first reproduction, interbirth interval, and weaning age, resemble those of large apes and differ significantly from those of cercopithecoids of equivalent or greater body mass.

We know of no published data on generation intervals in wild cercopithecoid monkeys. However, in a comparison of chimpanzees and wild baboons, patterns of female age-related birth rates are very similar if baboon age is scaled at half that of chimpanzee age (Altmann et al. 2010: figure 2). In this paper (figure 1), male birth rates are shown to peak at around eight to nine years and female rates at around fifteen years, together suggesting a male-female averaged generation interval around half that of chimpanzees. It is reasonable to conclude that all the apes, including gibbons, have slower life histories and longer generation intervals than Old World monkeys.

What can be said about generation intervals in hominoids at an earlier stage in their evolution? First, as observed by Wrangham (personal communication 2015), female generation intervals are similar in chimpanzees, orangutans, and human forager females, respectively 25.2, 26.7, and 25.6 years (for the three available genealogically determined non-forager societies it is 27 to 29). This suggests that the last common ancestors of living hominids, along with extinct members of that radiation, including early hominins, may have had female generation intervals around 25 years. (This assumes that gorillas had a smaller and more *Pan*-like ancestor, and have reduced their interbirth intervals as body size and diet changed from a more chimp-like ancestral condition.) Second, it is possible to draw some conclusions from the fact that aggregate change does not differ markedly among the extant hominoid lineages. For example, using four cercopithecine species as outgroups, Moorjani et al. (2016a) showed that gibbons and orangutans each showed 11 percent, gorillas 3 percent, and chimps 1 percent more change than modern humans, indicating that mutation rates in hylobatid and pongine lineages have been barely more than 10 percent faster than in humans, aggregated over time since separation.

The only possible way of estimating generation intervals in fossil hominoid species is to use maturational patterns in tooth enamel. Although Smith (2013) found that correlations among tooth maturation rates, eruption times, and life history parameters such as generation interval in closely related species can be low, Kelley and Schwartz (2010) report that ages of M1 eruption in *Gorilla, Pan,* and *Pongo* are "highly concordant with the comparative life histories of these great apes." Some information is available for first molar eruption ages in Miocene apes. Earlier Miocene *Ekembo nyanzae* and *Afropithecus turkanensis* and later Miocene *Sivapithecus parvada, Ouranopithecus macedoniensis,* and *Lufengpithecus lufengensis* have developmental patterns resembling those of chimpanzees (Kelley and Smith 2003; Smith et al. 2004; Smith 2008). As noted above, there is little evidence of significantly different aggregate genetic change among living hominoids as different in body size as gibbons and gorillas since the time of the crown hominoid common ancestor. If dental development is an approximate indicator of the pace of life history, some Miocene hominoids were more similar in life histories to living apes than to cercopithecoids, suggesting that hominoids at least since the Miocene were characterized by slow life history trajectories.

Ségurel et al. (2014) showed that shorter-generation species have higher mutation rates per unit time than those with longer generations (also see Amster and Sella 2016). Generation-time effect is therefore one contributor both to the relatively small differences in aggregate genetic change along related hominoid lineages, including the very small difference along the human lineage compared to the chimpanzee lineage, and to the much faster rates present in cercopithecoids (Moorjani et al. 2016a). Such effects reflect combinations of age of male puberty, mean generation interval, the ratio of male to female generation interval, rates of spermatogenesis, male and female mutation rates, and degree of male mutation bias, the ratio of male to female mutation rates (Moorjani et al. 2016a, 2016b; Amster and Sella 2016). These important parameters are still inadequately understood in humans, let alone in chimpanzees, and are effectively unknown in other primate species. Disentangling their likely contributions will be challenging, and clearly calls for expanding work on more individuals from wild primate populations using new methods of rapid whole genome sequencing.

Mutation Rates

A critical parameter is mutation rate, either in per year or per generation units. There are two ways of determining such rates: the so-called phylogenetic method, which depends on the fossil record to provide at least one calibration point, and a more recent approach that is not dependent on fossils, now referred to as the pedigree method.

One of the most frequently used phylogenetic mutation rates is based mostly on the analysis of Nachman and Crowell (2000) to estimate rates of per generation mutations (μ_g). Their model used a range of values taken from the paleontological literature to estimate *Pan-Homo* separation time, T, ranging from 4.5 to 6.0 million years. They also assumed two different *Pan-Homo* ancestral effective population sizes, N_e (either 10^4 or 10^5), for each value of T, and a generation interval of 20 years. The average overall per base per generation mutation rate was estimated as $\sim 2.5 \times 10^{-8}$; using the generation interval of 20 years yielded a per base per year mutation rate of $\sim 1.15 \times 10^{-9}$ (with a more realistic 25 years, this becomes $\sim 1.0 \times 10^{-9}$). Others, for example Scally et al. (2012), used a different fossil-based calibration, an estimated macaque-human separation date, and derived a similar value. Of course, such estimates assume that split times can be reliably determined from the fossil record. As we discuss in the following section, this is at best problematic. Also, any such estimates must inevitably fall after the true separation time. Most recent "clock" publications use or refer to a phylogenetic mutation rate of $\sim 1.0 \times 10^{-9}$ per base per year (for example, Hobolth et al. 2011; Prüfer et al. 2012, Scally et al. 2012; Prado-Martinez et al. 2013; Moorjani et al. 2016b).

A more recent focus on generating fossil free estimates of mutation rates is the pedigree method, which involves sequencing parent-offspring trios to high coverage and determining maternal and paternal mutational contributions to offspring. Challenges include avoiding false negatives and false positives, and ensuring that all novel mutations are counted over the genomic region analyzed (Moorjani et al. 2016a, 2016b; Amster and Sella 2016; Gao et al. 2016). As expected from different behaviors of their germlines, mutation rates in females vary only moderately with age while they are strongly correlated with paternal age (Goldmann et al. 2016).

Some of these studies report separate estimates for different categories of mutation, CpG ("spontaneous"), or non-CpG (mostly replication germline errors). In a classic analysis, Hwang and Green (2004) studied the extent to which different substitution types deviate from clocklike behavior in mammals, and their underlying factors. In the case of primates, two kinds of mutations are probably most important: (1) germline replication errors, especially in males where, in contrast to females, germline cells undergo continuous postpubertal replication; and (2) mutations occurring spontaneously and not linked to replication, many such involving change by deamination of methylated cytosine to thymine. Hwang and Green (2004) concluded that the most clocklike mutations were non replication-dependent CpG mutations, where cytosine in a cytosine-phosphate-guanine sequence mutates by deaminating spontaneously to thymine. Recently, Moorjani et al. (2016a) studied the extent of variation across primates and confirmed that many substitution types deviate from clock-like behavior, and thus may not always be suitable for dating evolutionary events, while, as shown by Hwang and Green (2004), CpG transitions occur in a quite clock-like manner. CpG mutations make up a small fraction of total genomic change, but with whole genome sequencing, enough human data are currently available for analysis; they have significantly higher mutation rates than the whole genome. Thus, events in hominine evolution are currently most reliably dated, with appropriate caveats (Moorjani et al. 2016b; Amster and Sella 2016; Gao et al. 2016), using CpG transitions (Moorjani et al. 2016a).

Moorjani et al. (2016b) and Scally (2016) summarize over a dozen different pedigree studies, sample sizes, and parental ages varying, and report a range of rates from $1.1-1.3 \times 10^{-8}$ per base per generation; depending on generation intervals used, this translates into a rate of approximately 0.5×10^{-9} per base per year. An additional approach used a well-sequenced and well-dated ancient genome from a 45,000-year-old *Homo sapiens* from Siberia (Fu et al. 2014). The sequence difference between the dated fossil and living human genomes yielded two mutation rate estimates, one of $0.44-0.63 \times 10^{-9}$ per base per year (Fu et al. 2014: S90), and a second of 0.43×10^{-9} (Fu et al. 2014: S82), with some preference for the lower range of values. This would yield ~1.3×10^{-8} per base per generation using 29 years per generation. Finally, sequence data of ancient DNA from Middle Pleistocene (ca. 400 Ka) early Neanderthals from

Spain (Meter et al. 2016) was compatible with a mutation rate of ~0.5 × 10⁻⁹ per base per year.

Estimating Speciation Times

Any estimate of speciation times requires assumptions about three parameters: per generation mutation rate, generation interval (thus determining mutation rate per year), and ancestral population genetic diversity. As already mentioned, relative to a macaque outgroup, the aggregate change along crown hominoid lineages varies no more than a little over 10 percent. Therefore, based on whole genome sequences, we can reasonably use a constant yearly mutation rate for at least a "first pass" at timing the hominoid radiation, with appropriate adjustments for hylobatids, *Pongo,* and *Gorilla.* As CpG-based estimates become more available, these will provide better estimates of separation times.

Assuming constant yearly mutation rates, the interspecific genetic "distances" will yield divergence times that are proportional to geological time. We are more interested in separation or speciation times, and these require estimates of the contribution of ancestral population diversity to determine its contribution to total diversity, in turn necessary for estimates of separation time. This is important because, for example, the separation time is only 57 percent of divergence time for *Pan-Homo,* while for *Pongo-Homo* it is 69 percent, and for *Macaca-Homo* 82 percent (Mailund et al. 2014).

Separation time estimates for the speciation events of most immediate interest to us, *Pan-Homo* and *Gorilla-Pan/Homo,* vary depending on parameter values and mutational classes used, ranging between 5.6 and 7.9 Ma (Prado-Martinez et al. 2013; Mailund et al. 2014; Moorjani et al. 2016a). We consider the Moorjani et al. (2016a) and Amster and Sella (2016) estimates for *Homo-Pan* (7.9 Ma, range 6.5 to 9.3) and *Homo-Gorilla* (10.8 Ma, range 9.4 to 12.2) speciation the best currently available point estimates, because they only use CpG mutations, those least subject to generation interval effects.

Estimating divergence times for *Pongo* and *Hylobates* lineages is more difficult for a variety of reasons. Using a single pedigree mutation rate yields estimates of separation times that most would consider to be far too old (e.g., Scally et al. 2012; Prado-Martinez et al. 2013; Amster and Sella 2016). Given that mutation rates have clearly evolved (Moorjani et al. [2016a] calculate radically different rates between hominoids, cercopithecoids, and ceboids), one option (Scally et al. 2012) is to adjust earlier rates to "fit" the fossil record. However, that would depend on the availability of a reliable record, which we doubt. As previously mentioned, among crown hominoid lineages, estimated fixation rates have varied at most 11 percent, suggesting that any adjustment for earlier mutation rates might be modest. Current best estimates, poor though they may be, place the separation of hylobatid and pongine lineages somewhere in the earlier Miocene, between ca. 23 and 16 Ma, with hominines radiating in the Late Miocene.

With these estimates in mind, we next examine the Miocene fossil record in terms of both the Late Oligocene/Early Miocene record for the diversification of hominoids and hominids (*Pongo* and *Hylobates* lineages) and the Late Miocene fossil record for evidence of hominine diversification (*Pan-Homo* and *Gorilla* lineages).

The Relevance of Miocene Hominoids to Reconstructing the LCA

It is reasonable to assume that all crown hominoid lineages originally diversified in Africa, in the earlier Miocene (23–14 Ma) for *Pongo* and *Hylobates*, and for gorillas, chimpanzees, humans, and their respective common ancestors in the African later Miocene (12–5 Ma). There are several reasons for caution when we use current evidence about Miocene apes to help us understand the evolution of crown hominoids, including reconstructing the LCA. Barely half of known Miocene ape species (summarized in Figure 2.4) have even a modest fossil record, geographical and temporal sampling is patchy (see below), and evaluating the relative likelihoods of homology is problematic. Consequently, there is little consensus on many aspects of Miocene ape phylogeny.

In the following discussion, we use "crown hominoid" to mean all taxa (e.g., species and higher taxa) descended from the LCA of the living members of the clade (we similarly define "crown cercopithecoids" or "crown catarrhines"). "Stem hominoids" refers to all taxa other than crown hominoids that are more closely related to crown hominoids than to living members of other groups (e.g., stem or crown cercopithecoids). Competent scholars disagree over which Miocene apes are hominoids (e.g., see Begun et al. 1997; Harrison 2010b; McNulty 2010; Alba et al. 2015), so we therefore use "hominoids" to refer

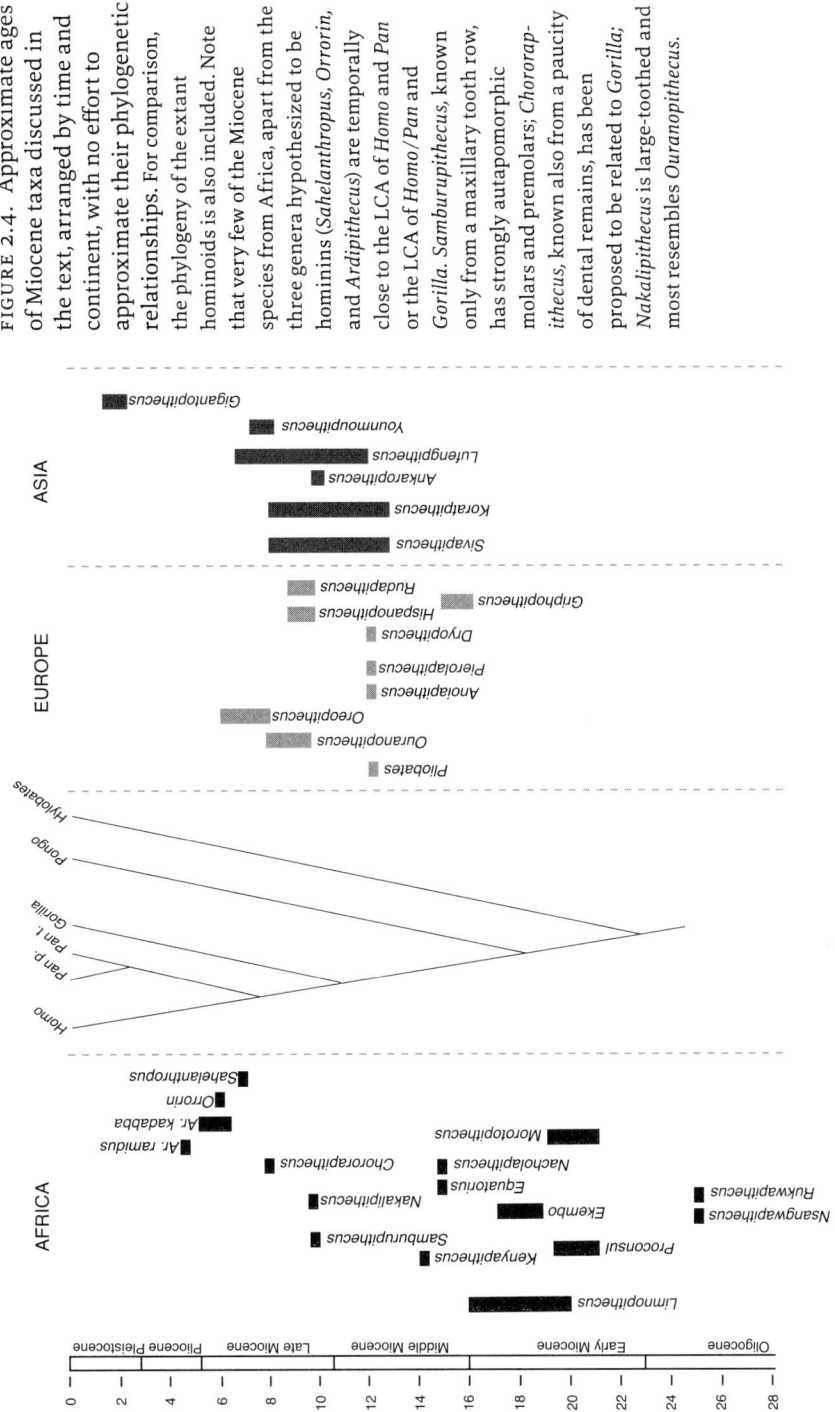

FIGURE 2.4. Approximate ages of Miocene taxa discussed in the text, arranged by time and continent, with no effort to approximate their phylogenetic relationships. For comparison, the phylogeny of the extant hominoids is also included. Note that very few of the Miocene species from Africa, apart from the three genera hypothesized to be hominins (*Sahelanthropus*, *Orrorin*, and *Ardipithecus*) are temporally close to the LCA of *Homo* and *Pan* or the LCA of *Homo/Pan* and *Gorilla*. *Samburupithecus*, known only from a maxillary tooth row, has strongly autapomorphic molars and premolars; *Chororapithecus*, known also from a paucity of dental remains, has been proposed to be related to *Gorilla*; *Nakalipithecus* is large-toothed and most resembles *Ouranopithecus*.

to the clade comprising all stem and crown hominoids, and "apes" to refer to taxa that might be either hominoids or stem catarrhines, with no agreement on stem- or crown-group membership.

The Miocene Epoch (see Figure 2.4) is formally subdivided into three subequal intervals (Early Miocene, 23–16 Ma; Middle Miocene, 16–11.6 Ma; Late Miocene, 11.6–5.3 Ma), but more informally we discuss the earlier Miocene prior to 14 Ma separately from the later Miocene (from 14 to 5.3 Ma). During the first half of the Miocene, global climates were warm, with less equatorial to polar contrast than is seen later (Zachos et al. 2001; Pound et al. 2012). A range of habitats from tropical forest to open grassland was present in Africa, with the proportion of forested habitats greater than in later times (Jacobs 2004; Couvreur et al. 2008; Bonnefille 2010; Feakins et al. 2013; Linder 2014). The 17 to 15 Ma interval was the warmest of the past 30 million years (Flower and Kennett 1994: Zachos et al. 2001), and since then global climates have become progressively and episodically cooler. Global cooling around 14 Ma (Zachos et al. 2001) resulted in more open habitats, but forests remained a significant component of global systems. Tectonic activity, particularly rifting and uplift, increased during the Middle Miocene, causing important paleogeographic and paleoclimatic changes (Popov et al. 2004; Zhang et al. 2014).

The Earlier Miocene African Record and Possible Hylobatid and Pongine Ancestors

Wherever their original source, catarrhines differentiated in Afro-Arabia beginning in the Eocene. By 19 Ma, cercopithecoids are clearly recorded in the Early Miocene African fossil record (Miller 1999; Miller et al. 2009). It has been suggested that the two catarrhine superfamilies might have become distinct as early as the late Oligocene (>24 Ma), but Harrison (2010a, 2010b) and Stevens et al. (2013) interpret *Kamoyapithecus hamiltoni* (Leakey et al. 1995), a possible hominoid, as a stem catarrhine. Others have suggested that by ~25 Ma there is evidence of a putative cercopithecoid *Nsungwepithecus gunnelli* and a putative hominoid *Rukwapithecus fleaglei* (Stevens et al. 2013), but both interpretations are questionable in our view, and further data are necessary to draw firmer phylogenetic conclusions.

Throughout the Early and Middle Miocene, African apes were far more diverse than cercopithecoids, with up to twenty-five plausible ape genera so far recorded (e.g., Harrison 2010b; McNulty 2010), and, at most, only seven cercopithecoid genera (Miller et al. 2009). Beginning in the Late Miocene, ape diversity declined, eventually reaching its current level of five living hominoid genera (or eight, depending on the status of the four hylobatid clades, now most frequently referred to as genera) by sometime during the Pleistocene, while cercopithecoids become steadily more taxically diverse. (Similar reciprocal patterns are seen in other groups, for example, perissodactyls versus artiodactyls, and bovids versus tragulids.) These relative shifts in diversity can be linked to global, regional, and local changes in the balance of forest to nonforest habitats, and to the varying dependence of crown hominoids on the availability of ripe fruit.

Of the twenty-five genera described from the first two-thirds of the African Miocene (McNulty 2010), twenty-one are from a restricted area in Kenya (mostly) and Uganda, the remainder coming from South Africa (Senut et al. 1997), Namibia (Conroy et al. 1992), and Saudi Arabia (Andrews et al. 1978). A fuller sampling of Africa would probably reveal greater diversity, perhaps considerably greater.

Miocene ape body sizes range from about 3 kg (smaller than gibbons) to 80 kg (the size of female gorillas). Relative tooth size (for taxa whose body size is known), as well as tooth morphology, varies considerably. A majority of species for which data are reliably reported have relatively thick enamel, although thinner enamel is also found (Andrews et al. 1991; Smith et al. 2003; Alba et al. 2012). Molar crown formation times vary, although, as discussed earlier, for many larger species formation times fall within the range of living great apes (Kelley and Smith 2003; Mahoney et al. 2007). Undistorted complete crania are not known (for example, Wu et al. 1983; Leakey et al. 1988a, 1988b; Ogihara et al. 2006; Kelley and Gao 2012; Kullmer et al. 2013), but enough is preserved or can be reconstructed for some taxa to demonstrate that their morphology is distinct from that of crown hominoids (Rae 1997; Alba et al. 2015).

Postcranial remains are rare, although four African genera are well to reasonably well sampled: *Ekembo* (previously *Proconsul*) *nyanzae* and *E. heseloni* (~17 Ma; McNulty et al. 2015; Ward 1993; DeSilva et al. 2010; Walker 1997), *Nacholapithecus kerioi* (~15 Ma, Nakatsukasa and Kunimatsu 2009), *Equatorius africanus* (~15 Ma, Ward et al. 1999), and *Morotopithecus bishopi* (~21 Ma, Gebo et al. 1997; MacLatchy 2004; Sanders and Bodenbender 1994; Nakatsukasa 2008), also referred to as *Afropithecus turkanensis* and dated more recently at 17–18 Ma by Harrison (2010b). The first three are best described as arboreal pronograde quadrupeds, albeit with different limb proportions and vertebral morphology. Relative to *Ekembo, Nacholapithecus* has postcranial features (vertebrae, limb proportions) that suggest a somewhat different positional repertoire than *Ekembo. Morotopithecus* shows adaptations in the lumbar spine for greater dorsostability than is seen in other currently known Early Miocene apes, its glenoid shape reflects a shoulder joint adapted for a range of motion including some suspension, and its femur shows adaptations for cautious pronograde behaviors similar to those of other Miocene genera (but see Senut et al. 2000). A distal tibia from Napak in Uganda (~19 to 20 Ma, Rafferty et al. 1995) is equivalent in size to a female gorilla or a male orangutan, and its morphology implies ape-like orthograde vertical climbing behaviors (DeSilva 2008; DeSilva et al. 2010). A talus from Songhor, Kenya, similar in age to Napak, suggests similar behaviors (DeSilva 2008). Rafferty et al. (1995), DeSilva (2008), and McNulty et al. (2015) refer these specimens to *Proconsul major*, while Senut et al. (2000) use *Ugandapithecus major*.

The brief summary above by no means does justice to the morphological variability (mostly postcranial) in the earlier Miocene, sampled almost entirely from eastern Africa, but provides a background for discussing phylogenetic relationships. What is the evidence that this material samples stem or even crown hominoids rather than stem catarrhines? What evidence is there for separation of hylobatids from hominids, and hence can crown hominids be reliably identified? And is there any information to be gleaned from this earlier Miocene record about the *Pan-Homo* LCA?

There is little consensus among scholars concerning the systematics of these Miocene catarrhine non-cercopithecoids. Perhaps the best represented cranially, dentally, and postcranially of the relevant Early and early Middle Miocene genera are *Ekembo nyanzae, E. heseloni* (Walker 1997; Harrison 2010b; McNulty et al. 2015), and *Nacholapithecus kerioi* (Nakatsukasa and Kunimatsu 2009). In their excellent recent discussion of *Proconsul* systematics, McNulty et al. (2015) include *Proconsul* and *Ekembo* in Hominoidea, Family *incertae sedis* in the formal diagnoses, but refer to them more neutrally in the discussion as crown catarrhines. Ward et al. (1991), Ward (1993), and Walker (1997) described a postcranium for *Ekembo* exhibiting a few features indicating either stem hominoid or stem hominid relationships. The hominoid-like morphologies of *Ekembo heseloni* and *Nacholapithecus kerioi* sacral and caudal vertebrae (Nakatsukasa 2004; Nakatsukasa et al. 2007) imply that they lacked external tails (but see Russo and Shapiro 2013). Despite a very small number of putative synapomorphies, they might be considered stem hominoids. *Morotopithecus bishopi* (Sanders and Bodenbender 1994; Gebo et al. 1997; MacLatchy 2004; Nakatsukasa 2008) shows greater similarity in the lumbar region and scapula to crown hominoids, and MacLatchy (2004) considered it a hominoid. In contrast, Harrison (2010b) and Susanna et al. (2014) suggest that the vertebral similarities are homoplasies.

Begun et al. (1997) performed a cladistic analysis involving 240 hard tissue characters. The most parsimonious cladogram placed "Proconsul" (including *Ekembo*), *Afropithecus,* and *Kenyapithecus* (which would include species now assigned to *Equatorius*) as stem hominoids. A more recent study using 191 characters and including more taxa (Stevens et al. 2013) largely concurs as far as Early Miocene taxa are concerned, as does the analysis of Alba et al. (2015) using 319 characters. As is invariably the case with such cladistic analyses, none of the analyzed fossil taxa possess all or even a significant majority of the characters. Particular placements should therefore be questioned on methodological grounds if no other, and it is worth noting that all three analyses recover a sister-taxon relationship of *Pan* and *Gorilla*.

Harrison (2010b: 444) synonymized *Morotopithecus* with *Afropithecus* and considered it as "merely a large orthograde proconsulid that developed its own unique adaptations in the vertebral column in parallel with those of extant hominoids (see also Nakatsukasa, 2008)." He continued to favor stem catarrhine status for almost all the Early and early Middle Miocene East African genera, implying therefore that the hominoid sacral-caudal similarities of *Ekembo* and *Nacholapithecus* were also homoplasies. He interpreted ~14 Ma *Kenyapithecus* (s.s. not including ~15–16 Ma *Equatorius*) as a probable stem hominid (less probably stem hominoid), making it the oldest African hominid, necessarily implying that hylobatids were already separate. Although the small-toothed species of *Limnopithecus, Dendropithecus,* and *Micropithecus* have sometimes been interpreted as early hylobatids, the current consensus is that they are stem catarrhines (Harrison 1981, 2010b; Stevens et al. 2013) or stem hominoids (Alba et al. 2015). Many discussions of hylobatid evolution assume that gibbons have always been small; Reichard and Barelli (2008) discuss alternative reasons to support larger size in stem hylobatids (see also Groves 1972).

In summary, there is no consensus on either the taxonomy or the phylogenetic relationships of these Early and early Middle Miocene East African catarrhines. We regard the evidence as currently insufficient to definitively support the presence of either hominids (pongines, hominines) or hylobatids in the available samples. However, the similarity to extant apes in some features of the Moroto vertebrae, Napak tibia, and Songhor talus, in contrast to comparable specimens from the younger Rusinga and Mfangano deposits, suggests that crown hominoids may have begun to diversify by around 20 Ma.

The African Later Miocene Record and the Diversification of Hominines

If our target times are reasonable for lineage separation of 9–12 million years for gorillas and 6–9 million years for chimpanzees and humans, then the most relevant record for reconstructing the LCA is the African later Miocene. This period is mostly dominated by fossils from Kenya and Ethiopia, with one highly suspect specimen of uncertain age from Niger. Despite the argument of Begun (2009) and Begun et al. (2012) that the hominines have a European rather than African origin, we agree with Cote (2004), Suwa et al. (2007), Kunimatsu et al. (2007), Pickford et al. (2008), and Harrison (2010b), that an African origin is more probable given current paleontological evidence.

A small number of Late Miocene African hominid species have been described in the last two decades, including *Samburupithecus kiptalami* from the Namurungule Formation in Kenya, dated at ~9.5 Ma (Ishida and Pickford 1997; Sawada et al. 1998); *Nakalipithecus nakayamai* (Kunimatsu et al. 2007) from Nakali in Kenya, dated at just under 10 Ma; and *Chororapithecus abyssinicus,* originally dated at 10 to 10.5 Ma from the Chorora Formation in Ethiopia (Suwa et al. 2007) but redated to ~8 Ma (Suwa et al. 2015). All are large-toothed but are in our view indeterminate as to relationships. *Samburupithecus kiptalami* (Ishida and Pickford 1997) is known only from a maxillary tooth row, with molar and premolar morphology being markedly distinct. Although Harrison (2010b: 460) accepts *Samburupithecus* as a tentative hominine, he raises the possibility that it is a "late surviving proconsulid," as does Begun (2010). Suwa et al. (2007) believe *Chororapithecus abyssinicus* to be related to but not directly ancestral to *Gorilla,* having somewhat thicker enamel and less developed shearing crests than the living ape. However, we agree with Harrison (2010b: 461) that, "given the paucity of the material and the high probability of functional convergence, it would seem that such a relationship cannot be currently convincingly demonstrated." Harrison (2010b) also mentions a possible close phylogenetic relationship with *Samburupithecus*. *Nakalipithecus nakayamai* (Kunimatsu et al. 2007) is also large toothed, and both Kunimatsu et al. (2007) and Harrison (2010b) raise the possibility of a phylogenetic relationship with *Ouranopithecus* (Koufos and de Bonis 2006), a Late Miocene Euro-Asian probable hominid with large and thickly enameled cheek teeth known from Greece, Bulgaria, and Turkey. More precise relationships of the two genera cannot be currently determined.

Isolated molars from the Ngorora Formation at Baringo in Kenya (Hill and Ward 1988; Pickford and Senut 2005), dated around 12.5 Ma, are hominoid, perhaps hominid, but indeterminate as to other relationships. A first or second upper molar matches those of female gorillas for size, and Hill and Ward (1988) thought the morphology similar to modern chimpanzee second molars. Kunimatsu et al. (2007) noted similarity in cusp configuration to molars of modern gorillas, but with lower cusps and more open occlusal foveae than in *Nakalipithecus,* and reportedly thinner enamel than in *N. nakayamai*. Relationships of the chimpanzee-sized lower molar also from the Ngorora Formation (Pickford and Senut 2005) are, as noted, indeterminate. Pickford and Senut (2005) describe three teeth or tooth fragments from the Lukeino Formation, Baringo Basin, Kenya, dated to 5.9 Ma, suggesting relationships to the gorilla lineage in the case of a fragmentary molar and heavily worn incisor. We believe the relationships to be indeterminate. A smaller lower molar represents a taxon clearly different from the hominin *Orrorin* recovered from the same formation. Finally, Pickford et al. (2008) report a possible hominoid mandibular fragment from Niger, dated based on a small nonprimate faunal sample to between 11 and 8 Ma. They describe it as a "best match" for *Pan,* but it is too fragmentary for any firm taxonomic or phylogenetic attributions.

Except for the dubious Niger fragment, these Late Miocene hominoids are from Kenya and Ethiopia only, and probably from habitats that ranged from predominantly forested to more mosaic, with some open habitats close to forests (Hill et al. 2002; Jacobs 2004; Kunimatsu et al. 2007; Bonnefille 2010; Uno et al. 2011; Feakins et al. 2013). Habitat heterogeneity is characteristic of much of Africa throughout the Miocene, although tropical rainforests were always abundant spatially and temporally.

There is a more modest level of ape diversity in the Late Miocene of Africa than in the earlier Miocene. If we add what we consider to be the presently unsampled gorilla lineage and that of the chimpanzee-human LCA, which may well have inhabited central African tropical rainforest, we count perhaps eight Late Miocene genera, a number that has increased in the past decade and is likely to be augmented with ongoing fieldwork. There being no plausible fossil evidence for the LCA or the gorilla lineage, nothing in the current record argues for or against our hypothesized separation estimate of 6–12 Ma for hominine diversification.

The Eurasian Late Miocene

We assume that hylobatid and orangutan lineages differentiated initially in Africa, at some stage dispersed out of (and disappeared from) Africa, and radiated in Asia. We should not assume that they were the lone precercopithecoid catarrhine dispersals prior to the latest Miocene, nor that dispersals were synchronous with each other or with those of other mammal groups. Dispersals have often been linked to a few hypothesized Miocene terrestrial links between Afro-Arabia and Europe and Asia, but the scale of paleogeographical reconstructions, even those of Popov et al. (2004), is too coarse to capture every possible connection. The fossil record is also too sparse in both time and space to possibly be a comprehensive record of dispersal timing and frequency.

Europe

The oldest European catarrhine, a partial thickly enameled probable ape tooth from Engelswies in southern Germany that may document the oldest hominoid in Europe, has been assigned to *Griphopithecus* (Heizmann and Begun 2001), although this assignment should be tentative given the minimal material. It is variously dated at 17.0 to 17.1 Ma by Böhme et al. (2011), 17.0 to 16.5 Ma by Heizmann and Begun (2001), or 16.5 to 16.2 Ma by Casanovas-Vilar et al. (2011, and personal communication). Turkey is now geographically Asian, but in the Miocene it was geologically and faunally part of southeastern Europe. From Pasalar in Turkey, *Griphopithecus* and *Kenyapithecus* (Kelley et al. 2008) are variously dated at 16.5 Ma (Heizmann and Begun 2001), 14.9 to 13.7 Ma (Casanovas-Vilar et al. 2011), or 13.8 to 13.6 Ma (Böhme et al. 2011). The former genus is also present at the slightly younger Turkish site of Çandir (Güleç and Begun 2003). There is agreement that the Pasalar specimens are hominoids, with a consensus that they are probably stem hominids

(Güleç and Begun 2003). Dentally, they resemble *Equatorius, Nacholapithecus,* and *Kenyapithecus* from sites in East Africa (Nachola, Baringo, Maboko, Fort Ternan) dated between ~16 and <14 Ma (Harrison 2010b).

The European hominoid record does not pick up again until ~13 Ma (Casanovas-Vilar et al. 2011), during a brief interval in which parts of Western Europe were forested with the kinds of fruit-bearing trees preferred by living hominoids. Whether the ~3 Ma "gap" after Engelsweis is real or reflective of an inadequate fossil record is unclear, but between ~13 and ~7 Ma the European record is good, particularly in Spain, France, Germany, Hungary, Greece, and Italy. It is unlikely that all proposed genera are valid, and only *Dryopithecus* (Mòya-Solà et al. 2009b), *Pierolapithecus* (Mòya-Solà et al. 2004), *Anoiapithecus* (Mòya-Solà et al. 2009a), *Pliobates* (Alba et al. 2015), *Hispanopithecus* (Mòya-Solà and Köhler 1996), *Rudapithecus* (Kordos and Begun 2002), *Ouranopithecus* (Koufos and de Bonis 2006), and *Oreopithecus* (Harrison 1986) are represented by more than teeth. All have faces that differ in completeness and preservation quality, and more reconstruction is necessary before firm conclusions can be drawn about similarities and differences. In the case of *Pierolapithecus, Pliobates, Hispanopithecus, Rudapithecus,* and *Oreopithecus,* the postcranium also differs in regional completeness and preservation quality (Harrison 1986; Kordos and Begun 2002; Alba et al. 2015), but all show to different degrees greater similarities to crown hominoids than is the case for most earlier African hominoids (or perhaps for the Pasalar *Griphopithecus* (Ersoy et al. 2008), although the postcranial sampling of the latter is minimal). Both *Pierolapithecus* (Mòya-Solà et al. 2004) and *Hispanopithecus* (Susanna et al. 2014) have features of the lumbar vertebrae, including more dorsal placement of transverse processes, that resemble those of crown hominoids and indicate adaptations for dorsostability (Alba et al. 2015). Reasonable estimates of ulnar and femoral lengths of *Hispanopithecus* (Mòya-Solà and Köhler 1996), suggest that its interlimb proportions fall between those of *Pan* and *Pongo*. The hand of *Hispanopithecus* has long and curved phalanges, consistent with suspensory behaviors, but its short metacarpals relative to phalangeal length differ from crown hominoids, and the distal metacarpal-proximal phalangeal joints, while resembling those of other Miocene hominoids, differ from those of all crown hominoids (Susman 1979; Almécija et al. 2007).

A case has been made that some or all of these later European taxa are stem hominoids, stem hominids, stem hominines, or stem pongines

(Mòya-Solà and Köhler 1996; Kordos and Begun 2002; Alba et al. 2015). Both formal and informal cladistic analyses are contradictory and ambiguous, in our view largely because the less than complete hard-tissue material representing fossils is difficult to atomize into characters and character states. We opine that most are probably stem hominoids and/or stem hominids because we see no plausible synapomorphies with either hominines or pongines. Of great interest is *Oreopithecus,* an intriguing Late Miocene ape from Italy and Sardinia (Harrison 1986; Sarmiento 1987). Postcranial resemblances are to highly suspensory arboreal crown hominoids, but cranially and dentally *Oreopithecus* is strongly autapomorphic and resembles no living or extinct ape.

In Turkey, after Pasalar and the slightly younger Çandir locality, there is a substantial temporal gap of almost 4 Ma before the next youngest hominoid, *Ankarapithecus* from Sinap, dated at just less than 10 Ma. *Ankarapithecus* is represented by two relatively complete facial specimens (Alpagut et al. 1996; Begun and Güleç 1998; Kappelman et al. 2003). It resembles several European Miocene taxa, such as *Dryopithecus, Hispanopithecus, Rudapithecus,* and *Ouranopithecus,* as well as both African apes and *Pongo.* There are disagreements as to relationships: Alpagut et al. (1996) and Kappelman et al. (2003) regard it as a stem hominid, Begun and Güleç (1998) as pongine. Kappelman et al. (2003) note the absence of shared cranial similarities with orangutans, and the sparse postcranial remains, which differ from contemporaries in Europe, resemble earlier Miocene hominoids in suggesting a more pronograde posture.

Asia

Indeterminate small catarrhines found in South Asia (Pakistan) at 16.8 Ma (Barry et al. 1986; Bernor et al. 1988; Harrison 2005) have been identified as pliopithecoids (a stem catarrhine superfamily) or hominoids, and Harrison (2005) identifies ~17 to 18 Ma catarrhines in East Asia (China) as pliopithecoids.

Younger South Asian, Southeast Asian, and East Asian hominoids are often collectively linked phylogenetically as pongines, varying in their inferred closeness to *Pongo* (Harrison 2010a), but there are several necessary caveats to those interpretations. *Sivapithecus* species are known from Pakistan, India, and Nepal, dated from just under 13 to around 8 Ma (see Kelley 2005; DeSilva et al. 2010; Morgan et al. 2015). A partial face of the youngest species, *Sivapithecus sivalensis* from the Potwar Plateau in Pakistan (Pilbeam 1982), now dated at around 9.3 Ma, shows a number of close facial and palatal resemblances to *Pongo* that have been widely interpreted as synapomorphies, yet there are no *Sivapithecus sivalensis* dental or mandibular similarities to the orangutan (Brown 1997; Kelley 2005). For example, *Pongo* lacks a symphyseal attachment area for an anterior digastric muscle, while it is present in *Sivapithecus* (Brown 1997), as well as in *Ankarapithecus* (Kappelman et al. 2003). The relationships of an older, damaged palatal specimen of *Sivapithecus indicus* (12.3 Ma) are disputed, with some advocating similarities to *Pongo* (Raza et al. 1983; Morgan et al. 2015) and others to *Ankarapithecus* (Begun and Güleç 1998), depending on whether its damaged palate resembles the orangutan pattern. Unassociated postcranial specimens from Pakistan assigned to *Sivapithecus* reflect pronograde behaviors and show minimal similarity to any crown hominoids, least of all to *Pongo* (Rose 1983, 1984, 1986; Pilbeam et al. 1990; Madar et al. 2002; DeSilva et al. 2010; Morgan et al. 2015). One exception is the distal humerus, which shows resemblances to all hominids (Pilbeam et al. 1990). In some features, such as the proximal manual phalangeal joint, *Sivapithecus* differs from all crown hominoids and resembles other Miocene taxa (Rose 1986). The older *Sivapithecus* species, *S. indicus*, like the younger *S. sivalensis,* is somewhat smaller than living hominids, with males being the size of female *Pan* and *Pongo* (~40 kg) and females the size of male *Papio* (~25 kg) (DeSilva et al. 2010; Morgan et al. 2015). The postcranially similar ~10 Ma *Sivapithecus parvada* is orangutan-like in size, with putative males ~70–80 kg, females ~30–40 kg (Kelley 1988; Morgan et al. 2015).

Further to the east and southeast in Myanmar and Thailand, dental and mandibular remains, ranging in age from ~13 to ~8 Ma, have been assigned to three species of *Khoratpithecus* (Chamainee et al. 2003, 2004; Jaeger et al. 2011). Mandibles of the Late Miocene *Khoratpithecus ayeyarwadyensis* (Myanmar) and *K. piriyai* (Thailand) lack a digastric fossa, thus providing a putative synapomorphy with *Pongo* (Chamainee et al. 2004; Jaeger et al. 2011). The Miocene faunas of Pakistan, India, Nepal, Myanmar, and Thailand share several large mammal species and genera, and differ from those of equivalent age in China (Chavasseau et al. 2006; Chavasseau 2008). The Chinese hominoids ranging in age from ~12 to ~6 Ma (Li et al. 2015) have been assigned to *Lufengpithecus*. Dentally, *Lufengpithecus* resembles the orangutan in some aspects of molar occlusal morphology (Kelley and Gao 2012). Schwartz (1997), Harrison (2010a), and Li et al. (2015) group *Sivapithecus, Khoratpithecus,* and

Lufengpithecus together as pongines, but Kelley and Gao (2012) show that the facial morphology of *Lufengpithecus* clearly differs from *Pongo*. Like *Ankarapithecus* and the European later Miocene apes, *Lufengpithecus* is probably best described as a stem hominid.

These South, Southeast, and East Asian hominoids present interesting challenges for inferring phylogenetic relationships: *Lufengpithecus* has orangutan-like teeth but there are no skeletal similarities. *Sivapithecus sivalensis* has orangutan-like facial-palatal similarities in Late Miocene, but there are no mandibular or postcranial similarities. *Khoratpithecus* has orangutan-like inferior symphyseal morphology, but no dental similarities. More complete and associated material may resolve these apparent paradoxes, and recovering material that is 5 to 7 Ma in areas where ancestral orangutans are likely to have lived (Indonesia?) will be essential.

Finally, likely hylobatid teeth from Yuanmou (Harrison 2010a), dated at ~7 to ~8 Ma (Li et al. 2015), fall within the size range of extant hylobatids, and could document one of the crown lineages or the hylobatid common ancestor or a close relative. The crown radiation probably began earlier than this during the Late Miocene (Carbone et al. 2014, adjusted using plausible mutation rate and generation interval).

Miocene Summary

This far from comprehensive review of the Miocene apes raises several important issues. First, we should not underestimate the limitations of the existing fossil samples. There are no undistorted and complete crania, few crania are associated with postcrania, and very few have even moderately complete postcrania. Taxonomic and phylogenetic discussions, which are dominated by dental evidence, are often conducted on the unstated assumption that cranio-dental morphology is less susceptible to homoplasy than postcranial, even though several studies prove this assumption is unwarranted (Sanchez-Villagra and Williams 1998; Young 2005; Williams 2007; Cardini and Elton 2008). Indeed, Williams (2007) concluded that postcrania may be slightly less prone to homoplasy than crania and teeth.

Fossil sampling for all mammals or other vertebrates, not just primates, in the Miocene is inadequate both temporally and geographically. For example, Africa west of the Rift Valley between the Tropics of Cancer and Capricorn is unsampled, northeastern Africa is quite scarcely sampled, and the Arabian Peninsula is hardly represented. Within Asia there is no record from Iran except for the far northwestern corner, nor for peninsular India or much of Southeast Asia, including Indonesia.

As noted above, discussions about taxonomy and phyogenetics based on the ape fossil record are heavily dependent on tooth morphology, including enamel thickness, which presents another cause for caution. First, it is unclear whether any morphological similarities to crown hominoids can be confirmed as homologies. Second, we are unaware of any evidence that tooth morphology can be used to reconstruct ape phylogenetic relationships, particularly at the lower taxonomic levels necessary here. We especially note the failure of Hartman (1987, 1988) to recover the correct relationships of living hominoids using molar occlusal morphology. It may be, as Smith (2008) suggested, that future research on detailed aspects of tooth development, particularly as revealed by phase contrast X-ray synchrotron microtomography (Smith et al. 2014), will better inform us about phylogenetic relationships; these would need to be tested using known relationships among living taxa. Third, enamel thickness is often discussed as though it had significant taxonomic or phylogenetic value, thick molar enamel being frequently cited as plesiomorphic for hominins because of its presence in a plurality of Miocene apes. For example, Alba et al. (2010) support this position by mapping enamel thickness onto a phylogenetic tree of Miocene apes, although the tree is but one of several possible trees. However, Schwartz (2000) showed that enamel thickness in living hominids carries no phylogenetic signal, and this is also apparent for other groups. For example, if thickness variation among *Macaca* species (Kato et al. 2013) is mapped onto a known phylogeny of the genus, there is no relationship with phylogeny. We are unaware of any current study showing that enamel thickness provides any reliable phylogenetic signal.

In contrast, enamel thickness does provide useful information about the physical properties of food (preferred, fallback, or both). Alba et al. (2010) and Kato et al. (2013) show significant variation in enamel thickness that can be plausibly linked to diet among, respectively, hylobatid and macaque species. In macaques, enamel thickness may vary with the degree of seasonal variation in physical properties of diet (Kato et al. 2013). Ungar (1996) reviewed tooth wear in a few Miocene European apes and showed that variation in wear patterns, reflecting diets, correlated to an extent with enamel thickness, and Xia et al. (2015) recently reaffirmed the use of tooth microwear to assess physical properties of ingested food. *Hispanopithecus* and *Rudapithecus* teeth with moderately thin enamel (Alba et al. 2011) show wear indicative of soft fruit eating, while wear on thickly enameled teeth of *Ouranopithecus* (Alba et al. 2011) reflects the consumption of harder objects. However, Nelson (2005) showed that *Sivapithecus*, which has thick enamel (Alba et al. 2011), had wear reflecting primary frugivory, with some harder objects. There is no one-to-one association here of enamel thickness and diet as inferred from microwear.

Finally, we note a fondness for *Ekembo* (or *Proconsul*) *nyanzae* in discussions of postcranial evolution in later hominoids, particularly hominins. This is surprising in that other Miocene species, for example, the geologically older material from Moroto, Napak, and Songhor, exhibit more postcranial resemblances to crown hominoids (MacLatchy 2004; DeSilva 2008; DeSilva et al. 2010), which should be considered in any hypotheses concerning hominoid and hominin postcranial evolution.

In his review of the Miocene ape record and its value for reconstructing the LCA, McNulty (2010) concluded (and we quote again, partially because this is perhaps representative of the current consensus among experts):

> a detailed knowledge of ape evolution is so important to human origins research (Andrews and Harrison, 2005). In the absence of a robust fossil record of gorillas and chimpanzees, it is only through a ... bottom-up approach—using the morphological diversity generated through 20 million years of ape evolution—that we can build reasonable hypotheses about the ancestral form which gave rise to the first fossil humans.

We should not forget that until molecular data became available, gorillas and chimpanzees were considered monophyletic and their numerous cranial and postcranial similarities treated as homologies; indeed, they were sometimes classified in the same genus, *Pan* (for example, Simpson 1963; Tuttle 1992). This brief review shows that most of the Miocene ape record has little to offer concerning reconstruction of the LCA (see Figure 2.4). We are skeptical that the Miocene ape record, or that of likely early hominins discussed below, requires reinterpreting *Pan-Gorilla* similarities as largely homoplasies.

We have no definitive record in Africa of hylobatid, pongine, gorillin, or panin lineages, with the exception of a handful of Middle Pleistocene *Pan* teeth (McBrearty and Jablonski 2005) and a possible *Pan* proximal femur (DeSilva et al. 2006). *Pongo* teeth are known from the early Pleistocene of

China (Harrison et al. 2014), and the oldest definite hylobatid remains (teeth) from the Chinese latest Miocene are probably temporally and cladistically close to the origin and diversification of the crown group (Harrison 2016).

In summary, there is no unequivocal evidence for the early evolution of any crown hominoid clade, with the sole exception of the widely accepted relationship of South Asian Late Miocene *Sivapithecus* to *Pongo*. We have noted other eastern and southeastern Asian possible orang relatives *(Khoratpithecus, Lufengpithecus)*. *Sivapithecus* is of interest in that the Late Miocene (~9.3 Ma) palate and face of *Sivapithecus sivalensis* resembles *Pongo,* while teeth, mandibles, and postcrania do not (Pilbeam 1982; Rose 1984, 1986, 1989; Pilbeam et al. 1990; Madar et al. 2002; Harrison 2010b; DeSilva et al. 2010; Morgan et al. 2015). This conflict leaves us with a choice of where to invoke homoplasy, and as we noted earlier—and despite the reviews of Sanchez-Villagra and Williams (1998), Young (2005), and Williams (2007)—homoplasy remains for many colleagues stubbornly located in the postcranium, not least in this case. A *Sivapithecus-Pongo* sister relationship would imply that the postcranial similarities of the crown hominoids, not least in the axial skeleton, cannot all be homologous. But this in turn raises other issues. Should the status of *Morotopithecus, Nacholapithecus, Ekembo, Oreopithecus, Pliobates, Pierolapithecus, Hispanopithecus,* or *Rudapithecus* be questioned because their varying hominoid-like postcranial features might be homoplasies? Some are likely hominoids, others hominids, but none can currently be linked to crown lineages.

However, as far as the LCA is concerned, we should not forget that, regardless of the position of *Sivapithecus* or of the extent of pongine-hominine postcranial homoplasy, a *Pan-Gorilla* sister relationship is still recovered in a range of formal phylogenetic analyses using both cranial-dental and postcranial characters (for example, Begun et al. 1997; Young 2003, 2005; Stevens et al. 2013; Alba et al. 2015). This supports the homology of much of their postcranial similarities (Pilbeam 2004; Williams 2011).

Hominin Fossil Evidence Regarding the LCA—The Skull

Although the fossil record of Miocene apes is of limited utility for making and testing inferences about the nature of the LCA, there is no disagreement that four proposed species of hominins from the late Miocene and early Pliocene are critical sources of evidence: *Sahelanthropus tchadensis, Orrorin tugenensis, Ardipithecus kadabba,* and *Ardipithecus ramidus.* Although these four species are currently assigned to three separate genera, it is possible, perhaps likely, that once they are better sampled and more carefully compared they will be subsumed within a single genus (for which *Ardipithecus* has priority); indeed, Haile-Selassie et al. (2004) have suggested this for the first three species, while Wood and Boyle (2016) assert that species differences have only moderate support.

Orrorin and *Ardipithecus* are from well-dated sediments in East Africa. *Orrorin tugenensis* was recovered from the Lukeino Formation of the Tugen Hills, Kenya, dated to between 6.0 and 6.2 Ma (Senut et al. 2001; Pickford and Senut 2001). *Ardipithecus kadabba* was first described by Haile-Selassie (2001) from the Middle Awash, Ethiopia, and is dated to 5.2–5.8 Ma. Further material dated at 5.6 to 5.8 Ma was recovered from the Middle Awash (Haile-Selassie et al. 2004). Simpson et al. (2015) described more material from the Adu-Asa Formation, Ethiopia, including four specimens dated to ca. 5.4 Ma and a single lower premolar, tentatively identified as *Ar. kadabba,* dated at ca. 6.3 Ma. *Ar. ramidus,* also from the Middle Awash in Ethiopia, is from sites dated to 4.3–4.5 Ma (White et al. 1994, 2009, 2015; Semaw et al. 2005), and is probably represented by the > 4.2 Ma Tabarin mandible from the Chemeron Formation at Baringo, Kenya.

Sahelanthropus tchadensis, which was discovered in lacustrine sediments in central Chad rather than from the volcanically active East African rift system, has been more difficult to date securely. Vignaud et al. (2002) initially reported a biochronological age for the Toros-Menalla sites of between 6 and 7 Ma, noting that the overall best match for the fauna was with the base of the Lothagam Nawata Formation, with a maximum age of 7.4 Ma (McDougall and Feibel 1999). A problem with this correlation, however, is that the faunas of Chad and Kenya are sufficiently different to indicate sampling from distinct faunal provinces (Lihoreau et al. 2006), rendering faunal-based correlations problematic. Lebetard et al. (2008, 2010) used a relatively new ^{10}Be/^{9}Be radiometric dating technique to estimate the site's age between 6.8 and 7.2 Ma, probably closer to 7.2 Ma. Suwa et al. (2015: 20) argued that an age of around 7 Ma is problematic, because the "summary dates are based on weighted means of highly variable individual analyses." In the same paper, Suwa et al. (2015: 21) suggested a biochronological placement between 6.5 and 7.4 Ma, equivalent to the lower levels of the Lothagam Narwata Formation (McDougall and Feibel 1999). We note that if only the small number of samples with low relative uncertainty are used (Lebetard et al. 2010), estimated dates (Lebetard et al. 2008) support an age of between 7.1 and 7.2 Ma.

These late Miocene and early Pliocene taxa raise two related, major questions. First, are they hominins? Second, what do they imply about the LCA of chimpanzees and humans? Both questions need to be addressed within an appropriate cladistic framework. Testing the hypothesis that they are hominins requires assessing whether the features they share with later undisputed hominins from the genus *Australopithecus* are homologies or homoplasies. On the one hand, researchers have argued that *Sahelanthropus, Orrorin,* and *Ardipithecus* are hominins because they share several derived characters with *Australopithecus,* including cranial and postcranial features indicative of bipedalism, reduced canine size, and expanded, thicker-enameled postcanine teeth (e.g., White et al. 1994, 2015; Brunet et al. 2002, 2005; Guy et al. 2005; Zollikofer et al. 2005; Lieberman 2011). On the other hand, other researchers have questioned whether for some or all of these taxa the purported shared derived features might be homoplasies (e.g., Senut et al. 2001; Andrews and Harrison 2005; Wood and Harrison 2011).

An even more contentious issue is what the earliest hominins imply about the nature of the LCA. If these taxa are early hominins, then we would expect their anatomy to be a mosaic of primitive features shared with the LCA, derived features shared with later hominins, and unique derived features. It therefore follows that we can make reasonable inferences about the LCA on the basis of the probably primitive features these hominins share with outgroups such as chimpanzees and gorillas. Simply stated, the features these hominins do not share with australopiths but do share with chimps and gorillas are very likely also to be shared with the LCA, unless these features evolved independently.

As noted above, however, analyses of the features early hominins share with other taxa have yielded conflicting interpretations. On the one hand, some researchers have argued that the apparently primitive features of *Sahelanthropus, Orrorin,* and *Ardipithecus* support the hypothesis of a somewhat chimp-like LCA. On the other hand, several researchers have argued that the anatomy of these fossils is unlike chimpanzees or gorillas, and that the LCA more closely resembled various Miocene hominoids (for review of these, see McNulty 2010). For example, Alméjica et al.'s (2013) analysis of the *Orrorin* femur led to the suggestion that the LCA was a generalized arboreal Miocene ape, largely orthograde (hence short-backed) but not specifically suspensory.

Further, White and colleagues (2015) argued that the anatomy of *Ardipithecus* as well as *Sahelanthropus* and *Orrorin* is "surprisingly" unlike chimps or gorillas, but instead was a "generalized" African ape that was relatively large-bodied, long-backed, pronograde, plantigrade, and a slow, careful climber.

With these concerns in mind, we therefore review what inferences can be made about the nature of the LCA using fossil data from *Sahelanthropus, Orrorin,* and *Ardipithecus* (also termed earliest hominins), in comparison with other taxa including later hominins (principally *Australopithecus*), extant apes, and, when relevant, Miocene apes. The basic question we ask is: What do the features evident in the earliest hominins, when viewed in their appropriate phylogenetic context, imply about the anatomy of the LCA? To repeat the logic articulated previously, these taxa potentially inform our understanding of the LCA in two ways. First, they provide a glimpse of morphological diversity around the time that the human and chimpanzee lineages separated (most likely between 6 and 9 Ma). More importantly, they can be used to *test* the predictions of the parsimony-based inference that the LCA was somewhat chimp-like and chimp-sized. According to this top-down logic, any features that the earliest hominins share with australopiths and not *Pan* or *Gorilla* are derived in hominins, hence less likely to be present in the LCA, and any features the earliest hominins share with chimpanzees and gorillas (after accounting for size) are more likely to be primitive features shared with the LCA. Although what follows is not an explicit cladistic analysis, we ask several key questions about morphological features in the teeth, cranium, upper limb, axial skeleton, and lower limb. First, how do we characterize the presumably primitive features present in *Ardipithecus, Sahelanthropus,* and *Orrorin* that are not shared with australopiths? Second, to what extent are these features shared with chimps, gorillas, or other taxa?

Dentition

Teeth are the best-preserved and most common component of the early hominin fossil record. Everyone who has studied early hominin teeth has recognized their many similarities to the African great apes, especially *Pan,* but several studies have argued that because the dentition of *Pan* and *Gorilla* is relatively specialized for frugivory, the LCA had more generalized teeth (Andrews and Harrison 2005; Schroer and Wood 2013; White et al. 2015). As noted above, from a phylogenetic perspective, the term "specialized" is synonymous with "derived," and the key questions to ask are whether the derived features of *Pan* and *Gorilla* evolved independently, and whether the presumably primitive dental features these taxa do not share with australopiths resemble the teeth of AGAs or other taxa.

Anterior Dentition

Great ape incisors tend to be large relative to the posterior dentition. The upper central incisor is especially broad and spatulate in *Pan* as well as *Pongo*; upper lateral incisors are also spatulate in *Pan* as in hominins, but more caniniform in *Gorilla*, *Pongo,* and Miocene apes. Australopith incisors, especially the upper centrals, are smaller and less spatulate, but the incisors of *Au. anamensis* and *Au. afarensis* are nonetheless intermediate in size between those of the African great apes and later hominins such as *Au. africanus* and *Paranthropus* (White et al. 2006; Ward et al. 2010; Lieberman 2011). Figure 2.5 shows lower incisor dimensions. *Sahelanthropus* has a chimp-like upper central incisor that is wide, spatulate, long-rooted, and within the size range of chimpanzees (Brunet et al. 2002). Details of the single known *Orrorin* upper central incisor are hard to glean from what is published, but its size is at the lower end of the chimp range, and its anatomy is described as relatively chimp-like (Senut et al. 2001). An *Ar. kadabba* I_2 is described as "distinctively narrower than" lateral incisors in *Pan,* and "metrically and morphologically comparable" to later hominins (Haile-Selassie 2001: 178). In contrast, *Ar. ramidus* incisors appear to be more derived, within or below the size range of *Au. afarensis* and *Au. anamensis* (Suwa et al. 2009b: S23).

Canines (see Figure 2.5) have been a major focus of research on early hominins. In general, *Pan* and *Gorilla* have large, sexually dimorphic canines, particularly uppers, that scale with positive allometry relative to body mass (Wood 1979); have a triangular shape with low shoulders and a groove on the mesial aspect; and form a honing complex with the P_3, leaving a distinct wear facet on the distal margin. Australopith upper canines are derived in several key respects: they are much smaller and less projecting; they have higher, flared shoulders, giving them a "diamond-shaped" profile; they lack a shearing facet with the P_3; they are more spatulate in shape; and they typically have a rounded basal tubercle that forms a cingulum. Of the early hominin canines so far identified, an upper canine attributed to *Orrorin* is at the lower end of the chimp size range, and generally primitive in terms of shape, with a triangular outline, a mesial groove, and low shoulders (Senut et al.

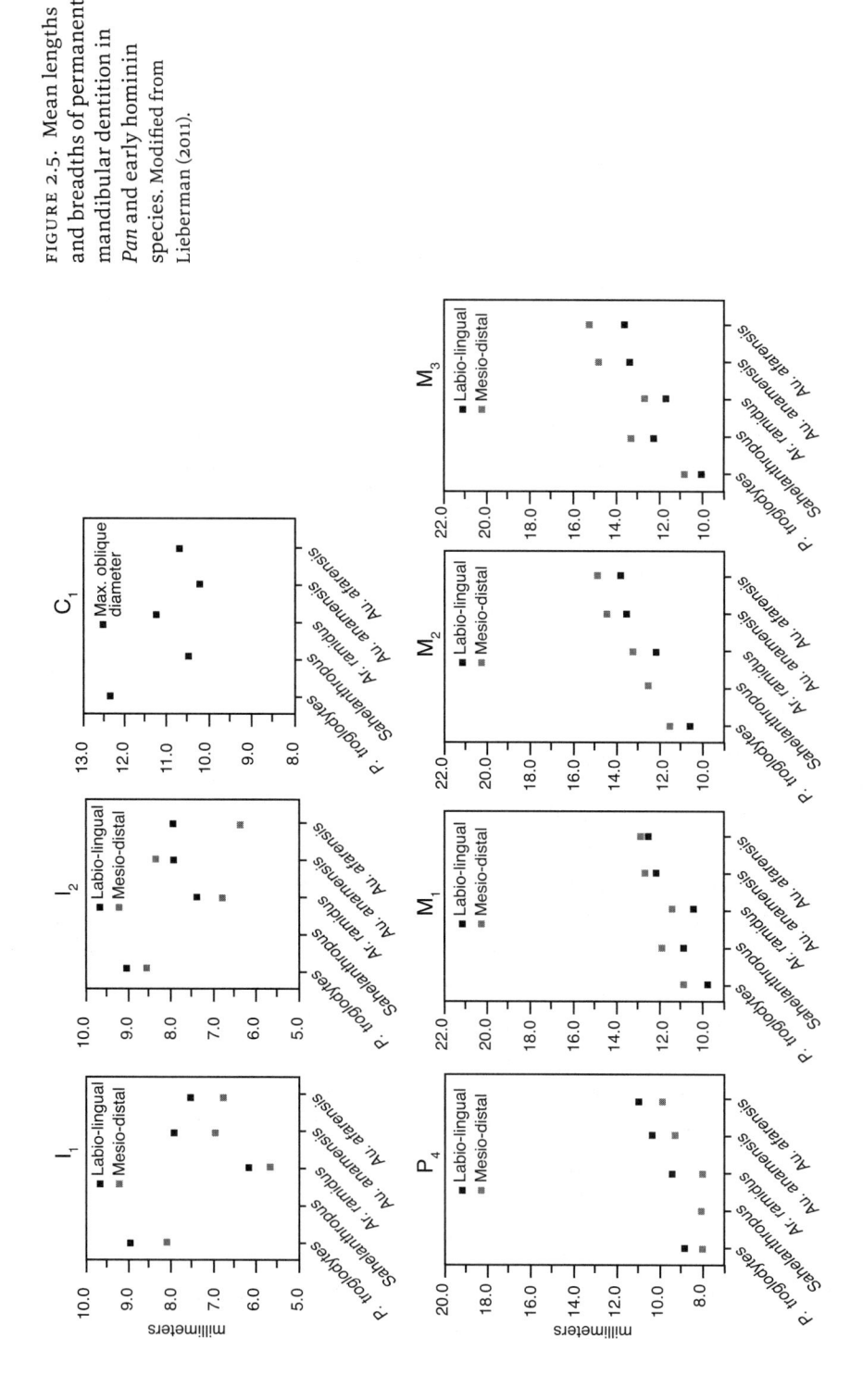

FIGURE 2.5. Mean lengths and breadths of permanent mandibular dentition in *Pan* and early hominin species. Modified from Lieberman (2011).

2001). The upper canines of *Sahelanthropus* and *Ar. kaddabba* have low shoulders but are more diamond-shaped than those of female *Pan* and *Gorilla*, and lack evidence for a C-P$_3$ honing complex (Brunet et al. 2002; Haile-Selassie et al. 2004); lower canines attributed to *Sahelanthropus, Ar. kadabba* and *Ar. ramidus* somewhat resemble female chimps in size and triangular shape, with primitive features such as a basal tubercle (Brunet et al. 2005; Haile-Selassie et al. 2004; Suwa et al. 2009b). In contrast, upper canines attributed to *Ar. ramidus* are decidedly more like those of *Australopithecus,* with high flaring shoulders that yield a diamond-shaped profile, and they are also less dimorphic than those of African apes (Suwa et al. 2009b; Plavcan et al. 2010). Like *Pan* and *Gorilla, Sahelanthropus* has robust canine roots, but like australopiths these roots are only slightly curved (Emonet et al. 2013).

Considering the anterior dentition as a whole, numerous features, especially in the canines, support the hypothesis that the earliest proposed hominins, especially *Ardipithecus,* are indeed hominins. Apart from these shared derived features, there is little evidence to contradict the parsimony-based inference that the incisors and canines of the LCA were generally chimp-like with large spatulate upper central incisors, triangular upper canines approximately the size of female chimps that honed against the lower premolars, and robust tooth roots (see Figure 2.5). All the proposed earliest hominins share some clear derived features shared with australopiths, such as reduced dimensions of the lateral and lower incisors and the upper canines, but their lower canines resemble those of chimps, and the features of the upper canines that are not shared derived with australopiths are also strikingly chimp-like in morphology.

Postcanine Dentition

It has long been appreciated that *Pan* and *Gorilla* share many postcanine features including generally bunodont crowns, thin enamel (slightly thinner in gorillas), and a molar crown size order of M$_1 \approx$ M$_2 <$ M$_3$. Gorillas, however, differ from chimps in several respects, often interpreted to reflect their more folivorous diet and larger body mass: they have absolutely larger crowns and more crown relief (larger talonids and paracones); in addition, chimps have more peripherally located cusps with an expanded mesial fovea (Aiello and Dean 1990; Uchida 1992; Pilbrow 2006). In contrast, the derived condition in *Australopithecus* includes large crowns (both absolutely and relatively), very thick enamel crowns, generally low crown relief, an absence of peripherally located cusps, flared crown bases, and a molar crown size order of $M_1 < M_2 < M_3$ (Wood and Abbott 1983; Wood et al. 1983; Wood 1991; Hlusko 2002).

In terms of size, *Orrorin's* postcanines are comparatively small (Senut et al. 2001), but as Figure 2.5 shows, almost all mandibular premolar and molar crown dimensions in *Sahelanthropus* and *Ardipithecus* are intermediate between chimps and australopiths (Brunet et al. 2002, 2005; Suwa et al. 2009b). *Sahelanthropus* and *Ardipithecus* also have tooth crown enamel thickness values that are intermediate between australopiths and either chimps or gorillas (White et al. 1994; Brunet et al. 2002, 2005; Suwa et al. 2009b; Skinner et al. 2015). Additionally, in both *Sahelanthropus* and *Ardipithecus,* the molar crown size order is $M_1 < M_2 > M_3$, and they have broad M_1s and M_2s in contrast with the more elongated crown shape of chimps and gorillas. However, patterns of nonmetric postcanine variation (e.g., the relative size and position of cusps and the presence of accessory features) of these species are complex and variable from species to species, making it difficult to evaluate what features are primitive or derived, let alone to predict the morphology of the LCA (for details, see Schroer and Wood 2013).

Considering the postcanine dentition as a whole, early hominins share numerous derived features with australopiths related to postcanine expansion, notably in terms of crown size and enamel thickness. Large, thickly enameled postcanine crowns are also common in Miocene apes such as *Chororapithecus, Samburupithecus,* and *Ouranopithecus,* although others have thinner enamel such as *Proconsul africanus, Otavipithecus,* and *Hispanopithecus* (Ishida and Pickford 1997; Suwa et al. 2007; Alba et al. 2010). These features are often variable among closely related species within the same genus, possibly because they reflect selection for chewing fallback foods with differing physical properties (Kato et al. 2013). Even so, selection for larger, thicker-enameled postcanine crowns is a clear trend in australopiths, culminating in the robust australopiths. It is reasonable to hypothesize that increased enamel thickness originated in stem hominins relative to the LCA in response to selection for processing tougher foods in less forested environments with less fruit, perhaps in conjunction with adaptations for bipedalism (Lieberman 2011). Apart from these derived aspects of premolar and molar anatomy, there is little about early hominin postcanines that stands out as being different from the African apes, apart from modest variations in enamel thickness and occlusal relief, as well as more centrally located cusps.

Cranium

Early hominin cranial material has figured importantly in debates about the nature of the LCA. Before the discovery of *Ardipithecus, Orrorin,* and *Sahelanthropus,* it was widely acknowledged that the crania of gracile australopiths, especially *Au. afarensis,* share many similarities with *Pan* and *Gorilla,* especially in the neurocranium and face. Considerable debate regarding the interpretation of these similarities has been stimulated by the relatively complete cranium, "Toumaï" (holotype TM-266-01-60-1, hereafter referred to as TM-266) attributed to *Sahelanthropus* (Brunet et al. 2002), and by the more fragmentary cranium, "Ardi" (ARA-VP-6/500), attributed to *Ar. ramidus* (White et al. 1994)—both of which were reconstructed virtually (Zollikofer et al. 2005; Suwa et al. 2009a) (see Figure 2.6). It has been argued that TM-266 and ARA-VP-6/500 share numerous derived features with later hominins (White et al. 1994, 2009; Suwa et al. 2009a; Brunet et al. 2002, 2005; Zollikofer et al. 2005; Guy et al. 2005), but some researchers have suggested these similarities might be homoplasies (Andrews and Harrison 2005; Wood and Harrison 2011), while others have argued that they most closely resemble gorillas (Wolpoff et al. 2002, 2006). Suwa et al. (2009a) and White et al. (2009, 2015) accept the hominin status of *Ardipithecus* and *Sahelanthropus,* but argue that the substantial differences in cranial shape between TM-266 and ARA-VP-6/500 compared to *Gorilla* and *Pan,* along with other evidence that the AGAs are specialized relative to Miocene apes, indicate that the LCA's craniofacial anatomy was unlike either chimpanzees or gorillas.

As with the dentition, we address two questions about the cranial remains attributed to *Sahelanthropus* and *Ardipithecus* (primarily TM-266 and ARA-VP-6/500). First, how do we characterize the features in the crania of these putative early hominins that are not shared with australopiths and thus likely to be primitive? Second, to what extent are these cranial features shared with chimpanzees, gorillas, or other taxa?

Neurocranium

Both TM-266 and ARA-VP-6/500 have relatively well-preserved cranial vaults, with the major exception of the posterior portion of the occipital in ARA-VP-6/500. A geometric morphometric (GM) comparison of the TM-266 reconstruction with apes and australopiths (see Figure 2.7) indicates that the

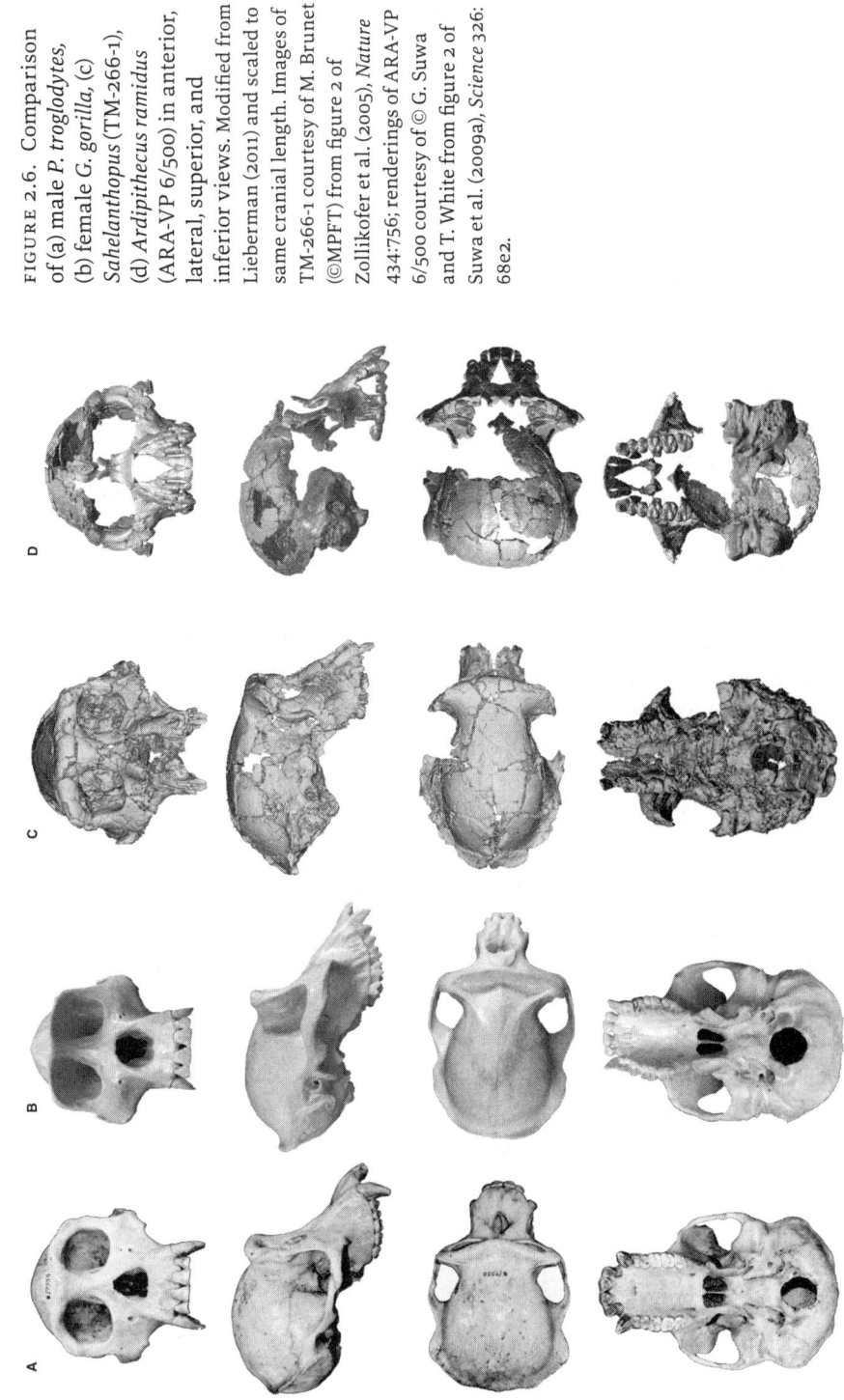

FIGURE 2.6. Comparison of (a) male *P. troglodytes*, (b) female *G. gorilla*, (c) *Sahelanthopus* (TM-266-1), (d) *Ardipithecus ramidus* (ARA-VP 6/500) in anterior, lateral, superior, and inferior views. Modified from Lieberman (2011) and scaled to same cranial length. Images of TM-266-1 courtesy of M. Brunet (©MPFT) from figure 2 of Zollikofer et al. (2005), *Nature* 434:756; renderings of ARA-VP 6/500 courtesy of © G. Suwa and T. White from figure 2 of Suwa et al. (2009a), *Science* 326: 68e2.

long, low superior contour of the vault in TM-266 is most similar to that of chimps (average of male and females), but also shares some similarities with female gorillas and *Au. afarensis* (Guy et al. 2005). Like *Pan*, the TM-266 vault differs from *Gorilla* in being more inferiorly rotated—the difference between *Pan* and *Gorilla* being the consequence of allometry (Berge and Penin 2004). As Figure 2.6 highlights, however, the *Sahelanthropus* vault resembles female gorillas more than chimps in being low and flat anterior to the apex of the vault, a difference between *Pan* and *Gorilla* that is unrelated to scaling (Berge and Penin 2004). Although a GM analysis of ARA-VP-6/500 has not been published, the anterior portion of the vault is rounded and more like *Pan* than *Gorilla*, reflecting a lower position of glabella and other parts of the upper face relative to the frontal squama. In posterior view, both TM-266 and ARA-VP-6/500 have bell-shaped contours with maximum width at the

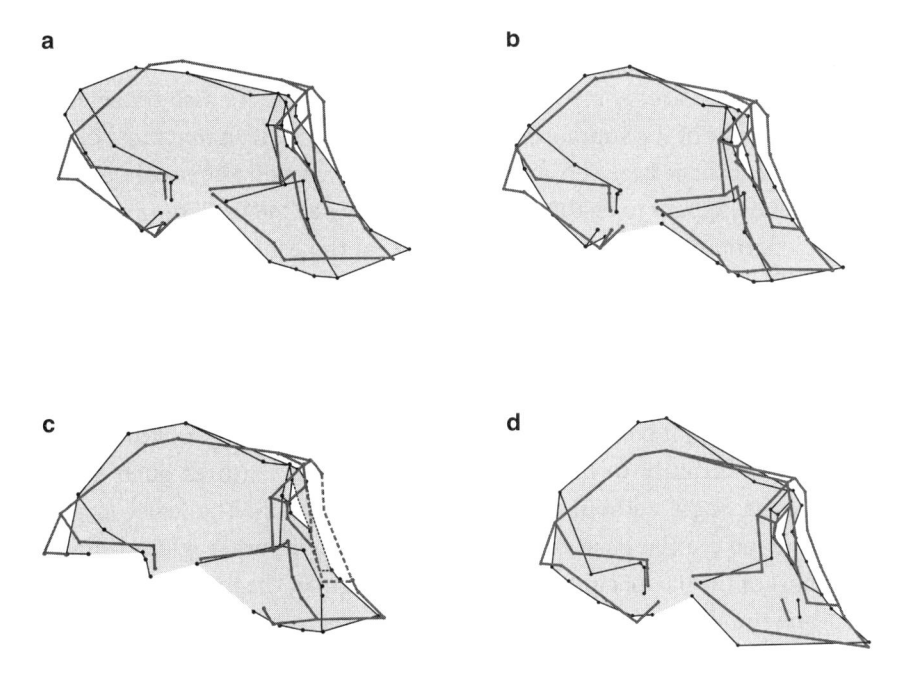

FIGURE 2.7. Geometric morphometric comparison in lateral view of *Sahelanthropus* (TM-266) in red outline with (a) adult female *G. gorilla* (pooled consensus), (b) adult *P. troglodytes* (pooled consensus of males and females), (c) *Au. afarensis* (AL-444-2), and (d) *Au. africanus* (STS 5). Note that, overall, TM-266 is most similar to *P. troglodytes*. Modified from Guy et al. (2005).

base of the vault, similar to *Gorilla, Pan,* and *Australopithecus*. Like gorillas, chimps (but not bonobos), and *Au. afarensis,* TM-266 and ARA-VP-6/500 have compound temporal crests (Brunet et al. 2002; Kimbel et al. 2004; Zollikofer et al. 2005; Suwa et al. 2009a).

Endocranial volume (ECV) estimates are 360–370 cm^3 for TM-266 (Zollikofer et al. 2005), and approximately 300 cm^3 (range: 280–350 cm^3) for ARA-VP-6/500 (Suwa et al. 2009a). As absolute values, these ECVs fall within the *Pan* range, but below the *Gorilla* range; if body mass in the 40-kg range is estimated for *Sahelanthropus* and *Ardipithecus* (Grabowski et al. 2015) and if these ECVs represent species averages, then relative brain size (encephalization quotients) for the two taxa are 2.0 and 1.7, respectively, similar to *Pan* but higher than in *Gorilla* (Lieberman 2011: 193). Given body size increases that we argue occurred in the *Gorilla* lineage since the hominine LCA, it is reasonable to infer that absolute and relative brain size increases occurred only in later hominins, starting with *Australopithecus*.

Overall, the size and shape of the neurocranium in *Ardipithecus* and *Sahelanthropus* is most similar to *Pan,* but the anterior portion of the vault in *Sahelanthropus* shares some similarities to *Gorilla* that are also reflected in the morphology of the supraorbital torus (see below). Given that most differences in vault shape between gorillas and chimps, as well as between chimps and bonobos, appear to be the result of scaling (Shea 1983b; Berge and Penin 2004; Lieberman et al. 2007), we conclude that the general shared configuration of the neurocranium in *Sahelanthropus, Ardipithecus, Pan,* and *Gorilla* is most probably primitive.

Cranial Base

The TM-266 cranial base is remarkably complete externally, but too poorly preserved internally to measure many important features such as the cranial base angle without substantial reconstruction. The less complete ARA-VP6/500 cranial base chiefly preserves the temporal regions (bilaterally), most of the basioccipital, and the left margin of the foramen magnum including the occipital condyle (see Figure 2.6).

Whereas the neurocrania of TM-266 and ARA-VP-6/500 are largely primitive, the posterior and central portions of the cranial bases of these putative early hominins share numerous derived features with later hominins, especially australopiths. As in australopiths, the anterior margin of the foramen magnum (basion) in both TM-266 and ARA-VP-6/500 is positioned nearly in the same coronal plane as the carotid foramina, indicating a relatively anteriorly positioned foramen magnum compared to apes (Dean and Wood 1981, 1982; White et al. 1994; Brunet et al. 2002; Guy et al. 2005; Suwa et al. 2009a; Kimbel et al. 2014). Although it is not possible to measure the orientation of the foramen magnum in ARA-VP-6/500, it appears to be oriented more horizontally relative to the face as in TM-266, and unlike the more posteriorly oriented foramen magnum of *Gorilla* and *Pan troglodytes* (Zollikofer et al. 2005; Suwa et al. 2009a); the foramen magnum in *Pan paniscus* is slightly more anteriorly positioned than in the other African apes, but has a similar orientation (Luboga and Wood 1990). The preserved portions of both TM-266 and ARA-VP-6/500 suggest that, like later hominins, they have relatively short, trapezoidally shaped basioccipitals and a relatively broad posterior cranial base that involves a petrous temporal that is shorter and relatively more coronally oriented than in any extant ape species (Dean and Wood 1982; Guy et al. 2005; Kimbel et al. 2014). In both hominins, the tympanic portion of the temporal is also mediolaterally short relative to cranial base width (Brunet et al. 2002; Kimbel et al. 2014).

ARA-VP-6/500 lacks a posterior (nuchal) portion of the cranial base, but this region is well preserved in TM-266 and includes additional derived features shared between *Sahelanthropus* and australopiths, probably related to bipedalism. Most notably, TM-266's nuchal plane is slightly shorter than in *Pan* and significantly shorter than in *Gorilla,* but comparable to *Au. afarensis;* its orientation is estimated to be 33°, thus more horizontally oriented than in *Pan* and in the range of gracile australopiths (Kimbel et al. 2004; Guy et al. 2005). In addition, TM-266, like australopiths, has a downwardly lipped nuchal crest.

Anatomically, the anterior cranial base extends from the middle of the sphenoid (the division between the pre- and postchordal portions of the basicranium) to the foramen caecum. Because neither ARA-VP-6/500 nor TM-266 preserves much internal cranial base anatomy, it is not possible to measure the length of the anterior cranial base or its internal angle relative to the posterior cranial base in either taxon. However, Suwa and colleagues (2009a) hypothesize that ARA-VP-6/500 has a more flexed cranial base compared to chimps (but not bonobos). They further hypothesize that chimps as well as gorillas are derived relative to bonobos (and the LCA) in having a relatively long, extended anterior cranial base. A major basis for the former inference is the distance between the foramen ovale and the pterygopalatine fossa, which approximates the length of the nasopharynx as well as the length of sphenoid lateral to the cranial base midline. Suwa et al. (2009a) show that this distance is longer both absolutely and relative to cranial base width in chimps compared to bonobos, a difference they hypothesize to be an adaptation in chimps to increase lower facial projection for enhancing gape during male-male competition. Although the anterior cranial base is almost completely missing in ARA-VP-6/500, they infer resemblance to bonobos, in which reduced facial prognathism is largely a function of a more flexed cranial base that necessarily rotates the whole face ventrally beneath the anterior cranial fossa because of constrained integration between the face and anterior cranial base (McCarthy and Lieberman 2001).

Although the cranial base in ARA-VP-6/500 as well as TM-266 may be more flexed than in chimps and more similar to bonobos, two issues need to be considered. First, the differences between chimps and bonobos are largely the result of heterochrony: adult bonobos have a configuration similar to subadult chimpanzees (Shea 1983a, 1983b; Cobb and O'Higgins 2004; Mitteroecker et al. 2005; Lieberman et al. 2007). Second, the cranial base angle (CBA1, basion-sella-foramen caecum) is a highly variable feature influenced by multiple ontogenetic interactions between the brain and the face relative to the cranial base. As a rule, CBA1 is more flexed in taxa with relatively large brains, and more extended in taxa with relatively anteroposteriorly long faces (Ross and Ravosa 1993; Spoor 1997; Lieberman 2000; McCarthy 2001; Lieberman et al. 2008). These interactions help explain why CBA1 averages 154° in *G. gorilla* and 156° in *P. troglodytes* but 148° in *P. paniscus* (McCarthy 2001), the last reflecting its relatively shorter face and paedomorphic similarity to adolescent *P. troglodytes,* whose CBA1 averages 147°–151° (Lieberman and McCarthy 1999). Efforts to measure cranial base angles in australopith fossils report that CBA1 tends to be more extended in species with anteroposteriorly long faces such as *Au. aethiopicus* (156°, KNM-WT 17000), but more flexed in australopiths with shorter, more retracted faces such as *Au. africanus* (147°, STS 5) and *Au. boisei* (135°, OH 5; 142°, KNM-ER 406) (Spoor 1997; Lieberman 2011). Based on these patterns, we predict an intermediate cranial base in *Sahelanthropus* and *Ardipithecus* because of the combination of relatively small ECVs (especially in ARA-VP-5/600) and anteroposteriorly shorter faces (see below). But to recap, CBA1 cannot be measured in ARA-VP-5/600, and any estimate in TM-266 will require substantial further reconstruction. In addition, there are several indications that the TM-266 cranial base is not as flexed as those of bonobos. For one, more flexed cranial bases (as in bonobos) tend to decrease rather than increase upper facial projection relative to the anterior cranial fossa, shortening rather than lengthening the supraorbital torus, as is evident in TM-266 (Lieberman 2000, 2011). Also, the nasopharynx in Toumaï (as judged by the distance from basion to the back of the palate) appears to be relatively long. In short, while bonobos are derived relative to the other African great apes in having a more flexed and shorter anterior cranial base, largely because of heterochrony, it is unclear without more data that this configuration is shared with *Sahelanthropus* or represents the primitive condition in hominines.

Finally, features of the lateral cranial base in ARA-VP-6/500 and TM-266 (see Figure 2.6) are generally like those of African apes, including a shallow mandibular fossa, a narrow tympanic tube, and a long, flat preglenoid region (White et al. 1994; Suwa et al. 2009a; Kimbel et al. 2014). *Au. anamensis* and *Au. afarensis* also share these primitive features (Kimbel et al. 1984, 2004; Ward et al. 2001).

Considering the cranial base as a whole, TM-266 and ARA-VP-6/500 share numerous derived features of the posterior cranial base with australopiths, they have generally chimpanzee-like lateral portions of the cranial base, and there is little evidence to suggest that *Ardipithecus* and *Sahelanthropus* have anterior cranial bases that differ markedly from those of *Pan* and *Gorilla*. One possible exception to this would be more cranial base flexion, which, if present, likely results from a relatively shorter face.

Face

Much of the face is preserved in TM-266 and ARA-VP-5/600 (although both are substantially reconstructed, the latter from numerous small fragments), and partial mandibles have been attributed to *Sahelanthropus, Ar. kadabba, Orrorin,* and *Ar. ramidus* (White et al. 1994; Pickford and Senut 2001; Brunet et al. 2002, 2005; Semaw et al. 2005). Given the paucity of taxonomically diagnostic features in the mandible, we focus here on the cranial portions of the face in TM-266 and ARA-VP-5/600, which share a number of derived features with australopiths but also retain some presumably primitive features that are potentially informative about the morphology of the LCA.

Compared with australopiths and the African great apes, the most obviously derived aspects of the TM-266 and ARA-VP-5/600 faces are anteroposterior length and position relative to the rest of the cranium, superoinferior length, and orientation in the sagittal plane (see Figure 2.6). A GM analysis, illustrated in Figure 2.7, comparing TM-266 with apes and other hominins, indicates that landmarks on the anterior aspect of the upper and middle face are more anteriorly positioned relative to the rest of the cranium in *Sahelanthropus* than in *Gorilla*, *Pan*, and even *Au. afarensis* (Guy et al. 2005). As Figure 2.7 shows, upper and midfacial projection in *Sahelanthropus* generates a relatively vertical face, an anteroposteriorly long supraorbital torus, and a more forwardly positioned zygomatic arch (see below). No GM analysis has been published for ARA-VP-5/600, but Suwa et al. (2009a: figure 4) report that measures of upper facial projection (but not midfacial projection) are also derived in this skull compared to the African great apes. Facial height (nasion to prosthion) is also shorter in TM-266, and especially shorter in ARA-VP-5/600 relative to *Pan* and *Gorilla* (Guy et al. 2005; Suwa et al. 2009a). Another derived feature that TM-266 and ARA-VP-5/600 reportedly share with later hominins is reduced subnasal prognathism, likely due to a relatively shorter premaxilla (Brunet et al. 2002; Guy et al. 2005; Suwa et al. 2009a). However, subnasal prognathism is variable in *Au. afarensis* (Kimbel et al. 2004) and *Au. africanus* (Lockwood 1999), and the anterior portion of the TM-266 premaxilla is heavily damaged, making it difficult to reconstruct with confidence. Like other later hominins, both TM-266 and ARA-VP-5/600 have little to no diastema between C_1 and P_3.

Although TM-266 and ARA-VP-5/600 lack most of the craniofacial features related to extreme masticatory robusticity that characterize australopiths, several features of the zygomatic arch in TM-266 and ARA-VP-5/600 may be shared derived with australopiths. In both TM-266 and ARA-VP-5/600, the anterior margin of the zygomatic process of the maxilla emerges above M^1, while this process arises above M^1-M^2 in apes, but above M^1-P^4 or further forward in australopiths (Figure 2.6). Therefore, as in other hominins, the zygomatic in *Sahelanthropus* and *Ardipithecus* is more forwardly placed in relation to the TMJ and the postcanine teeth. In addition, the zygomatic process of the maxilla is positioned more laterally relative to the orbits in TM-266 (but not ARA-VP-5/600) than in *Pan* and *Gorilla*, resulting in a slightly diamond-shaped face in frontal view also characteristic of australopiths (Rak 1983) (Figure 2.6). Anterior and lateral expansion of the zygomatic arches increases the mechanical advantage of the masseter in the sagittal and transverse planes, and thus may be adaptations, in conjunction with larger postcanine teeth with slightly thicker enamel (see above), for producing and resisting higher chewing forces (Lieberman 2011).

Although the faces of *Sahelanthropus* and *Ardipithecus* share many derived features with later hominins, these taxa otherwise share features with the African great apes, also extensively documented among gracile australopiths, that are likely to be primitive. For example, like *Pan, Gorilla, Au. Anamensis,* and *Au. afarensis* (Ward et al. 2001; White et al. 2006; Kimbel et al. 2004), TM-266 and ARA-VP-5/600 have U-shaped palates with long, parallel postcanine rows that curve slightly inward at the M^{3}s. Like *Pan* and *Gorilla,* they also have sizable frontal sinuses, and a wide and rugose frontozygomatic (Suwa et al. 2009a). The lateral margins of the piriform aperture in TM-266 and ARA-VP-5/600, as in *Pan* and *Gorilla,* are elevated and continuous with marked canine jugae. And, except for a slightly posteriorly receding symphysis in *Ardipithecus,* most aspects of the mandibles attributed to *Sahelanthropus* and *Ardipithecus* resemble those of *Pan,* including a lack of robusticity and canines that project above the postcanine occlusal plane (Brunet et al. 2002, 2005; Suwa et al. 2009a), which are likely to be shared primitive.

What does the mosaic of primitive and derived features evident in the faces of *Sahelanthropus* and *Ardipithecus* imply for the morphology of the LCA? As discussed above, White et al. (2015) accept the hominin status of *Sahelanthropus* and *Ardipithecus* but consider their craniofacial morphology to be essentially primitive for two reasons. First, because they assume that *Pan* and *Gorilla* are independently derived, they consider the many adaptations for masticatory robusticity in the *Sahelanthropus* and *Ardipithecus* faces to be primitive rather than derived. Second, White et al. (2015) propose that the more prognathic faces of chimps and gorillas are derived adaptations for having a wide gape to accommodate large canines used during male-male competition (Hylander 2013), and thus consider the reduced prognathism of bonobos to be shared primitive features with *Sahelanthropus* and *Ardipithecus,* hence also the ancestral condition of the LCA.

We agree that gorillas and chimps are different from known Miocene apes, but argued above that there are no currently known Miocene apes that are clearly and demonstrably related to hominins. *Sahelanthropus* and *Ardipithecus* share many apparent derived facial features with later hominins that may be adaptations for chewing mechanically demanding foods, and the above review highlights that a large percentage of the primitive features evident in the faces of TM-266 and ARA-VP-5/600 are shared not only with *Pan* and *Gorilla*, but also with early *Australopithecus*, notably *Au. anamensis* and *Au. afarensis*. In addition, while *Sahelanthropus* and *Ardipithecus* share with some Miocene apes several features related to masticatory robustness (e.g., *Sivapithecus, Ouranopithecus, Samburupithecus*), none of the latter taxa can be demonstrably related phylogenetically to the LCA. These features are also subject to homoplasy, as exemplified by the likely convergent similarities between the aforementioned *Ouranopithecus* and *Australopithecus* (Kunimatsu et al. 2007; but see De Bonis and Koufos 1994). Finally, the similarities in prognathism observed between bonobos and TM-266 and ARA-VP-5/600 are unlikely to be homologous because, as discussed above, bonobos have less prognathism than adult (but not subadult) chimpanzees due to paedomorphism and a relatively smaller face, which leads to a more flexed cranial base (Lieberman et al. 2007). Further, more evidence is needed to reliably conclude that the cranial base of either *Sahelanthropus* or *Ardipithecus* is more flexed than in *Pan* and *Gorilla* (see above). Moreover, reduced prognathism in these putative hominins appears to be a consequence of substantial elongation of the middle and upper face, resulting in greater projection of the upper face relative to the lower face (Guy et al. 2005; Lieberman 2011: 425–428). Testing the effects of upper facial projection on early hominin facial morphology will require better-preserved fossils. If correct, the similarities observed between bonobos and the earliest putative hominins may have different developmental origins. If this is the case, long and prognathic faces combined with a relatively extended CBA1 are likely shared primitive features between chimps and gorillas that are also present in LCA_{C-H}.

Hominin Fossil Evidence Regarding the LCA—The Postcranium

Many debates about the LCA focus on the hominin postcranium, chiefly over when, how, and why early hominins became habitual bipeds. Testing alternative hypotheses about the origins of bipedalism requires resolving disagreements about the locomotor repertoire of the LCA, especially whether it was predominantly orthograde or pronograde, and whether or not it was a knuckle-walker. As discussed above, when *Pan* and *Gorilla* were thought to be monophyletic, with hominins more distantly related, the most parsimonious inferences were that the LCA was not a knuckle-walker and instead had a less-derived locomotor repertoire, either orthograde or pronograde. Molecular evidence that humans and chimps are monophyletic combined with the numerous similarities between chimps and gorillas suggest, however, that the LCA's locomotor repertoire was like that of the African great apes. According to this logic, if the LCA did not combine suspensory orthograde climbing with knuckle-walking terrestrial locomotion, then knuckle-walking and the many postcranial similarities between gorillas and chimpanzees would have to be homoplasies. In contrast, approaches that reconstruct the LCA using early hominin and Miocene hominoid fossils have led to a variety of conflicting interpretations and inferences, ranging from chimp- to monkey-like.

As argued above, one limitation of using the Miocene fossil record to make inferences about the locomotor repertoire of the LCA is the phylogenetic relevance of the available data for the hominine clade. Although much attention has been focused on postcrania attributed to *Proconsul, Pierolapithecus, Oreopithecus,* and *Dryopithecus,* none of these apes are closely related to the *Pan* and *Gorilla,* limiting their relevance. Instead, the most pertinent fossil postcranial data for evaluating the LCA's anatomy come from putative early hominins, primarily the partial skeleton AR-VP-6/500 of *Ar. ramidus* (White et al. 1994, 2009, 2015). *Ar. ramidus,* however, is dated to between 4.5 and 4.3 Ma, and thus postdates the divergence of chimps and humans by at least 1.5 Ma, probably twice that. A few older postcranial remains, primarily from the upper limb, have been attributed to *Ar. kadabba* (Haile-Selassie 2001). In addition, the single 6.0 Ma fossil locality that yielded *O. tugenensis* includes two partial femurs, a partial humeral shaft, and a proximal manual phalanx (Senut et al. 2001). No postcrania have been officially attributed to *S. tchadensis,* but an as yet undescribed femur was apparently associated with the *Sahelanthropus* cranium (Beauvilain and Watté 2009).

As with the skull, to evaluate alternative hypotheses about the LCA we address several questions about the postcranial remains attributed to *Ardipithecus* and to a lesser extent *Orrorin* in relation to later hominins, the AGAs, and (with caveats) Miocene hominoids. First, what are the primitive features of these putative early hominins that are not shared derived with australopiths? Second, to what extent are these primitive features shared with *Pan* and *Gorilla* and with other potentially relevant taxa? Finally, are the postcranial similarities between *Pan* and *Gorilla* (including those related to knuckle-walking) homologies or homoplasies?

Axial Skeleton

There is intense interest in the axial skeleton of the LCA because of this region's well-characterized variation, whose functional importance and developmental bases are reasonably well understood (see Burke et al. 1995; O'Rahilly and Müller 2003; Aulehla and Pourquié 2010; Mallo et al. 2010; Galis et al. 2014). Although there are many aspects of axial variation to consider, we address two related variables that have been the focus of most debate about the LCA. First, there is disagreement over whether the LCA had three or four lumbar vertebrae like the "short-backed" African great apes, or whether it was "long-backed," with five or more lumbar vertebrae like most humans, several Miocene apes such as *Ekembo (Proconsul),* and other nonhominoid primates (Williams 2011). A second major point of disagreement is the position of the "transitional" vertebra, the element with more coronally oriented facets (thoracic-like) cranially and more sagittally oriented facets (lumbar-like) caudally, thus marking the transition between a more cranial region in which intervertebral rotation is facilitated and a more caudal region in which flexion and extension are facilitated. To avoid terminological confusion, we follow Williams (2012) and Hauesler et al. (2011) in classifying only vertebrae with rib facets as thoracic, without facets as lumbars; thoracolumbar vertebrae are pretransitional if their zygapophyseal orientations are like thoracics (more coronally oriented) and posttransitional if their zygapophyseal orientations are more sagitally oriented, regardless of presence or absence of ribs.

Ironically, a major reason for debates over the number of lumbar and posttransitional vertebrae in the LCA and the earliest hominins is that we know essentially nothing about the axial skeleton in any early hominin species. No vertebrae have yet been recovered for *Salehanthropus* or *Orrorin,* and the *Ar. ramidus* partial skeleton, ARA-VP-6/500, preserves only one crushed cervical and one crushed thoracic vertebra, along with two other indeterminate vertebral fragments (White et al. 2009). Even so, Lovejoy and colleagues (2009c) reconstructed *Ar. ramidus* with six lumbar vertebrae (see below). In contrast, average numbers of lumbar vertebrae (posttransitional vertebral numbers in parentheses) are 3.7 (4.0) in *P. troglodytes,* 3.5 (3.8) in *P. paniscus,*

3.6 (3.6) in lowland (western) gorillas, and 3.1 (2.6) in mountain gorillas (Pilbeam 2004; Williams 2011, 2012). Broadening the comparative perspective within hominoids, the average number of lumbar and posttransitional vertebrae is 4.0 (3.8) in *Pongo*, 5.2 (5.6) in *Hylobates lar,* and modal values of generally 7 (9) in cercopithecoids (Williams 2012). Turning to hominins, there are obviously no estimates of the number of posttransitional vertebrae in early hominins, but the derived condition in *Australopithecus* appears to be five lumbar and six posttransitional vertebrae, as evident from *Au. sediba* (Williams et al. 2013) and two *Au. africanus* specimens (Haeusler et al. 2002; Williams 2012; Williams et al. 2013). Unfortunately, the lumbar column of *Au. afarensis* is inadequately sampled (Johanson et al. 1982). Early *H. erectus*, as represented by the Nariokotome specimen, appears to have five lumbar and six posttransitional vertebrae (Haeusler et al. 2011). The few *H. neanderthalensis* lumbar columns known do not differ from *H. sapiens*, in which the most frequent vertebral combination is 7:12:5:5:4 (Trinkaus 1983; Arensberg 1991; Pilbeam 2004).

Based on the principle of parsimony, the LCA was most likely short-backed, with three or four lumbar vertebrae, at most four posttransitional vertebrae; if not, lumbar reduction would have to have evolved independently in gorillas, chimps, bonobos, and orangutans. This inference is also supported by great similarity between *Pan troglodytes* and *Gorilla gorilla* in relative frequencies of vertebral patterns in the two species, which very likely represents a common ancestral pattern for hominines (Pilbeam 2004; Williams 2011). Measures of within-species variation, however, are much higher in *G. gorilla* and both *Pan* species than in *G. beringei*, humans, and cercopithecoids, suggesting less stabilizing selection on the vertebral column in western lowland gorillas, chimps, and bonobos (Pilbeam 2004; Williams 2011). High levels of within-species variation, evident in all apes except *G. beringei*, indicate low selection pressure (Galis et al. 2014), and argue against homoplasy as the explanation for the short-backed similarities of all apes (Williams 2011).

As noted above, Lovejoy (2009c), McCollum et al. (2010), and White et al. (2015) have argued against the parsimony-based inference of a short-backed LCA with three or four lumbar vertebrae, and instead have proposed that the LCA had five or more lumbar vertebrae. One basis for this argument is the inference that because *Pan paniscus* has more precaudal vertebrae than other hominoids, it resembles a hypothetical hominoid ancestor with a long lumbar spine (McCollum et al. 2010). However, from a developmental perspective, it is more appropriate to describe vertebral numbers and assess homologies based on somite number and how somites are specified rather than on vertebral counts (O'Rahilly et al. 1990; Burke et al. 1995; O'Rahilly and Müller 2003; Pilbeam 2004; Mallo et al. 2010). In this respect, *Pan paniscus* has slightly *fewer* lumbar vertebrae than *Pan troglodytes* (3.5 as opposed to 3.7), slightly more thoracic vertebrae (13.5 as opposed to 13.1), and the bulk of the difference in precaudal numbers is due to difference in the sacral count (6.4 in *Pan paniscus* as opposed to 5.7 in *Pan troglodytes*) (Pilbeam 2004; Williams 2011). Consequently, the number of thoracic plus lumbar vertebrae totals differs between chimps and bonobos by a modest 0.3 vertebrae. This difference is not *meristic,* that is, due to the "loss" of a fractional vertebra, but rather is *homeotic,* due to shifts in ultimate identity of somites.

The other bases for Lovejoy and colleagues' (2009c) argument that the LCA had a long lumbar spine come from interpretations of the ARA-VP-6/500 pelvis and from lumbar counts in several Miocene species. As noted above, no lumbar vertebrae are preserved in ARA-VP-5/600, but a long lumbar spine is surmised from the pelvis's short iliac blades and reduced postauricular region, which are thought to indicate a relatively wide sacrum and a lack of lumbar entrapment by the iliac blades, thus facilitating more lumbar mobility, including a facultative lordosis (Lovejoy et al. 2009c). As discussed below, the presence of lumbar entrapment is not itself evidence for a long lumbar spine. In addition, it is problematic to infer that the LCA had a long lumbar spine from the Miocene ape fossil record. Lumbar counts are available for only three Miocene taxa from Africa: *Ekembo nyanzae, Nacholapithecus kerioi,* and *Morotopithecus bishopi.* Both *Ekembo nyanzae* and *Nacholapithecus kerioi* have six or possibly seven lumbar vertebrae, and both have at least one and possibly two posttransitional thoracic vertebrae (Ward 1993; Nakatsukasa et al. 2007; Williams 2012). However, as discussed above, these taxa are probable stem hominoids dated to 17 Ma or more (Nakatsukasa and Kunimatsu 2009; Harrison 2010b; Stevens et al. 2013; McNulty et al. 2015), and thus only distantly related to the African great apes. It is thus unsurprising that the lumbar vertebrae of *Ekembo* and *Nacholapithecus* differ from those of all crown hominoids in several important respects, including their greater relative length, and in having transverse processes that are located more on the vertebral bodies and not on the pedicle as in crown hominoids. *Ekembo* and *Nacholapithecus* represent basal stem hominoids with somewhat different kinds of pronograde positional repertoires, both lacking any features linking them to crown hominoids in general or to particular crown clades. We conclude that they have little relevance for making inferences about the vertebral morphology of the LCA_{G-CH} or the LCA_{C-H}.

Morotopithecus bishopi is different in that one lumbar vertebra (UMP 67.28) shares functionally important similarities with crown hominoids, notably transverse processes situated more dorsally, with relatively shorter vertebral bodies lacking keels (Ward 1993; Sanders and Bodenbender 1994; MacLatchy 2004). Nakatsukasa (2008) noted that other previously undescribed vertebrae from Moroto that are likely to be from the same individual as UMP 67.28 suggest that *M. bishopi* may well have resembled *Ekembo nyanzae* and *Nacholapithecus kerioi* in having at least six lumbar vertebrae. A more dorsostable spine and a positional repertoire different from either *Ekembo nyanzae* or *Nacholapithecus kerioi* is indicated, although the extent to which the derived features of the *M. bishopi* vertebral column can be interpreted as supporting crown hominoid or hominid status, and therefore relevance to the LCA_{C-H} continues to be debated (MacLatchy 2004; Nakatsukasa 2008; Harrison 2010b). However, in discussions of hominoid vertebral column evolution in general, and that of hominines in particular, *M. bishopi* deserves some level of priority over *Ekembo nyanzae* (or over the rarely mentioned *Nacholapithecus kerioi*).

Non-African Miocene hominoids have even less to contribute to our understanding of the LCA's lumbar spine. Two Spanish Miocene taxa, *Hispanopithecus laietanus* (Mòyà-Sola and Köhler 1996; Susanna et al. 2014), and *Pierolapithecus catalaunicus* (Mòyà-Sola et al. 2004), dated between 12 and 9.5 Ma (Casanovas-Vilar et al. 2011), preserve vertebrae. Although the number of lumbar or posttransitional vertebrae cannot be estimated definitively in either taxon, lumbar morphology and some other postcranial features imply torsos that are broader and shorter than in earlier Miocene African apes, with the possible exception of *Morotopithecus* (Hammond et al. 2013). The vertebral column of the Late Miocene ape *Oreopithecus bambolii* has been described by Harrison (1986), Sarmiento (1987), and Russo and Shapiro (2013), and is the only Miocene ape with close postcranial resemblance to crown hominoids, with Russo and Shapiro (2013) recording five lumbar and six sacral vertebrae. Its phylogenetic position, however, is uncertain: Harrison (1986) and Sarmiento (1987) place it as a crown hominoid or stem hominid, but its cranium and dentition are sufficiently autapomorphic that more precise allocation is not possible.

In summary, reconstruction of a long lumbar spine in the LCA requires the improbable convergent evolution of short lumbar spines in all the extant great apes. There is no direct evidence that early hominins such as *Ardipithecus* had long lumbar columns, and the Miocene ape postcranial record, especially the morphology of *Ekembo (Proconsul)*, contributes no significant, direct information about lumbar number in the LCA$_{C-H}$. For the past 150 years, the many similarities in the vertebral column of *Pan* and *Gorilla* were interpreted to be homologies (Huxley 1863; Gregory 1927; Osborn 1927; Straus 1949; Schultz 1963; Simpson 1963; Tuttle 1969; Groves 1986). Accordingly, we see no plausible arguments against the hypothesis that, at least for the vertebral column, the similarities of *Pan* and *Gorilla* in particular, and crown apes in general, are largely homologous, and that hominins are derived from an LCA with a short lumbar column similar to those in extant African apes.

Pelvis

The os coxae of *Ar. ramidus* are known primarily from the ARA-VP-6/500 partial skeleton, which includes a nearly complete but heavily damaged left pelvis, a portion of the right ilium, and a fragmentary distal sacrum. The published reconstruction of the ilium (Lovejoy et al. 2009d) indicates that it shares several key derived features with later hominins, notably australopiths. In contrast to the tall, narrow, and coronally oriented ilium of AGAs and other hominoids, the reconstructed ARA-VP-6/500 ilium is superoinferiorly short, anteroposteriorly long, laterally flared, and oriented primarily in the parasagittal plane. Additional derived features of the ilium include a prominent, anteriorly positioned anterior superior iliac spine (ASIS), and a protuberant and anteriorly positioned anterior inferior iliac spine (AIIS). According to Lovejoy (2009d), the iliac isthmus (the portion just superior to the acetabulum) is also derived and similar to later hominins in being short, broad, and sagittally oriented. Overall, the ARA-VP-5/600 ilium strongly suggests that *Ardipithecus* was a biped, because it had a basin-shaped upper pelvis with laterally positioned, parasagittally oriented small gluteal muscles that would have functioned effectively as abductors as well as medial rotators of the thigh during stance, key adaptations for controlling the center of mass of the upper body during bipedal locomotion (Saunders et al. 1953). In other words, by 4.5 Ma, selection for bipedalism had already substantially modified the hominin ilium.

Although the *Ardipithecus* ilium is generally australopith-like and thus provides little evidence with which to make inferences about the LCA's upper pelvis, the relationship of the ilium to the lower spine in early hominins and the LCA has generated considerable speculation. As described above, African great apes have a short, straight lumbar region with a mediolaterally narrow sacrum and ilia that extend cranially relative to the sacroiliac joint. Lovejoy and colleagues (Lovejoy et al. 2009d; Lovejoy and McCollum 2010) hypothesize that this configuration evolved independently in the great apes, especially chimps and gorillas, to "entrap" the lower lumbar vertebrae, limiting movement of the lumbar transverse processes relative to the iliac blades, and thus stiffening the lower back during orthograde climbing. Although ARA-VP-5/600 has short iliac blades, its sacrum is not well enough preserved to measure mediolateral width, and no lumbar vertebrae have been found. Nonetheless, Lovejoy and colleagues infer that the reduced postauricular region of ARA-VP-5/600 indicates a wide sacrum, and they contend that its supposed lack of other suspensory adaptations (see below) make it likely that *Ardipithecus*, hence the LCA, resembled monkeys in having more than five or six nonentrapped posttransitional vertebrae that would have been capable of a facultative lordosis (Lovejoy et al. 2009d). Apart from the indirect assessment of sacral width and lumbar number in ARA-VP-5/600, there are several reasons to question this argument. First, if the configuration of the lumbar vertebrae relative to the ilium in the LCA were monkey-like, then a short lumbar spine with a relatively narrow sacrum would have had to evolve independently at least four times in the crown hominoid genera. If, as McCollum et al. (2010) argue, it evolved independently even in the two *Pan* species, similar logic might suggest independent evolution in the dozen or so hylobatid species and in the two species of *Pongo* and *Gorilla*. In addition, the features used to infer lumbar entrapment are adaptations to stiffen the lower spine are not only untested but also uncertain. Although bipedal chimps do not lordose the lumbar spine to position the trunk above the hips, gorillas typically position the trunk more vertically above the hips during bipedal posture despite having entrapped lumbar vertebrae (Crompton et al. 2008). The hypothesis that lumbar entrapment constrains lordosis needs to be tested *in vivo* in gorillas and other apes. Finally, although the Miocene stem hominoid *Ekembo* had a pelvis and long posttransitional region reflecting adaptations for quadrupedal pronograde arboreal locomotion, unlike any crown hominoid (Rose 1983; Ward

1993; Walker 1997; Morgan et al. 2015), it is only distantly related to the African ape-human clade

In terms of the rest of the pelvis, the ischium and pubic bones are relatively independent from the ilium developmentally (Capellini et al. 2011) and morphologically (Grabowski et al. 2012; Lewton 2012). Thus it should not be surprising that the ARA-VP-5/600 ischium, unlike the ilium, has almost no derived hominin features; rather, it most strongly resembles that of African apes (Lovejoy et al. 2009d: 71e1). As in chimps, the *Ardipithecus* ischial body is relatively long and positions the ischial tuberosity further away from the acetabulum with a more dorsal (posterior) orientation than in humans and australopiths. In quadrupeds, this configuration permits the hip extensors to optimally produce torque during flexed positions during climbing, but at the cost of producing less torque during fully extended positions characteristic of bipedal locomotion (McHenry 1975; Stern and Susman 1983). However, if *Ardipithecus* was bipedal with human-like pelvic tilt, then its hip extensors could have produced nearly human-like hip extension moments while retaining the capacity to produce powerful chimp-like extensor moments during vertical climbing. The ARA-VP-5/600 pubis is also ape-like with a superinferiorly long body, but its pubic symphysis is broad relative to its length, a feature shared with other hominins (Lovejoy et al. 2009d: figure 3a).

In short, the *Ardipithecus* pelvis preserves a mosaic of derived iliac features shared with hominins that were likely adaptations for bipedal locomotion, along with primitive ischial and pubic features that are mostly shared with *Pan* and *Gorilla*. Arguments that the LCA did not have an African ape-like pelvis rely almost entirely on questionable inferences about anatomical features that are not preserved in the skeleton, and on the assumption that the pelvic morphologies evident in Miocene hominoids such as *Ekembo* (Ward 1993), *Pierolapithecus* (Hammond et al. 2013), and *Sivapithecus* (Morgan et al. 2015), none of which are closely related to hominines, are more likely to represent the condition of the LCA than any extant ape.

Femur

Published early hominin femora that can be used to make inferences about femoral anatomy in the LCA include three partial specimens attributed to *O. tugenensis,* including one fossil (BAR 1002'00) that preserves most of the proximal femur, plus two partial proximal femora of *Ar. ramidus* (ARA-

VP-5/600 and ARA-VP-1/701), neither of which preserves the head, neck, or greater trochanter.

Although the *Ardipithecus* femora lack most of the proximate anatomical regions that tend to be most useful for making taxonomic attributions, Lovejoy and colleagues (2009d) highlight two features that ARA-VP-1/701 shares with australopiths: a third trochanter (also known as a gluteal tuberosity) and a rugose hypotrochanteric fossa. Lovejoy et al. (2002) hypothesize that both these features are associated with hypertrophy of the ascending tendon of the gluteus maximus, and they are absent in *Pan* and *Gorilla,* whose gluteus maximus insertion is more distomedial and separated from the origin of the vastus lateralis by an elevated boss on the shaft (a lateral spiral pilaster) that is absent in hominins. However, instead of interpreting these features as shared derived with australopiths, Lovejoy and colleagues (2009d) argue they are shared primitive, because a strong gluteal tuberosity just inferior to the greater trochanter is also present in several Miocene hominoids including *Ekembo (Proconsul)* (Ward 1993), *Nacholopithecus* (Nakatsukasa et al. 2003), and *Dryopithecus* (Mòyà-Sola et al. 2009b). If so, then the configuration evident in chimpanzees and gorillas evolved convergently, possibly as an adaptation to relocate the insertion of the gluteus maximus more distomedially (Lovejoy et al. 2002). Although this hypothesis cannot be disproved, *Proconsul, Nacholopithecus,* and *Dryopithecus* are not closely related to hominines (see above), and the first two have been reconstructed as pronograde, nonsuspensory apes (the positional repertoire of *Dryopithecus* being unclear). Without fossil evidence for stem hominines, it is equally plausible to reconstruct the LCA_{G-CH} as a suspensory, orthograde ape lacking a third trochanter and a rugose hypotrochanteric fossa.

Differences of opinion about how to relate Miocene ape fossils to hominin and hominid femoral anatomy also apply to interpretations of the *Orrorin* femora, especially BAR 1002'00. Initial descriptions of this femur argued that it lacked any shared primitive features with AGAs, but instead shared numerous derived features with later hominins, including an intertrochanteric line, an elongated femoral neck, a third trochanter, an *obturator externus* groove, and an anteroposteriorly compressed femoral neck with a thin superior cortex (Senut et al. 2001; Pickford et al. 2002). Reanalyses of the fossil's femoral neck geometry corroborated its hominin-like morphology (Galik et al. 2004; Bleuze 2012; but see Kuperavage et al. 2010). A detailed multivariate analysis also concluded that BAR 1002'00 most closely resembled australopiths in having a long and anteroposteriorly compressed neck, a small femoral head, and a relatively wide proximal shaft (Richmond and Jungers 2008). However, Almécija and colleagues' (2013) GM comparison of the fossil to a large comparative sample including several Miocene taxa (e.g., *Ekembo, Hispanopithecus, Dryopithecus,* and *Equatorius*) yielded a different conclusion. Although Almécija et al. (2013) considered the BAR 1002'00 femur to be that of a bipedal hominin, their analysis placed the femur as phenetically intermediate between *Australopithecus* and Miocene hominoids rather than intermediate between extant humans and the African great apes. Mirroring some of the findings reported by Lovejoy et al. (2009d) for *Ardipithecus,* major similarities between BAR 1002'00 and Miocene apes include the presence of a protruding third trochanter; the shape, size, and orientation of the lesser trochanter; lateral protrusion of the base of the greater trochanter; the breadth of the proximal shaft; and the size and shape of the femoral head. Almécija et al. (2013) therefore concluded that the similarities among extant great apes, including orangutans, are convergent and most likely reflect selection for vertical climbing as a component of orthograde positional behaviors.

Although the *Orrorin* femur, like that of *Ardipithecus,* does share similarities with gibbons and early Miocene hominoids, we disagree that these phenetic similarities are strong evidence for a non-African ape-like LCA. As noted above, none of the pronograde Miocene apes included in the Almécija et al. (2013) analysis are closely related to hominines. Moreover, their analysis (Almécija et al. 2013: supplementary table 4) indicates that, phenetically, BAR 1002'00's closest overall resemblance is to other hominins rather than *Hylobates,* the exception being *Ekembo,* which they calculate as most similar to *Orrorin* and second most similar to AL-333 (hylobatids, in turn, most resemble *Equatorius, Ekembo,* and *Hispanopithecus*). In fact, it is more parsimonious to infer that extant great apes evolved from an as-yet-unknown orthograde common ancestor than to infer that not only *Gorilla* and *Pan* but also *Pongo* and *Hylobates* independently acquired adaptations for suspensory locomotion. Further, the effects of selection for bipedalism on femoral shape are unclear, and not always distinctive, as evinced by the overall similarity of hominins, including humans, to gibbons rather than chimps and gorillas in certain principal components, as reported by Almécija et al. (2013).

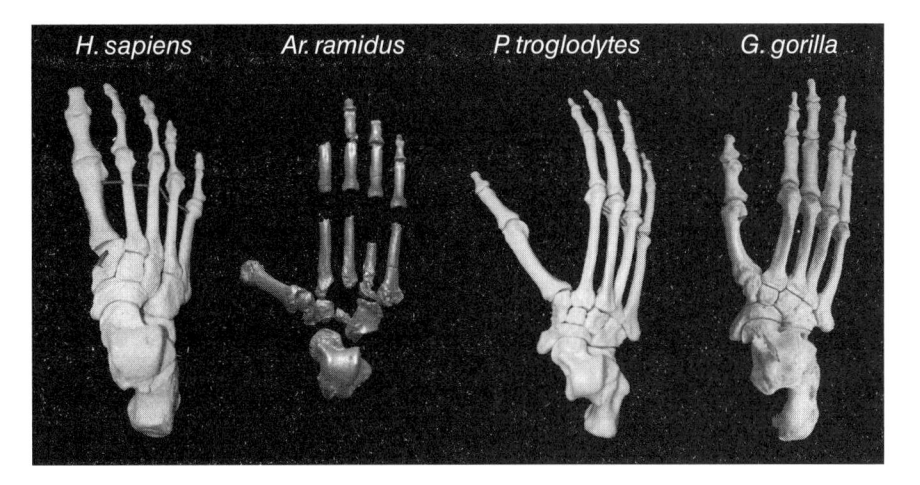

FIGURE 2.8. Dorsal view comparison of the foot of *H. sapiens, Ardipithecus ramidus* (ARA-VP-6/500), *Pan troglodytes,* and *G. gorilla.* ARA-VP-6/500 has been mirror-imaged. Rendering of ARA-VP 6/500 courtesy of © G. Suwa and T. White from figure 1b of Lovejoy et al. (2009a), *Science* 326: 72e2.

The Foot

Foot fossils are rare in the fossil record, and almost everything we know about the early hominin foot comes from the partial skeleton of *Ar. ramidus,* ARA-VP-5/600 (Lovejoy et al. 2009a). As shown in Figure 2.8, ARA-VP-5/600 includes the talus, cuboid, medial, and intermediate cuneiforms, a partial navicular, metatarsals I, II, III, and V, and several phalanges. There is also a proximal pedal phalanx (AME-VP-1/71) ascribed to *Ar. kadabba* (Haile-Selassie 2001). A younger (3.4 Ma) partial foot from Burtele, Ethiopia, however, exhibits several similarities to ARA-VP-5/600 (Haile-Selassie et al. 2012), differing in several respects from what we know of the feet of *Au. afarensis, Au. Africanus,* and *Au. sediba* (Harcourt Smith and Aiello 2004; Clarke and Tobias 1995; Kidd and Oxnard 2005; McHenry and Jones 2006; Zipfel et al. 2011). Given the anatomical and functional complexity of the foot, we focus separately on the rearfoot, midfoot, and forefoot.

Rearfoot

The only rearfoot element currently known for early hominins is the ARA-VP-5/600 talus. According to the detailed description of this bone (Lovejoy et al. 2009a), the ARA-VP-5/600 talus is generally chimp-like, with a high talocrural joint angle (the talar axis angle) of 14.5°, similar to that of chimps (15.5°±2.9) and quadrupedal Old World Monkeys (13.2°±2.2), but lower than gorillas (17.8°±2.7) or orangs (18.4°±3.5) and higher than humans (10.4°±2.3) and australopiths (7.4°±1.4). Substantial angulation of this joint, clearly a primitive feature, indicates a high degree of foot inversion (DeSilva 2009), suggesting that *Ardipithecus*, like chimpanzees, walked with a relatively varus knee. Like *Pan* and *Gorilla*, and unlike later hominins, the ARA-VP-5/600 talus also has an obliquely oriented groove for the flexor hallucis longus. Although the *Ardipithecus* talus therefore resembles those of chimpanzees, Lovejoy and colleagues (2009a) report that that it does have a few derived hominin features, notably a prominent tubercle for the attachment of the anterior talofibular ligament (probably to confer extra stability against hyperinversion), and minimal expression of the cotylar fossa. This fossa, which conjoins with the tibial malleolus, is pronounced in the African great apes and Miocene hominoids like *Ekembo (Proconsul)* (Walker 1997), probably to stabilize the joint in extreme dorsiflexion.

Midfoot

Features of the ARA-VP-5/600 midfoot are mostly primitive on the medial side and more derived on the lateral side. In terms of the lateral tarsals, Lovejoy et al. (2009a) report that the length of the cuboid in ARA-VP-5/600 relative to a geometric mean of capitate and talus measurements is 1.4, and thus similar to Old and New World monkeys (ca. 1.5), shorter than humans (1.8), and longer than in *Pan* and *Gorilla* (ca. 1.2). Although ARA-VP-5/600's relative cuboid length is interpreted as shared primitive with monkeys, the geometric mean used as a proxy for body mass has been shown to overestimate body mass in ARA-VP-5/600 by 59 percent—as 51 kg rather than 32 kg, thus likely underestimating its relative length (Grabowski et al. 2015). Further, even if humans and monkeys have relatively longer cuboids than the African great apes, diagnosing this trait as primitive would require cuboid length to have shortened independently at least three times in the great apes. An equally or more likely scenario is that there was selection for a longer cuboid in bipedal hominins to increase the out-lever arm of the plantarflexors during toe-off. Additional derived features of the cuboid are the shapes of the facets that articulate with the proximal heads of MT4 and MT5. These facets tend to be sinusoidal in *Pan* and *Gorilla* as well as Old World Monkeys to permit high angles of plantarflexion, but they are relatively flat in ARA-VP-5/600 and in humans, presumably to restrict joint mobility during toe-off (Lovejoy et al. 2009a).

Another likely derived feature of the lateral midfoot inferred from the ARA-VP-5/600 cuboid concerns the os peroneum (OP). This sesamoid bone in the peroneus longus tendon (also called the fibularis longus) articulates with a facet on the inferomedial aspect of the cuboid. Lovejoy et al. (2009a) describe the OP facet in ARA-VP-5/600 as broad, shallow, and within the peroneus longus groove, and argue that the OP is common in Old World Monkeys, lacking in extant great apes, and only partly calcified in humans; they also note that the facet is displaced proximal and lateral to the groove in just humans. As a result, they consider the presence of an OP in humans, *Ardipithecus,* and monkeys to be primitive, and the OP's absence to be a derived feature in AGAs. Since the OP increases the moment arm of the peroneus longus, which everts and plantarflexes the foot and possibly also stiffens the longitudinal arch (Kokubo et al. 2012), Lovejoy et al. (2009a) argue that the OP and its facet was lost in African great apes as an adaptation to permit "increased midtarsal laxity" during climbing.

There are two problems with this argument. First, comparative studies indicate that an OP is present in the peroneus longus tendon in most monkeys and gibbons, but in less than 30 percent of adult humans, and that partially calcified OPs are sometimes present in *Pan* and *Gorilla* (Manners-Smith 1908; Morton 1922; Oxnard and Lisowski 1980; Le Minor 1987; Muehleman et al. 2009). Second, despite almost always having an OP, gibbons and Old World Monkeys do not have a stiffened midfoot incapable of considerable flexion (Veereke and Aerts 2008; Patel 2010; DeSilva 2010). More detailed studies are necessary, but the expanded OP facet in ARA-VP-5/600 is unlikely to be a useful feature for inferring either phylogenetic relationships or midfoot mobility. We agree with Lovejoy et al. (2009a), however, that the location of the OP facet in ARA-VP-5/600 within the peroneus longus groove is probably primitive; in humans and early *Homo* (OH 8), the facet is located proximolaterally to the groove, reorienting the tendon more obliquely, thus increasing its moment arm to resist flexion in the midfoot between the tarsals and the metatarsals (Lewis 1989).

Although comparisons of the mostly human-like cuboid of ARA-VP-5/600 with other primates do not support the contention that LCA's lateral midfoot was primitive relative to African great apes, there is little disagreement that the more primitive, medial side of its midfoot shares many similarities to *Pan* and *Gorilla.* Lovejoy and colleagues (2009a) emphasize that the orientation of the MT1 joint on the medial cuneiform of ARA-VP-5/600 indicates a high degree of abduction, and that the spiral concavity of the proximal MT1 facet, also typical of *Pan* and *Gorilla,* would have facilitated conjunct rotation on the hemicylindrical facet of the medial cuneiform (Lewis 1989). Further, although only a small fragment of the ARA-VP-5/600 navicular is preserved, it appears to lack the anteroposterior elongation that is typical of *H. sapiens,* but is also argued to be longer than in *Pan* (Lovejoy et al. 2009a). Clearly, this hypothesis needs to be assessed metrically and with better material.

Metatarsals and Phalanges

Descriptions of the metatarsals and phalanges of *Ardipithecus* indicate that they share some derived features with hominins, but are otherwise very similar to those of the African great apes, especially *Pan.* In spite of having a very chimp-like, abducent first ray (see above), the lateral metatarsals of ARA-VP-5/600 are described as sharing a number of derived features with later hominins (Lovejoy et al. 2009a). In terms of the MT2, its proximal articular surface is relatively large compared to *Pan* and *Gorilla,* with invaginations on the dorsal-superior margin. These invaginations are interpreted as traces of substantial buttressing of the joint, and thus evidence for a human-like center of pressure during toe-off that was centered on the second MTP joint. As in other bipedal hominins, the ARA-VP-5/600 MT3 has a domed, dorsally oriented distal head, a deeply angled gutter on the dorsal aspect between the head and the shaft, a relatively straight shaft, and a relatively tall proximal articular surface (Latimer and Lovejoy 1990; Haile-Selassie et al. 2012). Altogether, these features are likely derived adaptations for toe-off during bipedal locomotion, because they would have facilitated greater hyperextension of the MTP joint than in *Pan* and *Gorilla,* helped prevent a midtarsal break, and improved resistance to high sagittal bending forces (Griffin et al. 2010; Proctor 2013; DeSilva et al. 2015; Fernández et al. 2015).

The pedal phalanges of ARA-VP-5/600 are generally ape-like in terms of their length and curvature, although the fourth and fifth proximal phalanges have strong dorsal canting of the proximal articular surface, along with domed distal heads (Lovejoy et al. 2009a) typical of later hominins (e.g., Griffin and Richmond 2010; Ward et al. 2011). A fourth proximal phalanx of *Ar. kadabba,* AME-VP-1/71, also has a dorsally canted proximal surface combined with strong curvature (Haile-Selassie 2001).

Forelimb

Forelimb remains of pre-*Australopithecus* hominins are known from specimens of *O. tugenensis* (Senut et al. 2001; Gommery and Senut 2006), *Ar. kadabba* (Haile-Selassie 2001; Haile-Selassie et al. 2009), and two partial skeletons of *Ar. ramidus* (ARA-VP-5/600, which includes the lower arm and most of both hands, and ARA-VP-7/2, a relatively complete forelimb), along with numerous unassociated specimens (Lovejoy et al. 2009b). Based on these remains, Lovejoy et al. (2009b) and White et al. (2015) argue that *Ardipithecus*, like later hominins, lacked specializations evident in *Pan* and *Gorilla* for both knuckle-walking and suspensory climbing. They further conclude that these hominins retain mostly primitive features for pronograde, nonsuspensory arboreal locomotion evident in monkeys and many Miocene apes, that hominins did not have a knuckle-walking ancestry, and that knuckle-walking and suspensory climbing features evolved independently in *Pan* and *Gorilla*.

As with arguments involving other parts of the *Ar. ramidus* skeleton, there are competing hypotheses. *Ar. ramidus*, a biped, was clearly not a knuckle-walker, and is dated to 1.5–3 million years after hominins are hypothesized to have diverged from the LCA, making the absence of relevant adaptations for knuckle-walking understandable. In addition, as we have noted previously, these conclusions require improbable levels of homoplasy. It bears repeating that when *Pan* and *Gorilla* were believed monophyletic, there was near universal agreement that their many forelimb similarities were homologies, with any differences interpreted as consequences of differences in body mass. Larson (1998) did argue persuasively for some degree of convergence in extant hominoid forelimb morphology (in part to explain the lack of upper limb features shared between *Pongo* and *Sivapithecus*), but Young's (2003) reanalysis of Larson's data showed, as did the aforementioned analyses of Begun et al. (1997), Stevens et al. (2013), and Alba et al. (2015), that the extant great apes clustered together, with branching sequence congruent with the molecular phylogeny. The extant hominids uniquely resemble each other in the forelimb in both hard and soft tissue anatomy (Groves 1986; Gibbs et al. 2002; Diogo and Wood 2011). Electromyographical studies of the forelimb have also shown that patterns of activity and recruitment are similar across great apes, with *Pan* and *Gorilla* patterns being most similar (Tuttle and Basmajian 1974, 1978a, 1978b; Tuttle et al. 1992; Larson and Stern 2013).

As with other anatomical regions, we evaluate alternative hypotheses about forelimb anatomy in the LCA by reviewing similarities and differences in the earliest hominin forelimb with australopiths, the African great apes, and other taxa including Miocene apes. Evaluating these features in their correct phylogenetic context is necessary to address competing claims, either that the LCA's forelimb resembled those of *Pan* and *Gorilla,* or that LCA and early hominins shared a primitive configuration of the forelimb unlike any of the great apes, whose many similarities are convergent (Lovejoy et al. 2009b).

Clavicle

A recent 3D geometric morphometric analysis of extant primate clavicles (Squyres and DeLeon 2015) shows a combination of phylogenetic and functional patterning, with overall resemblances between *Hylobates* and *Ateles,* as well as between *Pan* and *Homo,* implying functions similar enough, even though embedded in different behavioral repertoires, to select for or retain similar morphology.

A mostly lateral fragment of a left clavicle of *Ar. kadabba* (STD-VP-2/893) described by Haile-Selassie et al. (2009) from the Asa-Koma Member of the Adu-Asa Formation, Middle Awash, Ethiopia, and dated between 5.5 and 5.8 Ma, shows a mosaic of predominantly human- with some chimp-like features, and matches clavicles of *Au. afarensis*. The specimen is too fragmentary to conclude that its well-developed conoid tubercle implies a well-reinforced clavicular-scapular joint. Larson's review (2007) notes that the small australopith clavicular sample, while lacking thorough analysis, suggests an ape-like position of the shoulder.

Scapula

No scapulae have yet been recovered from *Sahelanthropus, Orrorin,* or *Ardipithecus,* but it has long been appreciated that there are many similarities between the scapulae of AGAs and australopiths, especially *Au. afarensis* and *Au. sediba* (for details, see Larson 2007; Haile-Selassie et al. 2010; Churchill et al. 2013). These similarities have been emphasized in several recent morphometric analyses. Focusing on the nearly complete Dikika scapula (DIK 1-1), Alemseged et al. (2006) and Green and Alemseged (2012) not only confirmed the ape-like shape of the *Au. afarensis* scapula but also found it to be morphologically more like gorillas than chimpanzees. A larger, subsequent analysis by Young et al. (2015) that included crown hominoids, hominins, and Old World monkeys found that scapular shape variation among bonobos, chimps, and almost all hominins falls along a single principal component of variation (explaining 47 percent of the variance), with gorillas being separated along a second component (explaining 19 percent of the variance). They therefore inferred that the LCA most likely had a scapula that was more like those of *Pan* than *Gorilla*. That said, there is variation in scapular shape among australopiths, as evident from the less cranially oriented spine of a partial adult scapula (KSD-VP-1/1) attributed to *Au. afarensis* from Woranso-Mile, Ethiopia (Haile-Selassie et al. 2010; but see Almécija 2016 for a cautionary note).

Although scapular shape differs between chimps and gorillas, in both species it reflects a compromise between a shoulder selected for support during quadrupedal knuckle-walking while also being effective in suspension and vertical climbing (Rose 1991). Relative to chimpanzees, gorilla scapulae are set more cranially, are relatively wider dorsoventrally, and have glenoids oriented less cranially and more ventrally than those of chimpanzees, in part because dorsoventral broadening is an inevitable structural correlate of orienting the glenoid more ventrally. It is likely that these differences reflect contrasts in body mass, involving selection for greater shoulder joint stability during knuckle-walking in the larger gorilla while accommodating less frequent arboreal vertical climbing among adults (Doran 1997).

Unlike *Pan* and *Gorilla, Au. afarensis* was a habitual biped, and neither a quadruped nor a knuckle-walker. Its scapular shape reflects selection for a slightly more ventrally oriented glenoid than in *Gorilla* relative to the vertebral border (21° compared with 25° for the ape). This morphology probably reflects some arboreal forelimb suspensory behaviors, along with other currently unclear functions.

Scapulae in Miocene apes are rare, and all are fragmentary and distorted. The least badly preserved is that of *Nacholapithecus kerioi* (Senut et al. 2004); that of *Ekembo (Proconsul) heseloni* is even less well preserved (Rose 1993; Senut et al. 2004). Even so, enough anatomy is preserved to observe that they most resemble *Colobus* or male *Nasalis*. There is no evidence that the forelimbs of these Miocene hominoids were adapted for vertical climbing or suspension, nor any reason to believe that these meager remains contribute anything to our understanding of the LCA (unless all the extant hominoids indeed independently evolved adaptations for suspensory locomotor behavior).

Humerus

Pre-*Australopithecus* humeri include a distal half of a diaphysis lacking epiphyses of *O. tugenensis* (BAR 1004'00, Senut et al. 2001), dated between 5.6 and 6.1 Ma; and the distal third of a left humerus shaft with trochlea, olecranon fossa, and some adjacent anatomy (STD-VP-3/78, Haile-Selassie et al. 2009), along with a shaft fragment (STD-VP-2/101, or 2/120 in Haile-Selassie 2001) of *Ar. kadabba* from the Asa-Koma Member of the Adu-Asa Formation, Middle Awash, Ethiopia, dated 5.5 to 5.8 Ma. Seven distal shaft fragments and two complete or reasonably complete *Ar. ramidus* humeri are known: ARA-VP-1/4, a humeral shaft lacking epiphyses, and ARA-VP-7/2, a humerus lacking some parts of the distal articular surfaces (White et al. 1994; Lovejoy et al. 2009b).

The *Ar. ramidus* ARA-VP-1/4 humerus contrasts with those of *Pan* and *Gorilla* and resembles later hominins in having a shallow bicipital groove and minimal head torsion (resembling hylobatids in this feature) along with an elevated, rugose deltopectoral crest. Lovejoy et al. (2009b: 70e6–70e7) reject the inference that this configuration is evidence for powerful arm muscles typical of suspensory locomotion, because they claim that the same morphology is a primitive trait evident in modern humans, cercopithecoids, *Ekembo heseloni, Sivapithecus indicus,* and *Au. afarensis,* less developed in great apes, and absent in gibbons. They further argue that deltopectoral rugosity is a "primitive trait dictated primarily by positional information rather than loading" that is retained in hominins but independently derived in "suspensory, vertical-climbing, and knuckle-walking apes, perhaps reflecting increased intermuscular fusion, as has occurred in gibbons, which also use suspension and exhibit almost no deltopectoral cortical surface manifestations (type 1 and/or 2)."

We have two comments about this argument. First, as noted by Begun (2010), the hypothesis that humeral anatomy in *Ar. ramidus* is the primitive condition for hominoids requires interpreting the many similarities among extant hominoids as homoplasies. We can frame a more formal phylogenetic framework as follows: add *Equatorius* and *Griphopithecus* to *Ekembo* as probable stem hominoids, and *Dryopithecus* and *Rudapithecus* as probable stem hominids (Begun 1992; Begun et al. 1997; Stevens et al. 2013; Alba et al. 2015), and use *Gorilla, Pongo,* and *Hylobates* as successive outgroups to *Pan*-hominins.

Viewed in this context, a more plausible interpretation is that the extant ape pattern is plesiomorphic for crown hominids, possibly for crown hominoids, with pre-*Australopithecus* hominins convergent on the Miocene stem hominoid pattern for reasons awaiting plausible biological explanation.

In addition, we question the classification of these features as so-called type 1 and 2 traits with a strong phylogenetic signal that indicates that the ancestors of hominins never relied on suspension, vertical climbing, or knuckle-walking. As defined by Lovejoy et al. (2009b: 70e7), type 1 traits are those "whose morphogenesis is the direct consequence of pattern formation; usually (but not always) subjected to direct selection;" and type 2 traits are "genetic but are pleiotropic to, or result from, hitchhiking on type 1 traits and are not themselves subject to selection." Apart from the lack of any data to test these hypothesized categorizations in the proximal humerus (or indeed anywhere), all morphology is "the direct consequence of pattern formation." This categorization fails to help the admittedly challenging task of turning complex three-dimensional hard tissue shapes into phylogenetically useful analytical categories.

As noted by Lovejoy et al. (2009b), the distal humeral morphology of *Ar. ramidus* and possibly *Ar. kadabba* (STD-VP-3/78, Haile-Selassie et al. 2009) resembles that of other crown hominoids, particularly in the trochlea, spherical capitulum, and deep zona conoidea. These features are also present in *Sivapithecus* (Pilbeam et al. 1990), *Rudapithecus* (Begun 1992), and *Oreopithecus* (Harrison 1986), but not in other known Miocene apes.

Ulna

The proximal ulna is represented by the third of a shaft attributed to *Ar. kadabba* (STD-VP-2/101, Haile-Selassie et al. 2009, or 2/120 in Haile-Selassie 2001) dated between 5.5 and 5.8 Ma, and for *Ar. ramidus* by ARA-VP-6/500-501 and ARA-VP-7/2-C-501 (Lovejoy et al. 2009b). The olecranon process is abbreviated as in all crown and some Miocene hominoids, *Hispanopithecus laeitanus* (Alba et al. 2012), *Rudapithecus* (Begun 1992), and *Oreopithecus* (Harrison 1986). The trochlear notch faces anteriorly rather than being retroflexed. Although Lovejoy et al. (2009b) note that a foreshortened olecranon is shared among hominins, the African great apes, and a number of Miocene hominoids that exhibit ulnar withdrawal at the wrist, they argue that a proximally oriented trochlear notch is an adaptation for habitual suspension because it is also found in *Pongo,* and that retroflexion of the trochlear notch is not a primitive trait. Instead, they argue that anteriorly facing trochlear notches are primitive adaptations for careful climbing and bridging, and that "its presence in *Ar. ramidus* further supports the hypothesis that hominids [hominins as used in this chapter] have never been adapted to either suspension or vertical climbing and that human and ape proximal ulnar morphology converged for different reasons" (Lovejoy et al. 2009b: 70e6).

In order to evaluate this argument, it is necessary to appreciate that ulnar withdrawal refers to lack of direct contact between the styloid process of the ulna and the proximal carpals, anatomy shared by crown hominoids reflecting the ability to ulnar deviate and supinate the wrist, important adaptations to suspension and climbing (Lewis 1989). Ulnar withdrawal in *Ar. ramidus* is therefore not surprising. The only Middle Miocene hominoid in which proximal carpal morphology shows no ulnar styloid contact with the antebrachial joint is *Pierolapithecus* (Mòyà-Solà et al. 2004), for which no ulna is known. The Late Miocene *Oreopithecus* ulna, although crushed, probably had no ulnar-antebrachial contact, according to Sarmiento (1987); it had an abbreviated olecranon (Harrison 1986; Sarmiento 1987).

As for the orientation of the trochlear notch relative to the shaft (measured as an angle), an approximating method was used in *Ar. ramidus* (Lovejoy et al. 2009b: figures S31–S32). Although the values published by Drapeau et al. (2005) and Harrison (1986) were generated differently, they are sufficiently similar to conclude that angular values are similar in humans and *Pan*, indicating a somewhat retroflexed notch; angular values are higher than these in *Gorilla* and *Pongo*, but lower in *Symphalangus* and still lower in *Hylobates*, where they are comparable to cercopithecoids and ceboids. *Oreopithecus* is more retroflexed than *Pan*. In contrast, as far as can be judged from not strictly comparable analyses, *Au. afarensis* resembles the siamang, as does *Ar. ramidus*. While Lovejoy et al. (2009b) use trochlear notch orientation to argue that *Ar. ramidus* orientation was not suspensory, its value is actually closer to that of suspensory hylobatids.

In our opinion, we lack enough information to make secure functional and phylogenetic inferences from trochlear orientation. *Homo*, which makes and uses tools, resembles knuckle-walking/suspensory *Pan* in this feature, whereas orthograde brachiating hylobatids resemble pronograde quadrupedal monkeys. Phylogenetic interpretation is also challenging, since the extant outgroups to *Pan-Homo*, *Gorilla*, and *Pongo* have higher angular values than *Pan-Homo*, while those of the largest hylobatid, *Symphalangus*, are only slightly lower (with *Hylobates* being lower still). One plausible possibility is that early hominins are derived relative to a more chimpanzee-like ancestor, with later hominins convergent on *Pan*.

Radius

One complete and slightly damaged radius (ARA-VP-6/500-039) and one complete distal radius (ARA-VP-7/2-B) are known for *Ar. ramidus*. Neither has been thoroughly described, although both show a greater distal articular surface / shaft angle, as in *Pan* and *Gorilla,* than do later hominins, a character argued to be part of the knuckle-walking adaptive complex (Richmond and Strait 2000; Orr 2005), but which Lovejoy et al. (2009b: 70e5) conclude to be a primitive character unrelated to knuckle-walking. Lovejoy et al. (2009b) also argue that the distal radius of *Ar. ramidus* indicates that broadening of the radiocarpal joints in African apes and hominins is convergent, but for different functional reasons: knuckle-walking in apes and tool-using or -making in hominins. As they state: "Scaphoid expansion and palmar broadening almost certainly underlie reduced radiocarpal joint angulation in later humans. Morphometrically based suppositions attributing these various characters to a history of knuckle-walking or suspension in *Australopithecus* have been critiqued previously on theoretical grounds and are now moot, because anatomical evidence indicates that *Ar. ramidus* was never reliant on either."

We do not dispute that the expansion and palmar broadening of the scaphoid in both knuckle-walkers and non-knuckle-walkers indicates that these traits cannot be used in isolation to infer knuckle-walking (Orr 2005), and we agree that they are probably primitive for hominines. However, it does not necessarily follow that hominines did not have a knuckle-walking ancestry. Instead of asserting that radiocarpal joints broadened in parallel in African apes and humans, a phylogenetic analysis shows that *Pan* and *Gorilla* are derived relative to all outgroups, hence it is as probable that humans are convergent on the African ape-like ancestral hominine morphotype.

Carpals

The carpus of *Ar. ramidus* is represented by many specimens, the best being an essentially complete wrist of ARA-VP-6/500. Extant hominines share early centrale-scaphoid fusion (evolved in parallel in lemuroids, Kivell and Begun 2007), as do fossil hominins (Kivell and Begun 2007), including *Ar. ramidus*

(Lovejoy et al. 2009b). Where known, Miocene hominoid carpals (*Ekembo, Equatorius, Nacholapithecus, Oreopithecus, Pierolapithecus,* and *Rudapithecus*) have an unfused scaphoid-centrale (Kivell and Schmitt 2009).

As Orr (2005) cogently argued, morphological similarities among closely related taxa such as *Pan, Gorilla,* and *Homo* (as well as fossil hominines) are to be expected. Identifying those due to phylogeny and those due to function (particularly positional behavior, in this case specifically knuckle-walking) is challenging. The thorough descriptions of Lewis (1989) document many carpal similarities among the hominines, especially for the capitate and hamate. Chimpanzees and gorillas differ from humans in some features that can be plausibly interpreted as adaptations for knuckle-walking, while many differences between *Pan* and *Gorilla* can be linked to the greater body mass of the latter.

Almécija and colleagues' (2015a) study of hamate variation in extant primates noted these complex phylogeny/function interactions. Hylobatids show more shape changes from the hypothetical hominoid last common ancestor than do hominids, and those within hominines *Gorilla* and *Homo* have changed more than *Pan*. Daver et al. (2014) also used 3D GM analyses to explore hamate variation in extant primates, including two *Au. afarensis* hamates. As with the Almécija et al. (2015a) analysis, the great apes are similar (particularly *Pan* and *Pongo*) and overlap with *Homo* in both principal components and canonical variates analyses. According to Daver (2009), the australopiths, both identified as *Au. afarensis,* differ from each other, with the younger (~3.0 Ma) more similar to *Homo* and the older (~3.5 Ma) more similar to African great apes.

Tocheri et al. (2007) and Orr et al. (2013) use 3D GM to explore patterns of capitate variation within and between species of Hominidae, interpolating fossil specimens, including two *Au. afarensis* and an *Au. africanus*. In this analysis, *Au. afarensis* hamates fall between humans and *Pan,* while *Au. africanus* resembles *Pan* and *Pongo*. Orr et al. (2013: 125) note an interesting developmental feature of the hominid capitate, involving the cartilaginous precursor cells that develop as a separate ossification center before forming the Mc3 styloid process in humans: "This cluster of cartilaginous precursor cells occasionally fuses to the capitate (3.5% of cases) or trapezoid (0.5%) rather than the Mc3, or it is maintained as a separate ossicle (2%; O'Rahilly 1953). In non-hominin primates, the styloid precursor cells merge with the cartilaginous anlagen of the capitate, such that it forms the dorsoradial corner of the bone. That condition characterizes all capitates and most Mc3s attributed to *Australopithecus* ... and *Ardipithecus*" (Lovejoy et al. 2009b). These and many other studies confirm the overall carpal similarities of hominids, especially the hominines, describing the derived features of later hominins and the varying morphologies of the australopiths and *Ar. ramidus* intermediate between *Homo,* and in particular the African great apes. As noted above, Orr (2005) reviewed carpal morphology in hominines and early hominins to evaluate evidence for or against a hominin knuckle-walking ancestry. He did so using a convergence test of adaptation in a very broad comparative framework by comparing *Pan* and *Gorilla* knuckle-walking with that of the New World terrestrial giant anteater, *Myrmecophaga tridactyla* (order Xenarthra, family Myrmecophagidae). Orr (2005: 639) concluded that of the frequently listed African ape knuckle-walking adaptations (distally extended dorsal ridge of the distal radius, proximal expansion of the nonarticular surface of the dorsal capitate, pronounced articular ridge on the dorsal aspects of the load-bearing metacarpal heads, metacarpal heads that are wider dorsally than volarly), only one feature, proximal expansion of the nonarticular surface of the dorsal capitate, did not also occur in some digitigrade cercopithecoids. Orr, however, noted that, given the phylogenetic context, the dorsal ridge of the distal radius in earlier hominins might be an African great ape adaptation.

Regarding *Ar. ramidus* carpals (Lovejoy et al. 2009b), other than the distal radius dorsal ridge there are no possible indicators of a knuckle-walking ancestry; indeed, given that *Ar. ramidus* is not a knuckle-walker, its presence in both a knuckle-walker and non-knuckle-walker shows that it cannot be a definitive functional marker. Nor does *Ar. ramidus* show proximal expansion of the nonarticular surface of the dorsal capitate. However, there are no features in the *Ar. ramidus* wrist that preclude or refute a knuckle-walking ancestry. To reiterate, the absence of knuckle-walking features in the descendant does not mean the ancestor could not have been a knuckle-walker.

Metacarpals and Phalanges

Pre-*Australopithecus* hand specimens include for *O. tugenensis* a manual phalangeal fragment (BAR 349'00, Senut et al. 2001) dated between 5.6 and 6.1 Ma, and a distal pollical phalanx (BAR 1901'01, Gommery and Senut 2006) dated between 5.65 and 5.62 Ma, and for *Ar. kadabba,* two partial phalanges (ALA-VP-2/11, and DID-VP-1/80, Haile-Selassie et al. 2009) from the Asa-Koma

Member of the Adu-Asa Formation, Middle Awash, Ethiopia, dated 5.5 to 5.8 Ma. The metacarpals and phalanges of *Ar. ramidus* are represented by many specimens, the best being that of the partial skeleton ARA-VP-6/500 (Lovejoy et al. 2009b). They differ from all extant apes, and most resemble *Au. afarensis* in morphology and digit proportions.

As with the arm, the hand also presents a combination of features in which hominoids contrast with non-hominoids, with the contrasts among hominoids being plausibly explained by the different positional paths taken by hylobatids, orangutans, the AGAs, and humans (Lewis 1989; Almécija et al. 2015b). One example of a hominoid hand feature in which all crown hominoids are similar, despite wide contrasts in body mass and positional behavior, involves the metacarpal-phalangeal joint epiphyses (Susman 1979). This we interpret as a crown hominoid synapomorphy, and a feature rarely found outside the living hominoids. *Ateles* (Rose 1986) and *Oreopithecus* (Mòyà-Solà et al. 1999) are exceptions.

Using thumb to fourth digit length ratios in living and fossil catarrhines, Almécija et al. (2015b) reconstructed the hand proportions in the hominine and hominin common ancestors. Estimates in living and selected fossil taxa were based on normalized ratios, in which proportions ("shape") were established by dividing digit lengths by the cube root of body mass. Using these data to calculate fourth phalangeal / thumb length ratios in hominines, the value for *Homo sapiens* is 1.3, for the more terrestrial *Gorilla beringei,* 1.8, and the more arboreal *Gorilla gorilla,* 1.9, while *Pan troglodytes* and *P. paniscus* have slightly higher ratios, 2.1. Among fossil hominins, the ratios for *H. neanderthalensis* and *Australopithecus sediba* are 1.24, and for *Ardipithecus ramidus* 1.9. Almécija et al. (2015b) conclude that the relatively high thumb to digit lengths ratio in humans and gorillas, in contrast to the lower ratio in chimpanzees, characterized both hominine and hominin ancestors. They argue further that support for this inference comes from the ratio in "the plesiomorphic ape" *Ekembo heseloni* of 1.6, resembling that seen in some cercopithecoids (1.5 for *Mandrillus leucophaeus,* 1.8 for *Macaca sylvanus*). A note of caution is warranted in that the Miocene ape *Nacholapithecus* differs from *Ekembo* in pedal phalangeal length, and this may well apply to hand morphology as well, and that as a stem hominoid, *Ekembo* is a quite distant outgroup.

An alternative conclusion is that the *Pan-Gorilla* ancestor resembled *Pan,* and that the biomechanical and kinematic requirements of increased body mass selected for *relatively* shorter thumb and (particularly) phalangeal lengths in *Gorilla,* yielding a reduction in ratio from 2.1 to 1.9. The hand proportions in the equivalently sized *Pan* and *Ar. ramidus* differ similarly (2.1 to 1.9), the difference reflecting the shorter fourth digit in the hominin. This is explicable for the latter as an adaptation for either a different pattern of hand use in arboreal positional behaviors, finer manipulative behaviors, or both. In addition, Almécija et al. (2015b: 9) used two different estimates of body mass for *Ar. ramidus,* 50.8 and 35.8 kg, and as noted above, the latter is more likely to be correct (Grabowski et al. 2015). Significantly, the lower weight estimate predicts a relative thumb length for *Ar. ramidus* similar to that of *Pan troglodytes* (24 vs. 23 units, respectively), and relative fourth ray length of 43 to 50, respectively, rather than 38 to 50 with a heavier *Ar. ramidus.*

Was the LCA a Knuckle-Walker?

As noted earlier, Groves (1986), Diogo and Wood (2011), and more recently Diogo et al. (2017) have argued that *Pan* has changed relatively little from parsimony-based reconstructions of the hominine LCA in terms of forelimb anatomy. Although early hominins were certainly not knuckle-walkers, the lack of adaptations for this unique mode of locomotion is not evidence that knuckle-walking ancestry cannot be plausibly reconstructed for the LCA. Instead, given the many other lines of evidence for a AGA-like LCA, we conclude that a knuckle-walking ancestor remains the most likely reconstruction.

Conclusion

Many lines of evidence, including comparative molecular and anatomical data from extant apes as well as fossil evidence of both hominoids and hominins, can be used to reconstruct when the LCA of chimpanzees and humans lived and what it looked like. Viewed as a whole, these data do not support arguments that the LCA differed substantially from chimpanzees and gorillas, and that *Pan* and *Gorilla* are too specialized and independently derived to resemble the LCA of humans and extant apes (White et al. 2009, 2015). We also disagree that African great ape-like reconstructions of the LCA are based on a referential, metaphorical model. Instead, both comparative (top-down) and fossil-based (bottom-up) lines of evidence, when viewed in their appropriate phylogenetic context, suggest that the LCA was morphologically somewhat similar to chimpanzees and gorillas, and probably more chimp-sized than gorilla-sized. To infer otherwise requires improbable levels of homoplasy in every region of the skeleton. To be sure, the LCA was not exactly like a chimpanzee, but comparative evidence (especially when body mass effects are considered) and principles of parsimony strongly suggest that neither chimpanzees nor gorillas have changed that much morphologically from the LCA_{G-CH} or the LCA_{C-H}. Although it is axiomatic that there has been as much molecular evolution in the gorilla and chimpanzee lineages as in the hominin lineage, there are many other examples of primate radiations of a similar age including hylobatids and macaques that have also undergone minimal morphological change.

The hypothesis that the LCA was like the African great apes also bolsters rather than contradicts the widely appreciated fact that gorillas and chimpanzees are highly derived not only relative to other extant apes but also to most Miocene apes. Although some have argued that the specialized nature of *Pan* and *Gorilla* relative to most Miocene apes makes them improbable ancestors for hominins, most Miocene apes are too distantly related to the hominines to be of much utility in reconstructing the LCA. It is a matter of fact that the hominin lineage not only evolved from a relatively derived clade of hominoids; it is also more closely related to just one of these genera, *Pan*, despite the African great apes being so similar morphologically that almost no one doubted their monophyly until the molecular evidence became overwhelming and irrefutable.

In short, if one accepts parsimony as a useful principle for reconstructing evolutionary events, then the most probable hypothesis remains the null hypothesis, that the LCA of humans and chimps generally resembled *Pan* and *Gorilla* in many aspects of habitat and of anatomy, and was like *Pan* in terms of size. The strongest challenge to this hypothesis has been the reconstruction of *Ar. ramidus* as being too unlike *Pan* and *Gorilla* to have evolved from a chimp- or gorilla-like LCA (White et al. 2009, 2015). As the above review has elucidated, we propose an alternative reconstruction of this early hominin. We did so by arguing, first, that a comparative phylogenetic approach is essential. Species in speciose genera are "time machines" in the sense that they all share, unchanged from their LCA, features defining the genus; in the case of the hominine clade, we argue (as have many others) that *Gorilla* and *Pan* are relatively conservative. We then reviewed the effects of differences in body mass on the morphology of the two very similar African great ape genera, showing, as have others, that allometry, as well as the kinematic effects of differences in mass, explain many skeletal differences between the two genera (in ways identical to analyses of numerous other species groups). These analyses support the hypothesis that a smaller ancestor of *Gorilla* would have resembled *Pan*.

Our review of the early hominin fossil record, mostly *Ar. ramidus*, notes similarities to the other living hominines, while acknowledging the derived nature of much of its anatomy. We agree that *Ar. ramidus* was bipedal and not a knuckle-walker like *Pan* and *Gorilla*, but we do not agree that its anatomy makes a knuckle-walking ancestor implausible or impossible. Nor did we find in its anatomy or in that of extant hominoids support for an exceptional level of hominoid postcranial homoplasy.

Our brief review of the genetic record and molecular clocks suggests that a best estimate of the divergence of the *Pan-Homo* lineages is between 6 and 9 Ma, and of the divergence of the *Gorilla-Pan/Homo* lineage is 9–12 Ma, thereby providing guides to the Miocene fossil record. As active research continues to better understand mutation rates, estimates of divergence times will continue to become less variable and more reliable. We were unable to find plausible fossil evidence for the divergence of hylobatids and pongines, expected in the Early Miocene of Africa, nor of the hominine radiation in the African Late Miocene. We recognize that a continuing challenge is understanding the evolution of pongines in Asia, and understand that a better and less ambiguous fossil record there will undoubtedly be important in clarifying hominoid evolution in general and that of hominines in particular. This is a case where progress clearly depends on greatly improving the Miocene fossil ape record, especially for the African Late Miocene, but also for Southeast Asia.

This is not the place to review the growing literature on homoplasy, a confounding challenge for all evolutionary biologists interested in comparative and phylogenetic approaches. Like bad weather, homoplasy is a common source of complaint but difficult to rectify. We are well aware that we have criticized the alternative LCA model for invoking excessive amounts of homoplasy, while at the same time invoking it when we believe it warranted. Perhaps, as Washburn (1973) wrote over four decades ago, it would be helpful to play the "human evolution game," by comparing several theories in terms of probability (actually, likelihood) in order to force us to assess the comparative likelihoods of two or more hypotheses given the same data. Do we believe our favored hypothesis is twice as likely as the other, five times, ten times, more? As Washburn so wisely noted (1973: 67): "Once very definite positions have been taken in public, an individual becomes emotionally involved, and change becomes difficult."

With Washburn's warning in mind, we propose that Mark Twain's famous and much-quoted observation might apply (metaphorically, of course) to reports of the demise of a *Pan*-like LCA: "The report of my death was an exaggeration." We continue to view *Pan* as a conservative genus, and hypothesize that the LCA when discovered (and recognized) will share many similarities with *Pan troglodytes* and *Pan paniscus*. A recent paper (Diogo et al. 2017) provides further support for these conclusions, based on an extensive set of myological data covering muscle morphology, attachments, and innervation, generating 166 head, neck, and forelimb (HNL) characters with two or more character states. Cladistic analyses (Diogo and Wood 2011) generated a primate tree that recovered the relationships of major groups as well as those to the generic level (including the sister group relationship of *Pan troglodytes* and *Homo sapiens*), fully congruent with that based on genomic data. Diogo et al. (2017) add data from *Pan paniscus*. Notably, of more than 120 HNL muscles, since the *Pan-Homo* split, *Pan troglodytes* showed only four minor changes, all plesiomorphic reversions, two prior to and two after the split from *Pan paniscus*. Reports on equivalent analyses of hindlimb data have not yet been published, but Diogo et al. (2017) report just one more difference of bonobos from humans than for chimpanzees. Overall, the *Pan* clade has evolved myologically at a much lower rate than humans since their common ancestor, with bonobos showing very marginally less change than chimpanzees. Based on muscles, this interesting study has, however, no relevance to hypotheses discussed earlier in this chapter concerning skeletal paedomorphism in bonobos, especially cranial, nor to cognition or behavior (Wobber et al. 2010). We continue to consider it more likely than not that the common ancestor resembled chimpanzees rather than bonobos in characters relevant to discussions of organismal evolution.

However, until there is agreement about the nature of the LCA, we can expect alternative hypotheses about the selective forces that led to the divergence of the hominin and chimpanzee lineages, including the origins of bipedalism in the former. Such differences of opinion will also fuel different interpretations of the roles of habitat structure and habitat change, locomotor and positional behavior, food and foraging behavior, social organization, and intra- and interspecific competition. These ought to be sources of fruitful exchange and discussion, and let us hope they will be.

Acknowledgments

For comments and suggestions, we thank John Barry, Terence Capellini, Jeremy DeSilva, Larry Flynn, Terry Harrison, Nick Holowka, Michèle Morgan, Herman Pontzer, Victoria Tobolsky, Bernard Wood, and two anonymous reviewers for Harvard University Press.

References

Adams, D. C., A. Cardini, L. R. Monteiro, P. O'Higgins, and F. J. Rohlf. 2011. Morphometrics and phylogenetics: Principal components of shape from cranial modules are neither appropriate nor effective cladistic characters. *Journal of Human Evolution* 60: 240–243.

Aiello, L., and M. C. Dean. 1990. *An Introduction to Human Evolutionary Anatomy*. London: Academic Press.

Alba, D. M., S. Almécija, I. Casanovas-Vilar, J. M. Méndez, and S. Mòyà-Solà. 2012. A Partial skeleton of the fossil great ape *Hispanopithecus laietanus* from Can Feu and the mosaic evolution of crown-hominoid positional behaviors. *PLoS ONE* 7: e39617.

Alba, D. M., S. Almécija, D. DeMiguel, J. Fortuny, M. P. de los Rios, M. Pina, J. M. Robles, and S. Mòyà-Solà. 2015. Miocene small-bodied ape from Eurasia sheds light on hominoid evolution. *Science* 350: 6260.

Alba, D. M., J. Fortuny, and S. Mòyà-Solà. 2010. Enamel thickness in the Middle Miocene great apes *Anoiapithecus, Pierolapithecus* and *Dryopithecus*. *Proceedings of the Royal Society B* 277: 2237–2245.

Alemseged, Z., F. Spoor, W. H. Kimbel, R. Bobe, D. Geraads, D. Reed, and J. G. Wynn. 2006. A juvenile early hominin skeleton from Dikika, Ethiopia. *Nature* 443: 296–301.

Almécija, S. 2016. Pitfalls reconstructing the last common ancestor of chimpanzees and humans. *Proceedings of the National Academy of Sciences* 113: E943–E944.

Almécija, S., D. M. Alba, S. Mòyà-Solà, and M. Köhler, M. 2007. Orang-like adaptations in the fossil hominoid *Hispanopithecus laietanus:* First steps towards great ape suspensory behaviours. *Proceedings of the Royal Society B* 274: 2375–2384.

Almécija, S., C. M. Orr, M. W. Tocheri, B. A. Patel, and W. L. Jungers. 2015a. Exploring phylogenetic and functional signals in complex morphologies: The hamate of extant anthropoids as a test-case study. *Anatomical Record* 298: 212–229.

Almécija, S., J. B. Smaers, and W. L. Jungers. 2015b. The evolution of human and ape hand proportions. *Nature Communications* 6: 7717.

Almécija, S., M. Tallman, D. M. Alba, M. Pina, S. Mòyà-Solà, and W. L. Jungers. 2013. The femur of *Orrorin tugenensis* exhibits morphometric affinities with both Miocene apes and later hominins. *Nature Communications* 4: 2888.

Alpagut, B., P. Andrews, M. Fortelius, J. Kappelman, I. Temizsoy, H. Çelebi, and W. Lindsay. 1996. A new specimen of *Ankarapithhecus meteai* from the Sinap Formation of central Anatolia. *Nature* 382: 349–351.

Altmann, J., L. Gesquiere, J. Galbany, P. O. Onyango, and S. C. Alberts. 2010. Life history context of reproductive aging in a wild primate model. *Annals of the New York Academy of Sciences* 1204: 127–138.

Amster, G., and G. Sella. 2016. Life history effects on the molecular clock of autosomes and sex chromosomes. *Proceedings of the National Academy of Sciences* 113: 1588–1593.

Andrews, P., W. R. Hamilton, and P. J. Whybrow. 1978. Dryopithecines from the Miocene of Saudi Arabia. *Nature* 274: 249–251.

Andrews, P., and T. Harrison. 2005. The last common ancestor of apes and humans. In D. E. Lieberman, R. J. Smith, and J. Kelley, eds., *Interpreting the Past: Essays on Human, Primate and Mammal Evolution in Honor of David Pilbeam*. The Hague: Brill Academic.

Andrews, P., and L. Martin. 1987. Cladistic relationships of extant and fossil hominoids. *Journal of Human Evolution* 16: 101–118.

Andrews, P., L. Martin, L. Aiello, and A. E. Scandrett. 1991. Hominoid dietary evolution [and discussion]. *Philosophical Transactions of the Royal Society of London B* 334: 199–209.

Arensberg, B. 1991. The vertebral column, thoracic cage and hyoid bone. In O. Bar Yosef and B. Vandermeersch, eds., *Le Squelette Moustérien de Kébara*. Paris: Éditions du CNRS.

Asfaw, B., T. White, O. Lovejoy, B. Latimer, S. Simpson, and G. Suwa. 1999. *Australopithecus garhi*: A new species of early hominid from Ethiopia. *Science* 284: 629–635.

Atchley, W. R., and B. K. Hall. 1991. A model for development and evolution of complex morphological structures. *Biological Reviews of the Cambridge Philosophical Society* 66: 101–157.

Aulehla, A., and O. Pourquié. 2010. Signaling gradients during paraxial mesoderm development. *Cold Spring Harbor Perspectives in Biology* 2: a000869.

Barry, J. C., L. L. Jacobs, and J. Kelley. 1986. An early Middle Miocene catarrhine from Pakistan with comments on the dispersal of catarrhines into Eurasia. *Journal of Human Evolution* 15: 501–508.

Beauvilain, A., and J.-P. Watté. 2009. Was Toumaï *(Sahelanthropus tchadensis)* buried? *Anthropologie* 47: 1–6.

Begun, D. R. 1992. Phyletic diversity and locomotion in primitive European hominids. *American Journal of Physical Anthropology* 87: 311–340.

Begun, D. R. 2009. Dryopithecins, Darwin, de Bonis, and the European origin of the African apes and human clade. *Geodiversitas* 31: 789–816.

Begun, D. R. 2010. Miocene hominids and the origins of the African apes and humans. *Annual Review of Anthropology* 39: 67–84.

Begun, D. R., and E. Guleç. 1998. Restoration of the type and palate of *Ankarapithecus meteai:* Taxonomic and phylogenetic implications. *American Journal of Physical Anthropology* 105: 279-314.

Begun, D. R., M. C. Nargolwalla, and L. Kordos. 2012. European Miocene hominids and the origin of the African ape and human clade. *Evolutionary Anthropology* 21: 10–23.

Begun, D. R., C. V. Ward, and M. D. Rose. 1997. Events in hominoid evolution. In D. R. Begun, C. V. Ward, and M. D. Rose, eds., *Function, Phylogeny, and Fossils*. New York: Plenum.

Berge, C., and X. Penin. 2004. Ontogenetic allometry, heterochrony, and interspecific differences in the skull of African apes, using tridimensional Procrustes analysis. *American Journal of Physical Anthropology* 124: 124–138.

Bernor, R. L., L. J. Flynn, T. Harrison, S. T. Hussain, and J. Kelley. 1988. *Dioysopithecus* from southern Pakistan and the biochronology and biogeography of early Eurasian catarrhines. *Journal of Human Evolution* 17: 339–358.

Biewener, A. A. 1989. Scaling body support in mammals: Limb posture and muscle mechanics. *Science* 245: 45–48.

Bleuze, M. 2012. Proximal femoral diaphyseal cross-sectional geometry in *Orrorin tugenensis. Homo* 63: 153–66.

Böhme, M., H. A. Aziz, J. Prieto, V. Bachtadse, and G. Schweigert. 2011. Biomagnetostratigraphy and environment of the oldest Eurasian hominoid from the Early Miocene of Engelsweis (Germany). *Journal of Human Evolution* 61: 332–339.

Bonnefille, R. 2010. Cenozoic vegetation, climate changes and hominid evolution in tropical Africa. *Global and Planetary Change* 72: 390–411.

Bookstein, F. L. 1994. Can biometrical shape be a homologous character? In B. K. Hall, ed., *Homology: The Hierarchical Basis of Comparative Biology*. New York: Academic Press.

Breuer, T., M. Breuer-Ndoundou Hockemba, C. Olejniczak, R. J. Parnell, and M. Stokes. 2009. Physical maturation, life-history classes and age estimates of free-ranging western gorillas: Insights from Mbeli Bai, Republic of Congo. *American Journal of Primatology* 71: 106–119.

Brown, B. 1997. Miocene hominoid mandibles: Functional and phylogenetic perspectives. In D. R. Begun, C. V. Ward, and M. D. Rose, eds., *Function, Phylogeny, and Fossils*. New York: Plenum.

Brunet, M., F. Guy, D. Pilbeam, D. E. Lieberman, A. Likius, H. T. Mackaye, M. S. Ponce de León, C. P. Zollikofer, and P. Vignaud. 2005. New material of the earliest hominid from the Upper Miocene of Chad. *Nature* 434: 752–755.

Brunet M., F. Guy, D. Pilbeam, H. Taisso Mackaye, A. Likius, D. Ahounta, A. Beauvilain, C. Blondel, H. Bocherens, J.-R. Boisserie, L. De Bonis, Y. Coppens, J. Dejax, C. Denys,

P. Duringer, V., Eisenmann, G. Fanone, P. Fronty, D. Geraads, T. Lehmann, F. Lihoreau, A. Louchart, A. Mahamat, G. Merceron, G. Mouchelin, O. Otero, P. Pelaez Campomanes, M. Ponce de Leon, J.-C. Rage, M. Sapanet, M. Schuster, J. Sudre, P. Tassy, X. Valentin, P. Vignaud, L. Viriot, A. Zazzo, and C. Zollikofer. 2002. A new hominid from the Upper Miocene of Chad, Central Africa. *Nature* 418: 145–151.

Burgess, R., and Z. Yang. 2008. Estimation of hominoid ancestral population sizes under Bayesian coalescent models incorporating mutation rate variation and sequencing errors. *Molecular Biology and Evolution* 25: 1979–1994.

Burke, A., C. E. Nelson, B. A. Morgan, and C. Tabin. 1995. *Hox* genes and the evolution of vertebrate axial morphology. *Development* 121: 333–346.

Caccone, A., and J. R. Powell. 1989. DNA divergence among hominoids. *Evolution* 43: 925–942.

Capellini, T. D., K. Handschuh, L. Quintana, E. Ferretti, G. Di Giacomo, S. Fantini, G. Vaccari, S. L. Clarke, A. M. Wenger, G. Bejerano, J. Sharpe, V. Zappavigna, and L. Selleri. 2011. Control of pelvic girdle development by genes of the Pbx family and Emx2. *Developmental Dynamics* 240: 1173–1189.

Carbone, L., R. A. Harris, S. Gnerre, K. R. Veeramah, B. Lorente-Galdos, J. Huddleston, T. J. Meyer, J. Herrero, C. Roos, B. Aken, F. Anaclerio, et al. 2014. Gibbon genome and the fast karyotype evolution of small apes. *Nature* 513: 195–201.

Cardini, A., and S. Elton. 2008. Does the skull carry a phylogenetic signal? Evolution and modularity in guenons. *Biological Journal of the Linnean Society* 93: 813–834.

Cartmill, M. 1992. New views on primate origins. *Evolutionary Anthropology* 1: 105–111.

Cartmill, M. 1994. A critique of homology as a morphological concept. *American Journal of Physical Anthropology* 94: 115–123.

Casanovas-Vilar, I., D. M. Alba, M. Garacés, J. M. Robles, and S. Mòyà-Solà. 2011. Updated chronology for the Miocene hominoid radiation in western Eurasia. *Proceedings of the National Academy of Sciences* 108: 5554–5559.

Chamainee, Y., D. Jolly, M. Benammi, P. Tafforeau, D. Duzer, I. Moussa, and J.-J. Jaeger. 2003. A Middle Miocene hominoid from Thailand and orangutan origins. *Nature* 422: 61–65.

Chamainee, Y., V. Suteethorn, P. Jintasakui, C. Vidthayanon, B. Marandat, and J.-J. Jaeger. 2004. A new orang-utan relative from the Late Miocene of Thailand. *Nature* 427: 439–441.

Chavasseau, O. 2008. Les faunes miocènes de grands mammifères d'Asie du Sud-Est: biochronologie et biogéographie. Ph.D. dissertation, University of Poitiers, France.

Chavasseau, O., Y. Chaimanee, S. T. Tun, A. N. Soe, J. C. Barry, B. Marandat, J. Sudre, L. Marivaux, S. Ducrocq, and J.-J. Jaeger. 2006. Chaungtha, a new Middle Miocene mammal locality from the Irrawddy Formation, Myanmar. *Journal of Asian Earth Sciences* 28: 354–362.

Churchill, S. E., T. W. Holliday, K. J. Carlson, T. Jashashvili, M. E. Macias, S. Mathews, T. L. Sparling, P. Schmid, D. J. de Ruiter, and L. R. Berger. 2013. The upper limb of *Australopithecus sediba*. *Science* 340: 1233477.

Clarke, R. J., and P. V. Tobias. 1995. Sterkfontein member 2 foot bones of the oldest South African hominid. *Science* 269: 521–524.

Cobb, S. N., and P. O'Higgins. 2004. Hominins do not share a common postnatal facial ontogenetic shape trajectory. *Journal of Experimental Zoology* 302: 302–321.

Cock, A. G. 1966. Genetical aspects of metrical growth and form in animals. *Quarterly Review of Biology* 41: 131–190.

Cole, T. M. 1992. Postnatal heterochrony of the masticatory apparatus in *Cebus apella* and *Cebus albifrons*. *Journal of Human Evolution* 23: 253–282.

Collard, M., and B. A. Wood. 2000. How reliable are human phylogenetic hypotheses? *Proceedings of the National Academy of Sciences* 97: 5003–5006.

Conroy, G. C., M. Pickford, B. Senut, J. Van Couvering, and P. Mein. 1992. *Otavipithecus namibiensis*, first Miocene hominoid from southern Africa. *Nature* 356: 144–148.

Cote, S. M. 2004. Origins of the African hominoids: An assessment of the paleobiogeographical evidence. *Comptes Rendus Palevol* 3: 323–340.

Couvreur, T. L. P., L. W. Chatrou, M. S. M. Sosef, and J. E. Richardson. 2008. Molecular phylogenetics reveal multiple tertiary vicarance origins of the African rain forest trees. *BMC Biology* 6: 54.

Crompton, R. H., E. E. Vereecke, and S. K. Thorpe. 2008. Locomotion and posture from the common hominoid ancestor to fully modern hominins, with special reference to the last common panin/hominin ancestor. *Journal of Anatomy* 212: 501–543.

Dainton, M., and G. A. Macho. 1999. Did knuckle walking evolve twice? *Journal of Human Evolution* 36: 171–94.

Darwin, C. 1971. *The Descent of Man and Selection in Relation to Sex*. London: Charles Murray.

Daver, G. 2009. The articular wrist complex in Miocene and Plio-Pleistocene African hominoids: Anatomofunctional and morphometric comparative approach. *Bulletins et Mémoires de la Société d'Anthropologie de Paris* 21: 233–239.

Daver, G., F. Détroit, G. Berillon, S. Prat, and D. Grimaud-Hervé. 2014. Fossil hominins, quadrupedal primates and the origin of human bipedalism: A 3D geometric morphometric analysis of the Primate hamate. *Bulletins et Mémoires de la Société d'Anthropologie de Paris* 26: 121–128.

Dean, M. C., and B. A. Wood. 1981. Metrical analysis of the basicranium of extant hominoids and *Australopithecus*. *American Journal of Physical Anthropology* 54: 63–71.

Dean, M. C., and B. A. Wood. 1982. Basicranial anatomy of Plio-Pleistocene hominids from East and South Africa. *American Journal of Physical Anthropology* 59: 157–174.

De Bonis, L., and G. Koufos. 1994. Our ancestors' ancestor: *Ouranopithhecus* is a Greek link in human ancestry. *Evolutionary Anthropology* 3: 75–81.

Delson, E., N. Eldredge, and I. Tattersall. 1977. Reconstruction of hominid phylogeny: A testable framework based on cladistic analysis. *Journal of Human Evolution* 6: 263–278.

De Manuel, M., M. Kuhlwilm, P. Frandsen, V. C. Sousa, T. Desai, J. Prado-Martinez, and J. Hernandez-Rodriguez. 2016. *Science* 354: 477–481.

DeSilva, J. M. 2008. Vertical climbing adaptations in the anthropoid ankle and midfoot: Implications for locomotion in Miocene catarrhines and Plio-Pleistocene hominins. Ph.D. dissertation, University of Michigan, Ann Arbor.

DeSilva, J. M. 2009. Functional morphology of the ankle and the likelihood of climbing in early hominins. *Proceedings of the National Academy of Sciences* 106: 6567–6572.

DeSilva, J. M. 2010. Revisiting the "midtarsal break." *American Journal of Physical Anthropology.* 141: 245–258.

DeSilva, J. M., R. Bonne-Annee, Z. Swanson, C. M. Gill, M. Sobel, J. Uy, and S. V. Gill. 2015. Midtarsal break variation in modern humans: Functional causes, skeletal correlates, and paleontological implications. *American Journal of Physical Anthropology* 156: 543–552.

DeSilva, J. M., M. E. Morgan, J. C. Barry, and D. Pilbeam. 2010. A hominoid distal tibia from the Miocene of Pakistan. *Journal of Human Evolution* 58: 147–154.

DeSilva, J., E. Shoreman, and L. MacLatchy. 2006. A fossil hominoid proximal femur from Kikorongo Crater, southwestern Uganda. *Journal of Human Evolution* 50: 687–695.

Diogo, R., and B. A. Wood. 2011. Soft-tissue anatomy of the primates: Phylogenetic analyses based on the muscles of the head, neck, pectoral region and upper limb, with notes on the evolution of these muscles. *Journal of Anatomy* 219: 273–359.

Diogo, R., J. L. Molnar, and B. Wood. 2017. Bonobo anatomy reveals stasis and mosaicism in chimpanzee evolution, and supports bonobos as the most appropriate extant model for the common ancestor of chimpanzees and humans. *Scientific Reports* 7: 608.

Doran, D. M. 1997. Ontogeny of locomotion in mountain gorillas and chimpanzees. *Journal of Human Evolution* 32: 323–344.

Drapeau, M. S. M., C. V. Ward, W. H. Kimbel, D. C. Johanson, and Y. Rak. 2005. Associated cranial and forelimb remains attributed to *Australopithecus afarensis* from Hadar, Ethiopia. *Journal of Human Evolution* 48: 593–642.

Edwards, S. V., and P. Beerli. 2000. Perspective: Gene divergence, population divergence, and the variance in coalescence time in phylogeographic studies. *Evolution* 54: 1839–1854.

Emonet, E. G., L. Andossa, H. Taïsso Mackaye, and M. Brunet. 2013. Subocclusal dental morphology of *Sahelanthropus tchadensis* and the evolution of teeth in hominins. *American Journal of Physical Anthropology* 153: 116–123.

Ersoy, A., J. Kelley, P. Andrews, and B. Alpagut. 2008. Hominoid phalanges from the Middle Miocene site of Pasalar, Turkey. *Journal of Human Evolution* 54: 518–529.

Feakins, S. J., N. E. Levin, H. M. Liddy, A. Sieracki, T. I. Eglinton, and R. Bonnefille. 2013. Northeast African vegetation change over 12 m.y. *Geology* 41: 295–298.

Fenner, J. N. 2005. Cross-cultural estimation of the human generation interval for use in genetics-based population divergence studies. *American Journal of Physical Anthropology* 128: 415–423.

Fernández, P. J., S. Almécija, B. A. Patel, C. M. Orr, M. W. Tocheri, and W. L. Jungers. 2015. Functional aspects of metatarsal head shape in humans, apes, and Old World monkeys. *Journal of Human Evolution* 86:136–46.

Filler, A. 2007. *The Upright Ape*. Franklin Lakes, NJ: Career Press.

Fleagle, J. G. 1985. Size and adaptation in primates. In W. L. Jungers, ed., *Size and Scaling in Primate Biology*. New York: Plenum.

Flower, B. P., and J. P. Kennett. 1994. The Middle Miocene climatic transition: East Antarctic ice sheet development, deep ocean circulation and global carbon cycling. *Palaeogeography, Palaeoclimatology, Palaeoecology* 108: 537–555.

Frankham, R. 1995. Effective population size / adult population size ratios in wildlife: A review. *Genetic Research Cambridge* 66: 95–107.

Fu, Q., H. Li, P. Moorjani, F. Jay, S. M. Slepchenko, A. A. Bondarev, P. L. F. Johnson, A. Aximu-Petri, K. Prüfer, C. de Filippo, M. Meyer, N. Zwyns, D. C. Salazar-García, Y. V. Kuzmin, S. G. Keates, P. A. Kosintsev, D. I. Razhev, M. P. Richards, N. V. Peristov, M. Lachmann, K. Douka, T. F. G. Higham, M. Slatkin, J.-J. Hublin, D. Reich, J. Kelso, T. B. Viola, and S. Pääbo. 2014. Genome sequence of a 45,000-year-old modern human from western Siberia. *Nature* 514: 445–450.

Galik, K., B. Senut, M. Pickford, D. Gommery, J. Treil, A. J. Kuperavage, and R. B. Eckhardt. 2004. External and internal morphology of the BAR 1002'00 *Orrorin tugenensis* femur. *Science* 305: 1450–1453.

Galis, F., D. R. Carrier, J. van Alphen, S. D. van der Mije, T. J. M. Van Dooren, J. A. J. Metz, and C. M. A. ten Broek. 2014. Fast running restricts evolutionary change of the vertebral column in mammals. *Proceedings of the National Academy of Sciences* 111: 11401–11406.

Gao, Z., M. J. Wyman, G. Sella, and M. Przeworski. 2016. Interpreting the dependence of mutation rates on age and time. *PLoS Biology* 14: e1002355.

Gebo, D. L. 1996. Climbing, brachiation, and terrestrial quadrupedalism: Historical precursors of hominid bipedalism. *American Journal of Physical Anthropology* 101: 55–92.

Gebo, D. L., L. MacLatchy, R. Kityo, A. Deino, J. Kingston, and D. Pilbeam. 1997. A hominoid genus from the Early Miocene of Uganda. *Science* 276: 401–404.

Gibbs, S., M. Collard, and B. A. Wood. 2000. Soft-tissue characters in higher primate phylogenetics. *Proceedings of the National Academy of Sciences* 97: 11130–11132.

Gibbs, S., M. Collard, and B. Wood. 2002. Soft-tissue anatomy of the extant hominoids: A review and phylogenetic analysis. *Journal of Anatomy* 200: 3–49.

Giles, E. 1956. Cranial allometry in the great apes. *Human Biology* 28: 43–58.

Goldmann, J. M., W. W. S. Wong, M. Pinelli, T. Farrah, D. Bodian, A. B. Stittrich, et al. 2016. Parent-of-origin-specific signatures of *de novo* mutations. *Nature Genetics* 48: 935–943.

Gommery, D., and B. Senut. 2006. The terminal thumb phalanx of *Orrorin tugenensis* (Upper Miocene of Kenya). *Geobios* 39: 372–384.

Goodman, M. 1963. Man's place in the phylogeny of the primates as reflected in serum proteins. In S. L. Washburn, ed., *Classification and Human Evolution*. Chicago: Aldine.

Gould, S. J. 1966. Allometry and size in ontogeny and phylogeny. *Biological Reviews of the Cambridge Philosophical Society* 41: 587–640.

Gould, S. J. 1975. Allometry in primates, with emphasis on scaling and the evolution of the brain. *Contributions to Primatology* 5: 244–292.

Grabowski, M. W., J. D. Polk, and C. C. Roseman. 2012. Divergent patterns of integration and reduced constraint in the human hip and the origins of bipedalism. *Evolution* 65: 1336–1356.

Grabowski, M. W., K. G. Hatala, W. L. Jungers, and B. G. Richmond. 2015. Body mass estimates of hominin fossils and the evolution of human body size. *Journal of Human Evolution* 85: 75–93.

Green, D. J., and Z. Alemseged. 2012. *Australopithecus afarensis* scapular ontogeny, function, and the role of climbing in human evolution. *Science* 338: 514–517.

Gregory, W. K. 1927. How near is the relationship of man to the chimpanzee-gorilla stock? *Quarterly Review of Biology* 2: 549–560.

Grehan, J. R., and J. H. Schwartz. 2009. Evolution of the second orangutan: phylogeny and biogeography of hominid origins. *Journal of Biogeography* 36: 1823–1844.

Griffin, N. L., and B. G. Richmond. 2010. Joint orientation and function in great ape and human proximal pedal phalanges. *American Journal of Physical Anthropology* 141: 116–123.

Griffin, N. L., K. D'Août, B. Richmond, A. Gordon, and P. Aerts. 2010. Comparative *in vivo* forefoot kinematics of *Homo sapiens* and *Pan paniscus*. *Journal of Human Evolution* 59: 608–619.

Groves, C. P. 1972. Systematics and phylogeny of gibbons. In D. M. Rumbaugh, ed., *Gibbon and Siamang*, vol. 1: *Evolution, Ecology, Behavior, and Captive Maintenance*. Basel: Karger.

Groves, C. P. 1986. Systematics of the great apes. In D. Swindler, ed., *Primate Biology*, vol. 1: *Systematics, Evolution, and Anatomy*. New York: Alan R. Liss.

Groves, C. P. 1988. The evolutionary ecology of the Hominoidea. *Anuario de Psicología* 39: 87–98.

Guleç, E., and D. R. Begun. 2003. Functional morphology and affinities of the hominoid mandible from Candir. *Courier Forschungsinstitut Senckenberg* 240: 89–111.

Gundling, T. 2005. *First in Line*. New Haven, CT: Yale University Press.

Guy, F., D. E. Lieberman, D. Pilbeam, M. Ponce de León, A. Likius, H. T. Mackaye, P. Vignaud, C. P. Zollikofer, and M. Brunet. 2005. Morphological affinities of the *Sahelanthropus tchadensis* (Late Miocene hominid from Chad) cranium. *Proceedings of the National Academy of Sciences* 102: 18836–18841.

Haile-Selassie, Y. 2001. Late Miocene hominids from the Middle Awash, Ethiopia. *Nature* 412: 178–181.

Haile-Selassie, Y., B. M. Latimer, M. Alene, A. L. Deino, L. Gibert, S. M. Melillo, B. Z. Saylor, G. R. Scott, and C. O. Lovejoy. 2010. An early *Australopithecus afarensis* post-

cranium from Woranso-Mille, Ethiopia. *Proceedings of the National Academy of Sciences* 107: 12121–12126.
Haile-Selassie, Y., B. Z. Saylor, A. Deino, N. E. Levin, M. Alene, and B. M. Latimer. 2012. A new hominin foot from Ethiopia shows multiple Pliocene bipedal adaptations. *Nature* 483: 565–569.
Haile-Selassie, Y., G. Suwa, and T. D. White. 2004. Late Miocene teeth from Middle Awash, Ethiopia, and early hominid dental evolution. *Science* 303: 1503–1505.
Haile-Selassie, Y., G. Suwa, and T. D. White, T. D. 2009. Hominidae. In Y. Haile-Selassie and G. Woldegabriel, eds., *Ardipithecus kadabba: Late Miocene Evidence from the Middle Awash, Ethiopia*. Berkeley: University of California Press.
Hammond, A. S., D. M. Alba, S. Almécija, and S. Mòyà-Solà. 2013. Middle Miocene *Pierolapithecus* provides a first glimpse into early hominid pelvic morphology. *Journal of Human Evolution* 64: 658–666.
Harcourt-Smith, W. E., and L. C. Aiello. 2004. Fossils, feet and the evolution of human bipedal locomotion. *Journal of Anatomy* 204: 403–416.
Harrison, T. 1981. New finds of small fossil apes from the Miocene locality at Koru in Kenya. *Journal of Human Evolution* 10: 129–137.
Harrison, T. 1986. A reassessment of the phylogenetic relationships of *Oreopithecus bambolii* Gervais. *Journal of Human Evolution* 15: 541–583.
Harrison, T. 2005. The zoogeographic and phylogenetic relationships of early catarrhine primates in Asia. *Anthropological Science* 113: 43–51.
Harrison, T. 2010a. Apes among the tangled branches of human origins. *Science* 327: 532–534.
Harrison, T. 2010b. Dendropithecoidea, Proconsuloidea, and Hominoidea. In L. Werdelin and W. J. Sanders, eds., *Cenozoic Mammals of Africa*. Berkeley: University of California Press.
Harrison, T. 2016. The fossil record and evolutionary history of hylobatids. In U. H. Reichard et al., eds., *Evolution of Gibbons and Siamang*. New York: Springer.
Harrison, T., C. Jin, Y. Zhang, Y. Wang, and M. Zhu. 2014. Fossil *Pongo* from the Early Pleistocene *Gigantopithecus* fauna of Chongzuo, Guanxi, southern China. *Quaternary International* 354: 59–67.
Hartman, S. E. 1987. Stereophotogrammetric analysis of occlusal morphology of extant hominoid molars: phonetics and function. *American Journal of Physical Anthropology* 80: 145–166.
Hartman, S. E. 1988. A cladistic analysis of hominoid molars. *Journal of Human Evolution* 17: 489–502.
Hartwig-Scherer, S. 1992. Skeletal growth allometry and the human-chimpanzee clade. *Anthropologie et Préhistoire* 106: 37–44.
Hartwig-Scherer, S. 1993. Allometry in hominoids: A comparative study of skeletal growth trends. Ph.D. dissertation, University of Zurich.
Hartwig-Scherer, S., and R. D. Martin. 1992. Allometry and prediction in hominoids: A solution to the problem of intervening variables. *American Journal of Physical Anthropology* 88: 37–57.

Hauesler, M., S. A. Martelli, and T. Boeni. 2002. Vertebral numbers of the early hominid lumbar spine. *Journal of Human Evolution* 43: 621–643.

Hauesler, M., R. Schiess, and T. Boeni. 2011. New vertebral and rib material point to modern bauplan of the Nariokotome *Homo erectus* skeleton. *Journal of Human Evolution* 61: 575–582.

Heizman, E. P. J., and D. R. Begun. 2001. The oldest Eurasian hominoid. *Journal of Human Evolution* 41: 463–481.

Helgason, A., B. Hrafnkelsson, J. R. Gulcher, R. Ward, and K. Stefánsson. 2003. A populationwide coalescent analysis of Icelandic matrilineal and patrilineal genealogies: Evidence for a faster evolutionary rate of mtDNA lineages than Y chromosomes. *American Journal of Human Genetics* 72: 1370–1388.

Hey, J., and C. A. Machado. 2003. The study of structured populations: New hope for a difficult and divided science. *Nature Reviews Genetics* 4: 535–543.

Hill, A. 1985. Early hominid from Baringo, Kenya. *Nature* 315: 222–224.

Hill, A., M. Leakey, J. D. Kingston, and S. Ward. 2002. New cercopithecoids and a hominoid from 12.5Ma in the Tugen Hills succession, Kenya. *Journal of Human Evolution* 42: 75–93.

Hill, A., and S. Ward. 1988. Origin of the Hominidae: The record of African large hominoid evolution between 14 My and 4 My. *Yearbook of Physical Anthropology* 31: 49–83.

Hlusko, L. 2002. Identifying metameric variation in extant hominoid and fossil hominid mandibular molars. *American Journal of Physical Anthropology* 118: 86–97.

Hobolth, A., J. Y. Dutheil, J. Hawks, M. H. Schierupo, and T. Mailund. 2011. Incomplete lineage sorting patterns among human, chimpanzee, and orangutan suggest recent orangutan speciation and widespread selection. *Genome Research* 21: 349–356.

Huxley, T. H. 1863. *Evidence as to Man's Place in Nature*. London: Williams and Norgate.

Hwang, D. G., and P. Green. 2004. Bayesian Markov chain Monte Carlo sequence analysis reveals varying neutral substitution patterns in mammalian evolution. *Proceedings of the National Academy* 101: 3994–14001.

Hylander, W. L. 2013. Functional links between canine height and jaw gape in catarrhines with special reference to early hominins. *American Journal of Physical Anthropology* 150: 247–259.

Inouye, S. E. 1992. Ontogeny and allometry of African apes manual rays. *Journal of Human Evolution* 23: 107–138.

Inouye, S. E. 1994. Ontogeny of knuckle-walking hand postures in African apes. *Journal of Human Evolution* 26: 459–485.

Ishida, H., and M. Pickford. 1997. A new Late Miocene hominoid from Kenya: *Samburupithecus kiptalami* gen. et sp. nov. *Comptes rendus de l'Académie des Sciences Paris, Sciences de la terre et des planets* 325: 823–829.

Jacobs, B. F. 2004. Palaeobotanical studies from tropical Africa: Relevance to the evolution of forest, woodland and savannah biomes. *Philosophical Transactions of the Royal Society of London B* 359: 1573–1583.

Jaeger, J.-J., A. N. Soe, O. Chavasseau, P. Coster, E.-G. Emonet, F. Guy, R. Lebrun, A. Maung, A. A. Khyaw, H. Shwe. S. T. Tun, K. L. Oo, M. Rugbumrung, H. Bocherens, M. Benammi, K. Chaivanich, P. Tafforeau, and Y. Chaimanee. 2011. First hominoid from the Late Miocene of the Irraddy Formation (Myanmar). *PLoS ONE* 6: e17065.

Johanson, D. C., C. O. Lovejoy, W. H. Kimbel, T. D. White, S. C. Ward, M. E. Bush, B. M. Latimer, and Y. Coppens. 1982. Morphology of the Pliocene partial hominid skeleton (A.L. 288-1) from the Hadar Formation, Ethiopia. *American Journal of Physical Anthropology* 57: 403–451.

Jungers, W. L. 1984. Aspects of size and scaling in primate biology with special reference to the locomotor skeleton. *Yearbook of Physical Anthropology* 27: 73–97.

Jungers, W. L., and M. S. Cole. 1992. Relative growth and shape of the locomotor skeleton in lesser apes. *Journal of Human Evolution* 23: 93–105.

Jungers, W. L., and J. G. Fleagle. 1980. Postnatal growth allometry of the extremities in *Cebus albifrons* and *Cebus apella*: A longitudinal and comparative study. *American Journal of Physical Anthropology* 53: 471–478.

Jungers, W. L., and S. E. Hartman. 1988. Relative growth of the locomotor skeleton in orang-utans and other large-bodied hominoids. In J. H. Schwartz, ed., *Orang-Utan Biology*. Oxford: Oxford University Press.

Kappelman, J., B. G. Richmond, E. R. Seiffert, A. M. Maga, and T. M. Ryan. 2003. Hominoidea (Primates). In M. Fortelius, J. Kappelman, S. Sen, and R. L. Bernor, eds., *Geology and Paleontology of the Miocene Sinap Formation, Turkey*. New York: Columbia University Press.

Kato, A., N. Tang, C. Borries, A. M. Papakyrikos, K. Hinde, E. Miller, Y. Kunimatsu, E. Hirasaki, D. Shimizu, and T. M. Smith. 2013. Intra- and interspecific variation in macaque molar enamel thickness. *American Journal of Physical Anthropology* 155: 447–459.

Keith, A. 1923. Man's posture: its evolution and disorders. *British Medical Journal* 1: 451–454, 499–502, 545–548, 587–590, 624–626, 669–672.

Kelley, J. 1988. A new large species of *Sivapithecus* from the Siwaliks of Pakistan. *Journal of Human Evolution* 17: 305–324.

Kelley, J. 2005. Twenty-five years contemplating *Sivapithecus* taxonomy. In D. E. Lieberman, R. J. Smith, and J. Kelley, eds., *Interpreting the Past: Essays on Human, Primate and Mammal Evolution in Honor of David Pilbeam*. Boston: Brill Academic.

Kelley, J., P. Andrews, and B. Alpagut. 2008. A new hominoid species from the Middle Miocene site of Pasalar, Turkey. *Journal of Human Evolution* 54: 455–479.

Kelley, J., and F. Gao. 2012. Juvenile hominoid cranium from the late Miocene of southern China and hominoid diversity in Asia. *Proceedings of the National Academy of Sciences* 109: 6882–6885.

Kelley, J., and G. T. Schwartz. 2010. Dental development and life history in living African and Asian apes. *Proceedings of the National Academy of Sciences* 107: 1035–1040.

Kelley, J., and T. M. Smith. 2003. Age at first molar emergence in early Miocene *Afropithecus turkanensis* and life-history evolution in Hominoidea. *Journal of Human Evolution* 44: 307–329.

Kidd, R., and Oxnard, C. 2005. Little foot and big thoughts: A re-evaluation of the Stw573 foot from Sterkfontein, South Africa. *Homo* 55: 189–212.

Kimbel, W. H., Y. Rak, and D. L. Johanson. 2004. *The Skull of Australopithecus afarensis.* Oxford: Oxford University Press.

Kimbel, W. H., G. Suwa, B. Asfaw, Y. Rak, and T. D. White. 2014. *Ardipithecus ramidus* and the evolution of the human cranial base. *Proceedings of the National Academy of Sciences* 111: 948–953.

Kimbel, W. H., T. D. White, and D. C. Johanson. 1984. Cranial morphology of *Australopithecus afarensis:* A comparative study based on a composite reconstruction of the adult skull. *American Journal of Physical Anthropology* 64: 337–88.

King, M.-C., and A. C. Wilson. 1975. Evolution at two levels in humans and chimpanzees. *Science* 188: 107–116.

Kivell, T. L., and D. R. Begun. 2007. Frequency and timing of scaphoid-centrale fusion in hominoids. *Journal of Human Evolution* 52: 321–340.

Kivell, T. L., and D. Schmitt. 2009. Independent evolution of knuckle-walking in African apes shows that humans did not evolve from a knuckle-walking ancestor. *Proceedings of the National Academy of Sciences* 106: 14241–14246.

Kokubo, T., T. Hashimoto, T. Nagura, T. Nakamura, Y. Suda, H. Matsumoto, and Y. Toyama. 2012. Effect of the posterior tibial and peroneal longus on the mechanical properties of the foot arch. *Foot and Ankle International* 33: 320–325.

Kordos, L., and D. R. Begun. 2002. Rudabánya: A Late Miocene subtropical swamp deposit with evidence of the origin of the African apes and humans. *Evolutionary Anthropology* 11: 45–67.

Koufos, G. D., and L. de Bonis. 2006. New material of *Ouranopithecus macedoniensis* from Late Miocene of Macedonia (Greece) and study of its dental attrition. *Geobios* 39: 223–243.

Kullmer, O., S. Benazzi, D. Schulz, P. Gunz, L. Kordos, and D. R. Begun. 2013. Dental arch restoration using tooth macrowear patterns with application to *Rudapithecus hungaricus,* from the Late Miocene of Rudabánya, Hungary. *Journal of Human Evolution* 64: 151–160.

Kunimatsu, Y., M. Nakatsukasa, Y. Sawada, T. Sakai, M. Hyodo, H. Hyodo, T. Itaya, H. Nakaya, H. Saegusa, A. Mazurier, M. Saneyoshi, H. Tsujikawa, A. Yamamoto, and E. Mbua. 2007. A new Late Miocene great ape from Kenya and its implications for the origins of African great apes and humans. *Proceedings of the National Academy of Sciences* 104: 19220–19225.

Kuperavage, A. J., H. J. Sommer, and R. B. Eckhardt. 2010. Moment coefficients of skewness in the femoral neck cortical bone distribution of BAR 1002'00. *Homo* 61: 244–252.

Langergraber, K. E., K. Prüfer, C. Rowney, C. Boesch, C. Crockfordd, K. Fawcette, E. Inouef, M. Inoue-Muruyama, J. C. Mitani, M. N. Muller, M. M. Robbins, G. Schubert, T. Stoinski, B. Viola, D. Watts, R. M. Wittig, R. W. Wrangham, K. Zuberbühler, S. Pääbo, and L. Vigilant. 2012. Generation times in wild chimpanzees and gorillas suggest earlier divergence times in great ape and human evolution. *Proceedings of the National Academy of Sciences* 109: 15716–15721.

Larson, S. G. 1998. Parallel evolution in the hominoid trunk and forelimb. *Evolutionary Anthropology* 6: 87–99.

Larson, S. G. 2007. Evolutionary transformation of the hominin shoulder. *Evolutionary Anthropology* 16: 172–187.

Larson, S. G., and J. T. Stern Jr. 2013. Rotator cuff muscle function and its relation to scapular morphology in apes. *Journal of Human Evolution* 65: 391–403.

Latimer, B. M., and C. O. Lovejoy. 1990. Metatarsophalangeal joints of *Australopithecus afarensis*. *American Journal of Physical Anthropology* 83: 13–23.

Leakey, R. E., M. G. Leakey, and A. C. Walker. 1988a. Morphology of *Afropithecus turkanensis* from Kenya. *American Journal of Physical Anthropology* 76: 289–307.

Leakey, R. E., M. G. Leakey, and A. C. Walker. 1988b. Morphology of *Turkanapithecus kalaolensis* from Kenya. *American Journal of Physical Anthropology* 76: 277–288.

Leakey, M. G., P. S. Ungar, and A. Walker. 1995. A new genus of large primate from the Late Oligocene of Lothidok, Turkana District, Kenya. *Journal of Human Evolution* 28: 519–531.

Lebatard, A.-E., D. L. Bourlès, R. Braucher, M. Arnold, P. Duringer, M. Jolivet, A. Moussa, P. Deschamps, C. Roquin, J. Carcaillet, M. Schuster, F. Lihoreau, A. Likius, H. T. Mackaye, P. Vignaud, and M. Brunet. 2010. Application of the authigenic $^{10}Be/^{9}Be$ dating method to continental sediments: Reconstruction of the Mio-Pleistocene sedimentary sequence in the early hominid fossiliferous areas of the northern Chad Basin. *Earth and Planetary Sciences Letters* 297: 57–70.

Lebatard, A.-E., D. L. Bourlès, P. Duringer, M. Jolivet, R. Braucher, J. Carcaillet, M. Schuster, N. Arnaud, P. Monié, F. Lihoreau, A. Likius, H. T. Mackaye, P. Vignaud, and M. Brunet. 2008. Cosmogenic nuclide dating of *Sahelanthropus tchadensis* and *Australopithecus bahrelghazali*: Mio-Pliocene hominids from Chad. *Proceedings of the National Academy of Sciences* 105: 3226–3231.

Le Minor, J. M. 1987. Comparative anatomy and significance of the sesamoid bone of the peroneus longus muscle *(os peroneum)*. *Journal of Anatomy* 151: 85–99.

Lewis, O. J. 1989. *Functional Morphology of the Evolving Hand and Foot*. Oxford: Oxford University Press.

Lewton, K. L. 2012. Evolvability of the primate pelvic girdle. *Evolutionary Biology* 39: 126–139.

Li, S., C. Deng, W. Dong, L. Sun, L. Liu, H. Qin, J. Yin, X. Ji, and R. Zhu. 2015. Magnetostratigraphy of the Xiaolongtan Formation bearing *Lufengpithecus keiyuanensis* in Yunnan, southwestern China: Constraint on the initiation time of the southern segment of the Xianshuihe–Xiaojiang fault. *Tectonophysics* 655: 213–226.

Lieberman, D. E. 1999. Homology and hominid phylogeny: Problems and potential solutions. *Evolutionary Anthropology* 7: 142–151.

Lieberman, D. E. 2000. Ontogeny, homology, and phylogeny in the Hominid craniofacial skeleton: The problem of the browridge. In P. O'Higgins, and M. J. Cohn, eds., *Development, Growth and Evolution*. London: Academic Press.

Lieberman, D. E. 2011. *The Evolution of the Human Head*. Cambridge, MA: Harvard University Press.

Lieberman, D. E., J. Carlo, M. Ponce de Leon, and C. P. Zollikofer. 2007. A geometric morphometric analysis of heterochrony in the cranium of chimpanzees and bonobos. *Journal of Human Evolution* 52: 647–662.

Lieberman, D. E., B. Hallgrímsson, W. Liu, T. E. Parsons, and H. A. Jamniczky. 2008. Spatial packing, cranial base angulation, and craniofacial shape variation in the mammalian skull: Testing a new model using mice. *Journal of Anatomy* 212: 720–35.

Lieberman, D. E., and R. C. McCarthy. 1999. The ontogeny of cranial base angulation in humans and chimpanzees and its implications for reconstructing pharyngeal dimensions. *Journal of Human Evolution* 36: 487–517.

Lieberman, D. E., C. F. Ross, and M. J. Ravosa. 2001. The primate cranial base: Ontogeny, function, and integration. *American Journal of Physical Anthropology* S31: 117–169.

Lihoreau, F., J.-R. Boisserie, L. Viriot, Y. Coppens, A. Likius, H. T. Mackaye, P. Tafforeau, P. Vignaud, and M. Brunet. 2006. Anthracothere dental anatomy reveals a Late Miocene Chado-Libyan bioprovince. *Proceedings of the National Academy of Sciences* 103: 8763–8767.

Linder, H. P. 2014. The evolution of African plant diversity. *Frontiers in Ecology and Evolution* 2: 1–14.

Lockwood, C. 1999. Sexual dimorphism in the face of *Australopithecus africanus*. *American Journal of Physical Anthropology* 108: 97–127.

Lovejoy, C. O., M. J. Cohn, and T. D. White. 1999. Morphological analysis of the mammalian postcranium: A developmental perspective. *Proceedings of the National Academy of Sciences* 96: 13247–13252.

Lovejoy, C. O., B. Latimer, G. Suwa, B. Asfaw, and T. D. White. 2009a. Combining prehension and propulsion: The foot of *Ardipithecus ramidus*. *Science* 326: 72e1–72e8.

Lovejoy, C. O., and M. A. McCollum. 2010. Spinopelvic pathways to bipedality: Why no hominids ever relied on a bent-hip-bent-knee gait. *Philosophical Transactions of the Royal Society of London B* 365: 3289–3299.

Lovejoy, C. O., R. S. Meindl, J. C. Ohman, K. G. Heiple, and T. D. White. 2002. The Maka femur and its bearing on the antiquity of human walking: Applying contemporary concepts of morphogenesis to the human fossil record. *American Journal of Physical Anthropology* 119: 97–133.

Lovejoy, C. O., S. W. Simpson, T. D. White, B. Asfaw, and G. Suwa. 2009b. Careful climbing in the Miocene: The forelimbs of *Ardipithecus ramidus* and humans are primitive. *Science* 326: 70e1–70e8.

Lovejoy, C. O., G. Suwa, S. W. Simpson, J. H. Matternes, and T. D. White. 2009c. The great divides: *Ardipithecus ramidus* reveals the postcrania of our last common ancestors with African apes. *Science* 326: 100–106.

Lovejoy, C. O., G. Suwa, L. Spurlock, B. Asfaw, and T. D. White. 2009d. The pelvis and femur of *Ardipithecus ramidus:* The emergence of upright walking. *Science* 326: 71e1–71e6.

Luboga, S. A., and B. A. Wood. 1990. Position and orientation of the foramen magnum in higher primates. *American Journal of Physical Anthropology* 81: 67–76.

MacLatchy. L. 2004. The oldest ape. *Evolutionary Anthropology* 13: 90–103.

Madar, S. I., M. D. Rose, J. Kelley, L. MacLatchy, and D. Pilbeam. 2002. New *Sivapithecus* postcranial specimens from the Siwaliks of Pakistan. *Journal of Human Evolution* 42: 705–752.

Mahoney, P., T. M. Smith, G. T. Schwartz, C. Dean, and J. Kelley. 2007. Molar crown formation in the Late Miocene Asian hominoids, *Sivapithecus parvada* and *Sivapithecus indicus*. *Journal of Human Evolution* 53: 61–68.

Mailund, T., K. Munch, and M. H. Schierup. 2014. Lineage sorting in apes. *Annual Review of Genetics* 48: 519–535.

Mallo, M., D. M. Wellik, and J. Deschamps. 2010. *Hox* genes and regional patterning of the vertebrate body plan. *Developmental Biology* 344: 7–15.

Manners-Smith, T. 1908. A study of the cuboid and os peroneum in the primate foot. *Journal of Anatomy and Physiology* 42: 397–414.

Matsumara, S., and P. Forster. 2008. Generation time and effective population size in Polar Eskimos. *Proceedings of the Royal Society B* 275: 1501–1508.

McBrearty, S., and N. G. Jablonski. 2005. First fossil chimpanzee. *Nature* 437: 105–108.

McCarthy, R. C. 2001. Anthropoid cranial base architecture and scaling relationships. *Journal of Human Evolution* 40: 41–66.

McCarthy, R. C., and D. E. Lieberman. 2001. The posterior maxillary (PM) plane and anterior cranial architecture in primates. *Anatomical Record* 264: 247–60.

McCollum, M. A., B. A. Rosenman, G. Suwa, R. S. Meindl, and C. O. Lovejoy. 2010. The vertebral formula of the last common ancestor of African apes and humans. *Journal of Experimental Zoology, Part B: Molecular Development and Evolution* 314: 123–34.

McDougall, I., and C. S. Feibel. 1999. Numerical age control for the Miocene–Pliocene succession at Lothagam, a hominoid-bearing sequence in the northern Kenya Rift. *Journal of the Geological Society of London* 156: 731–745.

McHenry, H. M. 1975. Fossils and the mosaic nature of human evolution. *Science* 190: 425–31.

McHenry, H. M., and A. L. Jones. 2006. Hallucial convergence in early hominids. *Journal of Human Evolution* 50: 534–539.

McNulty, K. P. 2010. Apes and tricksters: The evolution and diversification of humans' closest relatives. *Evolution: Education and Outreach* 3: 322–332.

McNulty, K. P., D. R. Begun, J. Kelley, and F. K. Manthi. 2015. A systematic revision of *Proconsul* with the description of a new genus of early Miocene hominoid. *Journal of Human Evolution* 84: 42–61.

Meter, M., J.-L. Arsuega, C. de Filippo, S. Nagel, A. Aximu-Petri, B. Nickel, I. Martinez, A. Gracia, et al. 2016. Nuclear DNA sequences from the Middle Pleistocene Sima de los Huesos hominins. *Nature* 531: 504–507.

Miller, E. R. 1999. Faunal correlation of Wadi Moghara, Egypt: Implications for the age of *Prohylobates tandyi*. *Journal of Human Evolution* 36: 519–533.

Miller, E. R., B. R. Benefit, M. L. McCrossin, J. M. Plavcan, M. G. Leakey, A. N. El-Barkooky, M. A. Hamdan, M. A. Gawad, S. M. Hassan, and E. L. Simons. 2009. Systematics of early and Middle Miocene Old World monkeys. *Journal of Human Evolution* 57: 195–211.

Mitteroecker, P., P. Gunz, and F. L. Bookstein. 2005. Heterochrony and geometric morphometrics: A comparison of cranial growth in *Pan paniscus* versus *Pan troglodytes*. *Evolution & Development* 7: 244–258.

Moorjani, P., C. E. G. Amorim, P. F. Arndt, and M. Przeworski. 2016a. Variation in the molecular clock of primates. *Proceedings of the National Academy of Sciences* 113: 10607–10612.

Moorjani, P., Z. Gao, and M. Przeworski. 2016b. Human germline mutation and the erratic evolutionary clock. *PLoS Biology* 14: e2000744.

Moorjani, P., S. Sankararaman, Q. Fu, M. Przeworski, N. Patterson, and D. Reich. 2016c. A genetic method for dating ancient genomes provides a direct estimate of human generation interval in the last 45,000 years. *Proceedings of the National Academy of Sciences* 113: 5652–5657.

Morgan, G. J. 1998. Emile Zuckerkandl, Linus Pauling, and the molecular evolutionary clock, 1959–1965. *Journal of the History of Biology* 31: 155–178.

Morgan, M. E., K. L. Lewton, J. Kelley, E. Otárola-Castillo, J. C. Barry, L. J. Flynn, and D. Pilbeam. 2015. A partial hominoid innominate from the Miocene of Pakistan: Description and preliminary analyses. *Proceedings of the National Academy of Sciences* 112: 82–87.

Morton, D. J. 1922. Evolution of the human foot. Part 1. *American Journal of Physical Anthropology* 5: 305–336.

Mòya-Solà, S., D. M. Alba, S. Almécija, I. Casanovas-Vilar, M. Köhler, S. De Esteban-Trivigno, J. M. Robles, J. Galindo, and J. Fortuny. 2009a. A unique Middle Miocene European hominoid and the origins of the great ape and human clade. *Proceedings of the National Academy of Sciences* 106: 9601–9606.

Mòyà-Solà, S., and M. Köhler. 1996. A *Dryopithecus* skeleton and the origins of great-ape locomotion. *Nature* 379: 156–159.

Mòya-Solà, S., M. Köhler, D. M. Alba, I. Casanovas-Vilar, and J. Galindo. 2004. *Pierolapithecus catalaunicus,* a new Middle Miocene great ape from Spain. *Science* 306: 1339–1344.

Mòya-Solà, S., M. Köhler, D. M. Alba, I. Casanovas-Vilar, J. Galindo, J. M. Robles, L. Cabrera, M. Garcés, S. Almécija, and E. Beamud. 2009b. First partial face and upper dentition of the Middle Miocene hominoid *Dryopithecus fontani* from Abocador de Can Mata (Vallès-Penedès Basin, Catalonia, NE Spain): Taxonomic and phylogenetic implications. *American Journal of Physical Anthropology* 139: 126–145.

Mòya-Solà, S., M. Köhler, and L. Rook. 1999. Evidence of hominid-like precision grip capability in the hand of the Miocene ape *Oreopithecus. Anthropology* 96: 313–317.

Muehleman, C., J. Williams, and M. L. Bareither. 2009. A radiologic and histologic study of the *os peroneum:* Prevalence, morphology, and relationship to degenerative joint disease of the foot and ankle in a cadaveric sample. *Clinical Anatomy* 22: 747–754.

Nachman, M. W., and S. L. Crowell. 2000. Estimate of the mutation rate per nucleotide in humans. *Genetics* 156: 297–304.

Nakatsukasa, M. 2004. Acquisition of bipedalism: the Miocene hominoid record and modern analogues for bipedal protohominids. *Journal of Anatomy* 204: 385–402.

Nakatsukasa, M. 2008. Comparative study of Moroto vertebral specimens. *Journal of Human Evolution* 55: 581–588.

Nakatsukasa, M., and Y. Kunimatsu. 2009. *Nacholapithecus* and its importance for understanding hominoid evolution. *Evolutionary Anthropology* 18: 103–119.

Nakatsukasa, M., Y. Kunimatsu, Y. Nakano, and H. Ishida. 2007. Vertebral morphology of *Nacholapithecus kerioi* based on KNM-BG 35250. *Journal of Human Evolution* 52: 347–369.

Nakatsukasa, M., Y. Kunimatsu, Y. Nakano, T. Takano, and H. Ishida. 2003. Comparative and functional anatomy of phalanges in *Nacholapithecus kerioi,* a Middle Miocene hominoid from northern Kenya. *Primates* 44: 371–412.

Nei, M., and A. K. Roychoudhury. 1974. Genic variation within and between the three major races of man, Caucasoids, Negroids, and Mongoloids. *American Journal of Human Genetics* 26: 421–443.

Nelson, S. V. 2005. Habitat requirements and the extinction of the Miocene ape, *Sivapithecus.* In D. E. Lieberman, R. J. Smith, and J. Kelley, eds., *Interpreting the Past: Essays on Human, Primate and Mammal Evolution in Honor of David Pilbeam.* Boston: Brill Academic.

Nunn, C. L. 2011. *The Comparative Method in Evolutionary Anthropology and Biology.* Chicago: University of Chicago Press.

Ogihara, N., M. Nakatsukasa, Y. Nakano, and H. Ishida. 2006. Computerized restoration of nonhomogeneous deformation of a fossil cranium based on bilateral symmetry. *American Journal of Physical Anthropology* 130: 1–9.

O'Rahilly, R. 1953. A survey of carpal and tarsal anomalies. *Journal of Bone & Joint Surgery* 35: 626–642.

O'Rahilly, R., and F. Müller. 2003. Somites, spinal ganglia, and centra. *Cells Tissues Organs* 173: 75–92.

O'Rahilly, R., F. Müller, and D. B. Meyer. 1990. The human vertebral column at the end of the embryonic period proper, 4: The sacrococcygeal region. *Journal of Anatomy* 168: 95–111.

Orr, C. M. 2005. Knuckle-walking anteater: A convergence test of adaptation for purported knuckle-walking features of African Hominidae. *American Journal of Physical Anthropology* 128: 639–658.

Orr, C. M., M. W. Tocheri, S. E. Burnett, R. D. Awe, E. W. Saptomo, T. Sutikna, Jatmiko, S. Wasisto, M. J. Morwood, and W. L. Jungers. 2013. New wrist bones of *Homo floresiensis* from Liang Bua (Flores, Indonesia). *Journal of Human Evolution* 64: 109–129.

Osborn, H. F. 1927. Recent discoveries relating to the origin and antiquity of man. *Proceedings of the American Philosophical Society* 66: 373–389.

Oxnard, C. E., and F. P. Lisowski. 1980. Functional articulation of some hominoid foot bones: Implications for the Olduvai (Hominid 8) foot. *American Journal of Physical Anthropology* 52: 107–117.

Patel, B. A. 2010. The interplay between speed, kinetics, and hand postures during primate terrestrial locomotion. *American Journal of Physical Anthropology* 141: 222–234.

Pickford, M. 2005. Orientation of the foramen magnum in late Miocene to extant African apes and hominids. *Anthropologie* 63: 191–198.

Pickford, M., and B. Senut. 2001. The geological and faunal context of Late Miocene hominid remains from Lukeino, Kenya. *Comptes Rendus de l'Académie des Sciences, Sciences de la terre et des planets* 332: 145–152.

Pickford, M., and B. Senut. 2005. Hominoid teeth with chimpanzee- and gorilla-like features from the Miocene of Kenya: Implications for the chronology of ape-human divergence and biogeography of Miocene hominoids. *Anthropological Science* 113: 95–102.

Pickford, M., B. Senut, D. Gommery, and J. Treil. 2002. Bipedalism in *Orrorin tugenensis* revealed by its femora. *Comptes Rendus Palevol* 1: 1–13.

Pickford, M., B. Senut, J. Morales, and J. Braga. 2008. First hominoid from the Late Miocene of Niger. *South African Journal of Science* 104: 337–339.

Pilbeam, D. 1982. New hominoid skull material from the Miocene of Pakistan. *Nature* 295: 232–234.

Pilbeam, D. 1986. Hominoid evolution and hominid origins. *American Anthropologist* 88: 295–312.

Pilbeam, D. R. 2002. Perspectives on the Miocene Hominoidea. In W. Hartwig, ed., *The Primate Fossil Record*. Cambridge: Cambridge University Press.

Pilbeam, D. 2004. The anthropoid postcranial axial skeleton: Comments on development, variation, and evolution. *Journal of Experimental Zoology (Molecular and Developmental Evolution)* 302B: 241–267.

Pilbeam, D. R., and S. J. Gould. 1974. Size and scaling in human evolution. *Science* 186: 892–901.

Pilbeam, D. R., M. D. Rose, J. C. Barry, and S. M. I. Shah. 1990. New *Sivapithecus* humeri from Pakistan and the relationship of *Sivapithecus* and *Pongo*. *Nature* 348: 237–239.

Pilbeam, D. R., and N. M. Young. 2004. Hominoid evolution: Synthesizing disparate data. *Comptes Rendus Palevol* 3: 303–319.

Pilbrow, V. 2006. Population systematics of chimpanzees using molar morphometrics. *Journal of Human Evolution* 51: 646–662.

Plavcan, J. M., C. V. Ward, and F. L. Paulus. 2010. Canine tooth crown size in *Australopithecus anamensis*. *Journal of Human Evolution* 57: 2–10.

Popov, S. V., F. Rögl, A. Y. Rozanov, F. F. Steininger, I. G. Shcherba, and M. Kovacs, eds. 2004. *Lithological-Paleogeographic Maps of the Paratethys*. Frankfurt am Main: Courier Forschungsinstitut Senckenberg.

Pound, M. J., A. M. Haywood, U. Salzmann, J. B. Riding, D. J. Lunt, and S. J. Hunter. 2012. A Tortonian (Late Miocene, 11.61–7.25 Ma) global vegetation reconstruction. *Palaeogeography, Palaeoclimatology, Palaeoecology* 300: 29–45.

Prado-Martinez, J., P. H. Sudmant, J. M. Kidd, H. Li, J. L. Kelley, B. Lorente-Galdos, K. R. Veeramah, A. E. Woerner, T. D. O'Connor, G. Santpere, A. Cagan, et al. 2013. Great ape genetic diversity and population history. *Nature* 499: 471–475.

Preuschoft, H., B. Demes, M. Meier, and H. F. Bär. 1985. Biomechanical principles of the maxilla of long-snouted vertebrates. *Zeitschrift für Morphologie und Anthropologie* 76: 1–24.

Proctor, D. J. 2013. Proximal metatarsal articular surface shape and the evolution of a rigid lateral foot in hominins. *Journal of Human Evolution* 65: 761–769.

Prüfer, K., K. Munch, I. Hellmann, K. Akagi, J. R. Miller, B. Walenz, S. Koren, G. Sutton, C. Kodira, R. Winer, J. R. Knight, et al. 2012. The bonobo genome compared with the chimpanzee and human genomes. *Nature* 486: 527–532.

Rae, T. C. 1997. The early evolution of the hominoid face. In D. R. Begun, C. V. Ward, and M. D. Rose, eds., *Function, Phylogeny, and Fossils*. New York: Plenum.

Rafferty, K. L., A. Walker, C. B. Ruff, M. D. Rose, and P. J. Andrews. 1995. Postcranial estimates of body weight in *Proconsul*, with a note on a distal tibia of *P. major* from Napak, Uganda. *American Journal of Physical Anthropology* 97: 391–402.

Rak, Y. 1983. *The Australopithecine Face*. New York: Academic Press.

Raza, S. M., J. C. Barry, D. Pilbeam, M. D. Rose, S. M. I. Shag, and S. Ward. 1983. New hominoid primates from the Middle Miocene Chinji Formation, Potwar Plateau, Pakistan. *Nature* 306: 52–54.

Reichard, R. H., and C. Barelli. 2008. Life history and reproductive strategies of Khao Yai *Hylobates lar:* Implications for social evolution in apes. *International Journal of Primatology* 29: 823–844.

Richmond, B. G., D. R. Begun, and D. S. Strait. 2001. Origin of human bipedalism: The knuckle-walking hypothesis revisited. *Yearbook of Physical Anthropology* 44: 70–105.

Richmond, B. G., and W. L. Jungers. 2008. *Orrorin tugenensis* femoral morphology and the evolution of hominin bipedalism. *Science* 319: 1662–1665.

Richmond, B. G., and D. S. Strait. 2000. Evidence that humans evolved from a knuckle-walking ancestor. *Nature* 404: 382–385.

Rose, M. D. 1983. Locomotor anatomy of Miocene hominoids. In D. L. Gebo, ed., *Postcranial Adaptation in Nonhuman Primates*. DeKalb: Northern Illinois University Press.

Rose, M. D. 1984. Hominoid postcranial specimens from the Middle Miocene Chinji Formation, Pakistan. *Journal of Human Evolution* 13: 503–516.

Rose, M. D. 1986. Further hominoid postcranial specimens from the Late Miocene Nagri Formation of Pakistan. *Journal of Human Evolution* 15: 333–367.

Rose, M. D. 1989. New postcranial specimens of catarrhines from the Middle Miocene Chinji Formation, Pakistan: Descriptions and a discussion of proximal humeral functional morphology in anthropoids. *Journal of Human Evolution* 18: 131–162.

Rose, M. D. 1991. The process of bipedalization in hominids. In Y. Coppens, and B. Senut, eds., *Origine(s) de la Bipédie chez les Hominidés*. Paris: Éditions du Centre Nationale de la Recherche Scientifique.

Rose, M. D. 1993. Locomotor anatomy of Miocene hominoids. In D. G. Gebo, ed., *Postcranial Adaptation in Nonhuman Primates*. DeKalb: Northern Illinois University Press.

Ross, C. F., and M. J. Ravosa. 1993. Basicranial flexion, relative brain size, and facial kyphosis in nonhuman primates. *American Journal of Physical Anthropology* 91: 305–324.

Russo, G. A., and L. J. Shapiro. 2013. Morphological correlates of tail length in the catarrhine sacrum. *Journal of Human Evolution* 61: 223–232.

Ruvolo, M. 1997. Molecular phylogeny of the hominoids: Inferences from multiple independent DNA sequence data sets. *Molecular Biology and Evolution* 14: 248–265.

Sanchez-Villagra, M. R., and B. A. Williams. 1998. Levels of homoplasy in the evolution of the mammalian skeleton. *Journal of Mammalian Evolution* 5: 113–126.

Sanders, W. J., and B. E. Bodenbender. 1994. Morphometric analysis of lumbar vertebra UMP 67-28: Implications for spinal function and phylogeny of the Miocene Moroto hominoid. *Journal of Human Evolution* 26: 203–237.

Sarich, V. M. 1983. Retrospective on hominoid macromolecular systematics. In R. L. Ciochon and R. S. Corruccini, eds., *New Interpretations of Ape and Human Ancestry*. New York: Plenum.

Sarich, V. M. 1999. Mammalian systematics: Twenty-five years among their albumins and transferrins. In F. S. Szalay, M. Novacek, and M. McKenna, eds., *Mammal Phylogeny: Placentals*. New York: Springer-Verlag.

Sarich, V. M., and A. C. Wilson. 1967a. Immunological time scale for hominid evolution. *Science* 158: 1200–1203.

Sarich, V. M., and A. C. Wilson. 1967b. Rates of albumin evolution in primates. *Science* 158: 142–148.

Sarmiento, E. E. 1987. The phylogenetic position of *Oreopithecus* and its significance in the origin of the Hominoidea. *American Museum Novitates* 2881: 1–44.

Sarmiento, E. E. 1994. Terrestrial traits in the hands and feet of gorillas. *American Museum Novitates* 3091: 1–56.

Saunders, J. B., V. T. Inman, and H. D. Eberhart. 1953. The major determinants in normal and pathological gait. *Journal of Bone and Joint Surgery* 35: 543–558.

Sawada, Y., M. Pickford, T. Itaya, T. Makinouchi, M. Tateishi, K. Kabeto, S. Ishida, and H. Ishida. 1998. K-Ar ages of Miocene Hominoidea (*Kenyapithecus* and *Samburupithecus*) from Samburu Hills, Northern Kenya. *Comptes rendus de l'Académie des Sciences Paris, Sciences de la terre et des planets* 326: 445–451.

Scally, A. 2016. The mutation rate in human evolution and demographic inference. *Current Opinion in Genetics & Development* 41: 36–43.

Scally, A., J. Y. Dutheil, L. W. Hillier, G. E. Jordan, I. Goodhead, J. Herrero, A. Hobolth, T. Lappalainen, T. Mailund, T. Marques-Bonet, S. McCarthy, et al. 2012. Insights into hominid evolution from the gorilla genome sequence. *Nature* 483: 169–175.

Schroer, K., and B. A. Wood. 2013. Evolution of hominin postcanine macromorphology: A comparative meta-analysis. In R. Scott and J. Irish, eds., *Anthropological Perspectives on Tooth Morphology*. Cambridge: Cambridge University Press.

Schultz, A. H. 1963. Age changes, sex differences, and variability as factors in the classification of primates. In S. L. Washburn, ed., *Classification and Human Evolution*. Chicago: Aldine.

Schwartz, G. T. 2000. Taxonomic and functional aspects of the patterning of enamel thickness distribution in extant large-bodied hominoids. *American Journal of Physical Anthropology* 111: 221–244.

Schwartz, J. H. 1997. *Lufengpithecus* and hominoid phylogeny: Problems in delineating and evaluating phylogenetically relevant characters. In D. R. Begun, C. V. Ward, and M. D. Rose, eds., *Function, Phylogeny, and Fossils*. New York: Plenum.

Ségurel, L., M. J. Wyman, and M. Przeworski. 2014. Determinants of mutation rate variation in the human germline. *Annual Review of Genomics and Human Genetics* 15: 47–70.

Semaw, S., S. W. Simpson, J. Quade, P. R. Renne, R. F. Butler, W. C. McIntosh, N. Levin, M. Dominguez-Rodrigo, and M. J. Rogers. 2005. Early Pliocene hominids from Gona, Ethiopia. *Nature* 433: 301–305.

Senut, B. 2007. The earliest putative hominids. In W. Henke, L. Tattersall, and T. Hardt, eds., *Handbook of Paleoanthropology*. New York: Springer.

Senut, B., M. Nakatsukasa, Y. Kunimatsu, Y. Nakano, T. Takano, H. Tsujiawa, D. Shimizu, M. Kagaya, and H. Ishida. 2004. Preliminary analysis of *Nacholapithecus* scapula and clavicle from Nachola, Kenya. *Primates* 45: 97–104.

Senut, B., M. Pickford, D. Gommery, and Y. Kunimatsu. 2000. Un nouveau genre d'hominoïde du Miocène inférieur d'Afrique orientale: *Ugandapithecus major* (Le Gros Clark & Leakey, 1950). *Comptes rendus de l'Académie des Sciences Paris, Sciences de la terre et des planets* 331: 227–233.

Senut, B., M. Pickford, D. Gommery, P. Mein, K. Cheboi, and Y. Coppens. 2001. First hominid from the Miocene (Lukeino Formation, Kenya). *Comptes Rendus de l'Académie des Sciences, Sciences de la terre et des planets* 332: 137–144.

Senut, B., M. Pickford, and D. Wessels. 1997. Panafrican distribution of Lower Miocene Hominoidea. *Comptes rendus de l'Académie des Sciences Paris, Sciences de la terre et des planets* 325: 741–746.

Shea, B. T. 1981. Relative growth of the limbs and trunk in the African apes. *American Journal of Physical Anthropology* 56: 179–201.

Shea, B. T. 1983a. Size and diet in the evolution of African ape craniodental form. *Folia Primatologica* 40: 32–68.

Shea, B. T. 1983b. Allometry and heterochrony in the African apes. *American Journal of Physical Anthropology* 62: 275–89.

Shea, B. T. 1985. Ontogenetic allometry and scaling: A discussion based on the growth and form of the skull in African apes. In W. L. Jungers, ed., *Size and Scaling in Primate Biology*. New York: Plenum.

Shea, B. T., R. E. Hammer, R. L. Brinster, and M. R. Ravosa. 1990. Relative growth of the skull and postcranium in giant transgenic mice. *Genetics Research.* 56: 21–34.

Simons, E. L., and D. Pilbeam. 1965. Preliminary revision of the Dryopithecinae (Pongidae, Anthropoidea). *Folia Primatologia* 3: 81–152.

Simpson, G. G. 1963. The meaning of taxonomic statements. In S. L. Washburn, ed., *Classification and Human Evolution*. Chicago: Aldine.

Simpson, S., L. Kleinsasser, J. Quade, N. E. Levin, W. C. McIntosh, N. Dunbar, S. Semaw, and M. J. Rogers. 2015. Late Miocene hominin teeth from the Gona Paleoanthropological Research Project area, Afar, Ethiopia. *Journal of Human Evolution* 81: 68–82.

Skinner, M. M., Z. Alemseged, C. Gaunitz, and J. J. Hublin. 2015. Enamel thickness trends in Plio-Pleistocene hominin mandibular molars. *Journal of Human Evolution* 85: 35–45.

Smith, R. J., and W. L. Jungers. 1997. Body mass in comparative primatology. *Journal of Human Evolution* 32: 523–559.

Smith, T. M. 2008. Incremental dental development: Methods and applications in hominoid evolutionary studies. *Journal of Human Evolution* 54: 205–224.

Smith, T. M. 2013. Teeth and human life-history evolution. *Annual Review of Anthropology* 42: 191–208.

Smith, T. M., L. B. Martin, and M. G. Leakey. 2003. Enamel thickness, microstructure and development in *Afropithecus turkanensis*. *Journal of Human Evolution* 44: 283–306.

Smith, T. M., L. B. Martin, D. J. Reid, L. de Bonis, and G. D. Koufos. 2004. An examination of dental development in *Graecopithecus freybergi (Ouranopithecus macedoniensis)*. *Journal of Human Evolution* 46: 551–577.

Smith, T. M., D. J. Reid, A. J. Olejniczak, P. Tafforeau, J.-J. Hublin, and M. Toussaint. 2014. Dental development in and age at death of the Scladina 1-4A juvenile Neanderthal. In M. Toussaint and D. Bonjean, eds., *The Scladina I-4A Juvenile Neandertal (Andenne, Belgium) Palaeoanthropology and Context*. Liège: Études et Recherches Archéologiques de l'Université de Liège.

Spoor, C. F. 1997. Basicranial architecture and relative brain size of Sts 5 *(Australopithecus africanus)* and other Plio-Pleistocene hominids. *South African Journal of Science* 93: 182–186.

Squyres, N., and V. B. DeLeon. 2015. Clavicular curvature and locomotion in anthropoid primates: A 3D geometric morphometric analysis. *American Journal of Physical Anthropology* 158: 257–268.

Stern, J. T., Jr., and R. L. Susman. 1983. The locomotor anatomy of *Australopithecus afarensis*. *American Journal of Physical Anthropology* 60: 279–317.

Stevens, N. J., E. R. Seiffert, P. M. O'Connor, E. M. Roberts, M. D. Schmitz, C. Krause, E. Gorscak, S. Ngasala, T. I. Hieronymus, and J. Temu. 2013. Palaeontological evidence for an Oligocene divergence between Old World monkeys and apes. *Nature* 497: 611–614.

Stoinski, T. S., B. Perdue, T. Breuer, and M. P. Hoff. 2013. Variability in the developmental life history of the genus *Gorilla*. *American Journal of Physical Anthropology* 152: 165–172.

Straus, W. L., Jr. 1949. The riddle of man's ancestry. *Quarterly Review of Biology* 24: 200–223.

Susanna, I., D. M. Alba, S. Almécija, and S. Mòyà-Solà. 2014. The vertebral remains of the late Miocene great ape *Hispanopithecus laietanus* from Can llobateres 2 (Vallès-Penedès Basin, NE Iberian Peninsula). *Journal of Human Evolution* 73: 15–34.

Susman, R. L. 1979. Comparative and functional morphology of hominoid fingers. *American Journal of Physical Anthropology* 50: 215–236.

Suwa, G., B. Asfaw, R. T. Kono, D. Kubo, C. O. Lovejoy, and T. D. White. 2009a. The *Ardipithecus ramidus* skull and its implications for hominid origins. *Science* 326: 68e1–7.

Suwa, G., Y. Beyene, H. Nakaya, R. L. Bernor, J.-R. Boisserie, F. Bibi, S. H. Ambrose, K. Sano, S. Katoh, and B. Asfaw. 2015. Newly discovered cercopithecid, equid and other mammalian fossils from the Chorora Formation, Ethiopia. *Anthropological Science* 123: 19–39.

Suwa, G., R. T. Kono, S. Katoh, B. Asfaw, and Y. Beyene. 2007. A new species of great ape from the late Miocene epoch in Ethiopia. *Nature* 448: 921–924.

Suwa, G., R. T. Kono, S. W. Simpson, B. Asfaw, C. O. Lovejoy, and T. D. White. 2009b. Paleobiological implications of the *Ardipithecus ramidus* dentition. *Science* 326: 94–99.

Thorpe, S. K. S., J. M. McClymount, and R. H. Crompton. 2014. The arboreal origins of human bipedalism. *Antiquity* 88: 906–926.

Tocheri, M. W., C. M. Orr, S. M. Larson, T. Sutikna, Jatmiko, E. W. Saptomo, R. A. Due, T. Djubiantono, M. J. Morwood, and W. L. Jungers. 2007. The primitive wrist of *Homo floresiensis* and its implications for hominin evolution. *Science* 317: 1743–1745.

Tremblay, M., and H. Vézina. 2000. New age estimates of intergenerational intervals for the calculation of age and origins of mutations. *American Journal of Human Genetics* 66: 651–658.

Trinkaus, E. 1983. *The Shanidar Neandertals*. New York: Academic Press.
Tuttle, R. H. 1969. Knuckle-walking and the problem of human origins. *Science* 166: 953–61.
Tuttle, R. H. 1981. Evolution of hominid bipedalism and prehensile capabilities. *Philosophical Transactions of the Royal Society of London B* 292: 89–94.
Tuttle, R. H. 1992. Electromyography of pronators and supinators in great apes. *American Journal of Physical Anthropology* 87: 215–226.
Tuttle, R. H., and J. V. Basmajian. 1974. Electromyography of forearm musculature in *Gorilla* and problems related to knuckle walking. In F. A. Jenkins, ed., *Primate Locomotion*. New York: Academic Press.
Tuttle, R. H., and J. V. Basmajian. 1978a. Electromyography of pongid shoulder muscles II: Deltoid, rhomboid, and "rotator cuff." *American Journal of Physical Anthropology* 49: 47–56.
Tuttle, R. H., and J. V. Basmajian. 1978b. Electromyography of pongid shoulder muscles III: Quadrupedal positional behavior. *American Journal of Physical Anthropology* 49: 57–70.
Tuttle, R. H., J. R. Hollowed, and J. V. Basmajian. 1992. Electromyography of pronators and supinators in great apes. *American Journal of Physical Anthropology* 87: 215–226.
Uchida, A. 1992. Intra-species variation among the great apes: Implications for taxonomy of fossil hominids. Ph.D. dissertation, Harvard University.
Ungar, P. S. 1996. Dental microwear of European Miocene catarrhines: Evidence for diets and tooth use. *Journal of Human Evolution* 31: 335–366.
Uno, K. T., T. E. Cerling, J. M. Harris, Y. Kunimatsu, M. G. Leakey, M Nakatsukasa, and H. Nakaya. 2011. Late Miocene to Pliocene carbon isotope record of differential diet change among East African herbivores. *Proceedings of the National Academy of Sciences* 108: 6509–6514.
Veereke, E. E., and P. Aerts. 2008. The mechanics of the gibbon foot and its potential for elastic energy storage during bipedalism. *Journal of Experimental Biology* 211: 3661–3670.
Vignaud, P., P. Duringer, H. T. Mackaye, A. Likius, C. Blondel, J.-R. Boisserie, L. de Bonis, V. Eisenmann, M.-E. Etienne, D. Geraadsk, F. Guy, T. Lehmann, F. Lihoreau, N. Lopez-Martinez, C. Mourer-Chauviré, O. Otero, J.-C. Rage, M. Schuster, L. Viriot, A Zazzo, and M. Brunet. 2002. Geology and palaeontology of the Upper Miocene Toros-Menalla hominid locality, Chad. *Nature* 418: 152–155.
Walker, A. 1997. Proconsul. In D. R. Begun, C. V. Ward, and M. Rose, eds., *Function, Phylogeny and Fossils: Miocene Hominoid Evolution and Adaptations*. New York: Plenum.
Wall, J. D. 2003. Estimating ancestral population sizes and divergence times. *Genetics* 163: 395–404.
Ward, C. V. 1993. Torso morphology and locomotion in *Proconsul nyanzae*. *American Journal of Physical Anthropology* 92: 291–328.
Ward, C. V., A. Walker, and M. F. Teaford. 1991. *Proconsul* did not have a tail. *Journal of Human Evolution* 21: 215–220.

Ward, C. V., M. G. Leakey, and A. Walker. 2001. Morphology of *Australopithecus anamensis* from Kanapoi and Allia Bay, Kenya. *Journal of Human Evolution* 41: 255–368.

Ward, C. V., M. J. Plavcan, and F. K. Manthi. 2010. Anterior dental evolution in the *Australopithecus anamensis–afarensis* lineage. *Philosophical Transactions of the Royal Society of London B* 365: 3333–3344.

Ward, C. V., W. H. Kimbel, and D. C. Johanson. 2011. Complete fourth metatarsal and arches in the foot of *Australopithecus afarensis*. *Science* 331: 750–753.

Ward, S., B. Brown, A. Hill, L. Kelley, and W. Downs. 1999. *Equatorius*: a new hominoid genus from the Middle Miocene of Kenya. *Science* 285: 1382–1386.

Washburn, S. L., ed. 1963. *Classification and Human Evolution*. Chicago: Aldine.

Washburn, S. L. 1967. Behaviour and the origin of man. *Proceedings of the Royal Anthropological Institute of Great Britain and Ireland* 3: 21–27.

Washburn, S. L. 1968. The study of human evolution. Condon Lectures, Oregon State System of Higher Education, Eugene.

Washburn, S. L. 1973. Human evolution: Science or game? *Yearbook of Physical Anthropology* 17: 67–70.

Wayne, R. K. 1986. Cranial morphology of domestic and wild canids: The influence of development on morphological change. *Evolution* 40: 243–61.

Weinert, H. 1927. Die kleinste Interorbitalbreite als stammesgeschichtliches Merkmal. *Zeitschrift für Morphologie und Anthropologie* 26: 450–488.

White, T. D., B. Asfaw, Y. Beyene, Y. Haile-Selassie, C. O. Lovejoy, G. Suwa, and G. WoldeGabriel. 2009. *Ardipithecus ramidus* and the paleobiology of early hominids. *Science* 326: 75–86.

White, T. D., C. O. Lovejoy, B. Asfaw, J. P. Carlson, and G. Suwa. 2015. Neither chimpanzee nor human, *Ardipithecus* reveals the surprising ancestry of both. *Proceedings of the National Academy of Sciences* 112: 4877–4884.

White, T. D., G. Suwa, and B. Asfaw. 1994. *Australopithecus ramidus*, a new species of early hominid from Aramis, Ethiopia. *Nature* 371: 306–312.

White, T. D., G. WoldeGabriel, B. Asfaw, S. Ambrose, Y. Beyene, R. L. Bernor, J. R. Boisserie, B. Currie, H. Gilbert, Y. Haile-Selassie, W. K. Hart, L. J. Hlusko, F. C. Howell, R. T. Kono, T. Lehmann, A. Louchart, C. O. Lovejoy, P. R. Renne, H. Saegusa, E. S. Vrba, H. Wesselman, and G. Suwa. 2006. Asa Issie, Aramis and the origin of *Australopithecus*. *Nature* 44: 883–889.

Williams, B. 2007. Comparing levels of homoplasy in the primate skeleton. *Journal of Human Evolution* 52: 480–489.

Williams, S. A. 2010. Morphological integration and the evolution of knuckle-walking. *Journal of Human Evolution* 58: 432–440.

Williams, S. A. 2011. Variation in anthropoid vertebral formulae: Implications for homology and homoplasy in hominoid evolution. *Journal of Experimental Zoology Part B: Molecular and Developmental Evolution* 318: 134–147.

Williams, S. A. 2012. Placement of the diaphragmatic vertebra in catarrhines: Implications for the evolution of dorsostability in hominoids and bipedalism in hominins. *American Journal of Physical Anthropology* 148: 111–122.

Williams, S. A., K. R. Ostrofsky, N. Frater, S. E. Churchill, P. Schmid, and L. R. Berger. 2013. The vertebral column of *Australopithecus sediba*. *Science* 340: 123996.

Wobber, V., R. Wrangham, and B. Hare. 2010. Application of the heterochrony framework to the study of behavior and cognition. *Communicative & Integrative Biology* 3:4, 337–339.

Wolpoff, M. H., J. Hawks, B. Senut, M. Pickford, and J. Ahern. 2006. An ape or the ape: Is the Toumaï cranium TM 266 a hominid? *PaleoAnthropology* 2006: 36–50.

Wolpoff, M. H., B. Senut, M. Pickford, and J. Hawks. 2002. Palaeoanthropology: *Sahelanthropus* or "*Sahelpithecus*"? *Nature* 419: 581–582.

Wood, B. A. 1979. An analysis of tooth and body size relationship in five primate taxa. *Folia Primatologica* 31: 187–211.

Wood, B. A. 1991. *Koobi Fora Research Project IV: Hominid Cranial Remains from Koobi Fora*. Oxford: Clarendon Press.

Wood, B. 2010. Reconstructing human evolution: Achievements, challenges, and opportunities. *Proceedings of the National Academy of Sciences* 107: 8902–8909.

Wood, B. A., and S. A. Abbott. 1983. Analysis of the dental morphology of Plio-Pleistocene hominids I: Mandibular molars: Crown area measurements and morphological traits. *Journal of Anatomy* 136: 197–219.

Wood, B. A., S. A. Abbott, and S. H. Graham. 1983. Analysis of the dental morphology of Plio-Pleistocene hominids II: Mandibular molars—study of cusp areas, fissure pattern and cross-sectional shape of the crown. *Journal of Anatomy* 137: 287–314.

Wood, B., and E. K. Boyle. 2016. Hominin taxic diversity: Fact or fantasy? *Yearbook of Physical Anthropology* 159: S37–S78.

Wood, B. A., and T. Harrison. 2011. The evolutionary context of the first hominins. *Nature* 470: 347–352.

Wood Jones, F. 1916. *Arboreal Man*. London: Edward Arnold.

Wood Jones, F. 1923. *The Ancestry of Man*. Brisbane: Gillies.

Wrangham, R. W., and D. R. Pilbeam. 2001 African apes as time machines. In B. M. F. Galdikas, N. E. Briggs, L. K. Sheeran, G. L. Shapiro, and J. Goodall, eds., *All Apes Great and Small*. Berlin: Springer.

Wu, R., Q. Xu, and Q. Lu. 1983. Morphological features of *Ramapithecus* and *Sivapithecus* and their phylogenetic relationships. *Acta Anthropologica Sinica* 2: 9–14.

Xia, J., J. Zheng, D. Huang, Z. R. Tian, L. Chen, Z. Zhou, P. S. Ungar, and L. Qian. 2015. New model to explain tooth wear with implications for microwear formation and diet reconstruction. *Proceedings of the National Academy of Sciences* 112: 10669–10672.

Yoder, A. D., and Z. Yang. 2004. Divergence dates for Malagasy lemurs estimated from multiple gene loci: Geological and evolutionary context. *Molecular Ecology* 13: 757–773.

Young, N. M. 2003. A reassessment of living hominoid postcranial variability: Implications for ape evolution. *Journal of Human Evolution* 45: 441–464.

Young, N. M. 2005. Estimating hominoid phylogeny from morphological data: Character choice, phylogenetic signal and postcranial data. In D. E. Lieberman, R. J.

Smith, J. Kelley, eds., *Interpreting the Past: Essays on Human, Primate and Mammal Evolution in Honor of David Pilbeam*. Boston: Brill Academic.

Young, N. M. 2008. A comparison of the ontogeny of shape variation in the anthropoid scapula: Functional and phylogenetic signal. *American Journal of Physical Anthropology* 136: 247–264.

Young, N. M., T. D. Capellini, N. T. Roach, and Z. Alemseged. 2015. Fossil hominin shoulders support an African ape-like last common ancestor of humans and chimpanzees. *Proceedings of the National Academy of Sciences* 112: 11829–11834.

Zachos, J., M. Pagani, L. Sloan, E. Thomas, and K. Billups. 2001. Trends, rhythms, and aberrations in global climate 65 Ma to present. *Science* 292: 686–693.

Zhang, Z., G. Ramstein, M. Schuster, C. Li, C. Contoux, and Q. Yan. 2014. Aridification of the Sahara desert caused by Tethys Sea shrinkage during the Late Miocene. *Nature* 513: 401–404.

Zipfel, B., J. M. DeSilva, R. S. Kidd, K. J. Carlson, S. E. Churchill, and L. R. Berger. 2011. The foot and ankle of *Australopithecus sediba*. *Science* 333: 1470–1420.

Zollikofer, C. P., M. S. Ponce de León, D. E. Lieberman, F., Guy, D. Pilbeam, A. Likius, H. T. Mackaye, P. Vignaud, and M. Brunet. 2005. Virtual cranial reconstruction of *Sahelanthropus tchadensis*. *Nature* 434: 755–759.

Zuckerkandl, E., and L. Pauling. 1965. Evolutionary divergence and convergence in proteins. In U. Bryson, and H. Vogel, eds., *Evolving Genes and Proteins*. New York: Academic Press.

3

Equal, Similar, but Different

Convergent Bonobos and Conserved Chimpanzees

BRIAN HARE *and* RICHARD W. WRANGHAM

> If the deduction that the bonobo is highly derived with respect to the other chimpanzees is correct, then its similarities to humans become even more significant.
>
> —COLIN GROVES (1981: 366)

A critical goal for human evolutionary biology is understanding when and how traits evolved in our ancestral lineage during the 6.5–9.3 million years since our split with the ancestors of chimpanzees and bonobos (Pilbeam and Lieberman, this volume). Comparisons among humans and living apes are essential in this task, because they allow us to discriminate between traits that we share through common descent and those that have been modified since our split from *Pan* (Figure 3.1). Central to interpreting these types of comparisons is determining which living ape, if any, most closely resembles our last common ape ancestor or LCA (Wrangham and Pilbeam 2001). Here we suggest, based on the most recent quantitative comparisons of behavior, physiology, development, and cognition, that in many respects chimpanzees *(Pan troglodytes)* most closely approximate our last common ancestor with chimpanzees and bonobos. We also suggest that bonobo *(Pan paniscus)* and human psychology are similar in ways that indicate that cognitive evolution in the two species was guided by parallel mechanisms (Hare et al. 2012).

FIGURE 3.1. Bonobos and chimpanzees both represent our closest living relatives and the two living representatives of the genus *Pan*. Embedded in an overall pattern of phenotypic similarity between the two species are important differences that can help reveal how apes are shaped by natural selection. (a) Adult male bonobo and (b) adult male chimpanzee each using long calls in response to the calls of neighboring males.

Phylogenetic comparisons with both living and extinct apes are crucial for reconstructing human evolution. Unfortunately, cognition, temperament, personality, and all other psychological features do not fossilize. Nor have they typically been linked to traits that can be measured in fossils. This means that for the study of psychological evolution our information is mostly restricted to comparisons of humans with bonobos and chimpanzees, which fortunately provide informative lessons. For example, all three species readily follow the gaze of another individual, but only humans explicitly attribute false beliefs to others (e.g., Hare et al. 2001; Tomasello et al. 2007; Herrmann et al. 2010; but see Krupenye et al. 2016 for implicit attribution of false belief by chimpanzees). We can conclude that the LCA used gaze-following to predict and manipulate the behavior of others, but it did not understand that others could have false beliefs. This lack of understanding makes sense, given that in humans the attribution of false beliefs is tightly linked to language development. Unlike gaze following, therefore, attribution of false beliefs is a derived feature of human evolution that evolved in the last 6.5–9.3 million years. Fossils cannot give us this level of resolution on questions of cognitive evolution (Hare 2011). When bonobos and chimpanzees share the same trait, comparisons with humans are thus directly illuminating.

By contrast, when bonobos and chimpanzees exhibit different traits, a further step is required to understand the evolutionary implications of the three-way comparison. The problem is that when one of the two species is more similar to humans, the similarity can be due either to shared ancestry or to convergence, and it is often not obvious which explanation is relevant. For example, direct quantitative comparisons show that chimpanzees are better at causal reasoning, while bonobos are more skillful in social tasks (Herrmann et al. 2010). To understand the implications of such observations for human evolution, we need to know if the relevant abilities evolved through shared ancestry or convergence.

Two Equal Cousins Both Similar and Different

Robert Yerkes, the so-called father of American primatology (Haraway 1993), did not realize that the infant bonobo and chimpanzee living with him for a year (i.e., Chim and Panzee) were different species (de Waal and Lanting 1998). His failure is understandable. The genomes of the two species differ by a mere 0.4 percent (Prüfer et al. 2012), and the two species are in many ways very similar in their morphology, feeding adaptations, cognition, life history, and social systems (Table 3.1).

Yet there are important differences that have persisted as long as ~1 million years. Genomic comparisons indicate that, unlike other primates that show graded hybridization zones during their early evolution, the two species evolved in part due to allopatric speciation. The Congo River provides an ancient and significant geographical barrier, since apes cannot swim (the formation of this river was previously considered to be 1.5–2.5 million years ago (Thompson 2003), but new evidence suggest a more ancient origin, 34 million years ago (Takemoto et al. 2015). Only limited genetic admixture between bonobos south of the Congo River and chimpanzees to the north of the river occurred since their separation (de Manuel et al. 2016). This effect

TABLE 3.1. Traits that are similar in bonobos, chimpanzees, and humans (foragers). The similarities allow for the inference that humans inherited these traits through common descent and that their origin is ancient. The traits that clearly differ in humans are likely derived.

	Bonobo	Chimpanzees	LCA	Foragers
Life History				
Long-lived	Yes	Yes	Yes	Yes
Slow reproduction	Yes	Yes	Yes	Yes
One offspring	Yes	Yes	Yes	Yes
Feeding Ecology				
Frugivorous	Yes	Yes	Yes	Omnivorous
Predictable unpredictability	Yes	Yes	Yes	Yes
Contest/scramble competition	Yes	Yes	Yes	Yes
Social Behavior				
Highly social	Yes	Yes	Yes	Yes
Fission-fusion	Yes	Yes	Yes	Yes
Male philopatry	Yes	Yes	Yes?	No
Morphology				
Quadrupedal knuckle-walker	Yes	Yes	Yes	No
Arboreal/suspensory adaptations	Yes	Yes	Yes	No
Large gut	Yes	Yes	Yes	No
Large brain	Yes	Yes	Yes	Yes
Cognition				
Social and manual tool use	Yes	Yes	Yes	Yes
Complex social field	Yes	Yes	Yes	Yes
Flexible inferential skills	Yes	Yes	Yes	Yes
High self-control	Yes	Yes	Yes	Yes
Flexible gestural communication	Yes	Yes	Yes	Yes

can be considered small and likely occurred during extreme climatic periods, when the Congo River's flow was reduced. The lack of gene flow allowed each species to evolve unique genetic and phenotypic features. Comparisons to humans suggest that about 3 percent of the human genome is more similar to either the bonobo or chimpanzee genome than the two sister species are to each other, while about a quarter of human genes have some elements more similar to one species of *Pan* over the other—but again in approximately equal proportions (Prüfer et al. 2012). Behavioral and cognitive comparisons likewise show striking differences between the two species of *Pan*, including many traits in which one *Pan* species is markedly more like humans (Table 3.3; Hare et al. 2012; Hare and Yamamoto 2015).

TABLE 3.2. Behaviors in bonobos and chimpanzees more similar to humans than to each other.

	Bonobo	Chimpanzee	Human Foragers
Extractive foraging[1]	Only captivity	Frequent	Frequent
Hetero/homosexual/infant nonconceptive sexual behavior[2]	Frequent	Absent	Frequent
Lethal aggression between groups[3]	Absent	Present	Present
Mother's importance to adult offspring[4]	High	Low	High
Infanticide/female coercion[5]	Absent	Present	Present
Levels of adult play[6]	High	Low	High
Cooperative hunting[7]	Absent	Present	Present
Sharing between strangers[8]	Present	Absent	Present
Male-male alliances[9]	Absent	Frequent	Frequent
Female gregariousness[10]	High	Low	High

1. Hohmann and Fruth 2003a; Gruber et al. 2010; Furuichi et al. 2015; Hopkins et al. 2015.
2. Kano 1992; Hashimoto and Furuichi 2006; Hohmann and Fruth 2000; Hare et al. 2007; Woods and Hare 2011; Ryu et al. 2015; Clay and de Waal 2015.
3. Wrangham 1999; Wilson et al. 2014.
4. De Lathouwers and Van Elsacker 2006; Surbeck et al. 2011; Schubert et al. 2013.
5. Hohmann and Fruth 2002; Surbeck et al. 2011.
6. Palagi and Paoli 2007; Wobber et al. 2010b.
7. Ihobe 1992; Watts and Mitani 2002; Surbeck and Hohmann 2008.
8. Tan and Hare in press; Tan and Hare 2013; Tan et al. 2015.
9. Kano 1992; Wrangham 1999.
10. Furuichi 2011; Stevens et al. 2015. "Even with the central role bonobos can play in testing hypotheses regarding ape and human evolution, bonobo research lags far behind work with chimpanzees. Searching ISI Web of Science and Google Scholar for 'bonobo' and 'chimpanzee' reveals that the bonobo makes up only 3% and 9% of the total citations indexed for both species collectively. This is far from the 50% that phylogeny alone would predict" (Hare and Yamamoto 2015).

Given these profiles of overall similarities mixed with differences that vary by species in their overlap with humans, how do we decide whether bonobos or chimpanzees are more representative of the LCA with respect to any particular trait? Three main hypotheses have been suggested:

1. **The mosaic hypothesis:** No living ape is more representative. The LCA's traits represented a mosaic of those found among bonobos and chimpanzees (Prüfer et al. 2012).
2. **The bonobo-like hypothesis:** The LCA most closely resembled bonobos (Zihlman and Cramer 1978; de Waal and Lanting 1998).
3. **The chimpanzee-like hypothesis:** The LCA most closely resembled chimpanzees (Wrangham and Pilbeam 2001).

The mosaic hypothesis claims that the relationship between traits in the living apes is not sufficiently consistent to suggest that one species is more representative of the LCA. It thus represents a null hypothesis. Prüfer et al. (2012) seem to suggest that comparisons of the bonobo, chimpanzee, and human genomes support this hypothesis, because they did not find any patterns indicating that one species of *Pan* has been under stronger selection or is more similar to humans than the other. In support, Table 3.2 shows that specific behavioral and cognitive traits are found for which each species of *Pan* is more similar to humans than to the other (Hare and Yamamoto 2015).

The bonobo-like hypothesis has been suggested based partly on qualitative comparisons of behavior and cognition. For example, within-group relationships show similar trends in tolerance between bonobos and humans (de Waal and Lanting 1998). Morphological similarities between bonobos and australopithecenes have also been used as a basis for inferring behavioral similarities between bonobos and a prehuman stage of the ancestral human lineage (Zihlman 1978). The bonobo-like hypothesis has not yet been carefully elaborated, however. In particular, no conceptual scheme has been proposed that purports to account for patterns of trait change in bonobos, chimpanzees and humans.

The chimpanzee-like hypothesis conforms to evidence that bonobos are highly derived phenotypically relative to other nonhuman apes (Wrangham and Pilbeam 2001). This hypothesis has led to quantitative tests that examine its core developmental and cognitive predictions (Hare et al. 2012; Wilkins et al. 2014). Below we review the hypothesis and the research that was designed to test it. We also highlight how this recent work contradicts the predictions of the two competing hypotheses. We conclude by suggesting where these hypotheses must be tested further.

Bonobos and Chimpanzees Behaving Socially

Both bonobos and chimpanzees are highly social, knuckle-walking, fruit-eating, large-brained apes. They live in fission-fusion communities composed of multiple breeding females and males, potentially including more than one hundred individuals. They form friendships, groom, fight, and reconcile within their groups. Males of both species are larger than females, and females have sexual swellings and mate promiscuously. In both species, females tend to emigrate and males are philopatric (Kano 1992; Boesch et al.

2002; Muller and Mitani 2005). Given the similarities between the two species, differences become more striking. Ethologists first began to notice these differences in the earliest studies of their social behavior—particularly involving the expression of aggression. Chimpanzees are characterized by clear dominance hierarchies within groups and hostile relationships between groups. Males are dominant, coerce females, commit infanticide, strive for alpha status, are highly territorial and xenophobic, and display coalitionary aggression that can be lethal (Muller and Mitani 2005; Wrangham et al. 2006; Muller et al. 2011; Wilson et al. 2014; Feldblum et al. 2014). Chimpanzee mothers support their offspring and are aggressive toward immigrants. They may form friendships with other mothers that can be used in aggressive coalitions, but alliance flexibility has not been reported (Kahlenberg et al. 2008; Pusey et al. 2008, 2013). In contrast, bonobos show much less intense forms of aggression within and between their groups. Bonobo males do strive for status, but do not form coalitions with other males or attain alpha status within their own groups (Kano 1992). Mothers are typically a male's most important social partner—with grandmothers potentially impacting male social status (Surbeck et al. 2011; Schubert et al. 2013). Bonobo females form coalitions with each other and are intolerant of male coercion (Kano 1992; Wrangham 2002; Hohmann and Fruth 2003b). Bonobos do not show the pattern of lethal aggression observed in chimpanzees, and have even been seen to socially interact and travel with neighboring groups (Furuichi 2011). Female immigrants are prized social partners for resident female bonobos, and they receive intense attention but relatively little physical aggression (Moscovice et al. 2015; Ryu et al. 2015; Sakamaki et al. 2015).

These stark differences in social behaviors related to aggression and affiliation again illustrate what is at stake in more precisely estimating which species is more representative of the LCA. One set of explanations and tests will be necessary to reconstruct human evolution if the LCA was a male-dominated, xenophobic species, and a totally different set if it was a more female-centered, relatively xenophilic species. A focus on this contrast leads to a unifying explanation for how these traits evolved.

Process Matters: The Evolution of Development

Bonobos were the last great ape discovered, first recognized in a museum. A small skull labeled as "juvenile" found inside a box on inspection was found to have fused cranial sutures. This suggested the skull was adult and must belong to a kind of chimpanzee that was different from *Pan troglodytes* (Schwarz 1929; Coolidge 1933). The developmental difference led not only to the identification of bonobos but also to the chimpanzee-like hypothesis.

Because ontogenetic stages are temporary, they are often given less attention than adult forms. However, changes in the timing of development provide natural selection with one of the richest sources of variation to act on (West-Eberhard 2003). Relatively small changes in developmental trajectories can have large effects on adult phenotypes, and in many cases selection acts strongest early in development (Hill et al. 2001; West-Eberhard 2003; Morimoto et al. 2014). Furthermore, changes in seemingly unrelated traits may be explained through a relationship formed via a common developmental pathway. Comparative analysis of development can therefore provide a powerful way to understand species differences.

While it may not be widely appreciated that bonobos were recognized as a species based on the fact that their skulls resembled those of juvenile chimpanzees, they are well known for having high levels of play and diverse patterns of sexual behavior in adulthood (Figure 3.2; Palagi and Cordoni 2012). These behavioral tendencies are unique compared to other apes, and exemplify a suite of bonobo traits that share two intriguing features. First, their expression in adult bonobos resembles their expression in juvenile chimpanzees. Second, their differences in expression between bonobos and chimpanzees resemble the differences in expression between domesticated animals and their wild ancestors (Wrangham and Pilbeam 2001; Hare et al. 2012; Wilkins et al. 2014). Thus, like bonobos compared to chimpanzees, domesticated animals have reductions in the size of crania, faces, and teeth, together with a variety of novel prosocial behaviors and accompanying physiological changes, compared to the adult forms of their wild progenitors. These observations suggest that the behavioral, physiological, and morphological changes observed both in bonobos and in domesticated animals are strongly influenced by heterochronic change. According to this idea, selection for a reduction in aggression is the driving force in generating a syndrome of traits associated with domestication. The domestication

FIGURE 3.2. Relative to bonobos, nonconceptive sexual behavior in chimpanzees is extremely rare in infants and adult same-sex pairings but does occasionally occur in juveniles. Bonobo nonconceptive sexual behavior emerges early and lasts throughout life: (a) Two adult female bonobos rub clitorises while eating fruit. (b) An adult male bonobo fellates another adult male bonobo (who is laughing) during an intense rough-and-tumble play session.

syndrome results from selection favoring individuals who retain juvenile characteristics into adulthood, including a reduced propensity for reactive aggression compared to the ancestral adult (Trut et al. 2009).

Following the logic developed for domesticated animals, Wrangham and Pilbeam (2001) therefore proposed that bonobo behavioral evolution is explicable by considering development. As with the initial discovery of bonobos, they began with morphology. They noted that while gorillas and chimpanzees share the same ontogenetic pattern for the trajectory of both their head and body, in bonobos the head is relatively small. They pointed out qualitative similarities between the recurring pattern of changes observed between domesticated and wild progenitors across species and the differences seen in bonobos and chimpanzees. They hypothesized that the observed convergence in traits of the "domestication syndrome" between bonobos and domesticated animals is a result of socioecological pressures that selected bonobos to be less aggressive, as in domesticated animals. The implication was that bonobos evolved from a chimpanzee-like ancestor when they were isolated on the left bank of the Congo River, thanks to selection for reduced aggression. Essentially, the list of qualitative differences in Table 3.2 that seem to suggest a mosaic pattern of evolution began to make sense using development as the organizing principle for their origin.

The Psychology of Bonobo Self-Domestication

Recently, Hare et al. (2012) formalized the "self-domestication" hypothesis (SDH) for bonobo evolution, and reviewed the body of work testing the main predictions of Wrangham and Pilbeam (2001). The SDH hypothesis relies heavily on the work of Dmitry Belyaev and colleagues, who experimentally selected a population of silver foxes *(Vulpes vulpes)* to be interested and prosocial towards humans, as opposed to aggressive and fearful toward them. The experimental line of foxes was compared to a line of control foxes selected randomly for their behavior toward humans; otherwise, the two fox populations were maintained identically. When compared to the control line of foxes, the experimental foxes showed the majority of physiological, morphological, and behavioral elements observed in the domestication syndrome. This included feminized crania, smaller teeth, floppy ears, shortened tails, more play and sexual behavior as adults, as well as prosocial behavior toward humans (Trut et al. 2009). The experimental foxes also became more skilled at using human gestures, similar to the way dogs are more skilled than wolves—even though the foxes were not selected based on their communicative abilities (Hare et al. 2005). Developmental comparisons of the two populations suggest that many of the differences are a result of heterochrony (cf. Gariepy et al. 2001). For example, the experimental foxes have extended socialization periods (Trut 1999). Selection against aggression and for prosociality thus appears to have acted on individual differences in developmental timing to produce a population of foxes that has not only become attracted to humans but also shows a suite of correlated by-products, from morphology to social cognition, none of which were under direct selection (Trut et al. 2009).

The silver fox experiment reveals the power of selection for prosocial behavior on morphology, behavior, physiology, and even cognition. It also suggests the possibility that natural selection acting on wild animal populations could drive a similar type of self-domestication (Hare et al. 2012). Because of the qualitative similarities documented between bonobos and domestic animals, quantitative tests were designed to examine the hypothesis that bonobos are self-domesticated. The majority of these a priori quantitative tests of the SDH involve comparing bonobo and chimpanzee cognition, since the SDH was proposed based largely on observational studies that led to clear predictions about where cognitive differences should evolve in response to selection against aggression (Wrangham and Pilbeam 2001; Hare et al. 2012). Key cognitive predictions were that bonobos should show greater prosociality than chimpanzees, as well as more flexible social skills relating to cooperation and communication—similar to what has been documented when comparing cognition between wild types and domesticated animals (e.g., Hare et al. 2002, 2010; Miklosi et al. 2003; Topal et al. 2009; Hernádi et al. 2012). Furthermore, if bonobos are juvenilized as a result of their reduced aggression being due to heterochronic change, they should show delays in their cognitive and behavioral development relative to chimpanzees (Wobber et al. 2010).

Prosocial Bonobos Prefer Strangers

The most obvious behavioral novelties common to all domestic animals are an increase in prosociality and reduction in the propensity for aggression. Specifically, domesticated animals are attracted to and interested in interacting with humans, while their wild-type ancestors are fearful or even aggressive toward people (Hare et al. 2012). Similarly, bonobos are more prosocial to conspecifics than chimpanzees. By contrast, natural selection did not act on their reaction to humans (Tan and Hare in press).

The relative increase in prosociality among bonobos is illustrated by the fact that, compared to chimpanzees, they co-feed more readily. Thus, when two bonobos are released into a room containing food in a single location, both tend to be able to eat. By contrast, chimpanzees under the same conditions rarely co-feed, because the dominant individual normally monopolizes the food. This result holds even for chimpanzees chosen for having a highly tolerant relationship with each other, whereas almost any bonobo pairs successfully co-feed. The high level of tolerance shown by bonobos allows them to cooperate more effectively than chimpanzees, even when chimpanzees are more experienced in the task (Hare et al. 2007). The different responses of the two species are in part hormonally mediated. In tests of food sharing, male bonobos show an increase in cortisol (but not testosterone) in anticipation of entering the room containing food, whereas male chimpanzees show an increase in testosterone (but not cortisol). The hormone differences thus indicate that, in response to the same socially stressful condition, bonobos become socially anxious, whereas chimpanzees prime themselves for competition (Wobber et al. 2010).

Parallel kinds of species difference in behavior occur in response to crowding. In crowded enclosures of captive groups, bonobos increase sociosexual and grooming behavior, whereas chimpanzees avoid aggression by decreasing social interactions (Aureli and de Waal 1997; Hare et al. 2007; Paoli et al. 2007). Thus, bonobos appear to use social interactions to diffuse tension, while chimpanzees must inhibit social interactions to accomplish the same outcome.

Like chimpanzees, bonobos sometimes share food in the wild (Yamamoto 2015). Experiments in captivity help understand the motivation and cognition behind this type of sharing. Chimpanzees share proactively only when they themselves cannot access the food (Warneken et al. 2007; Melis et al. 2010). However, bonobos share not only when they cannot reach the food, but also when, if they wanted to, they could keep it all for themselves. For example, one experiment allows bonobos to voluntarily share food by opening a one-way door to release another bonobo into the room with them. Bonobos with food repeatedly open the door, even when the other bonobo does not beg for food, and has no opportunity to exhibit reciprocal generosity (Hare and Kwetuenda 2010). Strikingly, bonobos will open the one-way door to share food with a stranger—a bonobo from a different social community. Even more remarkable, when given the choice of proactively sharing with a stranger or groupmate, bonobos prefer to share with a stranger (Bullinger et al. 2013; Tan and Hare 2013). The sharing tendencies of bonobos thus conform to the expectations of the self-domestication hypothesis by showing that there is a marked increase in prosociality in bonobos compared to chimpanzees.

Despite the unusual generosity of bonobos, it is important to note that their sharing is also constrained. Similar to chimpanzees, bonobos have not been observed to share monopolizable food that is in their possession unless their sharing leads to an affiliative social interaction (Jaeggi et al. 2010; Tan and Hare 2013; Tan et al. 2015). In addition, while chimpanzees readily share objects upon request, bonobos prefer to share food over objects (Warneken et al. 2007; Yamamoto et al. 2012; Krupenye et al. submitted).

Prosocial Temperament Leads to Social Cognition

An important cognitive result of domestication is more flexible cooperative communication (Hare 2011; but also see Lewejohann et al. 2010). Dogs are more skilled at spontaneously understanding human gestures than wolves (Hare et al. 2002, 2010). Dogs show similar flexibility in reading human pointing gestures, as seen in human infants (Hare and Tomasello 2005). Dogs are also much more likely than a wolf to make eye contact with a human when faced with an unsolvable task; they also making socially mediated errors only seen in human infants (Miklosi et al. 2003; Topal et al. 2009). Exogenous oxytocin may enhance the ability of dogs to use human gestures (Oliva et al. 2015). Dogs' unusual skill at using human gestures has been attributed to selection against aggression and for more prosocial behaviors toward humans. The strongest test of this idea coming from a comparison between Belyaev's silver foxes (selected to be attracted to and nonaggressive toward humans) and the control population bred randomly in relation to their response to humans. The experimental line of foxes was as skilled at using human gestures as dog puppies, and more skilled than control foxes (Hare et al. 2005). A similar relationship between temperament and social cognition has been observed in a variety of domesticated and nondomesticated species (e.g., Petit et al. 1992; Lewejohann et al. 2010; Hernándi et al. 2012; Michelleta and Waller 2012; Bray et al. 2015; Brust and Guenther 2015).

The self-domestication hypothesis therefore predicts that bonobos will outperform chimpanzees in tasks involving cooperative communication. This prediction has been upheld. When directly compared on a battery of temperamental tests given to bonobos, chimpanzees, orangutans *(Pongo pygmaeus),* and humans, bonobos are most similar to human children (who are shy and observant relative to the other ape species) (Herrmann et al. 2011). Similarly, across a wide range of cognitive tasks, bonobos are more sensitive than chimpanzees to social cues such as gaze direction (Herrmann et al. 2010; but see MacLean and Hare 2013). Moreover, when bonobos and chimpanzees are compared using an eye tracker, only bonobos spend significant time examining the eyes of social stimuli such as conspecific and human faces. By contrast, chimpanzees spent the majority of each trial examining the mouths of the same set of stimuli. The response of bonobos in this case is much more similar to that observed in normally functioning children, while the response of the chimpanzees resembles that of humans on the autism spectrum (Kano et al. 2015). In corroboration of this interpretation, the 2D4D ratios (ratio of the length of the second and fourth digits of the hand, a proxy for neonatal androgen exposure) observed in chimpanzees suggest high neonatal androgen levels, while bonobos suggest levels more similar to normally functioning humans. High neonatal androgen levels and concomitant 2D4D levels have been linked to autism spectrum disorder and aggressive tendencies in humans (McIntyre et al. 2009; Liu et al. 2012).

Chimpanzees are certainly highly sophisticated socially, but this first generation of direct comparisons strongly suggests that bonobos are more human-like than chimpanzees in terms of their sensitivity to subtle social cues. The mechanisms underlying these differences remain to be explored, but there are intriguing hints. A positive relationship between oxytocin and levels of mutual gaze in dogs suggests that the oxytocin system could be implicated. Dogs that make more eye contact with their owners have elevated levels of oxytocin, as do their owners (Nagasawa et al. 2009, 2015; Romero et al. 2014). When dogs are given oxytocin exogenously, they make more eye contact, and their owners' oxytocin level increases (Nagasawa et al. 2015). Increased prosociality in bonobos may be partly mediated by a similar oxytocin feedback loop that is less sensitive in chimpanzees. This idea is supported by previously observed genetic differences in bonobos related to the expression of oxytocin and vasopressin (Hammock and Young 2005; Staes et al. 2014; see also Garai et al. 2014). However, this idea remains largely untested. Variability in chimpanzee social behavior has been linked to urinary oxytocin levels (Wittig et al. 2014), but no similar tests have yet been conducted with bonobos.

Juvenilization: A Mechanism for Change

Heterochronic change remains the leading explanation for the domestication syndrome (Wilkins et al. 2014). If correct, the domestication syndrome should be accompanied by developmental shifts that lead to expanded windows of development. This means that some combination of early emerging traits combined with juvenile traits is preserved into adulthood (Wobber et al. 2011; Hare 2017). As expected, recent quantitative comparisons reveal that cognition is developmentally delayed in bonobos relative to chimpanzees (see Table 3.3 for a summary). For example, given that infant bonobos (with mothers) receive less aggression than infant chimpanzees, Wobber et al. (2010b) predicted that bonobos would show a delay in the development of social inhibition. The premise was that it is crucial for an infant chimpanzee to quickly understand whom they can and cannot forage near in order to avoid aggression and injury, whereas the high level of feeding tolerance toward infant bonobos makes this type of inhibition less vital. In a series of cognitive tasks, bonobos indeed showed delayed development of social inhibition in comparison to age-matched chimpanzees. Bonobos likewise showed evidence of developmental delay in spatial memory. When a cross-sectional sample was compared, older chimpanzees showed improvement in remembering where hidden treats were located in space, while bonobos did not; adult bonobos performed at levels similar to juvenile chimpanzees (Rosati and Hare 2012a).

Such differences might be partly hormonally mediated, given that male bonobos have been reported to show no increase in androgens around puberty. By contrast, male chimpanzees exhibit the expected pubertal rise in androgens (Wobber et al. 2014; but see Behringer et al. 2014). Furthermore, individual differences in chimpanzee androgen levels were strongly associated with performance on the spatial navigation task, whereas androgen levels and spatial cognition were not correlated in bonobos (Wobber and Herrmann 2015). Finally, bonobos also show delayed offset of the expression of thyroid hormone relative to chimpanzees, which might be linked to lower levels of aggression (Beringher et al. 2014). Other suspected cases of developmental delay include visual perspective taking (MacLean and Hare 2012) and reputation formation (Herrmann et al. 2013). Further experiments will test these ideas.

TABLE 3.3. Cognitive and behavioral traits for which bonobos have been found to be juvenilized relative to chimpanzees, and the relevant citation for quantitative comparisons that demonstrate delayed development in bonobos.

Juvenilized Traits in Bonobos	Descriptions of Delayed Development in Comparison to Chimpanzees	References
Play behavior	Adult bonobos show juvenile levels of play as adults.	Palagi and Paoli 2007; Hare et al. 2007; Wobber et al. 2010; Palagi and Cordoni 2012
Nonconceptive sexual behavior	Adult bonobos show higher levels of nonconceptive social sexual behavior than chimpanzees.	Hare et al. 2007 (although see Woods and Hare 2011)
Xenophilia and prosociality	Adult bonobos show juvenile levels of attraction to strangers (i.e., similar to immigrant females), and are prosocial rather than aggressive toward strangers in dyadic interactions.	Hare and Kwetuenda 2010; Tan et al. 2013
Maternal dependence	Bonobo infants remain dependent on mother for longer; adult males remain socially dependent on mothers throughout life.	De Lathouwers and Van Elsacker 2006; Surbeck et al. 2011
Food sharing	Adult bonobos show juvenile levels of food sharing observed in chimpanzees.	Hare et al. 2007; Wobber et al. 2010
Social inhibition	Bonobo infants and juveniles show delays in development of social inhibition.	Wobber et al. 2010
Spatial memory	Bonobos show delayed development in their ability to remember locations of hidden food.	Rosati and Hare 2012a
Physical cognition	Bonobos show delayed development in understanding of the physical properties of the world (i.e., connectivity, transposition, etc.).	Wobber et al. 2014

Bonobos thus show delay in various behaviors and types of cognitive performance relative to chimpanzees, whereas no cognitive task has yet been found in which chimpanzees show delay relative to bonobos. Bonobos also show social skills that emerge extremely early in comparison to chimpanzees. Even the youngest infant bonobos show a range of sexual behaviors not observed in infant chimpanzees (Woods and Hare 2011), and female bonobos appear to reach sexual maturity earlier than female chimpanzees (Beringher et al. 2014; Ryu et al. 2015). Infants use their early emerging sociosexual behavior to increase social tolerance in dyadic interactions with others. Domesticated species similarly tend to exhibit early reproduction (Wilkins et al. 2014).

Given that a range of differences between bonobos and chimpanzees represent a slower rate of development in bonobos, they could in theory have resulted from either paedomorphism (slowed development in bonobos) or peramorphism (accelerated development in chimpanzees). However, peramorphism in chimpanzees is an improbable mechanism, because chimpanzee cranial development is similar to gorillas *(Gorilla gorilla)* and other great apes (Wrangham and Pilbeam 2001; Hare et al. 2012), whereas among the great apes, bonobos are the exception in having a relatively juvenilized cranium (Lieberman et al. 2007; Durrleman et al. 2012). Since cognitive development is assumed to be causally linked to the development of the brain and cranium, the evidence for bonobo crania being paedomorphic supports the notion of bonobo cognition likewise being paedomorphic, as expected under the self-domestication hypothesis.

Future work may show that the developmental predictions of the self-domestication hypothesis can account for a variety of features in bonobos that are difficult to explain as independent ecological adaptations, such as the retention of an infantile white tail tuft into adulthood. Wilkins et al. (2014) recently proposed that the domestication syndrome is the result of a delay in the migration of melanocytes early in the development of the neural crest. These cells are responsible for the formation of morphological and physiological features altered through domestication (i.e., teeth, cartilage, melanin, and so on), as well as influencing brain development through their effects on growth factors. Hare et al. (2012) raise the possibility that a similar developmental mechanism might account for the convergence between bonobos and other domestic animals. The neural crest hypothesis points to future tests that could reveal the heritable mechanism that was affected to create the domestication syndrome across a wide variety of species, including bonobos.

Cognitive Support for Reduced Scramble Competition

The main scenarios proposed to explain the behavioral divergence of bonobos and chimpanzees are based on ecological differences (Wrangham 1986; Furuichi 2009). Over approximately a million years of separation on either side of the Congo River, contrasts in ecology are posited to have shaped the two species into their modern forms (Wrangham 1993; Furuichi 2009). Species-specific psychological predispositions evolved accordingly, such that the social behavior of the two species is now expected to differ even when they live in equivalent environments (Horuichi 2004; Serckx et al. 2015).

A critical species difference in social organization is that bonobo parties (temporary subgroups) are more stable than those in chimpanzees (Furuichi 2009). This gives bonobo mothers a predictable ability to form alliances with each other and with their adult sons (Furuichi 2011). Such alliances deter aggression by adult males, and are therefore inferred to have reduced the benefits to males of having a high propensity for aggression (Wrangham 1993; Furuichi 2011).

Using current ecological differences as a proxy for past selection pressures, the tendency for bonobos to form more stable parties is hypothesized to have evolved in a context of reduced scramble competition for fallback foods compared to chimpanzees. One possibility is that bonobo forests differ floristically from chimpanzee forests. For example, bonobo forests might have larger fruit trees or a higher density of food patches (Furuichi 2011). Against this, however, there is little evidence for consistent differences in these respects between bonobo and chimpanzee habitats (Sommer et al. 2011).

Another possibility is that the reduction in feeding competition in bonobos is due to the lack of gorillas within the species range of bonobos (Wrangham 1986; Wrangham and Peterson 1997). Gorillas rely heavily on the stems and leaves of terrestrial herbs for their food supply. These items are also eaten by chimpanzees and bonobos as fallback foods when preferred fruits are less available. The absence of gorillas south of the Congo River is therefore argued to afford bonobos a greater reliance on terrestrial herbs as fallback foods. As a result, bonobos can forage together in relatively stable parties, as occurs among gorillas. Novel quantitative tests of this hypothesis remain difficult, since there is no fossil record from the Congo Basin to test the distribution of gorillas over evolutionary time, and cropping rates of terrestrial herbs on either side of the Congo River have not been studied in detail.

A cognitive approach can help in testing ecological predictions (Rosati et al. 2014). If bonobos and chimpanzees were shaped over the past by ecological differences that mirror those thought to exist today, we should see species differences in foraging strategies that reflect those differences. Previous comparisons between closely related species illustrate such links among feeding strategies, decision making, and spatial memory (Santos and Rosati 2015). For example, frugivorous lemurs are more successful than more folivorous lemur species in a set of spatial memory tasks (Rosati et al. 2014). Likewise, in a stationary context, insectivorous tamarins *(Saguinus oedipus)* prefer small immediate rewards, whereas gummivorous marmosets *(Callithrix jacchus)* prefer larger delayed rewards (Rosati et al. 2006). However, if travel is required to obtain the same rewards, the monkeys reverse their preferences (Stevens et al. 2015). Thus the feeding ecology to which a species is adapted is expected to influence their foraging cognition.

Accordingly, if consistent differences in the nature of the Pleistocene food supply shaped bonobo and chimpanzee social tendencies, they are expected to have influenced their cognitive preferences also (Hare et al. 2012). To test for this possibility, bonobos and chimpanzees have been quantitatively compared across a range of foraging tasks designed to measure how they remember locations and value time and risk. Compared to bonobos, chimpanzees tend to prefer larger rewards, and they tolerate longer delays in waiting for them (Rosati et al. 2007; Rosati and Hare 2013). Similarly, in assessments of preference for a fixed or variable reward, bonobos prefer a small fixed reward, while chimpanzees prefer the option of a more risky, but potentially larger, variable reward (Heilbronner et al. 2008; Rosati and Hare 2016; see also Haun et al. 2011). Memory tasks reveal parallel results: chimpanzees tend to outperform bonobos in finding more hidden food items, across longer delays (Rosati and Hare 2012a; Rosati 2015). These differences between bonobos and chimpanzees thus conform to the ecological model of reduced scramble competition. They may again be mediated by androgens, since spatial performance was linked to basal testosterone levels in chimpanzees but not in bonobos (Rosati 2015; Wobber and Herrmann 2015).

Overall, therefore, research on developmental and cognitive comparisons between bonobos and chimpanzees tends to support the self-domestication hypothesis. Bonobos show a suite of traits that are seen in a range of species that have undergone juvenilization in the course of selection against aggression. Cognitive comparisons support the idea that this process occurred due to natural selection caused by reduced scramble competition. Neither the mosaic nor the bonobo-like ancestor model predicts or can explain the range of developmental differences reviewed here. Chimpanzees likely serve as the better living representative of the LCA, while the more derived bonobo gives us a powerful way to think about the evolution of the most derived ape of all.

Human Self-Domestication

It has been repeatedly suggested that humans are domesticated. Stephen J. Gould famously argued that humans are globally paedomorphic, with this developmental pattern being crucial to our species' evolution (Gould 1980). While the idea of humans being globally juvenilized has been rejected, the idea that humans are domesticated has gained new life (e.g., Leach 2003). The primary impetus for this is the comparative work explored above. Observing the constraints on chimpanzee cooperation and the unusual social skills of dogs led to the suggestion that temperamental evolution is a powerful mechanism to shift social problem-solving abilities (Hare and Tomasello 2005; Hare et al. 2007; Hare 2011; Bray et al. 2015). In support of this idea, children with shy temperament tend to develop sophisticated theory of mind skills earlier in development. This provides the first direct evidence that, even in humans, there may be an important link between temperament and social cognition that is crucial for unique forms of human cognition (Wellman et al. 2011; Lane et al. 2013; also see Rodrigues et al. 2009). Direct comparisons of cognitive development across a range of cognitive tasks in bonobo, chimpanzee, and human infants have also revealed that human infants have a set of early emerging social cognitive abilities. These same abilities are thought to be the cognitive foundation for participating in human culture (Herrmann et al. 2007). These findings support the idea that a shift in development led to the human condition (Wobber et al. 2014). Finally, there is even evidence from comparisons of fossil humans that late in our species evolution we underwent selection against aggression that may have shaped not only our morphology but more importantly our whole way of life (Nelson et al. 2011; Cieri et al. 2014). Initial comparisons of gene expression development in humans and chimpanzees also suggests heterochronic shifts. Many genes that are switched off early in the development of the chimpanzee prefrontal cortex are still switched on decades later in humans (Somel et al. 2009). This initial work suggests that developing predictions for a new generation of tests to examine human domestication will be an exciting area of research in coming years, and central to any test will likely be our more derived and potentially convergent relative, the bonobo (Hare 2017).

Past and Future Limitations

While a first generation of experimental studies offers support for the main cognitive predictions regarding the self-domestication hypothesis, not every study comparing cognition in the two species of *Pan* has revealed differences. Many of the similarities are observed in cognitive skills not predicted by the SDH to be altered. For example, bonobos and chimpanzees perform similarly on nonsocial inhibitory control in detour-reaching tasks (Vlamings et al. 2010; MacLean et al. 2014). The two species also both use play and grooming as social currency in similar ways (Schroepfer-Walker et al. 2015). Most significantly, on a standardized battery of cognitive tasks, bonobos and chimpanzees were identical in most ways except in areas that are predicted by the self-domestication hypothesis (i.e., bonobos being more skilled with social cues and chimpanzees being more skilled in some nonsocial tasks; Herrmann et al. 2010).

However, there have also been studies that predicted species differences but did not detect them. Although it was expected that bonobos would be more averse than chimpanzees to choosing ambiguous options, the two species show equal aversion to ambiguity (Rosati 2010; Krupenye et al. 2015). Likewise, bonobos are no more interested in joint object play than chimpanzees, and are no more skilled at using human gestures to find hidden food (MacLean and Hare 2013, 2015a). These findings should refine the self-domestication hypothesis, and act as examples of how quantitative comparisons that powerfully test the predictions of the SDH are now possible and can potentially allow for falsification.

To date, large-scale quantitative comparisons between bonobos and chimpanzees have been conducted at African sanctuaries designed to rescue, rehabilitate, and release bushmeat orphans ($n > 20$ per species). This raises the issue of how representative these individuals might be, given the acute trauma they experienced. Direct comparison of mother-reared and orphan bonobos and chimpanzees showed no differences in cognition, behavior, or hormonal profiles (Wobber and Hare 2011). Moreover, a number of successful releases of orphans back into the wild have been conducted (including orphans from Lola ya Bonobo; Andre et al. 2008). This suggests that the populations involved in recent quantitative comparisons are representative of the cognitive flexibility in the species to which they belong.

The self-domestication hypothesis suggests tests that may help explain both bonobo and human evolution. However, questions regarding evolutionary process in our lineage remain difficult to address. A small number of living species within the ape clade means comparisons between apes typically lack quantitative power. Biologists have developed sophisticated quantitative phylogenetic techniques to test evolutionary hypotheses that are not possible using a handful of species (MacLean et al. 2012). While this lack of power has made it difficult to resolve the extent to which the different species of *Pan* resemble the last common ancestor, it also highlights the strength of a broader comparative developmental approach that generated the self-domestication hypothesis.

Summary

Where fossils are silent, comparisons of the living apes are crucial to testing evolutionary ideas regarding the origin of traits. Interpretations of these comparisons rest heavily on which living ape is considered more representative of our LCA. A priori morphological, physiological, behavioral, and cognitive tests of the self-domestication hypothesis support the chimpanzee-like hypothesis. Bonobos not only are more prosocial and flexible in their social skills but also show signs of juvenilization across the phenotype. They also show early emerging sexual behavior related to their ability to maintain high social tolerance. This expanded window of development both earlier and later suggests that bonobos are more derived than chimpanzees from our last common ancestor, in line with a similar conclusion based on morphology. If support for the self-domestication hypothesis grows, it will not only clarify which traits are derived in humans, but will also generate hypotheses regarding the process(es) that shaped these traits. The self-domestication hypothesis has already led to a number of tests regarding the evolution of human cooperation and communication as a result of selection against within-group aggression. This means bonobos and chimpanzees will be needed equally to continue unlocking the secrets of human evolution.

Acknowledgments

We thank James Brooks and Eliot Cohen for help with our reference section.

References

Andre, C., C. Kamate, P. Mbonzo, D. Morel, and B. Hare. 2008. The conservation value of Lola ya Bonobo sanctuary. In T. Furuichi and J. Thompson, eds., *The Bonobos: Behavior, Ecology, and Conservation,* 303–322. New York: Springer.

Aureli, F., and F. B. M. de Waal. 1997. Inhibition of social behavior in chimpanzees under high-density conditions. *American Journal of Primatology* 41: 213–228.

Behringer, V., T. Deschner, C. Deimel, J. M. G. Stevens, and G. Hohmann. 2014. Age-related changes in urinary testosterone levels suggest differences in puberty onset and divergent life history strategies in bonobos and chimpanzees. *Hormones and Behavior* 66: 525–533.

Behringer, V., T. Deschner, R. Murtagh, J. M. G. Stevens, and G. Hohmann. 2014. Age-related changes in thyroid hormone levels of bonobos and chimpanzees indicate heterochrony in development. *Journal of Human Evolution* 66: 83–88.

Boesch, C., G. Hohmann, and L. Marchant, eds. 2002. *Behavioural Diversity in Chimpanzees and Bonobos.* Cambridge: Cambridge University Press.

Bray, E. E., E. L. MacLean, and B. A. Hare. 2015. Increasing arousal enhances inhibitory control in calm but not excitable dogs. *Animal Cognition* 18: 1317–1329.

Brust, V., and A. Guenther. 2015. Domestication effects on behavioural traits and learning performance: Comparing wild cavies to guinea pigs. *Animal Cognition* 18: 99–109.

Bullinger, A. F., J. M. Burkart, A. P. Melis, and M. Tomasello. 2013. Bonobos, *Pan paniscus,* chimpanzees, *Pan troglodytes,* and marmosets, *Callithrix jacchus,* prefer to feed alone. *Animal Behaviour* 85: 51–60.

Cieri, R. L., S. E. Churchill, R. G. Franciscus, J. Z. Tan, and B. Hare, B. 2014. Craniofacial feminization, social tolerance, and the origins of behavioral modernity. *Current Anthropology* 55: 419–443.

Clay, Z., and F. B. M. de Waal. 2015. Sex and strife: Post-conflict sexual contacts in bonobos. *Behaviour* 152: 313–334.

Coolidge, H. J. 1933. Pan paniscus. Pigmy chimpanzee from south of the Congo River. *American Journal of Physical Anthropology* 18: 1–59.

De Lathouwers, M., and L. Van Elsacker. 2006. Comparing infant and juvenile behavior in bonobos *(Pan paniscus)* and chimpanzees *(Pan troglodytes):* A preliminary study. *Primates* 47: 287–293.

de Manuel, M., et al. 2016. Chimpanzee genomic diversity reveals ancient admixture with bonobos. *Nature* 354: 477–481.

de Waal, F. B. M. 1987. Tension regulation and nonreproductive functions of sex in captive bonobos *(Pan paniscus). National Geographic Research* 3: 318–335.

de Waal, F. B., and F. Lanting. 1998. *Bonobo: The Forgotten Ape.* Berkeley: University of California Press.

Durrleman, S., X. Pennec, A. Trouve, N. Ayache, and J. Braga. 2012. Comparison of the endocranial ontogenies between chimpanzees and bonobos via temporal regression and spatiotemporal registration. *Journal of Human Evolution* 62: 74–88.

Feldblum, J. T., E. E. Wroblewski, R. S. Rudicell, B. H. Hahn, T. Paiva, M. Cetinkaya-Rundel, A. E. Pusey, and I. C. Gilby. 2014. Sexually coercive male chimpanzees sire more offspring. *Current Biology* 24: 2855–2860.

Furuichi, T. 2009. Factors underlying party size differences between chimpanzees and bonobos: A review and hypotheses for future study. *Primates* 50: 197–209.

Furuichi, T. 2011. Female contributions to the peaceful nature of bonobo society. *Evolutionary Anthropology* 20: 131–142.

Furuichi, T., K. Koops, H. Ryu, C. Sanz, T. Sakamaki, D. Morgan, and N. Tokuyama. 2015. Why do wild bonobos not use tools like chimpanzees do? *Behaviour* 152: 425–460.

Garai, C., T. Furuichi, Y. Kawamoto, H. Ryu, and M. Inoue-Murayama. 2014. Androgen receptor and monoamine oxidase polymorphism in wild bonobos. *Meta Gene* 2: 831–843.

Gariepy, J. L., D. J. Bauer, and R. B. Cairns. 2001. Selective breeding for differential aggression in mice provides evidence for heterochrony in social behaviours. *Animal Behaviour* 61: 933–947.

Gould, S. J. 1977. *Ontogeny and Phylogeny.* Cambridge, MA: Harvard University Press.

Groves, C. 1981. (Comments) Bonobos: Generalized hominid prototypes or specialized insular dwarfs? *Current Anthropology* 22: 363–375.

Gruber, T., Z. Clay, and K. Zuberbuhler. 2010. A comparison of bonobo and chimpanzee tool use: Evidence for a female bias in the *Pan* lineage. *Animal Behaviour* 80: 1023–1033.

Hammock, E. A. D., and L. J. Young. 2005. Microsatellite instability generates diversity in brain and sociobehavioral traits. *Science* 308: 1630–1634.

Haraway, D. 1993. Teddy Bear patriarchy. In N. B. Dirks, G. Eley, and S. B. Ortner, eds., *Culture/Power/History,* 49–95. Princeton, NJ: Princeton University Press.

Hare, B. 2007. From nonhuman to human mind: What changed and why? *Current Directions in Psychological Science* 16: 60–64.

Hare, B. 2009. What is the effect of affect on bonobo and chimpanzee problem solving? In A. Berthoz, ed., *Neurobiology of Umwelt: How Living Beings Perceive the World,* 89–102. New York: Springer.

Hare, B. 2011. From hominoid to hominid mind: What changed and why? *Annual Review of Anthropology* 40: 293–309.

Hare, B. 2017. Survival of the friendliest: *Homo sapiens* evolved via selection for prosociality. *Annual Review of Psychology* 68: 155–186.

Hare, B., J. Call, and M. Tomasello. 2001. Do chimpanzees know what conspecifics know? *Animal Behaviour* 61: 139–151.

Hare, B., M. Brown, C. Williamson, and M. Tomasello. 2002. The domestication of social cognition in dogs. *Science* 298: 1634–1636.

Hare, B., and S. Kwetuenda. 2010. Bonobos voluntarily share their own food with others. *Current Biology* 20: R230–R231.

Hare, B., A. P. Melis, V. Woods, S. Hastings, and R. Wrangham. 2007. Tolerance allows bonobos to outperform chimpanzees on a cooperative task. *Current Biology* 17: 619–623.

Hare, B., L. Plyusnina, N. Ignacio, O. Schepina, A. Stepika, R. Wrangham, and L. Trut. 2005. Social cognitive evolution in captive foxes is a correlated by-product of experimental domestication. *Current Biology* 15: 226–230.

Hare, B., A. Rosati, J. Breaur, J. Kaminski, J. Call, and M. Tomasello, M. 2010. Dogs are more skilled than wolves with human social cues: A response to Udell et al. (2008) and Wynne et al. (2008). *Animal Behaviour* 79: e1ee6.

Hare, B., and M. Tomasello. 2005. Human-like social skills in dogs? *Trends in Cognitive Sciences* 9: 439–444.

Hare, B., V. Wobber, and R. Wrangham. 2012. The self-domestication hypothesis: Evolution of bonobo psychology is due to selection against aggression. *Animal Behaviour* 83: 573–585.

Hare, B., and S. Yamamoto. 2015. Moving bonobos off the scientifically endangered list. *Behaviour* 152: 247–258.

Hashimoto, C., and T. Furuichi. 2006. Comparison of behavioral sequence of copulation between chimpanzees and bonobos. *Primates* 47: 51–55.

Haun, D. B. M., C. Nawroth, and J. Call. 2011. Great apes' risk-taking strategies in a decision making task. *PLoS ONE* 6: e28801.

Heilbronner, S. R., A. G. Rosati, J. R. Stevens, B. Hare, and M. D. Hauser. 2008. A fruit in the hand or two in the bush? Divergent risk preferences in chimpanzees and bonobos. *Biology Letters* 4: 246–249.

Hernádi, A., A. Kis, B. Turcsán, and J. Topál. 2012. Man's underground best friend: Domestic ferrets, unlike the wild forms, show evidence of dog-like social-cognitive skills. *PLoS One* 7: e43267.

Herrmann, E., J. Call, M. V. Hernandez-Lloreda, B. Hare, and M. Tomasello. 2007. Humans have evolved specialized skills of social cognition: The cultural intelligence hypothesis. *Science* 317: 1360–1366.

Herrmann, E., B. Hare, J. Call, and M. Tomasello. 2010. Differences in the cognitive skills of bonobos and chimpanzees. *PLoS ONE* 5: e12438.

Herrmann, E., B. Hare, J. Cissewski, and M. Tomasello. 2011. A comparison of temperament in nonhuman apes and human infants. *Developmental Science* 14: 1393–1405.

Herrmann, E., S. Keupp, B. Hare, A. Vaish, and M. Tomasello. 2013. Direct and indirect reputation formation in nonhuman great apes *(Pan paniscus, Pan troglodytes, Gorilla, Pongo pygmaeus)* and human children *(Homo sapiens)*. *Journal of Comparative Psychology* 127: 63–75.

Hill, K., C. Boesch, J. Goodall, A. Pusey, J. Williams, and R. Wrangham. 2001. Mortality rates among wild chimpanzees. *Journal of Human Evolution* 40: 437–450.

Hohmann, G. 2001. Association and social interactions between strangers and residents in bonobos *(Pan paniscus)*. *Primates* 42: 91–99.

Hohmann, G., and B. Fruth. 2000. Use and function of genital contacts among female bonobos. *Animal Behaviour* 60: 107–120.

Hohmann, G., and B. Fruth. 2002. Dynamics in social organization of bonobos *(Pan paniscus)*. In C. Boesch, G. Hohmann, and L. Marchant, eds., *Behavioral Diversity in Chimpanzees and Bonobos*, 138–150. Cambridge: Cambridge University Press.

Hohmann, G., and B. Fruth. 2003a. Culture in bonobos? Between-species and within-species variation in behavior. *Current Anthropology* 44: 563–571.

Hohmann, G., and B. Fruth. 2003b. Intra- and inter-sexual aggression by bonobos in the context of mating. *Behaviour* 140: 1389–1413.

Hohmann, G., and B. Fruth. 2003c. Lui Kotal: A new site for field research on bonobos in the Salonga National Park. *Pan Africa News* 10: 25–27.

Hopkins, W. D., J. Schaeffer, J. L. Russell, S. L. Bogart, A. Meguerditchian, and O. Coulon. 2015. A comparative assessment of handedness and its potential neuroanatomical correlates in chimpanzees *(Pan troglodytes)* and bonobos *(Pan paniscus)*. *Behaviour* 152: 461–492.

Horiuchi, S. 2004. A competition model within and across groups explaining the contrast between the societies of chimpanzees and bonobos. *Population Ecology* 46: 65–70.

Idani, G. 1991. Social relationships between immigrant and resident bonobo *(Pan paniscus)* females at Wamba. *Folia Primatologica* 57: 83–95.

Ihobe, H. 1992. Male-male relationships among wild bonobos *(Pan paniscus)* at Wamba, Republic of Zaire. *Primates* 33: 163–179.

Jaeggi, A. V., J. M. G. Stevens, and C. P. Van Schaik. 2010. Tolerant food sharing and reciprocity is precluded by despotism among bonobos but not chimpanzees. *American Journal of Physical Anthropology* 143: 41–51.

Kahlenberg, S. M., M. E. Thompson, M. N. Muller, and R. W. Wrangham. 2008. Immigration costs for female chimpanzees and male protection as an immigrant counterstrategy to intrasexual aggression. *Animal Behaviour* 76: 1497–1509.

Kano, F., S. Hirata, and J. Call. 2015. Social attention in the two species of *Pan:* Bonobos make more eye contact than chimpanzees. *PLoS ONE* 10: e0129684.

Kano, T. 1992. *The Last Ape: Pygmy Chimpanzee Behavior and Ecology*. Stanford, CA: Stanford University Press.

Krupenye, C., F. Kano, S. Hirata, J. Call, and M. Tomasello. 2016. Great apes anticipate that other individuals will act according to false beliefs. *Science* 354: 110–114.

Krupenye, C., A. Rosati, and B. Hare. 2015. Bonobos and chimpanzees exhibit economic framing effects. *Biology Letters.* 11: 20140527.

Krupenye, C., J. Tan, and B. Hare. Submitted. Bonobos share food but not toys and tools.

Kuroda, S. 1989. Developmental retardation and behavioral characteristics of pygmy chimpanzees. In P. G. Heltne and L. A. Marquardt, eds., *Understanding Chimpanzees*, 184–193. Cambridge, MA: Harvard University Press.

Lane, J. D., H. M. Wellman, S. L. Olson, A. L. Miller, L. Wang, and T. Tardif. 2013. Relations between temperament and theory of mind development in the United States and China: Biological and behavioral correlates of preschoolers' false-belief understanding. *Developmental Psychology* 49: 825–836.

Leach, H. M. 2003. Human domestication reconsidered. *Current Anthropology* 44: 349–368.

Lewejohann, L., T. Pickel, N. Sachser, and S. Kaiser. 2010. Wild genius—domestic fool? Spatial learning abilities of wild and domestic guinea pigs. *Frontiers in Zoology* 7: 9.

Lieberman, D. E., J. Carlo, M. P. de Leon, and C. P. E. Zollikofer. 2007. A geometric morphometric analysis of heterochrony in the cranium of chimpanzees and bonobos. *Journal of Human Evolution* 52: 647–662.

Liu, J. H., J. Portnoy, and A. Raine. 2012. Association between a marker for prenatal testosterone exposure and externalizing behavior problems in children. *Development and Psychopathology* 24: 771–782.

MacLean, E. L., and B. Hare. 2012. Bonobos and chimpanzees infer the target of another's attention. *Animal Behaviour* 83: 345–353.

MacLean, E. L., and B. Hare. 2013. Spontaneous triadic engagement in bonobos *(Pan paniscus)* and chimpanzees *(Pan troglodytes)*. *Journal of Comparative Psychology* 127: 245–255.

MacLean, E. L., and B. Hare. 2015a. Bonobos and chimpanzees exploit helpful but not prohibitive gestures. *Behaviour* 152, 493–520.

MacLean, E. L., and B. Hare. 2015b. Dogs hijack the human bonding pathway. *Science* 348: 280–281.

MacLean, E. L., B. Hare, C. L. Nunn, E. Addessi, F. Amici, R. C. Anderson, F. Aureli, J. M. Baker, A. E. Bania, A. M. Barnard, N. J. Boogert, E. M. Brannon, E. E. Bray, J. Bray, L. J. N. Brent, J. M. Burkart, J. Call, J. F. Cantlon, L. G. Cheke, N. S. Clayton, M. M. Delgado, L. J. DiVincenti, K. Fujita, E. Herrmann, C. Hiramatsu, L. F. Jacobs, K. E. Jordan, J. R. Laude, K. L. Leimgruber, E. J. E. Messer, A. C. D. Moura, L. Ostojic, A. Picard, M. L. Platt, J. M. Plotnik, F. Range, S. M. Reader, R. B. Reddy, A. A. Sandel, L. R. Santos, K. Schumann, A. M. Seed, K. B. Sewall, R. C. Shaw, K. E. Slocombe, Y. J. Su, A. Takimoto, J. Z. Tan, R. Tao, C. P. van Schaik, Z. Viranyi, E. Visalberghi, J. C. Wade, A. Watanabe, J. Widness, J. K. Young, T. R. Zentall, and Y. N. Zhao. 2014. The evolution of self-control. *Proceedings of the National Academy of Sciences* 111: E2140–E2148.

MacLean, E. L., L. J. Matthews, B. A. Hare, C. L. Nunn, R. C. Anderson, F. Aureli, E. M. Brannon, J. Call, C. M. Drea, N. J. Emery, D. B. M. Haun, E. Herrmann, L. F. Jacobs, M. L. Platt, A. G. Rosati, A. A. Sandel, K. K. Schroepfer, A. M. Seed, J. Z. Tan, C. P. van Schaik, and V. Wobber. 2012. How does cognition evolve? Phylogenetic comparative psychology. *Animal Cognition* 15: 223–238.

Mailund, T., A. E. Halager, M. Westergaard, J. Y. Dutheil, K. Munch, L. N. Andersen, G. Lunter, K. Prüfer, A. Scally, A. Hobolth, and M. H. Schierup. 2012. A new isolation with migration model along complete genomes infers very different divergence processes among closely related great ape species. *PLoS Genetics* 8: e1003125.

McIntyre, M. H., E. Herrmann, V. Wobber, M. Halbwax, C. Mohamba, N. de Sousa, R. Atencia, D. Cox, and B. Hare. 2009. Bonobos have a more human-like second-to-fourth finger length ratio (2D:4D) than chimpanzees: A hypothesized indication of lower prenatal androgens. *Journal of Human Evolution* 56: 361–365.

Melis, A. P., F. Warneken, and B. Hare. 2010. Collaboration and helping in chimpanzees. In E. V. Lonsdorf, S. R. Ross, and T. Matsuzawa, eds., *The Mind of the Chimpanzee: Ecological and Experimental Perspectives,* 278–393. Chicago: University of Chicago Press.

Micheletta, J., and B. M. Waller. 2012. Friendship affects gaze following in a tolerant species of macaque, *Macaca nigra. Animal Behaviour* 83: 459–467.

Miklosi, A., E. Kubinyi, J. Topal, M. Gacsi, Z. Viranyi, and V. Csanyi. 2003. A simple reason for a big difference: Wolves do not look back at humans, but dogs do. *Current Biology* 13: 763–766.

Morimoto, N., M. S. P. de Leon, and C. P. E. Zollikofer. 2014. Phenotypic variation in infants, not adults, reflects genotypic variation among chimpanzees and bonobos. *PLoS ONE* 9.

Moscovice, L. R., T. Deschner, and G. Hohmann. 2015. Welcome back: Responses of female bonobos *(Pan paniscus)* to fusions. *PLoS ONE* 10: e0127305.

Muller, M. N., S. M. Kahlenberg, M. E. Thompson, and R. W. Wrangham. 2007. Male coercion and the costs of promiscuous mating for female chimpanzees. *Proceedings of the Royal Society B: Biological Sciences* 274: 1009–1014.

Muller, M. N., and J. C. Mitani. 2005. Conflict and cooperation in wild chimpanzees. *Advances in the Study of Behavior* 35: 275–331.

Muller, M. N., M. E. Thompson, S. M. Kahlenberg, and R. W. Wrangham. 2011. Sexual coercion by male chimpanzees shows that female choice may be more apparent than real. *Behavioral Ecology and Sociobiology* 65: 921–933.

Nagasawa, M., T. Kikusui, T. Onaka, and M. Ohta. 2009. Dog's gaze at its owner increases owner's urinary oxytocin during social interaction. *Hormones and Behavior* 55: 434–441.

Nagasawa, M., S. Mitsui, S. En, N. Ohtani, M. Ohta, Y. Sakuma, T. Onaka, K. Mogi, and T. Kikusui. 2015. Oxytocin-gaze positive loop and the coevolution of human-dog bonds. *Obstetrical and Gynecological Survey* 70: 450–451.

Nelson, E., C. Rolian, L. Cashmore, and S. Shultz. 2011. Digit ratios predict polygyny in early apes, *Ardipithecus,* Neanderthals and early modern humans but not in *Australopithecus. Proceedings of the Royal Society B: Biological Sciences* 278: 1556–1563.

Oliva, J. L., J. L. Rault, B. Appleton, and A. Lill. 2015. Oxytocin enhances the appropriate use of human social cues by the domestic dog *(Canis familiaris)* in an object choice task. *Animal Cognition* 18: 991–991.

Palagi, E., and G. Cordoni. 2012. The right time to happen: Play developmental divergence in the two *Pan* species. *PLoS ONE* 7: e52767.

Palagi, E., and T. Paoli. 2007. Play in adult bonobos *(Pan paniscus):* Modality and potential meaning. *American Journal of Physical Anthropology* 134: 219–225.

Paoli, T., G. Tacconi, S. M. B. Tarli, and E. Palagi. 2007. Influence of feeding and short-term crowding on the sexual repertoire of captive bonobos *(Pan paniscus). Annales Zoologici Fennici* 44: 81–88.

Petit, O., C. Desportes, and B. Thierry. 1992. Differential probability of coproduction in 2 species of macaque *(Macaca tonkeana, Macaca mulatta). Ethology* 90: 107–120.

Prüfer, K., K. Munch, I. Hellmann, K. Akagi, J. R. Miller, B. Walenz, S. Koren, G. Sutton, C. Kodira, and R. Winer. 2012. The bonobo genome compared with the chimpanzee and human genomes. *Nature* 486: 527–531.

Pusey, A., C. Murray, W. Wallauer, M. Wilson, E. Wroblewski, and J. Goodall. 2008. Severe aggression among female *Pan troglodytes schweinfurthii* at Gombe National Park, Tanzania. *International Journal of Primatology* 29: 949–973.

Pusey, A. E., and K. Schroepfer-Walker. 2013. Female competition in chimpanzees. *Philosophical Transactions of the Royal Society B: Biological Sciences* 368: 20130077.

Rodrigues, S. M., L. R. Saslow, N. Garcia, O. P. John, and D. Keltner. 2009. Oxytocin receptor genetic variation relates to empathy and stress reactivity in humans. *Proceedings of the National Academy of Sciences* 106: 21437–21441.

Romero, T., M. Nagasawa, K. Mogi, T. Hasegawa, and T. Kikusui. 2014. Oxytocin promotes social bonding in dogs. *Proceedings of the National Academy of Sciences* 111: 9085–9090.

Rosati, A. G. 2015. Context influences spatial frames of reference in bonobos *(Pan paniscus). Behaviour* 152: 375–406.

Rosati, A. G., and B. Hare. 2012a. Chimpanzees and bonobos exhibit divergent spatial memory development. *Developmental Science* 15: 840–853.

Rosati, A. G., and B. Hare. 2012b. Decision making across social contexts: Competition increases preferences for risk in chimpanzees and bonobos. *Animal Behaviour* 84: 869–879.

Rosati, A. G., and B. Hare. 2013. Chimpanzees and bonobos exhibit emotional responses to decision outcomes. *PLoS ONE* 8: e63058.

Rosati, A. G., and B. Hare. 2016. Reward currency modulates human risk preferences. *Evolution and Human Behavior* 37: 159–168.

Rosati, A. G., K. Rodriguez, and B. Hare. 2014. The ecology of spatial memory in four lemur species. *Animal Cognition* 17: 947–961.

Rosati, A. G., J. R. Stevens, B. Hare, and M. D. Hauser. 2007. The evolutionary origins of human patience: Temporal preferences in chimpanzees, bonobos, and human adults. *Current Biology* 17: 1663–1668.

Rosati, A. G., J. R. Stevens, and M. D. Hauser. 2006. The effect of handling time on temporal discounting in two New World primates. *Animal Behaviour* 71: 1379–1387.

Ryu, H., D. A. Hill, and T. Furuichi. 2015. Prolonged maximal sexual swelling in wild bonobos facilitates affiliative interactions between females. *Behaviour* 152: 285–311.

Sakamaki, T., I. Behncke, M. Laporte, M. Mulavwa, H. Ryu, H. Takemoto, N. Tokuyama, S. Yamamoto, and T. Furuichi. 2015. Intergroup transfer of females and social relationships between immigrants and residents in bonobo *(Pan paniscus)* socie-

ties. In T. Furuichi, J. Yamagiwa, and F. Aureli, eds., *Dispersing Primate Females,* 127–164. New York: Springer.
Santos, L. R., and A. G. Rosati. 2015. The evolutionary roots of human decision making. *Annual Review of Psychology* 66: 321–347.
Schroepfer-Walker, K., V. Wobber, and B. Hare. 2015. Experimental evidence that grooming and play are social currency in bonobos and chimpanzees. *Behaviour* 152: 545–562.
Schubert, G., L. Vigilant, C. Boesch, R. Klenke, K. Langergraber, R. Mundry, M. Surbeck, and G. Hohmann. 2013. Co-residence between males and their mothers and grandmothers is more frequent in bonobos than chimpanzees. *PLoS ONE* 8: e83870.
Schultz, A. H. 1941. The relative size of the cranial capacity in primates. *American Journal of Physical Anthropology* 28: 273–287.
Schwarz, E. 1929. Das Vorkommen des Schimpansen auf den linken Congo-Ufer. *Revue de Zoologie et de Botanique Africaines* 16: 425–426.
Serckx, A., H. S. Kuhl, R. C. Beudels-Jamar, P. Poncin, J. F. Bastin, and M. C. Huynen. 2015. Feeding ecology of bonobos living in forest-savannah mosaics: Diet seasonal variation and importance of fallback foods. *American Journal of Primatology* 77: 948–962.
Somel, M., H. Franz, Z. Yan, A. Lorenc, S. Guo, T. Giger, J. Kelso, B. Nickel, M. Dannemann, S. Bahn, M. J. Webster, C. S. Weickert, M. Lachmann, S. Paabo, and P. Khaitovich. 2009. Transcriptional neoteny in the human brain. *Proceedings of the National Academy of Sciences* 106: 5743–5748.
Sommer, V., J. Bauer, A. Fowler, and S. Ortmann. 2011. Patriarchal chimpanzees, matriarchal bonobos: Potential ecological causes of a *Pan* dichotomy. In V. Sommer and C. Ross, eds., *Primates of Gashaka: Socioecology and Conservation in Nigeria's Biodiversity Hotspot,* 417–449. New York: Springer.
Staes, N., J. M. G. Stevens, P. Helsen, M. Hillyer, M. Korody, and M. Eens. 2014. Oxytocin and vasopressin receptor gene variation as a proximate base for inter- and intraspecific behavioral differences in bonobos and chimpanzees. *PLoS ONE* 9: e113364.
Stevens, J. M. G., E. de Groot, and N. Staes. 2015. Relationship quality in captive bonobo groups. *Behaviour* 152: 259–283.
Surbeck, M., and G. Hohmann. 2008. Primate hunting by bonobos at LuiKotale, Salonga National Park. *Current Biology* 18: R906–R907.
Surbeck, M., R. Mundry, and G. Hohmann. 2011. Mothers matter! Maternal support, dominance status and mating success in male bonobos *(Pan paniscus). Proceedings of the Royal Society B: Biological Sciences* 278: 590–598.
Takemoto, H., Y. Kawamoto, and T. Furuichi. 2015. How did bonobos come to range south of the Congo river? Reconsideration of the divergence of *Pan paniscus* from other *Pan* populations. *Evolutionary Anthropology* 24: 170–184.
Tan, J., A. Ariely, and B. Hare. In preparation. Bonobos demonstrate evolution in weak ties.

Tan, J., and B. Hare. 2013. Bonobos share with strangers. *PLoS ONE* 8: e51922.
Tan, J., and B. Hare. In press. Prosociality among non-kin in bonobos and chimpanzees compared. In B. Hare and S. Yamamoto, eds., *Bonobos: Unique in Mind Brain and Behaviour*. Oxford: Oxford University Press.
Tan, J., S. Kwetuenda, and B. Hare. 2015. Preference or paradigm? Bonobos show no evidence of other-regard in the standard prosocial choice task. *Behaviour* 152: 521–544.
Thompson, J. A. M. 2003. A model of the biogeographical journey from proto-*Pan* to *Pan paniscus*. *Primates* 44: 191–197.
Tomasello, M., B. Hare, H. Lehmann, and J. Call. 2007. Reliance on head versus eyes in the gaze following of great apes and human infants: The cooperative eye hypothesis. *Journal of Human Evolution* 52: 314–320.
Topal, J., G. Gergely, A. Erdohegyi, G. Csibra, and A. Miklosi. 2009. Differential sensitivity to human communication in dogs, wolves, and human infants. *Science* 325: 1269–1272.
Trut, L. 1999. Early canid domestication: The farm-fox experiment. *American Scientist* 87: 160–169.
Trut, L., I. Oskina, and A. Kharlamova. 2009. Animal evolution during domestication: The domesticated fox as a model. *Bioessays* 31: 349–360.
Vlamings, P. H. J. M., B. Hare, and J. Call. 2010. Reaching around barriers: The performance of the great apes and 3–5-year-old children. *Animal Cognition* 13: 273–285.
Warneken, F., B. Hare, A. P. Melis, D. Hanus, and M. Tomasello. 2007. Spontaneous altruism by chimpanzees and young children. *PLoS Biology* 5: 1414–1420.
Watts, D. P., and J. C. Mitani. 2002. Hunting behavior of chimpanzees at Ngogo, Kibale National Park, Uganda. *International Journal of Primatology* 23: 1–28.
Wellman, H. M., J. D. Lane, J. LaBounty, and S. L. Olson. 2011. Observant, nonaggressive temperament predicts theory-of-mind development. *Developmental Science* 14: 319–326.
West-Eberhard, M. J. 2003. *Developmental Plasticity and Evolution*. Oxford: Oxford University Press.
Wilkins, A. S., R. W. Wrangham, and W. T. Fitch. 2014. The "domestication syndrome" in mammals: A unified explanation based on neural crest cell behavior and genetics. *Genetics* 197: 795–808.
Wilson, M. L., C. Boesch, B. Fruth, T. Furuichi, I. C. Gilby, C. Hashimoto, C. L. Hobaiter, G. Hohmann, N. Itoh, K. Koops, J. N. Lloyd, T. Matsuzawa, J. C. Mitani, D. C. Mjungu, D. Morgan, M. N. Muller, R. Mundry, M. Nakamura, J. Pruetz, A. E. Pusey, J. Riedel, C. Sanz, A. M. Schel, N. Simmons, M. Waller, D. P. Watts, F. White, R. M. Wittig, K. Zuberbuhler, and R. W. Wrangham. 2014. Lethal aggression in *Pan* is better explained by adaptive strategies than human impacts. *Nature* 513: 414–417.
Wittig, R. M., C. Crockford, T. Deschner, K. E. Langergraber, T. E. Ziegler, and K. Zuberbuhler. 2014. Food sharing is linked to urinary oxytocin levels and bonding in related and unrelated wild chimpanzees. *Proceedings of the Royal Society B: Biological Sciences* 281: 20133096.

Wobber, V., and B. Hare. 2011. Psychological health of orphan bonobos and chimpanzees in African sanctuaries. *PLoS ONE* 6: e17147.

Wobber, V., B. Hare, J. Maboto, S. Lipson, R. Wrangham, and P. Ellison. 2010. Differential reactivity of steroid hormones in chimpanzees and bonobos when anticipating food competition. *Proceedings of the National Academy of Sciences* 107: 12457–12462.

Wobber, V., and E. Herrmann. 2015. The influence of testosterone on cognitive performance in bonobos and chimpanzees. *Behaviour* 152: 407–423.

Wobber, V., E. Herrmann, B. Hare, R. Wrangham, and M. Tomasello. 2014. Differences in the early cognitive development of children and great apes. *Developmental Psychobiology* 56: 547–573.

Wobber, V., R. Wrangham, and B. Hare. 2010a. Application of the heterochrony framework to the study of behavior and cognition. *Communicative and Integrative Biology* 3: 337–339.

Wobber, V., R. Wrangham, and B. Hare. 2010b. Bonobos exhibit delayed development of social behavior and cognition relative to chimpanzees. *Current Biology* 20: 226–230.

Woods, V., and B. Hare. 2011. Bonobo but not chimpanzee infants use socio-sexual contact with peers. *Primates* 52: 111–116.

Wrangham, R. W. 1986. Ecology and social relationships in two species of chimpanzee. In D. L. Rubenstein and R. W. Wrangham, eds., *Ecological Aspects of Social Evolution: Birds and Mammals,* 352–378. Princeton, NJ: Princeton University Press.

Wrangham, R. W. 1993. The evolution of sexuality in chimpanzees and bonobos. *Human Nature* 4: 47–79.

Wrangham, R. W. 1999. Evolution of coalitionary killing. *Yearbook of Physical Anthropology* 42: 1–30.

Wrangham, R. W. 2002. The cost of sexual attraction: Is there a trade-off in female *Pan* between sex appeal and received coercion? In C. Boesch, G. Hohmann, and L. Marchant, eds., *Behavioral Diversity in Chimpanzees and Bonobos,* 204–215. Cambridge: Cambridge University Press.

Wrangham, R. W., and D. Peterson. 1997. *Demonic Males: Apes and the Origins of Human Violence.* Boston: Houghton Mifflin Harcourt.

Wrangham, R. W., and D. R. Pilbeam. 2001. African apes as time machines. In B. M. F. Galdikas, N. E. Briggs, L. K. Sheeran, G. L. Shapiro, and J. Goodall, eds., *All Apes Great and Small.* Berlin: Springer.

Wrangham, R. W., M. L. Wilson, and M. N. Muller. 2006. Comparative rates of violence in chimpanzees and humans. *Primates* 47: 14–26.

Yamamoto, S. 2015. Non-reciprocal but peaceful fruit sharing in wild bonobos in Wamba. *Behaviour* 152: 335–357.

Yamamoto, S., T. Humle, and M. Tanaka. 2012. Chimpanzees' flexible targeted helping based on an understanding of conspecifics' goals. *Proceedings of the National Academy of Sciences* 109: 3588–3592.

Zihlman, A. L., and D. L. Cramer. 1978. Skeletal differences between pygmy *(Pan paniscus)* and common chimpanzees *(Pan troglodytes). Folia Primatologica* 29: 86–94.

PART II

Chimpanzees and the Evolution of Human Uniqueness

4

Introduction

Chimpanzees and Human Uniqueness

MARTIN N. MULLER

The opening chapters of this book reviewed the genetic evidence that chimpanzees and humans form a group to the exclusion of gorillas, the morphological evidence that chimpanzees and gorillas are extremely similar, and the inference that the last common ancestor of chimpanzees and humans was a fruit-eating, suspensory, knuckle-walking ape. Because large body size in gorillas is likely derived, and the major differences in social organization and behavior between chimpanzees and gorillas are the consequences of large body size in the latter, they further suggested that the LCA was morphologically and behaviorally like a chimpanzee. Supporting this characterization, unique aspects of bonobo morphology and behavior also appear derived.

The chapters in the second part of the book have two goals. First, they examine, in detail, the implications of this inference for understanding our own species. Does starting with a chimpanzee-like LCA help us to answer specific questions about human evolution? Second, they ask whether the initial inference is consistent with what is known about the evolution of particular traits. In other words, does a chimpanzee-like LCA make sense?

Each chapter in this section addresses the evolution of a specific trait, or cluster of traits, including life history parameters, diet, social relationships, cognition, and communication. Authors begin by reviewing similarities and differences between chimpanzees and humans. Whenever possible, human data are taken from studies of hunting and gathering populations. This is not because human foragers are viewed as somehow lacking culture, or variability, or history (Marks 2002). Rather, it reflects the fact that our species evolved in a foraging context, and despite considerable variation, foraging economies share multiple features, including a lack of material accumulation, routine food sharing, egalitarianism, and a sexual division of labor, that are likely to have been important in shaping human adaptations (Winterhalder 2001; Marlowe 2005). Foragers also maintain natural fertility patterns, which were the norm in preindustrial populations (Wood 1994).

Studying living foragers allows us to ask whether particular aspects of our species' life history or behavior are potentially ancient, or simply the result of the novel environments that we have recently created. For example, as Gurven and Gomes note in Chapter 5, it is sometimes assumed that the existence of menopause in humans requires no special explanation, because for most of our species' history women would rarely have lived past the age of fifty. Studies of extant foragers, who lack modern health care and live in generally challenging environments, falsify this assumption: grandmothers in these groups can enjoy long postreproductive lives (e.g., Blurton Jones 2016).

Human sleep patterns provide another example. Many researchers have hypothesized that artificial lighting, television, and computers have shortened the natural sleep duration in Western industrial populations, contributing to adverse health outcomes. Studies of foragers in Tanzania and Namibia cast doubt on this idea, as average sleep duration in these groups is no longer than what is observed in industrial populations, despite an absence of these technologies (Yetish et al. 2015; Samson et al. 2017).

For similar reasons, authors focus on data from wild chimpanzees over those in captivity. Captive animals enjoy optimal access to food, as well as routine medical care. They are additionally subject to artificial constraints on ranging, association, and group composition. Consequently, captive demography, behavior, and cognition are unlikely to accurately represent chimpanzees living in the environments to which they are adapted. Captive data are considered, however, when relevant data from the wild do not exist.

For traits that appear similar between chimpanzees and humans, authors address whether this is the result of homology (inheritance from a common

INTRODUCTION: CHIMPANZEES AND HUMAN UNIQUENESS

ancestor) or homoplasy (evolutionary convergence). Either process is possible, and ascertaining which is responsible for a particular similarity is always difficult. Reviewing data from other great apes, together with evidence from the hominin fossil record, when available, can be informative.

For traits that appear different between chimpanzees and humans, authors consider the likely selection pressures that could have pushed a chimpanzee-like ancestor toward a more human-like trait. They also ask when such changes were likely to have occurred in reference to the hominin fossil record. A major goal is to see whether chimpanzee data can help us to reconstruct the evolution of unique human traits by constraining hypotheses for their evolution.

Finally, authors evaluate the original inference that the last common ancestor of humans and chimpanzees was chimpanzee-like. Does the evolution of particular traits create problems for this idea?

Initial chapters focus on life history traits. The human life history is puzzling because it contains a mix of fast and slow elements. Humans grow slowly, have a long period of juvenile dependency, reproduce at a late age, and have a long life span, all traits that are normally associated with slow reproduction. As illustrated in Chapter 1, however, humans reproduce very quickly for a hominoid. Chapter 5 asks why humans enjoy lower mortality rates and longer lives than chimpanzees, including a postreproductive life span. Chapter 6 compares human and chimpanzee reproductive ecology to account for our fast reproduction.

The next four chapters address ecological and dietary adaptations in humans and chimpanzees. Chapter 7 examines the differences in locomotor ecology and energetics that result from human bipedalism, and Chapters 8 and 9 consider the evolution of the human dietary niche, which is distinct from that of chimpanzees in its emphasis on high-quality, easy to digest resources. Chapter 10 focuses on one such resource, comparing hunting and meat sharing in humans and chimpanzees.

Social behaviors in chimpanzees and humans show both surprising similarities and extraordinary differences. Chapter 11 traces the marked divergence between the chimpanzee and human mating systems, and Chapter 12 investigates the discrepant kinship structures that result from the human emphasis on long-term pair bonds. Chapter 13 reviews the similarities in intergroup conflict in chimpanzees and humans, and Chapter 14 suggests that these similarities led to parallels in patterns of sex-biased cooperation within groups. Chapter 15 examines the extensive cooperation between the sexes in human foraging and childrearing, which has no parallel in chimpanzee society, and is sometimes characterized as cooperative breeding. Chapter 16 considers sexual coercion in the two species, identifying shared patterns of sexual violence, as well as derived forms of coercion that are unique to humans.

The remaining chapters take up some of the complex traits that make humans unique in the animal world. Chapter 17 seeks the chimpanzee origins of human tool use and manufacture. Chapters 18 through 21 ask whether any of the building blocks of human culture, cognition, morality, and language can be found in chimpanzees, despite the significant gaps that clearly separate the species in all of these traits.

A major goal of this section is to provide a systematic and current account of the similarities and differences between human and chimpanzee behavior. Although some readers will doubtless find the argument for a chimpanzee-like LCA unconvincing, we hope that all will find the book useful as a comprehensive summary of our current understanding of our closest living relative.

References

Blurton Jones, N. *Demography and Evolutionary Ecology of Hadza Hunter-Gatherers.* Cambridge: Cambridge University Press.

Marks, J. 2002. *What It Means to Be 98% Chimpanzee.* Berkeley: University of California Press.

Marlowe, F. W. 2005. Hunter-gatherers and human evolution. *Evolutionary Anthropology* 14: 54–67.

Samson, D. R., A. N. Crittenden, I. A. Mabulla, A. Z. P. Mabulla, and C. L. Nunn. 2017. Hadza sleep biology: Evidence for flexible sleep-wake patterns in hunter-gatherers. *American Journal of Physical Anthropology* 162: 573–582.

Winterhalder, B. 2001. The behavioural ecology of hunter-gatherers. In C. Panter-Brick, R. H. Layton, and P. Rowley-Conwy, eds., *Hunter-Gatherers: An Interdisciplinary Perspective,* 12–38. Cambridge: Cambridge University Press.

Wood, J. W. 1994. *Dynamics of Human Reproduction: Biology, Biometry, Demography.* New York: Aldine de Gruyter.

Yetish, G., H. Kaplan, M. Gurven, B. Wood, H. Pontzer, P. R. Manger, C. Wilson, R. McGregor, and J. M. Siegel. 2015. Natural sleep and its seasonal variations in three pre-industrial societies. *Current Biology* 25: 1–7.

5

Mortality, Senescence, and Life Span

MICHAEL D. GURVEN *and* CRISTINA M. GOMES

"Little Mama," estimated to be over seventy years old, is one of the oldest chimpanzees ever recorded. Born in Africa, she is now a retiree living in a Florida theme park (Figure 5.1a). She is a rare exception, petite and healthy after receiving excellent care in the pet industry (Segal 2012). Madame Jeanne Calment, a French supercentenarian who died at the ripe age of 122, was the oldest human ever recorded (Figure 5.1b). She led a leisured life, was still riding her bicycle up until age 100, smoked cigarettes since age twenty-one, and had a good sense of humor ("I've never had but one wrinkle, and I'm sitting on it"). Although these two are hardly representative of their respective species, the gap in life span of a half century speaks to real biological differences in life span potential (Table 5.1).

The maximum human life expectancy has risen steadily by more than two years every decade over the past two centuries, a dramatic improvement that suggests new answers to old questions about species differences in programmed senescence and the existence of biologically determined maximal life-spans (Wachter and Finch 1997; Austad 1999; Oeppen and Vaupel 2003; Burger et al. 2012). Although much of the increase in life expectancy in the nineteenth century can be attributed to better sanitation,

FIGURE 5.1. Oldest recorded chimpanzee and human. (a) "Little Mama" at Florida's Lion Country Safari park is believed to be the oldest chimpanzee in captivity, between 60 and 74 years old. She has lived there since 1967. Photo courtesy of Andrew Halloran. (b) Jean Calment lived to 122 years. Here she celebrates her 121st birthday in 1996. Photo source: http://en.wikipedia.org/wiki/File:Jeanne-Calment-1996.jpg.

modern medicine, and improved diets (Riley 2001), there is strong evidence that the general pattern of a long life span is not unique to modern populations or even horticulturalists (cf. Lovejoy 1981; Washburn 1981), and that current increases in life span may be a consequence of plasticity in our evolved human life history. There is little support for Vallois's (1961: 222) infamous claim that among early humans, "few individuals passed forty years, and it is only quite exceptionally that any passed fifty," or the Hobbesian view of a "nasty, brutish and short" human life span (see also King and Jukes 1969; Weiss 1981).

Chimpanzees are the closest living genetic relatives to humans, and are likely to have a life history similar to our last common ancestor from 7–10 million years ago (Wrangham 1987, 2001). Compared to humans, our chimpanzee cousins have shorter lives, smaller brains, and bodies that grow and develop more rapidly (Isler and van Schaik 2009). Chimpanzee fertility is also lower than among human hunter-gatherers, owing primarily to longer intervals between births (Table 5.1). Closer attention to chimpanzee demography, diet, health, physiology, economics, and social behavior is therefore critical for understanding the evolution of human longevity and related life history traits. To date, however, there have been few comprehensive reviews of human hunter-gatherer life spans and survivorship that include detailed comparisons with chimpanzees.

TABLE 5.1. Comparison of life history traits among traditional humans living in natural fertility conditions and wild chimpanzees. Adapted from Gurven (2012).

Trait	Definition	Units	Humans	Chimpanzees	Percent Increase
Brain volume	Volume of the brain	cm^3	1,330[a,b]	330–430[a,b,c]	242%
Encephalization quotient (EQ)	Ratio of actual to predicted brain size (based on body size allometry)		7.4–7.8[c]	~2.2–2.5[c]	223%
Juvenile period	Weaning to menarche	years	12.9[d]	5[d]	158%
Adult life span	Life expectancy at age 15	years	37.7[e]	14[e]	148%
Maximum life span	Oldest observed individual	years	121[f]	54, 66[a,f,j]	83%
Fertility rate	Inverse of interbirth interval	#/yr	0.29[g]	0.21[h]	38%
Infant mortality rate	Probability of dying in first year	percent	23[e]	20[l]	15%
Juvenile survival	Probability of living to age 15	percent	57[e]	36[e]	58%
Interbirth interval	Time between successive births	months	41.3[g]	68.9[h,i]	−40%
Extrinsic mortality rate	Young adult mortality rate	percent/yr	1.1[e]	3.7[e]	−70%
Neonate mass	Mass at birth	kg	2.8[a,m]	1.7[a,m]	65%
Age at menarche	Birth to menstruation	years	15[d]	10[d]	50%
Age at first reproduction	Birth to reproduction	years	19.1[d,g]	12.8[k]	49%
Age at last reproduction	Birth to last reproduction	years	39[d,g]	27.7[k]	41%
Adolescence	Menarche to first reproduction	years	4.1[d]	3[d]	37%
Fetal growth rate	Growth rate: conception to birth	g/day	10.4[o]	7.6[o]	37%
Total fertility rate	Total number of live births	live births	6.1[g]	5[g,n]	22%
Gestation length	Conception to birth	days	269[a,p]	225[a,j]	20%
Body size	Average adult female mass	kg	47[d]	35[d]	34%

a. Not specific to natural fertility or wild populations; b. Schoenemann (2006); c. Roth and Dicke (2005); d. Walker et al. (2006); e. Gurven and Kaplan (2007); f. Finch (2007); g. Kaplan et al. (2000); h. Emery Thompson et al. (2007); I. Refers to interbirth interval when first infant survives to age four; IBI is 26.6 months otherwise when first infant dies; j. Bronikowski et al. (2011); k. Emery Thompson (2013); l. Hill et al. (2001); m. Lee et al. (1991); n. Based on completed fertility; TFR synthesized from age-specific fertility rates in Emery Thompson et al. (2007: table S1) is 6.9 live births; o. Fetal growth rate=neonate mass/gestation length; p. Wood (1994).

In this chapter, we first compare mortality rates and life span among humans and chimpanzees in the "wild" using the most complete available data on well-studied populations. Second, we describe the plasticity in life span in both species with improved conditions. Despite some similarities in early life mortality patterns, chimpanzees have shorter potential maximal life spans than humans. Humans everywhere have a substantial postreproductive life span, whereas chimpanzees mostly do not, and some robust females are capable of reproducing until the end of their adult lives. To help contextualize these species differences in survivorship, and presumably in aging, the third section summarizes what is known about causes of death both in the wild and under more modern conditions, in both young and old. We examine species differences in senescence rates, discuss the selection pressures that may have helped shape the distinct life histories of humans and chimpanzees, and evaluate leading hypotheses that might help explain why humans are longer-lived than chimpanzees. We link life history changes with species differences in social behavior, and conclude with suggestions for future directions.

Mortality

Humans

Hunter-gatherers with minimal exposure to modern medicine, and with traditional diets and activity regimes, are an important lens for understanding how selection helped shape the evolution of the human life course. Given that quality demographic data with reasonably accurate age and mortality estimation exist only for a handful of populations (Howell 1979; Blurton Jones et al. 1992; Hill and Hurtado 1996; Early and Headland 1998), we also consider simple horticulturalists without modern amenities as an additional source of data on mortality and senescence in preindustrial societies.

The age-specific probability of survival (l_x) from birth to adulthood shows a modest amount of variation across different populations of human hunter-gatherers, forager horticulturalists, and wild chimpanzees (Figure 5.2). The within-species variation is less marked than the between-species variation, for which humans have a higher probability of survival at all stages of life, with the exception of early infancy. Infant survival rates may be lower in

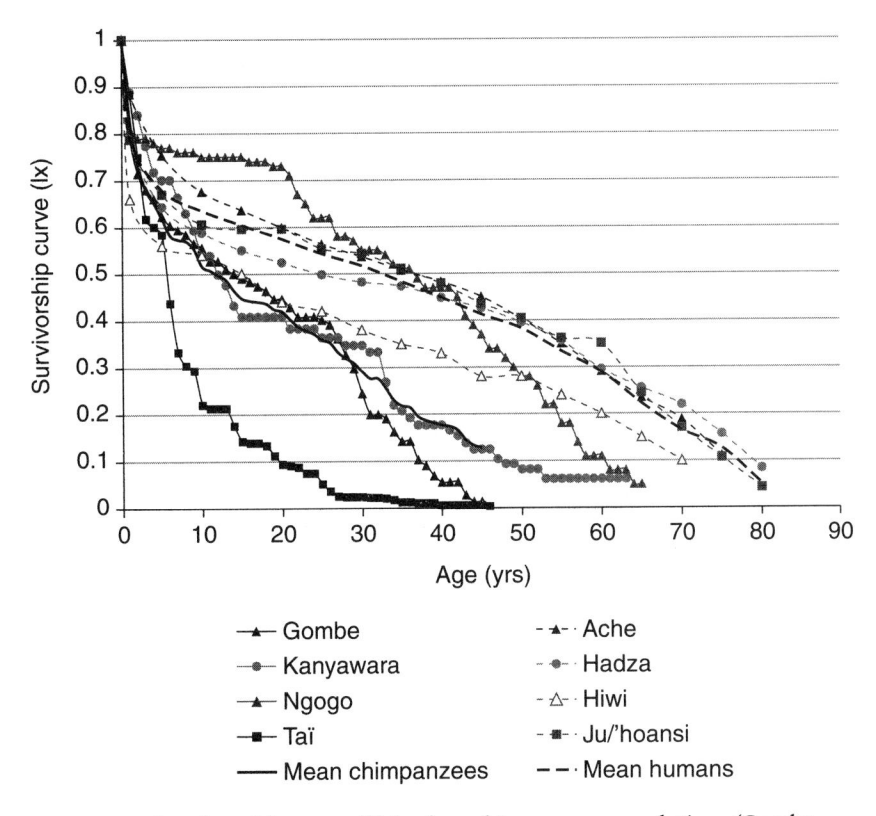

FIGURE 5.2. Survivorship curve (l_x) for four chimpanzee populations (Gombe, Kanyawara, Ngogo, and Taï), four traditional human populations (Aché, Hadza, Hiwi, and Ju/'hoansi), and the mean for chimpanzees (Bossou, Gombe, Kanyawara, Mahale, Ngogo, and Taï) and humans (Aché, Agta, Hadza, Hiwi, Ju/'hoansi, Tsimané, and Yanomamo). Both sexes are combined for each population. Sources: Gombe and Taï (Hill et al. 2001), Kanyawara (Muller and Wrangham 2014), Ngogo (Wood et al. 2017), Aché (Hill and Hurtado 1996), Agta (Early and Headland 1998), Hadza (Blurton Jones et al. 1992), Hiwi (Hill et al. 2007), Ju/'hoansi (Howell 1979), Tsimané (Gurven et al. 2007), and Yanomamo (Early and Peters 2000).

humans than chimpanzees, owing to birth complications or higher vulnerability of altricial neonates. Infant mortality rates among hunter-gatherer populations range from 14 to 40 percent, with a mean ± standard deviation of 27 ± 7 percent dying in the first year of life ($n = 15$) (Volk and Atkinson 2013). On average, 57 percent and 64 percent of children born survive to age fifteen among hunter-gatherers and forager-horticulturalists, respectively. Beyond age fifteen, adult mortality rates range from 1 to 1.5 percent per year until about age forty, when it increases exponentially. In spite of this variation, there does appear to be a clear premodern human pattern. There is remarkable similarity in age profiles of mortality risk over the life span. By age ten, the mortality hazard has slowed to 0.01, doubled to about 0.02 by age forty, doubled yet again before age sixty, and again by age seventy. Low mortality therefore persists until about age forty, when mortality accelerations become more evident. Overall, the mortality rate after age thirty doubles every seven to ten years (Finch 1994; Gurven and Kaplan 2007).

Data from extant foragers with little to no access to medical attention or modern foods, including the Ju/'hoansi, Aché, and Hadza, show that while at birth mean life expectancies range from thirty to thirty-seven years of life, women who survive to age forty-five can expect to live an additional twenty to twenty-two years (Blurton Jones et al. 2002; Gurven and Kaplan 2007). While there is significant variation across groups in life expectancy at early ages, there is significant convergence after about age thirty. With the exception of the Hiwi, who show over ten years less remaining during early ages and over five years less remaining during adulthood, and of the Hadza, whose life expectancy at each age is about two years longer than the rest at most adult ages, all other groups are hardly distinguishable. At age forty, the expected age at death is about sixty-three to sixty-six (i.e., twenty-three to twenty-six additional expected years of life), whereas by age sixty-five, expected age at death is only about seventy to seventy-six years of age. By that age, death rates become very high.

Human life span (whether measured as maximal life span or life expectancy) is longer than predicted for a typical mammal (or primate) of human body size, but not atypical given the larger than expected brain size of humans (Allman et al. 1993). Estimates based on regressions of various primate subfamilies and extant apes suggests a major increase in longevity between *Homo habilis* (52–56 yrs) to *Homo erectus* (60–63 yrs) occurring 1.7–2 million years ago, and further increases for *Homo sapiens* (66–72 yrs) (Judge and Carey 2000). Extrapolations for early *Homo sapiens* based on comparative analyses including both brain weights and body sizes among nonhuman primates similarly suggest a maximum life span between sixty-six and seventy-eight years (Hammer and Foley 1996). Although maximum life spans can be much larger than life expectancies (the average of which is often lowered due to high infant and child mortality), it is usually reported that Paleolithic humans had life expectancies of only fifteen to twenty years. This brief life span is believed to have persisted over thousands of generations (Cutler 1975; Weiss 1981) until less than ten thousand years ago, when early agriculture presumably caused a slight increase to about twenty-five years. Gage (2003) compiles over twelve reconstructed prehistoric life tables with similar life expectancies to form a composite life table with survivorship to age fifty (l_{50}) of about 2–9 percent and e_{45} values of about three to seven years.

There is a large paleodemographic literature concerning problematic age estimation in skeletal samples, and bias in bone preservation leading to underrepresentation of older individuals (see Buikstra and Konigsberg 1985; Walker et al. 1988; Buikstra 1997; Hoppa and Vaupel 2002). This literature is too large to discuss here, and we direct readers to recent treatments by O'Connell et al. (1999) and Kennedy (2003). Nonetheless, we point out some observations that further suggest problems with prehistoric life tables. Mortality rates in prehistoric populations are estimated to be lower than those for traditional foragers until about age two years. Estimated mortality rates then increase dramatically for prehistoric populations, so that by age forty-five they are over seven times greater than those for traditional foragers, even worse than the ratio of captive chimpanzees to foragers. Because these prehistoric populations cannot be very different genetically from the populations surveyed here, there must be systematic biases in the samples and/or in the estimation procedures at older ages, where presumably endogenous senescence should dominate as the primary cause of death. While excessive warfare could explain the shape of one or more of the prehistoric forager mortality profiles, it is improbable that these profiles represent the long-term prehistoric forager mortality profile. Such rapid mortality increase late in life would have severe consequences for our human life history evolution, particularly for senescence in humans. It is encouraging that recent treatments of prehistoric life tables show increasing similarity with those presented here based on ethnographic samples (Konigsberg and Herrmann 2006).

It is noteworthy that unlike chimpanzees (below), most hunter-gatherer and small-scale horticultural populations in the ethnographic record show positive population growth, on average 1 percent. Such growth could not have represented conditions over long stretches of our species history: in order to achieve population stationarity (i.e., zero growth), fertility would have to decline well below that ever observed in natural fertility populations (to a total fertility rate of four births per woman) or survivorship would need to decline below that ever observed (to $l_{15} = 0.41$) (Gurven and Kaplan 2007).

Either current conditions reflected in the demographic data are not representative of the past (i.e., warfare may have been more common, or fertility lower), or population dynamics over short periods of time may be better described as a "saw-tooth" pattern characterized by periods of growth followed by rapid population crashes (Hill and Hurtado 1996). It is difficult to evaluate these two possibilities in light of current knowledge about the past. While evidence suggests that climate varied widely throughout the Pleistocene and into the Holocene Epoch (Richerson et al. 2005), the extent to which past foragers typically experienced increasing, declining, or zero growth in past environments is unknown.

Chimpanzees

The largest dataset on mortality among free-living chimpanzees was compiled by Hill and colleagues (2001) to form a synthetic life table, based on data obtained from Gombe, Taï, Kanyawara (Kibale), Mahale, and Bossou field sites. We substituted the Kanyawara data in the former sample with an updated sample obtained by Muller and Wrangham (2014), and include a new Ngogo sample by Wood et al. (2017) from another part of the same Kibale forest. This increased the risk set by 93 percent to 7,214 risk years and deaths by 40 percent to 398. It is important to note that most wild chimpanzee populations except for Ngogo have been sampled during periods of stasis or decline, while hunter-gatherer populations are growing on average at about 1 percent (Gurven and Kaplan 2007).

Chimpanzees have a life expectancy of around fifteen years at birth, about half that of humans. Infant mortality ranges from 11 to 28 percent, with a mean±standard deviation of 17±7 percent in the first year of life, and 39±11 percent until age five. The mortality rate drops to about four between ages ten and fifteen. By age fifteen, the life expectancy is another twenty years, and by age thirty, the mortality rate is about 7 percent, with twelve additional years of life left. Remaining life expectancy if surviving to age forty-five is another six years. Life expectancy at birth is higher for females than males, 14.3 years versus 10.8 years, respectively.

Chimpanzee sites vary in mortality patterns with Taï showing the lowest survivorship and Ngogo showing the highest (9.9 percent average mortality rate per year over ages 10–35 yrs in Taï vs. 1.5 percent in Ngogo). Taï may be the most affected by anthropogenic factors, whereas Kanyawara reaches

FIGURE 5.3. An elderly chimpanzee male (estimated age fifty-six years) from Kanyawara. Photo by Martin N. Muller.

levels of survivorship comparable to humans between zero and fourteen years old. Mortality rates over this early life period are even lower in Ngogo than among humans (Figure 5.4). In Ngogo, life expectancy at birth is almost thirty-three years and infant mortality is 19 percent. In Kanyawara, life expectancy at birth is almost twenty years and infant mortality is 11 percent. These low rates are likely due to the low impact of disease transmission, predation, and habitat loss at both sites (Muller and Wrangham 2014). In addition, the rich resource base at Ngogo (three fruit trees: *Ficus mucuso, Chrysophyllum albidum,* and *Pterygota mildbraedii*) can support large populations (Wood et al. 2017), making it an ideal environment for population growth with abundant foods, little anthropogenic impact, and no natural predators. It is likely, however, that some deaths are missing from Ngogo, as their life table presents the probability of surviving from ages two to fifteen to be 95.8 percent! This reflects only five reported deaths in 1,169 chimp-risk years of observation. Another caveat about the representativeness of Ngogo is that the current age composition of their population does not mirror the shape of the survivorship curve (Martin Muller, pers. comm.).

(a)

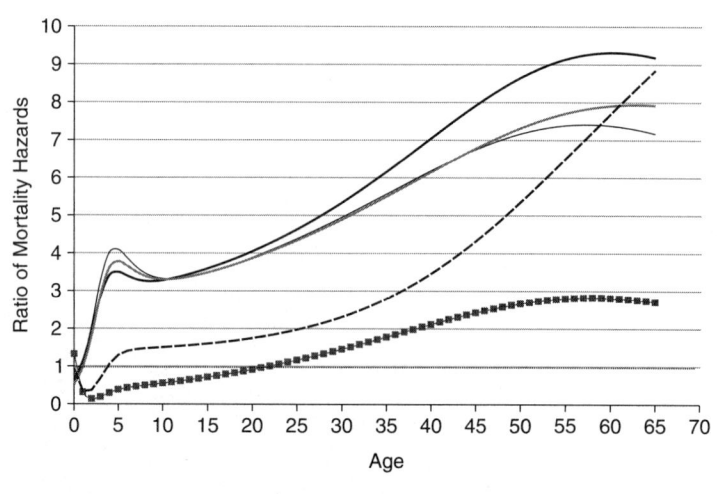

- Ratio of Wild chimps (HILL) to HGs
- Ratio of Wild Chimps (HILL+Kanyawara) to HGs
- Ratio of Captive Chimps to HGs
- Ratio of Ngogo to HGs
- Ratio of NEW Average Chimp to HGs

(b)

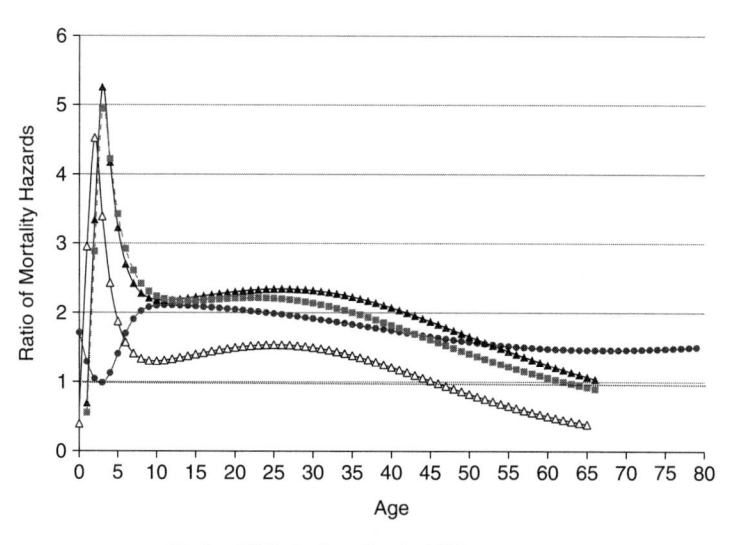

- Ratio of HGs to Acculturated HGs
- Ratio of Wild (HILL+Kany+Ngogo) to Captive Chimps
- Ratio of Wild Chimps (HILL) to Captives
- Ratio of New Chimp AVG to Captives

However, even if the Kanyawara and Ngogo mortality profiles are better representatives of chimpanzee life history in the absence of recent human interactions than the other sites, adult mortality still increases at a substantially higher rate than among humans. For example, remaining life expectancies at ages thirty, forty, and fifty are about ten, seven, and six years, respectively. Mortality rates are also increasing between the ages eleven and thirty-five at Kanyawara, and between ages twenty and thirty-five at Ngogo, the period when mortality rates in hunter-gatherers are flat. Although no life table yet exists for bonobos, one report based on a small sample suggests lower infant and juvenile mortality than among most chimpanzee groups (5 percent die in first year and 27 percent die before age six) (Furuichi et al. 1998). Overall, chimpanzees show a very different life course than human hunter-gatherers, with higher mortality and lower age-specific survival, especially during adulthood.

Plasticity in Captivity and Modern Environments

Improved conditions exhibit similar within-species effects on human and chimpanzee mortality profiles (Figure 5.4). Captive chimpanzees receive medical attention, abundant food, and protection from predation, and show large increases in survival rates (Dyke et al. 1995), though still substantially lower than the Ngogo pattern described above (see Figure 5.4). Infant and juvenile survival improves dramatically, from 35 percent surviving to age fifteen to 64 percent, similar to the human averages. The effects of captivity, however, diminish with age. The probability of reaching forty-five increases

FIGURE 5.4. Ratio of mortality hazards for chimpanzees and hunter-gatherers (HGs), illustrating (a) between-species and (b) within-species differences in age-specific mortality rates. Between-species comparisons include wild and captive chimpanzees versus traditional hunter-gatherers. Additional comparison shown between Ngogo chimpanzees and hunter-gatherers. Within-species comparisons include wild versus captive chimpanzees, and more traditional versus acculturated hunter-gatherers. Source: Wild chimpanzees (Gombe, Taï, Kanyawara, Mahale, Bossou, and Ngogo), Siler estimated based on composite of Hill et al. (2001), Kanyawara from Muller and Wrangham (2014), and Ngogo from Wood et al. (2017); captive chimpanzees (Siler estimated based on Dyke et al, 1995); traditional hunter gatherers (HG, Gurven and Kaplan 2007); acculturated hunter-gatherers (ibid.). Horizontal dotted line where the mortality ratio = 1 denotes equal mortality rates among compared populations.

from 5 percent in the wild to 20 percent in captivity, and remaining life expectancy at age forty-five is 4.6 and 7.2 years, respectively.

Among humans, the effects of improved conditions also seem to be greatest during childhood and middle adulthood, tapering off with age (Figure 5.4). Comparing mortality rates between hunter-gatherers and modern Americans, infant mortality is over thirty times greater among hunter-gatherers, and early child mortality is over one hundred times greater than encountered in the United States. Not until the late teens does the relationship flatten, with a more than tenfold difference in mortality. This difference is fivefold by age fifty, about fourfold by age sixty, and threefold by age seventy (Burger et al. 2012; Gurven and Kaplan 2007).

While captivity in chimpanzees improves survivorship at all ages, and early in life even matches or exceeds hunter-gatherer levels, differences are quite clear by age thirty-five. The twofold difference in mortality rates between captive chimps and hunter-gatherers accelerates steeply thereafter (Figure 5.4). By age forty-five, the expected future life span of chimpanzees in captivity is a third of the human expectation. Improved conditions for captive chimpanzees, though associated with lower mortality early in life than human hunter-gatherers, does not change the fundamental species differences in mortality rates in adulthood. From the age of lowest mortality rates, chimpanzee mortality rates increase, while human mortality rates remain relatively low and level for two decades before notably increasing. It appears that chimpanzees age much faster than humans and die earlier, even in protected environments.

Causes of Death

As shown above, chimpanzees, under the most favorable conditions in captivity, show much higher rates of adult mortality and a significantly shorter life span than foragers under the worst conditions (Gurven and Kaplan 2007). This is true in spite of the available evidence, which suggests that members of both species die from similar macro-causes, with the exception of predation (Table 5.2).

In traditional environments, the majority of deaths are due to infections and illness, representing 72 percent and 54 percent of all deaths in traditional humans and wild chimpanzees, respectively. Respiratory-related illnesses,

TABLE 5.2. Causes of death among humans and chimpanzees.

Cause	Traditional Humans n	Traditional Humans % Known	Wild Chimpanzees n	Wild Chimpanzees % Known	Industrialized Human Populations n	Industrialized Human Populations %	Captive Chimpanzees n	Captive Chimpanzees %
All illnesses	2,333	72.4	127	53.6	1,951,920	79.1	104	45.8
Infectious disease			85	35.9	170,521	4.6	13	5.7
Respiratory[a,b]	292	22.2	35	14.8	99,948	4.0	3	1.3
Gastrointestinal[a]	239	18.1	20	8.4	13,284	0.5	6	2.6
Fever[a]	107	8.1			699	0.0	4	1.8
Other infectious			30	12.7	56,590			
Chronic disease					1,425,687	57.8	91	40.1
Heart disease					784,454	31.8	69	30.4
Renal disease					51,084	2.1	22	9.7
Cancer					590,149	23.9		
Other illnesses[a]	317	24.1	42	17.7	355,712	14.4	133	58.6
Degenerative[c]	306	9.5	28	11.8				
Accidents	166	5.2	7	3.0	120,859	4.9	11	4.8
Violence	354	11.0	36	15.2	16,268	0.7		
Homicide[d]	164	6.0			16,259	0.7		
Warfare[d]	137	5.0			9	0.0		
Predation			20	8.4				
Feline-caused			10	4.2				
Human-caused			10	4.2				
Other causes of death	62	1.9	19	8.0	379,388	15.4	112	49.3
Total	3,221	100	237	100	2,468,435	100	227	100

Human data (n = 3,221) come from seven groups of hunter-gatherers and forager-horticulturalists (see Gurven and Kaplan 2007 for details). Wild chimpanzee sample (n = 237) is based on known reported deaths from Gombe (Williams et al. 2008), Mahale (Nishida et al. 2003), and Taï (Boesch and Boesch-Achermann 2000) populations. Data on industrialized human populations (n = 2,468,435) are based on Table 10 of the 2010 National Vital Statistical Reports for the United States (Murphy et al. 2013). Captive chimpanzee data (n = 227) compiled from Varki et al. (2009).

a. Illness breakdown does not exist for all human groups. These percentages are based on a risk set of 1,644 individuals and adjusted to sum to 72.4 percent.

b. Respiratory illness accounts for 48 percent of all illnesses in Gombe, 20 percent in Mahale, and 0 percent in Taï.

c. Degenerative illnesses overlap with chronic diseases, but greater specificity is lacking among most traditional human populations (but see Gurven and Kaplan 2007 for more details).

d. Information on violence-related deaths does not exist for all human groups. These percentages are based on a risk set of 2,272 individuals and adjusted to sum to 11.0 percent.

such as bronchitis, tuberculosis, pneumonias, and other viral infections, account for a fifth or more of illness-related deaths among humans, and 15 percent among wild chimpanzees. However, most infectious diseases are absent in many newly contacted Amazonian groups, because small, mobile populations cannot support these contagious vectors (Black 1975). Gastrointestinal illnesses account for 5–18 percent of deaths in traditional human societies. Diarrhea coupled with malnutrition is and remains one of the most significant causes of infant and early childhood deaths among forager populations. People living in tropical forest environments are especially vulnerable to helminthic parasites (Dunn 1968), which, although not usually lethal, can compromise growth and immune function. Few deaths related to gastrointestinal illness have been reported among wild chimpanzees, although one case of gastrointestinal anthrax was confirmed in the Taï chimpanzee population (Leendertz et al. 2004) and four chimpanzees in the Kasekela community at Gombe were reported to have gastrointestinal symptoms prior to death (Williams et al. 2008).

There is a notable lack of data on degenerative disease in traditional humans and wild chimpanzees, but they are probably very rare in both groups. Degenerative disease accounted for about 9 percent of adult deaths in a sample of hunter-gatherers and horticulturalists, with the highest representation among Northern Territory Aborigines. Neoplasms and possible heart disease each accounted for nine of the forty-nine deaths due to degenerative illness in adults over age sixty. It should be pointed out, however, that chronic illnesses as causes of death are the most difficult to identify, since more proximate causes are likely to be mentioned in verbal autopsies. In traditional humans, cases of degenerative deaths are confined largely to perinatal problems early in infancy, late-age cerebrovascular problems, as well as attributions of "old age" in the absence of any obvious symptom or pathology. Heart attacks and strokes appear to be rare, and do not account for these old-age deaths (see Eaton et al. 1988), which often occur when sleeping. Although some evidence of degenerative joint disease has been observed in Kibale chimpanzees, these were mild and unlikely to be lethal (Carter et al. 2008). Old chimpanzees, a potential high-risk group for degenerative disease, tend to "disappear," and, as with foragers, most of these deaths are attributed to senescence (Nishida et al. 2003). Neoplasms have been observed in captive chimpanzees, but reviews suggest lower incidence than in humans (and Old World monkeys) (Lowenstine et al. 2016); the higher proapoptotic gene expression (i.e., programmed cell death) observed in chimpanzees compared with humans is consistent with less carcinogenesis in chimpanzees (Arora et al. 2012).

Despite the expressed fear and cultural importance of dangerous predators, as represented by mythologies, stories, songs, and games, death by predation is rare among extant foragers. Grouping patterns, weapons, warning displays (e.g., fires), and other cultural means of avoiding predators may contribute to the reduced impact of predation on human survivorship (Wrangham et al. 2006). In contrast, predation could be an important cause of death among chimpanzees; however, scant data and the disappearance of predators from most chimpanzee habitats makes it difficult to determine whether the few predation reports (Boesch and Boesch-Achermann 2000; Furuichi 2000; Nakazawa et al. 2013) are rare anomalies, or are instead a typical source of mortality of past.

Intraspecies violent death appears to be a common feature of human and chimpanzee societies, accounting for 11 percent of 3,221 documented hunter-gatherer and horticultural deaths, and 15 percent of the 237 chimpanzee deaths (Table 5.2). However, Wrangham and colleagues (2006) report lower values for the latter, ranging between 1 percent and 3 percent. Infanticide is also commonly practiced in both. Those at greatest risk of being abandoned or killed in humans are the sickly, unwanted, those of questionable paternity, females, and those viewed as bad omens, such as twins (Milner 2000). Intergroup encounters are the most common context for violent deaths among chimpanzees (66 percent), with adult males the usual perpetrators (92 percent), and adult males (73 percent) and infants the most common victims (Wilson et al. 2014). Despite similar levels of violent deaths in chimpanzees and humans, the rate of intragroup violence (both lethal and nonlethal physical aggression) is remarkably higher in chimpanzees than in humans (Wrangham et al. 2006). It is likely, however, that violent deaths among humans decreased with increased state-level intervention and missionary influence in many small-scale groups around the world (e.g., Agta, Aché, Aborigines, Ju/'hoansi, Yanomamo).

Finally, the composition of accidental deaths varies across groups of traditional humans, including falls, river drownings, accidental poisonings, snake bites, burns, and getting lost. Together, accidental and violent deaths account for 4–43 percent (average 19 percent) of all deaths in traditional humans. In wild chimpanzees, accidents account for 3 percent of the reported deaths (Table 5.2), mostly due to falls from trees (Carter et al. 2008; Williams et al. 2008).

Neither hunter-gatherers nor wild chimpanzees appear to suffer from atherosclerosis or die from heart disease. It has often been remarked that few risk factors for heart disease and cardiovascular disease exist among active members of small-scale societies (Eaton et al. 1994). Obesity is rare, hypertension is low, cholesterol and triglyceride levels are low, and maximal oxygen uptake (VO_2max) is high. This is also likely to be the case among wild chimpanzees, for which evidence of the existence of atherosclerosis has never been reported. However, modern conditions have shifted the causes of death profile considerably. Lifestyle changes in industrialized human populations and captive chimpanzees compared to their traditional counterparts, which include the adoption of a pro-inflammatory diet, sedentary lifestyle, and a relatively aseptic environment, have led to a shift in morbidity from infectious to degenerative disease (Finch 2012). Infections decreased significantly with improvements in sanitation and control by vaccination and antibiotics in industrialized human societies and captive chimpanzee populations. The main cause of death among modern humans living in industrialized countries, and chimpanzees living in captivity, is now heart disease,[1] which accounts for more than 30 percent of deaths in both species (Table 5.2).

However, death by heart disease among humans is caused primarily by advanced coronary atherosclerosis, and is typically associated with high cholesterol, obesity, chronic inflammation, metabolic syndrome, and cigarette smoking. Among captive chimpanzees, heart failure is due to extensive interstitial myocardial fibrosis and arrhythmias (Varki et al. 2009). There is little evidence of arterial plaques in captive chimpanzees, despite their sedentary lifestyles, pro-atherogenic blood lipid profiles, occasional hypertension, and homozygosity for ApoE4 (a strong risk factor for atherosclerosis in humans). Finch (2012) suggests that humans are better adapted to chronic inflammation due to greater exposure to inflammation through changes in diet, technology, and pathogen exposure over evolutionary history (see below).

Senescence

The pace at which mortality rates double is a common empirical measure of demographic senescence, defined as $\ln 2 / \beta$, where β is the rate of increase when mortality grows exponentially with age. Exponential growth, or the Gompertz model, gives a reasonable fit to adult mortality patterns in a wide range of species, including humans and chimpanzees. Finch and colleagues (1990) report mortality rate doubling times (MRDTs) of seven to eight for a variety of recent human populations with low and high mortality. Despite the overall high mortality of hunter-gatherer populations, the adult mortality rate also doubles in seven years among Aché and nine years among Ju/'hoansi (Gurven et al. 2007). Hadza MRDT is just outside the reported range of other human populations, with MRDT of six years. The Hiwi MRDT shows rapid senescence (2.8 years). Several forager-horticulturalists and acculturated foragers show a similar MRDT of eight years, including two Yanomamo samples and settled Aché. The sample of forager-horticulturalists show MRDTs within the range of six to twelve. The acculturated foragers show a range of MRDT from seven to eleven. Overall, the highest-quality data among foragers shows a range of MRDT at six to ten. Chimpanzees show MRDT values that are roughly similar to that of human foragers, ranging between seven and nine years (De Magalhães 2006; Bronikowski et al. 2011). However, as described earlier, the onset of mortality rate doubling occurs at least ten to twenty years earlier among chimpanzees.

While humans appear to senesce more slowly than chimpanzees, it is an open question whether the pace of aging has slowed down in recent human history and among captive chimpanzees. Adult mortality has declined, but it is unclear whether the rate of functional, physiological decay has fallen in tandem. Aging is often tricky to define and measure. The crudest but most available method for making inferences about past aging patterns uses historical mortality data to measure age-related changes in mortality. For example, longitudinal analysis of European mortality data suggests that senescence has slowed over the past couple of centuries, where senescence is defined in several different ways (Gurven and Fenelon 2009). This observation is consistent with the notion that reductions in "extrinsic" age-independent mortality (e.g., infectious disease, accidents, and other nondegenerative causes) should lead to greater investments in repair and maintenance, thereby resulting in longer life span, as originally hypothesized by Williams (1957). However, MRDT tends to be lower in low mortality societies, which would conversely suggest a faster rate of aging. One explanation, called the "heterogeneity hypothesis," argues that in high mortality populations, only the robust survive to late ages, thereby giving the appearance at the population level of a slower rate of adult mortality increase (Vaupel et al.

1979; Hawkes et al. 2012). According to this view, low mortality populations would show greater heterogeneity in individual frailty among adult survivors, and so population-level mortality increase might seem faster. A simpler explanation rests on the peculiarity of MRDT and the manner in which it is estimated. It is possible that if the onset of exponential growth in mortality is pushed to later adult ages in low mortality populations, and if survivorship at the latest ages has improved less than at other adult ages, then estimates of MRDT will be lower and will give the appearance of a more rapid increase at the population level.

Unraveling species differences in aging will require moving beyond actuarial measures and instead focusing on changes in physiological condition and the selective forces impinging on their function. The similar role of infections as principal causes of death in chimpanzees and humans suggests that species differences in the ability to fight against and tolerate pathogens may be critical. The proximate pathways allowing humans to delay somatic aging are, however, not well understood, and no "magic bullet" biomarker has yet been discovered. Two biomarkers proposed to promote greater somatic maintenance and longevity include the steroid hormones estrogen and dehydroepiandrosterone sulfate (DHEAS) (Lane et al. 1997; Belvins and Coxworth 2013). Estrogen affects diverse tissue and cells, and plays an important role in the maintenance of many physiological systems (Lane et al. 1997; Roth et al. 2002; Kemnitz et al. 2006). Adrenal androgens such as DHEAS may be responsible for 75 percent of estrogen in women before menopause and close to 100 percent after menopause (Labrie 1991). Total DHEAS levels in women are three times higher than in age-matched female chimpanzees, and only after their late sixties do human concentrations fall to the highest chimpanzee level (Belvins and Coxworth 2013). However, rates of decline in DHEAS are slower than in human females.

At the cellular level, slower somatic aging in humans may be produced by a reduced rate of telomere attrition (Gomes et al. 2011; Hawkes and Coxworth 2013). Telomeres, which are the noncoding sequence at the end of chromosomes, protect chromosomes from deterioration and from fusion with neighboring chromosomes. With each cell division, telomeres become shorter, which decreases the organism's capacity to regenerate tissue. Thus, if telomere shortening were responsible for somatic aging, one might hypothesize that attrition rates in chimpanzees should be twice as fast than in humans (Hawkes and Coxworth 2013). However, chimpanzees and humans appear to show similar attrition rates, and chimpanzee telomeres are twice as long as those in humans. Indeed, telomere lengths may vary inversely with species-typical life spans (Gomes et al. 2011). Short telomeres combined with lower telomerase expression in humans may offer protection from runaway cell growth (cancer) in ways that are preferable to replicative aging. Species comparisons, even among primates, are therefore not yet clear.

Menopause and Postreproductive Life Span

Even for human populations living without health care, public sanitation, immunizations, or abundant and predictable food supply, up to one-third of the population is likely to live to age fifty, with an expected fifteen to twenty years remaining (Figure 5.5). With an average age of first reproduction of eighteen years, up to 40 percent of hunter-gatherer women could expect to reach the age at which a first grandchild would be born (36 yrs). For hunter-gatherers who survive to the age of reproduction, the average modal adult life span is about seventy-two years of age (range: 68–78; Gurven and Kaplan 2007). Existing paleontological evidence suggests that a postreproductive life span existed anywhere from 150,000–1.6 million years ago (Bogin and Smith 1996; Caspari and Lee 2004). Chimpanzees, on the other hand, have somatic aging rates similar to humans, and rarely survive their reproductive years (Goodall 1986; Emery Thompson et al. 2007; Jones et al. 2007). New evidence suggests that the rate of decline in ovarian follicular stock may even be faster in humans than chimpanzees (Cloutier et al. 2015), which further suggests that menopause is not a characteristic of chimpanzee life history; fertility decline is the ancestral trait common to humans and chimpanzees, whereas human postreproductive longevity is the derived trait. Indeed, attempts to model the evolution of menopause fail to show that the selective benefits of helping descendants would ever be sufficient to favor fertility cessation over extending reproduction to later ages (Hill and Hurtado 1991; Rogers 1993).

A more recent approach suggests that menopause would be favored when there are resource conflicts among women and their daughters-in-law; asymmetries in kinship—where women would be unrelated to their mothers-in-law's future offspring, but mothers-in-law are related to their grandchildren—give daughters-in-law the upper hand, and presumably priority reproduction (Cant and Johnstone 2008). This model, however, has fairly rigid assumptions

FIGURE 5.5. A Hadza grandmother cares for her grandson. Photo by Brian Wood.

(e.g., female dispersal and male philopatry, no coercion nor synergies in production or childcare, other kin relationships are ignored), and mixed empirical support (Lahdenperä et al. 2012; Mace and Alvergne 2012).

It is likely that extension of the female reproductive life span is not feasible due to trade-offs associated with the mammalian pattern of restricted oocyte production, where the complete, fixed supply of follicles is established in the second trimester of fetal development, and later subject to processes of gamete selection and decay (atresia) that seem oriented toward preserving embryo quality (Ellison 2001). Several theories have therefore been proposed to explain the extension of the human life span, rather than the evolution of menopause. The classic theory of senescence in evolutionary biology was first proposed by Medawar (1952), developed further by Williams (1957), and then formalized by Hamilton (1966). It proposes that as individuals age, they contribute less to biological fitness because less of their expected lifetime fertility remains. Consequently, natural selection acts more weakly to reduce mortality at older ages. The existence of substantial postreproductive life among humans therefore suggests that older individuals have "reproductive value" by increasing fitness through nonreproductive means.

George Williams was the first to suggest that beginning at ages forty-five to fifty, mothers may benefit more from investing their energy and resources in existing children rather than from producing new ones (1957). This idea became known thirty years later as the "Grandmother Hypothesis." One version of the Grandmother Hypothesis proposed by Kristen Hawkes and colleagues (Hawkes et al. 1998; Hawkes 2003) focuses on intergenerational transfers by older women. It proposes that older women can increase their inclusive fitness by enhancing offspring fertility and survivorship of grandchildren through provisioning or providing support to younger generations. Among foragers, the resources acquired by women are strength-intensive, disadvantaging young children and thereby increasing the value of older women's labor contributions. According to this view, extensions in the human life span are driven by selection on women, and the value of resource transfers from grandmothers to grandchildren. The initial inspiration for the Grandmother Hypothesis came from fieldwork done with Hadza foragers in Tanzania, where "hardworking" older women were observed to produce substantial quantities of food.

Peccei (2001) amends this view by pointing out that the long-term juvenile dependence among humans implies that adults who cease reproducing in their forties will not finish parenting until they are sixty or older (see also Lancaster and King 1985). The notion that most of the benefits to longevity derive from helping offspring rather than grandchildren has been called the "Mother Hypothesis."

An alternative view focuses on men. Marlowe argues that the extension of the life span is driven by selection on men, stressing the fact that men do not experience menopause and can have children into the seventh and eighth decades of life (2000). His argument, called the "Patriarch Hypothesis," is that as men age they accrue status and power that they use to obtain reproductive benefits. These benefits and the lack of a physiological menopause select for their greater longevity. Formal demographic models of life history evolution typically focus only on females, but two-sex demographic models where men tend to be older than their spouses may also lead to a pattern of delayed senescence after the age of fifty (Tuljapurkar et al. 2007). In the two-sex model, selection can favor survivorship for as long as men reproduce, lending additional support to the Patriarch Hypothesis. This model, however, requires extensive late-age male fertility more characteristic of polygynous societies, and/or mating patterns where fertile women mate with older men. Another model that does not require kin assistance (and that has been applied so far only to the arthropod *F. candida*) proposes that postreproductive life span can evolve as insurance against "life span indeterminacy," whereby greater variance in somatic and/or reproductive life spans selects for longer postreproductive life spans (Tully and Lambert 2011). The logic is that longer postreproductive life span reduces the risk of dying by chance before the cessation of reproduction. Both of these models are noteworthy in that neither requires extended parental or grandparental care.

The "Embodied Capital Model" suggests that timing of life events is best understood as an "embodied capital" investment process (Kaplan et al. 2000; Kaplan and Robson 2002; Gurven et al. 2006). Embodied capital is organized somatic tissue such as muscles, immune system components, and brains. In a functional sense, embodied capital includes strength, skill, knowledge, and other abilities. Humans are specialists in brain-based capital. High levels of knowledge and skill are required to exploit the suite of high-quality, difficult-to-acquire resources human foragers consume (Walker et al. 2002; Gurven et al. 2006). Those abilities require a large brain and a longtime commitment to development. This extended learning phase, during which productivity is low, is compensated for by higher productivity during the adult period. Since productivity increases with age, the time investment in skill acquisition and knowledge leads to selection for lowered mortality rates and greater longevity, because the returns on the investments in development occur at older ages. Thus, the long human life span coevolved with the lengthening of the juvenile period, increased brain capacities for information processing and storage, and intergenerational resource flows. Similarly, the "Reserve Capacity" hypothesis proposes that a supportive social system allowed mothers to wean their children earlier and to delay maturity, which allowed for a longer period of somatic investment. The larger reserve capacity resulting from a longer parental investment could result in prolonged longevity (Larke and Crews 2006; Bogin 2009).

Finally, the control-of-fire hypothesis complements these models by arguing that human use of fire for cooking helps increase the efficiency of provisioning by promoting food digestibility and energy, and by allowing early weaning through increased availability of weaning foods (Wrangham 2009; Wrangham and Carmody 2010). It also further reduces extrinsic mortality by detoxifying certain foods, helping to eliminate food-borne pathogens, and deterring predators.

Many of these evolutionary models and hypotheses are not mutually exclusive. They differ in their focus on women (Grandmothering, Mothering), men (Patriarch), or both sexes (Embodied Capital), their reliance on resource transfers as primary (Grandmothering, Mothering, Embodied Capital) or secondary (Patriarch) to life span extension, and whether slow development early in life and life span extension are coupled with economic surplus midlife and the skills-intensive nature of the human foraging niche requiring learning and instruction (Embodied Capital versus Grandmothering/Mothering).

All models except the Patriarch and Life Span Indeterminacy Hypotheses posit that future remaining (caloric) productivity, or productive value (V_x), can impact fitness even when reproductive value (R_x) (*sensu* Hamilton) is low or zero. In one study, V_x has been estimated as the sum of all future net caloric production, discounted by future mortality (Gurven et al. 2012). When comparing R_x and V_x in humans and chimpanzees, it is clear that they show similar profiles across the life course (Figure 5.6). Humans, however, show a huge surplus of caloric production in midlife, with declines occurring well beyond the reproductive years in both males and females. Given the ubiquity of food transfers among humans within and across generations,

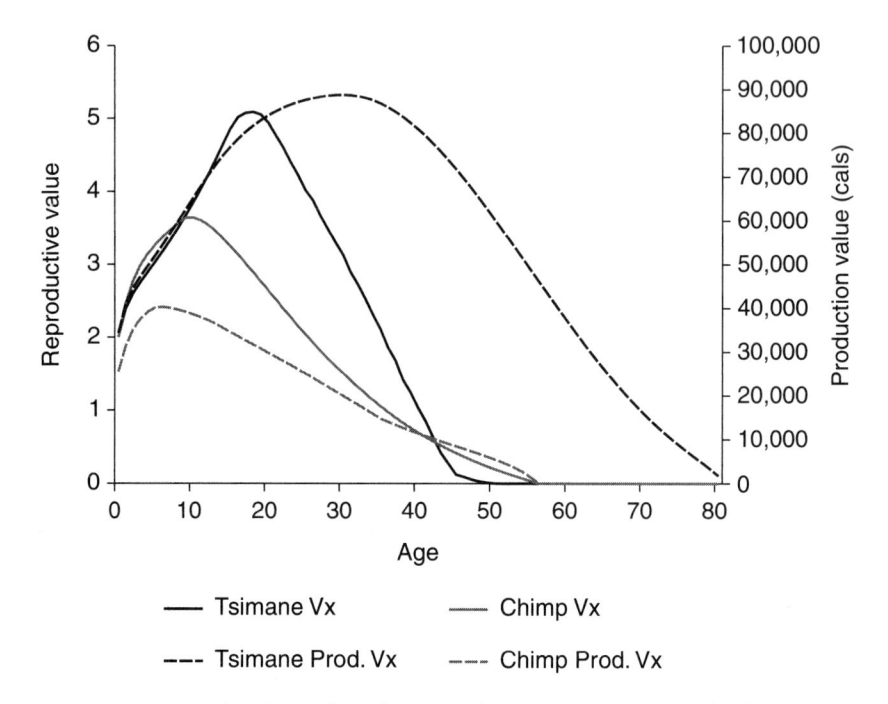

FIGURE 5.6. Reproductive and productive value among Tsimané and wild chimpanzees. Reproductive value (R_x) at age x reflects expected future remaining reproduction (i.e., $\sum_{i=x+1}^{\infty} l_i m_i$, where l_i is survival from birth to age i and m_i is annual fertility at age i). Productive value (V_x) is similar to reproductive value but replaces m_i with age-specific caloric production. Chimpanzee reproductive and productive values show similar trajectories with age; among humans, productive value peaks much later than reproductive value and remains substantial throughout much of adulthood, even after reproductive value is zero. Chimpanzee mortality and fertility data come from Hill et al. (2001) and Emery Thompson et al. (2007: table S1), respectively. Tsimané mortality data are from Gurven et al. (2007) and fertility from Gurven (unpublished data).

particularly to close kin, the high production value in late adulthood can greatly increase fitness impacts of older adults. Older adults make "transfers" of aid, advice, instruction, mediation, and other nonfood contributions that can also have important fitness consequences for kin (Gurven et al. 2012). Mortality rates in late adulthood coincide with rapid declines in V_x, and with lower numbers of potential recipients who are dependent kin (Gurven and Kaplan 2007).

These models were developed to explain the extended life span in the hominid lineage, relative to chimpanzees and other primates, but none directly address more recent changes in the human life span or the proximate mechanisms by which life span is extended. Since the early 1800s, human life expectancy has increased worldwide due to rapid declines in infant and child mortality; however, late-age mortality has continued to decline as well, and the modal age of adult death has increased by at least a decade (Vaupel 1997). Finch and colleagues have argued that genetic changes responding to alterations in infectious exposure, nutrition, and inflammatory immune responses over the course of hominin evolution are responsible for the lengthier life spans of humans, and the possibility for improved environmental conditions to continue lowering mortality rates (Finch 2012). A gradual reliance on scavenging, hunting, and cooking could alter the selective environment among hominins relative to forest-dwelling apes. Greater meat and fat consumption, pathogen exposure from scavenged meat, and noninfectious inflammagens (or harmful compounds called advanced glycation end products that speed up oxidative damage to cells and are implicated in worsening of many degenerative diseases) from cooked food would have selected for "meat-adaptive" genes. One of the more important of these includes apolipoprotein E alleles that are pro-inflammatory to heighten immune responses. While advantageous in high infection contexts, as in ancestral populations, pro-inflammatory genotypes unique to humans, in the low infection environments of the past century, have had adverse consequences on cardiovascular disease and brain aging (Finch and Sapolsky 1999; Finch and Stanford 2004). These genetic changes might explain why captive chimpanzees do not experience the same degenerative diseases as industrialized populations of humans (Varki et al. 2009).

Social Buffering and Extrinsic Mortality Reduction

While evidence and support are limited to pit the models above against each other, it is likely that low extrinsic mortality is a critical factor underlying the life history of long-lived species such as humans (Table 5.1). Among early humans, low juvenile-adult mortality was likely a prerequisite for further reductions in adult mortality and the further slowing of the life course. Whereas other long-lived species with low extrinsic mortality often inhabit microbe-free and predator-free microenvironments, the lower extrinsic mortality of early humans may have come from effective group defense against predation and from the nurturing of sick and injured individuals (Gurven et al. 2000; Sugiyama 2004).

The average forager female has about six births (Table 5.1), which places substantial burden on household feeding requirements. Allowances usually made for pregnant and lactating women, who reduce foraging efforts but nonetheless receive ample food, enable such a high level of human fertility with short interbirth intervals. However, the risk of food shortfalls occurs among males and females at all ages, as even adults in their peak production years cannot consistently meet the daily caloric needs of their large families. The human foraging niche leads to the possibility of greater risk of food shortfalls over the life course, but also includes a variety of other important risks that can impact fitness. Illness left untreated can lead to cascading morbidity and possibly death, and often impairs the ability to produce food or perform other important daily tasks. Death or divorce renders dependent children vulnerable to food shortage, disease, and lack of protection, and renders adults vulnerable to labor shortage. Conflict left unresolved, especially among kin, can result in fractured social and sharing networks, migration, fighting, and homicide. Theft and breakage of important tools, possessions, or other resources can potentially disrupt production and often incurs substantial costs to replace.

Human cooperation and sociality likely evolved to reduce risk in these fitness-relevant domains. Managing risk in the short term (e.g., daily food shortfalls) and in the long term (e.g., handling illness, feeding extra dependents, defense against predators and enemies) would result in lower baseline or "extrinsic" mortality. Baseline mortality in chimpanzees is about 70 percent higher than in humans (Table 5.1), due primarily to differences in predation and illness rates. Risk reduction was necessary for sustaining a foraging way of life, and is a central component of the evolved human life history. Although wild chimpanzees share food and other resources (Mitani and Watts 2001; Gomes and Boesch 2011), most food transfers are passive (Jaeggi and van Schaik 2011). Chimpanzees rarely live beyond age forty-five, the age when humans reach peak net economic productivity, and chimpanzee grandparents are rarely observed helping anyone (Goodall 1986). Chimpanzees also rarely help or care for sick individuals. Thus, humans may be unique in the breadth and volume of transfers and help given across different domains. But the question remains as to how *Homo sapiens* diverged from other hominins and why others did not follow the same path. As mentioned above, the control-of-fire hypothesis suggests that the strategic use of fire could have protected early hominins from predators. Additionally, early bipedalism, longer day ranges, and more efficient terrestrial locomotion among ancestral hominins likely pushed for greater hunting and gathering specialization, and a reliance on large packaged but relatively high-variance foods (Kaplan et al. 2000). During the drying of the Pleistocene and the expansion of African savannahs, these hominins would have been "pre-adapted" to better reap the gains from increased specialization on the mammal species, roots, and nuts that proliferated during this period.

Thus, we can speculate that the use of fire, economic gains from improved foraging behavior, and greater sociality may have helped lower extrinsic mortality risks in ancestral hominins. This initial lowering of extrinsic mortality could then select for further investment in reducing adult mortality rates. With a greater probability of reaching adulthood, selection would then have favored further specialization toward skill-intensive foraging subsistence strategies with delayed returns, and a higher rate of intergenerational transfers, as argued by the Embodied Capital Model (Kaplan and Robson 2002). Regardless of the original benefits of sociality, the mortality-lowering effect of social support should thus have pushed early hominins toward greater foraging specialization, extended development, and survivorship to produce and support kin at higher levels later in life.

Conclusion

There appears to be a characteristic life span for the human species, in which mortality decreases sharply from infancy through childhood, followed by a long period in which mortality rates remain essentially constant to about age forty, at which point mortality rises exponentially. Despite the potential for similar mortality levels in infancy and the juvenile period, chimpanzee mortality (except for Ngogo) tends instead to increase exponentially from its trough at age ten. There is a modal age of adult death of about seven decades for humans, whereas the mode does not exceed three decades in chimpanzees. Productivity and helping behavior can positively impact kin fitness before this time, after which senescence occurs rapidly and people die. We hypothesize that human bodies are designed to function well for about seven decades in the environment in which our species evolved.

There are differences in mortality rates among populations and among periods, especially in risks of violent death. However, those differences are small in a comparative cross-species perspective, and the similarity in mortality profiles of traditional peoples living in varying environments is impressive.

The evolved human life span is a core life history trait whose explanation frequently includes other derived or exaggerated traits. Compared to our chimpanzee cousins, humans not only have long lives, but large brains and bodies that grow and develop slowly (Isler and van Schaik 2009). Human diets are made up of high-quality, nutrient-dense foods that come in large packages, while cooperation, sociality, pair-bonds, divisions of labor and multigenerational resource transfers help underpin subsistence, parenting, and risk-management strategies. Whether these traits coevolved as a bundle or in sequence, and how they evolved, remain to be determined, but the answer will no doubt require greater attention to chimpanzee behavior and ecology. The Embodied Capital Model is the most comprehensive explanation for species differences to date, in that it ties together in one coherent framework the coevolution of long life, slow growth, encephalized brains, and high sociality—all as an outcrop of the shift to a more uniquely human foraging niche. However, it may not be the most parsimonious model. One comparative study involving fifty-seven bird and mammal species, however, provides broader support for the Embodied Capital Model by demonstrating positive relationships between cooperative foraging, greater sociality, and delayed foraging competency (Schuppli et al. 2012).

If we imagine the environments in which our ancestors evolved, environmental assaults and access to energy to combat those assaults are likely to have varied across time and locale. Such variation is likely to select for some phenotypic plasticity in allocations to defense and repair. At the same time, the hunting and gathering adaptation practiced by evolving humans appears along with a complex of long-term child dependence during which learning trumps productivity, and high productivity of adults, especially in middle age. Together, the costs of slowing senescence, preventing mortality, and the benefits of extended investment in descendants produced selection for a characteristic human life span, with some variance around the central tendency. The similar mortality profiles of eighteenth-century Sweden to hunting and gathering populations suggest that comparable age distributions of adult deaths occur under a relatively broad range of environmental conditions. The chimpanzee-human comparison does reveal, however, that species differences overwhelm differences in environmental conditions in determining mortality hazards throughout adulthood. This might suggest that some differences in our respective genomes have resulted in basic differences in rates of repair and tissue maintenance that manifest themselves in physiological deterioration at older ages.

Chimpanzees are likely to be a good ancestral model for testing hypotheses about the evolution of long life spans in humans. Future studies could benefit from controlled comparisons of biomarkers of aging among wild and captive chimpanzees, and among subsistence populations varying in diet, activity, and pathogen risk. Species comparisons of metabolism, immune function, repair mechanisms, and their genetic underpinnings will be especially instructive, and may provide insight into how and why the greater (or at least more specialized) investment in somatic maintenance that occurred over the course of hominin evolution resulted in human life span surpassing that of all other primates.

Endnotes

1. In the Yerkes captive chimpanzee colony, infectious disease was replaced by heart disease as the main cause of death when vaccination was introduced and sanitation improved in the early 1990s (Varki et al. 2009).

References

Allman, J., T. McLaughlin, and A. Hakeem. 1993. Brain weight and life-span in primate species. *Proceedings of the National Academy of Sciences* 90: 118–122.

Arora, G., R. Mezencev, and J. F. McDonald. 2012. Human cells display reduced apoptotic function relative to chimpanzee cells. *PLoS ONE* 7: e46182.

Austad, S. 1999. *Why We Age: What Science Is Discovering about the Body's Journey through Life.* New York: John Wiley.

Belvins, J., and J. Coxworth. 2013. Brief communication: Adrenal androgens and aging: Female chimpanzees *(Pan troglodytes)* compared with women. *American Journal of Physical Anthropology* 151: 643–648.

Black, F. L. 1975. Infectious disease in primitive societies. *Science* 187: 515–518.

Blurton Jones, N. G., K. Hawkes, and J. O'Connell. 2002. The antiquity of postreproductive life: Are there modern impacts on hunter-gatherer postreproductive lifespans? *Human Biology* 14: 184–205.

Blurton Jones, N. G., L. Smith, J. O'Connell, K. Hawkes, and C. L. Samuzora, C. L. 1992. Demography of the Hadza, an increasing and high density population of savanna foragers. *American Journal of Physical Anthropology* 89: 159–181.

Boesch, C., and H. Boesch-Achermann. 2000. *Chimpanzees of the Tai Forest: Behavioural Ecology and Evolution.* Oxford: Oxford University Press.

Bogin, B. 2009. Childhood, adolescence, and longevity: A multilevel model of the evolution of reserve capacity in human life history. *American Journal of Human Biology* 21: 567–577.

Bogin, B., and B. H. Smith. 1996. Evolution of the human life cycle. *American Journal of Human Biology* 8: 703–716.

Bronikowski, A. M., J. Altmann, D. K. Brockman, M. Cords, L. M. Fedigan, A. Pusey, T. Stoinski, W. F. Morris, K. B. Strier, and S.C. Alberts. 2011. Aging in the natural world: Comparative data reveal similar mortality patterns across primates. *Science* 331: 1325–1328.

Buikstra, J. E. 1997. Paleodemography: Context and promise. In R.R. Paine, ed., *Integrating Archaeological Demography: Multidisciplinary Approaches to Prehistoric Population,* 367–380. Carbondale, IL: Center for Archaeological Investigations.

Buikstra, J. E., and L. W. Konigsberg. 1985. Paleodemography: Critiques and controversies. *American Anthropologist* 87: 316–333.

Burger, O., A. Baudisch, and J. W. Vaupel. 2012. Human mortality improvement in evolutionary context. *Proceedings of the National Academy of Sciences* 109: 18210–18214.

Cant, M. A., and R. A. Johnstone. 2008. Reproductive conflict and the separation of reproductive generations in humans. *Proceedings of the National Academy of Sciences* 105: 5332–5336.

Carter, M. L., H. Pontzer, R. W. Wrangham, and J. K. Peterhans. 2008. Skeletal pathology in *Pan troglodytes schweinfurthii* in Kibale National Park, Uganda. *American Journal of Physical Anthropology* 135: 389–403.

Caspari, R., and S.-H. Lee. 2004. Older age becomes common late in human evolution. *Proceedings of the National Academy of Sciences* 101: 10895–10900.

Cloutier, C. T., J. E. Coxworth, and K. Hawkes. 2015. Age-related decline in ovarian follicle stocks differ between chimpanzees *(Pan troglodytes)* and humans. *Age* 37: 10.

Cutler, R. 1975. Evolution of human longevity and the genetic complexity governing aging rate. *Proceedings of the National Academy of Sciences* 72: 664–668.

De Magalhães, J. P. 2006. Species selection in comparative studies of aging and antiaging research. In P. M. Conn, ed., *Handbook of Models for Human Aging,* 9–20. London: Elsevier Academic.

Dunn, F. L. 1968. Epidemiological factors: Health and disease in hunter-gatherers. In R. B. Lee and I. DeVore, eds., *Man the Hunter,* 221–228. Chicago: Aldine.

Dyke, B., T. B. Gage, P. L. Alford, B. Senson, and S. Williams-Blangero. 1995. A model life table for captive chimpanzees. *American Journal of Primatology* 37: 25–37.

Early, J. D., and T. N. Headland. 1998. *Population Dynamics of a Philippine Rain Forest People: The San Ildefonso Agta.* Gainesville: University Press of Florida.

Early, J. D., and J. F. Peters. 2000. *The Xilixana Yanomami of the Amazon: History, Social Structure, and Population Dynamics.* Gainesville: University Press of Florida.

Eaton, S. B., M. J. Konner, and M. Shostak. 1988. Stone agers in the fast lane: Chronic degenerative diseases in evolutionary perspective. *American Journal of Medicine* 84: 739–749.

Eaton, S. B., M. C. Pike, R. V. Short, N. C. Lee, J. Trussell, R. A. Hatcher, J. W. Wood, C. M. Worthman, N. G. Blurton Jones, M. J. Konner, K. R. Hill, R. C. Bailey, and A. M. Hurtado. 1994. Women's reproductive cancers in evolutionary context. *Quarterly Review of Biology* 69: 353–367.

Ellison, P. T. 2001. *On Fertile Ground: A Natural History of Human Reproduction.* Cambridge, MA: Harvard University Press.

Emery Thompson, M. 2013. Reproductive ecology of female chimpanzees. *American Journal of Primatology* 75: 222–237.

Emery Thompson, M., J. H. Jones, A. E. Pusey, S. Brewer-Marsden, J. Goodall, D. Marsden, T. Matsuzawa, T. Nishida, V. Reynolds, Y. Sugiyama, and R. W. Wrangham. 2007. Aging and fertility patterns in wild chimpanzees provide insights into the evolution of menopause. *Current Biology* 17: 2150–2156.

Finch, C. 2007. *The Biology of Human Longevity.* San Diego: Academic Press.

Finch, C. E. 1994. *Longevity, Senescence, and the Genome.* Chicago: University of Chicago Press.

Finch, C. E. 2012. Evolution of the human lifespan, past, present, and future: Phases in the evolution of human life expectancy in relation to the inflammatory load. *Proceedings of the American Philosophical Society* 156: 9–44.

Finch, C. E., M. C. Pike, and M. Whitten. 1990. Slow mortality rate accelerations during aging in animals approximate that of humans. *Science* 249: 902–905.

Finch, C. E., and R. M. Sapolsky. 1999. The evolution of Alzheimer disease, the reproductive schedule and the apoE isoforms. *Neurobiology of Aging* 20: 407–428.

Finch, C. E., and C. B. Stanford. 2004. Meat adaptive genes and the evolution of slower aging in humans. *Quarterly Review of Biology* 79: 3–50.

Furuichi, T. 2000. Possible case of predation on a chimpanzee by a leopard in the Petit Loango Reserve, Gabon. *Pan Africa News* 7: 21–23.

Furuichi, T., G. I. Idani, H. Ihobe, S. Kuroda, K. Kitamura, A. Mori, T. Enomoto, N. Okayasu, C. Hashimoto, and T. Kano. 1998. Population dynamics of wild bonobos (*Pan paniscus*) at Wamba. *International Journal of Primatology* 19: 1029–1043.

Gage, T. B. 2003. The evolution of human phenotypic plasticity: Age and nutritional status at maturity. *Human Biology* 75: 521–537.

Gomes, C. M., and C. Boesch. 2011. Reciprocity and trades in wild West African chimpanzees. *Behavioral Ecology and Sociobiology* 65, 2183–2196.

Gomes, N. M., O. A. Ryder, M. L. Houck, S. J. Charter, W. Walker, N. R. Forsyth, S. N. Austad, C. Venditti, M. Pagel, J. W. Shay, and W. E. Wright. 2011. Comparative biology of mammalian telomeres: Hypotheses on ancestral states and the roles of telomeres in longevity determination. *Aging Cell* 10: 761–768.

Goodall, J. 1986. *The Chimpanzees of the Gombe: Patterns of Behavior.* Cambridge, MA: Harvard University Press.

Gurven, M. 2012. Human survival and life history in evolutionary perspective. In J. Mitani, J. Call, P. Kappeler, R. Palombit, and J. B. Silk, eds., *The Evolution of Primate Societies,* 293–314. Chicago: University of Chicago Press.

Gurven, M., W. Allen-Arave, K. Hill, and M. Hurtado. 2000. "It's a Wonderful Life": Signaling generosity among the Ache of Paraguay. *Evolution and Human Behavior* 21: 263–282.

Gurven, M., and A. Fenelon. 2009. Has the rate of actuarial aging changed over the past 250 years? A comparison of small-scale subsistence populations, and Swedish and English cohorts. *Evolution* 63: 1017–1035.

Gurven, M., and H. Kaplan. 2007. Longevity among hunter-gatherers: A cross-cultural comparison. *Population and Development Review* 33: 321–365.

Gurven, M., H. Kaplan, and M. Gutierrez. 2006. How long does it take to become a proficient hunter? Implications for the evolution of delayed growth. *Journal of Human Evolution* 51: 454–470.

Gurven, M., H. Kaplan, and A. Zelada Supa. 2007. Mortality experience of Tsimané Amerindians: Regional variation and temporal trends. *American Journal of Human Biology* 19: 376–398.

Gurven, M., J. Stieglitz, P. L. Hooper, C. Gomes, and H. Kaplan. 2012. From the womb to the tomb: The role of transfers in shaping the evolved human life history. *Experimental Gerontology* 47: 807–813.

Hamilton, W. D. 1966. The molding of senescence by natural selection. *Journal of Theoretical Biology* 12: 12–45.

Hammer, M., and R. Foley. 1996. Longevity, life history and allometry: How long did hominids live? *Journal of Human Evolution* 11: 61–66.

Hawkes, K. 2003. Grandmothers and the evolution of human longevity. *American Journal of Human Biology* 15: 380–400.

Hawkes, K., and J. E. Coxworth. 2013. Grandmothers and the evolution of human longevity: A review of findings and future directions. *Evolutionary Anthropology* 22: 294–302.

Hawkes, K., J. F. O'Connell, N. G. B. Jones, H. Alvarez, and E. L. Charnov. 1998. Grandmothering, menopause, and the evolution of human life-histories. *Proceedings of the National Academy of Sciences* 95: 1336–1339.

Hawkes, K., K. R. Smith, and J. K. Blevins. 2012. Human actuarial aging increases faster when background death rates are lower: A consequence of differential heterogeneity? *Evolution* 66: 103–114.

Hill, K., C. Boesch, J. Goodall, A. Pusey, J. Williams, and R. Wrangham. 2001. Mortality rates among wild chimpanzees. *Journal of Human Evolution* 40: 437–450.

Hill, K., and A. M. Hurtado. 1991. The evolution of reproductive senescence and menopause in human females. *Human Nature* 2: 315–350.

Hill, K., and A. M. Hurtado. 1996. *Ache Life History: The Ecology and Demography of a Foraging People.* New York: Aldine de Gruyter.

Hill, K., A. M. Hurtado, and R. S. Walker. 2007. High adult mortality among Hiwi hunter-gatherers: Implications for human evolution. *Journal of Human Evolution* 52: 443–454.

Hoppa, R. D., and J. W. Vaupel. 2002. *Paleodemography: Age Distributions from Skeletal Samples.* Cambridge: Cambridge University Press.

Howell, N. 1979. *Demography of the Dobe !Kung.* New York: Academic Press.

Isler, K., and C. P. van Schaik. 2009. The expensive brain: A framework for explaining evolutionary changes in brain size. *Journal of Human Evolution* 57: 392–400.

Jaeggi, A. V., and C. P. van Schaik. 2011. The evolution of food sharing in primates. *Behavioral Ecology and Sociobiology* 65: 2125–2140.

Jones, K., L. Walker, D. Anderson, A. Lacreuse, S. L. Robson, and K. Hawkes. 2007. Depletion of ovarian follicles with age in chimpanzees: Similarities to humans. *Biology of Reproduction* 77: 247–251.

Judge, D. S., and J. R. Carey. 2000. Postreproductive life predicted by primate patterns. *Journal of Gerontology: Biological Sciences* 55A: B201–209.

Kaplan, H., K. Hill, J. B. Lancaster, and A. M. Hurtado. 2000. A theory of human life history evolution: Diet, intelligence, and longevity. *Evolutionary Anthropology* 9: 156–185.

Kaplan, H. S., and A. J. Robson. 2002. The emergence of humans: The coevolution of intelligence and longevity with intergenerational transfers. *Proceedings of the National Academy of Sciences* 99: 10221–10226.

Kemnitz, J., E. Roecker, A. Haffa, J. Pinheiro, I. Kurzman, J. Ramsey, and E. MacEwan. 2006. Serum dehydroepiandrosterone sulfate concentrations across the life span of laboratory housed rhesus monkeys. *Journal of Medical Primatology* 29: 330–337.

Kennedy, G. E. 2003. Palaeolithic grandmothers? Life history theory and early *Homo. Journal of the Royal Anthropological Institute* 9: 549–572.

King, J. L., and T. H. Jukes. 1969. Non-Darwinian evolution. *Science* 164: 788–798.

Konigsberg, L. W., and N. P. Herrmann. 2006. The osteological evidence for human longevity in the recent past, In k. Hawkes and R. R. Paine, eds., *The Evolution of Human Life History,* 267–306. Santa Fe, NM: School of American Research Press.

Labrie, F. 1991. Intracrinology. *Molecular and Cellular Endocrinology* 78: C113–118.

Lahdenperä, M., D. O. Gillespie, V. Lummaa, and A. F. Russell. 2012. Severe intergenerational reproductive conflict and the evolution of menopause. *Ecology Letters* 15: 1283–1290.

Lancaster, J. B., and B. J. King. 1985. An evolutionary perspective on menopause, In V. Kerns and J. K. Brown, eds., *In Her Prime: A View of Middle-Aged Women,* 13–20. Garden City, NJ: Bergen and Garvey.

Lane, M. A., D. K. Ingram, S. S. Ball, and G. S. Roth. 1997. Dehydroepiandrosterone sulfate: A biomarker of primate aging slowed by calorie restriction. *Journal of Clinical Endocrinology & Metabolism* 82: 2093–2096.

Larke, A., and D. E. Crews. 2006. Parental investment, late reproduction, and increased reserve capacity are associated with longevity in humans. *Journal of Physiological Anthropology* 25: 119–131.

Lee, P. C., P. Majluf, and I. J. Gordon. 1991. Growth, weaning and maternal investment from a comparative perspective. *Journal of Zoology* 225: 99–114.

Leendertz, F. H., H. Ellerbrok, C. Boesch, E. Couacy-Hymann, K. Matz-Rensing, R. Hakenbeck, C. Bergmann, P. Abaza, S. Junglen, Y. Moebius, L. Vigilant, P. Formenty, and G. Pauli. 2004. Anthrax kills wild chimpanzees in a tropical rainforest. *Nature* 430: 451–452.

Lovejoy, C. O. 1981. The origin of man. *Science* 211: 341–350.

Lowenstine, L. J., R. McManamon, and K. A. Terio. 2016. Comparative pathology of aging great apes. *Veterinary Pathology* 53: 250–276.

Mace, R., and A. Alvergne. 2012. Female reproductive competition within families in rural Gambia. *Proceedings of the Royal Society B: Biological Sciences,* 279: 2219–2227.

Marlowe, F. W. 2000. The patriarch hypothesis: An alternative explanation of menopause. *Human Nature.* 11: 27–42.

Medawar, P. B. 1952. *An Unsolved Problem in Biology.* London: Lewis.

Milner, L. S. 2000. *Hardness of Heart/Hardness of Life: The Stain of Human Infanticide.* Lanham, MD: University Press of America.

Mitani, J. C., and D. P. Watts. 2001. Why do chimpanzees hunt and share meat? *Animal Behaviour* 61: 915–924.

Muller, M. N., and R. W. Wrangham. 2014. Mortality rates among Kanyawara chimpanzees. *Journal of Human Evolution* 66: 107–114.

Murphy, S. L., J. Xu, and K. D. Kochanek. 2013. Deaths: Final data for 2010. *National Vital Statistics Reports* 61: 1–117.

Nakazawa, N., S. Hanamura, E. Inoue, M. Nakatsukasa, and M. Nakamura. 2013. A leopard ate a chimpanzee: First evidence from East Africa. *Journal of Human Evolution* 65: 334–7.

Nishida, T., N. Corp, M. Hamai, T. Hasegawa, M. Hiraiwa-Hasegawa, K. Hosaka, K. D. Hunt, N. Itoh, K. Kawanaka, A. Matsumoto-Oda, J. C. Mitani, M. Nakamura, K. Norikoshi, T. Sakamaki, L. Turner, S. Uehara, and K. Zamma. 2003. Demography, female life history, and reproductive profiles among the chimpanzees of Mahale. *American Journal of Primatology* 59: 99–121.

O'Connell, J. F., K. Hawkes, and N. G. Blurton Jones. 1999. Grandmothering and the evolution of *Homo erectus*. *Journal of Human Evolution* 36: 461–485.

Oeppen, J., and J. W. Vaupel. 2003. Broken limits to life expectancy. *Science* 296: 1029–1031.

Peccei, J. S. 2001. Menopause: Adaptation or epiphenomenon? *Evolutionary Anthropology* 10: 43–57.

Richerson, P. J., R. L. Bettinger, and R. Boyd. 2005. Evolution on a restless planet: Were environmental variability and environmental change major drivers of human evolution? In F. M. Wuketits and F. j. Ayala, eds., *Handbook of Evolution*, vol. 2: *The Evolution of Living Systems*. Weinheim: Wiley-Blackwell.

Riley, J. 2001. *Rising Life Expectancy: A Global History*. Cambridge: Cambridge University Press.

Rogers, A. 1993. Why menopause? *Evolutionary Ecology* 7: 406–420.

Roth, G., and U. Dicke. 2005. Evolution of the brain and intelligence. *Trends in Cognitive Sciences* 9: 250–257.

Roth, G. S., M. A. Lane, D. K. Ingram, J. A. Mattison, D. Elahi, J. D. Tobin, D. Muller, and E. J. Metter. 2002. Biomarkers of caloric restriction may predict longevity in humans. *Science* 297: 811.

Schoenemann, P. T. 2006. Evolution of the size and functional areas of the human brain. *Annual Review of Anthropology* 35: 379–406.

Schuppli, C., K. Isler, and C. P. Van Schaik. 2012. How to explain the unusually late age at skill competence among humans. *Journal of Human Evolution* 63: 843–850.

Segal, K. 2012. Meet one of the oldest chimpanzees in captivity. CNN.com, http://www.cnn.com/2012/04/21/us/oldest-chimpanzee-in-captivity/.

Sugiyama, L. S. 2004. Illness, injury, and disability among Shiwiar forager-horticulturalists: Implications of health-risk buffering for the evolution of human life history. *American Journal of Physical Anthropology* 123: 371–389.

Tuljapurkar, S., C. Puleston, and M. Gurven. 2007. Why men matter: Mating pattern drives evolution of post-reproductive lifespan. *PLoS ONE* 2: e785.

Tully, T., and A. Lambert. 2011. The evolution of postreproductive life span as an insurance against indeterminacy. *Evolution* 65: 3013–3020.

Vallois, H. V. 1961. The social life of early man: The evidence of skeletons, In S. L. Washburn, ed., *Social Life of Early Man*, 214–235. Chicago: Aldine de Gruyter.

Varki, N., D. Anderson, J. G. Herndon, T. Pham, C. J. Gregg, M. Cheriyan, J. Murphy, E. Strobert, J. Fritz, and J. G. Else. 2009. Heart disease is common in humans and chimpanzees, but is caused by different pathological processes. *Evolutionary Applications* 2: 101–112.

Vaupel, J. W. 1997. The remarkable improvements in survival at older ages. *Philosophical Transactions of the Royal Society of London B* 352: 1799–1804.

Vaupel, J. W., K. G. Manton, and E. Stallard. 1979. The impact of heterogeneity in individual frailty on the dynamics of mortality. *Demography* 16: 439–454.

Volk, A. A., and J. A. Atkinson. 2013. Infant and child death in the human environment of evolutionary adaptation. *Evolution and Human Behavior* 34: 182–192.

Wachter, K. W., and C. Finch. 1997. *Between Zeus and Salmon: The Biodemography of Longevity*. Washington, DC: National Academies Press.

Walker, P. L., J. R. Johnson, and P. M. Lambert. 1988. Age and sex biases in the preservation of human skeletal remains. *American Journal of Physical Anthropology* 76: 183–188.

Walker, R., M. Gurven, K. I. M. Hill, A. Migliano, N. Chagnon, R. De Souza, G. Djurovic, R. Hames, A. M. Hurtado, and H. Kaplan. 2006. Growth rates and life histories in twenty-two small-scale societies. *American Journal of Human Biology* 18: 295–311.

Walker, R., K. Hill, H. Kaplan, and G. McMillan. 2002. Age-dependency in skill, strength and hunting ability among the Ache of eastern Paraguay. *Journal of Human Evolution* 42: 639–657.

Washburn, S. 1981. Longevity in primates. In J. McGaugh and S. Kiesler, eds., *Aging: Biology and Behavior.* New York: Academic Press.

Weiss, K. M. 1981. Evolutionary perspectives on human aging, In P. Amoss and S. Harrell, eds., *Other Ways of Growing Old,* 11–29. Stanford, CA: Stanford University Press.

Williams, G. C. 1957. Pleitropy, natural selection and the evolution of senescence. *Evolution* 11: 398–411.

Williams, J. M., E. V. Lonsdorf, M. L. Wilson, J. Schumacher-Stankey, J. Goodall, and A. E. Pusey. 2008. Causes of death in the Kasekela chimpanzees of Gombe National Park, Tanzania. *American Journal of Primatology* 70: 766–777.

Wilson, M. L., C. Boesch, B. Fruth, T. Furuichi, I. C. Gilby, C. Hashimoto, C. L. Hobaiter, G. Hohmann, N. Itoh, K. Koops, J. N. Lloyd, T. Matsuzawa, J. C. Mitani, D. C. Mjungu, D. Morgan, R. Mundry, M. N. Muller, M. Nakamura, J. Pruetz, A. E. Pusey, J. Riedel, C. Sanz, A. M. Schel, N. Simmons, M. Waller, D. P. Watts, F. White, R. Wittig, K. Zuberbühler, and R. W. Wrangham. 2014. Lethal aggression in *Pan* is better explained by adaptive strategies than human impacts. *Nature* 513: 414–417.

Wood, J. W. 1994. *Dynamics of Human Reproduction: Biology, Biometry and Demography.* New York: Aldine de Gruyter.

Wood, B. M., D. P. Watts, J. C. Mitani, and K. E. Langergraber. 2017. Favorable ecological circumstances promote life expectancy in chimpanzees similar to that of human hunter-gatherers. *Journal of Human Evolution* 105: 41–56.

Wrangham, R. 2009. *Catching Fire: How Cooking Made Us Human.* New York: Basic Books.

Wrangham, R. W. 1987. The significance of African apes for reconstructing human social evolution, In W. G. Kinzey, ed., *The Evolution of Human Behavior: Primate Models,* 51–71. New York: State University of New York Press.

Wrangham, R. W. 2001. Out of the *Pan,* into the fire: How our ancestors' evolution depended on what they ate, In F. B. M. de Waal, ed., *Tree of Origin: What Primate Behavior Can Tell Us about Human Social Evolution,* 121–143. Cambridge, MA: Harvard University Press.

Wrangham, R. W., and R. Carmody. 2010. Human adaptation to the control of fire. *Evolutionary Anthropology* 19: 187–199.

Wrangham, R. W., M. L. Wilson, and M. N. Muller. 2006. Comparative rates of violence in chimpanzees and humans. *Primates* 47: 14–26.

6

Fertility and Fecundity

MELISSA EMERY THOMPSON *and* PETER T. ELLISON

Humans are among the most abundant mammals on the planet, with a world population topping seven billion by 2012. This is a remarkable feat for a species with such a long life cycle. Humans typically give birth to a single offspring, nurturing it intensively through a period of growth and maturation slower than that of any other primate. Nevertheless, even in the most challenging environments, humans achieve a faster reproductive rate than that of our closest ape relatives. Thus, the traditional human family, centered around dependent children of varying ages and an elaborate network of caregivers (see Jaeggi et al., this volume), has no rival among the apes.

Underlying our fertility patterns are complex physiological mechanisms regulating the timing of conception and the allocation of energy to reproduction across the life course. Reproductive ecologists, in attempting to explain variation in fertility across human populations, have discovered that our species exhibits a strikingly conservative approach to reproduction. Human women reach reproductive maturity relatively late, stop reproducing early, and, during their peak reproductive years, experience a high rate of conception failure (Ellison 2001). Menstrual cycles are so susceptible to energetic stress that, even in the developed world where calories are no farther than the corner grocery store, ovulation may be inhibited should one choose to jog there.

In this chapter, we attempt to unravel the evolutionary history that led to the peculiar human pattern of reproduction by examining comparative data on the reproductive physiology and life history of chimpanzees (Figure 6.1). While gross comparisons of human and chimpanzee life history features have been discussed for decades (Schultz 1969; Charnov and Berrigan 1993; Kaplan et al. 2000; Blurton Jones 2001), those comparisons can only reveal *what* has changed in the evolution of the human species. Here, we refocus the question on *how* these changes may have come about. By examining the mechanisms that govern variation in fertility and fecundity within each species, we hope to discover the provenance of human reproductive adaptations. Critical to this issue is the recognition that the social context of reproduction has changed dramatically during the evolution of our

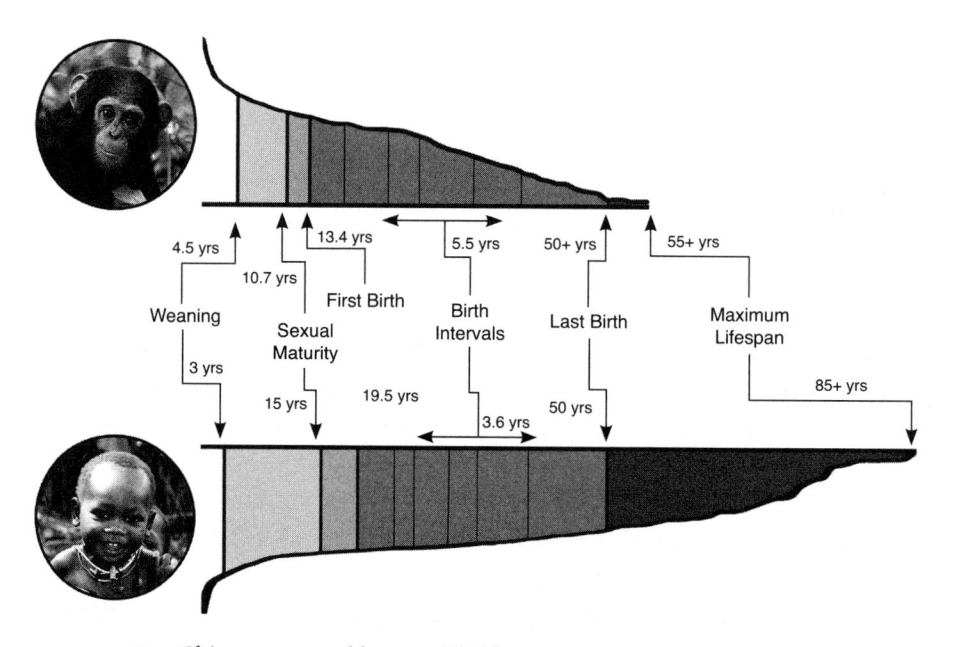

FIGURE 6.1. Chimpanzee and human life history comparison. Segment lengths demarcate major reproductive transitions, with average parameters for wild chimpanzees and human hunter-gatherers. Segment height indicates proportional female survivorship at each age. Mean birth interval is indicated; this is calculated from those intervals where the previous infant survived until the mother's next pregnancy. Variable reproductive segment lengths in the timeline follow variation in age-specific fertility, thus incorporating all births regardless of infant survival. Key sources, chimpanzees: Clark (1977); van der Rijt-Plooij and Plooij (1987); Wallis (1997); Nishida et al. (2003); Emery Thompson et al. (2007a); Emery Thompson (2013b). Key sources, humans: Howell (1979); Wood (1994); Hill and Hurtado (1996); Pennington (2001); Blurton Jones (2016). Photos: Suzi Eszterhas (chimpanzee), Brian Wood (Hadza child).

species. Thus, we ask to what extent the exceptional reproductive pattern in humans arises from uniquely derived reproductive adaptations in our species or from the socioecological context into which they are placed.

Life History Theory

Life history features, such as timing of maturity and birth rate, are outcomes of underlying trade-offs in the allocation of resources toward a range of somatic processes (Gadgil and Bossert 1970; Stearns 1989, 1992; Roff 1992). The basic explanatory framework in life history theory poses that organisms have a limited amount of energy to spend and must develop strategies for how to allocate energy in ways that maximize reproductive success in a given environment. If an organism allocates energy toward reproduction, it sacrifices investment in growth and survival-enhancing activities, such as immune function, energy storage, and cellular repair. Thus, faster reproductive rates are typically associated with earlier sexual maturation and shorter life spans. These features, in turn, are influenced by external factors, such as predation and disease, which alter the probability that a slow strategy will pay off. Slower reproductive rates are usually associated with more extensive parental care, as this maximizes the probability that the few offspring born will survive and thrive.

In comparison with other mammals, humans and the other great apes conform well to the predictions of standard life history models, exhibiting a comparatively slow rate of reproduction, slow growth, and long life span in association with large body size, minimal predation risk, and extensive offspring care (Charnov and Berrigan 1993; Ross 1998; Gurven and Gomes, this volume; Jaeggi et al., this volume). Humans exhibit the extremes of each of these features, with the exception of their birth rate, which is faster than in any of the other apes. This incongruity suggests that humans experience a less dramatic trade-off between offspring production and offspring care, or between reproduction and survival, than do the other apes.

The terms "fertility" and "fecundity" are used (or misused) inconsistently across fields. Here, we define "fertility" as the production of live offspring and "fecundity" as the physiological capacity to reproduce. These terms are particularly important to distinguish when considering humans, because behavioral factors may limit fertility without affecting fecundity. Additionally, demographers often consider fertility rates on a timescale that is removed from fecundity. For example, a woman's fertility may be evaluated retrospectively, after she is no longer fecund.

Fertility

Average interbirth intervals in wild chimpanzees range from 5.2 years to 6.6 years, excluding intervals in which the first infant died before four years of age (Emery Thompson 2013b). Human hunter-gatherers experience birth intervals of 3.1 (Aché) to 4.1 (Ju/'hoansi) years (Pennington 2001). The human total fertility rate (TFR), calculated as the average number of children born to women who have completed their reproductive years, is approximately 6 in most hunter-gathering societies (Pennington 2001). Rates as low as 2.6 (Efe) and as high as 8.0 (Aché) are reported, though the low estimates are likely skewed by failure to identify infants that die soon after birth (Pennington 2001). Relatively few chimpanzees live out their reproductive years (Hill et al. 2001; Emery Thompson et al. 2007a), making it difficult to derive TFR, as such. During her lifetime, the average wild chimpanzee female gives birth to only three or four offspring (Emery Thompson 2013b).

While there is clearly a large difference in the reproductive rates of chimpanzees and humans under natural conditions, there is also notable intraspecific variability and extensive overlap between the species. Among wild chimpanzees, interbirth intervals as short as two years, with the previous offspring surviving, have been reported (Emery Thompson et al. 2007a). In captivity, chimpanzees who raise their own infants achieve birth intervals as short as 1.7 years, with an average of approximately four years between births (Littleton 2005), similar to the reproductive rate of human hunter-gatherers. Captive chimpanzee females can give birth to eight or more offspring in their lifetimes (Littleton 2005; Roof et al. 2005). Human fertility also increases when energetic stress is low. Historical records from Europe and the United States indicate birth intervals of approximately two years were typical (Eijkemans et al. 2014). Hutterites, who practice natural fertility in nutritionally rich environments, have birth intervals averaging only thirteen months (Tietze 1957). In the modern United States, median interpregnancy interval is twenty months, and the lowest risk of adverse birth outcomes occurs with intervals of eighteen to twenty-three months (Zhu 2005).

Fecundity

Menstrual Cycle

The mechanisms governing ovulation and conception, reflected in hormonal fluctuations during the menstrual cycle, are remarkably similar in humans and other apes, despite considerable variation across primates (Figure 6.2; Graham 1981; Shideler and Lasley 1982; Bentley 1999; Shimizu et al. 2003). Primates generally exhibit a peak in estradiol production, indicative of a mature oocyte, just prior to ovulation, followed by a prolonged postovulatory (luteal) elevation of progesterone (Dixson 2013). Apes and humans also have a pronounced postovulatory elevation in estradiol (Nadler et al. 1985),

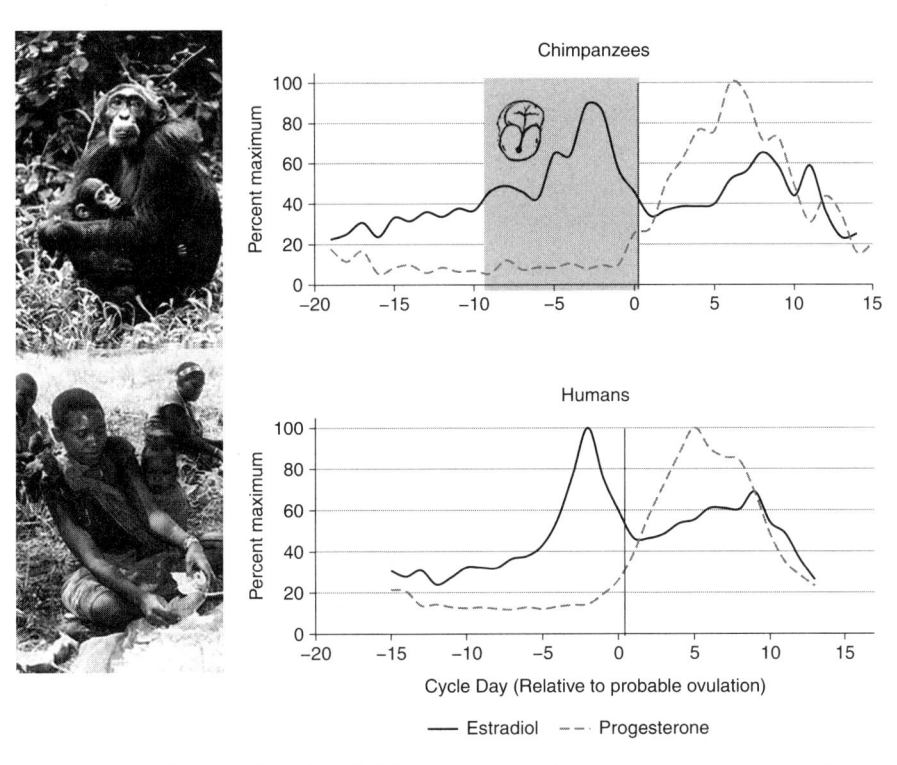

FIGURE 6.2. Menstrual cycles of chimpanzees and humans. Cycle data acquired, standardized, and averaged across multiple sources. Chimpanzee swelling phase denoted by shaded area. Chimpanzees: Deschner et al. (2003); Emery and Whitten (2003); Shimizu et al. (2003). Humans: Lipson and Ellison (1996); Baird et al. (1997). Photographs: Suzi Eszterhas (chimpanzee), Martin Muller (Hadza woman).

which is reduced or absent in monkeys (Graham 1981). Luteal phase estradiol appears to upregulate the action of progesterone (Fritz et al. 1987; Maas et al. 1992), and thus it is likely that both hormones are necessary to achieve the endometrial thickness needed for invasive forms of implantation. The timing and magnitude of peaks in pituitary gonadotropins (LH and FSH), which regulate ovarian steroid production, are also more similar in humans and chimpanzees than in cercopithecines (Ross et al. 1970; Howland et al. 1971; Graham 1981; Nadler et al. 1985).

Menstrual cycle lengths in women average twenty-five to thirty days (Treloar et al. 1967), whereas chimpanzee cycle lengths average thirty-four to forty days (Wallis 1997; Deschner et al. 2003; Emery Thompson 2005b). The longer chimpanzee cycle, which features a lengthened follicular phase, appears to be derived, as orangutan and gorilla cycles are approximately the same length as those of humans (Bentley 1999). Unlike these other species, chimpanzees exhibit a pronounced estrogen-dependent swelling of the sexual skin for a substantial portion of the follicular phase (~10–12 days: Emery and Whitten 2003; Emery Thompson 2013b). Selection on extended estrous advertisement in chimpanzees may have increased cycle length (Wrangham 2002), as species with sexual swellings generally exhibit lengthened and highly variable follicular phases (Zuckerman 1937).

Regulation of Fecundity

Human ovarian cycles exhibit an unusual characteristic: most of them do not lead to conception. Even among Western women in the best of health, cycles routinely have deficits in the production of estradiol and progesterone, affecting the likelihood of ovulation, fertilization, and implantation (Ellison et al. 1993b; Lipson and Ellison 1996; Ellison 2003). Chimpanzees exhibit similar variation across cycles, and that variation predicts the probability of conception (Emery and Whitten 2003; Emery Thompson 2005b). In both species, energetic condition is an important predictor of cycle fecundity. In humans, this is evident in higher ovarian hormone production by women in well-nourished Western populations relative to those in nutritionally stressed traditional populations (Ellison et al. 1993a, 1993b).

Within both modern and traditional populations, ovarian function covaries with various dimensions of energetic condition, including energy balance (i.e., weight change), seasonal food abundance, and energy expenditure (Ellison and Lager 1986; Ellison et al. 1989; Bentley et al. 1998, 1999). For example, weight loss during the monsoon season led to significantly reduced progesterone production in energetically stressed Nepali women (Panter-Brick et al. 1993). In a controlled study of American women, a vigorous running regimen led to reduced fecundity, particularly in those women who experienced weight loss as a result (Bullen et al. 1985). Among Polish farmers, the high workloads of the harvest led to reduced ovarian function despite sufficient energy intake (Jasienska and Ellison 1998). Similar effects have been reported in chimpanzees. Well-fed captive chimpanzees produce higher levels of ovarian steroids than wild chimpanzees (Emery Thompson 2005a). Wild chimpanzees experience increased fecundity, associated with a shorter waiting time to conception, during periods of high fruit availability (Emery Thompson and Wrangham 2008a). Within chimpanzee communities, females with higher body mass or better access to resources have higher ovarian hormone levels and faster birth rates (Pusey et al. 1997; Emery Thompson et al. 2007b; Jones et al. 2010). C-peptide of insulin levels, indicative of energy balance, are strong positive predictors of ovarian steroid production (Emery Thompson et al. 2014).

While ovarian dysfunction can be accompanied by oligomenorrhea, reduced fecundity may occur within the context of regular menstrual cycles in both chimpanzees and humans. While frequent nonconceptive cycling is not limited to the apes (e.g., baboons: Smuts and Nicolson 1989), it is a relatively rare phenomenon, raising the question of why conception is so difficult. Most primates reproduce seasonally as a strategy to manage the availability of energy for offspring production. Among seasonal breeders, the predominant strategy is to align the high-cost lactation period with the highest anticipated resource availability (Brockman and van Schaik 2005). This strategy is less viable for larger primates, such as the great apes, which face higher costs of offspring production and many years of infant dependency. For such species, a single season of abundance is not sufficient to support offspring nutrition, and lactation must be sustainable through periods of energy shortage. Thus, the probability of conception in larger primates is sensitive to maternal energetic condition (Brockman and van Schaik 2005; Emery Thompson 2013a). Low fecundability in the face of energy shortage may be a critical adaptation to maximize reproductive success in long-lived species by reducing failure rates (Ellison 1990). A conception delay of even several months is a relatively small cost compared to the loss of time and energy spent to produce an infant that does not survive, particularly if low adult mortality portends better opportunities to reproduce in the future. Reproductive success accrues very slowly for female apes, and infants require maternal care for an extended period of time, conditions that may lead mothers to prioritize survival over reproduction when resources are scarce.

Waiting Time to Conception

The waiting time from first exposed cycle to conception can vary according to a number of factors, including the frequency and timing of intercourse, rates of early (undetected) fetal loss, quality of male and female gametes, and ovarian function. In developed nations, most women conceive within four to ten months (Wood 1994). Waiting times in Ju/'hoansi hunter-gatherers average approximately 7.5 months (Bentley 1985). Waiting times in East African chimpanzees *(P. t. schweinfurthii)* are similar, averaging four to twelve months (Wallis 1997; Emery Thompson 2013b). In the one West African chimpanzee *(P. t. verus)* population with applicable data, the average waiting time is more than two years (Deschner and Boesch 2007). Curiously, this increase in waiting time does not equate to longer interbirth intervals than in East African chimpanzees; instead, West African chimpanzee females resume cycling sooner after a prior birth and then experience a longer conception delay. While it is difficult to find data on the number of cycles to conception in other nonseasonal primates, the little data available make it reasonable to conclude that chimpanzees and humans have among the lowest conception rates of any primate. Most of the few waiting times reported for orangutans, which have a much longer interbirth interval, are actually shorter (1–4 mos; Knott et al. 2009).

Based on the comparative data, the regulation of fecundity in humans and chimpanzees is remarkably similar, from the basics of ovarian regulation, to its responsiveness to energetic shortage, to the long delay before conception. The difference in fertility between the species cannot readily be attributed to differences in the physiology of conception.

Implantation, Pregnancy, and Parturition

Implantation

Great apes, including humans, differ from other primates in particular features of placental structure. Of key interest is that the ape trophoblast invades deep into the myometrial layer of the uterine wall, enabling more extensive vascular remodeling and access to the maternal circulation (Carter and Pijnenborg 2011; Pijnenborg et al. 2011). The presence of deep invasion in both species, and in the gorilla, refutes the hypothesis that this feature was linked to bipedalism (i.e., to relieve pressure on the vena cava: Rockwell et al. 2003) but is consistent with the hypothesis that deep placental invasion is necessary to support the oxygen and energy needs of a large fetal brain (Goodman 1961; Pijnenborg et al. 2011). The timing of invasion in chimpanzees appears, based on two museum specimens, to be delayed relative to humans, suggesting different timing of key steps in fetal development (Carter and Pijnenborg 2011; Pijnenborg et al. 2011). If, indeed, the invasive trophoblast is critical for the fetal brain, earlier invasion in humans may facilitate accelerated fetal brain growth (Pijnenborg et al. 2011).

The placenta of a chimpanzee weighs approximately 290 g (Soma 1990). This is roughly half the absolute size of the placenta in humans, and moderately lower in proportion to the fetal weight (chimpanzees: 0.165: Soma 1990; humans across populations 0.200: Ruangvutilert et al. 2002). The larger size of the human placenta is consistent with the greater energy transfer needed to accommodate the increased fetal brain and body size of human neonates.

Pregnancy Maintenance

Humans and other apes share qualitatively similar patterns of hormonal fluctuations during pregnancy (Reyes et al. 1975; Hobson et al. 1976; Shimizu et al. 2003). Of key interest is corticotropin-releasing hormone (CRH), a placentally derived peptide that plays a major role in the timing of fetal maturation and the initiation of parturition (McLean and Smith 2001). In humans and other apes, pregnancy features an exponential increase in CRH and its associated binding protein (Smith et al. 1999). Monkeys do not exhibit the same pattern of CRH release, and CRH is undetectable in prosimians (Smith et al. 1993). In chimpanzees and humans, CRH production is correlated with elevations of estrogens (chiefly, estriol), progesterone, and cortisol, suggesting similar pathways for hormonal regulation during pregnancy in both species (Smith et al. 1999). This pattern is largely similar in gorillas, though cortisol elevations are much less pronounced than in humans and chimpanzees.

Chorionic gonadotropin (CG) is essential for early pregnancy maintenance, and has been suggested to play a more extensive role in placental formation, metabolism, and maintenance (Cole 2009). Peaks of CG occur with near-identical timing in the first and third trimester of chimpanzees and humans, though absolute levels of CG during most of pregnancy are lower in chimpanzees (Clegg and Weaver 1972; Reyes et al. 1975; Faiman et al. 1981). CG peaks are differently timed in gorillas, and the third trimester peak appears blunted or absent in monkeys (Faiman et al. 1981). The functional significance of this difference is not clear. Profiles of prolactin, a hormone that prepares the mother for lactogenesis, are also similar in pregnant chimpanzees and humans, increasing gradually over gestation, whereas those of monkeys remain low until late in pregnancy (Faiman et al. 1981). Inhibin A and B, which suppress FSH during pregnancy in humans and some other primates, are present in low levels during chimpanzee pregnancy, suggesting that the feedback mechanisms for suppressing follicular development differ (Kondo et al. 2001).

Parturition

Median duration of gestation in captive chimpanzees is 232 days from insemination to live birth (range 203–244, $n=12$ normal, singleton pregnancies with known conception dates: Martin et al. 1978; Shimizu et al. 2003). Estimates from the wild, using the date of last maximal swelling before pregnancy as the date of conception, are similar (mean 225 days, range 208–235: Wallis 1997). The comparable figure, measured from ovulation to birth in humans, is 268 days (range 247–284, $n=113$: Jukic et al. 2013), about 16 percent longer. Birth weights of captive chimpanzees average 1.8 kg (0.7–3.3 kg, $n=436$ normal, singleton births: Fessler et al. 2005). Human infants average 3.0–3.5 kg at birth across populations, with both developed populations and hunter-gatherers falling in this range (Western: Wilcox 2001; Ju/'hoansi: Lee 1979). At thirty-three weeks, when a chimpanzee would be born, the human fetus weighs about the same as the chimpanzee, and fetal growth accelerates shortly thereafter (Lubchenco et al. 1966; Dobbing and Sands 1978). The chimpanzee neonate is born with a brain that is less than half the size of the human neonate (151 g versus 375 g), but closer to its adult size (40 percent versus 29 percent adult size: DeSilva and Lesnik 2006, 2008).

It has long been argued that the length of gestation, and more importantly the size of the brain at birth, has been constrained by narrowing of the human pelvis following the adaptation to bipedalism (Washburn 1960; Wittman and Wall 2007). But, when assessed relative to maternal body mass, human gestation is actually proportionally longer than in chimpanzees or gorillas (Dunsworth et al. 2012). Recent analyses indicate that human gestation is instead limited by the energetic demands of the growing fetal brain, which begins to outpace maternal metabolic capacity (Dunsworth et al. 2012). Nevertheless, the pelvis is small relative to the fetal infant's brain and widest in its transverse orientation. This severely constrains the mechanics of parturition, requiring the infant's head and shoulders to rotate, and raising the risk of birth complications (Wittman and Wall 2007). The near universality of birth assistance in human populations may be a compensating behavioral adaptation to reduce maternal and perinatal mortality (Rosenberg and Trevathan 1995, 2002). The presence of social support, even without direct assistance, has been shown to reduce the length of labor substantially (from 15.5 hrs to 7.7 hrs: Klaus et al. 1986; 19.3 hrs to 8.8 hrs: Sosa et al. 1980), as well as to reduce the rate of labor complications (from 59 percent to 27 percent: Klaus et al. 1986). In contrast, the chimpanzee pelvis allows ample room for the fetal head and does not necessitate fetal rotation (Rosenberg and Trevathan 1995, 2002). Labor in chimpanzees ranges from forty minutes to eight hours (Nissen and Yerkes 1943). Chimpanzee infants present facing either up or down and complicated labor is rare, even though mothers accomplish this task alone (Martin 1981; Nissen and Yerkes 1943). Ruff's (1995) analysis of pelvic morphology in fossil hominids suggests that australopithecines, and probably early *Homo*, would have been able to sustain birth without the complex rotation required in modern humans.

Fetal Loss

Early fetal losses are generally undetected in chimpanzees, and approaches to detecting and monitoring pregnancies vary across facilities, underestimating true rates of pregnancy loss. Captive facilities report a spontaneous abortion rate of 13 percent and a stillbirth rate of 8 percent from 1,255 pregnancies (Roof et al. 2005). The rate of pregnancy loss increases with age, as it does in humans (Roof et al. 2005).

Although the data on embryonic and fetal loss in humans is more extensive than for chimpanzees, it remains problematic. National and international registries note the likelihood of incomplete data, even for stillbirth ("late fetal loss"), usually defined as occurring after twenty-eight weeks gestation. Data on early fetal loss, embryonic loss, and preimplantation loss are even more difficult to obtain. Against this background, the following trends can be noted.

In the United States, fetal death rates after twenty weeks gestation have declined from 25.0 per thousand live births in 1942 to 6.2 per thousand in 2006 (MacDorman et al. 2012). Roughly half of this loss occurs between twenty and twenty-eight weeks gestation and half after twenty-eight weeks. Similar rates are reported for other Western, industrialized nations (Woods 2008). In the developing world, reported stillbirth rates are generally much higher, 15 to 65 per thousand live births, and are positively correlated with infant mortality rates (Woods 2008). If stillbirth rates in these countries are approximately half of total fetal loss rates, as in the United States, then total fetal loss rates from twenty weeks gestation would run from 30 to 130 per thousand live births. While these numbers are not derived in exactly the same way as for the chimpanzee studies cited above, and are based on a much larger sample of births, they suggest a much lower percentage of recognized pregnancies are lost in humans (~2.9–11.5 percent versus 21 percent).

Studies of early embryonic loss, from implantation at around three weeks after the last menstrual period, based on the detection of hCG by ultrasensitive assays, indicate total loss rates of 30 percent in U.S. women (Wilcox et al. 1988). Similar rates have been reported from Bangladesh using similar methods (Holman 1996). Some suggest that the average rates of conception among fecund couples engaging in unprotected intercourse (~20 percent per month) and for *in vitro* fertilization (~16 percent per embryo transfer) are reflective of even higher rates of loss per preimplantation zygote (Benagiano et al. 2010).

Postpartum Amenorrhea

Postpartum (or lactational) amenorrhea, the period of ovarian quiescence following the birth of an infant, is the longest and most variable component of the interbirth interval, making it a key determinant of fertility (Wood 1994). The period of postpartum amenorrhea is noticeably longer in most traditional hunter-gatherer populations, who breastfeed for three years or longer (Lee 1980; Konner 2005), than in Western populations, who typically breastfeed for less than one year, if at all. Within populations, women who breastfeed more intensively (i.e., more or longer bouts) experience longer periods of postpartum amenorrhea (Wood 1994). These data support the hypothesis that breastfeeding regulates the interbirth interval (McNeilly et al. 1994; McNeilly 2001). The evolutionary significance of this relationship is fairly intuitive: mothers avoid having to breastfeed a new infant until the previous one has been at least partially weaned. However, the mechanisms of how breastfeeding regulates postpartum fecundity have been unclear.

A key player in postpartum amenorrhea is the pituitary hormone prolactin, which triggers breastmilk production in response to infant suckling. Mothers who breastfeed more intensively experience frequent pulses of prolactin and maintain a higher average level than mothers who breastfeed less intensively (Tyson 1977a, 1977b; Gross and Eastman 1979; Stern et al. 1986). These observations initially supported the hypothesis that the intensity of nursing activity might regulate the interbirth interval (Tyson 1977a). However, subsequent work indicates that prolactin is either not directly involved in ovarian suppression or is insufficient to explain postpartum infecundity (McNeilly 1993; Wood 1994). For example, plasma prolactin concentrations failed to predict the timing of postpartum amenorrhea in a sample of intensively breastfeeding British women (Tay et al. 1996).

Subsequent research casts doubt on the assumption that breastfeeding alone is responsible for postpartum subfecundity. A study conducted in the United States and the Philippines found that, while the intensity of breastfeeding was generally associated with amenorrhea, even intensive breastfeeding was insufficient to prevent ovulation beyond six months postpartum (Gray et al. 1990). In the Gambia, prolactin concentrations in intensively breastfeeding women were dependent on maternal energetic factors, with women producing higher prolactin levels during the season of high workload and resource scarcity (Lunn et al. 1980). Nutritional supplementation also reduced prolactin secretion during lactation (Lunn et al. 1980). In the same study, lactating women in the United Kingdom produced substantially lower levels of prolactin than did even supplemented Gambia women, though it was unclear whether this was due to reduced nursing frequency or improved maternal energy balance. The Argentinian Toba, who have a recent hunting and gathering history, provide a clearer case. Toba women practice the frequent nursing pattern of their forebears despite having settled into semiurban environments of low workload and high resource availability (Valeggia and Ellison 2004). They nevertheless have short periods of lactational amenorrhea (~10 mos) and often wean their previous infants only after the next pregnancy is recognized (Valeggia and Ellison 2004). Thus, while it is clear that lactation plays an important role in postpartum amenorrhea, nursing intensity alone may be insufficient to suppress ovulation.

Further study of the Toba led to an alternative hypothesis: that postpartum amenorrhea is determined by the mother's ability to afford the next pregnancy. This hypothesis implicates breastfeeding, not via its direct inhibitory effects on gonadal regulation, but via its energetic impacts on the mother (Ellison and Valeggia 2003; Valeggia and Ellison 2004). For mothers in poor energetic condition, the cost of producing breastmilk (estimated at ~500 kcal/day during exclusive breastfeeding: Prentice and Prentice 1988) represents a higher "metabolic load," consuming a greater proportion of the total maternal energy budget, than for mothers in good condition. In that situation, a new conception could seriously compromise the ability of a mother to successfully feed both offspring. In further support of the metabolic load hypothesis, cycle resumption in Toba women was closely predicted by increasing body weight, as well as by concentrations of C-peptide of insulin, indicative of increased energy balance (Ellison and Valeggia 2003; Valeggia and Ellison 2009). Insulin is known to have potent effects on steroidogenic cells in the ovaries, upregulating both progesterone and estrogen production (Willis et al. 1996; Greisen et al. 2001), making it a promising candidate as a mediator between maternal condition and postpartum fecundity. In the Toba, maternal C-peptide levels increased across the postpartum period, achieving and surpassing levels characteristic of cycling females approximately four months prior to the onset of cycling (Ellison and Valeggia 2003; Valeggia and Ellison 2009).

Information is scarce on prolactin levels in nursing chimpanzees. In one unusual report, captive chimpanzees that had their infants removed at birth experienced prolonged postpartum amenorrhea (7–26 mos) in association with frequent nipple self-stimulation (Graham et al. 1991). These females had elevated prolactin levels, and cycling was reliably restored by administration of prolactin antagonists. However, data from wild chimpanzees strongly support the metabolic load hypothesis. When C-peptide of insulin levels were assessed longitudinally in lactating chimpanzee mothers, the patterns were remarkably similar to those reported for the Toba (Emery Thompson et al. 2012). Before infants begin feeding on solid foods (~6 mos), mothers had low levels of C-peptide. As in the human sample, maternal C-peptide levels increased prior to the resumption of cycling, exhibiting a similar peak just prior to cycling onset (Figure 6.3). Importantly, C-peptide levels did not simply increase incrementally throughout the postpartum period. Some chimpanzees showed early evidence of metabolic recovery, and resumed cycling early, whereas others did not recover their energetic condition, or their fecundity, for many years (Emery Thompson et al. 2012).

Similar mechanisms appear to control the onset of postpartum fecundity in chimpanzees and humans. A notable difference is the time over which these mechanisms operate. Postpartum amenorrhea in the Toba lasted only ten months (Valeggia and Ellison 2004), whereas the chimpanzees required an average of 3.4 years to resume cycling (Emery Thompson et al. 2012). Again, there is considerable intraspecific variation. The Ju/'hoansi experience approximately two years of amenorrhea (Konner and Worthman 1980; Bentley 1985), and women in Westernized populations with less-intense nursing practices may resume cycling within six months (Wood 1994). Nursing chimpanzees in captivity experience about one year of amenorrhea (Nadler et al. 1981), while estimates from wild populations vary from two years in one West African population (Taï: Deschner and Boesch 2007) to 4.6 years in East African populations (Mahale: Nishida et al. 2003; all sites: Emery Thompson 2013a). The aforementioned energetic models for postpartum cycle resumption predict that much of both the inter- and intraspecific variation in amenorrhea could be attributed to variation in the ability of mothers to recover their energetic condition after the previous reproductive event. Environmental resource availability probably plays a key role in this variation, but social and behavioral factors that affect energy access could be even more important.

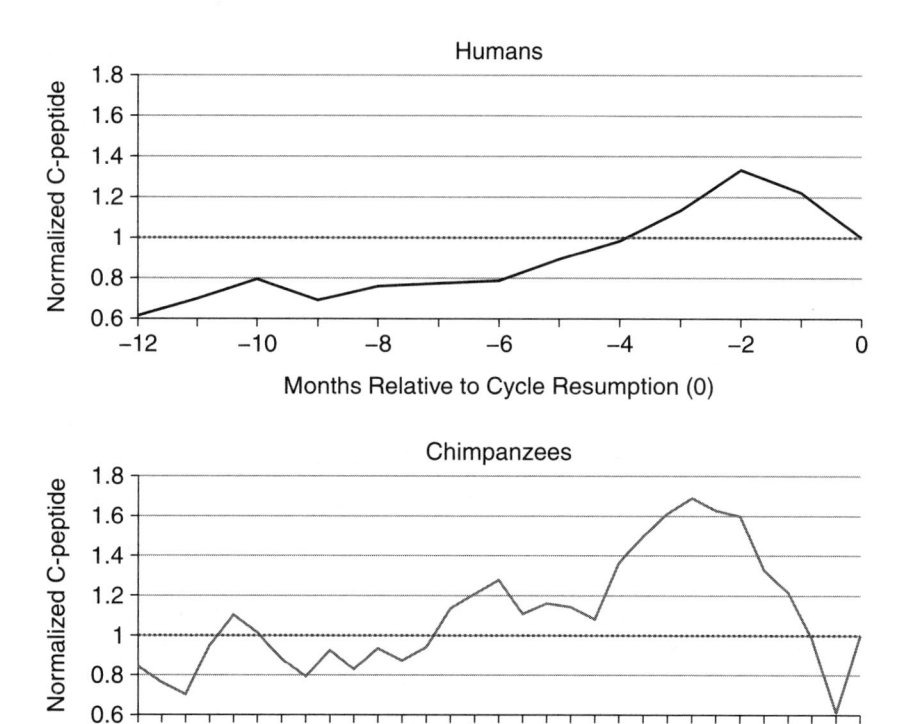

FIGURE 6.3. Energetics of postpartum amenorrhea. Illustrated by changes in urinary C-peptide of insulin levels relative to the mean levels of postpartum cycles. Redrawn from Ellison and Valeggia (2003) (humans, Toba) and Emery Thompson et al. (2012) (chimpanzees).

Reproductive Life Span

A final important component to fertility differences between species is the difference in the number of reproductive years. Chimpanzees reach menarche early (10.7 yrs: Wallis 1997; Nishida et al. 2003) relative to human hunter-gatherers (13–18 yrs: Howell 1979; Wood 1994; Hill and Hurtado 1996). Chimpanzees generally achieve first birth at approximately thirteen years (Emery Thompson 2013b). Age of first birth in human hunter-gatherers occurs at eighteen to twenty-three years (Pennington 2001). Though modernized populations experience markedly reduced age of menarche (~12–15 yrs: Wood 1994; Thomas et al. 2001), the practice of contraception often pushes first reproduction to a later age than in natural fertility populations.

Variation in the timing of menarche across human populations corresponds broadly with differences in energy availability, while within populations menarche is delayed under conditions of demonstrable energetic stress, as in young female athletes. This led to the hypothesis that female reproductive maturation was contingent on reaching a "critical fat" threshold that would allow for the maintenance of pregnancy and lactation (Frisch and Revelle 1970). While this idea continues to have considerable traction, it has not held up to empirical scrutiny (Cameron 1976; Scott and Johnston 1982, 1985). Instead, there is large body of evidence pointing to skeletal age as a key predictor of menarcheal age (Simmons and Greulich 1943; Eveleth and Tanner 1976; Marshall 1974). Mechanistically, maturation of the hypothalamic-pituitary-ovarian axis has important, independent effects on skeletal maturation and ovarian cycling (Wood 1994; Frank 1995). Functionally, the pelvis is the key mechanical constraint on reproduction, suggesting that selection may have acted to delay menarche until the skeleton could accommodate parturition (Ellison 1982). Pelvic dimensions are not similarly limiting for chimpanzee parturition. While both skeletal maturation and puberty are accelerated in captive chimpanzees versus wild chimpanzees, studies of chimpanzee skeletal maturation indicate a markedly different pattern than in humans, one that suggests that neither body fat nor skeletal age are good predictors of menarcheal age (Hamada et al. 1996, 1998, 2003).

Timing of menarche is only part of the story, because both humans and chimpanzees experience a prolonged period of subfecundity following menarche. In humans, social factors (e.g., marriage) often limit age of first reproduction beyond physiological limitations. Mean age of menarche and first birth are separated by about five years in hunter-gatherer populations (Pennington 2001), whereas chimpanzees experience delays of about two to three years in the wild (Emery Thompson 2013b). Adolescent human females experience irregular cycles, characterized by long (possibly overlapping) follicular phases, short luteal phases, and high rates of ovulatory failure (Montagu 1957; Wood 1994). While this has not been studied thoroughly in chimpanzees, long swelling duration and shorter post-swelling intervals among adolescents compared with adults suggest similar patterns (Wallis

1997). The factors predicting duration of subfecundity are unclear. A study of Finnish women found that early age of menarche predicted shorter periods of subfecundity, suggesting that similar fecundity-regulating mechanisms might govern both processes (Apter and Vikho 1983). However, the reverse was found in a large study of Bangladeshi women, and the demographic factors predicting age of menarche failed to predict subfecund intervals (Foster et al. 1986). In chimpanzees, female dispersal from one community to another makes it difficult to trace subfecundity in individuals. Yet females who disperse appear to have longer reproductive delays, most likely due to the social and energetic stresses of immigration (Nishida et al. 2003; Stumpf et al. 2009).

The end of reproductive life in humans is marked by menopause, when menstrual cycles cease due to depleted stocks of primordial oocytes ("follicles") in the ovaries (Wood 1994). This occurs at approximately age fifty in most populations (Wood 1994; Thomas et al. 2001). Rather than a single event in time, menopause is the result of an exponential loss in follicles throughout life, beginning when a female is still a fetus herself (Fortune 1994; Gosden et al. 2010). Despite differing ages of fertility onset, the age patterns of fertility decline in wild chimpanzees and human hunter-gatherers are nearly indistinguishable (Figure 6.4; Emery Thompson et al. 2007a; Hawkes and Smith 2010). Though the size of the initial follicular supply in chimpanzees is not known, the rate of depletion appears to be very similar to that of humans, at least until age thirty-five (Jones et al. 2007; Hawkes and Smith 2010). Thereafter, rates of follicular depletion in human ovaries appear to exceed those of chimpanzees, though the sample of aged chimpanzees is still scant (Cloutier et al. 2015). While some aged captive chimpanzees exhibit follicular exhaustion (Cloutier et al. 2015), there is also evidence of continued ovulation through age fifty, and sometimes beyond, in captive chimpanzees (Graham 1979; Lacreuse et al. 2008; Herndon et al. 2012). One report from a captive colony proposed an earlier menopausal age of thirty-five to forty years based on hormonal landmarks and sexual swelling frequency (Videan et al. 2006). However, these data were only consistent with decreased fecundity, rather than cessation of reproduction, at that age. Concentrations of LH in captive chimpanzees began declining before age thirty-five, while FSH increased moderately but significantly in the forties (Videan et al. 2006). This contrasts with the typical human menopausal pattern, characterized by sub-

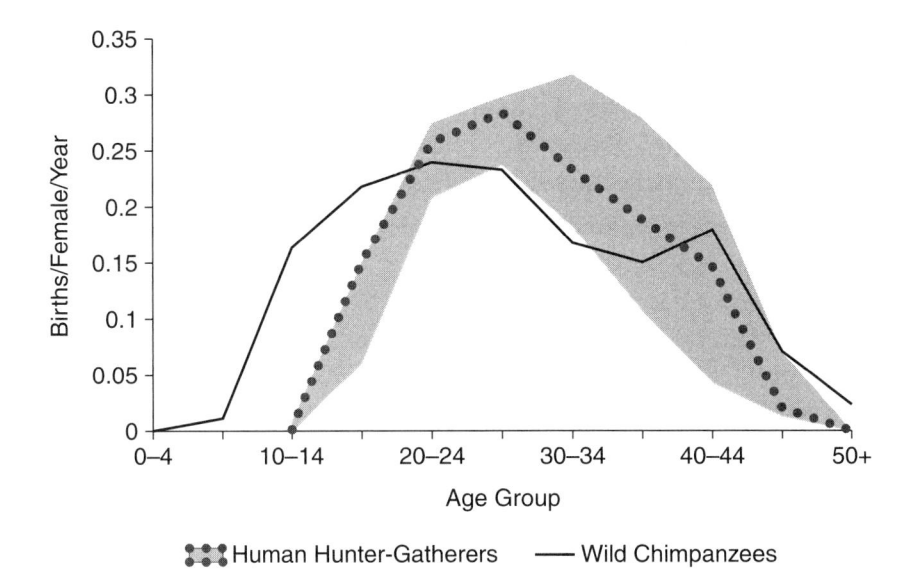

FIGURE 6.4. Age-specific fertility of wild chimpanzees compared with hunter-gatherer societies. Solid line (Emery Thompson et al. 2007a). Dotted line indicates the mean of three populations (Ju/'hoansi: Howell 1979; Aché: Hill and Hurtado 1996; Hadza: Blurton Jones 2016). The shaded area spans rates of the low-fertility Ju/'hoansi and high-fertility Aché. Hadza fertility rates approximate the mean.

stantial increases in both hormones beginning in perimenopause and persisting for decades after menopause (Chakravarti et al. 1976; Lenton et al. 1988; Fernandez et al. 1990).

Even in hunter-gathering populations, the majority of women who live to the age of sexual maturity live past the age of menopause (Pennington 2001). A wild female chimpanzee that lives to maturity can only expect to live, on average, to age thirty, and only about 1 percent live to age fifty (Hill et al. 2001). Death, rather than menopause, most often marks the end of reproductive life in chimpanzees. Given the earlier age of first reproduction, chimpanzees could experience a fertility rate comparable to that of humans, were they to live long enough. The most successful wild chimpanzee mother known, Fifi of the Gombe community in Tanzania, died at age forty-six after having given birth to nine infants, most of which survived (Jane Goodall Institute). Thus, the higher fertility of humans relative to chimpanzees may be more readily explained by changes in the biology of aging than the biology of reproduction (see Gurven and Gomes, this volume).

Variation in reproductive senescence is poorly understood for both species. Similarities in age of menopause between hunter-gatherer populations and industrialized societies (Wood 1994) suggest that the mechanisms that influence timing of menarche, both energetic (Thomas et al. 2001) and genetic (He et al. 2009), are independent of those that affect age of menopause. Indeed, the ages of menarche and menopause are correlated neither within populations (van Noord et al. 1997) nor between them (Thomas et al. 2001). Nevertheless, some populations are reported to experience menopause in their early forties (Goodman et al. 1985; Thomas et al. 2001), and substantial variation exists within populations. Factors such as smoking (Cramer et al. 1995), high parity (Jeune 1986), low parity (Gold 2001), low income (Stanford et al. 1987), and oral contraceptive use (de Vries et al. 2001) have been linked to accelerated menopause, but not consistently (Brambilla and McKinlay 1989; Whelan et al. 1990), and genetic factors appear to account for most of the individual variability (de Bruin et al. 2001; van Asselt et al. 2004). Among chimpanzees, one wild population (Bossou, Guinea) stands out for having both an early age of maturity, two years faster than other populations, and several apparently postreproductive females (Sugiyama 2004; Emery Thompson et al. 2007a).

In weighing the variance in age of menopause, it is important to consider that most studies rely on nonspecific symptoms of menopause (e.g., amenorrhea, infertility) rather than its specific cause (i.e., depleted ovarian follicles). Rates of sterility from other causes might be expected to increase with age, particularly in populations where sexually transmitted disease is prevalent or where high parity increases risk of puerperal sepsis (McFalls and McFalls 1984). Age is associated with chromosomal abnormalities in both sexes that may prevent conception or result in early fetal loss (Hook 1981; de la Rochebrochard and Thonneau 2002; Pellestor et al. 2003). Less often acknowledged is that the energetic trade-offs that compromise fecundity among young adults, as previously described in this chapter, may increase with age and give the appearance of early menopause. As general senescence, and associated health challenges, set in, individuals are less able to afford the costs of reproduction, even if they have follicular supplies remaining. This was demonstrated in a cross-population survey of wild chimpanzees, for which health status was a key predictor of low birth rates among older individuals (Emery Thompson et al. 2007a).

Reproduction in Context

The Receptive Period and Timing of Ovulation

Like many anthropoid primates, chimpanzee females exhibit a pronounced estrogen-dependent swelling of the anogenital region during their menstrual cycles. The swelling is at maximal tumescence for approximately twelve days in each cycle. While copulations are not strictly limited to the maximal swelling phase, the vast majority of copulations (>90 percent) between adult males and adult females occur during this period. The periovulatory period occurs reliably in the last few days of maximal swelling (Deschner et al. 2003; Emery Thompson 2005b), and females mate at greater frequencies at this time (Emery Thompson and Wrangham 2008b). This pattern appears to be driven by increased male mating interest rather than by increases in female proceptivity (Emery Thompson and Wrangham 2008b). Males may cue on small changes in swelling size that accompany the periovulatory period (Deschner et al. 2004). A preliminary study also suggests that vaginal fatty acids provide olfactory signals of impending ovulation (Matsumoto-Oda et al. 2002). However, for male chimpanzees, detection of ovulation is not an exact science. Inter-individual variation in swelling size exceeds intra-individual variation (Deschner et al. 2004), and wide variation in the duration of swelling may make it difficult to pinpoint ovulation without close monitoring. As a result, matings occur with high frequency throughout the swelling period, and it is hypothesized that exaggerated swellings evolved (in this species, as in others) to facilitate paternity confusion by females (Nunn 1999). Sexual swellings, therefore, both advertise and obscure ovulatory timing in chimpanzees.

All of the apes display genital swellings of some kind. Highly promiscuous bonobos have swellings that are larger and extend for a longer period of time than in common chimpanzees (Reichert et al. 2002). For gorillas, in which promiscuity is limited, swellings are small in size and confined to about three days during the periovulatory period (Nadler et al. 1979). Hylobatids also exhibit very small genital swellings encompassing ovulation (Barelli et al. 2007), and orangutans have genital swelling during pregnancy (Galdikas 1981). It is reasonable, therefore, to assume that some form of female sexual swelling was present in the last common ancestor of humans and chimpanzees. However, phylogenetic comparison across anthropoids indicates that sexual swellings are highly evolvable, having arisen and been lost several times, with even more numerous changes in form and function (Dixson 2013).

It is well documented that conception is only likely in humans when intercourse occurs within a six-day window ending on the day of ovulation (Wilcox et al. 1995). Compared to the striking sexual swellings of chimpanzees, however, human females demonstrate only very subtle, if any, externally detectable signals of periovulatory status. This contrast has led to a somewhat confusing literature on the function and evolution of concealed ovulation and extended sexual receptivity in our species (e.g., Lovejoy 1981). As described above, however, the exaggerated sexual swellings exhibited by chimpanzees are far from precise ovulatory advertisements and, in fact, may have become elaborated for the benefits of mating outside of the fertile period. By contrast, the fact that human copulations occur throughout the cycle need not necessarily indicate that ovulation is any more obscured than it is in chimpanzees, nor that ovulatory status has no influence on female sexual motivation. Thus, some authors have argued that humans cannot be claimed to have lost "estrus" because it was not a part of our recent evolutionary heritage (Dixson 2009), while others simply argue that it has never been lost (Gangestad and Thornhill 2008).

This issue may be more appropriately broken down into component questions that consider the roles of both male and female in driving copulation throughout the cycle. Do human females exhibit peaks in sexual desire or proceptivity at the time of ovulation? Are males able to detect ovulation, and how? And what drives the extension of sexuality across the cycle? Studies on the variation in copulation rates across the menstrual cycle produce mixed evidence for periovulatory peaks (Wood 1994; Dixson 2009). However, most studies use self-recollected day of last menstruation to estimate ovulation day, and few follow individual women throughout the cycle. A large study following sixty-eight women for 171 full menstrual cycles and using hormonal indicators of ovulation reported statistical peaks in the likelihood of intercourse during the periovulatory period (Wilcox et al. 2004). Even such positive evidence for cyclic effects on sexuality indicates low variability—daily probability of intercourse varied from 23.5 percent to 34.4 percent—compared to chimpanzees, whose daily copulation rates vary from 0 to 0.6 per hour (Figure 6.5). Recent investigations of a large sample indicate that variation in human sexual behavior can be at least partially attributable to cyclical variation in female sexual desire (Roney and Simmons

FERTILITY AND FECUNDITY 239

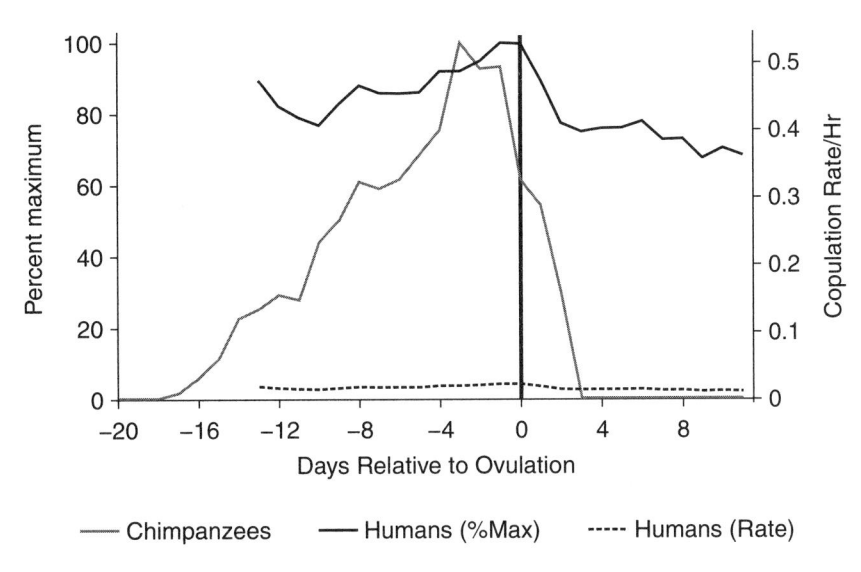

FIGURE 6.5. Distribution of sexual behavior across the cycle in chimpanzees and humans. Three-day running averages, aligned relative to ovulation (vertical line). Chimpanzee copulation rates from Emery Thompson (2005b), with ovulation day assigned to three days prior to detumescence (Deschner et al. 2003; Emery Thompson and Wrangham 2008b), adjusted for the proportion of females with maximal swelling each day. Sample human data are from Wilcox et al. (2004), originally reported as the proportion of women having intercourse each day (range 23.5–34.4). Data were plotted relative to chimpanzees as percentage maximum (solid black line), and rates estimated by conversion factor of just over one copulation per twelve-hour active period (dotted line).

2013). However, there is some evidence that males may identify periovulatory females independently of female behavioral cues. For example, male perceptions of female facial attractiveness change according to female menstrual phase (Roberts et al. 2004; Gangestad and Thornhill 2008), and men in a controlled environment appear to be able to detect differences in female body odor and other features associated with cycle phase (Haselton and Gildersleeve 2011). The significance of these findings has been downplayed due to small sample sizes and questionable relevance to humans in evolutionarily relevant contexts (Lancaster and Kaplan 2009; Henrich et al. 2010). However, such findings should not be surprising, because even if human females gain evolutionary benefits by concealing ovulation, males should experience correspondingly strong selection pressure to discern variance in female fertility. This form of sexually antagonistic coevolution is broadly comparable to what is observed in chimpanzees, though it is less clear that the extension of sexual behavior is functionally homologous between humans and chimpanzees.

Obscuring ovulatory timing in chimpanzees and other promiscuous primates is thought to further the goal of paternity confusion (Hrdy 1981). If the sexual swelling is graded in size or coupled with other cues, highly competitive or preferred males may still gain preferential access at the time of greatest fertilization potential (Nunn 1999). Extended sexuality in humans cannot easily be attributed to paternity confusion. On the contrary, maintenance of male investment within pair-bonds likely necessitates high confidence in paternity by the bonded male. Were paternity confidence gained by a reliable cue to ovulation, males might leave their partners to seek additional mating opportunities. It is, therefore, argued that females of many monogamous species exhibit extended sexuality in order to monopolize investment from their partners and prevent desertion (Rodriguez-Gironés and Enquist 2001). Consistent with this hypothesis, a recent study reported that women were most likely to solicit luteal-phase sex from their partners if they perceived that those partners were less invested in the relationship than they were (Grebe et al. 2013). Phylogenetic comparisons across primates suggest that "concealed ovulation" may be prerequisite for the evolution of monogamous mating systems (Sillen-Tullberg and Møller 1993).

Promiscuity

Despite similarities in the physiology of conception, the mating behavior of humans is a marked departure from that of chimpanzees. While both chimpanzees and humans live in social groupings consisting of multiple males and multiple females, chimpanzees do not form pair-bonds for mating or parenting. In the process of conceiving one infant, a typical chimpanzee female engages in hundreds of brief (~6 sec) copulations with nearly all males in her community (Stumpf et al. 2008). Such extreme promiscuity is consistent with the anti-infanticide strategy exhibited by many mammalian species (Palombit 2012; Hrdy 1979; Wolff and Macdonald 2004). By mating prolifically over several cycles, females provide male group members with evidence that they may have sired a future infant, reducing the possibility that males will compromise the infant's safety. Infanticide does occur in chimpanzees, though it most often occurs when a male has not had prior sexual access to the mother (Wilson and Wrangham 2003). In further support of the anti-infanticide hypothesis, female copulation rates increase when there are more potential partners, and the pattern of copulations quickly maximizes the number of unique partners (Watts 2007). It is, however, curious that bonobos display even greater female promiscuity than chimpanzees without any evidence of infanticide (Wrangham 2002).

Given the promiscuous behavior of females and the uncertainty about ovulatory timing, chimpanzee males have low paternity confidence. As a consequence, males offer little in the way of help with raising offspring, investing in future mating opportunities instead. Thus, the mother is responsible for all costs associated with infant care. Male chimpanzees do provide important group-wide benefits for female reproduction, such as access to food resources (Wilson et al. 2010). At Gombe, female reproductive rates increased with the size of the community home range, a consequence of costly male territorial efforts (Williams et al. 2004). However, on a day-to-day basis, association with males imposes costs of females. Females spend less time feeding (Wrangham and Smuts 1980; Pandolfi 2004) and suffer decreased energy balance when in the presence of large numbers of males (Emery Thompson et al. 2014). Given the importance of energy balance for ovarian function in cycling females, and for the resumption of cycling postpartum, it is likely that females pay a price in fecundity for their associations with males. Male chimpanzees also direct physical aggression toward females, especially during the sexual swelling period (Muller et al. 2009), and females who receive aggression from males have elevated stress hormone levels (Muller et al. 2007; Emery Thompson et al. 2010).

Offspring are born outside of marriage in nearly all human societies, though there is little evidence for promiscuous female mating on a chimpanzee scale. The testes, which are enlarged under conditions of sperm competition, are substantially smaller in humans than in chimpanzees and are smaller than expected for a promiscuous primate (Harcourt 1995; Dixson 2009; Muller and Pilbeam, this volume). Nevertheless, a large body of literature poses that, as in many other monogamously breeding species, limited extrapair mating is evolutionarily advantageous for human females, particularly if a current mate does not possess the genetic traits desirable in offspring (Thornhill and Gangestad 2008). Where genetic paternity determinations have been performed, up to 10 percent of paternities have been found to be incorrectly attributed. The figure varies with paternal confidence. When men are confident of their fatherhood, incorrect assignment is less than 2 percent, but when men are not confident, rates of incorrect assignment are over 25 percent (Anderson 2006).

Dietary Quality and Provisioning

The most important distinctions between the reproductive biology of chimpanzees and modern humans may not stem from their reproductive physiology, but rather from the pattern of energy flows to the reproductive process. Chimpanzee diets consist primarily of ripe fruit (~65 percent), supplemented by other plant parts and small amounts of insect and vertebrate prey (Hladik 1977; Newton-Fisher 1999). Meat from small prey is consumed irregularly and is unlikely to contribute significantly to chimpanzee energy budgets, particularly for females, who rarely hunt. Energy for gestation and lactation comes entirely from the individual female's foraging efforts, mediated by her own metabolic rate. Direct provisioning of young chimpanzees with foods other than breastmilk is rare. Even mothers rarely give solid food directly to their offspring, though they provide indirect assistance, such as helping infants and juveniles locate food in the forest, providing an observational model of foraging behavior and tool use, and tolerating feeding proximity. An infant chimpanzee slowly achieves metabolic independence from its mother, and relieves her of the metabolic burden of its support, as it gradually achieves self-sufficiency in foraging. A human mother also mediates the energy allocated to gestation and lactation via her own metabolism, but the energy inputs to her metabolism, in addition to her own efforts, come from a range of individuals beyond herself (Hawkes et al. 1997; Hill and Hurtado 2009; Reiches et al. 2009; Hooper et al. 2015). Given substantial evolution of cooperative and technological behavior in humans versus chimpanzees, community members can acquire larger and higher-quality foods that enable sharing. These foods may be altered through cooking and processing to further increase their energetic value (Carmody et al. 2011). The energetic burden of lactation is relieved not through the increasing ability of the infant to provide for itself, but by energy inputs from supplemental foods, which are directly or indirectly provided by individuals other than the mother (Sellen and Smay 2001). Hadza fathers, for example, bring home more meat and honey when they have unweaned offspring (Marlowe 2003). Similarly, Hadza grandmothers increase their foraging effort when they have young grandchildren, producing a positive effect on infant growth (Hawkes et al. 1997).

This pattern of energy flows to human reproduction has led to the concept of a "pooled energy budget" shared within human groups that cooperate in production and reproduction (Reiches et al. 2009; Kramer and Ellison 2010). Energy flows within this pooled budget not only through food sharing and the conversion of somatic energy into food through foraging and food processing, but through patterns of labor substitution that leverage the somatic energy of the most productively and reproductively capable members of the group (Gurven et al. 2012). In this way, human pooled energy budgets go far beyond what has been conventionally described as "cooperative breeding," and is more flexible than the caste structure of eusocial insects.

Pooled energy budgets have been documented for virtually all human subsistence strategies and levels of social organization. In subsistence foragers, alloparenting and food sharing are virtually universal (Hurtado et al. 1992; Fouts and Brookshire 2009; Meehan et al. 2013; Meehan and Roulette 2013). In societies with domestic dairy stock, milk, milk fat, and their derivatives provide energy-dense, easily digestible supplemental foods that can be introduced into a human infant's diet from an extremely early age (Gray 1995; Miller 2014). The availability of this dairy energy input into the human reproductive process depends, however, on a complex web of labor inputs and exchanges to which even relatively young children may contribute. In subsistence farming, labor exchanges have been documented to increase energy flows to the reproductive process (Kramer 2005).

This basic difference between the metabolic isolation of the chimpanzee female and the human reproductive female as the beneficiary of a large, flexible, and efficient pooled energy budget, is probably the most important source of the differences in reproductive patterns demonstrated by the two species. Even the emergence of menopause as a feature of human female life histories may ultimately derive from reductions in exogenous mortality and shifts in reproductive effort earlier in the life cycle made possible by pooled energy budgets (Hawkes et al. 1998; Kaplan et al. 2000; Reiches et al. 2009; Gurven et al. 2012).

Summary

One of the more remarkable features of the human life history is a high fertility rate, a conspicuous departure from the slow reproduction of our closest phylogenetic relatives. Consistent with their shared evolutionary history, humans and chimpanzees exhibit profound similarities in the physiology of reproduction, including shared features that are derived relative to other primate species. These include similarities in the menstrual cycle and the regulation of conception, structural and hormonal similarities in the maintenance of pregnancy, and similar mechanisms governing postpartum amenorrhea and the interbirth interval. Both species exhibit adaptations consistent with the high cost of producing large, encephalized offspring, including risk-averse strategies for allocation of reproductive effort. Thus, energy availability is critical for regulating variation in reproductive life histories of both species. Despite the relatively higher expense of human infants, human mothers appear to be less constrained by their reproductive systems than are chimpanzee mothers. Improved resource access and pooled energy budgets buffer the costs of reproduction for human mothers, whereas chimpanzee mothers are vulnerable to the competitive features of their social environments. In this manner, the social environment can be viewed as a fundamental component of the reproductive biology of these species.

References

Anderson, K. G. 2006. How well does paternity confidence match actual paternity? Evidence from worldwide nonpaternity rates. *Current Anthropology* 47: 513–520.

Apter, D., and R. Vikho. 1983. Early menarche, a risk factor for breast cancer, indicates early onset of ovulatory cycles. *Journal of Clinical Endocrinology and Metabolism* 57: 82–88.

Baird, D. D., A. J. Wilcox, C. V. Weinberg, F. Kamel, D. R. McConnaughey, P. R. Musey, and D. C. Collins. 1997. Preimplantation hormonal differences between the conception and non-conception menstrual cycles of 32 normal women. *Human Reproduction* 12: 2607–2613.

Barelli, C., M. Heistermann, C. Boesch, and U. H. Reichard. 2007. Sexual swellings in wild white-handed gibbon females *(Hylobates lar)* indicate the probability of ovulation. *Hormones and Behavior* 51: 221–230.

Benagiano, G., M. Farris, and G. Grudzinskas. 2010. Fate of fertilized human oocytes. *Reproductive BioMedicine Online* 21: 732–741.

Bentley, G. R. 1985. Hunter-gatherer energetics and fertility: A reassessment of the !Kung San. *Human Ecology* 13: 79–109.

Bentley, G. R. 1999. Aping our ancestors: Comparative aspects of reproductive ecology. *Evolutionary Anthropology* 7: 175–185.

Bentley, G. R., R. Aunger, A. M. Harrigan, M. Jenike, R. C. Bailey, and P. T. Ellison. 1999. Women's strategies to alleviate nutritional stress in a rural African society. *Social Science & Medicine* 48: 149–162.

Bentley, G. R., A. M. Harrigan, and P. T. Ellison. 1998. Dietary composition and ovarian function among Lese horticulturalist women of the Ituri Forest, Democratic Republic of Congo. *European Journal of Clinical Nutrition* 52: 261–270.

Blurton Jones, N. G. 2001. Some current ideas about the evolution of the human life history. In P. C. Lee, ed., *Comparative Primate Socioecology*, 140–166. Cambridge: Cambridge University Press.

Blurton Jones, N. G. 2016. *Demography and Evolutionary Ecology of Hadza Hunter-Gatherers*. Cambridge: Cambridge University Press.

Brambilla, D. J., and S. M. McKinlay. 1989. A prospective study of factors affecting age at menopause. *Journal of Clinical Epidemiology* 42: 1031–1039.

Brockman, D. K., and C. P. van Schaik. 2005. Seasonality and reproductive function. In D. K. Brockman and C. P. van Schaik, eds., *Seasonality in Primates: Studies of Living and Extinct Human and Non-Human Primates*, 269–305. Cambridge: Cambridge University Press.

Bullen, B. A., G. S. Skrinar, I. Z. Beitins, G. von Mering, B. A. Turnbull, and M. D. McArthur. 1985. Induction of menstrual disorders by strenous exercise in untrained women. *New England Journal of Medicine* 312: 1349–1353.

Cameron, N. 1976. Weight and skinfold variation at menarche and the critical body weight hypothesis. *Annals of Human Biology* 3: 279–282.

Carmody, R. N., G. S. Weintraub, and R. W. Wrangham. 2011. Energetic consequences of thermal and nonthermal food processing. *Proceedings of the National Academy of Sciences* 108: 19199–19203.

Carter, A. M., and R. Pijnenborg. 2011. Evolution of invasive placentation with special reference to non-human primates. *Best Practice & Research Clinical Obstetrics & Gynaecology* 25: 249–257.

Chakravarti, S., W. P. Collins, J. D. Forecast, J. R. Newton, D. H. Oram, and J. W. Studd. 1976. Hormonal profiles after the menopause. *British Medical Journal* 2: 784–787.

Charnov, E. L., and D. Berrigan. 1993. Why do female primates have such long lifespans and so few babies? or Life in the slow lane. *Evolutionary Anthropology* 1: 191–194.

Clark, C. B. 1977. A preliminary report on weaning among chimpanzees of the Gombe National Park, Tanzania. In S. Chevallier-Skolnikoff and F. E. Poirier, eds., *Primate Bio-Social Development*, 235–260. New York: Garland.

Clegg, M. T., and M. Weaver. 1972. Chorionic gonadotropin secretion during pregnancy in the chimpanzee *(Pan troglodytes)*. *Experimenal Biology & Medicine* 139: 1170–1174.

Cloutier, C. T., J. E. Coxworth, and K. Hawkes. 2015. Age-related decline in ovarian follicle stocks differ between chimpanzees *(Pan troglodytes)* and humans. *Age* 37: 10.

Cole, L. A. 2009. hCG and hyperglycosylated hCG in the establishment and evolution of hemochorial placentation. *Journal of Reproductive Immunology* 82: 112–118.

Cramer, D. W., B. L. Harlow, H. Xu, C. Fraer, and R. Barbieri. 1995. Cross-sectional and case-controlled analyses of the association between smoking and early menopause. *Maturitas* 22: 79–87.

de Bruin, J. P., H. Bovenhuis, P. A. H. van Noord, P. L. Pearson, J. A. M. van Arendock, E. R. te Velde, W. W. Kuurman, and M. Dorland. 2001. The role of genetic factors in age at natural menopause. *Human Reproduction* 16: 2014–2018.

de la Rochebrochard, E., and P. Thonneau. 2002. Paternal age and maternal age are risk factors for miscarriage; Results of a multicentre European study. *Human Reproduction* 17: 1649–1656.

Deschner, T., and C. Boesch. 2007. Can the patterns of sexual swelling cycles in female Taï chimpanzees be explained by the cost-of-sexual-attraction hypothesis? *International Journal of Primatology* 28: 389–406.

Deschner, T., M. Heistermann, K. Hodges, and C. Boesch. 2003. Timing and probability of ovulation in relation to sex skin swelling in wild West African chimpanzees, *Pan troglodytes verus. Animal Behaviour* 66: 551–560.

Deschner, T., M. Heistermann, K. Hodges, and C. Boesch. 2004. Female sexual swelling size, timing of ovulation, and male behavior in wild West African chimpanzees. *Hormones and Behavior* 46: 204–215.

DeSilva, J., and J. Lesnik. 2006. Chimpanzee neonatal brain size: Implications for brain growth in *Homo erectus. Journal of Human Evolution* 51: 207–212.

DeSilva, J. M., and J. J. Lesnik. 2008. Brain size at birth throughout human evolution: A new method for estimating neonatal brain size in hominins. *Journal of Human Evolution* 55: 1064–1074.

de Vries, E., I. den Tonkelaar, P. A. H. van Noord, Y. T. van der Schouw, E. R. te Velde, and P. H. M. Peeters. 2001. Oral contraceptive use in relation to age at menopause in the DOM cohort. *Human Reproduction* 16: 1657–1662.

Dixson, A. F. 2009. *Sexual Selection and the Origins of Human Mating Systems.* Oxford: Oxford University Press.

Dixson, A. F. 2013. *Primate Sexuality: Comparative Studies of the Prosimians, Monkeys, Apes, and Humans.* Oxford: Oxford University Press.

Dobbing, J., and J. Sands. 1978. Head circumference, biparietal diameter and brain growth in fetal and postnatal life. *Early Human Development* 2: 81–87.

Dunsworth, H. M., A. G. Warrener, T. Deacon, P. T. Ellison, and H. Pontzer. 2012. Metabolic hypothesis for human altriciality. *Proceedings of the National Academy of Sciences* 109: 15212–15216.

Eijkemans, M. J. C., F. van Poppel, D. F. Habbema, K. R. Smith, H. Leridon, and E. R. te Velde. 2014. Too old to have children? Lessons from natural fertility populations. *Human Reproduction* 29: 1304–1312.

Ellison, P. T. 1982. Skeletal growth, fatness, and menarcheal age: A comparison of two hypotheses. *Human Biology* 54: 269–281.

Ellison, P. T. 1990. Human ovarian function and reproductive ecology: New hypotheses. *American Anthropologist* 92: 933–952.

Ellison, P. T. 2001. *On Fertile Ground.* Cambridge, MA: Harvard University Press.

Ellison, P. T. 2003. Energetics and reproductive effort. *American Journal of Human Biology* 15: 342–351.

Ellison, P. T., and C. Lager. 1986. Moderate recreational running is associated with lowered salivary progesterone profiles in women. *American Journal of Obstetrics & Gynecology* 154: 1000–1003.

Ellison, P. T., S. F. Lipson, M. T. O'Rourke, G. R. Bentley, A. M. Harrigan, C. Panter-Brick, and V. J. Vitzthum. 1993a. Population variation in ovarian function. *The Lancet* 342: 433–434.

Ellison, P. T., C. Panter-Brick, S. F. Lipson, and M. T. O'Rourke. 1993b. The ecological context of human ovarian function. *Human Reproduction* 8: 2248–2258.

Ellison, P. T., N. R. Peacock, and C. Lager. 1989. Ecology and ovarian function among Lese women of the Ituri Forest, Zaire. *American Journal of Physical Anthropology* 78: 519–526.

Ellison, P. T., and C. R. Valeggia. 2003. C-peptide levels and the duration of lactational amenorrhea. *Fertility and Sterility* 80: 1279–1280.

Emery, M. A., and P. L. Whitten. 2003. Size of sexual swellings reflects ovarian function in chimpanzees *(Pan troglodytes). Behavioral Ecology and Sociobiology* 54: 340–351.

Emery Thompson, M. 2005a. Endocrinology and ecology of wild female chimpanzee reproduction. Ph.D. dissertation, Harvard University, Cambridge, MA.

Emery Thompson, M. 2005b. Reproductive endocrinology of wild female chimpanzees *(Pan troglodytes schweinfurthii):* Methodological considerations and the role of hormones in sex and conception. *American Journal of Primatology* 67: 137–158.

Emery Thompson, M. 2013a. Comparative reproductive energetics of human and nonhuman primates. *Annual Review of Anthropology* 42: 287–304.

Emery Thompson, M. 2013b. Reproductive ecology of wild female chimpanzees. *American Journal of Primatology* 75: 222–237.

Emery Thompson, M., J. H. Jones, A. E. Pusey, S. Brewer-Marsden, J. Goodall, D. Marsden, T. Matsuzawa, and R. W. Wrangham. 2007a. Aging and fertility patterns in wild chimpanzees provide insights into the evolution of menopause. *Current Biology* 17: 2150–2156.

Emery Thompson, M., S. M. Kahlenberg, L. C. Gilby, and R. W. Wrangham. 2007b. Core area quality is associated with variance in reproductive success among female chimpanzees at Kibale National Park. *Animal Behaviour* 73: 501–512.

Emery Thompson, M., M. N. Muller, and R. W. Wrangham. 2012. The energetics of lactation and the return to fecundity in wild chimpanzees. *Behavioral Ecology* 23: 1234–1241.

Emery Thompson, M., M. N. Muller, and R. W. Wrangham. 2014. Male chimpanzees compromise the foraging success of their mates in Kibale National Park, Uganda. *Behavioral Ecology and Sociobiology* 68: 1973–1983.

Emery Thompson, M., M. N. Muller, S. M. Kahlenberg, and R.W. Wrangham. 2010. Dynamics of social and energetic stress in wild female chimpanzees. *Hormones and Behavior* 58: 440–449.

Emery Thompson, M., and R. W. Wrangham. 2008a. Diet and reproductive function in wild female chimpanzees *(Pan troglodytes schweinfurthii)* at Kibale National Park, Uganda. *American Journal of Physical Anthropology* 135: 171–181.

Emery Thompson, M., and R. W. Wrangham. 2008b. Male mating interest varies with female fecundity in *Pan troglodytes schweinfurthii* of Kanyawara, Kibale National Park. *International Journal of Primatology* 29: 885–905.

Eveleth, P. B., and J. M. Tanner. 1976. *Worldwide Variation in Human Growth*. Cambridge: Cambridge University Press.

Faiman, C., F. I. Reyes, J. S. D. Winter, and W. C. Hobson. 1981. Endocrinology of pregnancy in the great apes. In C. E. Graham, ed., *Reproductive Biology of the Great Apes: Comparative and Biomedical Perspectives,* 45–68. New York: Academic Press.

Fernandez, B., J. L. Malde, A. Montero, and D. Acuña. 1990. Relationship between adenohypophyseal and steroid hormones and variations in serum and urinary melatonin levels during the ovarian cycle, perimenopause and menopause in healthy women. *Journal of Steroid Biochemistry* 35: 257–262.

Fessler, D. M. T., C. D. Navarrette, W. Hopkins, and M. K. Izard. 2005. Examining the terminal investment hypothesis in humans and chimpanzees: Associations among maternal age, parity, and birth weight. *American Journal of Physical Anthropology* 127: 95–104.

Fortune, J. E. 1994. Ovarian follicular growth and development in mammals. *Biology of Reproduction* 50: 225–232.

Foster, A., J. Menken, A. Chowdhury, and J. Trussell. 1986. Female reproductive development: A hazards model analysis. *Social Biology* 33: 183–198.

Fouts, H. N., and R. A. Brookshire. 2009. Who feeds children? A child's-eye-view of caregiver feeding patterns among the Aka foragers in Congo. *Social Science & Medicine* 69: 285–292.

Frank, G. R. 1995. The role of estrogen in pubertal skeletal physiology: Epiphyseal maturation and mineralization of the skeleton. *Acta Paediatrica* 84: 627–630.

Frisch, R. E., and R. Revelle. 1970. Height and weight at menarche and a hypothesis of critical body weights and adolescent events. *Science* 169: 397–398.

Fritz, M. A., P. K. Westfahl, and R. L. Graham. 1987. The effect of luteal phase estrogen antagonism on endometrial development and luteal function in women. *Journal of Clinical Endocrinology & Metabolism* 65: 1006–1013.

Gadgil, M., and W. H. Bossert. 1970. Life historical consequences of natural selection. *American Naturalist* 104: 1–24.

Galdikas, B. M. F. 1981. Orangutan reproduction in the wild. In C. E. Graham, ed., *Reproductive Biology of the Great Apes: Comparative and Biomedical Perspectives,* 281–300. New York: Academic Press.

Gangestad, S. W., and R. Thornhill. 2008. Human oestrus. *Proceedings of the Royal Society of London B* 282: 991–1000.

Gold, E. B. 2001. Factors associated with age at natural menopause in a multiethnic sample of midlife women. *American Journal of Epidemiology* 153: 865–874.

Goodman, M. 1961. The role of immunochemical differences in the phyletic development of human behavior. *Human Biology* 33: 131–162.

Goodman, M. J., A. Estioko-Griffin, P. B. Griffin, and J. S. Grove. 1985. Menarche, pregnancy, birth spacing and menopause among the Agta women foragers of Cagayan Province, Luzon, the Philippines. *Annals of Human Biology* 12: 169–177.

Gosden, R., E. Kim, B. Lee, K. Manova, and M. Faddy. 2010. Mammalian oocyte population throughout life. In M.-H. Verlhac and A. Villeneuve, eds., *Oogenesis: The Universal Process,* 387–401. Hoboken, NJ: John Wiley & Sons.

Graham, C. E. 1979. Reproductive function in aged female chimpanzees. *American Journal of Physical Anthropology* 50: 291–300.

Graham, C. E. 1981. Menstrual cycle of the great apes. In C. E. Graham, ed., *Reproductive Biology of the Great Apes: Comparative and Biomedical Perspectives,* 1–43. New York: Academic Press.

Graham, C. E., E. J. Struthers, W. C. Hobson, T. McDonald, C. Faiman, M. T. Buckman, and D. C. Collins. 1991. Postpartum infertility in common chimpanzees. *American Journal of Primatology* 24: 245–255.

Gray, R., O. M. Campbell, R. Apelo, S. S. Eslami, H. Zacur, R. M. Ramos, J. C. Gehret, and M. H. Labbock. 1990. Risk of ovulation during lactation. *Lancet* 11: 25–29.

Gray, S. J. 1995. Correlates of breastfeeding frequency among nomadic pastoralists of Turkana, Kenya: A retrospective study. *American Journal of Physical Anthropology* 98: 239–255.

Grebe, N. M., S. W. Gangestad, C. E. Garver-Apgar, and H. Thornhill. 2013. Women's luteal-phase sexual proceptivity and the functions of extended sexuality. *Psychological Science* 24: 2106–2110.

Greisen, S., T. Ledet, and P. Ovesen. 2001. Effects of androstenedione, insulin and luteinizing hormone on steroidogenesis in human granulosa luteal cells. *Human Reproduction* 16: 2061–2065.

Gross, B. A., and C. J. Eastman. 1979. Prolactin secretion during prolonged lactational amenorrhoea. *Australian and New Zealand Journal of Obstetrics & Gynaecology* 19: 95–99.

Gurven, M., J. Stieglitz, P. L. Hooper, C. Gomes, and H. Kaplan. 2012. From the womb to the tomb: The role of transfers in shaping the evolved human life history. *Experimental Gerontology* 47: 807–813.

Hamada, Y., K. Chatani, T. Udono, Y. Kikuchi, and H. Gunji. 2003. A longitudinal study on hand and wrist skeletal maturation in chimpanzees *(Pan troglodytes),* with emphasis on growth in linear dimensions. *Primates* 44: 259–271.

Hamada, Y., T. Udono, M. Teramoto, and I. Hayasaka, I. 1998. Development of the hand and wrist bones in chimpanzees. *Primates* 39, 157–169.

Hamada, Y., T. Udono, M. Teramoto, and T. Sugawara. 1996. The growth pattern of chimpanzees: Somatic growth and reproductive maturation in *Pan troglodytes*. *Primates* 37: 279–295.

Harcourt, A. H. 1995. Sexual selection and sperm competition in primates: What are male genitalia good for? *Evolutionary Anthropology* 4: 121–129.

Haselton, M. G., and K. Gildersleeve. 2011. Can men detect ovulation? *Current Directions in Psychological Science* 20: 87–92.

Hawkes, K., J. T. F. O'Connell, and N. G. Blurton Jones. 1997. Hadza women's time allocation, offspring provisioning, and the evolution of long postmenopausal life spans. *Current Anthropology* 38: 551–577.

Hawkes, K., J. T. F. O'Connell, N. G. Blurton Jones, H. Alvarez, and E. L. Charnov. 1998. Grandmothering, menopause, and the evolution of human life histories. *Proceedings of the National Academy of Sciences* 95: 1336–1339.

Hawkes, K., and K. R. Smith. 2010. Do women stop early? Similarities in fertility decline in humans and chimpanzees. *Annals of the New York Academy of Sciences* 1204: 43–53.

He, C., P. Kraft, C. Chen, J. E. Buring, G. Paré, S. E. Hankinson, S. J. Chanock, P. M. Ridker, D. J. Hunter, and D. I. Chasman. 2009. Genome-wide association studies identify loci associated with age at menarche and age at natural menopause. *Nature Genetics* 41: 724–728.

Henrich, J., S. J. Heine, and A. Norenzayan. 2010. The weirdest people in the world? *Behavioral and Brain Sciences* 33: 61–83.

Herndon, J. G., J. Paredes, M. E. Wilson, M. A. Bloomsmith, L. Chennareddi, and M. L. Walker. 2012. Menopause occurs late in life in the captive chimpanzee (*Pan troglodytes*). *Age* 34: 1145–1156.

Hill, K. R., C. Boesch, J. Goodall, A. E. Pusey, J. M. Williams, and R. W. Wrangham. 2001. Mortality rates among wild chimpanzees. *Journal of Human Evolution* 40: 437–450.

Hill, K. R., and A. M. Hurtado. 1996. *Ache Life History*. New York: Aldine de Gruyter.

Hill, K. R., and A. M. Hurtado. 2009. Cooperative breeding in South American hunter-gatherers. *Proceedings of the Royal Society of London B* 276: 3863–3870.

Hladik, C. M. 1977. Chimpanzees of Gabon and chimpanzees of Gombe: some comparative data on the diet. In T. H. Clutton-Brock, ed., *Primate Ecology: Studies of Feeding and Ranging Behaviour in Lemurs, Monkeys, and Apes*, 481–501. London: Academic Press.

Hobson, W. C., F. Coulston, C. Faiman, J. S. D. Winter, and F. I. Reyes. 1976. Reproductive endocrinology of female chimpanzees: a suitable model of humans. *Journal of Toxicology & Environmental Health* 1: 657–668.

Holman, D. J. 1996. Total fecundability and fetal loss in Bangladesh. Ph.D. dissertation, Pennsylvania State University.

Hook, E. B. 1981. Rates of chromosomal abnormalities at different maternal ages. *Obstetrics & Gynecology* 58: 282–285.

Hooper, P. L., M. Gurven, J. Winking, and H. S. Kaplan. 2015. Inclusive fitness and differential productivity across the life course determine intergenerational

transfers in a small-scale human society. *Proceedings of the Royal Society of London B* 282: 20142808.
Howell, N. 1979. *Demography of the Dobe !Kung.* New York: Academic Press.
Howland, B. E., C. Faiman, and T. M. Butler. 1971. Serum levels of FSH and LH during the menstrual cycle of the chimpanzee. *Biology of Reproduction* 4: 101–105.
Hrdy, S. B. 1979. Infanticide among animals: A review, classification, and examination of the implications for the reproductive strategies of females. *Ethology and Sociobiology* 1: 13–40.
Hrdy, S. B. 1981. *The Woman that Never Evolved.* Cambridge, MA: Harvard University Press.
Hurtado, A. M., K. R. Hill, H. S. Kaplan, and I. Hurtado. 1992. Trade-offs between female food acquisition and child care among Hiwi and Ache Foragers. *Human Nature* 3: 185–216.
Jane Goodall Institute. [No title.] http://www.janegoodall.ca/chimps-we-know-f-family.php.
Jasienska, G., and P. T. Ellison. 1998. Physical work causes suppression of ovarian function in women. *Proceedings of the Royal Society of London B* 265: 1847–1851.
Jeune, B. 1986. Parity and age at menopause in a Danish sample. *Maturitas* 8: 359–365.
Jones, J. H., M. L. Wilson, C. M. Murray, and A. Pusey. 2010. Phenotypic quality influences fertility in Gombe chimpanzees. *Journal of Animal Ecology* 79: 1262–1269.
Jones, K. P., L. C. Walker, D. Anderson, A. Lacreuse, S. L. Robson, and K. Hawkes. 2007. Depletion of ovarian follicles with age in chimpanzees: Similarities to humans. *Biology of Reproduction* 77: 247–251.
Jukic, A. M., D. D. Baird, C. R. Weinberg, D. R. McConnaughey, and A. J. Wilcox. 2013. Length of human pregnancy and contributors to its natural variation. *Human Reproduction* 28: 2848–2855.
Kaplan, H. S., K. R. Hill, J. Lancaster, and A. M. Hurtado. 2000. A theory of human life history evolution: Diet, intelligence, and longevity. *Evolutionary Anthropology* 9: 156–185.
Klaus, M. H., J. H. Kennell, S. S. Robertson, and R. Sosa. 1986. Effects of social support during parturition on maternal and infant morbidity. *British Medical Journal* 293: 585–587.
Knott, C. D., M. Emery Thompson, and S. A. Wich. 2009. The ecology of reproduction in wild orangutans. In S. A. Wich, S. S. Utami, T. Mitra Setia, and C. P. van Schaik, eds., *Orangutans: Ecology, Evolution, Behaviour, and Conservation,* 171–188. Oxford: Oxford University Press.
Kondo, M., T. Udono, W. Z. Jin, M. Funakoshi, K. Shimizu, M. Itoh, C. B. Herath, G. Watanabe, N. P. Groome, and K. Taya. 2001. Secretion of inhibin A and inhibin B throughout pregnancy and the early postpartum period in chimpanzees. *Journal of Endocrinology* 168: 257–262.
Konner, M. 2005. Hunter-gatherer infancy and childhood: The !Kung and others. In B. S. Hewlett and M. E. Lamb, eds., *Hunter-Gatherer Childhoods: Evolutionary, Developmental and Cultural Perspectives,* 19–64. New Brunswick, NJ: Transaction.

Konner, M. J., and C. M. Worthman. 1980. Nursing frequency, gonadal function, and birth spacing among !Kung hunter-gatherers. *Science* 207: 788–791.

Kramer, K. L. 2005. *Maya Children: Helpers at the Farm.* Cambridge, MA: Harvard University Press.

Kramer, K. L., and P. T. Ellison. 2010. Pooled energy budgets: Resituating human energy allocation trade-offs. *Evolutionary Anthropology* 19: 136–147.

Lacreuse, A., L. Chennareddi, K. G. Gould, K. Hawkes, S. R. Wijayawardana, J. Chen, K. A. Easley, and J. G. Herndon. 2008. Menstrual cycles continue into advanced old age in the common chimpanzee *(Pan troglodytes). Biology of Reproduction* 79: 407–412.

Lancaster, J. B., and H. S. Kaplan. 2009. The endocrinology of the human adaptive complex. In P. T. Ellison and P. G. Gray, eds., *Endocrinology of Social Relationships,* 95–119. Cambridge, MA: Harvard University Press.

Lee, R. B. 1979. *The !Kung San: Men, Women, and Work in a Foraging Society.* Cambridge: Cambridge University Press.

Lee, R. B. 1980. Lactation, ovulation, infanticide, and women's work: A study of hunter-gatherer population regulation. In M. N. Cohen, R. S. S. Malpass, and H. G. Klein, eds., *Biosocial Mechanisms in Population Regulation,* 321–348. New Haven, CT: Yale University Press.

Lenton, E. A., S. Sexton, S. Lee, and I. D. Cooke. 1988. Progressive changes in LH and FSH and LH: FSH ratio in women throughout reproductive life. *Maturitas* 10: 35–43.

Lipson, S. F., and P. T. Ellison. 1996. Comparison of salivary steroid profiles in naturally occurring conception and non-conception cycles. *Human Reproduction* 11: 2090–2096.

Littleton, J. 2005. Fifty years of chimpanzee demography at Taronga Park Zoo. *American Journal of Primatology* 67: 281–298.

Lovejoy, C. O. 1981. The origin of man. *Science* 211: 341–350.

Lubchenco, L. O., C. Hansman, and E. Boyd. 1966. Intrauterine growth in length and head circumference as estimated from live births at gestational ages from 26 to 42 weeks. *Pediatrics* 37: 403–408.

Lunn, P. G., S. Austin, A. M. Prentice, and R. G. Whitehead. 1980. Influence of maternal diet on plasma prolactin levels during lactation. *Lancet* 315: 623–625.

Maas, S., H. Jarry, A. Teichmann, W. Rath, W. Kuhn, and W. Wuttke. 1992. Paracrine actions of oxytocin, prostaglandin F2 alpha, and estradiol within the human corpus luteum. *Journal of Clinical Endocrinology & Metabolism* 74: 306–312.

MacDorman, M. F., S. E. Kirmeyer, and E. C. Wilson. 2012. Fetal and perinatal mortality, United States, 2006. *National Vital Statistics Reports* 60: 1–22.

Marlowe, F. 2003. A critical period for provisioning by Hadza men: Implications for pair bonding. *Evolution and Human Behavior* 24: 217–229.

Marshall, W. A. 1974. Interrelationships of skeletal maturation and sexual development and somatic growth in man. *Annals of Human Biology* 1: 29–40.

Marshall, W. A., and Y. de Limongi. 1976. Skeletal maturity and the prediction of age at menarche. *Annals of Human Biology* 3: 235–243.

Martin, D. E. 1981. Breeding great apes in captivity. In C. E. Graham, ed., *Reproductive Biology of the Great Apes: Comparative and Biomedical Perspectives*, 343–373. New York: Academic Press.

Martin, D. E., C. E. Graham, and K. G. Gould. 1978. Successful artificial insemination in the chimpanzee. *Symposia of the Zoological Society of London* 43: 249–260.

Matsumoto-Oda, A., R. Oda, Y. Hayashi, H. Murakami, N. Maeda, K. Kumazaki, K. Shimizu, and T. Matsuzawa. 2002. Vaginal fatty acids produced by chimpanzees during menstrual cycles. *Folia Primatologica* 74: 75–79.

McFalls, J. A., and M. H. McFalls. 1984. *Disease and Fertility.* New York: Academic Press.

McLean, M., and R. Smith. 2001. Corticotrophin-releasing hormone and human parturition. *Reproduction* 121: 493–501.

McNeilly, A. S. 1993. Lactational amenorrhea. *Endocrinology Metabolism Clinics of North America* 22: 59–73.

McNeilly, A. S. 2001. Lactational control of reproduction. *Reproduction, Fertility and Development* 13: 583–590.

McNeilly, A. S., C. C. K. Tay, and A. Glasier. 1994. Physiological mechanisms underlying lactational amenorrhea. *Annals of the New York Academy of Sciences* 709: 145–155.

Meehan, C. L., R. Quinlan, and C. D. Malcom. 2013. Cooperative breeding and maternal energy expenditure among Aka foragers. *American Journal of Human Biology* 25: 42–57.

Meehan, C. L., and J. W. Roulette. 2013. Early supplementary feeding among Central African foragers and farmers: a biocultural approach. *Social Science & Medicine* 96: 112–120.

Miller, E. M. 2014. Chronic undernutrition and traditional weaning foods are associated with fat stores in Ariaal infants of Northern Kenya. *American Journal of Physical Anthropology* 153: 286–296.

Montagu, M. F. A. 1957. *The Reproductive Development of the Female.* New York: Julian Press.

Muller, M. N., S. M. Kahlenberg, M. Emery Thompson, and R. W. Wrangham. 2007. Male coercion and the costs of promiscuous mating for female chimpanzees. *Proceedings of the Royal Society of London B* 274: 1009–1014.

Muller, M. N., S. M. Kahlenberg, and R. W. Wrangham. 2009. Male aggression against females and sexual coercion in chimpanzees. In M. N. Muller and R. W. Wrangham, eds., *Sexual Coercion in Primates: An Evolutionary Perspective on Male Aggression against Females*, 184–217. Cambridge, MA: Harvard University Press.

Nadler, R. D., C. E. Graham, D. C. Collins, and K. G. Gould. 1979. Plasma gonadotropins, prolactin, gonadal steroids and genital swelling during the menstrual cycle of lowland gorillas. *Endocrinology* 105: 290–296.

Nadler, R. D., C. E. Graham, D. C. Collins, and O. R. Kling. 1981. Postpartum amenorrhea and behavior of apes. In C. E. Graham, ed., *Reproductive Biology of the Great Apes: Comparative and Biomedical Perspectives,* 69–82. New York: Academic Press.

Nadler, R. D., C. E. Graham, R. E. Gosselin, and D. C. Collins. 1985. Serum levels of gonadotropins and gonadal steroids, including testosterone, during the menstrual cycles of the chimpanzee *(Pan troglodytes). American Journal of Primatology* 9: 273–284.

Newton-Fisher, N. E. 1999. The diet of chimpanzees in the Budongo Forest Reserve, Uganda. *African Journal of Ecology* 37: 344–354.

Nishida, T., N. Corp, M. Hamai, T. Hasegawa, M. Hiraiwa-Hasegawa, K. Hosaka, K. D. Hunt, N. Itoh, K. Kawanaka, A. Matsumoto-Oda, J. C. Mitani, N. Nakamura, K. Norikoshi, T. Sakamaki, L. Turner, S. Uehara, and Z. Zamma. 2003. Demography, female life history, and reproductive profiles among the chimpanzees of Mahale. *American Journal of Primatology* 59: 99–121.

Nissen, H. W., and R. M. Yerkes. 1943. Reproduction in the chimpanzee: Report on forty-nine births. *Anatomical Record* 86: 567–578.

Nunn, C. L. 1999. The evolution of exaggerated sexual swellings in primates and the graded-signal hypothesis. *Animal Behaviour* 58: 229–246.

Palombit, R. A. 2012. Infanticide: Male strategies and female counterstrategies. In J. C. Mitani, J. Call, P. M. Kappeler, R. A. Palombit, and J. B. Silk, eds., *The Evolution of Primate Societies,* 432–468. Chicago: University of Chicago Press.

Pandolfi, S. S. 2004. Ecological sex differences in gombe chimpanzees *(Pan troglodytes).* Ph.D. dissertation, Duke University, Durham, NC.

Panter-Brick, C., D. S. Lotstein, and P. T. Ellison. 1993. Seasonality of reproductive function and weight loss in rural Nepali women. *Human Reproduction* 8: 684–690.

Pellestor, F., B. Andréo, F. Arnal, C. Humeau, and J. Demaille. 2003. Maternal aging and chromosomal abnormalities: New data drawn from *in vitro* unfertilized human oocytes. *Human Genetics* 112: 195–203.

Pennington, R. 2001. Hunter-gatherer demography. In C. Panter-Brick, R. H. Layton, and P. Rowley-Conwy, eds., *Hunter-Gatherers: An Interdisciplinary Perspective,* 170–204. Cambridge: Cambridge University Press.

Pijnenborg, R., L. Vercruysse, and A. M. Carter. 2011. Deep trophoblast invasion and spiral artery remodelling in the placental bed of the chimpanzee. *Placenta* 32: 400–408.

Prentice, A. M., and A. Prentice. 1988. Energy costs of lactation. *Annual Review of Nutrition* 8: 63–79.

Pusey, A. E., J. M. Williams, and J. Goodall. 1997. The influence of dominance rank on the reproductive success of female chimpanzees. *Science* 277: 828–831.

Reichert, K. E., M. Heistermann, J. K. Hodges, C. Boesch, and G. Hohmann. 2002. What females tell males about their reproductive status: Are morphological and behavioural cues reliable signals of ovulation in bonobos *(Pan paniscus)? Ethology* 108: 583–600.

Reiches, M. W., P. T. Ellison, S. F. Lipson, K. C. Sharrock, E. Gardiner, and L. G. Duncan. 2009. Pooled energy budget and human life history. *American Journal of Human Biology* 21: 421–429.

Reyes, F. I., J. S. D. Winter, C. Faiman, and W. C. Hobson. 1975. Serial serum levels of gonadotrophins, prolactin, and sex steroids in the non-pregnant and pregnant chimpanzee. *Endocrinology* 96: 1447–1455.

Roberts, S. C., J. Havlicek, J. Flegr, M. Hruskova, A. C. Little, B. C. Jones, D. I. Perrett, and M. Petrie. 2004. Female facial attractiveness increases during the fertile phase of the menstrual cycle. *Proceedings of the Royal Society of London B* 271: 270–272.

Rockwell, L. C., E. Vargas, and L. G. Moore. 2003. Human physiological adaptation to pregnancy: Inter- and intraspecific perspectives. *American Journal of Human Biology* 15: 330–341.

Rodriguez-Gironés, M. A., and M. Enquist. 2001. The evolution of female sexuality. *Animal Behaviour* 61: 695–704.

Roff, D. A. 1992. *Evolution of Life Histories: Theory and Analysis*. New York: Chapman & Hall.

Roney, J. R., and Z. L. Simmons. 2013. Hormonal predictors of sexual motivation in natural menstrual cycles. *Hormones and Behavior* 63: 636–645.

Roof, K. A., W. D. Hopkins, M. K. Izard, M. Hook, and S. J. Schapiro. 2005. Maternal age, parity, and reproductive outcome in captive chimpanzees *(Pan troglodytes)*. *American Journal of Primatology* 67: 199–207.

Rosenberg, K., and W. Trevathan. 1995. Bipedalism and human birth: the obstetrical dilemma revisited. *Evolutionary Anthropology* 4: 161–168.

Rosenberg, K., and W. R. Trevathan. 2002. Birth, obstetrics and human evolution. *BJOG: An International Journal of Obstetrics & Gynaecology* 109: 1199–1206.

Ross, C. 1998. Primate life histories. *Evolutionary Anthropology* 6: 54–63.

Ross, G. T., C. M. Cargille, M. B. Lipsett, P. L. Rayford, J. R. Marshall, C. A. Strott, and D. Rodbard. 1970. Pituitary and gonadal hormones in women during spontaneous and induced ovulatory cycles. *Recent Progress in Hormone Research* 26: 1–62.

Ruangvutilert, P., V. Titapant, and V. Kerdphoo. 2002. Placental ratio and fetal growth pattern. *Journal of the Medical Association of Thailand* 85: 488–495.

Ruff, C. B. 1995. Biomechanics of the hip and birth in early *Homo*. *American Journal of Physical Anthropology* 98: 527–574.

Schultz, A. H. 1969. *The Life of Primates*. New York: Universe.

Scott, E. C., and F. E. Johnston. 1982. Critical fat, menarche, and the maintenance of menstrual cycles: A critical review. *Journal of Adolescent Health Care* 2: 249–260.

Scott, E. C., and F. E. Johnston. 1985. Science, nutrition, fat, and policy: Tests of the critical-fat hypothesis. *Current Anthropology* 2: 463–473.

Sellen, D. W., and D. B. Smay. 2001. Relationship between subsistence and age at weaning in "preindustrial" societies. *Human Nature* 12: 47–87.

Shideler, S. E., and B. L. Lasley. 1982. A comparison of primate ovarian cycles. *American Journal of Primatology* 3: 171–180.

Shimizu, K., C. Douke, S. Fujita, T. Matsuzawa, M. Tomonaga, M. Tanaka, K. Matsubayashi, and M. Hayashi. 2003. Urinary steroids, FSH and CG measurements for monitoring the ovarian cycle and pregnancy in the chimpanzee. *Journal of Medical Primatology* 32: 15–22.

Sillen-Tullberg, B., and A. P. Møller. 1993. The relationship between concealed ovulation and mating systems in anthropoid primates: A phylogenetic analysis. *American Naturalist* 141: 1–25.

Simmons, K., and W. W. Greulich. 1943. Menarcheal age and the height, weight, and skeletal age of girls age 7 to 17 years. *Journal of Pediatrics* 22: 518–548.

Smith, R., E. C. Chan, M. E. Bowman, W. J. Harewood, and A. F. Phippard. 1993. Corticotropin-releasing hormone in baboon pregnancy. *Journal of Clinical Endocrinology & Metabolism* 76: 1063–1068.

Smith, R. E., E. J. Wickings, M. E. Bowman, A. Belleoud, G. Dubreuil, J. J. Davies, and G. Madsen. 1999. Corticotropin-releasing hormone in chimpanzee and gorilla pregnancies. *Journal of Clinical Endocrinology & Metabolism* 84: 2820–2825.

Smuts, B., and N. Nicolson. 1989. Reproduction in wild female olive baboons. *American Journal of Primatology* 19: 229–246.

Soma, H. 1990. Placental implications for pregnancy complications in the chimpanzee (Pan troglodytes). *Zoo Biology* 9: 141–147.

Sosa, R., J. Kennell, M. Klaus, S. Robertson, and J. Urrutia. 1980. The effect of a supportive companion on perinatal problems, length of labor, and mother-infant interaction. *New England Journal of Medicine* 303: 7–10.

Stanford, J. L., P. Hartge, L. A. Brinton, R. N. Hoover, and R. Brookmeyer. 1987. Factors influencing the age at natural menopause. *Journal of Chronic Diseases* 40: 995–1002.

Stearns, S. C. 1989. Trade-offs in life history evolution. *Functional Ecology* 3: 259–268.

Stearns, S. C. 1992. *Evolution of Life Histories*. Oxford: Oxford University Press.

Stern, J. M., M. Konner, T. N. Herman, and S. Reichlin. 1986. Nursing behaviour, prolactin and postpartum amenorrhea during prolonged lactation in American and !Kung mothers. *Clinical Endocrinology* 25: 247–258.

Stumpf, R. M., M. Emery Thompson, and C. D. Knott. 2008. A comparison of female mating strategies in *Pan troglodytes* and *Pongo* spp. *International Journal of Primatology* 29: 865–884.

Sugiyama Y. 2004. Demographic parameters and life history of chimpanzees at Bossou, Guinea. *American Journal of Physical Anthropology* 124: 154–165.

Tay, C. C. K., A. F. Glasier, and A. S. McNeilly. 1996. Twenty-four hour patterns of prolactin secretion during lactation and the relationship to suckling and the resumption of fertility in breast-feeding women. *Human Reproduction* 11: 950–955.

Thomas, F., F. Renaud, E. Benefice, T. de Meeus, and J.-F. Guegan. 2001. International variability of ages at menarche and menopause: Patterns and main determinants. *Human Biology* 73: 271–290.

Thornhill, R., and S. W. Gangestad. 2008. *The Evolutionary Biology of Human Female Sexuality*. New York: Oxford University Press.

Tietze, C. 1957. Reproductive span and rate of reproduction among Hutterite women. *Obstetrical & Gynecological Survey* 12: 727–728.

Treloar, A. E., R. E. Boynton, B. G. Behn, and B. W. Brown. 1967. Variation of the human menstrual cycle through reproductive life. *International Journal of Fertility* 12: 77–126.

Tyson, J. E. 1977a. Neuroendocrine control of lactational infertility. *Journal of Biosocial Science* 9: 23–39.

Tyson, J. E. 1977b. Nursing and prolactin secretion: Principal determinants in the mediation of puerperal infertility. In P. G. Crosignani and C. Robyn, eds., *Prolactin and Human Reproduction,* 97–108. New York: Academic Press.

Valeggia, C. R., and P. T. Ellison. 2004. Lactational amenorrhoea in well-nourished Toba women of Formosa, Argentina. *Journal of Biosocial Science* 36: 573–595.

Valeggia, C. R., and P. T. Ellison. 2009. Interactions between metabolic and reproductive functions in the resumption of postpartum fecundity. *American Journal of Human Biology* 21: 559–566.

van Asselt, K. M., H. S. Kok, P. L. Pearson, J. S. Dubas, P. H. M. Peeters, E. R. te Velde, and P. A. H. van Noord. 2004. Heritability of menopausal age in mothers and daughters. *Fertility and Sterility* 82: 1348–1351.

van der Rijt-Plooij, H. H. C., and F. X. Plooij. 1987. Growing independence, conflict and learning in mother-infant relations in free-ranging chimpanzees. *Behaviour* 101: 1–86.

van Noord, P. A. H., J. S. Dubas, M. Dorland, H. Boersma, and E. te Velde. 1997. Age at natural menopause in a population-based screening cohort: the role of menarche, fecundity, and lifestyle factors. *Fertility and Sterility* 68: 95–102.

Videan, E. N., J. Fritz, C. Heward, and J. Murphy. 2006. The effects of aging on hormone and reproductive cycles in female chimpanzees *(Pan troglodytes)*. *Comparative Medicine* 56: 291–299.

Wallis, J. 1997. A survey of reproductive parameters in the free-ranging chimpanzees of Gombe National Park. *Reproduction* 109: 297–307.

Washburn, S. L. 1960. Tools and human evolution. *Scientific American* 203: 63–75.

Watts, D. P. 2007. Effects of male group size, parity, and cycle stage on female chimpanzee copulation rates at Ngogo, Kibale National Park, Uganda. *Primates* 48: 222–231.

Whelan, E. A., D. P. Sandler, D. R. McConnaughey, and C. R. Weinberg. 1990. Menstrual and reproductive characteristics and age at natural menopause. *American Journal of Epidemiology* 131: 625–632.

Wilcox, A. J. 2001. On the importance—and the unimportance—of birthweight. *International Journal of Epidemiology* 30: 1233–1241.

Wilcox, A. J., D. D. Baird, D. B. Dunson, D. R. McConnaughey, J. S. Kesner, and C. R. Weinberg. 2004. On the frequency of intercourse around ovulation: Evidence for biological influences. *Human Reproduction* 19: 1539–1543.

Wilcox, A. J., C. R. Weinberg, and D. D. Baird. 1995. Timing of sexual intercourse in relation to ovulation: Effects on the probability of conception, survival of the pregnancy, and sex of the baby. *New England Journal of Medicine* 333: 1517–1521.

Wilcox, A. J., C. R. Weinberg, J. F. O'Connor, D. D. Baird, J. P. Schlatterer, R. E. Canfield, E. G. Armstrong, and B. C. Nisula. 1988. Incidence of early loss of pregnancy. *New England Journal of Medicine* 319: 189–194.

Williams, J. M., G. W. Oehlert, J. V. Carlis, and A. E. Pusey. 2004. Why do male chimpanzees defend a group range? *Animal Behaviour* 68: 523–532.

Willis, D., H. Mason, C. Gilling-Smith, and S. Franks. 1996. Modulation by insulin of follicle-stimulating hormone and luteinizing hormone actions in human granulosa cells of normal and polycystic ovaries. *Journal of Endocrinology & Metabolism* 81: 302–309.

Wilson, M. L., M. Wells, S. M. Kahlenberg, and R. W. Wrangham. 2010. Food resources affect the timing of intercommunity interactions in the Kanyawara community of chimpanzees, Kibale National Park, Uganda. *American Journal of Physical Anthropology* 141: 246.

Wilson, M. L., and R. W. Wrangham. 2003. Intergroup relations in chimpanzees. *Annual Review of Anthropology* 32: 363–392.

Wittman, A. B., and L. L. Wall. 2007. The evolutionary origins of obstructed labor: Bipedalism, encephalization, and the human obstetric dilemma. *Obstetrical & Gynecological Survey* 62: 739–748.

Wolff, J. O., and D. W. Macdonald. 2004. Promiscuous females protect their offspring. *Trends in Ecology and Evolution* 19: 127–134.

Wood, J. W. 1994. *Dynamics of Human Reproduction: Biology, Biometry, Demography*. New York: Aldine de Gruyter.

Woods, R. 2008. Long-term trends in fetal mortality: Implications for developing countries. *Bulletin of the World Health Organization* 86: 460–466.

Wrangham, R. W. 2002. The cost of sexual attraction: Is there a trade-off in female *Pan* between sex appeal and received coercion? In C. Boesch, G. Hohmann, and L. F. Marchant, eds., *Behavioural Diversity in Chimpanzees and Bonobos*, 204–215. Cambridge: Cambridge University Press.

Wrangham, R. W., and B. Smuts. 1980. Sex differences in the behavioral ecology of chimpanzees in the Gombe National Park, Tanzania. *Journal of Reproduction and Fertility Supplement* 28: 13–31.

Zhu, B. P. 2005. Effect of interpregnancy interval on birth outcomes: findings from three recent US studies. *International Journal of Gynecology and Obstetrics* 89: S25–S33.

Zuckerman, S. 1937. The duration and phases of the menstrual cycle in primates. *Proceedings of the Zoological Society of London* A107: 315–329.

7

Locomotor Ecology and Evolution in Chimpanzees and Humans

HERMAN PONTZER

Few traits distinguish humans and chimpanzees as readily as locomotion. Humans are habitual, terrestrial bipeds, while chimpanzees primarily use a quadrupedal knuckle-walking gait when on the ground, and frequently climb and scramble in the trees. Both locomotor modes are rare. Bipedalism has evolved in only a handful of vertebrate groups, most notably the theropod dinosaurs and birds, and in very few mammalian orders (see Russo and Kirk 2013). Even among bipeds, human gait characteristics are unique (Gatesy and Biewener 1991; Roberts et al. 1998). Knuckle-walking is even less common than bipedalism, having evolved in only a handful of large bodied, semi-arboreal mammals like gorillas and giant anteaters (see Orr 2005). In this chapter I examine the common origins and divergent evolutionary trajectories of these two uncommon locomotor strategies.

Darwin, in the *Descent of Man*, noting the many anatomical similarities between humans and African apes, correctly deduced that our lineage began in Africa (Darwin 1871: 133). Humans and chimpanzees share a common, and evolutionarily conservative, primate bauplan. The radius and ulna, and the tibia and fibula, remain separated, the hands and feet retain five grasping rays (though these are less effective in the human foot), the number of thoracolumbar vertebral elements is similar (~17; Pilbeam 2004), and so on.

Humans, chimpanzees, and other extant apes are further united in the absence of a tail, which was lost among stem hominoids in the early Miocene (Nakatsukasa et al. 2004). Humans and chimpanzees also share a common set of postcranial muscles (Thorpe et al. 1999; Payne et al. 2006), and while there has been some suggestion that muscle tensile properties (Scholz et al. 2006; Bozek et al. 2014) or enervation (Walker 2006) might differ among *Homo* and *Pan,* vertebrate muscle is generally an evolutionarily conservative tissue (Biewener 2003).

Nonetheless, nearly every element of the musculoskeletal system has been reshaped by the divergent locomotor evolution of these two lineages (Figure 7.1). Humans exhibit the more derived musculoskeletal anatomy; major skeletal features associated with bipedalism are indicated in Figure 7.2. The most dramatic changes are evident in the pelvis (see Grabowksi et al. 2011): the iliac blades are shortened and wrap laterally and anteriorly, forming a bowl-shaped structure that supports the abdominal viscera and places the lesser gluteals (gluteus medius and minimus) in position to stabilize the trunk during single-leg support. The "S" curve of the human spine, with lumbar and cervical loridoses and thoracic kyphosis, positions the center of mass of the torso directly above the pelvis, aiding in balance and weight support. The vertical orientation of the spine is also reflected in the anterior placement of the foramen magnum. Humans have longer hind limbs than chimpanzees and other apes, in relation both to forelimb length and to body mass (Pontzer 2012). Human feet exhibit an adducted, non-grasping hallux and a springy but stiff midfoot with a plantar arch. Finally, our toes are short, further abrogating any grasping ability and reflecting instead the need to push off effectively with each step.

Chimpanzees share a suite of postcranial features with bonobos, gorillas, and orangutans, the other large-bodied nonhuman apes. Like other extant hominoids, chimpanzees exhibit a high intermembral index, with long arms and short legs. Their hands, and especially their fingers, are long, with curved phalanges, implicated in grasping and suspension. Similarly, the chimpanzee shoulder exhibits a cranially oriented glenoid fossa, reflecting the common use of overhead branches for support or safety (i.e., in case the supporting limb below fails) while in the trees (Hunt 1991). As in gorillas, the chimpanzee wrist does not permit extension beyond ~180 degrees, which is thought to be an adaptation for weight support during knuckle-walking. The columnar arrangement of the wrist is evident in the radiocarpal joint, in the scaphoid

LOCOMOTOR ECOLOGY AND EVOLUTION   261

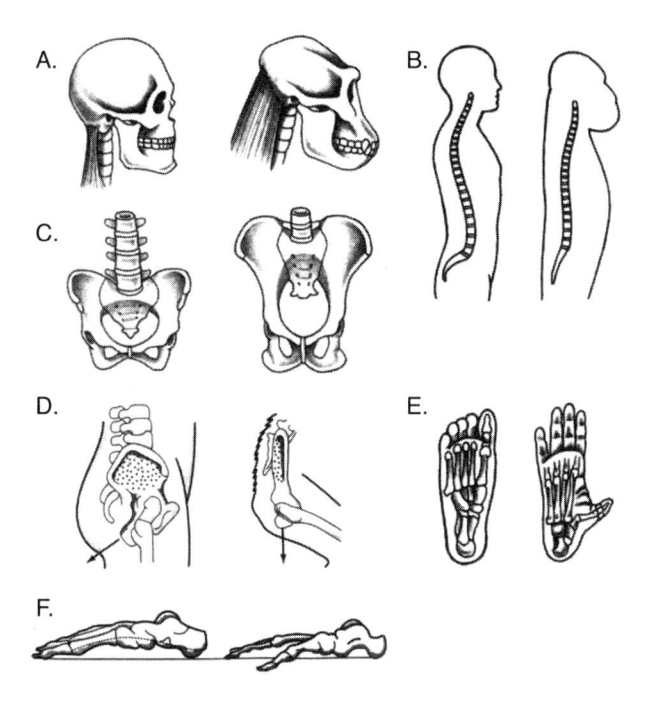

FIGURE 7.1. Overview of post-cranial differences between humans (left image in each pair) and chimpanzees. (a) The foramen magnum is positioned more anteriorly in humans, reflecting vertical orientation of the trunk and spine. (b) The spinal column in humans has a pronounced S-shape, with a thoracic kyphosis and lumbar loridosis. The lumbar vertebrae are also enlarged in humans, reflecting their role in weight bearing. (c and d) The human pelvis is bowl-shaped, with short, broad iliac blades that wrap transversely around the trunk. This places the origin of the lesser gluteal muscles (dotted area) in position to stabilize the trunk during single leg stance. The ischium in chimpanzees is long and projects caudally (black arrows), reflecting their crouched, bent-hip gait. (e) The hallux (first toe) of the human foot is enlarged and in line with the other rays, while chimpanzees and other primates have a grasping foot. (f) The human foot is relatively stiff, with a spring-like plantar arch (dotted line); the chimpanzee foot is compliant and lacks an arch. From figure 8.6 in Conroy and Pontzer (2012), with permission from W. W. Norton.

ridge that limits extension (Dainton and Macho 1999; Orr 2005; Kivell and Schmitt 2009). The metacarpal-phalangeal joint also bears signatures of knuckle-walking and stability; chimpanzees support their weight on the distal ends of the first phalanges of rays two through five. Spine and pelvis reflect vertical climbing and arboreal locomotion. Like other suspensory apes, chimpanzees are relatively orthograde, particularly in the canopy,

FIGURE 7.2. Human foragers regularly exploit arboreal resources, but face considerable risks from falling and often rely on technological solutions to gain access to the canopy. Here, Hadza men are seen using pegs to ascend a baobab tree (top left; photo: Martin Muller), and chopping into a smaller tree (ascended without pegs) to extract honey (bottom left; photo: Herman Pontzer). Chimpanzees move more adroitly in the canopy, with grasping feet enabling them to ascend safely (top right; photo: Ronan Donovan) and long, powerful arms and fingers providing dependable support in suspension (bottom right; photo: Ronan Donovan).

where bipedal, vertical trunk postures are common (Hunt 1991; Stanford 2006). The lumbar spine is short and functionally immobile, entrapped by the tall iliac blades and lower rib cage (Lovejoy et al. 2009b). The pelvis has a long, dorsally projecting ischium, which places the hamstrings in position for powerful extension of the hip during vertical climbing.

Locomotor Ecology and Performance

To make sense of these anatomical differences, we must look to the locomotor ecology and performance of humans and chimpanzees. Here, I will draw from studies of hunter-gatherer populations for assessments of human locomotor ecology, as living foragers are the best available ecological models for Pleistocene humans (Marlowe 2005). Physiological data on locomotor efficiency are drawn from lab studies. Comparisons with chimpanzees indicate that humans travel farther, faster, and more economically on the ground, but have also lost a significant measure of arboreal capability.

On the Ground

Humans and chimpanzees travel farther each day than other apes, and indeed farther than most other primates (Table 7.1). In both species, males range farther each day, on average, than females. Among human foragers, men typically forage for animal foods, while and women forage for plant foods, and men's mean day ranges (14.1 km/d) are roughly 50 percent greater than women's (9.5 km/d) (Marlowe 2005). In chimpanzees, males travel farther (3.6 km/d) than females (3.0 km/d) because males tend to forage in larger groups and engage more frequently in border patrols (Wrangham 2000). Home range size is somewhat dependent on the local ecology (Table 7.2), but is roughly an order of magnitude larger among human foragers compared to chimpanzees.

Few studies have examined walking speeds, but the available data indicate that humans also travel faster than chimpanzees. GPS data from Hadza hunter-gatherers indicate a mean walking speed of 3.8±0.3 km/hr for women and 4.4±0.6 km/hr for men (Pontzer et al. 2015). This is ~34 percent faster than mean walking speeds reported for chimpanzees in the Kanywara community in Kibale National Park, Uganda (males: 3.2, females: 2.8 km/hr:

Hunt 1989; Table 7.1). Combining these walking speeds and distances with species average day ranges, estimated mean time per day spent walking for human foragers (women: 9.5 km/d ÷ 3.8 km/hr = 2.5 hr/d; men: 14.1 ÷ 4.4 = 3.2 hr/d) is more than twice that of chimpanzees (females: 3.0 ÷ 2.8 = 1.1 hr/d; males: 3.6 ÷ 3.2 = 1.1 hr/d). Notably, human foragers often walk carrying heavy burdens (e.g., food, water, firewood, or infants), which can slow their walking speed considerably. Chimpanzee mothers carry their young as well, but speeds appear to be largely unaffected (Hunt 1989), presumably because these juveniles do not present a particularly large burden. Newborn chimpanzees are always carried, but weigh relatively little (~1.8 kg, about 5 percent of mother's mass), and offspring are carried less and less frequently as they approach weaning (~3 yrs, 8.5 kg, or 25 percent of mother's mass: AnAge database). Chimpanzees rarely carry food or other items long distances. By comparison, it is not uncommon for Hadza women to return to camp carrying ten kilograms of gathered foods along with their infant and other supplies (personal observation).

Ranging farther and faster reflects humans' superior endurance capabilities. Compared to most other mammals, and all other primates, humans have excellent aerobic capacity. They are able to walk or run twenty kilometers or more in a single effort, and can maintain these intense aerobic workloads indefinitely, day after day (Carrier 1984; Bramble and Lieberman 2004; Marlowe 2010). Nothing approaching this level of aerobic output has been reported for chimpanzees in the wild. Further, while no lab studies have expressly focused on ape aerobic abilities, observations of chimpanzees in treadmill studies suggest that their endurance capacities are quite limited compared to humans (Pontzer et al. 2014; Pontzer 2017).

Humans exhibit a suite of derived features that enhance our endurance abilities. Humans are hairless and sweat prodigiously, more than any other mammal, which protects against overheating during long periods of exercise (Carrier 1984; Ruxton and Wilkerson 2011; Lieberman 2015). Humans also carry more muscle mass in our hind limbs than do chimpanzees and other apes (Payne et al. 2006), which improves aerobic capacity and fatigue resistance during walking and running (Weibel et al. 2004). Improved cooling abilities and greater hind limb muscle mass would improve endurance during walking, running, and many other activities. Bramble and Lieberman (2004) have argued further that a number of derived human musculoskeletal features, including our long Achilles tendon, springy plantar arch, de-

TABLE 7.1. Comparison of locomotor ecology and performance for chimpanzees and humans.

Species	Sex	Mass (kg)[a]	Home Range (km²)[b]	Day Range (km/d)[c]	Walking Speed (m/s)[d]	Travel Cost (kCal/km)[e]	Climb (m/day)[f]	Time Arboreal (%24 hr. day)
Chimpanzees	Male	41.7 ±1.6	22	3.6	0.88	44.1	104	65–95[g]
	Female	34.4 ±2.0		3.0	0.77	36.4	99	
Humans	Male	55.2 ±9.2	175	14.1	1.22	Walk: 27.0 Run: 51.3	10–20*	<5*
	Female	47.3 ±6.9		9.5	1.05	Walk: 23.2 Run: 44.0	<5*	<1*

a. Chimpanzees: Carter et al. 2008 (n = 5 communities). Humans: Walker et al. 2006.
b. Chimpanzees: median calculated from data in Table 2, using means for each community. Humans: median for human foragers, from Marlowe 2005.
c. Chimpanzees: data from Table 2, using mean male and female values for Taï and excluding Middle community values. Humans: Marlowe 2005.
d. Chimpanzees: Hunt 1989. Humans: Pontzer et al. 2015 (Hadza foragers).
e. Chimpanzees: Pontzer et al. 2014. Humans: Rubenson et al. 2007.
f. Chimpanzees: Pontzer and Wrangham 2004.
g. Chimpanzees: using daytime values (Takemoto et al. 2004; Wrangham 1977) and assuming twelve hours per day in night nests.
* Humans: gross estimates for Hadza foragers (Brian M. Wood, personal communication).

coupled neck and shoulder girdle, and larger hip, knee, and ankle joints, are specifically related to selection for endurance running.

Human walking is also more economical (i.e., more distance is covered per unit of energy expended) than that of chimpanzees. The energy cost of chimpanzee walking (1.06 kcal/km per kg body mass: Pontzer et al. 2014) is twice that of human walking (0.49: Rubenson et al. 2007). The greater walking economy of humans may be an adaptation to longer day ranges (discussed below), and is primarily a result of humans' longer hind limbs and straight-legged gait. Chimpanzees have shorter hind limbs and use more crouched postures (i.e., more flexed hip and knee, sometimes referred to as a "bent-hip, bent-knee" gait) when walking, which increases the muscle activity needed for locomotion and consequently increases metabolic cost (Sockol et al. 2007; Pontzer et al. 2009). Notably, when humans run, we adopt more flexed hind limb postures (i.e., a more flexed hip and knee), and consequently our running costs (0.93 kcal/km per kg of body mass; Rubenson et al. 2007) are nearly as high as chimpanzee walking costs (Pontzer et al. 2009, 2014). Thus, human walking is substantially more economical than that of chimpanzees, but human running is not (Table 7.1).

In the Trees

It is plain to anyone who has observed chimpanzees in the wild that their climbing and arboreal scrambling abilities are superior to humans' (Figure 7.2). Chimpanzee elbow and wrist flexor muscles are nearly twice as large as humans' (Thorpe et al. 1999), and together with their long arms and long, curved fingers (Figure 7.1), these large forelimb flexor muscles enable powerful grasping and resistance to fatigue during climbing and suspension. The chimpanzee hind limb is also adapted for arboreality, with a grasping foot and pelvic morphology and lumbar morphology that stiffens the lower back and provides a large lever arm for the hip extensor muscles, enhancing vertical climbing capability.

Wild chimpanzees spend the majority of their lives in the trees, although time spent arboreally varies widely across seasons and sites (Table 7.2). Takemoto (2004) reported that chimpanzees in the Bossou community in Guinea spent an average of 76.6 percent of daylight hours arboreally, most of it resting (68.6 percent of arboreal observations). Time spent in the canopy ranged between 97.1 percent and 57.9 percent seasonally, with the remainder spent

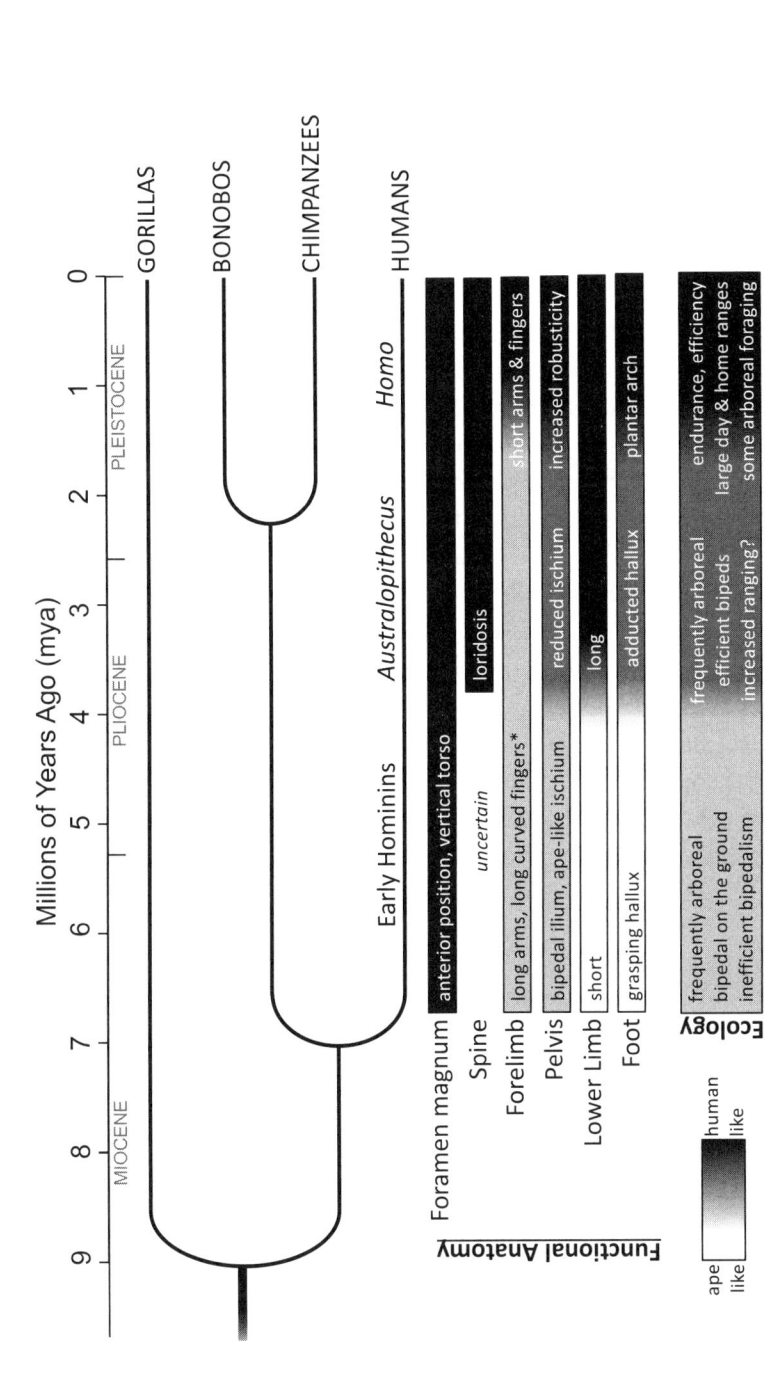

FIGURE 7.3. Overview of morphological and ecological change in the hominin lineage related to locomotion. "Ape-like" and "human-like" refer to function and performance; details of the anatomy may still differ. * Forelimbs in early hominins and *Australopithecus* lack knuckle-walking adaptations evident in chimpanzees.

terrestrially. Wrangham (1977) found less of the day spent arboreally in Gombe chimpanzees, with daytime observations of arboreal behavior ranging from 35.5 percent to 68.8 percent. Assuming that chimpanzees spend approximately twelve hours each night in their arboreal nests, time spent in the trees would account for ~65 percent to ~95 percent of each twenty-four-hour day. However, while chimpanzees spend most of their time in the trees, most of their locomotion (82–92 percent of observed bouts) is terrestrial (Carlson 2005). Chimpanzees climb approximately one hundred meters per day (Pontzer and Wrangham 2004), less than 5 percent of the distance they typically travel on the ground (Table 7.1).

While no documented human group approaches the level of arboreality seen in chimpanzees, it would be incorrect to characterize humans as entirely terrestrial. Many, perhaps most, human foragers depend on arboreal foraging for substantial portions of their diet. One common arboreal food is honey: human foragers (typically men) across a range of habitats in tropical and temperate latitudes regularly climb trees to harvest honey (Venkataraman et al. 2013; Marlowe et al. 2014). Among traditional Hadza foragers, living in dry savanna in northern Tanzania, honey accounts for 15 percent of the calories in the diet, ranging seasonally from <5 percent in the dry season to ~50 percent in the wet season (Marlowe et al. 2014). Arboreal honey harvesting accounts for the majority of foraging effort among men in some parts of the year for the Efe pygmies in the Ituri rainforest of Zaire (Bailey 1991). Arboreal foraging is not limited to honey. For example, Hadza women and children will occasionally scramble into low trees or bushes to harvest fruit (personal observation), and in some populations, such as the Aché of Paraguay, game animals are sometimes sought in the canopy (Hill and Hawkes 1983). The amount of time and effort that humans spend arboreally varies widely across populations and seasonally, making it difficult to calculate a useful or reliable human average. There also appears to be a consistent sex difference in tree climbing among human groups, unlike in chimpanzees, with men doing the large majority of arboreal foraging. Nonetheless, while humans spend much less time in the canopy than chimpanzees do, arboreal foraging among human hunter-gatherers must not be dismissed in ecological and evolutionary reconstructions, particularly in light of the dietary importance of arboreal foods and the mortality risk that climbing entails.

TABLE 7.2. Ranging characteristics for populations of chimpanzees and human hunter-gatherers.

Location	Community	Sex	Body Mass kg	Source	Home Range km²	Source	Day Range km/d	Source
Chimpanzees								
Gombe NP, Tanzania	Kasakela, Kahama	Male	39.0	1	Kasak.: 5.4–12.0	5	4.6	15
		Female	31.3		Kahama: 10		3.2	
Mahale, Tanzania	K and M	Male	42.0	2	K: 6.2–10.6	6	—	
		Female	35.2		M: 13.9–27.5			
Kibale NP, Uganda	Kanywara and Ngogo	Male	43.0[a,b]	3	Kany: 14.9–37.8	7, 8	2.4[a]	16
		Female	36.9[a,b]		Ngogo: 28.8–35.2	9	2.0[a]	
Taï Forest, Côte de Ivoire	North and South	Male			North: 13.9–26.4	10	N: 3.7, S: 4.3	10
		Female			South: 26.5		N: 3.6, S: 4.1	
	Middle	—			12.1		2.1	
Tomboronko, Senegal	Fongoli	—			63.0	11	Males: 3.3	17
Budongo, Uganda	Sonso	Male			6.8	12	—	
Humans								
Kalahari (desert)	Ju/'hoansi	Male	50.6	4	~475	13	14.9	18
		Female	42.2				9.1	
Paraguay (rainforest)	Aché	Male	59.8	4	~180	13	19.2	18
		Female	53.7				9.2	
Tanzania (savannah)	Hadza	Male	54.0	4	122	14	12.2	19
		Female	48.0				6.2	

a. Data from Kanyawara.
b. Estimated from skeletal samples.
Sources: 1: Pusey et al. 2005. 2: Uehara and Nishida 1987. 3: Carter et al. 2008. 4: Walker et al. 2006. 5: Williams et al. 2002. 6: Nakamura et al. 2013. 7: Chapman and Wrangham 1993. 8: Wilson et al. 2001. 9: Mitani et al. 2010. 10: Herbinger et al. 2001. 11: Pruetz 2006. 12. Newton-Fisher 2003. 13: Estimated from km²/individual (Leonard and Robertson 2000), assuming a local size of twenty-five individuals (Marlowe 2005). 14: Marlowe 2010: 261. 15: Wrangham 1977. 16: Pontzer and Wrangham 2004. 17: Wessling 2011 (wet season only, May–August). 18: Leonard and Robertson 1997. 19: Pontzer et al. 2015.

The evolution and maintenance of arboreal adaptations in chimpanzees appears to reflect selection for safety rather than climbing economy (Pontzer and Wrangham 2004). Laboratory studies measuring the metabolic cost of climbing indicate that human climbing economy is no different than other, arboreally adapted primates (monkeys, lemurs, lorises; Hanna et al. 2008). In fact, the available data indicate that the energy expended to climb is a fixed, linear function of body mass for all mammals, regardless of anatomical specialization (Taylor et al. 1972; Hanna et al. 2008). Thus, the loss of climbing adaptations in humans likely had no discernible effect on climbing economy.

The loss of suspensory adaptations in the forelimb, a grasping foot, and other arboreal morphology likely increased the risk of falling from the canopy. Falls while climbing are a significant source of morbidity and mortality among human foragers (particularly men; Venkataraman et al. 2013), even though the amount of time spent in the canopy is small compared to other primates. For example, falls accounted for 3.6 percent of known-cause mortality (7/193 deaths) among adult male Aka pygmies in a study by Hewlett and colleagues (1986), and 1.7 percent (4/238) of recorded deaths among adult male Agta foragers (Headland et al. 2011). Serious injuries from falls are even more common. Bennett and colleagues (1973) surveyed 127 Hadza men (age≥20 yrs, from their table 5) and reported twenty-one cases of serious injury, mostly to the lower limbs and back, due to falls from trees while foraging for honey, an incidence of ~17 percent. Falls are also an important source of injury and death among chimpanzees and other apes (Lovell 1990; Jurmain 1997; Carter et al. 2008), but their accident rate in the canopy (falls/hour) appears to be far lower than humans'. Despite their much higher exposure to the risk of falling (i.e., time spent in the canopy), deaths from injuries related to falls account for only 1.0 percent of known-cause deaths in wild chimpanzees [includes males and females: 2/86 at Gombe (Williams et al. 2008) and 0/116 at Mahale (Nishida et al. 2003)], not significantly different than the rate for Agta and Aka males ($p = 0.95$, Fisher's exact test). Notably, reports of injury and death from tree falls in humans are limited to males. Risk of falling in chimpanzees is presumably similar in males and females since both forage and sleep in the canopy, but further study is warranted to examine whether hunting, displaying, or fighting increases the risk of falling for males.

Divergent Locomotor Evolution in Humans and Chimpanzees

How and when did these differences evolve, and in what ecological context? Darwin, lacking a paleontological record of any substance, posited that bipedalism evolved under selection to free the hands for making and using stone tools (Darwin 1871). With the benefit of 150 years of fossil and archeological discovery, we can confidently say that Darwin was wrong: stone tools and improved manipulative abilities appear roughly four million years after the evolution of hominin bipedalism. However, this may be the sole point on which modern paleoanthropologists concur.

Current debate in hominin locomotor evolution begins at the beginning, with disagreement over the anatomy and locomotion of the last common ancestor (LCA) of *Homo* and *Pan*. Many, including Pilbeam and Lieberman in this volume, have argued that the LCA was a semi-arboreal knuckle-walker similar in its locomotor habits and anatomy to modern chimpanzees (Pilbeam 1996; Richmond and Strait 2000; Richmond et al. 2001; Orr 2005). This line of argument is certainly compelling given the long list of derived postcranial traits shared by *Gorilla* and *Pan,* and their locomotor similarities, as well as other lines of inference (Pilbeam and Lieberman, this volume). However, there is currently no broad consensus regarding the postcranial anatomy of the LCA. The hominoid fossil record from the late Miocene has yet to produce a clear candidate for the species or genus from which humans and chimpanzees diverged, and many Middle and Late Miocene hominoids, even those such as *Sivapithecus* that have clear affinities to living apes, lack the suspensory and knuckle-walking adaptations common among African apes today (Morgan et al. 2014). Further, osteological study of the wrist suggests some degree of homoplasy in the knuckle-walking adaptations evident in *Gorilla* and *Pan* (Dainton and Macho 1999; Kivell and Schmitt 2009). White and colleagues have taken this proposal further, arguing that the broad suite of postcranial similarities among *Pan* and *Gorilla* evolved independently, and that the LCA was a more generalized ape lacking specializations for knuckle-walking or suspension seen among extant apes (Lovejoy et al. 2009a, 2009b, 2009c; White et al. 2015).

Synthetic inferential approaches notwithstanding (Pilbeam and Lieberman, this volume), resolving the debate over the nature of the LCA may hinge upon the future recovery of fossils that can be confidently assigned to the *Pan* or *Gorilla* lineages. Here, I review human and chimpanzee locomotor evolution in the context of a chimpanzee-like LCA, and briefly discuss how interpretations of these evolutionary trajectories would change if the LCA lacked the knuckle-walking and suspensory adaptations present in modern chimpanzees. I begin with an overview of the fossil record before discussing ecological reconstructions.

Locomotor Evolution in the Hominin Lineage

Fossil discoveries over the past two decades have revealed a remarkable degree of diversity in locomotor anatomy both within and between hominin genera (Harcourt-Smith and Aiello 2004; Harcourt-Smith 2016). Here, I review broad evolutionary trends in hominin locomotor anatomy and function, focusing on taxa that are thought to represent forms ancestral to modern humans rather than representing derived sister groups (e.g., *Paranthropus*, *H. floresiensis*).

The oldest widely recognized hominin species (*Sahelanthropus tchadensis*, *Orrorin tugenensis*, and *Ardipithecus ramidus*) exhibit anatomical adaptations for bipedalism. Thus, bipedalism, along with blunted, less dimorphic canines, appears to be one of the first human-like traits to evolve in the hominin lineage. The oldest of these species, *Sahelanthropus tchadensis*, dated to between 6 and 7 million years ago (Ma), preserves only the skull and fragments of the postcranium (Brunet et al. 2002). Its diagnosis as a biped relies on the orientation and position of the foramen magnum, which is tucked under the skull and forward along the basicranium, indicating that the spinal column was habitually held upright (Zollikofer et al. 2005). *Orrorin tugenensis*, dated to 5.8 Ma, preserves the proximal portion of a femur along with other fragmentary remains (Senut et al. 2001). The arrangement of muscle markings at the base of the femoral neck (specifically the appearance and morphology of a groove imparted by the obturator externus muscle) and a human-like pattern of cortical bone thickness in the femoral neck suggest that this species was a biped (Galik et al. 2004). For both of these early hominins, little can be said about their locomotion other than that they were likely bipedal when on the ground.

By far the best known early hominin species is *Ardipithecus ramidus*, dated to 4.4 Ma, for which a largely complete skeleton and many other remains have been recovered (White et al. 2009). The forelimb of *Ardipithecus* indicates a significant portion of its time was spent in the trees: the forelimbs and fingers are long, and the phalanges are curved (Lovejoy et al. 2009a), its hind limbs are short relative to body mass, similar to chimpanzees and gorillas (Pontzer 2012), and, crucially, it retains a grasping foot with an abducted, ape-like hallux (Lovejoy et al. 2009b). However, the pelvis and skull appear to have been reshaped for the mechanical demands of bipedalism. As in *Sahelanthropus* and later hominins, the foramen magnum in *Ardipithecus* is positioned anteriorly, under the skull, indicating a vertical trunk. The pelvis, which was badly crushed *in situ,* appears to have short, broad iliac blades that wrap transversely around the trunk, placing the lesser gluteals in position to stabilize the trunk during single leg stance. Notably, the ischium is ape-like, long and caudally projecting, placing the hamstrings in position for powerful hip extension during vertical climbing. The overall impression is of a rather ape-like species (albeit with notable morphological differences relative to living apes) that was competent and comfortable in the trees and bipedal while on the ground. This mixed anatomical suite is consistent with the isotopic evidence, which indicates a semiforested habitat with opportunities for both arboreal and terrestrial foraging (Nelson 2013).

Species of the genus *Australopithecus,* dating from 1.9 to 4.1 Ma, retain several climbing adaptations in the forelimb but exhibit a more derived hind limb. Several partial *Australopithecus* skeletons and numerous other remains have been recovered over nearly a century of survey and excavation, lending greater confidence to locomotor reconstructions than is currently possible for earlier hominins. The australopith forelimb is long, with long, curved phalanges and a superiorly oriented glenoid fossa, indicative of regular and effective arboreal climbing and scrambling. However, the pelvis, hind limb, and foot are fully committed to terrestrial bipedalism: the pelvis is broad but functionally similar to modern humans, with a reduced ischium, the hind limb is long relative to body mass (Pontzer 2012), and the hallux is adducted, in line with the other toes. *Australopithecus afarensis* (3.4–3.1 Ma) may even have had a plantar arch (Ward et al. 2011), but this is debated (Harcourt-Smith and Aiello 2004). The australopith spine shows a lumbar lordosis similar to modern humans' (Whitcombe et al. 2007; Williams et al. 2013).

The genus *Homo,* dating from ~2.4 Ma to the present, is marked by the loss of climbing-related features in the forelimb: finger and forelimb length are reduced, and the glenoid fossa of the scapula is oriented laterally. Body size increases ca. 34 percent from *Australopithecus* (Pontzer 2012). The pelvis and hind limb joints in *Homo erectus* and later hominins are larger and more robust, reflecting their greater body size and possibly an increase in ranging. Bramble and Lieberman (2004) have argued that these and other features are adaptations specific to endurance running. Evidence for dietary change in early *Homo* to include more meat further suggests an increase in ranging, because foraging for meat typically requires daily travel distances four times longer than foraging for plant foods (Garland 1983; Carbone et al. 2005). This increase in ranging and physical activity is generally thought to have led to selection for hairlessness and sweating to improve thermoregulation, perhaps in *Homo erectus* (Bramble and Lieberman 2004).

Locomotor Evolution in Chimpanzees

The evolutionary history of locomotor change in the chimpanzee lineage is largely unknown. None of the fossil hominoids yet recovered from Late Miocene deposits are particularly chimpanzee-like, indicating one of two scenarios: either we have simply been unlucky thus far in recovering Late Miocene or Pliocene members of the *Pan* lineage, or the *Pan* lineage has experienced a substantial degree of anatomical change since the Late Miocene. With its fossil record essentially nonexistent, even the basic outline of locomotor evolution in *Pan* remains in dispute, with researchers debating whether all African apes share a knuckle-walking ancestor, or whether knuckle-walking evolved independently in *Pan* and *Gorilla* (Pilbeam and Lieberman, this volume). The similarity in postcranial anatomy between chimpanzees and bonobos indicates that their shared locomotor anatomy is at least 1–2 million years old, the date of their divergence as determined from their DNA (Hey 2010; Stone et al. 2010). Evidence for a chimpanzee-like LCA (Pilbeam and Lieberman, this volume) implies that chimpanzee locomotor anatomy arose much earlier, prior to the *Gorilla-Pan* divergence ~10 Ma, and has changed little over the past seven million years.

Evolution of Hominin and Chimpanzee Ranging Ecology

With the benefit of an ever-improving hominin fossil record, the broad outlines of hominin locomotor evolution and ecology are emerging (Figure 7.3).

As reconstructions of fossil hominin locomotor abilities improve, so does our understanding of the likely ecological pressures shaping hominin anatomy. Based on the postcranial evidence from *Ardipithecus,* it seems most likely that the earliest hominins frequented the trees to forage and possibly to sleep, similar to modern chimpanzees. Given their short legs, ape-like ischium, and grasping foot, it is unlikely that these early hominins walked as economically as modern humans (Pontzer et al. 2009, 2014). By eliminating the role of the forelimbs in locomotion, early bipedalism would also have reduced the volume of muscle available to do the work of walking, with two limbs supporting body weight rather than four. Muscle volume is strongly correlated with locomotor endurance: more muscle generally means a greater volume of mitochondria available for aerobic respiration and a higher maximum rate of energy throughput (i.e., VO_2max), resulting in a greater resistance to muscle fatigue (Weibel et al. 2004). Consequently, removing the forelimb muscles from their role in locomotion would likely have led to a corresponding decrease in walking and running endurance in early hominins (Pontzer 2017).

Reduced endurance and ape-like walking efficiency would argue against the "Savanna Hypothesis," which posits that bipedalism evolved in early hominins as an adaptation for increased ranging and efficient terrestrial travel in an increasingly open, savanna environment (e.g. Rodman and McHenry 1980; Isbell and Young 1996; Foley and Elton 1998). Limited endurance would also be inconsistent with provisioning models proposed by Lovejoy (1981, 2009), since traveling roundtrip to and from food sources would presumably require increased daily travel (at least for males) compared to the feed-as-you-go foraging of chimpanzees and other apes.

Instead, it seems more likely that bipedalism evolved in the context of relatively little terrestrial travel, and here a number of scenarios have been proposed. Several researchers, noting that chimpanzees and other apes commonly walk and stand bipedally while moving through the canopy, have suggested that hominin bipedalism could have evolved as an outgrowth of orthograde postures and bipedalism in the trees (Stanford 2006; Thorpe et al. 2007; Crompton et al. 2010). Observations of baboons (Jolly 1970; Wrangham 1980) and chimpanzees (Hunt 1991) using bipedalism while feeding on shrubby terrestrial vegetation has led others to propose that hominin bipedalism evolved in those contexts. Still others have suggested that early hominin bipedalism evolved as a wading adaption to exploit shallow-water resources, which apes and baboons do in some riparian and delta habitats (Verhaegan et al. 2002; Wrangham et al. 2009). And in captive and provisioned settings, chimpanzees sometimes use bipedal gaits to carry food away from the group and monopolize it, leading some to suggest that carrying was the key selection pressure for hominin bipedalism (Videan and McGrew 2002; Carvahlo et al. 2012).

All of these explanations for hominin bipedalism are plausible, and few, if any, are mutually exclusive. Further, these hypotheses do little to address Jolly's (1970) critique: Why would any of these behaviors require a wholesale change in postcranial anatomy and the loss of key arboreal traits if chimpanzees are already adept at them? Proponents of arboreal scenarios (Stanford 2006; Thorpe et al. 2007; Crompton et al. 2010) can at least point to gibbons as a living example of arboreal orthogrady leading to some preference for bipedalism when walking on the ground (Vereecke et al. 2006). Nonetheless, there remains little consensus as to why early hominin bipedalism evolved, and no clear path forward for distinguishing current, competing hypotheses. It may prove necessary (if difficult) to revisit these and other hypotheses for early hominin bipedalism with more rigid, quantitative tests of their proposed form-function-fitness relationships (Arnold 1983).

Whatever pressures shaped the adoption of bipedalism in the earliest hominins, adaptations for improved walking economy are evident by 3.7 Ma in *Australopithecus afarensis*, with longer hind limbs, adducted hallux, and reduced ischium (Pontzer et al. 2009, 2014; Pontzer 2012). Increased molar size and enamel thickness in australopiths also suggests a substantial change in diet and foraging ecology from earlier hominins, perhaps to lower-quality foods and drier, more open habitats. The co-occurrence of locomotor and dietary changes suggests that this ecological shift imposed longer day ranges and a greater emphasis on terrestrial travel. Further adaptations for efficiency and endurance in early *Homo* appear to coincide with an ecological shift to increased hunting and scavenging (Bramble and Lieberman 2004), although many of the critical features may have already been present in *Australopithecus* (Pontzer 2012). Regardless, dietary reliance on meat would likely have led to an increase in ranging effort and selection for improved endurance and efficiency in *Homo*. The loss of climbing adaptations in the forelimb also signals a decrease in arboreal foraging in *Homo* (Figure 7.3).

If the LCA was indeed chimpanzee-like in its locomotor habits and anatomy, a semiarboreal knuckle-walker with a long day range relative to other apes, then hominin bipedalism likely coincided with a reduction in ranging: the walking economy of early hominins does not appear to have been substantially better than in chimpanzees, and the loss of forelimb support would likely have reduced locomotor endurance. In turn, this scenario suggests that early hominins diverged in an environment with relatively high food availability or exploited foods that were more common (and perhaps therefore of lower quality), thereby reducing terrestrial travel between food patches. Ecological shifts at ~4 Ma with *Australopithecus,* and then ~2 Ma with *Homo,* would have led to selection for improved terrestrial efficiency and endurance. By contrast, locomotor ecology and anatomy in the chimpanzee lineage would have remained essentially unchanged during the past seven million years, with *Pan* restricted to environments suited to their particular balance of arboreal and terrestrial abilities.

Alternatively, if the LCA was a more primitive, generalized primate lacking the knuckle-walking specializations and relatively long daily travel distances evident in modern chimpanzees, then the adoption of bipedalism in early hominins need not signal a decrease in ranging. Daily travel distances in most primates are relatively modest, averaging only 1.5 km per day (Leonard and Robertson 1997). Early hominins could have presumably maintained such small day ranges even with a reduction in walking endurance. With bipedalism established as the preferred terrestrial mode of locomotion over two million years of early hominin evolution, selection for improved efficiency and endurance with *Australopithecus* and *Homo* would have shaped the bipedal solutions discussed above. Similarly, under this model, knuckle-walking in the chimpanzee lineage would presumably have evolved under selection for improved walking performance, as day ranges increased to the 3–4 km/day common in chimpanzees and bonobos today. This ecological and evolutionary shift would have occurred sometime between ~7 and 2 Ma, after the divergence from the hominin lineage but prior to the separation of bonobos and chimpanzees. Likewise, the independent evolution of knuckle-walking in gorillas could have been an adaptation to increased terrestriality (and reduced arboreality) in their lineage. Evaluating these scenarios will require fossil evidence from the *Pan* and *Gorilla* lineages.

Summary

Despite their close evolutionary relationship, humans and chimpanzees are remarkably different in their locomotor habits and anatomy. Humans are highly economical bipedal walkers and excellent endurance athletes, with arboreal capabilities that are severely limited compared to other primates. Chimpanzees are inefficient on the ground compared to humans, but retain formidable climbing abilities. Bipedalism is evident in the earliest hominins, but increased economy and endurance appear to have evolved much later. As hominin evolutionary history comes into increasingly better focus through over a century of fossil discoveries, our curiosity must inevitably shift to the *terra incognita* of the chimpanzee fossil record. Multiple lines of evidence indicate that hominins diverged radically from an LCA that was remarkably chimpanzee-like in its locomotor anatomy and ecology (Pilbeam and Lieberman, this volume), an evolutionary trajectory that is all the more remarkable when set against the apparent stasis in *Pan*. Nonetheless, there remain important questions of anatomical and ecological evolution in both humans and chimpanzees that will only be addressed directly with fossil evidence from the *Pan* lineage. Let us hope that a chimpanzee fossil record is preserved, and that we have the good fortune to uncover it.

Acknowledgments

I thank Martin Muller and Richard Wrangham for the invitation to contribute to this volume. Theodore Headland generously provided help with data from the Agta Demographic Database. Conversations and collaborations with David Raichlen, Daniel Lieberman, Brian Wood, and Richard Wrangham over the years have helped shape my thinking on human locomotor evolution and ecology, for which I am grateful. David Pilbeam and Martin Muller provided helpful comments that improved this paper.

References

Arnold, S. J. 1983. Morphology, performance, and fitness. *American Zoologist* 23: 347–361.

Bailey, R. C. 1991. *The Behavioral Ecology of Efe Pygmy Men in the Ituri Forest, Zaire*. Ann Arbor: Museum of Anthropology, University of Michigan.

Bennett, F. J., N. A. Barnicot, J. C. Woodburn, M. A. Periera, and B. E. Henderson. 1973. Studies on viral, bacterial, rickettsial and treponemal diseases in the Hadza of Tanzania and a note on injuries. *Human Biology* 45: 243–272

Biewener, A. A. 2003. *Animal Locomotion*. Oxford: Oxford University Press.

Bozek, K., Y. Wei, Z. Yan, X. Liu, J. Xiong, M. Sugimoto, M. Tomita, S. Pääbo, R. Pieszek, C. C. Sherwood, P. R. Hof, J. J. Ely, D. Steinhauser, L. Willmitzer, J. Bangsbo, O. Hansson, J. Call, P. Giavalisco, and P. Khaitovich. 2014. Exceptional evolutionary divergence of human muscle and brain metabolomes parallels human cognitive and physical uniqueness. *PLoS Biology* 12: e1001871.

Bramble, D. M., and D. E. Lieberman. 2004. Endurance running and the evolution of *Homo*. *Nature* 432: 345–352.

Brunet, M., G. Guy, D. Pilbeam, H. T. Mackaye, A. Likius, D. Ahounta, A. Beauvilain, C. Blondel, H. Bocherens, J. R. Boisserie, L. De Bonis, Y. Coppens, J. Dejax, C. Denys, P. Duringer, V. Eisenmann, G. Fanone, P. Fronty, D. Geraads, T. Lehmann, F. Lihoreau, A. Louchart, A. Mahamat, G. Merceron, G. Mouchelin, O. Otero, P. Pelaez Campomanes, M. Ponce de Leon, J. C. Rage, M. Sapanet, M. Schuster, J. Sudre, P. Tassy, X. Valentin, P. Vignaud, L. Viriot, A. Zazzo, and C. Zollikofer. 2002. A new hominid from the Upper Miocene of Chad, Central Africa. *Nature* 418: 145–151.

Carbone, C., G. Cowlishaw, N. J. B. Isaac, and J. M. Rowcliffe. 2005. How far do animals go? Determinants of day range in mammals. *American Naturalist* 165: 290–297.

Carlson, K. J. 2005. Investigating the Form-Function Interface in African Apes: Relationships Between Principal Moments of Area and Positional Behaviors in Femoral and Humeral Diaphyses. *American Journal of Physical Anthropology* 127: 312–334.

Carrier, D. R. 1984. The energetic paradox of human running and hominid evolution. *Current Anthropology* 25: 483–495.

Carter, M. L., H. Pontzer, R. W. Wrangham, and J. K. Peterhans. 2008. Skeletal pathology in *Pan troglodytes schweinfurthii* in Kibale National Park, Uganda. *American Journal of Physical Anthropology* 135: 389–403.

Carvalho, S., D. Biro, E. Cunha, K. Hockings, W. C. McGrew, B. G. Richmond, and T. Matsuzawa. 2012. Chimpanzee carrying behaviour and the origins of human bipedality. *Current Biology* 22: R180–181.

Chapman, C. A., and R. W. Wrangham. 1993. Range use of the forest chimpanzees of Kibale: Implication for the understanding of chimpanzee social organization. *American Journal of Primatology* 31: 263–273.

Conroy, G. C., and H. Pontzer. 2012. *Reconstructing Human Origins: A Modern Synthesis*. New York: W. W. Norton.

Crompton, R. H., W. I. Sellers, and S. K. S. Thorpe. 2010. Arboreality, terrestriality and bipedalism. *Philosophical Transactions of the Royal Society B* 365: 3301–3314.

Dainton, M., and G. A. Macho. 1999. Did knuckle-walking evolve twice? *Journal of Human Evolution* 36: 171–194.

Darwin, C. 1871. *Descent of Man and Selection in Relation to Sex*. New York: Barnes & Noble.

Foley, R. A., and S. Elton. 1998. Time and energy: The ecological context for the evolution of bipedalism. In E. Strasser, J. Fleagle, A. Rosenberger, and H. McHenry, eds., *Primate Locomotion: Recent Advances*, 419–433. New York: Plenum.

Galik, K., B. Senut, M. Pickford, D. Gommery, J. Treil, A. J. Kuperavage, and R. B. Eckhardt. 2004. External and internal morphology of the BAR 1002'00 *Orrorin tugenensis* femur. *Science* 305: 1450–1453.

Garland, T. 1983. Scaling the ecological cost of transport to body mass in terrestrial mammals. *American Naturalist* 121: 571–87.

Garland, T., Jr. 1999. Laboratory endurance capacity predicts variation in field locomotor behaviour among lizard species. *Animal Behaviour* 58: 77–83.

Gatesy, S. M., and A. A. Biewener. 1991. Bipedal locomotion: Effects of size, speed and limb posture in birds and humans. *Journal of Zoology* 224: 127–147.

Grabowski, M. W., J. D. Polk, and C. C. Roseman. 2011. Divergent patterns of integration and reduced constraint in the human hip and the origins of bipedalism. *Evolution* 65: 1336–1356.

Hanna, J. B., D. Schmitt, and T. M. Griffin. 2008. The energetic cost of climbing in primates. *Science* 320 (5878): 898.

Harcourt-Smith, W. E. 2016. Early hominin diversity and the emergence of the genus *Homo*. *Journal of Anthropological Science* 94: 19–27.

Harcourt-Smith, W. E., and L. C. Aiello. 2004. Fossils, feet and the evolution of human bipedal locomotion. *Journal of Anatomy* 204: 403–16.

Harrison, T. 2010. Apes among the tangled branches of human origins. *Science* 327: 532–534.

Headland, T. N., J. D. Headland, and R. Uehara. 2011. *Agta Demographic Database: Chronicle of a Hunter-Gatherer Community in Transition*. SIL Language and Culture Documentation and Description, version 2.0. Dallas: SIL, www.sil.org/resources/publications/entry/9299.

Herbinger, I., C. Boesch, and H. Rothe. 2001. Territory characteristics among three neighboring chimpanzee communities in the Taï National Park, Côte d'Ivoire. *International Journal of Primatology* 22: 143–167.

Hewlett, B. S., J. M. H. van de Koppel, and M. van de Koppel. 1986. Causes of death among Aka pygmies of the Central African Republic. In L. L. Cavalli-Sforza, ed., *African Pygmies*, 45–63. New York: Academic.

Hey, J. 2010. The divergence of chimpanzee species and subspecies as revealed in multi-population isolation-with-migration analyses. *Molecular Biology and Evolution* 27: 921–933.

Hill, K., and K. Hawkes. 1983. Neotropical hunting among the Aché of eastern Paraguay. In R. B. Hames and W. T. Vickers, eds., *Adaptive Responses of Native Amazonians*, 139–188. New York: Academic Press.

Hunt, K. D. 1989. Positional behavior in *Pan troglodytes* at the Mahale Mountains and the Gombe Stream National Parks, Tanzania. Ph.D. dissertation, University of Michigan, Ann Arbor.

Hunt, K. D. 1991. Mechanical implications of chimpanzee positional behavior. *American Journal of Physical Anthropology* 86: 521–536.

Isbell, L. A., and T. P. Young. 1996. The evolution of bipedalism in hominids and reduced group size in chimpanzees: Alternative responses to decreasing resource availability. *Journal of Human Evolution* 30: 389–397.

Jolly, C. J. 1970. The seed-eaters: A new model of hominid differentiation based on a baboon analogy *Man.* 5: 1–26.

Jurmain, R. 1997. Skeletal evidence of trauma in African apes, with special reference to the Gombe chimpanzees. *Primates* 38: 1–14.

Kivell, T. L., and D. Schmitt. 2009. Independent evolution of knuckle-walking in African apes shows that humans did not evolve from a knuckle-walking ancestor. *Proceedings of the National Academy of Sciences* 106: 14241–14246.

Leonard, W. R., and M. L. Robertson. 1997. Comparative primate energetics and hominid evolution. *American Journal of Physical Anthropology* 102: 265–281.

Leonard, W. R., and M. L. Robertson. 2000. Ecological correlates of home range variation in primates: Implications for hominid evolution. In *On the Move: How and Why Animals Travel in Groups,* ed. S. Boinski and P. A. Garber, 628–648. Chicago: University of Chicago Press.

Lieberman, D. E. 2015. Human locomotion and heat loss: An evolutionary perspective. *Comprehensive Physiology* 5: 99–117.

Lovejoy, C. O. 1981. The origin of man. *Science* 211: 341–350.

Lovejoy, C. O. 2009. Reexamining human origins in light of *Ardipithecus ramidus. Science* 326: 74e1–8.

Lovejoy, C. O., S. W. Simpson, T. D. White, B. Asfaw, and G. Suwa. 2009a. Careful climbing in the Miocene: The forelimbs of *Ardipithecus ramidus* and humans are primitive. *Science* 326: 70e1–8.

Lovejoy, C. O., G. Suwa, S. W. Simpson, J. H. Matternes, and T. D. White. 2009b. The great divides: *Ardipithecus ramidus* reveals the postcrania of our last common ancestors with African apes. *Science* 326: 100–106.

Lovejoy, C. O., G. Suwa, L. Spurlock, B. Asfaw, and T. D. White. 2009c. The pelvis and femur of *Ardipithecus ramidus:* The emergence of upright walking. *Science* 326: 71e1–6.

Lovell, N. C. 1990. *Patterns of Illness and Injury in Great Apes: A Skeletal Analysis.* Washington, DC: Smithsonian Institution Press.

Marlowe, F. W. 2005. Hunter-gatherers and human evolution. *Evolutionary Anthropology* 14: 54–67.

Marlowe, F. W. 2010. *The Hadza: Hunter-Gatherers of Tanzania.* Berkeley: University of California Press.

Marlowe, F. W., and J. C. Berbesque. 2009. Tubers as fallback foods and their impact on Hadza hunter-gatherers. *American Journal of Physical Anthropology* 140: 751–758.

Marlowe, F. W., J. C. Berbesque, B. Wood, A. Crittenden, C. Porter, and A. Mabulla. 2014. Honey, Hadza, hunter-gatherers, and human evolution. *Journal of Human Evolution* 71: 119–28.

Mitani, J. C., D. P. Watts, and S. J. Amsler. 2010. Lethal intergroup aggression leads to territorial expansion in wild chimpanzees. *Current Biology* 20: R507–R508.

Nakamura, M., N. Corp, M. Fujimoto, S. Fujita, S. Hanamura, H. Hayaki, K. Hosaka, M. A. Huffman, A. Inaba, E. Inoue, N. Itoh, N. Kutsukake, M. Kiyono-Fuse, T. Kooriyama, L. F. Marchant, A. Matsumoto-Oda, T. Matsusaka, W. C. McGrew, J. C. Mitani, H. Nishie, K. Norikoshi, T. Sakamaki, M. Shimada, L. A. Turner, J. V. Wakibara, and K. Zamma. 2013. Ranging behavior of Mahale chimpanzees: A 16 year study. *Primates* 54: 171–182.

Nakatsukasa, M., C. V. Ward, A. Walker, M. F. Teaford, Y. Kunimatsu, and N. Ogihara. 2004. Tail loss in *Proconsul heseloni*. *Journal of Human Evolution* 46: 777–784.

Nelson, S. V. 2013. Chimpanzee fauna isotopes provide new interpretations of fossil ape and hominin ecologies. *Proceedings of the Royal Society B: Biological Sciences* 280 (1773): 20132324.

Newton-Fisher, N. E. 2003. The home range of the Sonso community of chimpanzees from the Budongo Forest, Uganda. *African Journal of Ecology* 41: 150–156.

Nishida, T., N. Corp, M. Hamai, T. Hasegawa, M. Hiraiwa-Hasegawa, K. Hosaka, K. Hunt, N. Itoh, K. Kawanaka, A. Matsumoto-Oda, J. Mitani, M. Nakamura, K. Norikoshi, T. Sakamaki, L. Turner, S. Uehara, and K. Zamma. 2003. Demography, female life history, and reproductive profiles among the chimpanzees of Mahale. *American Journal of Primatology* 59: 99–121.

Orr, C. M. 2005. Knuckle-walking anteater: A convergence test of adaptation for purported knuckle-walking features of African Hominidae. *American Journal of Physical Anthropology* 128: 639–658.

Payne, R. C., R. H. Crompton, K. Isler, R. Savage, E. E. Vereecke, M. M. Gunther, S. K. Thorpe, and K. D'Août. 2006. Morphological analysis of the hindlimb in apes and humans, I: Muscle architecture. *Journal of Anatomy* 208: 709–724.

Pilbeam, D. 2004. The anthropoid postcranial axial skeleton: Comments on development, variation, and evolution. *Journal of Experimental Zoology, B: Molecular Development and Evolution* 302: 241–267.

Pilbeam, D. R. 1996. Genetic and morphological records of the Hominoidea and hominid origins: A synthesis. *Molecular Phylogenetics and Evolution* 5: 155–168.

Pontzer, H. 2007. Limb length and the scaling of locomotor cost in terrestrial animals. *Journal of Experimental Biology* 210: 1752–1761.

Pontzer, H. 2012. Ecological energetics in early *Homo*. *Current Anthropology* 53(S6): S346–S358.

Pontzer, H. 2017. Economy and endurance in human evolution. *Current Biology* 27: R613–R621.

Pontzer, H., D. A. Raichlen, and P. S. Rodman. 2014. Bipedal and quadrupedal locomotion in chimpanzees. *Journal of Human Evolution* 66: 64–82.

Pontzer, H., D. A. Raichlen, and M. D. Sockol. 2009. The metabolic cost of walking in humans, chimpanzees, and early hominins. *Journal of Human Evolution* 56: 43–54.

Pontzer, H., D. A. Raichlen, B. M. Wood, M. Emery Thompson, S. B. Racette, A. Z. P. Mabulla, and F. W. Marlowe. 2015. Energy expenditure and activity in Hadza hunter-gatherers. *American Journal of Human Biology* 27: 628–637.

Pontzer, H., and R. W. Wrangham. 2004. Climbing and the daily energy cost of locomotion in wild chimpanzees: Implications for hominoid locomotor evolution. *Journal of Human Evolution* 46: 315–333.

Pontzer, H., and R. W. Wrangham. 2006. The ontogeny of ranging in wild chimpanzees. *International Journal of Primatology* 27: 295–309.

Pruetz, J. D. 2006. Feeding ecology of savanna chimpanzees *(Pan troglodytes verus)* at Fongoli, Senegal. In C. Boesch, G. Hohmann, and M. M. Robbins, eds., *Feeding Ecology in Apes and Other Primates,* 161–182. Cambridge: Cambridge University Press.

Pusey, A. E., G. W. Oehlert, J. M. Williams, and J. Goodall. 2005. Influence of ecological and social factors on body mass of wild chimpanzees. *International Journal of Primatology* 26: 3–31.

Richmond, B. G., D. R. Begun, and D. S. Strait. 2001. Origin of human bipedalism: The knuckle-walking hypothesis revisited. *American Journal of Physical Anthropology* S33: 70–105.

Richmond, B. G., and D. S. Strait. 2000. Evidence that humans evolved from a knuckle-walking ancestor. *Nature* 404: 382–385.

Roberts, T. J., R. Kram, P. G. Weyand, and C. R. Taylor. 1998. Energetics of bipedal running, I: Metabolic cost of generating force. *Journal of Experimental Biology* 201: 2745–2751.

Rodman, P. S., and H. M. McHenry. 1980. Bioenergetics and the origin of hominid bepedalism. *American Journal of Physical Anthropology* 52: 103–106.

Rubenson, J., D. B. Heliams, S. K. Maloney, P. C. Withers, D. G. Lloyd, and P. A. Fournier. 2007. Reappraisal of the comparative cost of human locomotion using gait specific allometric analyses. *Journal of Experimental Biology* 210: 3513–3524.

Russo, G. A., and E. C. Kirk. 2013. Foramen magnum position in bipedal mammals. *Journal of Human Evolution* 65: 656–670.

Ruxton, G. D., and D. M. Wilkinson. 2011. Avoidance of overheating and selection for both hair loss and bipedality in hominins. *Proceedings of the National Academy of Sciences* 108: 20965–20969.

Scholz, M. N., K. D'Août, M. F. Bobbert, and P. Aerts. 2006. Vertical jumping performance of bonobo *(Pan paniscus)* suggests superior muscle properties. *Proceedings of the Royal Society B: Biological Sciences* 273: 2177–2184.

Senut, B., M. Pickford, D. Gommery, P. Mein, K. Cheboi, and Y. Coppens. 2001. First hominid from the Miocene (Lukeino Formation, Kenya). *Comptes Rendus de l'Académie des Sciences* 332: 137–144.

Sockol, M. D., D. A. Raichlen, and H. Pontzer. 2007. Chimpanzee locomotor energetics and the origin of human bipedalism. *Proceedings of the National Academy of Sciences* 30: 12265–12269.

Stanford, C. B. 2006. Arboreal bipedalism in wild chimpanzees: Implications for the evolution of hominid posture and locomotion. *American Journal of Physical Anthropology* 129: 225–231.

Stone, A. C., F. U. Battistuzzi, L. S. Kubatko, G. H. Perry, Jr., E. Trudeau, H. Lin, and S. Kumar. 2010. More reliable estimates of divergence times in *Pan* using complete mtDNA sequences and accounting for population structure. *Philosophical Transactions of the Royal Society B: Biological Sciences* 365: 3277–3288.

Takemoto, H. 2004. Relation to microclimate in the tropical forest. *American Journal of Physical Anthropology* 124: 81–92.

Taylor, C. R., S. L. Caldwell, and V. J. Rowntree. 1972. Running up and down hills: some consequences of size. *Science* 178 (4065): 1096–1097.

Taylor, C. R., and V. J. Rowntree. 1973. Running on two or on four legs: Which consumes more energy? *Science* 179: 186–187.

Thorpe, S. K. S., R. H. Crompton, M. M. Gunther, R. F. Ker, and R. M. Alexander. 1999. Dimensions and moment arms of the hind- and forelimb muscles of common chimpanzees *(Pan troglodytes)*. *American Journal of Physical Anthropology* 110: 179–199.

Thorpe, S. K., R. L. Holder, and R. H. Crompton. 2007. Origin of human bipedalism as an adaptation for locomotion on flexible branches. *Science* 316: 1328–1331.

Uehara, S., and T. Nishida. 1987. Body weights of wild chimpanzees *(Pan troglodytes schweinfurthii)* of the Mahale Mountains National Park, Tanzania. *American Journal of Physical Anthropology* 72: 315–321.

Venkataraman, V. V., T. S. Kraft, and N. J. Dominy. 2013. Tree climbing and human evolution. *Proceedings of the National Academy of Sciences* 110: 1237–1242.

Vereecke, E. E., K. D'Août, and P. Aerts. 2006. Locomotor versatility in the white-handed gibbon *(Hylobates lar):* A spatiotemporal analysis of the bipedal, tripedal, and quadrupedal gaits. *Journal of Human Evolution* 50: 552–567.

Verhaegen, M., P.-F. Puech, and S. Munro. 2002. Aquarboreal ancestors? *Trends in Ecology and Evolution* 17: 212–217.

Videan, E. N., and W. C. McGrew. 2002. Bipedality in chimpanzee *(Pan troglodytes)* and bonobo *(Pan paniscus):* Testing hypotheses on the evolution of bipedalism. *American Journal of Physical Anthropology* 118: 184–90.

Walker, A. 2009. The strength of great apes and the speed of humans. *Current Anthropology* 50: 229–234.

Walker, R. S., M. Gurven, K. Hill, A. Migliano, N. Chagnon, R. De Souza, G. Djurovic, R. Hames, A. M. Hurtado, H. Kaplan, K. Kramer, W. J. Oliver, C. Valeggia, and T. Yamauchi. 2006. Growth rates and life histories in twenty-two small-scale societies. *American Journal of Human Biology* 18: 295–311.

Ward, C. V., W. H. Kimbel, and D. C. Johanson. 2011. Complete fourth metatarsal and arches in the foot of *Australopithecus afarensis*. *Science* 331: 750–753.

Weibel, E. R., L. D. Bacigalupe, B. Schmitt, and H. Hoppeler. 2004. Allometric scaling of maximal metabolic rate in mammals: Muscle aerobic capacity as determinant factor. *Respiratory Physiology & Neurobiology* 140: 115–132.

Wessling, E. R. 2011. Rank-related differences in the travel patterns of savanna chimpanzees *(Pan troglodytes verus)* at Fongoli, Senegal. M.A. thesis, Iowa State University.

Whitcome, K. K., L. J. Shapiro, and D. E. Lieberman. 2007. Fetal load and the evolution of lumbar lordosis in bipedal hominins. *Nature* 450: 1075–1078.

White, T. D., B. Asfaw, Y. Beyene, Y. Haile-Selassie, C. O. Lovejoy, G. Suwa, and G. WoldeGabriel. 2009. *Ardipithecus ramidus* and the paleobiology of early hominids. *Science* 326 (5949): 75–86.

White, T. D., C. O. Lovejoy, B. Asfaw, J. P. Carlson, and G. Suwa. 2015. Neither chimpanzee nor human, Ardipithecus reveals the surprising ancestry of both. *Proceedings of the National Academy of Sciences* 112 (16): 4877–4884.

Williams, J. M., E. V. Lonsdorf, M. L. Wilson, J. Schumacher-Stankey, J. Goodall, and A. E. Pusey. 2008. Causes of death in the Kasekela chimpanzees of Gombe National Park, Tanzania. *American Journal of Primatology* 70: 766–777.

Williams, J. M., A. E. Pusey, J. V. Carlis, B. P. Farm, and J. Goodall. 2002. Female competition and male territorial behaviour influence female chimpanzees' ranging patterns. *Animal Behaviour* 63: 347–360.

Williams, S. A., K. R. Ostrofsky, N. Frater, S. E. Churchill, P. Schmid, and L. Berger. 2013. The vertebral column of *Australopithecus sediba*. *Science* 340: 1232996.

Wilson, M. L., M. D. Hauser, and R. W. Wrangham. 2001. Does participation in intergroup conflict depend on numerical assessment, range location, or rank for wild chimpanzees? *Animal Behaviour* 61: 1203–1216.

Wrangham, R., D. Cheney, R. Seyfarth, and E. Sarmiento. 2009. Shallow-water habitats as sources of fallback foods for hominins. *American Journal of Physical Anthropology* 140: 630–642.

Wrangham, R. W. 1977. Feeding behavior of chimpanzees in Gombe National Park, Tanzania. In T. H. Clutton-Brock, ed., *Primate Ecology*, 503–538. London: Academic Press.

Wrangham, R. W. 1980. Bipedal locomotion as a feeding adaptation in gelada baboons and its implication for hominid evolution. *Journal of Human Evolution* 9: 329–331.

Wrangham, R. W. 2000. Why are male chimpanzees more gregarious than mothers? A scramble competition hypothesis. In P. M. Kappeler, ed., *Primate Males, Causes and Consequences of Variation in Group Composition*. Cambridge: Cambridge University Press.

Zollikofer, C. P. E., M. S. Ponce de Leon, D. E. Lieberman, F. Guy, D. Pilbeam, A. Likius, H. T. Mackaye, P. Vignaud, and M. Brunet. 2005. Virtual cranial reconstruction of *Sahelanthropus tchadensis*. *Nature* 434: 755–759.

8

Evolution of the Human Dietary Niche

Initial Transitions

SHERRY V. NELSON *and* MARIAN I. HAMILTON

Modern human diets differ from those of chimpanzees by including more high-quality, nutrient-dense foods that are easy to digest. In this chapter, we trace early hominin transitions from a chimpanzee-like diet specializing in ripe fruit to exploitation of different resources and increasing dietary breadth. In characterizing the diet of early hominins, we pay particular attention to isotopic data, with new analyses that allow us to explore three questions: (1) What habitat supported the transition from ape to hominin? (2) Where within that habitat did early hominins feed? And (3) how was the early hominin dietary niche constrained by habitat and climate?

The Dietary Niche of Early Hominins

The first 4–5 million years of human evolution are marked by hominins that resemble great apes in several key features, including similar body and gut size, similar or only slightly enlarged brains, and comparable life history trajectories (Wood and Collard 1999; Dean et al. 2001; Zollikofer et al. 2005;

Suwa et al. 2009a; White et al. 2009). Given that the large bodies and brains of apes are metabolically expensive (McNab 1990; Aiello and Wheeler 1995), and that long periods of lactation require a large number of calories per reproductive event (Aiello and Key 2002), great apes require a high-quality diet compared with other primates. Apes meet this demand by specializing in ripe fruits, in rainforests where the dry season is no longer than four months. Within the same forest, chimpanzees feed more on ripe fruits than do frugivorous, or fruit-eating, monkeys, and their diets are of higher quality when compared for carbohydrates, fiber, and antifeedants (Conklin-Brittain et al. 1998; Wrangham et al. 1998). Mountain gorillas are an exception with respect to this diet, but they can survive on lower-quality foods because their extremely large guts allow for greater fiber intake. Likewise, among the lesser apes, only the largest siamangs feed extensively on leaves, and then only on young leaves (MacKinnon and MacKinnon 1978). A few chimpanzee populations inhabit drier regions outside of rainforests, but they live at very low densities, and whereas their home ranges are extensive during the wet season, during the dry season they are limited to gallery, or riverine, forests similar to typical ape habitats (Baldwin et al. 1982; McGrew et al. 1981).

Early hominins must have had similar energetic requirements, but their habitat reconstructions indicate that they lived in more open, drier, and more seasonal habitats compared with their ape cousins (Cerling et al. 2011b). For the tens of thousands of faunal specimens collected at hominin localities, none is an ape, further confirming that hominins exploited a different niche. Our challenge is to determine what plant foods, and hence what habitat, could have met early hominin energetic needs, replacing rainforest fruits, before digestive morphology and the archaeological record yield evidence for a transition to heavier reliance on animal source foods and processed items. We can make dietary inferences based on dental and cranial morphology, dental microwear, and more recently, stable isotope analyses.

Through early hominin evolution, premolars and molars increase in size from chimpanzee-like in some of the earliest hominins, to several times larger in the robust australopithecines (Wood and Lonergan 2008; Suwa et al. 2009b; Grine et al. 2012). Tooth enamel thickness also increases over time, from thin like that of chimpanzees, to hyper-thick in robust australopithecines (Suwa et al. 2009b, Skinner et al. 2015). Coincident with increases in tooth size, facial and mandibular morphology grow increasingly robust, suggesting a diet increasing in hardness, abrasion, or toughness. An analysis of bite force using tooth and facial morphology suggested that australopithecines were not chewing hard foods (Demes and Creel 1988), although recent finite element models of bony structure suggest that *Australopithecus (Au.) africanus* may have used its premolars to break large nuts and seeds (Strait et al. 2009).

Numerous dental microwear analyses of australopithecines using both scanning electron microscopy and confocal microscopy corroborate the evidence for a lack of hard objects in australopithecine diets. Dental microwear reflects the physical properties of foods most recently consumed. Microwear studies in living species have demonstrated significant differences among folivores, frugivores, grazers, and hard-object feeders. Analyses of microscopic pits and scratches, as well as occlusal surface fractal complexity, suggest that while some australopithecines may have relied on hard objects as fallback foods, they were not preferred foods (Ungar and Sponheimer 2011). Therefore, increasing megadonty and facial buttressing in australopithecines must relate to either a tough diet requiring high cyclical loads, or an abrasive diet. If a tough diet, then our challenge is to determine what tough foods would also have been of high quality, because fiber is considered an antifeedant (Conklin-Brittain et al. 1998). Potential abrasive elements include plants with high phytolith loads, grit, or dust (Lucas et al. 2008; Hummel et al. 2011; Rabenold and Pearson 2011).

Stable isotope analyses complement dental microwear reconstructions by addressing not only food sources, but also subsistence behavior, such as terrestrial versus arboreal feeding, as well as habitat reconstructions. Isotope values of an animal's tooth enamel reflect its diet and drinking habits during tooth formation (Luz and Kolodny 1985; Quade et al. 1992; Wang and Cerling 1994; Quade et al. 1995). Carbon ($\delta^{13}C$) isotopes are informative due to differences in carbon assimilation between different plants (Bender 1971; Vogel 1980; Farquhar et al. 1989). C_3 plants, which include almost all trees and shrubs and only grasses in areas experiencing cool, wet growing seasons, yield $\delta^{13}C$ values between −22‰ and −35‰. C_4 plants include exclusively grasses and sedges growing in habitats with warm season precipitation and yield $\delta^{13}C$ values between −10‰ and −15‰. Additionally, the degree of water and light stress a plant undergoes affects its carbon and oxygen ($\delta^{18}O$) isotopic values. Plants under irradiance stress must become water-use efficient to function at lower respiration rates, but still absorb CO_2 for plant growth. This process leads to enrichment of ^{13}C in plants in habitats where evaporative stresses are high (Ehleringer et al. 1986). Plants in evaporative habitats are enriched in ^{18}O because the lighter oxygen isotope evaporates more readily (Quade et al. 1995). Therefore, the lowest δ^{13}C and δ^{18}O values represent forest floors where plants are shaded and under less evaporative stress, while the highest values represent open habitats such as savannas or forest upper canopies.

An animal's tooth enamel reflects where it feeds in these environments. For example, in Kibale National Park, a chimpanzee habitat, carbon isotope values differentiate animals feeding in forest versus more open habitats, and within the forest, oxygen isotope values distinguish terrestrial versus arboreal feeders (Nelson 2013). Likewise, oxygen isotope data partition monkey species of Taï Forest, Côte d'Ivoire (another chimpanzee habitat), along a vertical gradient according to feeding height (Krigbaum et al. 2013). In forests with dense canopies, ^{13}C-depleted CO_2 produced by soil respiration and rotting leaf litter remains trapped under the canopy and lowers the δ^{13}C of CO_2 in local air, and this CO_2 is extensively recycled during photosynthesis, resulting in low δ^{13}C values of plants (Van der Merwe and Medina 1991; Cerling et al. 1997). These low δ^{13}C values are reflected in Ituri forest fauna of the Democratic Republic of the Congo and Taï Forest fauna of Côte d'Ivoire, two closed-canopy forests (Cerling et al. 2004; Krigbaum et al. 2013).

Here we explore differences between isotope values of apes versus early hominins in order to address the following questions: (1) What habitats supported the transition from ape to hominin? (2) Within those vegetation mosaics, where did early hominins forage? And (3) did early hominins become more generalist feeders relative to apes, and if so, what habitat or climate limitations did they continue to face? In order to address these questions, we compiled carbon and oxygen isotope data from modern apes, Miocene apes, early hominins, and associated faunas where available (Table 8.1).

Tooth enamel records the carbon isotopic signal from vegetation eaten, with an enrichment in δ^{13}C due to fractionation from metabolic processes. Experimental studies have shown that this fractionation factor differs across animals with different digestive systems (Passey et al. 2005; Crowley et al. 2010). Cerling et al. (1999) report a fractionation factor of 14‰ in an extensive analysis of mammals ranging in size from elephants to small antelopes, but they caution that animals that do not produce large quantities of methane might have fractionation factors closer to 13‰. Cerling et al. (2004) estimate a fractionation factor of 12.8‰ for primates in the Ituri forest based

TABLE 8.1. Hominoid isotope datasets.

Hominoid/Hominin	Locality	With Fauna	$\delta^{18}O$ Values	Sample Type	References
Rainforest chimp	Kibale National Park, Uganda	X	X	Tooth enamel	Nelson 2013
Rainforest chimp	Ituri Forest, DRC	X, 1 hominoid	X	Tooth enamel	Cerling et al. 2004
Rainforest chimp	Loango, Gabon			Hair	Oelze et al. 2014
Gallery forest chimp	Ishasha, DRC			Hair	Schoeninger et al. 1999
Gallery forest chimp	Ugalla, Tanzania			Hair	Schoeninger et al. 1999
Gallery forest chimp	Fongoli, Senegal			Hair	Sponheimer et al. 2006
Bonobo	Salonga, DRC			Hair	Oelze et al. 2011
Lowland gorilla	Loango, Gabon			Hair	Oelze et al. 2014
Mountain gorilla	Bwindi, Uganda			Feces	Blumenthal et al. 2012
Sivapithecus	Siwaliks, Pakistan	X	X	Tooth enamel	Nelson 2007
Griphopithecus	Pasalar, Turkey	X, 1 hominoid	X	Tooth enamel	Quade et al. 1995
Kenyapithecus	Ft. Ternan, Kenya	X, 0 hominoid	X	Tooth enamel	Cerling et al. 1997
Oreopithecus	Baccinello, Fiume Santo, Italy	X	X	Tooth enamel	Nelson and Rook 2016
Gigantopithecus	Sanhe, Longgudong, Juyuandong, China	X	X	Tooth enamel	Nelson 2014; Qu et al. 2014
Pongo	Sanhe, China	X	X	Tooth enamel	Qu et al. 2014
Orrorin	Lukeino, Kenya	X, 0 hominin	X	Tooth enamel	Roche et al. 2013
Ardipithecus	Aramis, Ethiopia	X	X	Tooth enamel	White et al. 2009
Au. anamensis	Turkana Basin			Tooth enamel	Cerling et al. 2013
Kenyanthropus	Turkana Basin			Tooth enamel	Cerling et al. 2013
Au. deyiremida	Woranso-Mille, Ethiopia	X	X	Tooth enamel	Levin et al. 2015
Au. afarensis	Hadar, Ethiopia	X	X	Tooth enamel	Bedaso et al. 2013; Wynn et al. 2013

TABLE 8.1. (continued)

Hominoid/Hominin	Locality	With Fauna	$\delta^{18}O$ Values	Sample Type	References
Au. bahrelghazali	Koro Toro, Chad	X	X, fauna only	Tooth enamel	Lee-Thorp et al. 2012; Zazzo et al. 2000
Au. africanus	Makapansgat, Sterkfontein, S. Africa			Tooth enamel	Sponheimer et al. 2013
Au. aiethiopicus	Turkana Basin			Tooth enamel	Cerling et al. 2013
Au. boisei	Koobi Fora, Kenya	X	X	Tooth enamel	Cerling et al. 2011a
Au. robustus	Swartkrans, S. Africa			Tooth enamel	Sponheimer et al. 2013
Au. sediba	Malapa, S. Africa			Tooth enamel	Henry et al. 2012
Early *Homo*	Koobi Fora, Kenya	X	X	Tooth enamel	Cerling et al. 2013

Datasets used in our analyses. All hominoids and hominins are represented by carbon isotope data. Those with faunal isotope and oxygen isotope data are marked.

on isotopic analyses of tooth enamel and plant foods. Prowse et al. (2004) use a fractionation factor of 13‰ for humans. Passey et al. (2005) report a fractionation factor of 13.3‰ for suids. Clementz et al. (2009) report a fractionation factor of 12.7‰ for carnivores, and Passey et al. (2005) report 11.5‰ for rodents. Therefore, we use an enrichment factor of 13‰ for primates, 13.3‰ for suids, 12.7‰ for carnivores, 11.5‰ for rodents, and 14‰ for all other mammals. Given that we compare samples from tooth enamel, hair, and feces, we first adjust specimen values for the fractionation between the tissues and the diet they represent. For tooth enamel, we subtract the appropriate fractionation factor as stated above for animals with different digestive systems in order to obtain dietary $\delta^{13}C$ values. For hair, we subtract 3.2‰ (Oelze et al. 2011; Sponheimer et al. 2006), and for feces we subtract 0.3‰ (Blumenthal et al. 2012).

Finally, in making comparisons across geologic time, we adjust $\delta^{13}C$ values to correct for shifts in atmospheric carbon isotope ratios. Atmospheric data from Scripps field site at Mauna Loa, Hawaii, indicate that as of 2008, the $\delta^{13}C$ value of the atmosphere is −8.2‰ (Keeling et al. 2010). Fauna from Kibale and the Ituri forest were collected in the 1990s, when the $\delta^{13}C$ value of the atmosphere was −7.9‰. For fossil faunas, Passey et al. (2009) estimated atmospheric CO_2 $\delta^{13}C$ values through geologic time based on changes measured in deep ocean carbonates (Zachos et al. 2001). When estimating dietary $\delta^{13}C$ values for fossil species, or when comparing different localities, we correct for atmospheric differences by subtracting from each sample the amount needed to make atmospheric CO_2 values comparable to the modern value of −8.2‰.

By comparing $\delta^{13}C$ values of faunas from different localities, we can capture differences in the vegetation mosaic of these different habitats (Figure 8.1). All early hominin localities yield higher $\delta^{13}C$ values than ape localities, both modern and Miocene (median = −18.4 versus −28.3‰, Mann-Whitney: $Z = 23.63$; $p < 0.001$), as well as greater variance in values (variance = 21.2 versus 5.8). Global vegetation isotope analyses indicate that for much of the world, higher $\delta^{13}C$ values are associated with lower mean annual precipitation (MAP) (Diefendorf et al. 2010; Kohn 2010). The differences in $\delta^{13}C$ values between hominin and ape localities suggest that even the earliest hominins inhabited significantly drier habitat than any ape, modern or extinct. Furthermore, the greater variance suggests that early hominins experienced a vegetation mosaic that included some C_4 grasses. All hominin localities yield $\delta^{13}C$ values higher than −22‰, the end member for C_3 plants (Bender 1971; Vogel 1980; Farquhar et al. 1989).

Within this drier and more varied mosaic, we next consider potential differences in subsistence behavior of the early hominins relative to apes. First, we compare $\delta^{13}C$ values across species, irrespective of their contemporaneous faunas (Figure 8.2). Among modern apes, three modern chimpanzee populations occupying gallery forests surrounded by savannas (Ishasha, Ugalla, and Fongoli) yield significantly higher $\delta^{13}C$ values compared with all rainforest chimpanzees, gorillas, and bonobos (median = −25.6 versus −28.7‰, $Z = 12.26$; $p < 0.001$), and they demonstrate less variance in their values (0.24 versus 0.54). While gallery forests make up a small fraction of the habitat mosaic in these locations, these chimpanzee populations make disproportionate use of the forests for food and nests (Baldwin et al. 1982; McGrew et al. 1988; Sept 1992; Hunt and McGrew 2002; Copeland 2009), and thus rely extensively on forests supported by groundwater near the river in spite of lower MAP and the presence of more open habitat. The higher $\delta^{13}C$ values of the gallery forest versus rainforest chimpanzees are consistent with lower MAP, and their lower variance suggests a more constrained feeding niche space.

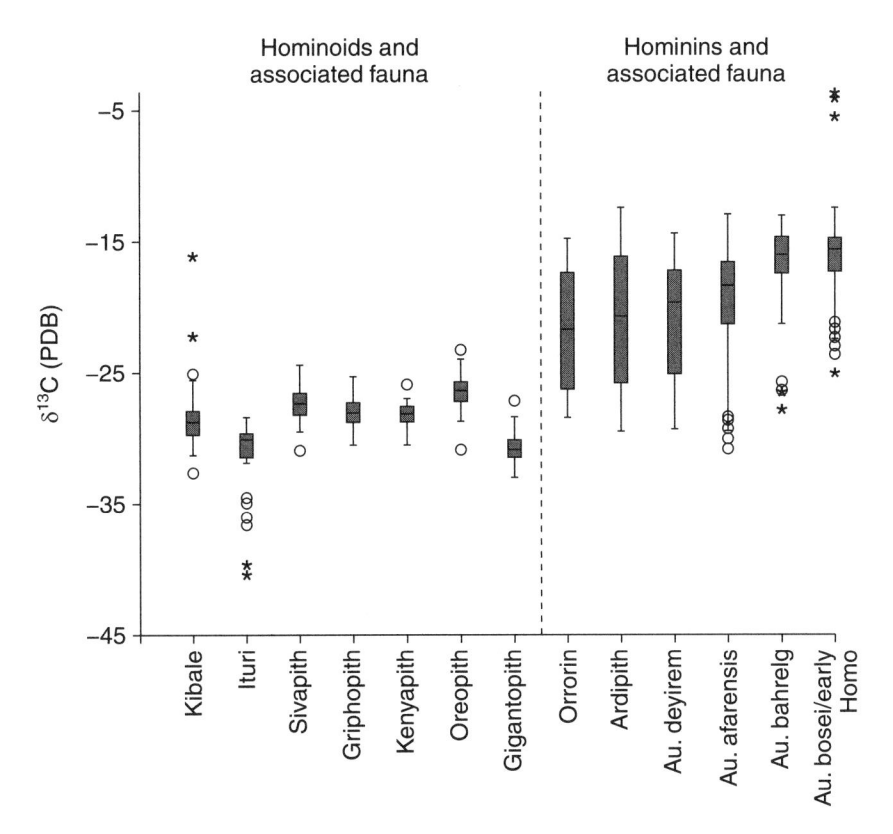

THE HUMAN DIETARY NICHE: INITIAL TRANSITIONS 293

FIGURE 8.1. Distribution of dietary carbon isotope values for modern and fossil hominoid and early hominin faunas. For each faunal set, the middle line represents the median. The bottom of the box is the 25th percentile, while the top of the box is the 75th percentile. T-bars extend to 1.5 times the height of the box or to the minimum and maximum values. Points are outliers, with asterisks being extreme outliers more than three times the height of the box.

Among fossil apes, including *Sivapithecus, Oreopithecus, Gigantopithecus,* and fossil orangutans, *Oreopithecus* is the only species to yield higher $\delta^{13}C$ values than rainforest apes (median = −25.7‰, $Z = 4.80$; $p < 0.001$), with values comparable to those of gallery forest chimpanzees, but with greater variance (1.65). The greater variance might suggest a different or more expansive feeding niche. Pollen, isotope, and geologic data indicate that *Oreopithecus* inhabited a swamp-like forest with subtropical elements despite lower MAP than Asian forests of comparable botanical composition (Harrison and Harrison 1989; Benvenuti et al. 1994; Nelson and Rook 2016). *Oreopithecus* thus

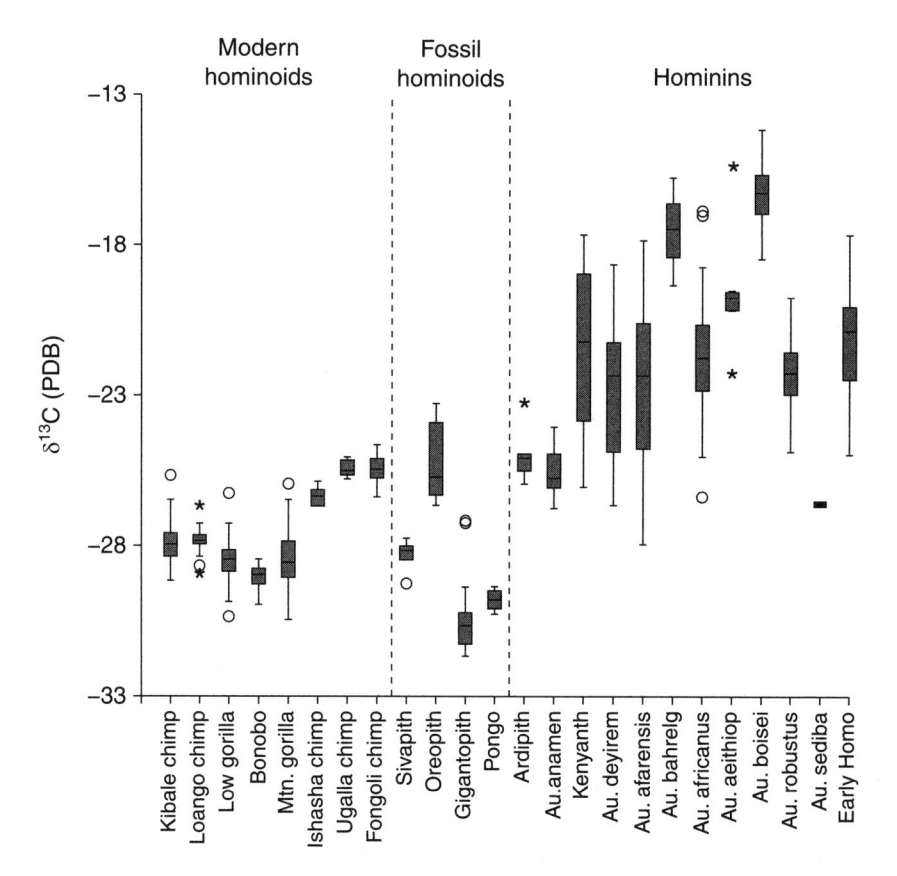

FIGURE 8.2. Distribution of dietary carbon isotope values for modern and fossil hominoids and early hominins. For each species or population, the middle line represents the median. The bottom of the box is the 25th percentile, while the top of the box is the 75th percentile. T-bars extend to 1.5 times the height of the box or to the minimum and maximum values. Points are outliers, with asterisks being extreme outliers more than three times the height of the box.

represents another ape that relied on vegetation supported by permanent water sources in spite of relatively low MAP.

Among early hominins, the two oldest species sampled, *Ardipithecus (Ar.) ramidus* and *Au. anamensis,* yield $\delta^{13}C$ values comparable to those of gallery forest chimpanzees and *Oreopithecus* (median −25.0 and −25.7‰; variance 0.64 and 0.63). Thus the carbon data are compatible with the hypothesis that gallery forest chimpanzees might be models for the last common ancestor of chimpanzees and humans, with gallery forests buffering the transition from rainforest to open woodlands and savannas (Moore 1996). More extensive isotope analyses of faunas or plants associated with gallery forest chimpanzees would greatly aid in these comparisons. All later hominins yield significantly higher $\delta^{13}C$ values (median = −21.3‰, Z = 6.31; p < 0.001) except for *Au. sediba,* which is represented by only two samples. These values provide evidence for the incorporation of some C_4 or aquatic vegetation in hominin diets, with notably high levels of C_4 or aquatic plants in *Au. bahrelghazali* and *Au. boisei* diets. We mention aquatic vegetation because some aquatic plants yield $\delta^{13}C$ values similar to those of C_4 plants, not because of their photosynthetic pathway but because carbon dioxide diffuses slowly across water barriers, driving $\delta^{13}C$ values high (Farquhar et al. 1989). Furthermore, Wrangham et al. (2009) have suggested that river deltas might have been key habitats for the earliest hominins given the presence of underground storage organs (tubers) and fruits as possible abundant and relatively high-quality food sources.

Interestingly, for species represented by more than just a few samples, variance in $\delta^{13}C$ values is much higher in post–*Au. anamensis* species compared with apes and earlier hominins, ranging between 1.24–7.01, with the lowest variance in two robust australopithecines, *Au. boisei* (1.24) and *Au. robustus* (1.65). The high variance of most australopithecine species suggests increasing niche breadth, with specialization in the robust australopithecines. To determine subsistence behavior, such as feeding on aquatic vegetation or arboreal foods, and to quantify niche breadth, we use a combination of carbon and oxygen isotopes (Figure 8.3). Within Kibale fauna, $\delta^{18}O$ values distinguish between forest terrestrial feeders, arboreal frugivores, and arboreal folivores, with terrestrial feeders exhibiting the lowest $\delta^{18}O$ values and arboreal folivores the highest (Nelson 2013). These isotope distinctions are likely the effect of lower irradiance and evaporative stresses on the forest floor versus the canopy, and greater evaporative stress in leaves versus fruits because of the role leaves play in transpiration. Kibale chimpanzee $\delta^{18}O$ values fall within the arboreal frugivore range, higher than those of terrestrial baboons and L'Hoest monkeys, but lower than folivorous colobus monkeys. Chimpanzees are therefore characterized by low $\delta^{13}C$ values reflective of tropical forest, but high $\delta^{18}O$ values relative to terrestrial species in the community, due to feeding in the canopy. *Sivapithecus,* a Miocene ape from Pakistan, also exhibits this combination of carbon and oxygen values (Nelson 2007), as do many giraffes. While combined low carbon

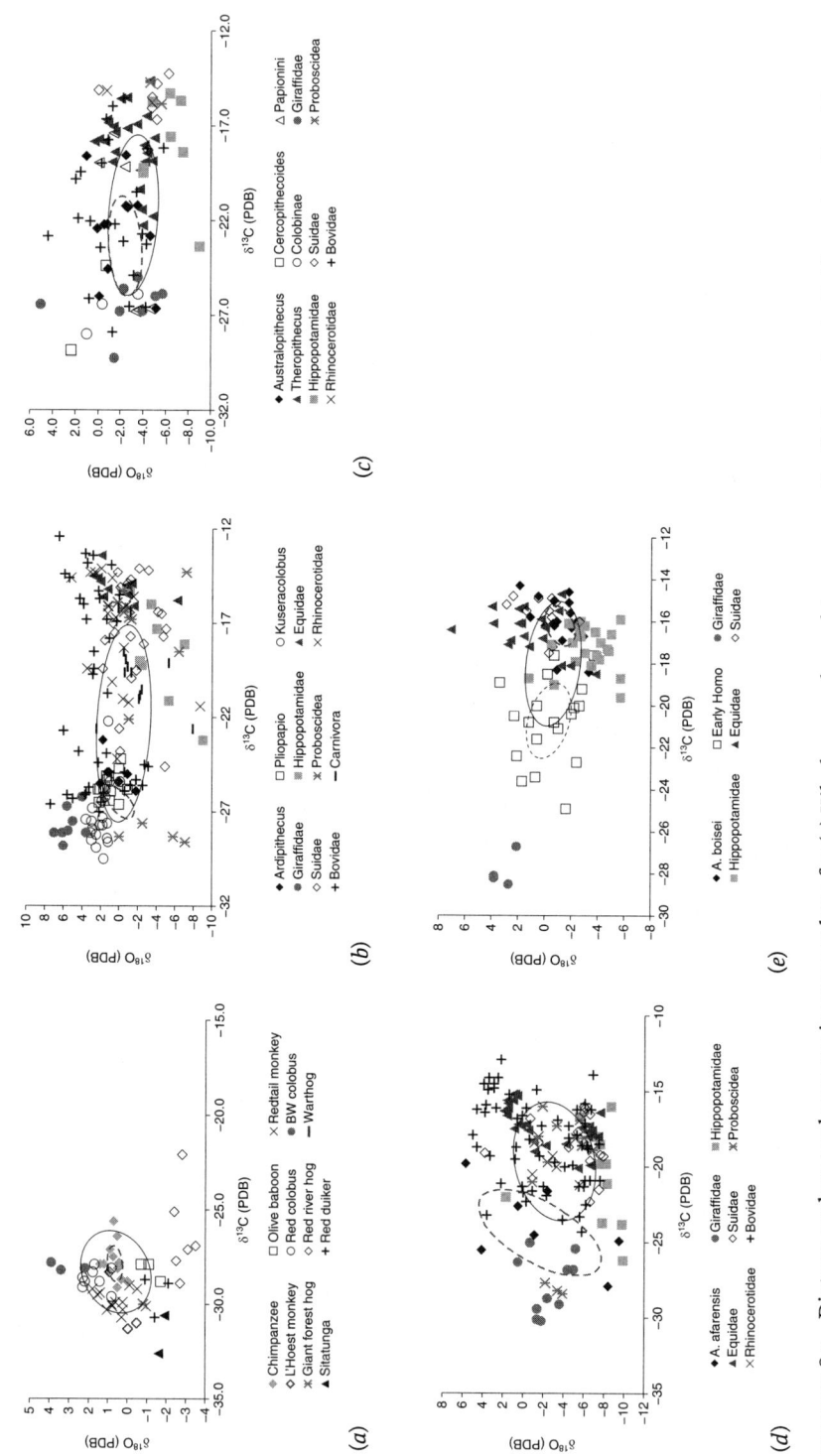

FIGURE 8.3. Dietary carbon and oxygen isotope values for (a) Kibale, a modern chimpanzee site; (b) Aramis, *Ardipithecus* locality; (c) Woranso-Mille, *Au. Deyiremeda*; (d) Hadar, *Au. afarensis*; and (e) Koobi Fora, *Au. boisei* and early *Homo*. Solid ellipses represent total available niche space of fauna. Dashed ellipses represent percentage of total available niche space occupied by modern chimpanzees and the five hominin taxa. For Woranso-Mille, the ellipse with long dashes is *Au. boisei*, and the ellipse with short dashes is early *Homo*.

and oxygen isotopes suggest feeding on forest floors and low carbon with high oxygen suggest feeding in forest canopies, high carbon and oxygen indicate the most evaporative or open habitats. Equids usually yield this combination of values. Finally, high carbon combined with low oxygen might indicate feeding on aquatic plants or underground storage organs. Both underground organs and aquatic plants should yield relatively low $\delta^{18}O$ values, because they are under the least evaporative stress underground or in water, but this remains to be shown empirically with modern vegetation. As stated previously, aquatic plants often yield high $\delta^{13}C$ values due to the water barrier (Farquhar et al. 1989). Roots also yield consistently higher $\delta^{13}C$ values than do leaves (von Fischer and Tieszen 1995; Hobbie and Werner 2004). Hippopotamuses usually yield the combination of high carbon and low oxygen isotope values. Among apes, *Oreopithecus* also yields this carbon-oxygen combination, suggesting that it took advantage of aquatic foods within its swamp habitat (Nelson and Rook 2016). Among hominins, *Au. boisei* yields a similar pattern, again suggesting that it fed on submerged plants in a wetland setting (Shipman and Harris 1988; Cerling et al. 2011a).

Carbon and oxygen isotope data are available for faunas and apes/hominins from the chimpanzee site Kibale, *Ar. ramidus* Aramis site [4.4 million years ago (Ma)], an australopithecine that may be *Au. deyiremeda* intermediate between *Au. anamensis* and *Au. afarensis* at Woranso-Mille, *Au. afarensis* in the Hadar formation, and both *Au. boisei* and early *Homo* at Koobi Fora (Figure 8.3). For temporal control, we restricted the *Au. deyiremeda* dataset to the 3.8–3.6 Ma fauna, the Hadar data to the Sidi Hakoma formation (3.4–3.2 Ma), and the Koobi Fora data to the KBS and Okote members (1.9–1.5 Ma). To determine proportions of different food sources using carbon and oxygen stable isotopes, we use Bayesian mixing models, which provide several advantages over classic statistical methods (Phillips 2001; Inger and Bearhops 2008). Basic or linear mixing models are limited in the number of potential sources that can be included (only $n-1$ the number of isotopic systems analyzed), and while they are able to calculate a range of possible solutions, they cannot estimate which solutions are more or less likely (Phillips 2001; Phillips and Gregg 2003). We address both of these shortcomings by using a Bayesian framework. In our models, potential food sources and the unknown diet of interest are defined isotopically by a mean and standard deviation. Here, we define potential sources for each hominin using contemporaneous nonhominin fauna with known dietary niches, broadly defined as "grazers" (e.g., equids, most bovids, modern warthog-like suids), "aquatic grazers" (e.g., hippopotamus), and "arboreal browsers" (e.g., giraffes, colobus monkeys). A unique combination of carbon and oxygen isotopic values characterizes each niche as described above. In Aramis and Kibale National Park, a fourth niche dimension, "terrestrial browser," is occupied by *Deinotherium*, Neotragini, and *Kolpochoerus* (Aramis) and modern olive baboons, L'Hoest monkeys, bushpigs, and red duikers (Kibale). The absence of these niches in later hominin sites could be either an artifact of fossilization bias or an indication of the absence of that niche in those environments, perhaps due to increasingly open habitats. We then include the carbon and oxygen measurements for the associated hominin as the unknown diet of interest. We conducted all analyses using the SIAR program in R (Parnell et al. 2010). We used a noninformative Dirichlet distribution prior, as is suggested for models without any observational dietary data. We then ran a series of 200,000 Markov chain Monte Carlo simulations (plus 50,000 burn-ins), each of which proposes a random vector of potential contributions of the food sources (faunal niches) to the unknown diet (the hominin), and determines the probability of that proportionate contribution vector resulting in the observed diet. The vector was accepted if its probability was higher than the previously proposed vector. Through this process, we iteratively approached the most likely combination of proportionate food source contributions (Parnell et al. 2010; Newsome et al. 2012). We could then calculate the mean proportionate contribution from these iterations along with a standard deviation. The more evenly distributed the proportionate source contributions were, the more generalized the hominin niche. Highly skewed proportionate source contributions indicate niche specialization.

There are inherent limitations to estimating niche breadth through mixing models. First, the models assume that all sources included represent some portion of the diet; therefore, the model will never return a result of 0 percent contribution. Second, the model assumes that all dietary sources are included. Omitting a potential source will result in skewed and potentially impossible proportionate attributions (Phillips and Gregg 2003; Phillips 2012). To ensure that contribution estimations were reasonable despite these limitations, we ran additional models using taxa with known niche spaces (such as the grazing and highly specialized *Theropithecus*) as the unknown diet of interest. Returns were consistent with its known grazing specialization. We repeated this process for the species of Kibale National Park, with the model again yielding results compatible with known diets.

Our Bayesian mixing models indicate that the oldest hominin represented by the data, *Ar. ramidus,* had a diet similar to that of chimpanzees with respect to proportions of arboreal versus terrestrial feeding (Figure 8.4). Within the terrestrial food sources, *Ar. ramidus* incorporated more C_4 vegetation, compatible with a more varied and open vegetation mosaic (White et al. 2009). *Au. deyiremeda* and *Au. afarensis,* the two oldest australopithecines in this study, were characterized by more even distributions across all resources, suggesting an increasingly generalized diet. *Au. deyiremeda* derived just under half of its resources from arboreal browsing and just under half from terrestrial grazing, with a small (11 percent) contribution of aquatic grazing resources. *Au. afarensis* further increased the proportionate spread of resources, indicating an increased trend toward generalism. This latter species derived just under half of all dietary contributions from arboreal browsing, with the remaining half of the diet split almost equally between terrestrial and aquatic grazing resources (27 percent and 25 percent, respectively). No Hadar or Woranso-Mille fauna represented the terrestrial browsing niche, demonstrating either a lack of availability of such a niche or a lack of sampled fauna from such a niche. If the latter, these missing data would result in an overestimation of the arboreal browsing or terrestrial grazing foods for these australopithecines. However, the terrestrial versus arboreal proportions of the diet are consistent with australopithecine postcranial morphologies that suggest a mosaic of bipedal and suspensory features (Wood and Collard 1999). *Theropithecus,* also from the Sidi Hakoma member in Hadar, illustrates the proportionate contributions expected for a strong specialist, with 67 percent of its diet derived from terrestrial grazing.

Analyses of the Okote and KBS formations from Koobi Fora indicate that *Au. boisei* and early *Homo* occupied different niches. Terrestrial and aquatic grazing resources contributed equally to the diet of *Au. boisei,* suggesting that *boisei* abandoned significant arboreal subsistence behaviors characteristic of earlier australopithecines in favor of occupying a grazing niche. While this is a more specialized niche than *Au. deyiremeda* and *Au. afarensis,* it was not a specialist in the same sense as *Theropithecus.* Early *Homo,* on the other hand, mirrored the pattern of a dietary generalist seen in early australopithecines, but with increasing aquatic grazing contributions at the expense of browsing resources. However, similar to data from Hadar, terrestrial browsing was not

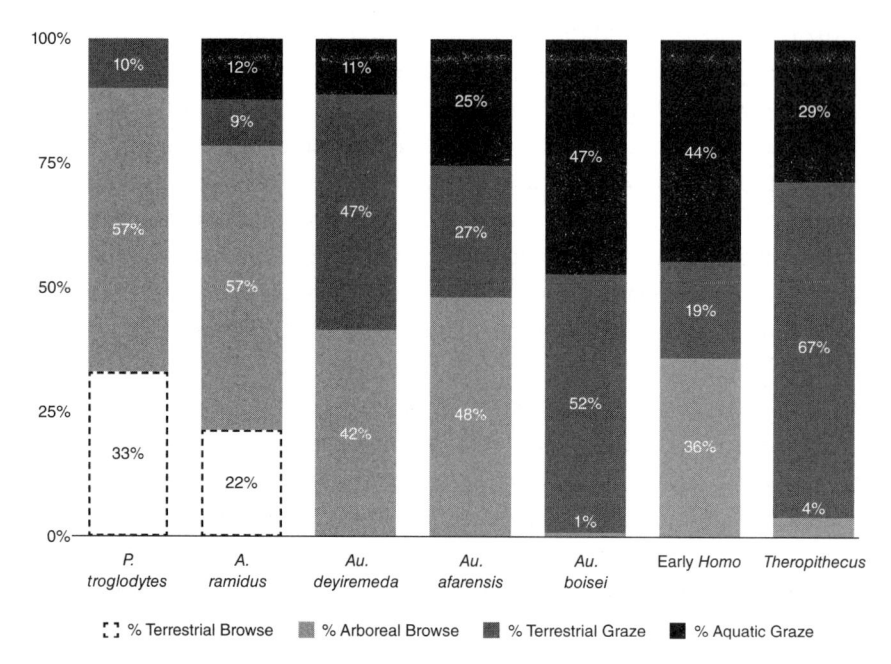

FIGURE 8.4. Proportionate contributions of diet sources for modern chimpanzees at Kibale, five hominin species, and *Theropithecus*, estimated through Bayesian mixing models.

represented in the fauna from Koobi Fora. Again, if terrestrial browsers simply were not included in the original isotope dataset, the missing data could have resulted in overestimation of graze or arboreal browse for early *Homo*.

To assess how much of the available vegetation mosaic each hominin used for foraging, we calculated the standard ellipse area of each hominin's isotopic niche space and compared that to the ellipse area of the entire contemporaneous fauna. Standard ellipse areas are similar to convex hull measurements in that both measure the amount of bivariate isotopic space occupied by a given taxon. However, standard ellipse areas are far less sensitive to changes in sample sizes compared with convex hulls (Layman et al. 2007; Jackson et al. 2011). By calculating the standard ellipse area for each hominin and dividing it by the area occupied by the entire faunal community, we can estimate the proportion of available niche space used by each species. Generalist species are expected to have higher percentages, while specialists' percentages are expected to be lower (Newsome et al. 2012).

Standard ellipse areas are calculated using the mean of each isotopic system (here, carbon and oxygen) to determine the location of the ellipse in isotopic space, and the covariance matrix between the isotopes (analogous to the standard deviation of bivariate data) to determine its size and shape (Jackson et al. 2011). A vague normal prior is assigned to the mean of each isotope, and a vague Inverse-Wishart prior is assigned to the covariance matrix (McCarthy 2007; Jackson et al. 2011). A series of Markov chain Monte Carlo simulations generates posterior estimates of the isotopic means and their covariance matrixes, which can then be used to draw a standard ellipse around the isotopic niche space, calculate its area, and estimate the uncertainty. We repeated this process using data from hominin species as well as all faunas from the different localities to derive percentage use of the available niche space from the ratios (Newsome et al. 2012).

Comparisons of hominin standard ellipse areas with the area of contemporaneous fauna are consistent with increasing hominin niche breadth through time (Figures 8.3 and 8.5). *Ar. ramidus* yielded a small niche breadth similar to modern chimpanzees (using 9 percent and 15 percent of available niche space). Among early australopithecines, niche space increased dramatically, up to 45 percent for *Au. deyiremeda* and 80 percent for *Au. afarensis*. It is possible that lack of terrestrial browsing fauna in the Hadar data might have driven this high estimate for *Au. afarensis*. *Au. boisei* returned to chimpanzee-like niche breadth parameters (16 percent), although with very different sources contributing. Early *Homo* maintained *Au. deyiremeda*–like niche space (46 percent).

These data are consistent with chimpanzee-like niche space for hominins such as *Ar. ramidus* prior to four million years ago, both in the types of resources contributing to their diet (Figures 8.2 and 8.4) and in the proportions of available resources exploited (Figure 8.5). Gallery forest chimpanzees might present a good analogue for some of the earliest hominins. Carbon and oxygen isotope analyses of these marginal chimpanzees and their faunas, as well as other early hominins, would greatly aid in testing this hypothesized model of the last common ancestor. After this initial transition, subsequent australopithecines greatly expanded their niche breadth. A steady increase in the contribution of aquatic resources at the expense of arboreal browsing could have contributed greatly to their expanding subsistence niches, supporting a wetland hypothesis (Wrangham et al. 2009) for suitable habitat and food resources that predate substantial exploitation of animal foods or

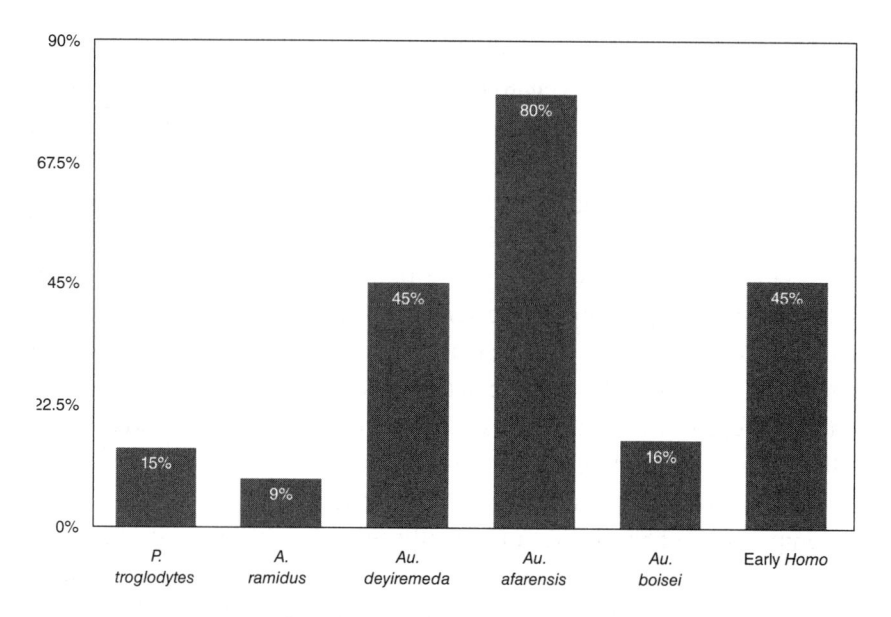

FIGURE 8.5. Percentage of total available dietary niche space occupied by modern chimpanzees and five hominin taxa. Percentages were calculated by dividing the standard ellipse area occupied by each hominin in bivariate isotope space by the total standard ellipse area of all fauna from the hominin's locality. Increased percentages correspond with increased dietary generalism.

improvement of plant and animal items through processing. *Au. boisei* appears to have specialized in wetland vegetation at Koobi Fora, although it was still more omnivorous than the highly specialized *Theropithecus*. Contemporaneous early *Homo* at Koobi Fora remained a dedicated generalist, continuing to utilize portions of all available niche spaces in the environment, including arboreal foods.

Finally, while the australopithecines expanded their habitat and feeding niches beyond those of modern apes, it is notable that the highest carbon isotope values among modern apes, Miocene apes, and early hominins are all associated with permanent water sources (Figure 8.2). In each case, high carbon isotope values likely indicate lower mean annual precipitation than that experienced by comparable species. Gallery forest chimpanzees make use of riverine forests to live in regions where the dry season is longer than those experienced by other great apes. In the Miocene, *Oreopithecus* appears to have exploited aquatic food sources in a swamp at a time when all other apes were driven to extinction in Europe due to cooling, aridification, greater seasonality of precipitation, and forest loss. Among australopithecines represented by oxygen isotope and faunal data, *Au. boisei* exploited the most open habitat, but through wetland resources. Considered together, our isotope analyses suggest that, like their ape cousins, early hominins still experienced habitat limitations driven by seasonal precipitation regimes. Australopithecines were not yet capable of expanding their range to more temperate regions.

Conclusions

Our review of chimpanzee, human, and ancestral human diets highlights several priorities for future research. With respect to the transition from ape to hominin, morphological, microwear, and isotopic analyses have provided insights into potential characteristics of early hominin plant food sources. However, we continue to be challenged with the question of what potential C_4 or aquatic vegetation requiring powerful mastication would have been of high enough quality to support large-bodied, large-brained, slow-growing hominins. Isotope data suggest that we should look to habitats with gallery forests or wetlands as potential sources. Extensive isotopic analyses of faunas and vegetation from these modern habitats, with a focus on both carbon and oxygen isotopes, would provide a direct means of comparison to fossil faunas and could address the vagaries of their vegetation mosaics. Furthermore, we need more extensive sampling of hominins and their faunas, particularly for the South African hominins, for too many of them are represented by only a few specimens, only carbon isotopes, or no contemporaneous fauna. With only a few samples, Bayesian analyses simply do not yield reliable results. Without oxygen isotopes and faunal samples, we fail to capture a more complete reconstruction of subsistence behavior. In addition to isotopic reconstructions, macronutrient analyses addressing carbohydrate, protein, and fiber content of non-ape foods in gallery forest and wetland settings could potentially identify high-quality resources. Finally, questions remain concerning the driving factor behind thick tooth enamel, and resolution of the effects of phytolith loads and grit will also aid us in narrowing our search for potential early hominin foods.

Summary

Early hominins faced the challenge of meeting hominoid high energetic requirements but in more open, drier, and more seasonal habitats compared with their ape ancestors. In this setting of limited ripe fruit availability, they exploited different foods from those of apes, and ultimately expanded their dietary niche breadth. We used carbon and oxygen stable isotopes of early hominins and their contemporaneous fauna to explore these dietary transitions. The earliest hominins yield carbon isotope values similar to those of chimpanzees living in gallery forests. Later australopithecines incorporated some C_4 or aquatic plants into their diets. Among those hominins with the highest carbon values, and thus most open habitats, oxygen isotopes suggest that they fed in wetlands. Thus, early hominins likely still experienced habitat limitations driven by seasonal precipitation regimes, and they alleviated the driest climates with expansive standing water.

To explore different proportions of food sources and niche breadth, we employed Bayesian mixing models. Our models indicate that the earliest hominins had diets similar to that of chimpanzees with respect to proportions of arboreal versus terrestrial feeding. Later gracile australopithecines were characterized by more even distributions across all resources, suggesting an increasingly generalized diet. Early *Homo*, like its predecessors, was a dietary generalist, but with increasing aquatic grazing contributions at the expense of browsing resources. Finally, within our Bayesian framework, we used standard ellipse areas to calculate dietary niche breadth. The earliest hominins resembled chimpanzees in breadth, while later gracile australopithecines and early *Homo* expanded their niches. With respect to the last common ancestor, our results suggest that the earliest hominins such as *Ar. ramidus* exploited a chimpanzee-like niche space, based upon isotope values that resemble those of gallery forest chimpanzees, proportions of arboreal versus terrestrial resources contributing to their diet, and a narrow niche breadth.

References

Aiello, L. C., and C. Key. 2002. Energetic consequences of being a *Homo erectus* female. *American Journal of Human Biology* 14: 551–565.

Aiello, L. C., and P. Wheeler. 1995. The expensive tissue hypothesis: The brain and the digestive system in human and primate evolution. *Current Anthropology* 36: 199–221.

Baldwin, P., W. McGrew, and C. Tutin. 1982. Wide-ranging chimpanzees at Mt. Asserik, Senegal. *International Journal of Primatology* 3: 367–385.

Bedaso, Z. K., J. G. Wynn, Z. Alemseged, and D. Geraads. 2013. Dietary and paleoenvironmental reconstruction using stable isotopes of herbivore tooth enamel from middle Pliocene Dikika, Ethiopia: Implication for Australopithecus afarensis habitat and food resources. *Journal of Human Evolution* 64: 21–38.

Bender, M. M. 1971. Variations in the $^{13}C/^{12}C$ ratios of plants in relation to the pathway of photosynthetic carbon dioxide fixation. *Phytochemistry* 10: 1239–1245.

Benvenuti, M., A. Bertini, and L. Rook. 1994. Facies analysis, vertebrate paleontology and palynology in the Late Miocene Baccinello-Cinigiano Basin (southern Tuscany). *Memorie della Societa Geologica Italiana* 48: 414–423.

Berna, F., P. Goldberg, L. K. Horwitz, J. Brink, S. Holt, M. Bamford, and M. Chazan. 2012. Microstratigraphic evidence of in situ fire in the Acheulean strata of Wonderwerk Cave, Northern Cape province, South Africa. *Proceedings of the National Academy of Sciences* 109: E1215–E1220.

Björck, I., H. Liljeberg, and E. Ostman. 2000. Low glycaemic-index foods. *British Journal of Nutrition* 83: S149–S155.

Blumenthal, S., K. Chritz, J. Rothman, and T. Cerling. 2012. Detecting intraannual dietary variability in wild mountain gorillas by stable isotope analysis of feces. *Proceedings of the National Academy of Sciences* 109: 21277–21282.

Cerling, T. E., J. M. Harris, S. H. Ambrose, M. G. Leakey, and N. Solounias. 1997. Dietary and environmental reconstruction with stable isotope analyses of herbivore tooth enamel from the Miocene locality of Fort Ternan, Kenya. *Journal of Human Evolution* 33: 635–650.

Cerling, T. E., J. M. Harris, and M. G. Leakey. 1999. Browsing and grazing in elephants: The isotope record of modern and fossil proboscideans. *Oecologia* 120: 364–374.

Cerling, T., J. Hart, and T. Hart. 2004. Stable isotope ecology in the Ituri forest. *Oecologia* 138: 5–12.

Cerling, T., F. Manthi, E. Mbua, L. Leakey, M. Leakey, R. Leakey, F. Brown, F. Grine, J. Hart, P. Kaleme, H. Roche, K. Uno, and B. Wood. 2013. Stable isotope-based diet reconstructions of Turkana Basin hominins. *Proceedings of the National Academy of Sciences* 110: 10501–10506.

Cerling, C., E. Mbua, F. Kirera, F. Manthi, F. Grine, M. Leakey, M. Sponheimer, and K. Uno. 2011a. Diet of *Paranthropus boisei* in the early Pleistocene of East Africa. *Proceedings of the National Academy of Sciences* 108: 9337–9341.

Cerling, T., J. Wynn, S. Andanje, M. Bird, D. Korir, N. Levin, W. Mace, A. Macharia, J. Quade, and C. Remien. 2011b. Woody cover and hominin environments in the past 6 million years. *Nature* 476: 51–56.

Clementz, M., K. Fox-Dobbs, P. Wheatley, P. Koch, and D. Doak. 2009. Revisiting old bones: Coupled carbon isotope analysis of bioapatite and collagen as an ecological and paleoecological tool. *Geological Journal* 44: 605–620.

Conklin-Brittain, N., R. Wrangham, and K. Hunt. 1998. Dietary response of chimpanzees and cercopithecines to seasonal variation in fruit abundance, II: Macronutrients. *International Journal of Primatology* 19: 971–998.

Copeland, S. 2009. Potential hominin plant foods in northern Tanzania: semi-arid savannas versus savanna chimpanzee sites. *Journal of Human Evolution* 57: 365–378.

Crowley, B., M. Carter, S. Karpanty, A. Zihlman, P. Koch, and N. Dominy. 2010. Stable carbon and nitrogen isotope enrichment in primate tissues. *Oecologia* 164: 611–626.

Dean, C., M. G. Leakey, D. Reid, F. Schrenk, G. T. Schwartz, C. Stringer, and A. Walker. 2001. Growth processes in teeth distinguish modern humans from *Homo erectus* and earlier hominins. *Nature* 414: 628–631.

Demes, B., and N. Creel. 1988. Bite force, diet, and cranial morphology of fossil hominids. *Journal of Human Evolution* 17: 657–670.

Diefendorf, A., K. Mueller, S. Wing, P. Koch, and K. Freeman. 2010. Global patterns in leaf ^{13}C discrimination and implications for studies of past and future climate. *Proceedings of the National Academy of Sciences* 107: 5738–5743.

Ehleringer, J. R., C. B. Field, Z. F. Lin, and C. Y. Kuo. 1986. Leaf carbon isotope and mineral-composition in subtropical plants along an irradiance cline. *Oecologia* 70: 520–526.

Farquhar, G. D., J. R. Ehleringer, and K. T. Hubick. 1989. Carbon isotope discrimination and photosynthesis. *Annual Review of Plant Physiology and Plant Molecular Biology* 40: 503–537.

Grine, F., M. Sponheimer, P. Ungar, J. Lee-Thorp, and M. Teaford. 2012. Dental microwear and stable isotopes inform the paleoecology of extinct hominins. *American Journal of Physical Anthropology* 148: 285–317.

Harrison, T., and T. Harrison. 1989. Palynology of the late Miocene *Oreopithecus*-bearing lignite from Baccinello, Italy. *Palaeogeography, Palaeoclimatology, Palaeoecology* 76: 45–65.

Henry, A., P. Ungar, B. Passey, M. Sponheimer, L. Rossouw, M. Bamford, P. Sandberg, D. de Ruiter, and L. Berger. 2012. The diet of *Australopithecus sediba*. *Nature* 487: 90–93.

Hobbie, E. A., and R. A. Werner. 2004. Intramolecular, compound-specific, and bulk carbon isotope patterns in C_3 and C_4 plants: A review and synthesis. *New Phytologist* 161: 371–385.

Hummel, J., E. Findeisen, K. Sudekum, I. Ruf, T. Kaiser, M. Bucher, M. Clauss, and D. Codron. 2011. Another one bites the dust: Faecal silica levels in large herbivores correlate with high-crowned teeth. *Proceedings of the Royal Society B* 278: 1742–1747.

Hunt, K., and W. McGrew. 2002. Chimanzees in the dry habitats of Assirik, Senegal, and Semliki Wildlife Reserve, Uganda. In C. Boesch, G. Hohmann, and L. Merchant, eds., *Behavioural Diversity in Chimpanzees and Bonobos*, 35–51. New York: Cambridge University Press.

Inger, R., and S. Bearhops. 2008. Applications of stable isotope analyses to avian ecology. *Ibis* 150: 447–461.

Jackson, A., R. Inger, A. Parnell, and S. Bearhops. 2011. Comparing isotopic niche widths among and within communities: SIBER—Stable Isotope Bayesian Ellipses. *Journal of Animal Ecology* 80: 595–602.

Keeling, R., S. Piper, A. Bollenbacher, and S. Walker. 2010. Monthly atmospheric $^{13}C/^{12}C$ isotopic ratios for 11 SIO stations. Trends: A compendium of data on global change. Carbon Dioxide Information Analysis Center, Oak Ridge National Laboratories. Oak Ridge, TN: U.S. Department of Energy.

Kohn, M. 2010. Carbon isotope compositions of terrestrial C_3 plants as indicators of (paleo)ecology and (paleo)climate. *Proceedings of the National Academy of Sciences* 107: 19691–19695.

Krigbaum, J., M. Berger, D. Daegling, and S. McGraw. 2013. Stable isotope canopy effects for sympatric monkeys at Taï Forest, Côte d'Ivoire. *Biology Letters.* 9: 20130466.

Layman, C., D. Arrington, C. Montana, and D. Post. 2007. Can stable isotope ratios provide for community-wide measures of trophic structure? *Ecology* 88: 42–48.

Lee-Thorp, J., A. Likius, H. Mackaye, P. Vignaud, M. Sponheimer, and M. Brunet. 2012. Isotopic evidence for an early shift to C_4 resources by Pliocene hominins in Chad. *Proceedings of the National Academy of Sciences* 109: 20369–20372.

Levin, N., Y. Haile-Selassie, S. Frost, and B. Saylor. 2015. Dietary change among hominins and cercopithecids in Ethiopia during the early Pliocene. *Proceedings of the National Academy of Sciences* 112: 12304–12309.

Lucas, P., P. Constantino, B. Wood, and B. Lawn. 2008. Dental enamel as a dietary indicator in mammals. *BioEssays* 30: 374–385.

Luz, B., and Y. Kolodny. 1985. Oxygen isotope variations in phosphates of biogenic apatites, IV: Mammal teeth and bones. *Earth and Planetary Science Letters* 75: 29–36.

MacKinnon, J., and K. MacKinnon. 1978. Comparative feeding ecology of six sympatric primates in West Malaysia. In D. J. Chivers and J. Herbert, eds., *Recent Advances in Primatology,* vol. 1: *Behaviour,* 305–321. London: Academic Press.

McCarthy, M. 2007. *Bayesian Methods for Ecology.* Cambridge: Cambridge University Press.

McGrew, W., P. Baldwin, and C. Tutin. 1981. Chimpanzees in a hot, dry open habitat: Mt. Asserik, Senegal, West Africa. *Journal of Human Evolution* 10: 227–244.

McGrew, W., P. Baldwin, and C. Tutin. 1988. Diet of wild chimpanzees *(Pan troglodytes verus)* at Mt. Assirik, Senegal. *American Journal of Primatology* 16: 213–226.

McNab, B. 1990. The physiological significance of body size. In J. Damuth and B. MacFadden, eds., *Body Size in Mammalian Paleobiology,* 11–23. Cambridge: Cambridge University Press.

Moore, J. 1996. Savanna chimpanzees, referential models of the last common ancestor. In W. McGrew, L. Marchant, and T. Nishida, eds., *Great Ape Societies,* 275–292. Cambridge: Cambridge Univeristy Press.

Nelson, S. 2007. Isotopic reconstructions of habitat change surrounding the extinction of *Sivapithecus*, a Miocene hominoid, in the Siwalik Group of Pakistan. *Palaeogeography, Palaeoclimatology, Palaeoecology* 243: 204–222.

Nelson, S. 2013. Chimpanzee fauna isotopes provide new interpretations of fossil ape and hominin ecologies. *Proceedings of the Royal Society B* 280: 20132324.

Nelson, S. 2014. The paleoecology of Early Pleistocene *Gigantopithecus blacki* inferred from isotopic analyses. *American Journal of Physical Anthropology* 155: 571–578.

Nelson, S., and L. Rook. 2016. Isotopic reconstructions of habitat change surrounding the extinction of *Oreopithecus*, the last European ape. *American Journal of Physical Anthropology* 160: 254–271.

Newsome, S., J. Yeakel, P. Wheatley, and M. Tinker. 2012. Tools for quantifying isotopic niche space and dietary variation at the individual and population level. *Journal of Mammalogy* 93: 329–341.

Oelze, V. M., B. T. Fuller, M. P. Richards, B. Fruth, M. Surbeck, J. J. Hublin, and G. Hohmann. 2011. Exploring the contribution and significance of animal protein in the diet of bonobos by stable isotope ratio analysis of hair. *Proceedings of the National Academy of Sciences* 108: 9792–9797.

Oelze, V. M, J. S. Head, M. M. Robbins, M. Richards, and C. Boesch. 2014. Niche differentiation and dietary seasonality among sympatric gorillas and chimpanzees in Loango National Park (Gabon) revealed by stable isotope analysis. *Journal of Human Evolution* 66: 95–106.

Parnell, A., R. Inger, S. Bearhops, and A. Jackson. 2010. Source partitioning using stable isotopes: Coping with too much variation. *PLoS ONE* 5: e9672.

Passey, B., L. Ayliffe, A. Kaakiner, Z. Zhang, J. Eronen, Y. Zhu, L. Zhou, C. Cerling, and M. Fortelius. 2009. Strengthened East Asian summer monsoons during a period of high-latitude warmth? Isotopic evidence from Mio-Pliocene fossil mammals and soil carbonates from northern China. *Earth and Planetary Science Letters* 277: 443–452.

Passey, B., T. Robinson, L. Ayliffe, C. Cerling, M. Sponheimer, D. Dearing, B. Roeder, and J. Ehleringer. 2005. Carbon isotope fractionation beween diet, breath CO_2, and bioapatite in different mammals. *Journal of Archaeological Science* 32: 1459–1470.

Phillippy, B. Q., J. M. Bland, and T. J. Evens. 2003. Ion chromatography of phytate in roots and tubers. *Journal of Agricultural and Food Chemistry* 51: 350–353.

Phillips, D. 2001. Mixing models in analyses of diet using multiple stable isotopes: A critique. *Oecologia* 127: 166–170.

Phillips, D. 2012. Converting isotope values to diet composition: The use of mixing models. *Journal of Mammalogy* 93: 342–352.

Phillips, D., and J. Gregg. 2003. Source partitioning using stable isotopes: Coping with too many sources. *Oecologia* 136: 261–269.

Prowse, T., H. Schwarcz, S. Saunders, L. Bondioli, and R. Macchiarelli. 2004. Isotope paleodiet studies of skeletons from the imperial Roman cemetery of Isola Sacara, Rome, Italy. *Journal of Archaeological Science* 31: 259–272.

Qu, Y., C. Jin, Y. Zhang, Y. Hu, X. Shang, and C. Wang. 2014. Preservation assessments and carbon and oxygen isotopes analysis of tooth enamel of *Gigantopithecus blacki* and its contemporary animals from Sanhe Cave, Chongzuo, China. *Quaternary International* 354: 52–58.

Quade, J., T. E. Cerling, P. Andrews, and B. Alpagut. 1995. Paleodietary reconstruction of Miocene faunas from Pasalar, Turkey, using stable carbon and oxygen isotopes of fossil tooth enamel. *Journal of Human Evolution* 28: 373–384.

Quade, J., T. E. Cerling, J. C. Barry, M. E. Morgan, D. R. Pilbeam, A. R. Chivas, J. A. Lee-Thorp, and N. J. Van der Merwe. 1992. A 16-Ma record of paleodiet using carbon and oxygen isotopes in fossil teeth from Pakistan. *Chemical Geology* 94: 183–192.

Rabenold, D., and O. Pearson. 2011. Abrasive, silica phytoliths and the evolution of thick molar enamel in primates, with implications for the diet of *Paranthropus boisei*. *PLoS ONE* 6: e28379.

Roche, D., L. Segalen, B. Senut, and M. Pickford. 2013. Stable isotope analyses of tooth enamel carbonate of large herbivores from the Tugen Hills deposits: Paleoenvironmental context of the earliest hominins. *Earth and Planetary Science Letters* 381: 39–51.

Schoeninger, M. J., J. Moore, and J. M. Sept. 1999. Subsistence strategies of two "savanna" chimpanzee populations: The stable isotope evidence. *American Journal of Primatology* 49: 297–314.

Sept, J. 1992. Was there no place like home? A new perspective on early hominid archaeological sites from the mapping of chimpanzee nests. *Current Anthropology* 33: 187–207.

Shipman, P., and J. Harris. 1988. Habitat preference and paleoecology of *Australopithecus boisei* in eastern Africa. In F. Grine, ed., *Evolutionary History of the "Robust" Australopithecines*, 343–381. New York: Aldine de Gruyter.

Skinner, M., Z. Alemseged, and C. Gaunitz. 2015. Enamel thickness trends in Plio-Pleistocene hominin mandibular molars. *Journal of Human Evolution* 85: 35–45.

Sponheimer, M., Z. Alemseged, T. Cerling, F. Grine, W. Kimbel, M. Leakey, J. Lee-Thorp, F. Manthi, K. Reed, B. Wood, and J. Wynn. 2013. Isotopic evidence of early hominin diets. *Proceedings of the National Academy of Sciences* 110: 10513–10518.

Sponheimer, M., J. Loudon, D. Codron, M. Howells, J. Pruetz, J. Codron, D. de Ruiter, and J. Lee-Thorp. 2006. Do "savanna" chimpanzees consume C_4 resources? *Journal of Human Evolution* 51: 128–133.

Strait, D., G. Weber, S. Neubauer, J. Chalk, B. Richmond, P. Lucas, M. Spencer, C. Schrein, P. Dechow, C. Ross, I. Grosse, B. Wright, P. Constantino, B. Wood, B. Lawn, W. Hylander, Q. Wang, C. Byron, and D. Slice. 2009. The feeding biomechanics and dietary ecology of *Australopithecus africanus*. *Proceedings of the National Academy of Sciences* 106: 2124–2129.

Suwa, G., B. Asfaw, R. Kono, D. Kubo, O. Lovejoy, and T. White. 2009a. The *Ardipithecus ramidus* skull and its implications for hominid origins. *Science* 326: 68e61–68e67.

Suwa, G., R. Kono, S. Simpson, B. Asfaw, O. Lovejoy, and T. White. 2009b. Paleobiologial implications of the *Ardipithecus ramidus* dentition. *Science* 326: 94–99.

Ungar, P., and M. Sponheimer. 2011. The diets of early hominins. *Science* 334: 190–193.
Van der Merwe, N. J., and E. Medina. 1991. The canopy effect, carbon isotope ratios and foodwebs in Amazonia. *Journal of Archaeological Science* 18: 249–259.
Vogel, J. C. 1980. *Fractionation of the Carbon Isotopes during Photosynthesis*. Berlin: Springer-Verlag.
von Fischer, J. C., and L. L. Tieszen. 1995. Carbon isotope characterization of four tropical forests in Luquillo, Puerto Rico. *Biotropica* 27: 138–148.
Wang, Y., and T. E. Cerling. 1994. A model of fossil tooth and bone diagenesis: Implications for paleodiet reconstruction from stable isotopes. *Palaeogeography Palaeoclimatology Palaeoecology* 107: 281–289.
White, T., B. Asfaw, Y. Beyene, Y. Haile-Selassie, O. Lovejoy, G. Suwa, and G. Woldegabriel. 2009. *Ardipithecus ramidus* and the paleobiology of early hominids. *Science* 326: 75–86.
Wood, B., and L. C. Aiello. 1998. Taxonomic and functional implications of mandibular scaling in early hominins. *American Journal of Physical Anthropology* 105: 523–538.
Wood, B., and M. Lonergan. 2008. The hominin fossil record: Taxa, grades and clades. *Journal of Anatomy* 212: 354–376.
Wood, B. A., and M. Collard. 1999. The human genus. *Science* 284: 65–71.
Wrangham, R., D. Cheney, R. Seyfarth, and E. Sarmiento. 2009. Shallow-water habitats as sources of fallback foods for hominins. *American Journal of Physical Anthropology* 140: 630–642.
Wrangham, R., N. Conklin-Brittain, and K. Hunt. 1998. Dietary response of chimpanzees and cercopithecines to seasonal variation in fruit abundance, I: Antifeedants. *International Journal of Primatology* 19: 949–969.
Wynn, J., M. Sponheimer, W. Kimbel, Z. Alemseged, K. Reed, and Z. Bedaso. 2013. Diet of *Australopithecus afarensis* from the Pliocene Hadar Formation, Ethiopia. *Proceedings of the National Academy of Sciences* 110: 10495–10500.
Zachos, J., M. Pagani, L. Sloan, E. Thomas, and K. Billups. 2001. Trends, rhythms, and aberrations in global climate 65 Ma to present. *Science* 292: 686–693.
Zazzo, A., H. Bocherens, M. Brunet, A. Beauvilain, D. Billoi, H. Mackaye, P. Vignaud, and A. Mariotti. 2000. Herbivore paleodiet and paleoenvironmental changes in Chad during the Pliocene using stable isotope ratios of tooth enamel carbonate. *Paleobiology* 26: 294–309.
Zollikofer, C., M. Ponce de Leon, D. Lieberman, F. Guy, D. Pilbeam, A. Likius, H. Mackaye, P. Vignaud, and M. Brunet. 2005. Virtual cranial reconstruction of *Sahelanthropus tchadensis*. *Nature* 434: 755–759.

9

Evolution of the Human Dietary Niche

Quest for High Quality

RACHEL N. CARMODY

Modern humans display morphological and physiological commitments to an exceptionally energy-rich diet that is easy to chew and digest. In this chapter, I review evidence supporting and challenging the two major hypotheses that have been proposed to explain the emergence of digestive adaptations in *Homo* that suggest a rapid increase in dietary quality: (1) enhanced exploitation of animal source foods, and (2) the improvement of available plant and animal foods through cooking and other forms of food processing. Such dietary data provide a foundation for understanding many aspects of chimpanzee and human behavior, morphology, and physiology that are the foci of other chapters in this volume.

Modern humans expend substantially more energy per kilogram of body mass than do our closest living relatives in *Pan*, a differential driven partly by our larger activity budgets (Leonard and Robertson 1997). Evidence for the emergence of high-energy endeavors like long-distance running (Bramble and Lieberman 2004) and shorter interbirth intervals (Aiello and Key 2002) suggest that the same was also true of hominins beginning with *Homo erectus* (*sensu lato*) versus earlier hominins. In addition, females of the genus *Homo* experienced energetically demanding increases in body size over time, despite their particular sensitivity to energy availability due to the high metabolic costs of gestation and lactation (Leonard and Robertson 1994; Ruff et al. 1997; Wood and Collard 1999; Aiello and Wells 2002).

Notably, although these physiological and behavioral adaptations necessitate a high caloric intake, modern and ancestral members of *Homo* show reduced investment in food ingestion. Modern humans spend substantially less time feeding compared with chimpanzees and compared with expectations based on phylogeny and body mass (Organ et al. 2011). Whereas chimpanzees spend four to six hours per day chewing, humans typically chew for less than one hour (Wrangham 2009). Compared with chimpanzees, humans also have less robust masticatory apparatuses, including smaller oral cavities, reduced molars, slighter mandibles, more gracile chewing muscles per unit body size, and genetic mutations in myosin that effectively limit bite force (Wood and Aiello 1998; Lieberman et al. 2004; Lucas 2004; Stedman et al. 2004; Wrangham 2009; Lieberman 2011; Eng et al. 2013). Bayesian reconstructions suggest that molar size reduction in early *Homo* (*H. habilis* and *H. rudolfensis*) can be explained by the overall rate of craniodental and body size evolution, but that later *Homo* (*H. erectus, H. neanderthalensis,* and *H. sapiens*) experienced an accelerated reduction suggestive of a major change in diet (Organ et al. 2011).

Similarly, modern humans also show reduced investment in food digestion. Humans living in industrialized countries expend an average of 6–9 percent of meal energy in digestion compared with the mammalian average of 13–16 percent (Westerterp 2004; Boback et al. 2007). Although human-chimpanzee comparisons are limited by small samples and variable measurement methods, modern humans appear to have smaller and less sacculated intestinal tracts that are approximately 60 percent of the expected weight/volume for our body mass (Aiello and Wheeler 1995). In particular, we exhibit reduced caeca and colons that limit our ability to retain and extract energy from foods that escape digestion in the small intestine, like resistant starch and fiber (Martin et al. 1985; Milton 1987). Ancestral hominins beginning with *H. erectus* may also have possessed small guts, judging from the tapered dimensions of the thoracic cavity and narrower pelvis compared with earlier hominins (Aiello and Wheeler 1995).

Higher energy budgets despite reduced digestive structures imply that, in contrast to earlier hominins and chimpanzees, humans and ancestral members of the genus *Homo* gained routine access to energy-rich yet easily masticated and digested food items over the course of our evolution. Human evolutionary biologists have advanced two major proposals for the dietary change underlying this transition. The first contends that ancestral hominins incorporated increasing amounts of high-calorie, easy-to-digest animal products into the diet. The second contends that ancestral hominins exploited a wide range of common resources, but transformed them into richer, easier-to-digest materials through advances in food processing, particularly cooking.

Animal Source Food Hypothesis

Compared with those of plants, the tissues of animals tend to be energy-rich due to their relatively high protein and fat contents (Speth 2010). The potential impact of these animal source foods has been articulated widely since the 1960s (Washburn and Lancaster 1968; Shipman and Walker 1989; Aiello and Wheeler 1995; Milton 1999; Stanford and Bunn 2001; Bunn 2007). Since that time, abundant archaeological, biological, and ethnographic evidence has pointed to their exploitation by ancestral hominins.

Three distinct well-preserved faunal assemblages at Kanjera, Kenya, dating to 2 Ma offer strong evidence that Oldowan hominins routinely acquired and processed small ungulate carcasses and had at least occasional access to larger game (Ferraro et al. 2013). Bones bearing cut marks and/or hammerstone damage were found throughout a stratified sequence of several hundred years, suggesting persistent carnivory of meat, organ, and marrow tissues. Analysis of the bones present suggests that these hominins acquired most small carcasses in a complete or relatively complete state, implying hunting as opposed to scavenging. By contrast, cranial remains were overrepresented among larger carcasses, indicating that hominins may have scavenged these items in pursuit of nutrient-rich brain tissues. More generally, the emergence of lithic technology and associated cut-marks on fossilized bones suggests at least occasional exploitation of animal foods at 2.6 Ma (Domínguez-Rodrigo et al. 2010), and possibly as early as 3.4 Ma (McPherron et al. 2010).

Various types of biological evidence likewise support a long history of animal source food exploitation by hominins. For instance, similar to obligate carnivores (Hedberg et al. 2007), humans exhibit diminished ability to synthesize the sulfonic acid taurine from precursor amino acids (Hayes and Sturman 1981; Chesney et al. 1998). Taurine is found primarily in animal tissue and has been argued to have been consumed sufficiently to reduce selective pressure for *in vivo* synthesis (Mann 2007). Likewise, humans lack the ability to efficiently synthesize long-chain omega-3 polyunsaturated fatty acids like docosahexaenoic acid (DHA) that are required for cell membrane structure and function, as well as fetal and postnatal brain development (Broadhurst et al. 2002; Brenna and Carlson 2014). Preformed DHA is uniquely available from animal foods (Broadhurst et al. 2002), suggesting that habitual access to such animal sources may have relaxed selective constraints on human encephalization (Kyriacou et al. 2016). However, it is possible that human DHA requirements are higher today than in the past owing to high Western consumption of omega-6 fatty acids, which inhibit a key enzyme involved in converting the α-linolenic acid available in some leafy plants into DHA (Simopoulos 2006). In addition, three species of taeniid tapeworms, host-specific parasites for which carnivores are definitive hosts, use humans as their primary host. Dating the genetic divergence of these human-associated species suggests routine consumption of animal source foods over the last one million years (Henneberg et al. 1998; Hoberg et al. 2001). Finally, anatomical changes at the shoulder that would have enabled hominins to throw projectiles with high speed and accuracy, presumably for the purpose of hunting game, first appear together in *Homo erectus* at ~2 Ma (Roach et al. 2013).

Ancestral exploitation of animal foods may explain why animal foods continue to represent a large percentage of the diet across most modern hunter-gatherer groups. Whereas chimpanzees typically obtain <5 percent of their diet from animal foods, hunted and fished resources were found to provide ≥50 percent of dietary energy for 73 percent of the 229 hunter-gatherer societies in Murdock's *Ethnographic Atlas* (Murdock 1967; Cordain et al. 2000). Among the Hadza, who are commonly used as models of foraging behavior in ancestral *Homo,* meat accounts for 32 percent of annual calories (Marlowe et al. 2014).

However, there are difficulties with the idea that an increase in animal source foods was uniquely responsible for the dietary transition in *Homo.* For one, the pursuit of animal foods has been argued to be a high-risk strategy in that it demands a large up-front energetic investment but is associated with low rates of success, even among hunter-gatherers with modern weaponry (Hawkes 1991). This has led to speculation that hunting was only made possible by hominins having access to a consistent alternative source of energy-rich foods that could provide a buffer against a stretch of unsuccessful hunts (Wrangham et al. 1999). While energetic investments might be lower if hominins relied at least partially on scavenging rather than hunting, the contribution of scavenging to the ancestral human diet remains unclear (Ragir et al. 2000; O'Connell et al. 2002; Speth 2010; Smith et al. 2015). Surveys of modern hunter-gatherer groups suggest wide variation in the proportion of animal carcasses that are acquired by scavenging, with scavenging accounting for 20 percent of carcasses exploited by the Hadza (O'Connell et al. 1988), at least 9 percent among the Ju/'hoansi (Yellen 1991), but just 0.3 percent among the Bofi and Aka (Lupo and Schmitt 2005). In these cases, scavenged meat is almost always cooked, raising the question of whether scavenging was made possible in humans by the control of fire. A recent report suggests that the rapid growth of bacterial pathogens on exposed surfaces of carcasses would have made it difficult for hominins to scavenge safely without focusing on marrow, employing strategies of carrion selection to minimize pathogen load, or cooking to kill pathogens (Smith et al. 2015). Perhaps this helps explain why chimpanzees rarely exploit scavenged meat (Goodall 1986; Ragir et al. 2000).

Second, it has been argued that seasonal depletions of body fat in game animals would have placed hominins at risk of "rabbit starvation," a condition of negative energy balance that can arise in mammalian omnivores when diets derive a large proportion of their calories from lean protein, due to the high costs of protein digestion and the limited capacity of the liver for urea synthesis (Noli and Avery 1988; Cordain et al. 2000; Speth 2010). As Speth and colleagues observe, this would have lowered energy gains during precisely those periods when subsistence resources were least available (Speth and Spielmann 1983; Speth 1989, 2010).

Third, a high intake of meat would have presented a masticatory challenge since hominins have low-crested bunodont molars that do not efficiently fracture compliant animal tissues (Zink et al. 2014). This has led some to challenge whether hominins could physically have ingested sufficient quantities of animal foods per day, given that chimpanzees seem to ingest a maximum of 400 calories per hour when chewing carcasses (Wrangham and Conklin-Brittain 2003). Further, even if chewing this quantity of meat were physically possible, it is unclear whether hominins would have had sufficient time remaining in the day for other fitness-maximizing activities (Wrangham 2009).

Fourth, human physiology is dependent on an efficient supply of glucose; this is particularly true of the brain, which accounts for 20–25 percent of adult basal energy expenditure (Fonseca-Azevedo and Herculano-Houzel 2012). Our adult glucose requirement of approximately 170 g per day is typically met by a combination of dietary carbohydrate as well as gluconeogenesis utilizing select amino acids, glycerol from dietary fats, and short-chain fatty acids produced during the microbial fermentation of carbohydrates and protein (Hardy et al. 2015). But gluconeogenesis alone is typically too inefficient to meet the metabolic needs of the brain, necessitating the breakdown of fat to produce ketones as an alternative energy substrate (Westman et al. 2007). Adaptations for more efficient gluconeogenesis are theoretically possible: the Inuit, whose traditional diet derived up to 96 percent of calories from animal sources (Nobmann et al. 1992), have enlarged livers with increased capacities for gluconeogenesis and the excretion of urea excretion, as well as high penetrance (68 percent) of a non-synonymous G→A transition in CPT1A, a gene encoding an enzyme involved in the digestion of long-chain fatty acids that are abundant in animal foods (Clemente et al. 2014). However, the majority of modern human populations appear to bear no such adaptations. Therefore, it is likely that ancestral hominins were physiologically limited in the proportion of the diet that could be derived from animal sources.

Finally, if humans were adapted to meet our expanded energy needs through the exploitation of animal foods, we might expect vegetarians to exhibit compromised energy status and perhaps an associated impairment of female reproductive capacity (Ellison 2003). However, this does not appear to be the case. A recent review found that median body mass indices for vegetarian women and men in the United States were 23.7 and 24.3, respectively, which is close to the median values of 24.8 and 25.3 among women and men eating typical American mixed diets (Carmody and Wrangham 2009). Reproductive-aged women consuming vegetarian diets exhibit no suppression of ovarian function compared with those consuming diets that include meat (Barr 1999), and there are no differences in the age at menarche of girls eating vegetarian versus omnivorous diets (Rosell et al. 2005).

Thus, the available evidence suggests that animal foods were critical components of the ancestral hominin diet, but that animal foods alone were unlikely to have supported the suite of dietary adaptations that arose in *Homo*.

Food Processing Hypothesis

Problems with increased animal foods as an independent solution led to new proposals that ancestral hominins may have met their higher energy needs by improving available plant and animal foods through food processing, especially cooking (Wrangham et al. 1999; Carmody and Wrangham 2009; Wrangham 2009). In theory, a reduction in food structural integrity elicited by the physical and chemical effects of cooking should facilitate chewing and raise the energetic value of foods by increasing their digestibility and lowering the metabolic costs of mastication and digestion. Moreover, by improving consistently available plant items like tubers, which are believed to have been a critical food resource in addition to meat for ancestral hominins (Hatley and Kappelman 1980; Hawkes et al. 1997; Laden and Wrangham 2005; Perry et al. 2007), cooking should confer an important increase in the predictability of the energy supply. The control of fire may also have improved access to available high-energy but high-risk foods, such as scavenged meat (Ragir 2000; Smith et al. 2015) and honey (Wrangham 2011). Higher and more predictable energetic returns should relax biological constraints, allowing for the coevolution of larger total energy budgets and smaller masticatory and digestive capacities.

Diverse evidence supports the idea that cooking would have improved energy availability across a range of ancestral hominin diets. In the case of plant foods, heat degrades the polysaccharides that are responsible for intercellular adhesion, heightening fracturability (Lillford 2001; Jarvis et al. 2003). This lowers the work of mastication (Lieberman 2011), reducing both force per chew and the number chewing cycles required before particles are combined into a bolus and swallowed (Engelen et al. 2005; Zink et al. 2014). Heat also gelatinizes intracellular starch, a process that breaks bonds between adjacent molecules and exposes their hydrogen bonding sites, causing the granule to swell with water and rupture (Svihus et al. 2005). Gelatinization greatly reduces the resistance of starch to hydrolysis by amylases (Holm et al. 1988; Björck et al. 2000; Peres and Oliva-Teles 2002; Svihus et al. 2005; Tester et al. 2006). In addition, cooking disrupts the integrity of plant cell walls, both allowing starch to come into contact with amylases and allowing the digestion products to escape the food matrix (Singh et al. 2013). These effects lead to consistent and significant increases in the bioavailability of cooked starches compared with raw starches, with benefits ranging from 12 percent for oats to 35 percent for green bananas (Carmody and Wrangham 2009). Indeed, it has been suggested that substantial increases in starch digestibility due to cooking would have rendered amylase production a rate-limiting factor in starch digestion (Hardy et al. 2015), thus creating selection pressure for high copy numbers of the genes encoding salivary amylase (Perry et al. 2007), and possibly pancreatic amylase (Carpenter et al. 2015), in the human lineage.

It has generally been assumed that dietary lipids are fully digested regardless of their state. However, recent research suggests that cooking increases the energy gained per calorie of lipid ingested through its impacts on plant cell microstructure (Groopman et al. 2015). In peanuts, a lipid-rich oilseed, cooking disrupted cell walls and denatured the oleosin layer of proteins that otherwise shield lipids from digestive lipases. This led to a greater lipid digestibility in mice fed cooked versus raw peanuts, whereas physical disruption of raw peanuts by blending had comparatively little effect. Such results are consistent with a growing body of literature suggesting that the calorie values assigned to several varieties of nut are overestimated due to inaccessibility of their cell-bound lipids (Baer et al. 2012; Novotny et al. 2012; Grundy et al. 2015).

Cooking also transforms animal foods, although the digestive implications are less well understood. The elasticity of raw meat makes it difficult to chew because biting forces are dissipated, preventing the propagation of fractures (Lieberman 2011). Heat denatures the proteins present in muscle fibers, causing contraction along the grain that leads to stiffening (McGee 2004; Tornberg 2005). In addition, the collagen surrounding each fascicle of muscle fibers generally remains too tough for mastication until heated to 60–70°C, when collagen begins to hydrolyze into gelatin (McGee 2004). While these effects should theoretically facilitate the mastication of meat, a recent study of the material properties of raw versus cooked meat samples found only partial support for this hypothesis (Zink et al. 2014). The effects of cooking on masticatory efficiency may ultimately depend on factors such as species, muscle type, fat content, and animal age (Purslow 1999; McGee 2004; Lepetit 2008; Zink et al. 2014).

In addition to influencing masticatory performance, the protein denaturation and collagen gelatinization that occur in meat due to cooking could also theoretically increase bioavailability and/or reduce the metabolic cost of food ingestion, digestion, absorption, and assimilation, a cumulative expenditure known as diet-induced thermogenesis. In the case of bioavailability, Evenepoel and colleagues (1998, 1999) found that the bioavailability of egg protein in human volunteers was 91–94 percent when cooked, compared with just 51–65 percent when raw, attributing this improvement to denaturation making cleavage sites more accessible to proteolytic enzymes (Davies et al. 1987; Bax et al. 2012). Heat-induced denaturation of protein is common to all forms of animal protein, but whether the bioavailability of meat protein is similarly improved with cooking remains to be empirically confirmed. In the case of diet-induced thermogenesis, Boback and colleagues (2007) found that Burmese pythons *(Python molurus)* exhibited 13 percent lower diet-induced thermogenesis after consuming cooked versus raw meat meals, attributing these effects to role that solubilization of collagen plays in increasing the surface area of the tissue exposed to gastric acids and enzymes, thereby reducing the work of gastric digestion. However, the digestive physiologies of pythons and humans differ substantially, and so the translational significance remains uncertain.

Beyond its role in improving energy gain from a given food, the control of fire may have afforded better access to resources that are problematic to acquire or consume. We have already considered the possibility that cooking was essential for safely exploiting the meat of scavenged carcasses (Ragir 2000; Smith et al. 2015). Honey from stinging honeybees *(Apis mellifera)* provides a second example of a high-energy but high-risk food resource that is often acquired using fire. Honey is a highly valued resource for chimpanzees as well as many hunter-gatherer groups (Marlowe et al. 2014). But whereas honey makes up a small fraction of dietary calories for chimpanzees, it is an important food for many hunter-gatherers (Figure 9.1). The difference in honey exploitation between chimpanzees and humans appears to be at least partly due to humans controlling fire, as smoke is routinely used to sedate the bees to prevent stinging (Wrangham 2011). As Marlowe and colleagues describe of the Hadza: "If there is a lot of honey in the hive they climb down and start a fire. They make a torch and carry it back up the tree

FIGURE 9.1. Chimpanzee honey consumption is limited by difficult-to-access hives and aggressive bees (left). Foragers like the Hadza access much more honey (right) by using smoke to pacify bees, and tools to chop into their hiding places. Kanyawara photo by Ronan Donovan. Hadza photo by Brian Wood.

to smoke the bees. According to one view, this induces panic, which causes the bees to consume great amounts of honey to store for the future. However, this makes the bees drowsy. Once the bees are sufficiently sated, one can reach in and grab the honeycomb or parts of it without getting stung so many times" (Marlowe et al. 2014: 124). Similar tactics are used routinely by other foraging groups (Wrangham 2011).

Although hominins likely accessed honey from *Apis mellifera* prior to the control of fire, aided perhaps by hand axes (Marlowe et al. 2014), the control of fire would have increased the efficiency of its harvest. Any improvement in efficiency would have been significant for energy gain, as honey is the most energy-dense wild food resource, providing >3,000 kcal/kg (Murray et al. 2001). A recent review suggested that fifteen of sixteen warm-climate foragers in the Standard Cross Cultural Sample include honey in their diets, the exception being the Badjau of the Philippines, who spend most of their time on boats (Marlowe et al. 2014). In the two populations where honey use has been studied best, honey represents a major source of dietary calories: 15 percent of annual calories among the Hadza (Marlowe et al. 2014) and up to 80 percent of daily calories during the honey season among the Mbuti (Ichikawa 1981). In addition to its high caloric value, increased access to honey would have had important consequences for the stability of energy gain across seasons, as honey is most available during the rainy season, when hunting is typically less productive (Marlowe et al. 2014).

Likewise, fire may lower the barrier to exploitation for a variety of other plant and animal foods. For instance, roasting reduces the toughness of the cortex of *Vigna* tubers, allowing the Hadza to peel them by hand (Dominy et al. 2008). Cooking also reduces plant secondary compounds such as tannins that taste bitter or astringent (Xu and Chang 2009), likely increasing their palatability. Indeed, among the 828 plant species estimated to be the most important in human diets worldwide, 40 percent are never eaten raw (Proches et al. 2008), suggesting that cooking plays an important role in their acceptance. Cooking also increases the suitability of many animal foods, including killing food-borne pathogens such as *Campylobacter* that are common in the guts of wild game (Ragir et al. 2000), and reducing some of the cyanobacterial and dinoflagellate toxins present in shellfish (Indrasena and Gill 2000).

Although access to fire likely increased the potential range of accessible foods, widespread adoption of cooking might also have precipitated a fall in dietary diversity (Wrangham and Carmody 2016). Optimal foraging theory suggests that consumers will exploit fallback items when food is scarce and exclude these items when food is abundant. If cooking could be used to increase the quality of common food resources, we might expect that ancestral hominins eating cooked food would show lower dietary diversity than their predecessors consuming raw food. In support of this idea, where chimpanzees and humans inhabit the same area, chimpanzees appear to eat a wider variety of plants than do humans. In Bossou, for example, chimpanzees were observed to consume 30.1 percent of identified plant species (200 of 664) compared with the 11.4 percent consumed by humans (76 of 664) (Sugiyama and Koman 1992). Moreover, the gut microbial communities of humans are substantially less diverse compared with those of wild apes, suggesting a narrower range of ingested compounds (Moeller et al. 2014). Thus, cooking may have influenced human dietary diversity in contrasting ways.

Yet perhaps the strongest evidence for a critical role for cooking in the human transition to a high-quality diet is that modern humans appear to fare poorly on raw diets, even when they incorporate animal foods. In the most extensive study, a questionnaire-based survey of 572 urban raw foodists, body mass index was inversely correlated with both the proportion of raw food in the diet and the length of time since the adoption of raw foodism (Koebnick et al. 1999). Encouraging the hypothesis that modern humans may be biologically committed to a diet that includes cooked items (Wrangham and Conklin-Brittain 2003), 50 percent of nonpregnant female subjects under forty-five years of age who consumed a 100 percent raw diet were amenorrheic, presumably as a result of being energy-limited (Ellison et al. 1993; Ellison 2003). Critically, the odds of energy deficiency or amenorrhea were similar for vegetarian and omnivorous raw foodists, implying that the inclusion of raw animal source foods alone did not measurably improve energy status.

Parsimony suggests that the value of these improvements would have been appreciated by ancestral hominins. All modern human societies cook (Wrangham 2009). Although great apes cannot cook, preference tests in which sanctuary chimpanzees, bonobos, gorillas, and orangutans were presented with food items either cooked or raw suggest that apes prefer most items cooked, including meat, sweet potato tuber, and carrots (Wobber et al. 2008; Warneken and Rosati 2015). In addition, chimpanzees appear to value cooked items sufficiently to resist consuming raw items until they can be cooked, as well as to give up possession of raw items in order for them to be cooked (Warneken and Rosati 2015). Together, such data suggest that the widespread adoption of cooking would have followed closely after the control of fire, and that the last common ancestor of humans and chimpanzees probably had the cognitive capacity to engage in cooking.

Nevertheless, the cooking hypothesis faces a critical challenge in terms of timing. The major transition toward higher energy budgets and reduced digestive capacity appears to have occurred roughly 1.9 Ma, yet the earliest widely accepted evidence for controlled fire dates only to 1.0 Ma at Wonderwerk Cave, South Africa (Berna et al. 2012). At the majority of sites with Lower Paleolithic evidence of fire, it is unclear whether burned materials are the result of human activity or natural phenomena such as bush fires, lightning strikes, or volcanic activity (Beaumont 2011; Roebroeks and Villa 2011). However, the deep cave site at Wonderwerk makes it difficult to challenge hominin-controlled fire as an explanation for the extensive distribution of well-preserved ashed plants and charred bone remains found in association with heated sediment and manuports bearing pot-lid fractures (Berna et al. 2012). Charred plant materials and heated microflints have also been found at the Lower Paleolithic site of Gesher Benot Ya'aqov in Israel, dated to 790 Ka on the basis of thermoluminescence-validated burned microdebitage (Goren-Inbar et al. 2004; Alperson-Afil 2008). But whether humans used fire habitually at this stage is unclear, as the archaeological record reveals a prominent intensification of fire use around 300–400 Ka (Roebroeks and Villa 2011; Shahack-Gross et al. 2014; Shimelmitz et al. 2014). The earliest direct evidence for cooking itself, in the form of a repeatedly used hearth containing burned bones and lithics at Qesem Cave, Israel, dates to just 300 Ka (Shahack-Gross et al. 2014), with additional well-accepted hearths around 250 Ka (James 1989; Brace 1995; Goldberg et al. 2001; Shipman 2009). Moreover, despite evidence that Neanderthals at least occasionally consumed cooked starches (Henry et al. 2011), the limited evidence of fire use in some Middle Paleolithic sites, especially during colder periods, raises the possibility that Neanderthals living as recently as 40 Ka may not have had the capacity to create fire even if they could utilize it (Sandgathe et al. 2011).

One potential explanation for the limited evidence of fire use by ancestral hominins is that traces of fire may vanish too quickly to be preserved in the archaeological record (Figure 9.2). For instance, the one-year-old remains of a Hadza cooking hearth that was used continuously for four months left only some oxidized soil and isolated bits of charcoal, with no traces of charred food items (Mallol et al. 2007). Likewise, Sergant et al. (2006) reported that despite abundant burned bone, shells, and other artifacts at Mesolithic sites in northwest Europe, direct evidence of the control of fire is very limited.

A possible alternative is that the widespread adoption of nonthermal processing, as opposed to cooking, contributed to the significant rise in dietary energy inferred for *H. erectus*. While the food preparation methods used by *H. erectus* are not known, it is reasonable to expect that nonthermal processing predated cooking. Stone tools have been confidently dated to 2.5–3.3 Ma (Semaw et al. 1997, 2003; de Heinzelin et al. 1999; Harmand et al. 2015), prior to the emergence of *H. erectus,* and tool-assisted butchery was reported to have occurred at 3.4 Ma in *Australopithecus afarensis* (McPherron et al. 2010). Cut-marks on animal bones and microwear on flakes indicate slicing (Keeley and Toth 1981; Semaw et al. 2003; Dominguez-Rodrigo et al. 2005; Pobiner et al. 2008). In addition, hammerstones and spheroids as well as unmodified

FIGURE 9.2. Little remains of this Hadza hearth, just weeks after the camp residents have moved on. Photo by Martin Muller.

stones and wooden implements were available for pounding (Panger et al. 2002; Wrangham 2007). Although there is no direct evidence of food pounding in the archaeological record dating back to *H. erectus,* evidence that chimpanzees *(P. troglodytes)* pound the stems of oil palms *(Elaeis guineensis)* (Yamakoshi and Sugiyama 1995) and mash fruits (Fernández-Carriba and Loeches 2001) suggests that simple processing behaviors are ancestral.

Nonthermal processing is expected to confer many of the same physical benefits as cooking through its effects on structural integrity. In the case of plant foods, nonthermal processing reduces structural integrity by disrupting cell walls, leading to the release of cellular water and an associated reduction in turgor pressure (Jarvis et al. 2003). For instance, a recent study found a 42 percent reduction in toughness in tubers following pounding with a hammerstone (Zink et al. 2014). Nonthermal processing also releases cell-bound nutrients, making them more available for digestion (Grundy et al. 2015). In the case of meat, nonthermal processing mechanically separates muscle fiber bundles from one another and from the surrounding collagen-rich matrix (McGee 2004), facilitating chewing. Zink and Lieberman (2016) recently modeled the physical effects of nonthermal processing for a hypothetical ancestral diet composed of two-thirds tubers and one-third meat, finding that simple techniques such as pounding and slicing could have increased chewing efficiency and decreased bite forces sufficiently to initiate selection for smaller masticatory structures.

In other ways, however, the effects of nonthermal processing are unlikely to match those of cooking, because nonthermal processing involves no chemical modifications. For instance, heat is necessary for the gelatinization of starches, denaturation of proteins, and breakdown of lipid microstructure, factors previously argued to influence bioavailability (Evenepoel et al. 1998, 1999; Tester et al. 2006; Carmody and Wrangham 2009; Groopman et al. 2015).

Recent feeding studies offer direct evidence that the energetic benefits of cooking exceed those of nonthermal processing. Mice eating sweet potato tubers or lean beef achieved significantly higher body mass outcomes when these items were fed in processed versus unprocessed form, controlling for intake and activity (Carmody et al. 2011). The positive energetic effects of cooking greatly exceeded those of pounding in both tuber and meat, a conclusion further supported by food preferences in fasted animals. Similarly, cooking consistently increased energy gain from peanuts, whereas blending did not (Groopman et al. 2015). The superior effects of cooking that were paralleled across foods suggest that the adoption of cooking would likely have led to energetic gains for ancestral hominins even if pounding was already in widespread use and regardless of whether plant or animal foods predominated in the diet. Moreover, because these studies used domesticated foods, the results likely underestimate the effects of cooking on wild items, such as ungulate meat with low fat content, wild tubers containing relatively resistant starch, or uncultivated nuts bearing high loads of antifeedant compounds.

Intriguingly, other forms of food processing that have barely been considered in the context of human evolution, such as natural fermentation, could be expected to modify the physical and chemical structure of foods in ways that facilitate chewing and make foods easier to digest. Fermentative bacteria and molds degrade plant cell walls, giving the human digestive tract access to cell-bound carbohydrates, proteins, and lipids (Potter and Hotchkiss 1998). In addition, the enzymatic splitting of cellulose, hemicellulose, and related polymers yields simple sugars that would otherwise have escaped digestion in the small intestine. Fermentation can also modify digestibility through indirect effects. For instance, many edible tubers, roots, grains, and legumes contain substantial quantities of the phosphorus-storage compound phytate (Phillippy et al. 2003). Phytase inhibits the activity of digestive enzymes such as trypsin, pepsin, alpha-amylase, and beta-galactosidase and decreases the bioavailability of dietary proteins by forming complexes (Reddy et al. 1989). These consequences can be prevented through fermentation, in which microbial 3-phytase hydrolyzes phytate to inorganic orthophosphate and a series of lower phosphoric esters (Reddy and Pierson 1994). Although the earliest direct archaeological evidence for fermentation in pottery jars dates only to 7,000 B.C. in China (McGovern et al. 2004), many human cultures ferment foods and beverages using methods such as caching or piling, which leave no obvious remains. The contributions of naturally fermented foods to the chimpanzee diet, the incorporation of fermentation practices by ancestral hominins, and the energetic significance of food fermentation in the modern human diet are areas ripe for further inquiry.

Ultimately, it is likely that both increased access to animal foods and developments in food processing technology supported the evolution of adaptations for high dietary quality in humans. However, observations that modern humans cannot thrive on raw omnivorous diets that have been heavily processed by nonthermal means, despite thriving well on cooked vegetarian diets, suggests that cooking played a pivotal role in shaping human biology. Emerging evidence of positive selection in genes influenced in their expression level by the consumption of a cooked diet, together with the finding that sequence changes in these genes arise before the split between the human and Neanderthal-Denisovan lineages, suggests that human adaptation to a cooked diet had likely begun by 275–765 Ka (Carmody et al. 2016).

Outlook

With respect to the dietary revolution in *Homo*, much remains to be discovered about the relative energetic benefits of animal foods, cooked foods, and foods processed by nonthermal means, including fermentation. Valuable information about the relative contributions of animal foods and food processing in explaining the human commitment to an energy-rich diet will come from controlled experiments that compare these alternatives head-to-head with respect to nutrient bioavailability, diet-induced thermogenesis, and impacts on host-microbial interactions in energy harvest. Focusing on the mechanisms underlying net energy gain will allow results from experiments using a given food substrate to offer predictions about impacts expected for other items; this ability to extrapolate will be important given the diversity of food resources and preparation methods used. However, especially helpful will be experiments focused on net energy gain, in which body mass and composition are evaluated in humans or animal models fed different diets. Net energy gain experiments will allow us to evaluate how nutrient bioavailability, diet-induced thermogenesis, and host-microbial interactions combine energetically, for these effects could covary in food-specific ways.

Additionally, work to elucidate several unknowns will help to clarify the evolution of the modern human dietary niche: What were the relative contributions of hunting and scavenging in the ancestral hominin diet? What fraction of the diet was composed of meat versus fat-rich tissues such as organs, brains, and marrow? To what extent did ancestral hominins exploit aquatic plants and marine animals? Is there a combination of raw foods that provides an adequate human diet under diverse environmental conditions? To what extent does cooking increase net energy gain after accounting for the energy invested in the process of building and tending a fire? Do clinical data on energy status and female fecundity uphold the idea that humans are physiologically committed to a cooked diet? In addition, future genomic studies may ultimately allow us to more carefully probe molecular signals of human adaptation to diets rich in animal foods, cooked items, and/or nonthermally processed items, shedding light on the biological properties under selection and the timing of these events in human history.

Finally, rapid advances in high-throughput sequencing and bioinformatics have made it possible to characterize the microbial communities of humans and great apes with unprecedented resolution. Studies to date suggest that gut microbial structure and function are exquisitely sensitive to diet (Ley et al. 2008; David et al. 2014; Carmody et al. 2015). Thus, efforts to assess shifts in the gut microbiome since the split between humans and chimpanzees could help reveal dietary adaptations in the human lineage. Doing so requires identifying aspects of gut microbial structure and function that are shared across culturally and geographically diverse human populations to the exclusion of African apes. As a first step toward this goal, recent work has shown that compositional change in gut microbial communities was slow during the divergence of African apes, with the distance between communities generally mirroring the genetic distance between hosts (Moeller et al. 2014). By contrast, human gut microbial communities exhibit evidence of having diverged from the ancestral state at an accelerated rate. Compared with other African apes, the human gut microbiota exhibited a fivefold enrichment in *Bacteroides,* a bacterial genus linked to diets rich in animal foods (Wu et al. 2011; David et al. 2014), and a fivefold reduction in *Methanobrevibacter,* an archaeon that renders the fermentation of undigested carbohydrates and protein more efficient by using the short-chain fatty acid end products for methanogenesis (Samuel and Gordon 2006). These data would appear to reinforce the idea that humans have specialized in a relatively high-quality diet rich in animal foods and low in indigestible materials (Moeller et al. 2014), but it remains unclear whether such results are driven by patterns of host-microbial coevolution or simply the current diets being consumed by the sampled populations. Future controlled experiments contrasting the functions of human versus nonhuman primate gut microbiomes under shared dietary scenarios, as well as interrogation of genomic signals of host-microbial functional complementarity, will help to discriminate these ultimate and proximate effects.

Energy gain through food consumption is the most basic objective of any organism, underlying virtually all aspects of its biology. Illuminating the origins of the human diet and the factors that have shaped it are thus prerequisites to knowing ourselves. Efforts to trace the evolution of the human diet deliver not only an academic understanding of our past, but also a framework for appreciating how cultural or temporal variations in diet influence human health today, and a means of anticipating how present dietary habits could shape our future.

Summary

Compared with chimpanzees, humans have large energy budgets but small structures for mastication and digestion, suggesting that we gained routine access to higher-quality food items at some point after our lineages diverged. Evidence of a step change in digestive adaptations with the rise of *Homo erectus* approximately 2 Ma have led to two major proposals for the dietary innovations underlying these changes: increased dependence on animal source foods and the advent of cooking and nonthermal food processing technologies. The animal source foods hypothesis is supported by diverse archaeological, biological, and ethnographic evidence of intensified reliance on animal products in *Homo*, but it is challenged by the inherent unreliability of hunting/scavenging success and physiological limits to protein utilization, coupled with encephalization-driven demand for efficient dietary sources of glucose. The food processing hypothesis is supported by processing-driven increases in macronutrient digestibility, reductions in digestive expenditures, and enhancements in access to high-quality resources that are problematic to acquire or consume, but it is challenged by a paucity of evidence for human-controlled fire before 1 Ma, coupled with the comparatively limited projected benefits of nonthermal processing afforded by early *Homo* lithic technology. Ultimately, increased reliance on animal foods and processing technologies likely contributed in parallel to dietary adaptations in the human lineage. Future research probing key unknowns, coupled with novel exploration of unique human-microbial interactions in nutrient digestion, will pave the way for new insights into our dietary past and how its legacy shapes our present and future.

References

Aiello, L. C., and C. Key. 2002. Energetic consequences of being a *Homo erectus* female. *American Journal of Human Biology* 14: 551–565.

Aiello, L. C., and J. C. K. Wells. 2002. Energetics and the evolution of the genus *Homo*. *Annual Review of Anthropology* 31: 323–338.

Aiello, L. C., and P. Wheeler. 1995. The expensive tissue hypothesis: the brain and the digestive system in human and primate evolution. *Current Anthropology* 36: 199–211.

Alperson-Afil, N. 2008. Continual fire-making by hominins at Gesher Benot Ya'aqov, Israel. *Quaternary Science Reviews* 27: 1733–1739.

Baer, D. J., S. K. Gebauer, and J. A. Novotny. 2012. Measured energy value of pistachios in the human diet. *British Journal of Nutrition* 107: 120–125.

Barr, S. I. 1999. Vegetarianism and menstrual cycle disturbances: Is there an association? *American Journal of Clinical Nutrition* 70: 549S–554S.

Bax, M. L., L. Aubry, C. Ferreira, J. D. Daudin, P. Gatellier, D. Rémond, and V. Santé-Lhoutellier. 2012. Cooking temperature is a key determinant of *in vitro* meat protein digestion rate: Investigation of underlying mechanisms. *Journal of Agricultural and Food Chemistry* 60: 2569–2576.

Beaumont, P. B. 2011. The edge: More on fire-making by about 1.7 million years ago at Wonderwerk Cave in South Africa. *Current Anthropology* 52: 585–595.

Berna, F., P. Goldberg, L. K. Horwitz, J. Brink, S. Holt, M. Bamford, and M. Chazan. 2012. Microstratigraphic evidence of in situ fire in the Acheulean strata of Wonderwerk Cave, Northern Cape province, South Africa. *Proceedings of the National Academy of Sciences* 109: E1215–E1220.

Björck, I., H. Liljeberg, and E. Ostman. 2000. Low glycaemic-index foods. *British Journal of Nutrition* 83: S149–S155.

Boback, S. M., C. L. Cox, B. D. Ott, R. Carmody, R. W. Wrangham, and S. M. Secor. 2007. Cooking and grinding reduces the cost of meat digestion. *Comparative Biochemistry and Physiology, Part A* 148: 651–656.

Brace, C. L. 1995. *The Stages of Human Evolution*, 5th ed. Englewood Cliffs, NJ: Prentice-Hall.

Bramble, D. M., and D. E. Lieberman. 2004. Endurance running and the evolution of *Homo*. *Nature* 432: 345–352.

Brenna, J. T., and S. E. Carlson. 2014. Docosahexaenoic acid and human brain development: Evidence that a dietary supply is needed for optimal development. *Journal of Human Evolution* 77: 99–106.

Broadhurst, C. L., Y. Q. Wang, M. A. Crawford, S. C. Cunnane, J. E. Parkington, and W. F. Schmidt. 2002. Brain-specific lipids from marine, lacustrine, or terrestrial food resources: Potential impact on early African *Homo sapiens*. *Comparative Biochemistry and Physiology, Part B* 131: 653–673.

Bunn, H. T. 2007. Meat made us human. In P. S. Ungar, ed., *Evolution of the Human Diet: The Known, the Unknown, and the Unknowable*, 191–211. New York: Oxford University Press.

Carmody, R. N., M. Dannemann, A. W. Briggs, B. Nickel, E. E. Groopman, R. W. Wrangham, and J. Kelso. 2016. Genetic evidence of human adaptation to a cooked diet. *Genome Biology and Evolution* 8: 1091–1103.

Carmody, R. N., G. K. Gerber, J. M. Luevano Jr., D. M. Gatti, L. Somes, K. L. Svenson, and P. J. Turnbaugh. 2015. Diet dominates host genotype in shaping the murine gut microbiota. *Cell Host & Microbe* 17: 72–84.

Carmody, R. N., G. S. Weintraub, and R. W. Wrangham. 2011. Energetic consequences of thermal and nonthermal food processing. *Proceedings of the National Academy of Sciences* 108: 19199–19203.

Carmody, R. N., and R. W. Wrangham. 2009. The energetic significance of cooking. *Journal of Human Evolution* 57: 379–391.

Carpenter, D., S. Dhar, L. M. Mitchell, B. Fu, J. Tyson, N. A. Shwan, F. Yang, M. G. Thomas, and J. A. Armour. 2015. Obesity, starch digestion and amylase: Association between copy number variants at human salivary (AMY1) and pancreatic (AMY2) amylase genes. *Human Molecular Genetics* 24: 3472–3480.

Chesney, R. W., R. A. Helms, M. Christensen, A. M. Budreau, X. B. Han, and J. A. Sturman. 1998. The role of taurine in infant nutrition. *Advances in Experimental Medicine and Biology* 442: 463–476.

Clemente, F. J., A. Cardona, C. E. Inchley, B. M. Peter, G. Jacobs, L. Pagani, D. J. Lawson, T. Antão, M. Vicente, M. Mitt, M. DeGiorgio, Z. Faltyskova, Y. Xue, Q. Ayub, M. Szpak, R. Mägi, A. Eriksson, A. Manica, M. Raghavan, M. Rasmussen, S. Rasmussen, E. Willerslev, A. Vidal-Puig, C. Tyler-Smith, R. Villems, R. Nielsen, M. Metspalu, B. Malyarchuk, M. Derenko, and T. Kivisild. 2014. A selective sweep on a deleterious mutation in CPT1A in arctic populations. *American Journal of Human Genetics* 95: 584–589.

Cordain, L., J. B. Miller, S. B. Eaton, N. Mann, S. H. A. Holt, and J. D. Speth. 2000. Plant-animal subsistence ratios and macronutrient energy estimations in worldwide hunter-gatherer diets. *American Journal of Clinical Nutrition* 71: 682–692.

David, L. A., C. F. Maurice, R. N. Carmody, D. B. Gootenberg, J. E. Button, B. E. Wolfe, A. V. Ling, A. S. Devlin, Y. Varma, M. A. Fischbach, S. B. Biddinger, R. J. Dutton, and P. J. Turnbaugh. 2014. Diet rapidly and reproducibly alters the human gut microbiome. *Nature* 505: 559–563.

Davies, K. J. A., S. W. Lin, and R. E. Pacifici. 1987. Protein damage and degradation by oxygen radicals, IV: Degradation of denatured protein. *Journal of Biological Chemistry* 262: 9914–9920.

de Heinzelin, J., J. D. Clark, T. White, W. Hart, P. Renne, G. WoldeGabriel, Y. Beyene, and E. Vrba. 1999. Environment and behavior of 2.5-million-year-old Bouri hominids. *Science* 284: 625–629.

Domínguez-Rodrigo, M., T. R. Pickering, and H. T. Bunn. 2010. Configurational approach to identifying the earliest hominin butchers. *Proceedings of the National Academy of Sciences* 107: 20929–20934.

Dominguez-Rodrigo, M., T. R. Pickering, S. Semaw, and M. Rogers. 2005. Cutmarked bones from Pliocene archaeological sites at Gona, Afar, Ethiopia: Implications for the function of the world's oldest stone tools. *Journal of Human Evolution* 48: 109–121.

Dominy, N. J., E. R. Vogel, J. D. Yeakel, P. Constantino, and P. W. Lucas. 2008. Mechanical properties of plant underground storage organs and implications for dietary models of early hominins. *Evolutionary Biology* 35: 159–175.

Ellison, P. T. 2003. Energetics and reproductive effort. *American Journal of Human Biology* 15: 342–351.

Ellison, P. T., C. Panter-Brick, S. F. Lipson, and M. T. O'Rourke. 1993. The ecological context of human ovarian function. *Human Reproduction* 8: 2248–2258.

Eng, C. M., D. E. Lieberman, K. D. Zink, and M. A. Peters. 2013. Bite force and occlusal stress production in hominin evolution. *American Journal of Physical Anthropology* 151: 544–557.

Engelen, L., A. Fontijn-Tekamp, and A. van der Bilt. 2005. The influence of product and oral characteristics on swallowing. *Archives of Oral Biology* 50: 739–746.

Evenepoel, P., D. Claus, B. Geypens, M. Hiele, K. Geboes, P. Rutgeerts, and Y. Ghoos. 1999. Amount and fate of egg protein escaping assimilation in the small intestine of humans. *American Journal of Physiology: Gastrointestinal and Liver Physiology* 277: 935–943.

Evenepoel, P., B. Geypens, A. Luypaerts, M. Hiele, and P. Rutgeerts. 1998. Digestibility of cooked and raw egg protein in humans as assessed by stable isotope techniques. *Journal of Nutrition* 128: 1716–1722.

Fernández-Carriba, S., and A. Loeches. 2001. Fruit smearing by captive chimpanzees: A newly observed food-processing behavior. *Current Anthropology* 42: 143–147.

Ferraro, J. V., T. W. Plummer, B. L. Pobiner, J. S. Oliver, L. C. Bishop, D. R. Braun, P. W. Ditchfield, J. W. Seaman III, K. M. Binetti, J. W. Seaman Jr., F. Hertel, and R. Potts. 2013. Earliest archaeological evidence of persistent hominin carnivory. *PLoS ONE* 8: e62174.

Fonseca-Azevedo, K., and S. Herculano-Houzel. 2012. Metabolic constraint imposes tradeoff between body size and number of brain neurons in human evolution. *Proceedings of the National Academy of Sciences* 109: 18571–18576.

Goldberg, P., S. Weiner, O. Bar-Yosef, Q. Xu, and J. Liu. 2001. Site formation processes at Zhoukoudian, China. *Journal of Human Evolution* 41: 483–530.

Goodall, J. 1986. *The Chimpanzees of Gombe: Patterns of Behavior*. Cambridge, MA: Harvard University Press.

Goren-Inbar, N., N. Alperson, M. E. Kislev, O. Simchoni, Y. Melamed, A. Ben-Nun, and E. Werker. 2004. Evidence of hominin control of fire at Gesher Benot Ya'aqov, Israel. *Science* 304: 725–727.

Groopman, E. G., R. N. Carmody, and R. W. Wrangham. 2015. Cooking increases net energy gain from a lipid-rich food. *American Journal of Physical Anthropology* 156: 11–18.

Grundy, M. M., T. Grassby, G. Mandalari, K. W. Waldron, P. J. Butterworth, S. E. Berry, and P. R. Ellis. 2015. Effect of mastication on lipid bioaccessibility of almonds in a randomized human study and its implications for digestion kinetics, metabolizable energy, and postprandial lipemia. *American Journal of Clinical Nutrition* 48: 109–121.

Hardy, K., J. Brand-Miller, K. D. Brown, M. G. Thomas, and L. Copeland. 2015. The importance of dietary carbohydrate in human evolution. *Quarterly Review of Biology* 90: 251–268.

Harmand, S., J. E. Lewis, C. S. Feibel, C. J. Lepre, S. Prat, A. Lenoble, X. Boes, R. L. Quinn, M. Brenet, A. Arroyo, N. Taylor, S. Clement, G. Daver, J.-P. Brugal, L. Leakey, R. A. Mortlock, J. D. Wright, S. Lokorodi, C. Kirwa, D. V. Kent, and H. Roche. 2015.

3.3-million-year-old stone tools from Lomekwi 3, West Turkana, Kenya. *Nature* 521: 310–315.

Hatley, T., and J. Kappelman. 1980. Bears, pigs, and Plio-Pleistocene hominids: A case for the exploitation of belowground food resources. *Human Ecology* 8: 371–387.

Hawkes, K. 1991. Showing off: Tests of an hypothesis about men's foraging goals. *Ethology and Sociobiology* 12: 29–54.

Hawkes, K., J. F. O'Connell, and N. G. Blurton Jones. 1997. Hadza women's time allocation, offspring provisioning and the evolution of long post-menopausal lifespans. *Current Anthropology* 38: 551–577.

Hayes, K. C., and J. A. Sturman. 1981. Taurine in metabolism. *Annual Review of Nutrition* 1: 401–425.

Hedberg, G. E., E. S. Dierenfeld, and Q. R. Rogers. 2007. Taurine and zoo felids: Considerations of dietary and biological tissue concentrations. *Zoo Biology* 26: 517–531.

Henneberg, M., V. Sarafis, and K. Mathers. 1998. Human adaptations to meat eating. *Human Evolution* 13: 229–234.

Henry, A. G., A. S. Brooks, and D. R. Piperno. 2011. Microfossils in calculus demonstrate consumption of plants and cooked foods in Neanderthal diets (Shanidar III, Iraq; Spy I and II, Belgium). *Proceedings of the National Academy of Sciences* 108: 486–491.

Hoberg, E. P., N. L. Alkire, A. de Queiroz, and A. Jones. 2001. Out of Africa: Origins of the *Taenia* tapeworms in humans. *Proceedings of the Royal Society of London Series B: Biological Sciences* 268: 781–787.

Holm, J., I. Lundquist, I. Björck, A. C. Eliasson, and N. G. Asp. 1988. Relationship between degree of gelatinisation, digestion rate *in vitro,* and metabolic response in rats. *American Journal of Clinical Nutrition* 47: 1010–1016.

Ichikawa, M. 1981. Ecological and sociological importance of honey to the Mbuti net hunters, eastern Zaire. *African Study Monographs* 1: 55–68.

Indrasena, W. M., and T. A. Gill. 2000. Thermal degradation of partially purified paralytic shellfish poison toxins at different times, temperatures, and pH. *Journal of Food Science* 65: 948–953.

James, S. R. 1989. Hominid use of fire in the Lower and Middle Pleistocene: A review of the evidence. *Current Anthropology* 30: 1–26.

Jarvis, M. C., S. P. H. Briggs, and J. P. Knox. 2003. Intercellular adhesion and cell separation in plants. *Plant Cell and Environment* 26: 977–989.

Keeley, L., and N. Toth. 1981. Microwear polishes on early stone tools from Koobi Fora, Kenya. *Nature* 293: 464–465.

Koebnick, C., C. Strassner, I. Hoffmann, and C. Leitzmann. 1999. Consequences of a long-term raw food diet on body weight and menstruation: Results of a questionnaire survey. *Annals of Nutrition and Metabolism* 43: 69–79.

Kyriacou, K., D. M. Blackhurst, J. E. Parkington, and A. D. Marais. 2016. Marine and terrestrial foods as a source of brain-selective nutrients for early modern humans in the southwestern Cape, South Africa. *Journal of Human Evolution* 97: 86–96.

Laden, G., and R. Wrangham. 2005. The rise of the hominids as an adaptive shift in fallback foods: Plant underground storage organs (USOs) and australopith origins. *Journal of Human Evolution* 49: 482–498.

Leonard, W. R., and M. L. Robertson. 1994. Evolutionary perspectives on human nutrition: The influence of brain and body size on diet and metabolism. *American Journal of Human Biology* 6: 77–88.

Leonard, W. R., and M. L. Robertson. 1997. Comparative primate energetics and hominid evolution. *American Journal of Physical Anthropology* 102: 265–281.

Lepetit, J. 2008. Collagen contribution to meat toughness: Theoretical aspects. *Meat Science* 80: 960–967.

Ley, R. E., M. Hamady, C. Lozupone, P. J. Turnbaugh, R. R. Ramey, J. S. Bircher, M. L. Schlegel, T. A. Tucker, M. D. Schrenzel, R. Knight, and J. I. Gordon. 2008. Evolution of mammals and their gut microbes. *Science* 320: 1647–1651.

Lieberman, D. E. 2011. *The Evolution of the Human Head*. Cambridge, MA: Harvard University Press.

Lieberman, D. E., G. E. Krovitz, F. W. Yates, M. Devlin, and M. S. Claire. 2004. Effects of food processing on masticatory strain and craniofacial growth in a retrognathic face. *Journal of Human Evolution* 46: 655–677.

Lillford, P. J. 2001. Mechanisms of fracture in foods. *Journal of Texture Studies* 32: 397–417.

Lucas, P. 2004. *Dental Functional Morphology: How Teeth Work*. Cambridge: Cambridge University Press.

Lupo, K. D., and D. N. Schmitt. 2005. Small prey hunting technology and zooarchaeological measures of taxonomic diversity and abundance: Ethnoarchaeological evidence from Central African forest foragers. *Journal of Anthropological Archaeology* 24: 335–353.

Mallol, C., F. W. Marlowe, B. M. Wood, and C. C. Porter. 2007. Earth, wind, and fire: Ethnoarchaeological signals of Hadza fires. *Journal of Archaeological Science* 34: 2035–2052.

Mann, N. 2007. Meat in the human diet: An anthropological perspective. *Nutrition & Dietetics* 64: S102–S107.

Marlowe, F. W., J. C. Berbesque, B. Wood, A. Crittenden, C. Porter, and A. Mabulla. 2014. Honey, Hadza, hunter-gatherers, and human evolution. *Journal of Human Evolution* 71: 119–128.

Martin, R. D., D. J. Chivers, A. M. MacLarnon, and C. M. Hladik. 1985. Gastrointestinal allometry in primates and other mammals. In W. I. Jungers, ed., *Size and Scaling in Primate Biology*, 61–89. New York: Plenum.

McGee, H. 2004. *On Food and Cooking: The Science and Lore of the Kitchen*. New York: Scribner.

McGovern, P. E., J. Zhang, J. Tang, Z. Zhang, G. R. Hall, R. A. Moreau, A. Nunez, E. D. Butrym, M. P. Richards, C. S. Wang, G. Cheng, Z. Zhao, and C. Wang. 2004. Fermented beverages of pre- and proto-historic China. *Proceedings of the National Academy* 101: 17593–17598.

McPherron, S. P., Z. Alemseged, C. W. Marean, J. G. Wynn, D. Reed, D. Geraads, R. Bobe, and H. A. Bearat. 2010. Evidence for stone-tool-assisted consumption of animal tissues before 3.39 million years ago at Dikika, Ethiopia. *Nature* 466: 857–860.

Milton, K. 1987. Primate diets and gut morphology: Implications for hominid evolution. In M. Harris and E. B. Ross, eds., *Food and Evolution: Towards a Theory of Human Food Habits,* 93–115. Philadelphia: Temple University Press.

Milton, K. 1999. A hypothesis to explain the role of meat-eating in human evolution. *Evolutionary Anthropology* 8: 11–21.

Moeller, A. H., Y. Li, E. Mpoudi Ngole, S. Ahuka-Mundeke, E. V. Lonsdorf, A. E. Pusey, M. Peeters, B. H. Hahn, and H. Ochman. 2014. Rapid changes in the gut microbiome during human evolution. *Proceedings of the National Academy of Sciences* 111: 16431–16435.

Murdock, G. P. 1967. Ethnographic atlas: A summary. *Ethnology* 6: 109–236.

Murray, S. S., M. J. Schoeninger, H. T. Bunn, T. R. Pickering, and J. A. Marlett. 2001. Nutritional composition of some wild plant foods and honey used by Hadza foragers of Tanzania. *Journal of Food Composition and Analysis* 14: 3–13.

Nobmann, E. D., T. Byers, A. P. Lanier, J. H. Hankin, and M. Y. Jackson. 1992. The diet of Alaska Native adults: 1987–1988. *American Journal of Clinical Nutrition* 55: 1024–1032.

Noli, D., and G. Avery. 1988. Protein poisoning and coastal subsistence. *Journal of Archaeological Science* 15: 395–401.

Novotny, J. A., S. K. Gebauer, and D. J. Baer. 2012. Discrepancy between the Atwater factor predicted and empirically measured energy values of almonds in human diets. *American Journal of Clinical Nutrition* 96: 296–301.

O'Connell, J. F., K. Hawkes, and N. Blurton Jones. 1988. Hadza scavenging: Implications for Plio/Pleistocene hominid subsistence. *Current Anthropology* 29: 356–363

O'Connell, J. F., K. Hawkes, K. D. Lupo, and N. G. Blurton Jones. 2002. Male strategies and Plio-Pleistocene archaeology. *Journal of Human Evolution* 43: 831–872.

Organ, C. L., C. L. Nunn, Z. Machanda, and R. W. Wrangham. 2011. Phylogenetic rate shifts in chewing time during the evolution of *Homo*. *Proceedings of the National Academy of Sciences* 108: 14555–14559.

Panger, M. A., A. S. Brooks, B. G. Richmond, and B. Wood. 2002. Older than the Oldowan? Rethinking the emergence of hominin tool use. *Evolutionary Anthropology* 11: 235–245.

Peres, H., and A. Oliva-Teles. 2002. Utilization of raw and gelatinized starch by European sea bass *(Dicentrarchus labrax)* juveniles. *Aquaculture* 205: 287–299.

Perry, G. H., N. J. Dominy, K. G. Claw, A. S. Lee, H. Fiegler, R. Redon, J. Werner, F. A. Villanea, J. L. Mountain, R. Misra, N. P. Carter, C. Lee, and A. C. Stone. 2007. Diet and the evolution of human amylase gene copy number variation. *Nature Genetics* 39: 1256–1260.

Phillippy, B. Q., J. M. Bland, and T. J. Evens. 2003. Ion chromatography of phytate in roots and tubers. *Journal of Agricultural and Food Chemistry* 51: 350–353.

Pobiner, B., M. Rogers, C. Monahan, and J. Harris. 2008. New evidence for hominin carcass processing strategies at 1.5 Ma, Koobi Fora, Kenya. *Journal of Human Evolution* 55: 103–130.

Potter, N. N., and J. H. Hotchkiss. 1998. *Food Science,* 5th ed. New York: Springer.

Proches, S., J. R. U. Wilson, J. C. Vamosi, and D. M. Richardson. 2008. Plant diversity in the human diet: Weak phylogenetic signal indicates breadth. *BioScience* 58: 151–159.

Purslow, P. P. 1999. The intramuscular connective tissue matrix and cell/matrix interactions in relation to meat toughness. 46th International Congress of Meat Science and Technology, Yokohama, Japan, 210–219.

Ragir, S. 2000. Diet and food preparation: Rethinking early hominid behavior. *Evolutionary Anthropology* 9: 153–155.

Ragir, S., M. Rosenberg, and P. Tierno. 2000. Gut morphology and the avoidance of carrion among chimpanzees, baboons, and early hominids. *Journal of Anthropological Research* 56: 477–512.

Reddy, N. R., and M. D. Pierson. 1994. Reduction in antinutritional and toxic components in plant foods by fermentation. *Food Research International* 27: 281–290.

Reddy, N. R., M. D. Pierson, S. K. Sathe, and D. K. Salunkhe. 1989. Phytates in cereals and legumes. Boca Raton, FL: CRC Press.

Roach, N. T., M. Venkadesan, M. J. Rainbow, and D. E. Lieberman. 2013. Elastic energy storage in the shoulder and the evolution of high-speed throwing in *Homo. Nature* 498: 483–486.

Roebroeks, W., and P. Villa. 2011. On the earliest evidence for habitual use of fire in Europe. *Proceedings of the National Academy of Sciences* 108: 5209–5214.

Rosell, M., P. Appleby, and T. Key. 2005. Height, age at menarche, body weight and body mass index in life-long vegetarians. *Public Health Nutrition* 8: 870–875.

Ruff, C. B., E. Trinkaus, and T. W. Holliday. 1997. Body mass and encephalization in Pleistocene *Homo. Nature* 387: 173–176.

Samuel, B. S., and J. I. Gordon. 2006. A humanized gnotobiotic mouse model of host-archaeal-bacterial mutualism. *Proceedings of the National Academy of Sciences* 103: 10011–10016.

Sandgathe, D. M., H. L. Dibble, P. Goldberg, S. P. McPherron, A. Turq, L. Niven, and J. Hodgkins. 2011. On the role of fire in Neandertal adaptations in Western Europe: Evidence from Pech de l'Azé IV and Roc de Marsal, France. *PaleoAnthropology* 2011: 216–242.

Semaw, S., P. Renne, J. Harris, C. Feibel, R. Bernor, N. Fesseha, and K. Mowbray. 1997. 2.5-million-year-old stone tools from Gona, Ethiopia. *Nature* 385: 333–336.

Semaw, S., M. J. Rogers, J. Quade, P. R. Renne, R. F. Butler, M. Dominguez-Rodrigo, D. Stout, W. S. Hart, T. Pickering, and S. W. Simpson. 2003. 2.6-million-year-old stone tools and associated bones from OGS-6 and OGS-7, Gona, Afar, Ethiopia. *Journal of Human Evolution* 45: 169–177.

Sergant, J., P. Crombé, and Y. Perdaen. 2006. The "invisible" hearths: A contribution to the discernment of Mesolithic non-structured surface hearths. *Journal of Archaeological Science* 33: 999–1007.

Shahack-Gross, R., F. Berna, P. Karkanas, C. Lemorini, A. Gopher, and R. Barkai. 2014. Evidence for the repeated use of a central hearth at Middle Pleistocene (300 ky ago) Qesem Cave, Israel. *Journal of Archaeological Science* 44: 12–21.

Shimelmitz, R., S. L. Kuhn, A. J. Jelinek, A. Ronen, A. E. Clark, and M. Weinstein-Evron. 2014. "Fire at will": The emergence of habitual fire use 350,000 years ago. *Journal of Human Evolution* 77: 196–203.

Shipman, P. 2009. Cooking debate goes off the boil. *Nature* 459: 1059–1060.

Shipman, P., and A. Walker. 1989. The costs of becoming a predator. *Journal of Human Evolution* 18: 373–392.

Simopoulos, A. P. 2006. Evolutionary aspects of diet, the omega-6/omega-3 ratio and genetic variation: Nutritional implications for chronic diseases. *Biomedicine & Pharmacotherapy* 60: 502–507.

Singh, J., L. Kaur, and H. Singh. 2013. Food microstructure and starch digestion. *Advances in Food Nutritional Research* 70: 137–179.

Smith, A. R., R. N. Carmody, R. J. Dutton, and R. W. Wrangham. 2015. The significance of cooking for early hominin scavenging. *Journal of Human Evolution* 84: 62–70.

Speth, J. A., and K. A. Spielmann. 1983. Energy source, protein metabolism, and hunter-gatherer subsistence strategies. *Journal of Anthropological Archaeology* 2: 1–31.

Speth, J. D. 1989. Early hominid hunting and scavenging: The role of meat as an energy source. *Journal of Human Evolution* 18: 329–343.

Speth, J. D. 2010. *The Paleoanthropology and Archaeology of Big-Game Hunting: Protein, Fat or Politics?* New York: Springer.

Stanford, C. B., and H. T. Bunn. 2001. *Meat-Eating and Human Evolution.* New York: Oxford University Press.

Stedman, H. H., B. W. Kozyak, A. Nelson, D. M. Thesier, L. T. Su, D. W. Low, C. R. Bridges, J. B. Shrager, N. Minugh-Purvis, and M. A. Mitchell. 2004. Myosin gene mutation correlates with anatomical changes in the human lineage. *Nature* 428: 415–418.

Sugiyama, Y., and J. Koman. 1992. The flora of Bossou: Its utilization by chimpanzees and humans. *African Study Monographs* 13: 127–169.

Svihus, B., A. K. Uhlen, and O. M. Harstad. 2005. Effect of starch granule structure, associated components and processing on nutritive value of cereal starch: A review. *Animal Feed Science and Technology* 122: 303–320.

Tester, R. F., X. Qi, and J. Karkalas. 2006. Hydrolysis of native starches with amylases. *Animal Feed Science and Technology* 130: 39–54.

Tornberg, E. 2005. Effects of heat on meat proteins: Implications on structure and quality of meat products. *Meat Science* 70: 493–508.

Warneken, F., and A. G. Rosati. 2015. Cognitive capacities for cooking in chimpanzees. *Proceedings of the Royal Society B: Biological Sciences* 282: 20150229.

Washburn, S. L., and C. S. Lancaster. 1968. The evolution of hunting. In R. B. Lee and I. DeVore, eds., *Man the Hunter,* 293–303. Cambridge, MA: Harvard University Press.

Westerterp, K. R. 2004. Diet induced thermogenesis. *Nutrition and Metabolism* 1: 5.

Westman, E. C., R. D. Feinman, J. C. Mavropoulos, M. C. Vernon, J. S. Volek, J. A. Wortman, W. S. Yancy, and S. D. Phinney. 2007. Low-carbohydrate nutrition and metabolism. *American Journal of Clinical Nutrition* 86: 276–284.

Wobber, V., B. Hare, and R. Wrangham. 2008. Great apes prefer cooked food. *Journal of Human Evolution* 55: 340–348.

Wood, B., and L. C. Aiello. 1998. Taxonomic and functional implications of mandibular scaling in early hominins. *American Journal of Physical Anthropology* 105: 523–538.

Wood, B. A., and M. Collard. 1999. The human genus. *Science* 284: 65–71.

Wrangham, R. 2009. *Catching Fire: How Cooking Made Us Human*. New York: Basic Books.

Wrangham, R., and R. Carmody. 2016. Influences of the control of fire on the energy value and composition of the human diet. In J. Lee-Thorp and M. A. Katzenberg, eds., *The Oxford Handbook of the Archaeology of Diet*. Oxford: Oxford University Press.

Wrangham, R. W. 2007. The cooking enigma. In P. S. Ungar, ed., *Evolution of the Human Diet: The Known, the Unknown, and the Unknowable,* 308–323. New York: Oxford University Press.

Wrangham, R. W. 2011. Honey and fire in human evolution. In J. Sept and D. Pilbeam, eds., *Casting the New Wide,* 91–109. Cambridge, MA: Peabody Museum.

Wrangham, R. W., and N. L. Conklin-Brittain. 2003. Cooking as a biological trait. *Comparative Biochemistry and Physiology, Part A* 136: 35–46.

Wrangham, R. W., J. H. Jones, G. Laden, D. Pilbeam, and N. L. Conklin-Brittain. 1999. The raw and the stolen: Cooking and the ecology of human origins. *Current Anthropology* 40: 567–594.

Wu, G. D., J. Chen, C. Hoffmann, K. Bittinger, Y. Y. Chen, S. A. Keilbaugh, M. Bewtra, D. Knights, W. A. Walters, R. Knight, R. Sinha, E. Gilroy, K. Gupta, R. Baldassano, L. Nessel, H. Li, F. D. Bushman, and J. D. Lewis. 2011. Linking long-term dietary patterns with gut microbial enterotypes. *Science* 334: 105–108.

Xu, B., and S. K. C. Chang. 2009. Phytochemical profiles and health-promoting effects of cool-season food legumes as influenced by thermal processing. *Journal of Agricultural and Food Chemistry* 57: 10718–10731.

Yamakoshi, G., and Y. Sugiyama. 1995. Pestle-pounding behavior of wild chimpanzees at Bossou, Guinea: A newly observed tool-using behavior. *Primates* 36: 489–500.

Yellen, J. E. 1991. Small mammals: !Kung San utilization and the production of faunal assemblages. *Journal of Anthropological Archaeology* 10: 1–26.

Zink, K. D., and D. E. Lieberman. 2016. Impact of meat and Lower Palaeolithic food processing techniques on chewing in humans. *Nature* 531: 500–503.

Zink, K. D., D. E. Lieberman, and P. W. Lucas. 2014. Food material properties and early hominin processing techniques. *Journal of Human Evolution* 77: 155–166.

10

From *Pan* to Man the Hunter

Hunting and Meat Sharing by Chimpanzees, Humans, and Our Common Ancestor

BRIAN M. WOOD *and* IAN C. GILBY

Humans eat more meat than any other anthropoid primate, attesting to a major shift in the diet of our hominin ancestors. Hunting and meat sharing are central to hypotheses explaining the evolution of several derived human traits, including large brains, long childhoods, small guts and teeth, complex cooperation, the sexual division of labor, cooperative breeding, and the expansion of *Homo* spp. around the world (Read 1914; Dart and Salmons 1925; Dart 1926, 1949, 1953; Washburn and Lancaster 1968; Laughlin 1968; Isaac 1978; Hill 1982; McGrew 1992b; Bickerton 2009; Gurven and Hill 2009; Isler and van Schaik 2014). Empirical tests of these hypotheses are challenging, however, as they require reconstruction of the behavior and diet of extinct species. Together with the fossil and archaeological records, studies of living apes and human foraging societies are essential for understanding how hominin behavior changed since our lineage split from that of the great apes. Here we use the behavior of chimpanzees and human hunter-gatherers to make inferences about hunting and meat sharing by their last common ancestor (LCA), and to inform our understanding of the causes and consequences of increased meat consumption in the human lineage.

Chimpanzees hunt vertebrates at all research sites across Africa (reviewed in Uehara 1997; Newton-Fisher 2014). Given the relative rarity of predation, however, data regarding frequency, seasonal patterns, hunting strategies, and meat sharing have primarily come from six long-term sites (Table 10.1). By contrast, the worldwide sample of human forager societies available for comparison is *much* larger. We focus on six African groups (Table 10.2), three of which (Aka, Efe, and Mbuti) hunt in forested habitats similar to those occupied by the chimpanzee populations in our sample. Our human societies also include those that occupy drier, more open habitats, including the Hadza, Ju/'hoansi, and Central Kalahari foragers (/Gui and //Gana, which we also refer to as /Gui-//Gana, following Tanaka 2014). While small, our human sample represents considerable environmental, genetic, and cultural diversity.

In both species, hunting behavior varies considerably by individual, population, and season. For example, among chimpanzees, the presence or absence of "impact hunters," who catalyze group hunting (Gilby et al. 2008, 2015), may lead to long-term variation in hunting frequency both within and among social groups. Absence of red colobus monkeys (*Procolobus* spp., chimpanzees' most frequent prey) at a particular site will make hunting relatively rare (e.g., Budongo forest, Uganda; Newton-Fisher et al. 2002). Elsewhere, high densities of fruiting trees can support very large communities and foraging parties (e.g., Ngogo; Potts et al. 2011), which facilitate hunting (Mitani and Watts 2001, 2005). Finally, depending on local ecology, hunting frequency may undergo short-term variation, either regularly (e.g., more hunting during the dry season at Gombe; Stanford et al. 1994a) or unpredictably (e.g., during periods when preferred fruit is abundant at Kanyawara; Gilby and Wrangham 2007).

Similarly, researchers describe large disparities in individual hunting skill in the Ju/'hoansi, /Gui-//Gana, Hadza, Efe, and Mbuti (Lee 1979; Ichikawa 1983; Bailey and Aunger 1989; Tanaka 2014; Wood and Marlowe 2013). Hunting success and the proportion of meat in the diet can vary widely, depending on the presence of certain hunters, the length of the observation period, and the occurrence of low-probability but high-yield large game kills (Hill and Kintigh 2009). As with chimpanzees, hunting may be seasonal, but because humans also hunt migratory birds and mammals, variation is likely to be more pronounced. Finally, reliance on tools and the potential for rapid changes in hunting technology provide further sources of variation within and among human groups. With these caveats in mind, we now explore broad species-level similarities and differences.

Similarities in Hunting and Sharing by Chimpanzees and Humans

Most Prey Weigh Less Than 10 kg

Since the first report of meat eating by chimpanzees at Gombe (Goodall 1963), predation has been observed or inferred in all known chimpanzee populations (Table 10.1; Newton-Fisher 2014). Summing across all field sites, chimpanzees have been documented to hunt at least forty vertebrate prey species, but the most frequent by far is the red colobus monkey (Struhsaker 2010), which accounts for 53 percent (Mahale: Nishida et al. 1992) to 88 percent (Ngogo: Watts and Mitani 2002b) of all kills at sites where the two species coexist (Uehara 1997; Mitani 2009; Newton-Fisher 2014). East African chimpanzees selectively prey upon immature red colobus (Takahata et al. 1984; Stanford et al. 1994a; Stanford 1998; Mitani and Watts 1999), so while adults (*Procolobus tephrosceles*) may weigh up to 13 kg (Kingdon 1997), most victims are much smaller (Figure 10.1). At Gombe, between August 1970 and April 1975, the median estimated weight of red colobus prey was 4 kg (calculated from Wrangham and Bergmann-Riss 1990). In thirty-five successful red colobus hunts between 1999 and 2002 at Gombe, median estimated carcass mass was 3 kg (calculated from Gilby 2004). Even at Taï, where almost half of the red colobus captured were adults (Boesch and Boesch-Achermann 2000), mean carcass size must have been less than 10 kg, as this is the maximum adult mass of the red colobus species found there (*Procolobus badius*; Kingdon 1997).

After arboreal monkeys, the next most frequent chimpanzee prey species are duiker, bushbuck, and bushpig. Bushbuck prey at Gombe are invariably young fawns (Goodall 1986), which are at most the size of adult blue duikers (~3.5–9 kg; Kingdon 1997). All five bushpigs killed by chimpanzees at Mahale between 1979 and 1982 were juveniles (Takahata et al. 1984), and of thirty-two bushpig kills observed at Gombe between 1972 and 1981, all but three victims were "small, still in their striped natal coats" (Goodall 1986: 276), which weigh approximately 1.5 kg (Wrangham and Bergmann-Riss 1990). Notably,

TABLE 10.1. Key chimpanzee study sites that have contributed to our understanding of long-term predation patterns.

Population (Study Group)	Location	Subspecies	Habitat	Start of Continuous Study	Mammalian Prey Species	Hunting Frequency Highest during:	% of Prey Red Colobus	Amount of Meat in Diet	Key Hunting and Meat Sharing Studies
Gombe National Park (Kasekela)	Tanzania, East Africa	*P. t. schweinfurthii*	Evergreen, semideciduous forest; woodland[1]	1960	11[2]	Dry season[3,4]	53%[5]–85%[6]	<5% of feeding time[2,7]; 22 g/day[4]	3–6, 8–17
Mahale National Park (M-Group)	Tanzania, East Africa	*P. t. schweinfurthii*	Semideciduous forest, woodland[18]	1966	17[19]	Dry season[20]	53%[21]		18–24
Kibale National Park (Kanyawara)	Uganda, East Africa	*P. t. schweinfurthii*	Moist deciduous forest, swamp, grassland, colonizing forest[25]	1989	5[26]	Periods of high preferred fruit abundance[27]	83%[26]		14, 16–17, 27–28
Taï Forest	Côte d'Ivoire, West Africa	*P. t. verus*	Evergreen, semideciduous forest[29]	1979	7[29–30]	Wet season[29]	81%[29]	25–180 g/day[29]	29–33

| Kibale National Park (Ngogo) | Uganda, East Africa | *P. t. schweinfurthii* | Evergreen rainforest, grassland, colonizing forest[34] | 1995 | 12[35] | Periods of high ripe fruit abundance[36,37] | 88%[38] | 35–43 |
| Fongoli | Senegal, West Africa | *P. t. verus* | Woodland, grassland, bamboo, gallery forest[44] | 2005 | 6[45] | Wet season (tool-assisted)[45] | 0% (not present)[45] | 44–45 |

1. Clutton-Brock and Gillett 1979; 2. McGrew 1992a; 3. Stanford et al. 1994a; 4. Gilby 2004; 5. Goodall 1986; 6. Stanford 1998; 7. Gombe Stream Research Centre, unpublished data; 8. Wrangham 1975; 9. Goodall 1963; 10. Teleki 1973; 11. Stanford et al. 1994b; 12. Gilby 2006; 13. Gilby et al. 2006; 14. Gilby et al. 2010; 15. Gilby et al. 2013; 16. Gilby et al. 2015; 17. Gilby et al. in revision; 18. Nishida et al. 1979; 19. Uehara 1997; 20. Hosaka et al. 2001; 21. Nishida et al. 1992; 22. Kawanaka 1982; 23. Takahata et al. 1984; 24. Uehara et al. 1992; 25. Chapman and Wrangham 1993; 26. Kibale Chimpanzee Project, unpublished data; 27. Gilby and Wrangham 2007; 28. Gilby et al. 2008; 29. Boesch and Boesch-Achermann 2000; 30. Boesch and Boesch 1989; 31. Boesch 1994b; 32. Boesch 1994b; 33. Gomes and Boesch 2009; 34. Struhsaker 1997; 35. Watts and Mitani 1997; 36. Mitani and Watts 2001; 37. Mitani and Watts 2005; 38. Watts and Mitani 2002b; 39. Langergraber et al. 2007; 40. Mitani and Watts 1999; 41. Watts and Mitani 2002a; 42. Mitani 2006; 43. Sobolewski et al. 2012; 44. Pruetz and Bertolani 2007; 45. Pruetz et al. 2015.

TABLE 10.2. African foragers with detailed studies of hunting and meat sharing.

Population	Study Location	Subsistence	Habitat, Average Yearly Rainfall	Key Hunting Technologies	Study Period(s)	Estimated Dietary Contribution of Hunted Meat (% daily kcal)	Key Sources
Efe	Ituri forest, Democratic Republic of Congo (DRC)	Hunting, honey collecting, fishing, plant gathering, trade with Lese farmers, wage labor	Primary evergreen lowland rainforest, secondary forest, and cultivated areas ~1,800 mm	Bow, arrow, axe, knife, poison, fire, spear, tree perch, dogs, beating stick, noisemaker, clothing	1980–1982[1,2]	~9%	1–2
Mbuti	Ituri forest, DRC	Hunting, honey collecting, fishing, plant gathering, trade with Bira, Nande, and Lese farmers, wage labor	Primary evergreen lowland rainforest, secondary forest, and cultivated areas ~1,800 mm	Net, bow, spear, arrow, trap, axe, knife, poison, fire, quiver, tree perch, dogs, beating stick, noise makers, clothing	1972–1975 1980–1985	Hunting yields are high, but diet not estimated owing to the trade and selling of meat	3–7
Aka	Southern Central African Republic, northeast Congo, western DRC	Hunting, gathering plants and insects, honey collecting, fishing, bushmeat trading, wage labor and trade with neighboring farmers, some manioc and plantain cultivation	Primary evergreen and deciduous forest, swamp forest, secondary forest, cultivated areas ~1,600 mm	Spears, nets, traps, crossbows, knives, poisoned arrows, guns, fire, clothing	1980–2000	In forest with limited trade: 20–40%, in village setting: 10–20%[8]	9–15

Group	Location	Subsistence	Habitat	Tools	Study period	Meat in diet	Refs
Hadza	Lake Eyasi area, northern Tanzania	Hunting, scavenging, honey collecting, gathering, occasional trade with neighboring pastoralists and farmers	Acacia-Commiphora-Baobab woodland and savanna ~650 mm	Bow, arrow, axe, knife, fire, fire drill, hunting blind, slings, rope, poison, clothing, weaverbird hook	1985–1989; 1996–ongoing	>50% (1985–1989) ~33% (1996–2007)	16–22
Ju/'hoansi	Northern Kalahari desert, northeast Namibia	Hunting, scavenging, trapping, gathering plants, laboring for and exchange with pastoralists, ranchers, and traders	Sandy desert dunes and flats, open scrub-savanna ~400 mm	Bow, arrow, quiver, hunting blind, spear, knife, axe, springhare probe, poison, rope snare, digging stick, fire, fire making kit, hunting bag, carrying skin, carrying net, rope, dogs, club	1960s–1970s	~29–40%[24,271,243]	23–26
/Gui and //Gana	Central Kalahari desert, Botswana	Hunting, trapping, gathering plants, occasional wage labor, exchange with pastoralists, ranchers, and traders	Sandy desert dunes and flats, open scrub-savanna ~400 mm	Bow, arrow, spear, club, quiver, poison, hunting blind, knife, springhare hook, snare, traps, digging stick, fire, hunting bag, carrying skin, carrying net, rope, dogs, horses	1958–1966; 1966–1974; 1982–1983	~20%	27–30

1. Bailey and Aunger 1989; 2. Bailey 1991; 3. Harako 1976; 4. Harako 1981; 5. Tanno 1976; 6. Ichikawa 1983; 7. Hart and Hart 1986; 8. Hewlett, personal communication; 9. Bahuchet 1985; 10. Bahuchet 1990; 11. Kitanishi 1995; 12. Kitanishi 1996; 13. Noss and Hewlett 2001; 14. Lewis 2002; 15. Balinga et al. 2006; 16. Hawkes et al. 1991; 17. Hawkes et al. 2001a; 18. Hawkes et al. 2001b; 19. Marlowe 2010; 20. Wood and Marlowe 2013; 21. Marlowe et al. 2014; 22. Wood and Marlowe 2014; 23. Yellen 1977; 24. Lee 1979; 25. Biesele and Barclay 2001; 26. Lee 2003; 27. Tanaka 1980; 28. Silberbauer 1981; 29. Osaki 1984 30 Tanaka 2014.

FIGURE 10.1. A chimpanzee at Gombe National Park captures a juvenile red colobus monkey. Photo by Ian Gilby.

one of the exceptions in Goodall's (1986) sample was "estimated at just over half adult size," suggesting it weighed at least 25 kg (Kingdon 1997).

In the 1960s at Gombe, chimpanzees regularly killed baboons at a chimpanzee provisioning station that brought the two species together at unusually high rates. In 1968–1969, the eight baboon victims of known age averaged twenty-seven weeks old (Teleki 1973), probably weighing between 2.0 and 3.1 kg (based on known weights of two male baboons of eighteen and forty-one weeks of age; Gombe Stream Research Centre, unpublished data). Two others were listed as "juvenile" and one as "infant." After provisioning was reduced and eventually discontinued, chimpanzees preyed upon baboons at much lower rates, but continued to target infants and juveniles exclusively (Gombe Stream Research Centre, unpublished data). Finally, other chim-

FROM *PAN* TO MAN THE HUNTER 347

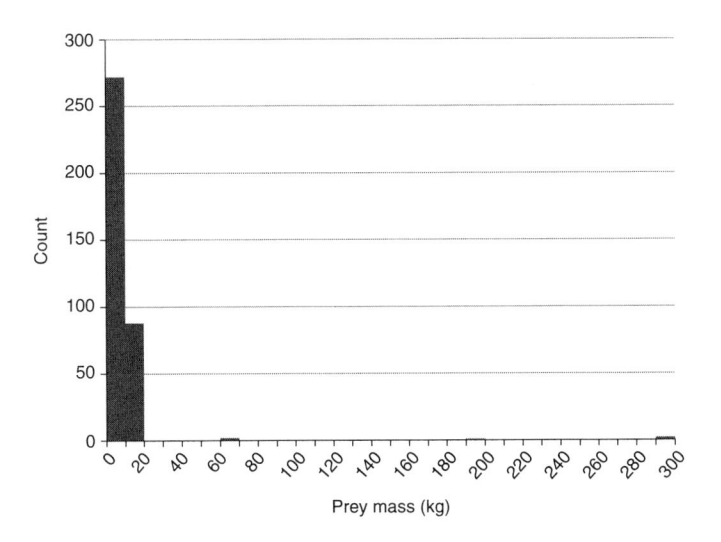

FIGURE 10.2. Body mass of prey killed by Mbuti. Data are from Tanno (1976), Harako (1981), and Ichikawa (1983). Body mass include total mass of carcasses brought into camp, including edible and inedible parts.

panzee prey include eggs, nestlings, rodents, and nocturnal primates (reviewed by Uehara 1997; Newton-Fisher 2014), all of which weigh less than one kilogram.

African foragers also typically capture prey weighing less than 10 kg, although they do occasionally obtain much larger prey (see below). Figure 10.2 shows the estimated body mass of 365 animals acquired by Mbuti foragers using nets ($n=320$), bows and arrows ($n=39$), spears ($n=4$), and via scavenging ($n=2$). The median body mass in this sample is 4.5 kg, and 70 percent of animals killed weighed less than 10 kg. Like chimpanzees, Mbuti hunters selectively target red colobus monkeys, partly because they live in large, easily located groups (Harako 1981). Among the Aka foragers, Noss and Hewlett (2001) report that 75 percent of the animals killed in net hunts were blue duikers. The average weight of all animals killed by the Efe was 4.6 kg (Bailey 1991).

Among the Hadza, between 2005 and 2009, 79 percent of the animals killed and brought to camp weighed less than 10 kg (Wood and Marlowe 2013). More small game were killed but consumed outside of camp (Wood and Marlowe 2014). In Lee's twenty-eight-day work diary of the Dobe Ju/'hoansi in 1964, 78 percent of the carcasses brought to camp were animals weighing less than 10 kg, and the median body mass of all recorded prey was 5.4 kg. Similarly, we calculated the median weight of prey killed by the Ju/'hoansi in Yellen's 1968 sample to be approximately 4 kg, with 56 percent weighing less than 10 kg (Yellen 1977). Using data from Silberbauer (1981), we estimate that 83 percent of prey killed by the/Gui-//Gana over the course of a year weighed less than 10 kg. In sum, the median body mass of prey killed by forest foragers and those in more open habitats seem to be rather similar, all within the 4–6 kg range.

Hunting Is Male-Biased

Among all primates that regularly hunt vertebrates, including chimpanzees, baboons (Harding 1975), and capuchins (Fedigan 1990), males hunt more frequently than females. Boesch and Boesch-Achermann (2000) estimate that male chimpanzees at Taï consume almost seven times more meat than adult females, a difference that can be detected in nitrogen and carbon isotopes extracted from hair and bone collagen (Fahy et al. 2013). This male bias is largely driven by predation upon red colobus monkeys, typically performed by adult males, which made the majority of kills at Taï (81.6 percent; calculated from Boesch and Boesch 1989), Ngogo (93 percent; Watts and Mitani 2002a) and Gombe (89.4 percent: Stanford et al. 1994a). Although females at Kasekela (Gombe) and Kanyawara encountered red colobus less often than males did, when present at a hunt, females were significantly less likely to participate (Gilby et al. 2017). At Kasekela, this may be because female hunters often immediately lose their kill to males (Gilby et al. 2017). Instead of focusing on red colobus, which are active and aggressive, females appear to follow a risk-averse hunting strategy, specializing in relatively low-cost prey. Females at Gombe captured approximately 60 percent of the sedentary prey items (e.g., bushbuck), and a killer was significantly more likely to be female if the prey was sedentary than if it was a red colobus monkey (Gilby et al. 2017). At Kanyawara, females were more likely to capture black and white colobus (Gilby et al. 2017), which are typically less active and aggressive than red colobus. At Mahale, nine of thirteen hunts by females targeted duiker or bushbuck (Takahata et al. 1984), and at Fongoli, females were significantly more likely to engage in tool-assisted capture of cavity-dwelling galagos than males were (Pruetz et al. 2015).

As with chimpanzees, adult males in African foraging societies are responsible for the majority of the hunting. For example, in Lee's (1979) work diary of the Dobe Ju/'hoansi, every kill was made by men and their dogs. However, women do hunt in some contexts. Further analysis of foods brought to seven Hadza camps studied by Wood and Marlowe (2013) shows that females acquired 3.2 percent of the total mass of animals that were brought to camp. The animals that Hadza females killed were small and relatively immobile: tortoises, infant bushpigs, hyrax, and nesting birds. Similar to the Hadza, /Gui and //Gana women occasionally kill birds, small mammals, and tortoises using their hands, clubs, or digging sticks (Tanaka 1980). More female hunting is apparent among Pygmy foragers. Aka and Mbuti women frequently participate in cooperative net hunts, helping drive game into linked nets stretching 500–800 m through the forest. While participating in such hunts, women typically take on less dangerous roles such as driving the animals, rather than capturing and killing them. Noss and Hewlett (2001) report that Aka women net-hunted on 18.1 percent of observation days, significantly more often than men did (11.6 percent). Including all types of hunts, Kitanishi (1995) reports that 0–20 percent of Aka women hunted per observation day, compared to 40–70 percent of men.

Beyond their actual participation in hunts, women provide various kinds of help that aids men's hunting. Hadza and Ju/'hoansi women help men track wounded animals (Biesele and Barclay 2001), and in all groups in our sample, women help carry meat from kill or scavenging sites. However, there appear to be bounds on the kinds of hunting practiced by African forager women—to our knowledge there are no reports of women hunting solitarily, hunting with projectiles, or killing large game.

Hunting Is Often Communal

When chimpanzees encounter a troop of red colobus, reactions range from indifference to immediate hunting by all present (Goodall 1986; Stanford et al. 1994a). A ubiquitous predictor of hunting probability is the number of adult male chimpanzees present at an encounter with red colobus monkeys—parties containing many males are more likely to hunt than those with fewer males (Boesch and Boesch-Achermann 2000; Hosaka et al. 2001; Mitani and Watts 2001; Gilby et al. 2008, 2015). By contrast, little is known about the relationship between chimpanzee party size and opportunistic hunts of solitary or cryptic prey (e.g., duiker, bushbuck, etc.), although we suspect that such hunts may be more likely to occur when parties are small and thus harder for the prey to detect. Indeed, at Kasekela, chimpanzee party size was smaller at kills of terrestrial or concealed prey than at kills of red colobus (Gilby et al. 2017).

In all African forager groups, there are cases in which hunters act alone, and others in which they work together in coordinated groups. Harako (1976) describes hunts in which members of multiple Mbuti residential bands (~30–60 people) join together to track, stalk, and spear elephants. Group hunts in which some individuals drive game toward others waiting with bows and arrows or nets are common among forest-living hunters (Aka, Mbuti, Efe), but rare in more open country foragers (Hadza, Ju/'hoansi, /Gui, and //Gana). Ichikawa (1983) argues that game drives are more common in forests because it is more difficult in such habitats to visually spot and then stalk animals, as hunters often do in more open country. /Gui men often hunt in pairs (Silberbauer 1981), as do Hadza men hunting at night (Marlowe 2010). During the day, Hadza men usually forage alone, and men in all other groups occa-

FIGURE 10.3. A group of Hadza men cooperatively hunting zebra. Photo by Brian Wood.

sionally do so, armed with spears or bows. Yellen (1977: 73) notes that Dobe Ju/'hoansi usually hunt alone or with a single partner, but that "all men in camp may cooperate to follow a wounded animal, help butcher it, and carry the meat back to camp."

Meat Is Shared Strategically

Significant human-chimpanzee differences in life history, cultural and technological sophistication, and social organization correspond with major species differences in the characteristics of social relationships, and particularly the role of meat sharing. Nevertheless, we can see in rough outline that within each species, meat is shared preferentially with species-specific "key social partners."

There are several lines of evidence that chimpanzees share meat strategically among allies. The strongest support comes from Ngogo, where rates of giving and receiving meat were positively correlated among partners, and there were positive associations between dyadic meat sharing rates and rates of grooming and coalitionary support (Mitani and Watts 2001; Mitani 2006). At Mahale, one alpha male (Ntologi) shared preferentially with frequent grooming partners, and supported these males in aggressive conflicts (Nishida et al. 1992). Similar correlations were found at the Yerkes Regional Primate Center (de Waal 1989), but not at Gombe, where sharing among males was correlated with neither grooming frequency nor time spent in close proximity (Gilby 2006). At Taï, Boesch (1994b) describes a "social mechanism limiting access to meat by non-hunters," proposing that sharing decisions are based on an individual's past contributions to collective action. The evidence for this conclusion is weak, however (see "Human Hunting Is More Collaborative," below), and more research is needed to rule out the alternative explanation that active hunters get more meat simply because they are more motivated to do so.

While correlational studies are consistent with the hypothesis that chimpanzees use meat in a system of generalized reciprocal exchange, few studies have directly tested whether sharing decisions are directly based on previous exchanges. High fission-fusion dynamics, the relative rarity of meat eating, and uncertainty over the appropriate time frame of exchange make observational studies of contingent reciprocity particularly challenging in the wild. In one captive study, de Waal (1997) found that food sharing was

FIGURE 10.4. Chimpanzees sharing meat. Photo by Ian Gilby.

more likely to occur if the recipient had groomed the donor within the preceding ninety minutes than if no grooming had occurred. Although the effect was rather small, these exchanges were partner-specific, and there was some evidence of turn-taking.

Among African foragers, the most fundamental way that hunters influence who will receive meat is by deciding with whom they live. Unlike chimpanzees, in which all males stay in their natal groups for their entire lives and members of neighboring communities are generally hostile to one another, African foragers form much more permeable residential groups. They move their residences within large territories and maintain contact with hundreds of individuals through temporary visits and the fission, fusion, and formation of new residential camps (Woodburn 1968b; Yellen 1977; Hill et al. 2014). As such, humans have more flexibility to choose with whom to associate, exchange information, and share food. Inter-camp movement is a critical means by which foragers regulate their social environments (Turnbull 1968; Woodburn 1968b; Lee 1979).

The meat sharing that occurs in African forager camps varies according to prey type, method of capture, and many other factors. When two or more individuals collaborate in a hunt, they are generally assured privileged access to the resulting meat. For example, among the Mbuti, individuals who lend their nets to others receive larger shares than do others in camp, including those who actually used the nets to capture the animals (Harako 1981). Among the Ju/'hoansi and the Hadza, two men often travel and work together when hunting, but usually only one man's arrow actually strikes and kills the animal. Nevertheless, both men subsequently enjoy privileged access to the carcass at the kill site, where they often eat raw liver and bone marrow, and cooked rib meat and parts of the head before carrying the animal back to camp (Yellen 1977). Meat sharing among the Efe is also dependent upon an individual's relative contribution to the kill (Bailey 1991). During cooperative hunts of duiker, the man whose arrow first hit the animal receives the largest share (hind quarters and liver), followed by others whose arrows struck the animal (front leg), and then any owner of a hunting dog (front leg and head). Finally, the older men receive first claims to organs and axial parts.

Studies of correlations between dyadic meat sharing totals show strong evidence for reciprocity in meat sharing among the Hadza and the Aka (Gurven 2004; Gurven and Hill 2009, Crittenden and Zes 2015). No such studies of reciprocity in meat transfers have been conducted among other African foragers. Among the Hadza, Ju/'hoansi, and Aka, husbands are expected to share meat with the parents of their wives; this bride-service determines men's residential choices, especially early in marriages (Hames and Draper 2004; Wood and Marlowe 2011). Among the set of people living in one camp, Hadza men have been shown to bias distributions of all the foods they acquire in ways that benefit their key social partners and dependents, including their wives, children, kin living in other households, and the kin of their wives (Wood and Marlowe 2013).

Differences among Chimpanzees and Humans

Humans Kill More Species

Among all study sites and in all years of observation, chimpanzees have been observed to prey upon only forty vertebrate species (reviewed in Newton-Fisher 2014). There is, however, considerable variation in prey diversity among sites. For example, seventeen prey species have been recorded at Mahale (Uehara 1997), compared to seven at Taï (Boesch and Boesch 1989; Boesch and Boesch-Achermann 2000), even though the availability of potential prey and observation efforts are similar at the two sites. The mean number of mammalian species consumed at each long-term site listed in Table 10.1 is less than ten (although at some sites, certain taxa are listed as one species, e.g., "rodents"). Additionally, prey profiles may change over time. For example, at Mahale, seven species were captured between 1966 and 1981, compared to twelve between 1983 and 1989 (Hosaka et al. 2001).

While chimpanzees specialize on only a few small arboreal and terrestrial species, mainly mammals, human foragers regularly hunt aquatic, terrestrial, subterranean, arboreal, and flying prey, including fish, reptiles, birds, and mammals. Hadza foragers alone kill more vertebrate species than all chimpanzee populations combined: in 242 observation days between 2005 and 2009, they killed forty-one different species (Wood and Marlowe 2013) and recognize several hundred species as potential prey (Marlowe 2010). The Mbuti and Ju/'hoansi hunt and kill at least fifty-three (Harako 1976) and eighty species (Lee 1979), respectively.

Humans Acquire Much More Meat via Scavenging

In general, chimpanzees appear reluctant to scavenge. The bacteria that accumulate rapidly in carcasses pose a greater hazard to chimpanzees than to human foragers, who kill such bacteria by cooking (Smith et al. 2015). Over an eleven-year period, the Ngogo chimpanzees were observed to scavenge only four times, even though they had the opportunity to do so every one hundred days (Watts 2008). Over about twenty years, there were only seven, ten, and two scavenging events at Taï (Boesch and Boesch-Achermann 2000), Gombe (Goodall 1986), and Kanyawara (Gilby et al. 2017), respectively. Chimpanzees at Mahale scavenged twice on fresh adult bushbuck, probably killed by leopards (Hasegawa et al. 1983). The first was, "not large for a normal adult bushbuck," and the second, an adult male, had already been defleshed. With a mass of at least 24 kg (adult females weigh 24–60 kg, males 30–80 kg; Kingdon 1997), these are the largest carcasses that chimpanzees have been reported to eat at any site (with the possible exception of the Gombe bushpig described earlier). However, the chimpanzees consumed only a small portion of the carcasses, perhaps because a leopard lurked nearby.

More often, chimpanzees "power scavenge"; that is, they seize carcasses directly from other predators (reviewed in Watts 2008). At Gombe, over fifty-three community-years of research in Kasekela and Mitumba, there were forty-eight cases in which chimpanzees seized prey (forty-seven bushbuck and one bushpig) from baboons. In one case described by Goodall (1986), adult female Melissa and her daughter Gremlin displayed together at a male baboon, throwing branches and waving their arms before taking the carcass. To our knowledge, baboons have never been observed to seize meat from chimpanzees.

/Gui-//Gana (Silberbauer 1981), Ju/'hoansi (Yellen 1977), and Hadza (Bunn et al. 1988; O'Connell et al. 1988; Hawkes et al. 1991) drive lions, leopards, caracals, cheetahs, and wild dogs off their kills. Among the Ju/'hoansi, scavenging contributed around 9 percent of the total prey mass acquired. Hadza data from the 1980s show that about 20 percent of all animal flesh was acquired through scavenging (Bunn et al. 1988; O'Connell et al. 1988). Hadza men acquired on average 1.3 kg/day through scavenging, compared to 4.9 kg/day from ambush and intercept hunting (Hawkes et al. 1991). Hyenas and other scavengers often locate animals that have died from Hadza and Ju/'hoansi poisoned arrows, and foragers subsequently drive such scavengers off the carcasses. There are far fewer reports of passive or power scavenging by African forest foragers, probably due to lower visibility and faster rates of decomposition. Harako (1981) describes an Mbuti forager appropriating a red colobus from a crowned eagle, and another case of hunters finding and scavenging a buffalo that died for unknown reasons. Lupo and Schmitt (2005) report that only 0.3 percent of the animals that Aka and Bofi foragers acquired were scavenged. Wrangham (personal communication) observed Efe scavenge a rotten red colobus monkey that was subsequently cooked and eaten.

Human Foragers Occasionally Kill Relatively Large Prey

By targeting prey considerably smaller than themselves, chimpanzees follow a pattern common among predators (Cohen et al. 1993). At Gombe, median adult male body mass is 39 kg (Pusey et al. 2005), and the largest recorded prey item captured there was a young bushpig estimated at 25 kg (Goodall 1986). At other sites, chimpanzees occasionally capture adult male black and white colobus, which may weigh >20 kg (Kingdon 1997, although Watts and

Mitani [2015] estimate that adult male black and white colobus at Kibale weigh only 9.9 kg). There are many larger animals in some chimpanzee habitats that they do not hunt, including buffalos, hippos, elephants, okapis, adult forest hogs, and adult bushpigs. The largest animals that chimpanzees kill are other adult chimpanzees (Wilson et al. 2014), although these are almost never eaten (but see Pruetz et al. 2017). Nevertheless, we believe that intraspecific killing among chimpanzees has important implications for the evolution of human hunting (see below).

By contrast, even though humans typically kill animals weighing less than 10 kg (see "Most Prey Weigh Less Than 10 kg," above), they can (and do) kill animals much larger than themselves. Okiek (Huntingford 1929), Aka (Kitanishi 1996), Mbuti (Harako 1976), and /Gui-//Gana (Silberbauer 1981) foragers killed African elephants. Hadza, Ju/'hoansi, and /Gui-//Gana have been observed to kill adult male giraffes, which average 1,865 kg (Kingdon 1997). Early twentieth-century accounts by Kohl-Larsen and Cooper (cited in Marlowe 2010) indicate that the Hadza once hunted hippopotamus and rhinoceros, which are no longer found in their area. Mbuti foragers also kill forest buffalos (up to 320 kg), okapis (210–250 kg), giant forest hogs (100–275 kg), and bushpigs (45–150 kg) (adult masses, Kingdon 1997). Such large animals are rare, and therefore opportunities are infrequent, but they result in enormous amounts of meat, which has important implications for food sharing.

Humans Acquire Much More Meat

Meat is a valuable source of protein, fat, iron, vitamin B_{12}, and other micronutrients (Tennie et al. 2009, 2014); however, its contribution to chimpanzee diets is still poorly understood. Nevertheless, by any measure, meat constitutes a small proportion of any individual's diet (e.g., less than 5 percent of feeding time at Gombe: Goodall 1986; McGrew 1992a). At Taï, males and females consume an average of 180 and 25 grams of meat per day, respectively (Boesch and Boesch-Achermann 2000), although there is considerable variation by season and individual.

By contrast, in terms of total calories, hunted meat is estimated to make up between 9 percent and more than 50 percent of the yearly diet of African foragers (Table 10.2). Marlowe (2005) estimates that among all warm-climate nonequestrian foragers, the average contribution is 25 percent. Each nuclear family of Mbuti net hunters acquired 5.3 kg of game per day (Tanno 1976). As is common among Central African foragers, the Mbuti traded a great deal of this meat with their agricultural neighbors, and were able to receive 12,000 calories of cassava in exchange for only 2,500–4,000 calories of meat. These data attest to the high potential yields from cooperative net hunting in rainforests, and illustrate that meat is usually prized over plant foods on a calorie-for-calorie basis (Hill 1988). Among the Hadza in the 1980s, Hawkes et al. (1991) report that men acquired 4.9 kg/hunter-day. Assuming that camps contained three nonhunters per hunter, this yielded 1.2 kg (~1,830 kcal) per hunter-day, clearly a very large fraction of their diet. Hadza data from 1995–2009 indicate a lower fraction, between 25–35 percent, depending on the season and camp location (Marlowe 2010). Tanaka (1980) estimates that /Gui and //Gana foragers acquired 0.3 kg of meat per person-day, lower than other African foragers, but more than chimpanzees.

Human Foragers Spend More Time Hunting

At Gombe, the average hunt (including failures) of red colobus lasts twenty-eight minutes (Stanford 1998). Using this value with raw data from Gilby et al. (2013), we calculate that between 1976 and 2007, focal male chimpanzees at Gombe spent an average of thirty-five minutes hunting per one hundred hours of observation, or 3.5 minutes per ten-hour day. By contrast, focal males in the same sample spent almost 4.9 hours feeding on plant matter during the average ten-hour day (Gombe Stream Research Centre, unpublished data). Chimpanzees have never been observed (or suspected) to hunt at night.

On average, while living in hunting camps, Mbuti men and women spend between 6.8 (Ichikawa 1983) and 7.5 hours/day (Tanno 1976) net hunting. In Bailey's (1991) study, Efe men spent an average of 2.7 hours/day in the forest searching for, pursuing, butchering, and carrying game. Hadza men spend 4.1 hours out of camp on average (Hawkes et al. 1991), but not all that time is spent hunting. Wood and Marlowe (2014) show that Hadza men spent 62.8 percent of their time out of camp in generalized search (for game, fruit, honey, etc.), and 9 percent following animal tracks, pursuing visually encountered animals, processing carcasses, or atop rock outcrops scanning the landscape for animals. Lee (1979) estimates that Ju/'hoansi men spent 29.1 hours per week in subsistence, tool making, and tool repair, and that 83 percent of this work effort (i.e., 3.4 hours/day) was hunting-related. Clearly, African foragers spend much more time hunting than do chimpanzees, although measures of chimpanzee hunting do not include travel time. In addition, humans hunt both during daylight hours and at night (Lee 1979; Hawkes et al. 2001b).

Human Foragers Use Many Complex Tools When Hunting

Some chimpanzees use tools while *consuming* meat. In twenty-six of twenty-eight kills at Taï, chimpanzees modified small sticks to extract bone marrow (Boesch and Boesch 1989). Sticks and leaves are also sometimes used to extract brain matter from monkey crania at Taï (Boesch and Boesch 1989) and Gombe (Teleki 1973). Critically, however, there is no evidence that chimpanzees have ever used tools to cut meat or break open bones, even at sites where they routinely use stone anvils to break open nuts. For a review of chimpanzee tool use, including discussion of the morphological constraints on tool manufacture, see Rolian and Carvalho (this volume).

Chimpanzees rarely use tools while hunting, with one notable exception. At Fongoli, chimpanzees forcibly probe tree cavities with sharpened sticks to flush out, disable, or kill galagos (Pruetz and Bertolani 2007; Pruetz et al. 2015). Over a ten-year period, thirty-five (of a possible forty-four) individuals in this community made 308 attempts to capture galagos in this manner (Pruetz et al. 2015). When infants and juveniles (which never succeeded) were excluded from the analyses, a kill was made in 10.3 percent of attempts. One case of similar behavior was reported at Mahale, when an adolescent female used a stick to flush a squirrel from a hollow tree cavity (Huffman and Kalunde 1993). Goodall (1986: 541) describes cases in which a Gombe chimpanzee "broke off a branch, pushed it into an opening, and moved it rapidly backward and forward." However, apart from ants, termites, and bees, nothing emerged. The high proportion of tool-assisted predation at Fongoli may be due to in part to the absence of red colobus monkeys, but it is unclear why it is so rare at other sites (Pruetz et al. 2015). One possibility is that galago density is greater in the savanna habitat where the Fongoli community ranges, or they might not occupy tree holes at other sites (e.g.,

Kanyawara; R. Wrangham, personal communication). Pruetz et al. (2015: 9) propose that tool-assisted predation "enables individuals who would be less likely to chase down larger vertebrate prey access to an energetically and nutritionally valuable food resource in a patchy savannah environment," a claim that has important implications for hunting by early hominins.

In sharp contrast to chimpanzees, humans use numerous types of multi-part tools when hunting (Table 10.2). These tools are used to kill prey outright or slowly via blood loss, poisoning, or sepsis. Among the Hadza, hunting tools are more technologically complex than are tools used for gathering (Marlowe 2010). This trend in tool complexity seems likely to apply to other African foragers. All African foragers occasionally capture small game by hand, without the use of deadly weapons, but even on such occasions, hunters are wearing clothing, carrying tools, and benefiting from technological aids to their foraging. Through the use of tools, humans can kill arboreal, terrestrial, subterranean, and aquatic prey, exploiting more predatory niches than any other predator on earth.

Language Facilitates Hunting and Sharing among Humans

Apart from pant-hoots, anticipatory food grunts, and specific "hunting calls" (Mitani and Watts 1999), which simply seem to advertise that a hunt is underway, there is no indication that chimpanzees deliberately communicate during a hunt. Even at Taï, where complex collaboration has been reported (Boesch 1994b, 2002; Boesch and Boesch-Achermann 2000), hunters do not appear to intentionally signal to one another.

By contrast, language is an enormously powerful tool that allows humans to accumulate, maintain, and use complex bodies of knowledge. Using language and mimicry, hunters can recall and represent past hunts, and imagine and plan future hunting scenarios. While hunting, African foragers use sophisticated repertoires of hunting calls, whistles, words, body language, and hand signals to communicate intent and coordinate actions. Language allows for the retention and pooling of collective memories, and the effective teaching of hunting skills. Language permits more effective planning of cooperative hunts, the management of meat distribution, and the resolution of problems associated with collective action and coordination (Bowles et al. 2010; Smith 2010).

Humans Travel Greater Distances in Search of Meat

It is difficult to ascertain the extent to which chimpanzee ranging patterns are affected by hunting. While the Gombe chimpanzees are more likely to encounter red colobus monkeys on days when they travel greater distances (Gilby et al. 2013), they do not seem to deliberately seek prey. Instead, encounters appear to occur by chance. In contrast, 41 percent of encounters with red colobus at Ngogo occurred on "hunting patrols," during which chimpanzees traveled quickly and quietly, apparently searching for monkeys (Mitani and Watts 1999). Similar behavior occurs at Taï, where there were "clear signs of hunting intention before any prey was seen or heard" in 50 percent of hunts (Boesch and Boesch 1989). Using published data from Gombe, Kanyawara, Taï, and Fongoli (males only), Pontzer (this volume) calculated a mean day range of 3.6 km/day for adult males and 3.0 km/day for adult females. Chimpanzees at Fongoli, which occupy a woodland-savanna habitat, travel much further in the dry season (J. Pruetz, personal communication), which was not included in this sample. However, there is no evidence that hunting dictates ranging patterns in this or other chimpanzee populations.

Humans spend more time traveling, and they do so faster than chimpanzees (Pontzer, this volume). As a result, human daily path lengths are longer. For example, Efe men are reported to have traveled on average 9.4 km per day (Bailey 1991), while Pontzer et al. (2012) found that on average, Hadza men traveled 11.4 km and women 5.8 km per day. Marlowe (2005) reports that among warm-climate nonequestrian foragers, males and females traveled on average 14.1 and 9.5 km per day, respectively. The larger day range of forager males is undoubtedly due to men's hunting, and the fact that men pursue foods that are more mobile, unpredictable, and scarce than female-acquired foods.

Silberbauer (1981) estimates that /Gui-//Gana hunters living together at one camp made use of a maximal foraging area that was 800 km^2 in size, which is more than nine times larger than the largest chimpanzee home range (85 km^2 at Fongoli: Wilson et al. 2014). /Gui-//Gana foragers also moved camps several times a year, and thus made use of an even larger total foraging area. Like chimpanzees, human foragers living in more productive environments use smaller ranges. The Mbuti studied by Harako (1981) used 100–200 km^2 territories in one year, while those studied by Tanno (1976) made use of 120–150 km^2.

Human Hunting Is Most Productive during the Dry Season; Chimpanzee Hunting Seasonality Varies by Site

At Ngogo, chimpanzee hunting frequency is correlated not with rainfall, but instead with ripe fruit availability (Mitani and Watts 2001), which enables the formation of large hunting parties (Mitani and Watts 2005). Similarly, at Kanyawara, hunting increases when preferred fruits are particularly abundant (providing surplus energy), and there is no regular, seasonal pattern (Gilby and Wrangham 2007). At Mahale and Gombe, however, hunting is strongly seasonal, peaking during the dry season (Hosaka et al. 2001; Stanford et al. 1994a), when large chimpanzee parties form (Takahata et al. 1984; Gilby et al. 2006). Hunting at Taï is also seasonal, reaching its maximum during the wettest months (Boesch and Boesch-Achermann 2000). Boesch and Boesch-Achermann (2000) attribute this increase to reduced alternative food sources, increased ease of capture (due to slippery branches) and the red colobus birth season. At Fongoli, 95.1 percent of tool-assisted predation occurred during the wet season (Pruetz et al. 2015).

In contrast to the variation observed among chimpanzee populations, all African foragers experience an increase in hunting productivity during the dry season. Among the Hadza, this appears to be because the movement of game is predictably restricted to fewer sources of water and forage during the dry season (Hawkes et al. 2001b; Wood and Marlowe 2013), making nighttime ambush hunting particularly effective. Hunting in general, and especially large game hunting, is more frequent and more successful for the Ju/'hoansi in the dry and late dry seasons (Lee 1979: 104). The Mbuti hunt during the dry season, while during the rainy seasons they reside in villages where they depend almost entirely upon farm products (Tanno 1976). Regarding the Mbuti, Tanno (1976: 115) notes that "the hours per day spent in net hunting are determined mainly by the rainfall . . . if it begins to rain while the hunting is in progress, they cease hunting and return to the camp." The Efe's hunting season is also the driest part of the year (Bailey 1991: 65–67).

Human Males in Their Forties and Fifties Acquire Significant Amounts of Meat

Among the Taï chimpanzees, learning to hunt is a long process, with the necessary skills acquired over at least twenty years, starting at the age of nine or ten (Boesch and Boesch-Achermann 2000). At both Gombe (Kasekela) and Kanyawara, male hunting probability follows an inverted U-shaped distribution, with males in the twenty-one to twenty-five year age category exhibiting the highest rates (Gilby et al. 2015). There was a significant decline in hunting rates after ages thirty and forty at Kasekela and Kanyawara, respectively. Similarly, success rates peaked between ages twenty-one and twenty-five at Kasekela before falling. However, at Kanyawara, hunting success rates continued to increase with age—though males older than thirty-six rarely hunted, when they did so they were more likely to succeed than their younger counterparts were. Importantly, there is considerable individual variation in both hunting propensity and skill—at Gombe, Frodo had made at least forty-two kills by the time he was fifteen years old (Stanford et al. 1994a), in contrast to the next most successful young hunter, Ferdinand, with twenty kills (data from Gilby et al. 2015). Over the course of the whole Gombe study, all other males made ten or fewer kills by age fifteen. Frodo continued to exhibit high participation and kill rates for his whole life (Gilby et al. 2015).

Because humans experience lower adult mortality than chimpanzees, groups of human foragers contain more males over the age of forty than do chimpanzee communities. Chimpanzees at Kanyawara have only a 14 percent chance of reaching age forty, and a 9 percent chance of reaching age fifty (Muller and Wrangham 2014). Among the Hadza, 40 percent of males reach age forty and 35 percent reach age fifty; if they survive to age forty, they can expect to live another twenty-three years (Blurton Jones 2016).

There are few quantitative records of men's hunting success by age among African foragers. Silberbauer (1981) notes that /Gui-//Gana hunters begin to kill large game around the age of eighteen, and are most effective and active between their late twenties and the age of thirty-five. At around age forty, men start to shift their efforts away from bow-and-arrow hunting and toward trapping and digging for springhares. Lee (1979) notes that Ju/'hoansi men's peak hunting success occurs between the ages of thirty and forty-five. Mbuti men appear to have peak success as spear hunters between ages thirty and fifty (Harako 1981). The most detailed data on hunting rates by age among African foragers come from the Hadza. In Table 10.3, we provide additional analysis of Wood and Marlowe's (2013) food returns data.

Hadza men aged forty and over contributed only 20 percent of the observation days in this sample but acquired 40 percent of the total meat brought to camp. The highest hunting returns (kg brought to camp per observation

TABLE 10.3. Rates of Hadza men acquiring and bringing meat to camp by age.

Age	n Individuals	Person Days Observed	n Small Game	n Large Game	Total kg Game	kg Game/Person Day
0–9	32	1,072 (27%)	9	0	0.85	0.00
10–19	14	737 (18%)	62	0	31.05	0.04
20–29	17	631 (16%)	63	7	1,060.12	1.68
30–39	19	744 (18%)	52	18	1,459.16	1.96
40–49	11	378 (9%)	22	8	521.26	1.38
50–59	8	292 (7%)	28	12	1,144.60	3.92
60–69	3	59 (1%)	3	1	43.60	0.74
70+	2	84 (2%)	3	0	0.33	0.00

Note: These data are derived from the sample described in Wood and Marlowe (2013), which includes seven camps observed between 2005 and 2009. Weights include all parts of animals brought into camp, including edible and inedible parts.

day) were among men age fifty to fifty-nine. Surely, if Hadza had an age structure more similar to chimpanzees, there would be much less meat in their diets. Given cross-cultural patterns in the age dependency of economic productivity, this is also likely to be the case among other African foragers. Gurven and Gomes (this volume) provide a discussion of the importance of intergenerational cooperation and food sharing in traditional human populations, and the role it might have played in the evolution of the human life span.

Human Hunters Ambush Prey

With the exception of searching for red colobus at Ngogo and Taï, chimpanzees appear to encounter prey by chance during routine travel. While it is possible that they may alter their travel routes to pass through prey-rich areas, there are no reports of any chimpanzees waiting for prey to arrive. By contrast, Mbuti (Harako 1976), Hadza (Hawkes et al. 1991), and Efe (Bailey 1991) hunters often search for fresh signs of prey, and then, based on such signs, select a location where the animal is likely to pass in the near future—often water sources, fruiting trees, game paths, or salt licks. Hunters then conceal themselves, either by using naturally occurring vegetation or by constructing a blind, stand, or pit, and wait quietly for animals to arrive. This requires foresight, calculation, and patience, and is commonly used for killing the largest game. Mbuti hunters use this technique to spear buffalo,

FIGURE 10.5. A Hadza couple cooperatively hunting a hyrax—the wife blocking possible points of escape while her husband spears the animal. Photo by Brian Wood.

okapi, giant forest hog, and bushpig (Harako 1981). Hadza men use it to shoot zebra, eland, buffalo, impala, and smaller game. This is a very effective strategy for the Hadza in the late dry season, when animal movements are more predictable, and while hunting at night (Hawkes et al. 2001b). Ju/ 'hoansi men also occasionally hunt at night from blinds (Lee 1979).

Human Hunting Is More Collaborative

There has been considerable interest in the degree to which chimpanzees "work together" during communal hunts of arboreal monkeys. Boesch (2002) argues that increasingly sophisticated cognitive ability is required for species to move along a continuum of cooperation from "similarity," when hunters simultaneously target the same prey without coordination, to "collaboration," in which they "perform different complementary actions" (Boesch and Boesch 1989). True collaboration involves observation, anticipation, and reaction to the actions of the prey and other predators in space and time. It has also been proposed that particular psychological traits, including impulse control (Stevens and Hauser 2004), have evolved to facilitate collaboration. Boesch (1994b) has argued that truly collaborative species have also evolved the ability to selectively form bonds with "trusted partners," which allows individuals to more reliably predict the outcome of collaborative actions, such as meat sharing after a kill. However, chimpanzees at most sites rarely hunt in a manner that is consistent with true collaboration (Mitani and Watts 2001; Gilby et al. 2006, 2015). The exception is Taï, where Boesch and Boesch-Achermann describe the frequent occurrence of hunts in which "drivers" funnel colobus monkeys past "blockers" (who restrain themselves from chasing) toward "ambushers" and "chasers" who ultimately make the kill (Boesch and Boesch 1989; Boesch 1994b, 2002; Boesch and Boesch-Achermann 2000). However, with the data presented, one cannot rule out the more parsimonious alternative that this apparent coordination is a by-product of the selfish efforts of several hunters, each attempting to make his own kill and reacting to the actions of others (Gilby and Connor 2010; Tomasello and Moll 2010; Gilby et al. 2015).

African foragers work together, share knowledge, tools, and food, and perform complementary actions that far exceed the complexity and effectiveness of the limited coordination observed among chimpanzees. Foragers collaborate in all their subsistence pursuits. The range of collaborative hunting activities is quite broad—multiple camps of Mbuti work together to track and spear elephants, with individuals playing different roles based on age and skill (Harako 1976); /Gui- //Gana men travel together and coordinate shots of arrows at their prey (Silberbauer 1981); Efe and Mbuti women drive duikers toward their husbands, who tend nets, with all the individually owned nets tied together as one unit for the group's collaborative hunt (Tanno 1976).

Human hunters are never quite alone—their behavior is structured by the fact that they can expect to find information, food, water, warmth, and shelter when they return home. Solitary human hunters also draw on coordinated assistance from others as needs arise. When solitary Ju/'hoansi, /Gui-//Gana, Hadza, and Mbuti men shoot animals with arrows, they return to their camps and recruit help from others to track wounded prey, butcher carcasses, and carry the meat back to their camp (Woodburn 1968a; Harako 1976; Yellen 1977; Lee 1979; Biesele and Barclay 2001).

Social Norms and Kinship Regulate Hunting and Sharing among Humans, While Chimpanzees Are Primarily Motivated by Immediate Self-Interest

Hunting in groups increases the probability that a given chimpanzee immediately obtains meat (Boesch 1994b; Gilby et al. 2008, 2010; Tennie et al. 2009). While subsequent sharing may secure reciprocated (delayed) benefits in the form of more meat, coalitionary support, grooming, or sex (de Waal 1989, 1997; Mitani 2006; Gomes and Boesch 2009), chimpanzees do not appear to base their hunting decisions on the potential for delayed exchanges—neither the presence of a preferred male social partner at Kanyawara (Gilby et al. 2008) nor sexually receptive females at Ngogo, Kanyawara, or Gombe (Mitani and Watts 2001; Gilby et al. 2006, 2008, 2015) improved the chances of a hunt occurring. Instead, the primary motivation for a chimpanzee to initiate or join a hunt seems to be to obtain meat for itself.

Most meat sharing by chimpanzees (e.g., 76.4 percent of 529 sharing events at Gombe; Gilby, unpublished data) is "passive," in which a possessor neither facilitates nor resists others' attempts to take pieces or feed from the same carcass. While there is growing evidence of preferential sharing with frequent grooming and coalition partners, the "sharing under pressure" (Wrangham 1975) or "harassment" (Stevens 2004) hypothesis also explains a significant proportion of chimpanzee meat-sharing events. Similar to the tolerated theft model proposed for humans (Blurton Jones 1984), the sharing under pressure hypothesis states that in the face of persistent begging, it is immediately beneficial for a possessor to share if doing so reduces the negative effects of harassment. It is critical to note that harassment does not equal "aggression"; instead, harassment need only impose a cost for the possessor to immediately benefit by sharing. Gilby (2006) found that (1) meat possessors at Gombe consumed meat less efficiently when faced with many beggars, (2) sharing with beggars encouraged their departure, and (3) sharing probability increased with the level of harassment. We emphasize that while there are many reasons why chimpanzees share meat, most of these transfers would not occur in the absence of harassment.

Like chimpanzees, African foragers face pressure to share. However, physical harassment and theft are rare. Instead, meat sharing is governed by social norms that establish guidelines for distribution, and which reduce chances of outright conflict or freeloading. First, the hunter who made the kill typically benefits the most. Hadza men keep heavier and more valuable shares of their kills than they give to others (Wood and Marlowe 2013, 2014). Among the /Gui-//Gana, the hunter who kills a large animal receives the back meat, including the sinew, which is used for a variety of tools. He also retains the skin, which is extremely valuable for clothing, carrying devices, sleeping surfaces, and working surfaces (Tanaka 1980). The same norm of hunters keeping the valuable skins from their kills is found among the Hadza (Wood and Marlowe 2013). Second, in all groups, hunters share preferentially with their wives (e.g., Hadza: Wood and Marlowe 2013). The institution of marriage is one means by which, relative to chimpanzees, human females come to play a more active role in managing the sharing of meat. In reference to the Aka, Noss and Hewlett (2001) note that women "control the division and distribution of game acquired captured on net hunts." Hadza men deliver small animals, fruit, and honey directly to their households, and their wives largely control the subsequent sharing that occurs within the family and with neighbors. By contrast, although male chimpanzees preferentially share meat with maternal brothers when there is the opportunity to do so, most meat-sharing events occur between unrelated males (Langergraber et al. 2007). Third, sharing of tools or contribution of labor to a kill obliges the owner to share. Fourth, meat-sharing norms that benefit older males are reported for the Efe, the Ju/'hoansi, the /Gui-//Gana, and the Hadza.

The Evolution of Human Hunting

While it is possible that some similarities in hunting and meat sharing arose independently in the human and chimpanzee lineages via convergent evolution, the most parsimonious explanation is that these are homologies; it is therefore likely that the LCA (1) hunted mostly small prey, (2) exhibited male-biased hunting, (3) often hunted in groups, (4) occasionally appropriated carcasses from other predators, and (5) shared meat with key social partners but also under pressure. In their reconstructions of ancestral states in hominoid evolution, Pickering and Domínguez-Rodrigo (2010) and Duda and Zrzavý (2013) similarly propose that the LCA already engaged in more hunting than is typical for a primate. Environmental changes affecting the availability of prey likely provided selective pressures resulting in early differences in hunting behavior within the hominin lineage. As the climate became cooler and drier, African woodland and savanna habitats expanded (Reed 1997; Bobe and Behrensmeyer 2004). To exploit these more open habitats, early hominins would have needed to increase their day range (Foley 1987), encountering more terrestrial prey species, particularly grazing herbivores. This would have lowered the frequency of encountering arboreal monkeys. However, since chimpanzees have been shown to respond to reduced numbers of preferred prey by switching to alternative species (Watts and Mitani 2015), early hominins surely did the same, incorporating easily acquired terrestrial prey such as infant antelopes and bovids, fledgling birds, and hyrax. Additionally, in woodland and savanna habitats, large herbivores tend to congregate near sources of water, consequently attracting large carnivores (Valeix et al. 2010). In order to exploit the expanding savanna, early hominins would have also needed to visit waterholes, presenting them with more frequent opportunities to capture prey and scavenge fresh kills, although they would also have experienced increased predation risk. The fact that chimpanzees engage in coalitionary killing of adult conspecifics (Wilson and Glowacki, this volume) suggests that even without the use of weapons, the LCA likely had the capacity to kill large-bodied, dangerous animals.

While paleoclimatic data provide hints about what life could have been like for early hominins between 7.9 Ma and 3.3 Ma, the archaeological record itself is silent, and therefore our ability to make inferences about their behavior is limited. It is possible that rates of hominin carnivory and scavenging increased during this period, but without supportive data, the most we can say is that meat comprised at least 1–5 percent of their total diet (comparable to that of chimpanzees).

The earliest stone tools appear at 3.3 Ma at the site of Lomekwi, Kenya (Harmand et al. 2015). It seems safe to assume that the hominins that manufactured such stone tools would also have fashioned simple wooden implements of a complexity similar to those used by chimpanzees today. We therefore suggest that by 3 Ma, hominins were using minimally modified sticks as clubs or short spears in various contexts including predator defense, power scavenging, hunting, and digging for animal or plant foods.

The site of Kanjera South, Kenya, has produced well-preserved stone tools and cut-marked bones that are approximately 2.0 million years old. These attest to hominin occupation of a grassland environment adjacent to Lake

Victoria (Ferraro et al. 2013). The counts and distributions of stone tool cutmarks and carnivore tooth marks on the faunal material indicate that by this point hominins regularly engaged in hunting and/or power scavenging, giving them access to large, fleshed carcasses. The species responsible for these early assemblages is not known, but an early member of our genus, either *Homo habilis* or *Homo ergaster,* is the most likely candidate. While it is not possible to estimate the median body mass of prey items killed by these hominins, the Kanjera South assemblage contains numerous wildebeest-sized prey, and a few even larger. Spatial and taphonomic analyses at this site indicate that hominins returned with fleshed carcasses to a central place, representing a notable change in social organization relative to that of chimpanzees and our reconstructed LCA. As discussed by Blumenschine et al. (1991), transporting food back to a central place can serve several functions: providing refuge from risks of predation (Isaac 1983), accessing cached stone tools to aid butchering (Potts 1984), and provisioning others.

These archaeological materials reflect several of the derived features of modern human hunting: technological sophistication, the occupation of a wider range of habitats, and the (at least occasional) incorporation of large prey. The power scavenging and/or hunting involved in acquiring these larger-bodied prey almost certainly involved cooperation, although the degree of collaboration is unknown. However, we can assume that at the very least, they exhibited the kind of loosely coordinated group action typical of chimpanzee hunts of red colobus.

While some have argued that only large carcasses provide sufficient evidence for food sharing (e.g., Binford 1981), studies of chimpanzees and human foragers illustrate that even very small game are regularly shared. Most likely, the meat derived from the kills at Kanjera South and Olduvai was being shared, but the specific details of such transfers and the nature of in-group social relationships involved remain a mystery. Given the technological sophistication and foraging capabilities of these hominins, we estimate that meat would have contributed a significant fraction of their diet, perhaps 5–20 percent of their total calories. Similarly, these hominins were probably acquiring other high-quality extracted foods (e.g., honey, tubers, nuts), leading to an overall increase in dietary quality and diversity relative to the LCA.

Given that the changes in social organization and foraging strategies suggested by the archeological record correspond closely to a marked increase in brain size in the genus *Homo,* several researchers have proposed that dietary shifts (including increased carnivory) led to, resulted from, or coevolved with larger brains and increased intelligence (Foley 2001; Foley and Lee 1991; Kaplan et al. 2000). In principle, more fossil and archaeological data could help resolve the timing of these shifts. At present, the fact that increased meat eating and early brain size expansion coincide relatively closely in time lends support to ideas that changes in diet and social structure coevolved with larger brains.

Nearly coincident in time with these examples of increased meat eating and encephalization of early *Homo* in Africa, there is evidence for the rapid dispersal of *Homo erectus* outside of Africa, north to the Republic of Georgia by 1.8 Ma (Gabunia et al. 2000), and east to the island of Java by 1.7 Ma. This range expansion is not likely to be a consequence of population growth, but rather was due to a fundamental shift in ranging and foraging capabilities (Anton et al. 2002). Comparative analyses show that primates with larger brains are better able to buffer themselves against highly seasonal environments (van Woerden et al. 2012), and among mammals, larger brains are correlated with greater chances of survival in novel habitats (Sol et al. 2008). In particular, hunting would have facilitated the expansion of hominins into and through colder habitats, where gathering would have been less productive.

The control of fire undoubtedly had several important impacts on hunting, meat eating, and many other features of hominin diet and socioecology (Wrangham 2009; Carmody, this volume). Considerable debate surrounds the identification of the first controlled fire (Gowlett and Wrangham 2013), but the earliest irrefutable evidence appears at Wonderwerk Cave in South Africa, approximately 1.0 Ma (Berna et al. 2012). A hominin (probably *Homo ergaster*) that could roast meat could kill the harmful bacteria that rapidly accumulate in carcasses postmortem (Smith et al. 2015). A fire-controlling hominin could thus increase its meat consumption without any change in the rate of killing prey or encountering carcasses killed by other predators. Cooking and tool-aided pounding and tenderizing of meat increase its digestibility (Carmody et al. 2011), and reducing the costs of digestion could free up time and energy for other pursuits. Beyond cooking, fire has myriad adaptive functions among living foragers, and the cross-cultural investigation by Scherjon et al. (2015) shows that fire is frequently used to drive and kill game, to make landscapes more attractive to game animals, and to clear vegetation so as to facilitate travel, including pursuit of prey. The antiquity of such behaviors is unknown, but hopefully future archaeological research will shed light on this important topic.

At least three significant characteristics of hunting among modern human foragers require language-based cumulative culture: (1) complex weapon manufacture, (2) extensive collaboration, and (3) sharing norms. If the LCA already hunted in groups, as we assume it did, then the advent of language would permit more effective information sharing among hunters. This would have increased prey encounter rates and reduced handling costs, ultimately leading to greater foraging efficiency. Archaeological materials from the last 500 Ka are suggestive of language-aided cultural innovations. Hafted spear points appear in South Africa (Kathu Pan) dated to 500 Ka (Wilkins et al. 2012). During the Middle Stone Age, spear points exhibit regional differences in style across Africa (McBrearty and Brooks 2000). In the Levant region, butchery marks found on the bones of large mammals become increasingly standardized after 190 Ka, perhaps indicating task specialization in the realm of butchery and meat sharing (Stiner et al. 2009).

Summary

In *The Descent of Man,* Darwin (1871) speculated that men who could "best defend and hunt for their families ... would succeed in rearing a greater average number of offspring." Data from modern human foraging populations show that better hunters do indeed have higher reproductive success (Smith 2004). Therefore, any genetic traits that underlie hunting ability remain under positive selection (in some populations), although of course not all such traits are genetic, and all require development, practice, and learning. Since success in hunting is known to lead to both higher reproductive success and prestige, Darwin's inference remains reasonable, and we can assume that over an extraordinarily long period of time, genetic and cultural adaptations have arisen owing to their fitness consequences in the hunting context.

The use of deadly weapons, extensive collaboration, and social norms are key components of hunting and sharing among modern African foragers, but the cultural and genetic factors underlying these traits would not have arisen *only* vis-à-vis their role in hunting. Adaptations that arose in other domains would have been co-opted and applied in hunting, and vice versa. A simple wooden hunting spear, for example, is but a lengthened digging stick, and traits underlying the efficient manufacture and use of either tool would influence its use in other domains. Likewise, new means of effectively defending against predators or enemies—concealment, detection, mobbing, or the use of weapons, could all be usefully employed in confrontational scavenging or in hunting. Traits that facilitate success in hunting are functionally linked to other important domains of behavior and social organization. Both humans and chimpanzees engage in coalitionary killing of adult conspecifics, and this suggests that even without the use of weapons, the human-chimpanzee LCA likely had the capacity to kill large and dangerous animals. The same neural architecture that gives humans the ability to symbolically represent, communicate, and reason about the behavior of other people is likely to also be useful for predicting the behavior of nonhuman predators and hunted prey.

The meat-sharing norms that humans exhibit today only make sense given a host of derived changes in the structure of human social life: the development of pair-bonds, increased dependence of juveniles on maternal, paternal, and grandparental investment, larger networks of social interaction, and longer lives (Jaeggi et al., this volume). Several researchers have argued that these traits coevolved with or depended on more human-like hunting (Kaplan et al. 2000; van Schaik and Burkart 2010). These ideas remain viable, but significant obstacles remain in our ability to conclusively test them owing to the difficulties of reconstructing past hunting, food sharing, or social organization. Regardless of the original function(s) of our species' technological skills, cognitive traits, and collaborative subsistence systems, surely one consequence of them is that all humans—not just hunter-gatherers—today exploit a more meat-centric dietary niche that remains relatively unexploited by chimpanzees or any other hominoid.

References

Anton, S. C., W. R. Leonard, and M. L. Robertson. 2002. An ecomorphological model of the initial hominid dispersal from Africa. *Journal of Human Evolution* 43: 773–785.

Bahuchet, S. 1985. *Les pygmées Aka et la Forét Centrafricaine*. Paris: SELAF.

Bahuchet, S. 1990. Food sharing among the pygmies of central africa. *African Study Monographs* 11: 27–53.

Bailey, R. C. 1991. *The Behavioral Ecology of Efe Pygmy Men in the Ituri Forest, Zaire*. Anthropological Papers No. 86, Museum of Anthropology, University of Michigan.

Bailey, R., and R. Aunger. 1989. Net hunters vs. archers: Variation in women's subsistence strategies in the Ituri forest. *Human Ecology* 17: 273–297.

Balinga, M., S. Moses, E. Fombod, T. Sunderland, S. Chantal, and S. Asahar. 2006. *A preliminary assessment of the vegetation of the Dzanga Sangha Protected Area Complex, Central African Republic*. Central Africa Regional Program for the Environment.

Berna, F., P. Goldberg, L. K. Horwitz, J. Brink, S. Holt, M. Bamford, and M. Chazan. 2012. Microstratigraphic evidence of *in situ* fire in the Acheulean strata of Wonderwerk Cave, Northern Cape province, South Africa. *Proceedings of the National Academy of Sciences* 109: E1215–E1220.

Bickerton, D. 2009. *Adam's Tongue: How Humans Made Language, How Language Made Humans*. New York: Hill and Wang.

Biesele, M., and S. Barclay. 2001. Ju/'hoan women's tracking knowledge and its contribution to their husbands hunting success. *African Study Monographs* 26: 67–84.

Binford, L. 1981. *Ancient Men and Modern Myths*. New York: Academic Press.

Blumenschine, R. J., A. Whiten, and K. Hawkes. 1991. Hominid carnivory and foraging strategies, and the socio-economic function of early archaeological sites [and discussion]. *Philosophical Transactions of the Royal Society B: Biological Sciences* 334: 211–221.

Blurton Jones, N. G. 1984. A selfish origin for human food sharing: tolerated theft. *Ethology and Sociobiology* 5: 1–3.

Blurton Jones, N. G. 2016. *Demography and Evolutionary Ecology of Hadza Hunter-Gatherers*. Cambridge: Cambridge University Press.

Bobe, R., and A. K. Behrensmeyer. 2004. The expansion of grassland ecosystems in Africa in relation to mammalian evolution and the origin of the genus *Homo*. *Palaeogeography, Palaeoclimatology, Palaeoecology* 207: 399–420.

Boesch, C. 1994a. Chimpanzees-red colobus monkeys: A predator-prey system. *Animal Behaviour* 47: 1135–1148.

Boesch, C. 1994b. Cooperative hunting in wild chimpanzees. *Animal Behaviour* 48: 653–667.

Boesch, C. 2002. Cooperative hunting roles among Taï chimpanzees. *Human Nature* 13: 27–46.

Boesch, C., and H. Boesch. 1989. Hunting behavior of wild chimpanzees in the Taï National Park. *American Journal of Physical Anthropology* 78: 547–573.

Boesch, C., and H. Boesch-Achermann. 2000. *The Chimpanzees of the Taï Forest: Behavioural Ecology and Evolution*. Oxford: Oxford University Press.

Bowles, S., E. A. Smith, and M. Borgerhoff Mulder. 2010. The emergence and persistence of inequality in premodern societies. *Current Anthropology* 51: 7–17.

Bunn, H. T., L. E. Bartram, and E. M. Kroll. 1988. Variability in bone assemblage formation from Hadza hunting, scavenging, and carcass processing. *Journal of Anthropological Archaeology* 7: 412–457.

Carmody, R. N., G. W. Weintraub, and R. W. Wrangham. 2011. Energetic consequences of thermal and nonthermal food processing. *Proceedings of the National Academy of Sciences* 108: 19199–19203.

Chapman, C. A., and R. W. Wrangham. 1993. Range use of the forest chimpanzees of Kibale: Implications for the understanding of chimpanzee social organization. *American Journal of Primatology* 31: 263–273.

Clutton-Brock, T. H., and J. B. Gillett. 1979. A survey of forest composition in the Gombe National Park, Tanzania. *African Journal of Ecology* 17: 131–158.

Cohen, J. E., S. L. Pimm, P. Yodzis, and J. Saldaña. 1993. Body sizes of animal predators and animal prey in food webs. *Journal of Animal Ecology* 62: 67–78.

Crittenden, A. N., and D. A. Zes. 2015. Food sharing among Hadza hunter-gatherer children. *PLoS ONE* 10: e0131996.

Dart, R. A. 1926. Taungs and its significance. *Natural History* 26: 315–327.

Dart, R. A. 1949. The predatory implemental technique of *Australopithecus*. *American Journal of Physical Anthropology* 7: 1–38.

Dart, R. A. 1953. The predatory transition from ape to man. *International Anthropological and Linguistic Review* 1: 201–218.

Dart, R. A., and A. Salmons. 1925. *Australopithecus africanus:* The man-ape of south africa. In L. Garwin and T. Lincoln, eds., *A Century of Nature: Twenty-One Discoveries that Changed Science and the World,* 10–20. Chicago: University of Chicago.

Darwin, C. 1871. *The Descent of Man, and Selection in Relation to Sex.* London: John Murray.

de Waal, F. B. M. 1989. Food sharing and reciprocal obligations among chimpanzees. *Journal of Human Evolution* 18: 433–459.

de Waal, F. B. M. 1997. The chimpanzee's service economy: Food for grooming. *Evolution and Human Behavior* 18: 375–386.

Duda, P., and J. Zrzavý. 2013. Evolution of life history and behavior in *Hominidae:* Towards phylogenetic reconstruction of the chimpanzee–human last common ancestor. *Journal of Human Evolution* 65: 424–446.

Fahy, G. E., M. Richards, J. Riedel, J.-J. Hublin, and C. Boesch. 2013. Stable isotope evidence of meat eating and hunting specialization in adult male chimpanzees. *Proceedings of the National Academy of Sciences* 110: 5829–5833.

Fedigan, L. M. 1990. Vertebrate predation in *Cebus capucinus:* Meat eating in a neotropical monkey. *Folia Primatologica* 54: 196–205.

Ferraro, J. V., T. W. Plummer, B. L. Pobiner, J. S. Oliver, L. C. Bishop, D. R. Braun, P. W. Ditchfield, J. W. Seaman III, K. M. Binetti, and J. W. Seaman Jr. 2013. Earliest archaeological evidence of persistent hominin carnivory. *PloS ONE* 8: e62174.

Foley, R. 1987. *Another Unique Species: Patterns in Human Evolutionary Ecology.* Harlow, UK: Longman Science and Technology.

Foley, R. A. 2001. The evolutionary consequences of increased carnivory in hominids. In C. B. Stanford and H. T. Bunn, eds., *Meat-Eating and Human Evolution,* 305–331. Oxford: Oxford University Press.

Foley, R. A., and P. C. Lee. 1991. Ecology and energetics of encephalization in hominid evolution. *Philosophical Transactions of the Royal Society B: Biological Sciences* 334: 223–232.

Gabunia, L., A. Vekua, D. Lordkipanidze, C. C. Swisher, R. Ferring, A. Justus, M. Nioradze, M. Tvalchrelidze, S. C. Anton, G. Bosinski, O. Joris, M. A. de Lumley, G. Majsuradze, and A. Mouskhelishvili. 2000. Earliest Pleistocene hominid cranial remains from Dmanisi, Republic of Georgia: Taxonomy, geological setting, and age. *Science* 288: 1019–1025.

Gilby, I. C. 2004. Hunting and meat sharing among the chimpanzees of Gombe National Park, Tanzania. Ph.D. dissertation, University of Minnesota.

Gilby, I. C. 2006. Meat sharing among the Gombe chimpanzees: Harassment and reciprocal exchange. *Animal Behaviour* 71: 953–963.

Gilby, I. C., and R. C. Connor. 2010. The role of intelligence in group hunting: Are chimpanzees different from other social predators? In E. V. Lonsdorf, S. R. Ross, and T. Matsuzawa, eds., *The Mind of the Chimpanzee: Ecological and Experimental Perspectives,* 220–233. Chicago: University of Chicago Press.

Gilby, I. C., L. E. Eberly, L. Pintea, and A. E. Pusey. 2006. Ecological and social influences on the hunting behaviour of wild chimpanzees *(Pan troglodytes schweinfurthii). Animal Behaviour* 72: 169–180.

Gilby, I. C., L. E. Eberly, and R. W. Wrangham. 2008. Economic profitability of social predation among wild chimpanzees: Individual variation promotes cooperation. *Animal Behaviour* 75: 351–360.

Gilby, I. C., M. Emery Thompson, J. D. Ruane, and R. W. Wrangham. 2010. No evidence of short-term exchange of meat for sex among chimpanzees. *Journal of Human Evolution* 59: 44–53.

Gilby, I. C., Z. P. Machanda, D. C. Mjungu, J. Rosen, M. N. Muller, A. E. Pusey, and R. W. Wrangham. 2015. "Impact hunters" catalyse cooperative hunting in two wild chimpanzee communities. *Philosophical Transactions of the Royal Society B: Biological Sciences* 370: 20150005.

Gilby, I. C., Z. P. Machanda, R. C. O'Malley, C. M. Murray, E. V. Lonsdorf, K. Walker, D. C. Mjungu, E. Otali, M. N. Muller, M. Emery Thompson, A. E. Pusey, and R. W. Wrangham, R. W. In revision. *Predation by female chimpanzees: Toward an understanding of sex differences in meat acquisition among early hominins.*

Gilby, I. C., M. L. Wilson, and A. E. Pusey. 2013. Ecology rather than psychology explains co-occurrence of predation and border patrols in male chimpanzees. *Animal Behaviour* 86: 61–74.

Gilby, I. C., and R. W. Wrangham. 2007. Risk-prone hunting by chimpanzees *(Pan troglodytes schweinfurthii)* increases during periods of high diet quality. *Behavioral Ecology and Sociobiology* 61: 1771–1779.

Gomes, C. M., and C. Boesch. 2009. Wild chimpanzees exchange meat for sex on a long-term basis. *PloS ONE* 4: e5116.

Goodall, J. 1963. Feeding behaviour of wild chimpanzees: A preliminary report. *Symposium of the Zoological Society of London* 10: 39–47.

Goodall, J. 1986. *The Chimpanzees of Gombe: Patterns of Behavior*. Cambridge, MA: Harvard University Press.

Gowlett, J. A. J., and R. W. Wrangham. 2013. Earliest fire in Africa: Towards the convergence of archaeological evidence and the cooking hypothesis. *Azania-Archaeological Research in Africa* 48: 5–30.

Gurven, M. 2004. To give and to give not: The behavioral ecology of human food transfers. *Behavioral and Brain Sciences* 27: 543–559.

Gurven, M., and K. Hill. 2009. Why do men hunt? A reevaluation of "Man the Hunter" and the sexual division of labor. *Current Anthropology* 50: 51–74.

Hames, R., and P. Draper. 2004. Women's work, child care, and helpers-at-the-nest in a hunter-gatherer society. *Human Nature* 15: 319–341.

Harako, R. 1976. The Mbuti as hunters: A study of ecological anthropology of the Mbuti pygmies. *Kyoto University African Studies* 10: 37–99

Harako, R. 1981. The cultural ecology of hunting behavior among Mbuti pygmies in the Ituri forest, Zaire. In R. Harding and G. Teleki, eds., *Omnivorous Primates: Gathering and Hunting in Human Evolution*, 499–555. New York: Columbia University Press.

Harding, R. S. 1975. Meat-eating and hunting in baboons. In R. Tuttle, ed., *Socioecology and Psychology of Primates*, 245–257. The Hague: Mouton.

Harmand, S., J. E. Lewis, C. S. Feibel, C. J. Lepre, S. Prat, A. Lenoble, X. Boës, R. L. Quinn, M. Brenet, and A. Arroyo. 2015. 3.3-million-year-old stone tools from Lomekwi 3, West Turkana, Kenya. *Nature* 521: 310–315.

Hart, T. B., and J. A. Hart. 1986. The ecological basis of hunter-gatherer subsistence in African rain forests: The Mbuti of eastern Zaire. *Human Ecology* 14: 29–55.

Hasegawa, T., M. Hiraiwa, T. Nishida, and H. Takasaki. 1983. New evidence on scavenging behavior in wild chimpanzees. *Current Anthropology* 24: 231–232.

Hawkes, K., J. F. O'Connell, and N. G. Blurton Jones. 1991. Hunting income patterns among the Hadza: Big game, common goods, foraging goals and the evolution of human diet. *Philosophical Transactions: Biological Sciences* 334: 243–251.

Hawkes, K., J. F. O'Connell, and N. G. Blurton Jones. 2001a. Hadza meat sharing. *Evolution and Human Behavior* 22, 113–142.

Hawkes, K., J. F. O'Connell, and N. G. Blurton Jones. 2001b. Hunting and nuclear families: Some lessons from the Hadza about men's work. *Current Anthropology* 42: 681–709.

Hill, K. 1982. Hunting and human evolution. *Journal of Human Evolution* 11: 521–544.

Hill, K. 1988. Macronutrient modifications of optimal foraging theory: An approach using indifference curves applied to some modern foragers. *Human Ecology* 16: 157–197.

Hill, K. R., and K. Kintigh. 2009. Can anthropologists distinguish good and poor hunters? Implications for hunting hypotheses, sharing conventions, and cultural transmission. *Current Anthropology* 50: 369–377.

Hill, K. R., B. M. Wood, J. Baggio, A. M. Hurtado, and R. T. Boyd. 2014. Hunter-gatherer inter-band interaction rates: Implications for cumulative culture. *PloS ONE* 9: e102806.

Hosaka, K., T. Nishida, M. Hamai, A. Matsumoto-Oda, and S. Uehara. 2001. Predation of mammals by the chimpanzees of the Mahale Mountains, Tanzania. In B. Galdikas, N. Briggs, L. Sheeran, G. Shapiro, and J. Goodall, eds., *All Apes Great and Small*, vol. 1: *African Apes*, 107–130. New York: Klewer Academic.

Huffman, M. A., and M. S. Kalunde. 1993. Tool-assisted predation on a squirrel by a female chimpanzee in the Mahale Mountains, Tanzania. *Primates* 34: 93–98.

Huntingford, G. W. B. 1929. Modern hunters: Some account of the Kamelilo-Kapchepkendi Dorobo (Okiek) of Kenya colony. *Journal of the Anthropological Institute of Great Britain and Ireland* 59: 333–378.

Ichikawa, M. 1983. An examination of the hunting-dependent life of the Mbuti pygmies, eastern Zaire. *African Study Monographs* 4: 55–76.

Isaac, G. 1978. The food-sharing behavior of protohuman hominoids. *Scientific American* 238: 90–108.

Isaac, G. L. 1983. Bones in contention: Competing explanations for the juxtaposition of Early Pleistocene artifacts and faunal remains. In J. Clutton-Brock and C. Grigson, eds., *Animals and Archaeology*, 3–19. Cambridge: Cambridge University Press.

Isler, K., and C. P. van Schaik. 2014. How humans evolved large brains: Comparative evidence. *Evolutionary Anthropology* 23: 65–75.

Kaplan, H. S., K. Hill, J. Lancaster, and A. M. Hurtado. 2000. A theory of human life history evolution: Diet, intelligence, and longevity. *Evolutionary Anthropology* 9: 156–185.

Kawanaka, K. 1982. Further studies on predation by chimpanzees of the Mahale Mountains. *Primates* 23: 364–384.

Kingdon, J. 1997. *The Kingdon Field Guide to African Mammals*. London: Academic Press.

Kitanishi, K. 1995. Seasonal changes in the subsistence activities and food intake of the Aka hunter-gatherers in northeastern Congo. *African Study Monographs* 16: 73–118.

Kitanishi, K. 1996. Variability in the subsistence activities and distribution of food among different aged males of the Aka hunter-gatherers in northeastern Congo. *African Study Monographs* 17: 35–57.

Langergraber, K., J. C. Mitani, and L. Vigilant. 2007. The limited impact of kinship on cooperation in wild chimpanzees. *Proceedings of the National Academy of Sciences* 104: 7786–7790.

Laughlin, W. S. 1968. An integrating biobehavior system and its evolutionary importance. In R. B. Lee and I. DeVore, eds., *Man the Hunter*, 304–320. New York: Aldine de Gruyter.

Lee, R. B. 1979. *The !Kung San: Men, Women, and Work in a Foraging Society*. Cambridge: Cambridge University Press.

Lee, R. B. 2003. *The Dobe Ju/'hoansi*. Belmont, CA: Wadsworth.

Lewis, J. 2002. Forest hunter-gatherers and their world: A study of the Mbendjele Yaka Pygmies of Congo-Brazzaville and their secular and religious activities and representations. Ph.D. dissertation, University of London.

Lupo, K. D., and D. N. Schmitt. 2005. Small prey hunting technology and zooarchaeological measures of taxonomic diversity and abundance: Ethnoarchaeological evidence from Central African forest foragers. *Journal of Anthropological Archaeology* 24: 335–353.

Marlowe, F. W. 2005. Hunter-gatherers and human evolution. *Evolutionary Anthropology* 14: 54–67.

Marlowe, F. W. 2010. *The Hadza: Hunter-Gatherers of Tanzania.* Berkeley: University of California Press.

Marlowe, F. W., J. C. Berbesque, B. M. Wood, A. Crittenden, C. Porter, and A. Mabulla. 2014. Honey, hunter-gatherers, and human evolution. *Journal of Human Evolution* 71: 119–128.

McBrearty, S., and A. S. Brooks. 2000. The revolution that wasn't: A new interpretation of the origin of modern human behavior. *Journal of Human Evolution* 39: 453–563.

McGrew, W. C. 1992a. *Chimpanzee Material Culture.* Cambridge: Cambridge University Press.

McGrew, W. C. 1992b. Two nonhuman primate models for the evolution of human food sharing: Chimpanzees and callitrichids. In H. Barkow, L. Cosmides, and J. Tooby, eds., *The Adapted Mind: Evolutionary Psychology and the Generation Culture,* 229–243. Oxford: Oxford University Press.

Mitani, J. C. 2006. Reciprocal exchange in chimpanzees and other primates. In P. M. Kappeler and C. P. van Schaik, eds., *Cooperation in Primates: Mechanisms and Evolution,* 101–113. Heidelberg: Springer-Verlag.

Mitani, J. C. 2009. Cooperation and competition in chimpanzees: Current understanding and future challenges. *Evolutionary Anthropology* 18: 215–227.

Mitani, J. C., and D. P. Watts. 1999. Demographic influences on the hunting behavior of chimpanzees. *American Journal of Physical Anthropology* 109: 439–454.

Mitani, J. C., and D. P. Watts. 2001. Why do chimpanzees hunt and share meat? *Animal Behaviour* 61: 915–924.

Mitani, J. C., and D. P. Watts. 2005. Seasonality in hunting by non-human primates. In D. K. Brockman and C. P. van Schaik, eds., *Seasonality in Primates: Studies of Living and Extinct Human and Non-Human Primates,* 215–240. Cambridge: Cambridge University Press.

Muller, M. N., and R. W. Wrangham. 2014. Mortality rates among Kanyawara chimpanzees. *Journal of Human Evolution* 66: 107–114.

Newton-Fisher, N. E. 2014. Chimpanzee hunting behavior. In W. Henke and I. Tattersal, eds., *SpringerReference,* vol. 2, 2nd ed, 1–19. Berlin: Springer-Verlag.

Newton-Fisher, N. E., H. Notman, and V. Reynolds. 2002. Hunting of mammalian prey by Budongo Forest chimpanzees. *Folia Primatologica* 73: 281–283.

Nishida, T., T. Hasegawa, H. Hayaki, Y. Takahata, and S. Uehara. 1992. Meat-sharing as a coalition strategy by an alpha male chimpanzee? In T. Nishida, W. C. McGrew, P. Marler, M. Pickford, and F. B. M. de Waal, eds., *Topics in Primatology,* vol. 1: *Human Origins,* 159–174. Tokyo: University of Tokyo Press.

Nishida, T., S. Uehara, and R. Nyundo. 1979. Predatory behavior among wild chimpanzees of the Mahale Mountains. *Primates* 20: 1–20.

Noss, A. J., and B. S. Hewlett. 2001. The contexts of female hunting in Central Africa. *American Anthropologist* 103: 1024–1040.

O'Connell, J. F., K. Hawkes, and N. Blurton Jones. 1988. Hadza hunting, butchering, and bone transport and their archaeological implications. *Journal of Anthropological Research* 44: 113–161.

Osaki, M. 1984. The social influence of change in hunting technique among the central Kalahari San. *African Study Monographs* 5: 49–62.

Pickering, T. R., and M. Domínguez-Rodrigo. 2010. Chimpanzee referents and the emergence of human hunting. *Open Anthropology Journal* 3: 107–113.

Pontzer, H., D. A. Raichlen, B. M. Wood, A. Z. Mabulla, S. B. Racette, and F. W. Marlowe. 2012. Hunter-gatherer energetics and human obesity. *PLoS ONE* 7: e40503.

Potts, R. 1984. Home bases and early hominids: Reevaluation of the fossil record at Olduvai Gorge suggests that the concentrations of bones and stone tools do not represent fully formed campsites but an antecedent to them. *American Scientist* 72: 338–347.

Potts, K. B., D. P. Watts, and R. W. Wrangham. 2011. Comparative feeding ecology of two communities of chimpanzees *(Pan troglodytes)* in Kibale National Park, Uganda. *International Journal of Primatology* 32: 669–690.

Pruetz, J. D., and P. Bertolani. 2007. Savanna chimpanzees, *Pan troglodytes verus,* hunt with tools. *Current Biology* 17: 412–417.

Pruetz, J. D., P. Bertolani, K. B. Ontl, S. Lindshield, M. Shelley, and E. G. Wessling. 2015. New evidence on the tool-assisted hunting exhibited by chimpanzees *(Pan troglodytes verus)* in a savannah habitat at Fongoli, Sénégal. *Royal Society Open Science* 2: 140507.

Pruetz, J. D., K. B. Ontl, E. Cleaveland, S. Lindshield, J. Marshack, and E. G. Wessling. 2017. Intragroup lethal aggression in West African chimpanzees *(Pan troglodytes verus):* Inferred killing of a former alpha male at Fongoli, Senegal. *International Journal of Primatology* 38: 31–57.

Pusey, A. E., G. W. Oehlert, J. M. Williams, and J. Goodall. 2005. Influence of ecological and social factors on body mass of wild chimpanzees. *International Journal of Primatology* 26: 3–31.

Read, C. 1914. On the differentiation of man from the anthropoids. *Man* 14: 181–186.

Reed, K. E. 1997. Early hominid evolution and ecological change through the African Plio-Pleistocene. *Journal of Human Evolution* 32: 289–322.

Scherjon, F., C. Bakels, K. MacDonald, and W. Roebroeks. 2015. Burning the land: An ethnographic study of off-site fire use by current and historically documented foragers and implications for the interpretation of past fire practices in the landscape. *Current Anthropology* 56: 299–326.

Silberbauer, G. 1981. Hunter/gatherers of the Central Kalahari. In R. Harding and G. Teleki, eds., *Omnivorous Primates: Gathering and Hunting in Human Evolution,* 455–498. New York: Columbia University Press.

Smith, A. R., R. N. Carmody, R. J. Dutton, and R. W. Wrangham. 2015. The significance of cooking for early hominin scavenging. *Journal of Human Evolution* 84: 62–70.

Smith, E. A. 2004. Why do good hunters have higher reproductive success? *Human Nature* 15: 343–364.

Smith, E. A. 2010. Communication and collective action: Language and the evolution of human cooperation. *Evolution and Human Behavior* 31: 231–245.

Sobolewski, M. E., J. L. Brown, and J. C. Mitani. 2012. Territoriality, tolerance and testosterone in wild chimpanzees. *Animal Behaviour* 84: 1469–1474.

Sol, D., S. Bacher, S. M. Reader, and L. Lefebvre. 2008. Brain size predicts the success of mammal species introduced into novel environments. *American Naturalist* 172: S63–S71.

Stanford, C. B. 1998. *Chimpanzee and Red Colobus*. Cambridge, MA: Harvard University Press.

Stanford, C. B., J. Wallis, H. Matama, and J. Goodall. 1994a. Patterns of predation by chimpanzees on red colobus monkeys in Gombe National Park, 1982–1991. *American Journal of Physical Anthropology* 94: 213–228.

Stanford, C. B., J. Wallis, E. Mpongo, and J. Goodall. 1994b. Hunting decisions in wild chimpanzees. *Behaviour* 131: 1–18.

Stevens, J. R. 2004. The selfish nature of generosity: Harassment and food sharing in primates. *Proceedings of the Royal Society of London Series B: Biological Sciences* 271: 451–456.

Stevens, J. R., and M. D. Hauser. 2004. Why be nice? Psychological constraints on the evolution of cooperation. *Trends in Cognitive Sciences* 8: 60–65.

Stiner, M. C., R. Barkai, and A. Gopher. 2009. Cooperative hunting and meat sharing 400–200 kya at Qesem Cave, Israel. *Proceedings of the National Academy of Sciences* 106: 13207–13212.

Struhsaker, T. T. 1997. *The Ecology of an African Rainforest*. Gainesville: University Presses of Florida.

Struhsaker, T. T. 2010. *The Red Colobus Monkeys: Variation in Demography, Behavior and Ecology of Endangered Species*. Oxford: Oxford University Press.

Takahata, Y., T. Hasegawa, and T. Nishida. 1984. Chimpanzee predation in the Mahale Mountains from August 1979 to May 1982. *International Journal of Primatology* 5: 213–233.

Tanaka, J. 1980. *The San, Hunter-Gatherers of the Kalahari: A Study in Ecological Anthropology*. Tokyo: University of Tokyo Press.

Tanaka, J. 2014. *The Bushmen: A Half-Century Chronicle of Transformations in Hunter-Gatherer Life and Ecology*. Kyoto: Kyoto University Press.

Tanno, T. 1976. The Mbuti net-hunters in the Ituri forest, eastern Zaire: Their hunting activities and band composition. *Kyoto University African Studies* 10: 101–135.

Teleki, G. 1973. *The Predatory Behavior of Wild Chimpanzees*. Lewisburg, PA: Bucknell University Press.

Tennie, C., I. C. Gilby, and R. Mundry. 2009. The meat-scrap hypothesis: Small quantities of meat may promote cooperation in wild chimpanzees *(Pan troglodytes)*. *Behavioral Ecology and Sociobiology* 63: 421–431.

Tennie, C., R. C. O'Malley, and I. C. Gilby. 2014. Why do chimpanzees hunt? Considering the benefits and costs of acquiring and consuming vertebrate versus invertebrate prey. *Journal of Human Evolution* 71: 38–45.

Tomasello, M., and H. Moll. 2010. The gap is social: Human shared intentionality and culture. In P. M. Kappeler and J. B. Silk, eds., *Mind the Gap: Tracing the Origins of Human Universals,* 331–349. Heidelberg: Springer.

Turnbull, C. 1968. The importance of flux in two hunting societies. In R. B. Lee and I. DeVore, eds., *Man the Hunter,* 132–137. New York: Aldine de Gruyter.

Uehara, S. 1997. Predation on mammals by the chimpanzee *(Pan troglodytes). Primates* 38: 193–214.

Uehara, S., T. Nishida, M. Hamai, T. Hasegawa, H. Hayaki, M. A. Huffman, K. Kawanaka, S. Kobayashi, J. C. Mitani, Y. Takahata, H. Takasaki, and T. Tsukahara. 1992. Characteristics of predation by the chimpanzees in the Mahale Mountains National Park, Tanzania. In T. Nishida, W. C. McGrew, P. Marler, M. Pickford, and F. B. M. de Waal, eds., *Topics in Primatology,* vol. 1: *Human Origins,* 143–158. Tokyo: University of Tokyo Press.

Valeix, M., A. J. Loveridge, Z. Davidson, H. Madzikanda, H. Fritz, and D. W. Macdonald. 2010. How key habitat features influence large terrestrial carnivore movements: Waterholes and African lions in a semi-arid savanna of north-western Zimbabwe. *Landscape Ecology* 25: 337–351.

van Schaik, C. P., and J. M. Burkart. 2010. Mind the gap: Cooperative breeding and the evolution of our unique features. In P. M. Kappeler and J. B. Silk, eds., *Mind the Gap: Tracing the Origins of Human Universals,* 477–496. Heidelberg: Springer.

van Woerden, J. T., E. P. Willems, C. P. van Schaik, and K. Isler. 2012. Large brains buffer energetic effects of seasonal habitats in catarrhine primates. *Evolution* 66: 191–199.

Washburn, S. L., and C. S. Lancaster. 1968. The evolution of hunting. In R. B. Lee and I. DeVore, eds., *Man the Hunter,* 293–303. New York: Aldine de Gruyter.

Watts, D. P. 2008. Scavenging by chimpanzees at Ngogo and the relevance of chimpanzee scavenging to early hominin behavioral ecology. *Journal of Human Evolution* 54: 125–133.

Watts, D. P., and J. C. Mitani. 2002a. Hunting and meat sharing by chimpanzees at Ngogo, Kibale National Park, Uganda. In C. Boesch, G. Hohmann, and L. Marchant, eds., *Behavioural Diversity in Chimpanzees and Bonobos,* 244–255. Cambridge: Cambridge University Press.

Watts, D. P., and J. C. Mitani. 2002b. Hunting behavior of chimpanzees at Ngogo, Kibale National Park, Uganda. *International Journal of Primatology* 23: 1–28.

Watts, D. P., and J. C. Mitani. 2015. Hunting and prey switching by chimpanzees *(Pan troglodytes schweinfurthii)* at Ngogo. *International Journal of Primatology* 36: 728–748.

Wilkins, J., B. J. Schoville, K. S. Brown, and M. Chazan. 2012. Evidence for early hafted hunting technology. *Science* 338: 942–946.

Wilson, M. L., C. Boesch, B. Fruth, T. Furuichi, I. C. Gilby, C. Hashimoto, C. L. Hobaiter, G. Hohmann, N. Itoh, K. Koops, J. N. Lloyd, T. Matsuzawa, J. C. Mitani, D. C. Mjungu,

D. Morgan, M. N. Muller, R. Mundry, M. Nakamura, J. D. Pruetz, A. E. Pusey, J. Riedel, C. Sanz, A. M. Schel, N. Simmons, M. Waller, D. P. Watts, F. White, R. M. Wittig, Z. Zuberbühler, and R. W. Wrangham. 2014. Lethal aggression in *Pan* is better explained by adaptive strategies than human impacts. *Nature* 513: 414–417.

Wood, B. M., and F. W. Marlowe. 2011. Dynamics of postmarital residence among the Hadza: A kin investment model. *Human Nature* 22: 128–138.

Wood, B. M., and F. W. Marlowe. 2013. Household and kin provisioning by Hadza men. *Human Nature* 24: 280–317.

Wood, B. M., and F. W. Marlowe. 2014. Toward a reality-based understanding of Hadza men's work: A response to Hawkes et al. (2014). *Human Nature* 25: 620–630.

Woodburn, J. 1968a. An introduction to Hadza ecology. In R. B. Lee and I. DeVore, eds., *Man the Hunter,* 49–55. New York: Aldine de Gruyter.

Woodburn, J. 1968b. Stability and flexibility in Hadza residential groupings. In R. B. Lee and I. DeVore, eds., *Man the Hunter,* 103–110. New York: Aldine de Gruyter.

Wrangham, R. 2009. *Catching Fire: How Cooking Made Us Human.* New York: Basic Books.

Wrangham, R. W. 1975. The behavioural ecology of chimpanzees in Gombe National Park, Tanzania. Ph.D. dissertation, Cambridge University.

Wrangham, R. W., and E. Bergmann-Riss. 1990. Rates of predation on mammals by Gombe chimpanzees, 1972–1975. *Primates* 31: 157–170.

Yellen, J. E. 1977. *Archaeological Approaches to the Present: Models for Reconstructing the Past.* New York: Academic Press.

11

The Evolution of the Human Mating System

MARTIN N. MULLER *and* DAVID R. PILBEAM

Mating systems can be classified using many criteria, but in primates, conspicuous features include the number of mates that a female takes during a single ovarian cycle, the duration and exclusivity of sexual relationships, and the manifestation of sexual competition and mate choice (Borgerhoff Mulder 1992; Dixson 1997). Although both human and chimpanzee mating strategies vary among individuals and populations, stark differences are evident between the two species in all of these features. A female chimpanzee can mate with more than two dozen males in a day, at times averaging five copulations per hour (Mitani, personal communication; Watts 2007). Consequently, over one ovarian cycle, a female can often expect to mate with all of the adult and adolescent males in her social group (Wrangham 2002). Male chimpanzees routinely fight over sexual access to cycling females, and prefer older mothers as mating partners (Muller and Wrangham 2004; Muller et al. 2006). Female choice appears directed toward the goal of mating with as many males as possible, potentially to confuse paternity and reduce the risk of male infanticide (Muller et al. 2011). No human society has ever been described that exhibits these characteristics.

Instead, the most salient feature of the human mating system is the ubiquity of the pair-bond, defined as an enduring, preferential, affiliative relationship between two individuals that includes a sexual component (Quinlan

2008; Young et al. 2011). Bernard Chapais (2008: 160) notes that the essence of pair-bonding lies in its "selective and relatively stable character." It is "essentially and simply, the opposite of sexual promiscuity" (Chapais 2008: 161).

The divergence between humans and chimpanzees in the mating realm is all the more striking given their similarity in basic social organization.[1] Both live in multimale, multifemale groups that exhibit fission-fusion association (i.e., in which segments of a larger community merge or split throughout the day). For human foragers, local band sizes vary from 13 to 250, with a median around 25 (Marlowe 2005). Bands are embedded in broader communities that average 250–500 individuals, but can be much larger (Layton et al. 2012). Chimpanzee communities fall in the range of human bands, varying from 7 to 200, with a median of 42 (Mitani, personal communication; Wilson et al. 2014). The human pattern of stable breeding bonds situated in a larger multimale, multifemale social group is unusual among primates, and unique among the living hominoids (Rodseth et al. 1991).

We begin this chapter by considering commonalities among human forager mating systems, and contrasting these with the chimpanzee mating system. We then attempt to trace the evolution of the human system, using evidence from both the fossil record and comparative studies of nonhuman primates. Important anatomical and physiological traits that are associated with mating systems are examined, including sexual dimorphism in body size, canine height, and vocal pitch, and anatomical features related to sperm competition.

Human and Chimpanzee Mating Systems

Duration and Exclusivity of Sexual Relationships

Under the definition above, pair-bonds can be polygynous, polyandrous, or monogamous, and need not be lifelong. Thus, although pair-bonding is a human universal, considerable variation exists, both within and between societies, in the number of bonds that an individual maintains, the average duration of those bonds, and their sexual exclusivity. The most common arrangement among foragers, however, involves monogamous pair-bonds. In a sample of thirty-one warm-climate, nonequestrian foragers, Marlowe (2010) reported that the median percentage of women with co-wives was 10 percent.

Most polygynously married women in the sample had only one co-wife. In a sample of 168 Hadza men, only 2.4 percent had ever been in a polygynous marriage (Blurton Jones 2016).

When polygynous marriages do occur in foragers, they are often short-term or transitional (Hill and Hurtado 1996). Among the Hadza, for example, such marriages tend not to last for longer than a few months, because as soon as one of the wives finds out about the other, the man is "thrown out" (Blurton Jones 2016). Across foraging societies, polygyny is less common when men's contribution to the diet is higher, suggesting an ecological constraint under which most men are unable to provide sufficient food to retain more than one wife at a time (Marlowe 2003).

Pair-bond stability shows considerable variation across the few foraging societies in which it has been surveyed (Blurton Jones et al. 2000). Howell (1979) estimated that 22 percent of Ju/'hoansi marriages lasted more than thirty years. Blurton Jones (2016) calculated that 60 percent of Hadza marriages reached their fifth anniversary, 40 percent their tenth, and about 16 percent continued to twenty-five years. Aché marriages are less stable; Hill and Hurtado (1996) reported that *all* first marriages in their sample ended in divorce. Across foragers, divorce consistently occurs most often during the first months or years of a marriage, becoming less likely with time and offspring (Ju/'hoansi: Howell 1979; Agta: Headland 1987; Aka: Hewlett 1992; Aché and Hiwi: Hurtado and Hill 1992; Hill and Hurtado 1996; Hadza: Blurton Jones et al. 2000; Blurton Jones 2016). In the Hadza, for example, the likelihood of divorce steadily declines after 3.5–4 years of marriage (Blurton Jones 2016). For both the Ju/'hoansi and the Hadza, the average number of lifetime marriages is two (Howell 1979; Blurton Jones 2016).

Divorce is more common when men have more reproductive opportunities and experience less mating competition—for example, when the sex ratio is skewed toward women (Blurton Jones et al. 2000). It is also more common when paternal care is less essential—for example, with increasing alloparental involvement (Quinlan and Quinlan 2007, 2008). Marriages are most stable in societies in which men and women make equal contributions to subsistence, probably because when men contribute little, pair-bonds do not benefit women, and when men control resources, they are more likely to engage in affairs or serial monogamy (Quinlan and Quinlan 2007, 2008).

Data on marital infidelity are difficult to acquire, because sexual promiscuity by married individuals is universally disfavored. However, affairs are cited cross-culturally as a leading cause of divorce (Betzig 1989; Hill and Hurtado 1996; Marlowe 2010; Blurton Jones 2016). Thus, the incidence of infidelity likely parallels that of divorce in many societies. Winking and colleagues (2007) found evidence for this relationship in the Tsimané, a Bolivian group that combines foraging with horticulture and wage labor. The frequency of Tsimané men's affairs was highest in the early years of marriage, and varied inversely with the number of dependent children. Unfortunately, such detailed data are not available from any foragers.

Chimpanzees, by contrast, show little evidence for any form of stable breeding bond. Most chimpanzee mating takes place in one of two group contexts (Tutin 1979): *Opportunistic mating* occurs with little male aggression or coercion, and typically involves females who are less likely to conceive (i.e., subfecund nulliparous females or mothers early in the follicular phase). *Possessive mating* occurs when males employ aggression to discourage contact between estrous females and rival males. Such aggression is most common when mothers are in the fertile (periovulatory) period (Muller and Mitani 2005; Emery Thompson and Wrangham 2008).

In both of these group settings, females mate broadly, and the proportion of community males that have mated with a female rises steeply with her total copulation number (Wrangham 2002). Females appear highly motivated to mate with multiple partners. For example, in Kanyawara, estrous females increased their solicitation rates of *all* lower-ranked adult males when the alpha male was absent (Muller et al. 2011). Copulations may not be divided evenly among males, either because of male coercion or female choice (see Muller, this volume). However, in most communities, by the time a female conceives she will have mated with most, if not all, of the resident males (with 400–3,000 total copulations per conception: Wrangham 2002).

Consortships constitute a third major context for chimpanzee mating (Tutin 1979). In a consortship, a male restricts access to a female by accompanying her to a peripheral part of the territory, where they may remain undetected for several days to more than a month. Although consortships can be initiated by male aggression, many observers have argued that female cooperation is ultimately necessary, because unwilling females can easily attract additional suitors by vocalizing (Tutin 1979; Goodall 1986; Boesch and Boesch-Achermann 2000; Nishida 2012). The inherent risks of this option are illustrated by Wild's (2009) description of a Kanyawara female who thwarted a male's attempt to force her into consortship by screaming, but was subse-

FIGURE 11.1. Chimpanzee mating in Kanyawara, Kibale National Park. The chimpanzee mating system is highly promiscuous, and during a single ovarian cycle a female sometimes mates with all of the subadult and adult males in her community. Mothers, like the female pictured above, are highly attractive to males. Younger females must sometimes be very sexually assertive to successfully solicit male interest. Photo by Martin N. Muller.

quently beaten severely, first by the frustrated male, and then by the males who found them together. Inferring that a female's silent presence in a consortship demonstrates her preference for the accompanying male is thus unjustifiable without further evidence.

Stanford (1996) characterized chimpanzee consortships as a potential form of incipient pair-bonding. Consistent with this idea, participation in a consortship might reflect a female's desire to produce offspring with a particular sire. Such was the conclusion of Boesch and Boesch-Achermann (2000), who reported that consortships at Taï only involved alpha males, or males who had previously been alpha. Alternatively, consortships might reflect a female strategy to confuse paternity by giving males who would not be competitive in the possessive mating context a chance at conception. This idea is consistent with data from Gombe, where 19 percent of conceptions took place during consortships, and low-ranking males were the most frequent beneficiaries (Wroblewski et al. 2009). Such hypotheses are difficult to evaluate, given the furtive nature of chimpanzee consortships, the potential for coercion, and the fact that females can mate in all three of the contexts described above during a single ovarian cycle (e.g., Boesch and Boesch-Achermann 2000; Wroblewski et al. 2009).

There is currently no evidence that consortships represent preferential, enduring relationships. No tendency has been found for specific male-female pairs to form consortships at higher rates than expected, nor do males develop a special interest in the offspring from such unions. This issue of paternity recognition has been investigated with genetic data from two long-term study sites. Wroblewski (2010) found no evidence that Gombe fathers associated preferentially with, or biased any behavior toward, their offspring. Similarly, Lehmann and colleagues (2006) reported that Taï fathers did not associate preferentially with, or groom or play more frequently with, their offspring than other infants. They found weak evidence that Taï fathers showed an increased duration of play with their offspring, but this effect was not found in all age classes. More recently, Murray and colleagues (2016) reported that Gombe males were more likely both to be found in parties with, and to interact with, their offspring than with unrelated infants. Although these findings were presented as strong evidence for paternal recognition, plausible alternatives exist. For example, mothers, not fathers, may have been responsible for patterns of association at the party level. And if infants were generally more attracted to high-ranking than low-ranking males, then they might have interacted more with their fathers simply because high-ranking males are more likely to sire infants (patterns observed in mountain gorillas: Rosenbaum et al. 2011, 2015). Thus, the evidence for paternal recognition in chimpanzees remains weak. In any case, increased association between fathers and infants at Gombe was only observed during the infant's first six months, and association between fathers and mother-infant pairs did not increase the male's chances of siring the mother's next offspring (Murray et al. 2016). At present, then, a convincing connection between chimpanzee consortships and pair bonds is lacking. Further work is clearly needed in this area, however, and surprises are possible.

Genetic paternity data from Ngogo recently revealed that male and female chimpanzees who shared similar ranges within the community territory were more likely to have offspring together than other dyads (Langergraber et al. 2013). The authors interpreted this as a form of stable breeding association, suggesting that "the gap between the social and mating systems of humans and their closest living relatives may not be as large as previously thought." This conclusion is surprising because it is based entirely on conception frequency and ranging data, without any indication of social bonding. It contrasts with firm evidence that within a community, adult females show no social preferences for any particular males to whom they are not related (Machanda et al. 2013). Current evidence indicates that the probability of conception and the spatial association of chimpanzee females with males merely reflect a shared preference for a particular feeding range, rather than for each other's company. And although females may be more likely to conceive with specific males, they are still mating promiscuously with many males across the community. Accordingly, until there is evidence of social bonding between female and male chimpanzees, we must differ from Langergraber et al. (2013) by concluding that chimpanzee social relationships show no meaningful similarity to human pair-bonds.

Intrasexual Competition

Sexual competition is another aspect of the mating system that takes different forms in humans and chimpanzees. Chimpanzee males routinely fight over mating opportunities, aggressively disrupting the copulations of their rivals (Muller and Mitani 2005). In one year at Kanyawara, adult males chased or attacked other adults approximately once every fourteen observation hours, with males and females targeted at equivalent rates (Muller and Wrangham 2004). In mating contexts, however, this rate of escalated aggression more than doubled, increasing further prior to ovulation.

Within forager groups, by contrast, men show comparatively low rates of physical aggression toward other men (Boehm 1999). Lee (1979), for example, recorded only thirty-four instances of hand-to-hand fighting over six years in a population of several hundred Ju/'hoansi. And Hill and Hurtado "never observed a scuffle between Aché men in seventeen years of work with them" (1996: 70). Published rates of nonlethal aggression among foragers are sparse, but Wrangham and colleagues (2006) produced estimates from Burbank's (1992) detailed data on a recently settled group of Australian aboriginals. Although Burbank characterized this group as having strikingly high rates of violence, attack rates were 1/384 and 1/182 those of wild chimpanzees, for males and females, respectively. Even considering that violence among modern hunter-gatherers is likely inhibited by the prospect of state intervention (Hill and Hurtado 1996; Hill et al. 2007), the gap between humans and chimpanzees in this domain is wide.

When violence does erupt within forager groups, however, it is regularly precipitated by male jealousy and sexual competition (Knauft 1987, 1991). Most aggression among the Hadza occurs either when a man discovers that his wife has had an affair, or when two men are competing for the same woman (Marlowe 2010; Woodburn 1979). Sexual competition is similarly the leading cause of within-group violence among the Ju/'hoansi (Howell 1979; Lee 1979) and the Hiwi (Hill et al. 2007). Prior to settlement, club fighting represented the only socially sanctioned form of within-group violence among Aché men (Hill and Hurtado 1996). Spontaneous club fights were rare, but "almost always" involved infidelity (Hill and Hurtado 1996). During organized club fights, men who engaged in sexual affairs with married women were likely to become targets.

And although violence among men within forager groups is infrequent, it is much more likely to turn lethal than in chimpanzees. Hill and Hurtado (1996) estimate that, prior to first contact, club fights were responsible for 8 percent of Aché male deaths over the age of fifteen, and within-group killing accounted for 11 percent of adult male deaths. Within-group homicide rates for foragers are scarce, but estimates for the Ju/'hoansi and the Hadza fall around twenty to sixty per 100,000 person-years (Blurton Jones 2016). This is similar to the rate of within-group killing of adults calculated for chimpanzees (Wilson et al. 2014), who, as noted, engage in physical aggression at dramatically higher rates.

The critical difference between the lethality of violence in chimpanzees and humans is likely technological. Projectile weapons used in hunting, such as spears and poisoned arrows, make killing potentially easy for men, whether from a distance or from ambush (Woodburn 1982; Boehm 1999; Chapais 2008). Chimpanzees, by contrast, find it very difficult to kill other adults without assistance. Indeed, all reported killings result from group attacks, rather than dyadic fights (Goodall 1986; Wilson et al. 2014). A number of researchers have attributed the remarkable human propensity to cooperate, and avoid conflict, to the development of weapons, which increased the costs of violence by making fatal injuries more likely (Boehm 1999; Okada and Bingham 2008; Phillips et al. 2014).

Do comparatively low levels of nonlethal aggression among forager men imply that physical competition over women was simply not an important factor in human evolution (e.g., Lancaster and Kaplan 2009)? The prevalence of within-group killings in the context of male sexual rivalry suggests otherwise. Although men rarely fight over immediate opportunities to mate, in the manner of chimpanzee males, the threat of violence probably affects men's mating prospects among many foragers. Hill and Hurtado (1996: 227), for example, report that older Aché men "discouraged the young men from marrying or having sex, and boys would become targets in club fights when they began to 'have' women openly." Marlowe (2010) noted that among the Hadza, violence and threats of violence occur most frequently among young men, when they are competing for young, single women. It is thus plausible that larger and stronger men might start their reproductive careers earlier. More formidable men might also be less likely to be cuckolded, or to lose mates to a rival. Unfortunately, the data needed to test such specific mechanisms are currently lacking. Two studies have reported positive associations between men's fertility and body size in foragers (Aché: Hill and Hurtado 1996; Ju/'hoansi: Kirchengast 2000), but there are many potential explanations for this pattern, some of which are unrelated to male aggression.

Female-female competition also takes different forms in chimpanzees and humans. Chimpanzee females exhibit aggression at much lower rates than chimpanzee males, but at dramatically higher rates than women in any human society (Muller 2002; Wrangham et al. 2006). Most aggression among chimpanzee females involves either immediate contests over food or struggles for long-term access to resource-rich feeding areas (reviewed in Pusey and Schroepfer-Walker 2013). There is no evidence that female chimpanzees compete aggressively over mates (Goodall 1986; Matsumoto-Oda 2007). Long-term data from Mahale indicate that chimpanzee estrous cycles overlap less than would be expected by chance, which may help to curtail competition for mating opportunities among females (Matsumoto-Oda 2007).

Data from foragers are again scarce, but women universally engage in direct, physical aggression at lower rates than men in the same societies (reviewed in Archer 2004; Aka foragers: Hess et al. 2010). When female-female aggression does occur, however, it routinely involves competition over mates and the critical resources they provide (Burbank 1987; Campbell 1999; Benenson 2013). Adolescent females are more likely to use aggression than older women, with disputes centering around three key issues: "management of sexual reputation, competition over access to resource-rich young men, and protecting heterosexual relationships from takeover by rival women" (Campbell 1995). The striking difference between humans and chimpanzees in female sexual rivalry is probably driven by the importance of pair-bonding and paternal investment in human reproduction, which leads to competition within both sexes for the most desirable partners (Stewart-Williams and Thomas 2013).

Mate Preferences

Men's mate preferences show clear differences from those of chimpanzee males, particularly with respect to female age. Multiple behavioral measures indicate that chimpanzee males actively choose older mothers as mating partners, in preference to either younger mothers or nulliparous females (Muller et al. 2006). At Kanyawara, males initiate copulations more often with older mothers in estrus, are more likely to join parties containing such mothers, and show higher rates of male-male aggression over access to them (Muller et al. 2006). Consequently, older mothers mate more frequently with the most dominant males.

In humans, by contrast, cross-cultural studies indicate that women's sexual attractiveness peaks before motherhood and declines with age (Symons 1995; Jones 1995, 1996). Many of the traits that men value in potential mates are cues of nubility (i.e., having recently started ovulatory cycling, but not yet become pregnant), with men seeming to prefer women younger than the age of peak fecundity (i.e., probability of successful conception) (Hadza: Marlowe 2004). This attention to reproductive value over immediate fecundity led Symons (1995) to propose that men's evolved psychology of sexual attraction is adapted to long-term pair-bonds. Such bonds promote a preference for youth, because men who choose young partners maximize their future reproductive opportunities with those partners.[2] In the promiscuous chimpanzee system, males instead focus their mating effort on the females with the highest probability of immediate conception. These are generally older mothers, who experience fewer cycles per conception than younger mothers or nulliparous females (Kibale Chimpanzee Project, unpublished data).

Other aspects of human life history that should favor a male preference for youth include high levels of paternal investment and a long female postreproductive life span (Symons 1995). Specifically, younger women are less likely to have costly offspring from previous partners, and are less likely to have undergone menopause. These are not relevant factors for chimpanzees, who show neither substantial direct paternal investment nor menopause (Emery Thompson et al. 2007).

The strength of female mate preferences differs conspicuously between humans and chimpanzees. Women show strong preferences for some men over others, and can be severely traumatized when their choices are circumvented by men's sexual aggression (Buss 1989; Thornhill and Thornhill 1990). Female chimpanzees, by contrast, appear eager to mate with a wide range of males, and rarely resist male sexual advances. At Gombe, Goodall (1986) reported that over a five-year period, females responded positively to 95.9 percent of male copulatory advances within one minute ($n = 1,475$). Instances in which they did not do so almost all involved a single sterile female, or females rebuffing the advances of male relatives. The comparative lack of choosiness among female chimpanzees reflects an underlying strategy of paternity confusion, which may reduce the risk of male infanticide (Palombit 2012).

Do female chimpanzees nevertheless maintain weak preferences for particular males? At Taï, females near ovulation mated more frequently with high-ranking males, or males who subsequently became high-ranking, possibly indicating a preference for dominant males (Stumpf and Boesch 2005). It is currently impossible, however, to exclude an alternative hypothesis that dominant males were simply more successful in excluding rivals from mating, and in constraining female promiscuity (Muller et al. 2011).

Cross-cultural evidence suggests that women prefer men with broad chests and shoulders, narrow hips, and muscular torsos (reviewed in Gangestad and Scheyd 2005; Dixson 2009). This is often interpreted as a preference for male formidability and dominance, which could benefit women who choose long-term mates capable of protecting them and their offspring (Snyder et al. 2011). However, in a foraging context, more muscular males might also be better hunters and providers (Hadza: Apicella 2014). Indirect genetic benefits to offspring are also possible, if these physical characteristics are indicators of genetic quality (Gangestad and Scheyd 2005). Evidence for the potential importance of genetic benefits comes from studies indicating that women's preferences for muscular men are stronger when considering short-term, rather than long-term, mates (Frederick and Haselton 2007).

Evolution of the Human Mating System

Assuming for the moment that the last common ancestor of chimpanzees and humans exhibited a chimpanzee-like mating system, how did a system centered around monogamous pair-bonds evolve? A number of influential scenarios posit a direct transition from promiscuity to monogamy (e.g., Lovejoy 1981; Kaplan et al. 2000, 2009). These models assume that the sexual, parental, and economic aspects of modern human pair-bonds evolved concurrently, with an increase in the dependency of offspring that necessitated paternal provisioning (reviewed by Grueter et al. 2012; Chapais 2013, this volume).

Chapais (2008, 2011, 2013, this volume) argues that for several reasons the transition is unlikely to have been so direct. First, polygyny is extremely common among primates, and is the rule in multilevel primate societies (i.e., those with distinct family units nested within larger bands), because whenever females group together, the potential for male monopolization is present. Polygyny is also widespread in human societies, even though most marriages in any given society may be monogamous. More importantly, phylogenetic analyses of pair-bonding and paternal care in both primates and other mammals indicate that monogamy has evolved frequently in the absence of paternal care. In the subset of species that exhibit both, paternal care probably followed the adoption of monogamy, rather than evolving with it (Grueter et al. 2012; Chapais 2013; Opie et al. 2013). Consequently, Chapais proposes an intermediate stage in the shift from promiscuity to monogamy, in which multimale-multifemale groups partition into multiple, polygynous, one-male units. This scenario is parsimonious, because it involves only one initial change—from short-term mating bonds to long-term bonds, as occurred with the evolution of the hamadryas and gelada baboon social systems from a promiscuous ancestor (Grueter et al. 2012; Chapais 2013). Further, and importantly, the model does not require all of the unique aspects of human life history and ecology to evolve simultaneously.

Testing such models for the divergence of the human and chimpanzee mating systems ultimately requires evidence from the fossil record, in the form of anatomical traits that are reliably associated with sexual competition and mate choice. The most commonly used traits are sexual dimorphism in body size and canine height, as extreme dimorphism in these indicates the presence of polygyny. In this section we assess the degree of human dimorphism in comparative perspective, attempt to trace the history of dimorphism in the human lineage, and evaluate functional explanations for human dimorphism.

Dimorphism in Body Size

Body size dimorphism shows remarkable variation across primates, which include everything from the monomorphic lemurs, callitrichids, and gibbons to the massively dimorphic baboons, mandrills, gorillas, and orangutans. Darwin (1871) was the first to argue that sexual selection could favor increased male size if large males were more effective than small ones in aggressive competition over mates. Although alternative hypotheses have been proposed, it is now clear that Darwin's is the best explanation for variation in dimorphism across the order (Mitani et al. 1996; Plavcan 2001, 2004, 2011; Muller and Emery Thompson 2012). For example, Plavcan and van Schaik (1997b) ranked a large sample of primates based on the observed frequency and intensity of male-male competition, and found a positive relationship between these "competition levels" and dimorphism in body mass. Mitani and colleagues (1996) found a similar relationship between dimorphism and the operational sex ratio (OSR), a measure of male mating competition representing the number of reproductively active males to females, accounting for temporal changes in the availability of females owing to factors such as pregnancy and lactation.

Table 11.1 shows estimates of sexual dimorphism in body mass for multiple human and chimpanzee populations. The ratio of male to female mass for chimpanzees is approximately 1.3, which is the average for a primate with a multimale/multifemale mating system (Plavcan 2012b). The mean ratio for human foragers is 1.19, which is higher than gibbons and some monogamous monkeys, but at the low end of the primate range (Plavcan 2012b). Ostensibly, this supports the idea that humans have an evolutionary history of reduced mating competition, in comparison with chimpanzees.

Placing human dimorphism in primate perspective is complicated, however, by the fact that humans exhibit substantial dimorphism in body composition that is not reflected in measures of body mass (Dixson 2009). Specifically, because women maintain larger fat stores than men, body mass

TABLE 11.1. Sexual dimorphism in body size and composition in African apes and human foragers.

Sample	M Body Mass	F Body Mass	M/F Mass	M Body Fat	F Body Fat	Reference
Hadza	50.9 ($n=13$)	43.4 ($n=17$)	1.17	13.5% ($n=13$)	21% ($n=17$)	Pontzer et al. 2012
Ju/'hoansi	47.9 ($n=79$)	40.19 ($n=74$)	1.19			Truswell and Hansen 1976
Aché	60 ($n=41$)	52 ($n=32$)	1.15			Hurtado and Hill 1987
Hiwi	56 ($n=62$)	48 ($n=65$)	1.17			Hurtado and Hill 1987
Forager mean ($n=30$ groups)	53.9	45.6	1.19			Marlowe and Berbesque 2012
Mahale chimpanzees	42 ($n=6$)	35.2 ($n=8$)	1.19			Uehara and Nishida 1987
Gombe chimpanzees	39	31.3	1.25			Pusey et al. 2005
Pan paniscus (captive)	42.7 ($n=7$)	34.3 ($n=6$)	1.24	<1% ($n=5$)	3.6% ($n=5$)	Zihlman and Bolter 2015
Pan paniscus	45 ($n=7$)	33.2 ($n=6$)	1.36			Jungers and Susman 1984
P. t. schweinfurthii	42.7 ($n=21$)	33.7 ($n=26$)	1.27			Smith and Jungers 1997
P. t. troglodytes	59.7 ($n=5$)	45.8 ($n=4$)	1.30			Smith and Jungers 1997
Gorilla	170.4 ($n=10$)	71.5 ($n=3$)	2.38			Smith and Jungers 1997

ratios conceal pronounced sex differences in muscularity and strength. This is not an issue for most nonhuman primates, because they exhibit relatively little sexual dimorphism in their levels of body fat (McFarland 1997) and have low fat stores (when living on natural diets in the wild; Dittus 2013). For example, isotope studies of wild baboons reported a mean body fat percentage of 1.9 percent for females ($n=13$; Altmann et al. 1993), and a dissection study of wild Tocque macaques reported a two-sex mean of 2.1 percent ($n=13$; Dittus 2013). Even in captivity, bonobos of both sexes appear to maintain very low levels of body fat (Zihlman and Bolter 2015: 6 females = 3.6 percent, 7 males = <1 percent).

Sexual dimorphism in musculature is pronounced in humans, particularly in the upper body. Men in Western populations maintain, on average, approximately 61 percent more skeletal muscle than women, including 75 percent more in the arms and 50 percent more in the legs (Lassek and Gaulin 2009). A review of European and American dissection data, for example, showed a ratio of male muscle mass/female muscle mass of 1.54, and a ratio of male fat-free mass to female fat-free mass of 1.4 ($n = 31$ men, 20 women; Clarys et al. 1999). A larger X-ray study of Americans reported similar ratios—1.6 for male muscle mass/female muscle mass, and 1.43 for male fat-free mass/female fat-free mass ($n = 96$ men, 174 women; Kim et al. 2004). Comparable measures of muscle dimorphism are not available from any foraging population. The Hadza ratio of male fat-free mass to female fat-free mass is lower, however, at 1.28, implying slightly less muscle dimorphism than Western populations ($n = 13$ men, 17 women; Pontzer et al. 2012).

Such differences in musculature produce large disparities in strength between men and women, even among foragers. The ratio of adult male hand-grip strength to adult female hand-grip strength for the Hadza is 1.6 ($n = 240$ men, 218 women; Marlowe 2010). The comparable figure for Aché in the twenty to twenty-nine year age group is 1.55 (Walker and Hill 2003).

How extreme is human muscle dimorphism, from a primate perspective? Few autopsy data are available from nonhuman primates, but Zihlman and Bolter (2015) reported a ratio of male muscle mass to female muscle mass of 1.72 in captive bonobos ($n = 6$ females, 7 males). A much smaller sample of captive lowland gorillas showed a ratio of 2.24 (Zihlman and McFarland 2000; $n = 2$ females, 2 males). One of the female gorillas in this sample was old and infirm, however, with relatively little muscle. Considering only the young, healthy female produces a ratio in gorillas of 1.87. This limited sample suggests that muscle dimorphism in humans, like body mass dimorphism, is slightly less than that seen in chimpanzees and bonobos, but more data are needed.

Dimorphism in Canine Height

Sexual dimorphism in canine size varies even more dramatically across primates than body mass, with some species showing male canine heights greater than four times those of females (Plavcan 2004). Canines are often

TABLE 11.2. Sexual dimorphism in maxillary canine height.

Sample	Maxillary Canine Height (M/F)
Gorilla gorilla	1.73
Pan troglodytes	1.43
Pan paniscus	1.38
Ardipithecus ramidus	1.12*
Australopithecus afarensis	1.19 ($n=4$)
Australopithecus africanus	1.13 ($n=5$)
Paranthropus robustus	1.25 ($n=5$)
Homo sapiens	1.15

Note: Estimates are from Plavcan and van Schaik (1997a), except for *A. ramidus*, which is from Suwa et al. (2009).

* The estimate for *A. ramidus* is based on maximum canine crown length. However, maxillary canine heights in the very small *ramidus* sample are similar in range and average size to those of other early hominins (Suwa 2009: figure S8).

employed as slashing weapons, and Darwin (1871) was the first to propose that mating competition could promote a sex difference in canine size, if males with larger canines were more formidable opponents. This general explanation has been strongly supported by a wide range of studies (Plavcan et al. 1995; Plavcan 2004; Leigh et al. 2008).

Humans exhibit low-crowned canines for their body size, and very low levels of dimorphism, with men maintaining maxillary canines, on average, around 10 percent more projecting than those of women (Plavcan 2012b). Chimpanzees, by contrast, display formidable canines that project well past their incisors and premolars, and maintain a ratio of male/female maxillary canine height of approximately 1.42 (Plavcan 2004). Table 11.2 presents comparative data from multiple populations.

The implications of reduced canine dimorphism for the evolution of the human mating system are not straightforward. Although extreme dimorphism in both canine height and body size invariably signals a polygynous mating system, a male-biased operational sex ratio, and a high degree of male-male competition, the absence of sexual dimorphism in these traits is not strongly linked with any particular mating system across primates (Plavcan 2004). Consequently, monomorphism is not evidence for a monogamous mating system, nor for a lack of aggressive competition among males (Lawler 2009; Plavcan 2012a), as is sometimes assumed (e.g., Sayers et al. 2012).

Canine dimorphism correlates modestly with body mass dimorphism across primates ($n = 79$, $r = 0.626$: Plavcan 2004), suggesting distinct evolutionary constraints on body size and canine height. Importantly, body size appears to be more limited in arboreal than terrestrial species, probably for reasons of safety and maneuverability in the canopy (Kappeler 1990; Lawler et al. 2005). Body size is also more constrained in females than in males, owing to the energetic costs of gestation and lactation. Because these factors are less relevant for canine height, some arboreal primates exhibit both extreme canine dimorphism and high levels of aggressive mating competition, but little body size dimorphism. Other species lack canine dimorphism because both sexes maintain large, weaponized canines that they use in aggressive competition (Plavcan 2004).

Dietary factors can also limit canine height. To be useful as weapons, upper and lower canines cannot overlap when the mouth is fully open. Consequently, maximal jaw gape places a constraint on canine length (Hylander 2013). Maximal gape, in turn, is constrained by the placement of the jaw muscles. Positioning these muscles anteriorly allows for greater bite force in the posterior dentition, but limits the opening of the mouth. Positioning these muscles posteriorly allows for wider gape—and longer canines—but decreases the mechanical efficiency of chewing (Hylander 2013). Reliance on difficult-to-chew foods, therefore, could promote a reduction in canine height, even in species exhibiting high levels of male-male competition. This may be an important factor in explaining the unique hominin pattern of dimorphism.

Sexual Dimorphism in the Fossil Record

There is general agreement that by 1.5 Ma, body size dimorphism in *Homo* species was equivalent to that of most living human populations, males being on average around 15 to 20 percent heavier and somewhat taller than females (Smith 1980; Grabowski et al. 2015). Prior to 1.5 Ma the story is less clear, not least because sexing the often-isolated specimens used in analyses is not possible. This can be partially addressed by comparing variation in a fossil sample, treated as equivalent to a "population," to that in extant species with known dimorphism.

The best-represented species prior to *Homo* is *Australopithecus afarensis*, for which there is currently no agreement on degree of body size dimorphism.

Reno et al. (2003) concluded that *afarensis* dimorphism was similar to that in *Homo sapiens,* while others (for example, Plavcan et al. 2005; Grabowski et al. 2015) inferred substantially greater dimorphism. One problem involves choosing the scaling relationship for predicting body mass from a skeletal metric, such as femoral head diameter. Bipedal humans show positive allometry, while nonhuman catarrhines scale geometrically, and this means that a given difference in skeletal measurements translates into different body mass difference depending on which allometric relationship is chosen. Most studies assume implicitly or explicitly that the australopithecines are "sufficiently" bipedal to warrant use of human scaling values. We favor the approaches of Plavcan et al. (2005), which maximizes the likelihood that the same individual is not counted more than once, and Grabowski et al. (2015), which estimates body weights from specimens representing clearly different individuals. These studies infer body mass dimorphism for *Au. afarensis* greater than that of humans and *Homo erectus* (males 15 to 20 percent heavier than females), and probably greater than that of chimpanzees (males 20 to 40 percent heavier), but much less than gorillas (males >100 percent heavier). The most recent study (Grabowski et al. 2015) estimates a pattern in which *afarensis* males were 50 to 60 percent heavier than females. Similar values obtain for other species with sufficient samples: *Au. africanus, Paranthropus robustus,* and *P. boisei* (Grabowski et al. 2015).

Because of small sample sizes, it is challenging to estimate degree of stature dimorphism in pre-*Homo* species, but an approach is possible given the recovery of a probable male partial skeleton dated at just under 3.6 Ma, *Au. afarensis,* KSD-VP-1/1 (Haile-Selassie et al. 2010), substantially older than the ca. 3.2 Ma female partial skeleton A.L. 288-1 ("Lucy") (Johanson et al. 1982). Each preserves enough of the femur and tibia to estimate lengths with reasonable confidence. Haile-Selassie and colleagues (2010: table S6) summarize these data, and comparative samples are included for *Homo sapiens, Pan troglodytes,* and *Gorilla gorilla* ($n = 25$ for each). When compared to the 95 percent confidence intervals of the comparands, the difference between the *Au. afarensis* femoral and tibial lengths substantially exceeds the confidence intervals for humans and chimpanzees, and falls at the 95 percent limits for gorillas. This suggests a degree of stature dimorphism exceeding that seen in humans.

By contrast, it appears that male canine height is reduced in all known hominins or putative hominins. The oldest probable hominin cranium,

Sahelanthropus tchadensis (Brunet et al. 2002), dated around 7 Ma (Lebetard et al. 2008), is likely to be male given its substantial supraorbital torus. Its canine had apical wear but was clearly small, metrically similar to those of younger pre-*Australopithecus* species. Similarly, a maxillary canine attributed to *Orrorin tugenensis* (Senut et al. 2001), dated a little under 6 Ma, shows low canine height, and resembles that of the somewhat younger *Ardipithecus kadabba* (Haile-Selassie et al. 2004; Simpson et al. 2015), dated between 5.8 and 5.4 Ma. Canine height falls between female *Pan troglodytes* and female *Pan paniscus,* and canine height is clearly reduced relative to males in either *Pan* species (Suwa et al. 2009: figure S8).

In chimpanzees and gorillas, dimorphism in canine height is associated with a canine honing complex, especially in males, in which the mesiobuccal lower premolar surface abrades the distal surface of the upper canine, sharpening (or honing) it. Simpson and colleagues (2015) note that there is modest change in the morphology of the upper canine-lower third premolar complex between the Late Miocene *Ar. kadabba* and Early Pliocene (4.4 Ma) *Ar. ramidus* (Suwa et al. 2009), with the earlier form showing some occlusal contact. However, there is no significant canine honing in pre-*ramidus* canines, size remains similar, and crown height is (perhaps rapidly) reduced by wear.

Plavcan and van Schaik (1997a) were able to sex a small sample of *Au. afarensis* maxillary canines, and a subset of these were sufficiently unworn that height could be measured. Dimorphism in canine length, breadth, and height was less than even *Pan paniscus*. Indeed, data summarized by Suwa et al. (2009: supplementary material) show that crown height and its variability vary little in pre-*Homo* hominins, while Spoor et al. (2007) and Suwa et al. (2009) show that canines were reduced still more in *Homo* species.

Why is canine height reduced so early in the hominin lineage? Three major hypotheses have been offered. First, as described previously, if early hominins showed dietary shifts toward difficult-to-chew foods, this would have selected for anterior placement of the jaw muscles, decreasing gape and constraining canine height (Hylander 2013). As Hylander notes, the critical test is with pre-*Australopithecus* hominins, but yields no clear answer. Canine height in putative males is already reduced, while cheek (chewing) tooth size and degree of facial projection are intermediate between chimpanzees and *Australopithecus afarensis.* Second, hominin males may have adopted alternative modes of competition that rendered weaponized canines superfluous. One possibility is that bipedalism freed the hands to employ tools as weapons.

This scenario is supported by the observation that male chimpanzees in some sites use sticks as bludgeons, but very clumsily, owing to their awkward bipedalism (Kahlenberg and Wrangham 2010; Muller, personal observation). Cartmill and Smith (2009) argue that handheld weapons would not have made canines redundant, because males wielding both clubs and large canines would be especially imposing. This argument, however, ignores the costs of producing canines. A third possibility is that aggressive competition between males was reduced early in the hominin lineage, making male weaponry obsolete (e.g., Lovejoy 1981). The considerable degree of body size dimorphism that persists in the lineage is inconsistent with this view, however.

Sexual dimorphism in facial size is generally correlated with body mass dimorphism across primates, though in many anthropoids it greatly exceeds body mass dimorphism (Plavcan 2003). Estimating facial dimorphism is challenging, and there are few systematic studies for hominins. However, several species appear to show highly dimorphic faces, despite more modest body size dimorphism. Lockwood (1999) compared facial measures in five putative males and eight putative females of *Australopithecus africanus* and concluded that variation exceeded that in *Pan troglodytes* and *Homo sapiens* (Zulu sample) and approached that of *Gorilla beringei* (Figure 11.2). Lockwood et al. (2007) showed that facial dimorphism in *Paranthropus robustus* was equivalent to that seen in *Gorilla,* with a pattern of extended male growth. Spoor et al. (2007) document variation in facial and particularly browridge robusticity in *Homo erectus* suggesting marked facial dimorphism in a species with reduced body mass dimorphism (Grabowski et al. 2015). Smith (1980) showed that Neanderthals were somewhat more facially robust and dimorphic than later *Homo sapiens,* consonant with the analyses of Cieri et al. (2014) suggesting successive reduction in facial robusticity and dimorphism between pre-80 Ma, post-80 Ma, and living *Homo sapiens.*

Why these species exhibited disproportional facial dimorphism is not currently clear. One possibility is that body mass dimorphism is being systematically underestimated (Plavcan 2003). Another is that bipedalism places constraints on male body size, but not facial size, and large faces make males appear more formidable. In either case, the existence of substantial facial dimorphism is inconsistent with the idea that male-male competition was significantly reduced early in the hominin lineage.

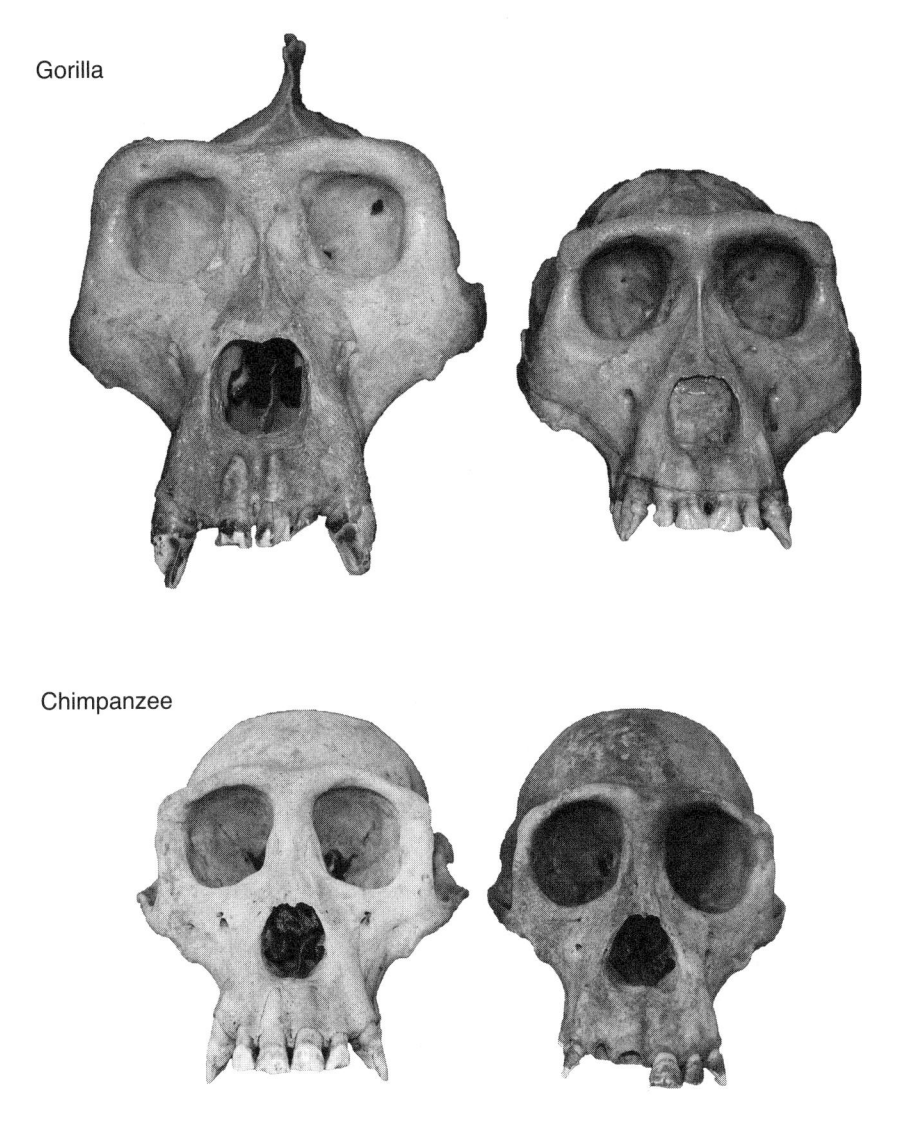

FIGURE 11.2. Facial size dimorphism in gorillas and chimpanzees. Facial size dimorphism correlates with body mass dimorphism across primates, but greatly exceeds it in many hominoids. *Australopithecus africanus* and *Paranthropus robustus* have both been estimated to have facial size dimorphism closer to that of gorillas than chimpanzees, despite lower estimated body mass dimorphism than gorillas. Large faces may help males look more impressive in sexual competition. (Males at left, females at right.)

Explaining the Modern Human Pattern of Dimorphism

The low level of canine dimorphism found in modern humans appeared early in the hominin lineage and, as discussed, likely reflects an increased reliance on tools as weapons, biomechanical constraints on jaw gape, or some combination of the two. How, then, do we explain the pronounced sexual dimorphism in strength and musculature that characterizes the species?

One prominent hypothesis is that, as in other primates, sexual selection has favored increased male musculature for its use in aggressive competition over mating opportunities (Lassek and Gaulin 2009; Puts 2010). Sell et al. (2012) list twenty-six sex differences in humans that "suggest male design for combat." These include greater height, weight, and upper body strength in men, together with faster reaction time and more accurate throwing. All are argued to be associated with the use of ancestral weaponry, such as clubs, handaxes, spears, and bows.

Lancaster and Kaplan (2009), by contrast, argue that dimorphism in human musculature is driven not by male competition, but by the sexual division of labor, with men showing adaptations to sustained and heavy work (see also Wolpoff 1976; Frayer 1980). Hunting in particular is a difficult and dangerous activity, requiring strength and endurance, that is undertaken primarily by men in most foraging societies (Apicella 2014; Murdock and Provost 1973; Lancaster and Kaplan 2009; Gurven and Hill 2009). All of the sex differences identified by Sell and colleagues (2012) as adaptations for combat are as plausibly interpreted as adaptations to big-game hunting employing the same ancestral weapons.

One potential test of these alternatives is to examine sex differences in the development of strength and musculature across the life span, to see how these align with investment in mating and foraging effort. If sex differences in musculature evolved primarily to support mating competition, then they should be most pronounced during life history phases when men are competing for mates. If such differences evolved primarily to support men's work, then they should peak when men's foraging effort is most intense.

The most detailed dataset on physical performance in foragers comes from a cross-sectional study of Aché men and women. Walker and Hill (2003) used a battery of tests to measure grip strength, upper body strength (maximum number of push-ups, pull-ups, and chin-ups), running speed (in the fifty-meter dash), and aerobic capacity in a large sample, across a wide age range. For men, all of these measures peaked in the early twenties and then showed linear declines with age. Women maintained their peak upper body strength throughout the life span, and showed more gradual declines than men in running speed and grip strength. Women's aerobic capacity showed a slight decline during the reproductive years, but returned to peak levels during the postreproductive span. Sex differences in physical performance were most pronounced during the early twenties.

Hill and Hurtado (1996) reported a median age at first birth of twenty-three for Ache men, supporting an association between men's peak strength and reproductive competition (Figure 11.3). By contrast, Aché men's foraging shows little relationship to strength across the life span. Hunting return rates peak around age forty, when men produce food at twice the rate of those at maximum strength (Walker et al. 2002).

Data from Hadza foragers are not as comprehensive, but measures of men's bow-pull and grip strength peak between the ages of twenty-five and thirty (Marlowe 2000, 2010), with men's median age at first marriage in the same window, at twenty-five to twenty-six (Blurton Jones 2016). As with the Aché, Hadza men's hunting returns and hunting reputation peak

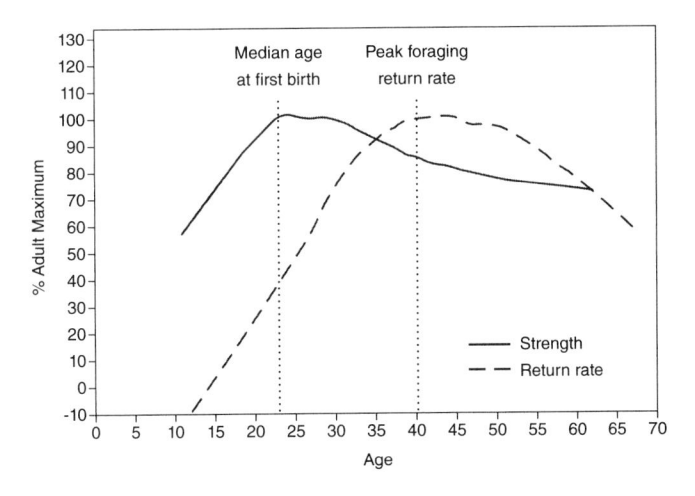

FIGURE 11.3. Aché men's strength and foraging returns across the life span. Aché men reach peak strength in their early twenties, after which they show a steady decline. Foraging returns, by contrast, peak around age forty, when strength has declined significantly. Maximum strength coincides with median age at men's first birth, suggesting an association with mating effort. Data from Walker et al. (2002) and Walker and Hill (2003).

later—around age forty (Marlowe 2000; Apicella 2014). Sexual dimorphism in strength thus appears to be more closely associated with reproductive competition than with foraging effort.[3]

A second test of these alternative hypotheses incorporates sexually dimorphic traits other than size and strength. Human dimorphism is not limited to muscle mass, but includes conspicuous sex differences in facial hair, body hair, larynx size, voice pitch, and pattern baldness (Dixson 2009). These traits are unlikely to be important for men's work or the sexual division of labor. Across primates, however, they are predictably linked with male mating competition.

The presence of striking secondary sexual adornments, such as beards, capes, crests, and cheek pads, for example, is strongly associated with a polygynous mating system in primates. Dixson and colleagues (2005) rated a sample of 124 primate species for dimorphism in visual traits using a six-point scale (with 0 indicating no difference between males and females, and 5 indicating that males possessed a prominent adornment completely absent in females). Distinct traits were scored separately, and the scores were summed, producing values from 0 (in monomorphic species) to 32 (in the mandrill). Across the sample, visual scores correlated positively with levels of dimorphism in body size. Both monogamous males and males living in multimale, multifemale mating systems showed low levels of visual dimorphism. Males in polygynous species exhibited high values, though in most cases it was not clear whether male-male competition or female choice was responsible (Dixson et al. 2005).

Chimpanzee males do not maintain any secondary sexual adornments that are not present in females (Figure 11.4). Humans, by contrast, show sex differences in facial hair, body hair, and pattern baldness that are more extreme than anything seen in any non-polygynous primate (Dixson et al. 2005). Although the functions of facial and body hair are not clear, they are unlikely to relate to male work. Cross-cultural research has found little support for the idea that women consistently prefer bearded faces (Muscarella and Cunningham 1996; Dixson and Vasey 2012; Dixson and Brooks 2013). In numerous studies, however, both men and women rate faces displaying full beards as more masculine, dominant, strong, and aggressive (Addison 1989; Muscarella and Cunningham 1996; Neave and Shields 2008; Dixson and Vasey 2012). At present, then, most evidence favors the hypothesis that men's beards "augment the effectiveness of human aggressive facial displays," ren-

FIGURE 11.4. Secondary sexual adornments in chimpanzees and humans. Although chimpanzee males are larger than females, and have longer canines, they do not maintain additional secondary sexual adornments. Female (left) and male (right) faces are consequently similar, though males have more pronounced supraorbital tori (browridges). Human men too have more pronounced browridges than women, but show notable dimorphism in facial hair and pattern baldness. Photos by Martin N. Muller.

dering their owners more intimidating (Dixson and Vasey 2012). This supports the idea that male-male aggression was an important factor in human evolution.

Sex differences in larynx size and vocal pitch are similarly consistent with a history of aggressive sexual competition among men. Across primates, polygynous species show increased levels of dimorphism in the vocal apparatus (including the larynx, laryngeal sacs, and hyoid bone) compared to those with monogamous or multimale, multifemale mating systems (Dixson 2009). Such dimorphism cannot be explained as a by-product of body size dimorphism, as it exceeds what would be predicted from dimorphism in body size alone. Instead, dimorphism in these structures supports the loud calls that males employ in mating competition—specifically to attract mates, repel sexual rivals, and defend territorial boundaries (Delgado 2006; Dixson 2009).

Chimpanzees show little dimorphism in larynx size or vocal pitch, possibly because both sexes use long calls (the pant-hoot vocalization) to contact others from a distance. Males call more frequently than females, however, and males with high testosterone levels call more frequently than males with low levels, suggesting that pant-hoots may play a role in male competition (Fedurek et al. 2016).

Humans, by contrast, maintain substantial dimorphism in the vocal tract and the vocal folds. These structures are, respectively, 15 percent and 60 percent longer in men, resulting in a lower fundamental vocal frequency, with almost no overlap between the male and female distributions (Puts et al. 2007, 2014). The lengthening and thickening of men's vocal folds occurs during puberty, under the influence of testosterone. As with the case of facial hair, there is no reason to think that vocal dimorphism relates to men's work or the sexual division of labor. Instead, decreasing vocal tract resonant frequencies likely helps to exaggerate perceived body size, making men sound more formidable to competitors (Fitch and Reby 2001).

Substantial evidence supports the idea that human vocal dimorphism was shaped by sexual selection. In multiple studies, men with more masculine voices have reported having more sexual partners than men with higher voices (reviewed in Puts et al. 2014). Among the Hadza, vocal pitch was negatively correlated with the number of children fathered, such that men with lower voices had more offspring (Apicella et al. 2007).

Both male competition and female choice may have selected for vocal dimorphism. Women generally show a preference for slightly lower than average voices in men, and rate men's voices as more attractive when they are experimentally masculinized, rather than feminized (Collins 2000; Feinberg et al. 2005, 2006; Saxton et al. 2006; Hodges-Simeon et al. 2010; Babel et al. 2014; Puts et al. 2014). Puts (2005) reported a stronger preference for deep voices in short-term mating contexts, and during the fertile phase of the ovarian cycle. A similar pattern was observed among the

Hadza, with only women who were breastfeeding not preferring lower-pitched men's voices (Apicella and Feinberg 2009).

Both men and women perceive men with lower voices as being more physically and socially dominant than men with higher voices, and experimentally lowering men's voices increases the perception of dominance (Puts et al. 2006, 2007; Jones et al. 2010; Wolff and Puts 2010). Several studies have found associations between dimorphic vocal traits and measures of body size or upper body strength, suggesting that masculine voices are honest signals of formidability (Evans et al. 2006; Puts et al. 2012; Hodges-Simeon et al. 2014; Pisanski et al. 2014). Puts and colleagues (2014) note that experimentally masculinizing men's voices increases perceived dominance more than it does attractiveness, and suggest that male competition may have been more important than female choice in driving vocal dimorphism.

Thus, from a broad primate perspective, multiple aspects of human sexual dimorphism are difficult to explain except as the result of an evolutionary history in which male-male aggression had clear fitness consequences. Sell and colleagues (2012) have further argued that aspects of human psychology are best explained by such a history. Specifically, both men and women are capable of quickly and accurately assessing men's physical strength from cues in the body, face, and voice, in both Western and non-Western populations (Fink et al. 2007; Sell et al. 2009, 2010). When making such assessments, stronger men consistently rate potential antagonists as less formidable than do weaker men (Fessler et al. 2014). And in various studies, men who are larger or physically stronger feel more entitled to deference from others, anger more quickly, and employ aggression more frequently than do smaller or weaker men (Archer and Thanzami 2007; Gallup et al. 2007; Sell et al. 2009, 2012; Muñoz-Reyes et al. 2012; Salas-Wright and Vaughn 2016), including among Aka foragers (Hess et al. 2010). Such links among male attitudes, behaviors, and formidability would not be expected in a species in which male aggression was unimportant (Sell et al. 2009, 2012).

Most discussions of aggression and sexual dimorphism focus on men competing for mates within groups. Is it possible, however, that intergroup conflict, rather than intragroup mating competition, was the driving force behind the sex differences reviewed above? As previously noted, foragers who have not been pacified by state intervention, or by more powerful farming or pastoralist neighbors, generally exhibit higher rates of killing between groups than within groups. This pattern ostensibly supports the notion that intergroup killing was the more important selective force.

The coalitionary nature of intergroup conflict, however, suggests a different conclusion. The most common form of warfare in foraging societies is the stealth raid, in which a group of men attempt to isolate and kill a member of the neighboring group at little or no risk to themselves (Wrangham and Glowacki 2012). Because this kind of interaction involves overwhelming power disparities, it is unlikely to favor increased male size. In support of this conclusion, across primates, species that frequently engage in coalitionary fights, such as chimpanzees, have smaller canines than expected, given their overall frequency and intensity of aggression (Plavcan et al. 2005).

Sperm Competition

Across primates, male aggression over access to mating opportunities is common when females live together in monopolizable groups. When multiple males are present in a group, however, scramble competition, in the form of postmating sperm competition, can also occur. If females normally mate with more than one male during a single ovarian cycle, then selection favors morphological and physiological adaptations in males to increase their odds of successful fertilization (reviewed in Dixson 2012). Although such adaptations do not leave fossil evidence, their presence or absence in living species is informative about an evolutionary history of promiscuous mating.

As previously discussed, female chimpanzees routinely mate with multiple males over the course of a single ovarian cycle. Consequently, males show clear physiological adaptations to sperm competition, including large testes for body size, a fast sperm production rate, substantial sperm reserves in the epididymis, and a large number of sperm per ejaculate (Harcourt et al. 1981; Dixson 2009, 2012). Sperm midpiece volume is large, reflecting the number of mitochondria available to power sperm movement, and sperm swimming speed and power are high (Anderson and Dixson 2002; Nascimento et al. 2008). Defective, pleiomorphic sperm, which result from errors in meiosis, are present at low levels (~4.5 percent: Seuanez et al. 1977).

Humans are sometimes characterized as exhibiting "intermediate" levels of sperm competition, based primarily on the observation that they have a testes to body size ratio smaller than that of promiscuous chimpanzees, but larger than that of polygynous gorillas (e.g., Shackelford and Goetz 2007).

Dixson (2009, 2012) identifies multiple problems with this characterization, including the fact that humans have smaller testes than predicted for a primate of their body size, and that variation in human testes size is considerable, with the ratio in some populations overlapping that of some gorilla samples. Most ape measurements also come from captive individuals, who appear to have smaller testes than their wild counterparts.

Additional comparative anatomical and physiological evidence is inconsistent with significant levels of human sperm competition. Humans maintain modest sperm reserves in the epididymis, have low numbers of sperm per ejaculate, and have sperm with a smaller midpiece volume than do gorillas (Dixson 2009). Humans produce abnormal, pleiomorphic sperm at a rate of 27 percent, almost identical to the gorilla rate (29 percent: Seuanez et al. 1977). Although sperm competition favors a short vas deferens (because sperm then have a smaller distance to travel during copulation) that is thickly muscled (to propel sperm with increased force), the human vas deferens is long and thinly muscled (Anderson et al. 2004).

Aspects of human penile morphology have also been argued to reveal an evolutionary history of sperm competition. Gallup and colleagues (2003), for example, argued that the prominent glans and pronounced coronal ridge are adaptations for displacing sperm from the female reproductive tract. This is inconsistent with comparative data, however, which show that this morphology is extremely common in polygynous primates—including the gorilla—that show little or no sperm competition (Dixson 2009). Smith (1984) argued that the human penis is exceptionally long, owing to selection, via sperm competition, for depositing sperm closer to the cervix during mating. Dixson (2009, 2012), however, shows that this conclusion was based on flawed data. When reliable measurements from larger samples of nonhuman primates are considered, human penis length is exactly what is expected for a primate of human body size. Dixson (2009) notes that human penile circumference is comparatively large, but no larger than expected from human vaginal circumference, which has likely increased to facilitate the birth of large-brained infants.

Finally, males in a number of promiscuous primates produce copulatory plugs, formed by proteins from the seminal vesicles (semenogelin 1 and 2) interacting with the prostatic enzyme vesiculase (Dixson 2009). Such plugs probably function to block the sperm of rival males. SEMG1 and SEMG2, the genes encoding these proteins, show evidence of strong positive selection in promiscuous species, like the chimpanzee, whereas they are nonfunctional pseudogenes in gorillas and at least one gibbon, which have one-male mating systems (Carnahan and Jensen-Seaman 2008). In humans, these genes are under neutral selection, perhaps indicating a more recent promiscuous ancestor in the human than in the gorilla lineage, but inconsistent with sperm competition as a recent selective force.

Conclusions and Summary

Human and chimpanzee mating systems differ markedly in the duration and exclusivity of sexual relationships, the manifestation of sexual competition, and patterns of mate choice. These differences evince adaptations to pair-bonding in humans, and a promiscuous mating system in chimpanzees. For example, when women fight with other women, it is normally over relationships with men and the critical paternal investment that men provide. Chimpanzee females, by contrast, never fight over males, who provide little or no investment. Males in both species fight over access to females, but such conflict is more persistent and pervasive in chimpanzees. Human men are most likely to compete aggressively when they are first starting their reproductive careers. Women prefer more formidable men, whereas chimpanzee females do not show strong mate preferences. Men are attracted to younger women, who have longer reproductive careers ahead of them. Chimpanzee males prefer older females, who have a greater immediate chance of conception. The history of promiscuous mating in chimpanzees has produced a suite of physiological adaptations to sperm competition that are absent in humans.

Human foragers show a strong tendency toward monogamous pair-bonding, though divorce rates can be high early in marriages, leading to serial monogamy. The evolutionary importance of pair-bonding is ostensibly supported by low levels of canine and body mass dimorphism in humans, together with the lack of adaptations for sperm competition. Humans maintain significant sexual dimorphism in strength and musculature, however, suggesting either (1) that aggressive mating competition (and likely some degree of polygynous pair-bonding) was important in our evolutionary history, or (2) that the sexual division of labor has selected for increased muscularity in men to meet the physical demands of hunting. Supporting the role of aggression, strength differences between the sexes are most pronounced during the life history phase when men are first competing over mating opportunities and forming pair-bonds, not when men's foraging effort is most intense. Humans also show high levels of sexual dimorphism in imposing traits, such as facial hair and vocal pitch. These are not related to foraging, but in other primates are associated with polygyny and aggressive mating competition.

The hominin fossil record supports some degree of polygyny in the australopithecines and paranthropines, as several species appear to have maintained body mass dimorphism exceeding that of chimpanzees, and facial size dimorphism approaching that of gorillas. Although canine height is reduced early in the hominin lineage, a lack of canine dimorphism is not associated with any particular mating system in living primates. Canine reduction may reflect biomechanical constraints on the jaw resulting from a change in diet or a shift to the use of handheld weapons in male-male competition, facilitated by bipedalism. The fossil evidence is consistent with the hypothesis that, in the hominin lineage, pair-bonds initially evolved in a polygynous context, and prior to intense paternal provisioning, the sexual division of labor, and derived aspects of human life history, such as slow maturation and extended juvenile dependency.

Endnotes

1. Here we follow Kappeler and van Schaik (2002) in defining social organization in primates as the "size, sexual composition and spatiotemporal cohesion of a society," and social structure as the "pattern of social interactions and the resulting relationships among the members of a society."
2. If long-term bonds promote a preference for youth, then one might expect to see evidence for this preference in other pair-bonded primates, such as hamadryas baboons or gorillas. Hamadryas males do often target juvenile females for their initial harem formation (Kummer 1968), but this is not strong evidence for preference. These relationships are initially characterized by maternal, rather than sexual, behavior by the male, and may simply result from the intense competition to control females in a highly polygynous mating system.
3. Apicella (2014: figure 2) appears to show a peak in men's upper-body strength in the late thirties, but this is an artifact of imposing a poor-fitting quadratic regression curve on the data. A Loess curve through the same data indicates a peak in strength in the mid to late twenties, as in Marlowe (2000, 2010), though men appear to maintain peak strength later than in Marlowe's data. (Apicella's measure is a composite of strength and arm circumference measures.)

Acknowledgments

We thank Christopher Boehm, Kristen Hawkes, Richard Wrangham, and an anonymous reviewer for helpful comments on this chapter.

References

Addison, W. E. 1989. Beardedness as a factor in perceived masculinity. *Perceptual and Motor Skills* 68: 921–922.

Altmann, J., D. Schoeller, S. A. Altmann, P. Muruthi, and R. M. Sapolsky. 1993. Body size and fatness of free-living baboons reflect food availability and activity levels. *American Journal of Primatology* 30: 149–161.

Anderson, M. J., and A. F. Dixson. 2002. Sperm competition: Motility and the midpiece in primates. *Nature* 416: 496.

Anderson, M. J., J. Nyholt, and A. F. Dixson. 2004. Sperm competition affects the structure of the mammalian vas deferens. *Journal of Zoology* 264: 97–103.

Apicella, C. L. 2014. Upper-body strength predicts hunting reputation and reproductive success in Hadza hunter-gatherers. *Evolution and Human Behavior* 35: 508–518.

Apicella, C. L., and D. Feinberg. 2009. Voice pitch alters mate-choice-relevant perception in hunter-gatherers. *Proceedings of the Royal Society B* 276: 1077–1082.

Apicella, C. L., D. Feinberg, and F. W. Marlowe. 2007. Voice pitch predicts reproductive success in male hunter-gatherers. *Biology Letters* 3: 682–684.

Archer, J. 2004. Sex differences in aggression in real-world settings: A meta-analytic review. *Review of General Psychology* 8: 291–322.

Archer, J., and V. L. Thanzami. 2007. The relation between physical aggression, size and strength, among a sample of young Indian men. *Personality and Individual Differences* 43: 627–633.

Babel, M., G. McGuire, and J. King. 2014. Towards a more nuanced view of vocal attractiveness. *PLoS ONE* 9: e88616.

Benenson, J. F. 2013. The development of human female competition: Allies and adversaries. *Philosophical Transactions of the Royal Society B* 368: 20130079.

Betzig, L. 1989. Causes of conjugal dissolution: A cross-cultural study. *Current Anthropology* 30: 654–676.

Blurton Jones, N. G. 2016. *Demography and Evolutionary Ecology of Hadza Hunter-Gatherers*. Cambridge: Cambridge University Press.

Blurton Jones, N. G., F. Marlowe, K. Hawkes, and J. F. O'Connell. 2000. Paternal investment and hunter-gatherer divorce rates. In L. Cronk, N. Chagnon, and W. Irons, eds., *Adaptation and Human Behavior: An Anthropological Perspective*, 69–90. New York: Aldine.

Boehm, C. 1999. *Hierarchy in the Forest: The Evolution of Egalitarian Behavior.* Cambridge, MA: Harvard University Press.

Boesch, C., and H. Boesch-Achermann. 2000. *The Chimpanzees of the Taï Forest: Behavioral Ecology and Evolution.* Oxford: Oxford University Press.

Borgerhoff Mulder, M. 1992. Reproductive decisions. In E. A. Smith and B. Winterhalder, eds., *Evolutionary Ecology and Human Behavior.* New York: Aldine.

Brunet, M., F. Guy, D. Pilbeam, H. T. Mackaye, A. Likius, D. Ahounta, A. Beauvilain, C. Blondel, H. Bocherens, J.-R. Boisserie, L. de Bonis, Y. Coppens, J. Dejax, C. Denys, P. Duringer, V. Eisenmann, G. Fanoné, P. Fronty, D. Geraads, T. Lehmann, F. Lihoreau, A. Louchart, A. Mahamat, G. Merceron, G. Mouchelin, O. Otero, P. P. Campomanes, M. Ponce de Leon, J.-C. Rage, M. Sapanet, M. Schuster, J. Sudre, P. Tassey, X. Valentin, P. Vignaud, L. Ciriot, A. Zazzo, and C. Zollikofer. 2002. A new hominid from the Upper Miocene of Chad, Central Africa. *Nature* 418: 45–151.

Burbank, V. K. 1987. Female aggression in cross-cultural perspective. *Cross-Cultural Research* 21: 70–100.

Burbank, V. K. 1992. Sex, gender, and difference: Dimensions of aggression in an Australian aboriginal community. *Human Nature* 3: 251–277.

Buss, D. M. 1989. Sex differences in human mate preferences: Evolutionary hypotheses tested in 37 cultures. *Behavioral and Brain Sciences* 12: 1–49.

Campbell, A. 1995. A few good men: Evolutionary psychology and female adolescent aggression. *Ethology and Sociobiology* 16: 99–123.

Campbell, A. 1999. Staying alive: Evolution, culture, and women's intrasexual aggression. *Behavioral and Brain Sciences* 22: 203–252.

Carnahan, S. J., and M. I. Jensen-Seaman. 2008. Hominoid seminal protein evolution and ancestral mating behavior. *American Journal of Primatology* 70: 939–948.

Cartmill, M., and F. H. Smith. 2009. *The Human Lineage.* Hoboken, NJ: Wiley-Blackwell.

Chapais, B. 2008. *Primeval Kinship: How Pair-Bonding Gave Birth to Human Society.* Cambridge, MA: Harvard University Press.

Chapais, B. 2011. The evolutionary history of pair-bonding and parental collaboration. In C. Salmon and T. K. Shackelford, eds., *The Oxford Handbook of Evolutionary Family Psychology.* Oxford: Oxford University Press.

Chapais, B. 2013. Monogamy, strongly bonded groups, and the evolution of human social structure. *Evolutionary Anthropology* 22: 52–65.

Cieri. R. L., S. E. Churchill, R. G. Franciscus, T. Tan, and B. Hare. 2014. Craniofacial feminization and the origins of behavioral modernity. *Current Anthropology* 55: 419–443.

Clarys, J. P., A. D. Martin, M. J. Marfell-Jones, V. Janssens, D. Caboor, and D. T. Drinkwater. 1999. Human body composition: A review of adult dissection data. *American Journal of Human Biology* 11: 167–174.

Collins, S. A. 2000. Men's voices and women's choices. *Animal Behaviour* 60: 773–780.

Darwin, C. 1871. *The Descent of Man, and Selection in Relation to Sex*. London: Murray.

Delgado, R. A. 2006. Sexual selection in the loud calls of male primates: Signal content and function. *International Journal of Primatology* 27: 5–25.

Dittus, W. P. 2013. Arboreal adaptations of body fat in wild toque macaques *(Macaca sinica)* and the evolution of adiposity in primates. *American Journal of Physical Anthropology* 152: 333–344.

Dixson, A. F. 1997. Evolutionary perspectives on primate mating systems. *Annals of the New York Academy of Sciences* 807: 42–61.

Dixson, A. F. 2009. *Sexual Selection and the Origins of Human Mating Systems*. Oxford: Oxford University Press.

Dixson, A. F. 2012. *Primate Sexuality*, 2nd ed.. Oxford: Oxford University Press.

Dixson, B. J., and R. C. Brooks. 2013. The role of facial hair in women's perceptions of men's attractiveness, health, masculinity and parenting abilities. *Evolution and Human Behavior* 34: 236–241.

Dixson, A., B. Dixson, and M. Anderson. 2005. Sexual selection and the evolution of visually conspicuous sexually dimorphic traits in male monkeys, apes, and human beings. *Annual Review of Sex Research* 16: 1–19.

Dixson, B. J., and P. L. Vasey. 2012. Beards augment perceptions of men's age, social status, and aggressiveness, but not attractiveness. *Behavioral Ecology* 23: 481–490.

Emery Thompson, M., J. H. Jones, A. E. Pusey, S. Brewer-Marsden, J. Goodall, D. Marsden, T. Matsuzawa, T. Nishida, V. Reynolds, Y. Sugiyama, and R. W. Wrangham. 2007. Aging and fertility patterns in wild chimpanzees provide insights into the evolution of menopause. *Current Biology* 17: 2150–2156.

Emery Thompson, M., and R. W. Wrangham. 2008. Male mating interest varies with female fecundity in *Pan troglodytes* of Kanyawara, Kibale National Park. *International Journal of Primatology* 29: 885–905.

Evans, S., N. Neave, and D. Wakelin. 2006. Relationships between vocal characteristics and body size and shape in human males: An evolutionary explanation for a deep male voice. *Biological Psychology* 72: 160–163.

Fedurek, P., K. E. Slocombe, D. K. Enigk, M. Emery Thompson, R. W. Wrangham, and M. N. Muller. 2016. The relationship between testosterone and long-distance calling in wild male chimpanzees. *Behavioral Ecology and Sociobiology* 70: 659–672.

Feinberg, D. R., B. C. Jones, M. J. Law Smith, F. R. Moore, L. M. DeBruine, R. E. Cornwell, S. G. Hillier, and D. I. Perret. 2006. Menstrual cycle, trait estrogen level, and masculinity preferences in the human voice. *Hormones and Behavior* 49: 215–222.

Feinberg, D. R., B. C. Jones, A. C. Little, D. M. Burt, and D. I. Perrett. 2005. Manipulations of fundamental and formant frequencies influence the attractiveness of human male voices. *Animal Behaviour* 69: 561–568.

Fessler, D. M. T., C. Holbrook, and M. M. Gervais. 2014. Men's physical strength moderates conceptualizations of prospective foes in two disparate societies. *Human Nature* 25: 393–409.

Fink, B., N. Neave, and H. Seydel. 2007. Male facial appearance signals physical strength to women. *American Journal of Human Biology* 19: 82–87.

Fitch, W. T., and D. Reby. 2001. The descended larynx is not uniquely human. *Proceedings of the Royal Society B* 268: 1669–1675.

Frayer, D. W. 1980. Sexual dimorphism and cultural evolution in the Late Pleistocene and Holocene of Europe. *Journal of Human Evolution* 9: 399–415.

Frederick, D., and M. Haselton. 2007. Why is muscularity sexy? Tests of the fitness indicator hypothesis. *Personality and Social Psychology Bulletin* 33: 1167–1183.

Gallup, A., D. White, and G. Gallup. 2007. Handgrip strength predicts sexual behavior, body morphology, and aggression in male college students. *Evolution and Human Behavior* 28: 423–429.

Gallup, G. G., R. L. Burch, M. L. Zappieri, R. A. Parvez, M. L. Stockwell, and J. A. Davis. 2003. The human penis as a semen displacement device. *Evolution and Human Behavior* 24: 277–289.

Gangestad, S. W., and G. J. Scheyd. 2005. The evolution of physical attractiveness. *Annual Review of Anthropology* 34: 523–548.

Goodall, J. 1986. *The Chimpanzees of Gombe: Patterns of Behavior*. Cambridge, MA: Harvard University Press.

Grabowski, M., K. G. Hatala, W. L. Jungers, and B. G. Richmond. 2015. Body mass estimates of hominin fossils and the evolution of human body size. *Journal of Human Evolution* 85: 75–93.

Grueter, C. C., B. Chapais, and D. Zinner. 2012. Evolution of multilevel social systems in nonhuman primates and humans. *International Journal of Primatology* 33: 1002–1037.

Gurven, M., and K. Hill. 2009. Why do men hunt? A re-evaluation of "Man the Hunter" and the sexual division of labor. *Current Anthropology* 50: 51–74.

Haile-Selassie, Y., B. M. Latimer, M. Alene, A. L. Deino, L. Gibert, S. M. Melillo, B. Z. Saylor, G. R. Scott, and C. O. Lovejoy. 2010. An early *Australopithecus afarensis* postcranium from Woranso-Mille, Ethiopia. *Proceedings of the National Academy of Sciences* 107: 12121–12126.

Haile-Selassie, Y., G. Suwa, and T. D. White, T. D. 2004. Late Miocene teeth from Middle Awash, Ethiopia, and early hominid dental evolution. *Science* 303: 1503–1505.

Harcourt, A. H., P. H. Harvey, S. G. Larson, and R. Short. 1981. Testis weight, body weight and breeding system in primates. *Nature* 293: 55–57.

Headland, T. N. 1987. Kinship and social behavior among Agta Negrito hunter-gatherers. *Ethnology* 26: 261–280.

Hess, N., C. Helfrecht, E. Hagen, A. Sell, and B. Hewlett. 2010. Interpersonal aggression among Aka hunter-gatherers of the Central African Republic: Assessing the effects of sex, strength, and anger. *Human Nature* 21: 330–354.

Hewlett, B. S. 1992. Husband-wife reciprocity and the father-infant relationship among Aka pygmies. In B. S. Hewlett, ed., *Father-Child Relations: Cultural and Biosocial Contexts,* 153–176. New York: Aldine.

Hill, K., and M. Hurtado. 1996. *Ache Life History*. New York: Aldine.

Hill, K., A. M. Hurtado, and R. S. Walker. 2007. High adult mortality among Hiwi hunter-gatherers: Implications for human evolution. *Journal of Human Evolution* 52: 443–454.

Hodges-Simeon, C. R., M. Gurven, R. Cardenas, D. A. Puts, and J. C. Gaulin. 2014. Vocal fundamental and formant frequencies are honest signals of threat potential in peripubertal males. *Behavioral Ecology* 25: 984–988.

Hodges-Simeon, C. R., D. A. Puts, and J. C. Gaulin. 2010. Perceptions of dominance and attractiveness depend on different parameters in the voice. *Human Nature* 21: 406–427.

Howell, N. 1979. *Demography of the Dobe !Kung*. New York: Aldine.

Hurtado, A. M., and K. R. Hill. 1987. Early dry season subsistence ecology of Cuiva (Hiwi) foragers of Venezuela. *Human Ecology* 15: 163–187.

Hurtado, A. M., and K. Hill. 1992. Paternal effect on offspring survivorship among Ache and Hiwi hunter-gatherers: Implications for modeling pair-bond stability. In B. S. Hewlett, ed., *Father-Child Relations: Cultural and Biosocial Contexts*, 31–55. New York: Aldine.

Hylander, W. L. 2013. Functional links between canine height and jaw gape in catarrhines with special reference to early hominins. *American Journal of Physical Anthropology* 150: 247–259.

Johanson, D. C., C. O. Lovejoy, W. H. Kimbel, T. D. White, S. C. Ward, M. E. Bush, B. M. Latimer, and Y. Coppens. 1982. Morphology of the Pliocene partial hominid skeleton (A.L. 288-1) from the Hadar Formation, Ethiopia. *American Journal of Physical Anthropology* 57: 403–451.

Jones, B. C., D. R. Feinberg, L. M. DeBruine, A. C. Little, and J. Vukovic. 2010. A domain-specific opposite-sex bias in human preferences for manipulated voice pitch. *Animal Behaviour* 79: 57–62.

Jones, D. 1995. Sexual selection, physical attractiveness and facial neoteny: Cross-cultural evidence and implications. *Current Anthropology* 36: 723–748.

Jones, D. 1996. An evolutionary perspective on physical attractiveness. *Evolutionary Anthropology* 5: 97–109.

Jungers, W. L., and R. L. Susman. 1984. Body size and skeleton allometry in African apes. In R. L. Susman, ed., *The Pygmy Chimpanzee*, 131–177. New York: Plenum.

Kahlenberg, S. M., and R. W. Wrangham. 2010. Sex differences in chimpanzees' use of sticks as play objects resemble those of children. *Current Biology* 20: R1067–R1068.

Kaplan, H., K. Hill, J. Lancaster, and A. M. Hurtado. 2000. A theory of human life history evolution: Diet, intelligence, and longevity. *Evolutionary Anthropology* 9: 156–185.

Kaplan, H. S., P. L. Hooper, and M. Gurven. 2009. The evolutionary and ecological roots of human social organization. *Philosophical Transactions of the Royal Society B* 364: 3289–3299.

Kappeler, P. 1990. The evolution of sexual size dimorphism in prosimian primates. *American Journal of Primatology* 21: 201–214.

Kappeler, P. M., and C. P. van Schaik. 2002. Evolution of primate social systems. *International Journal of Primatology* 23: 707–740.

Kim, J., S. Heshka, D. Gallagher, D. P. Kotler, L. Mayer, J. Albu, W. Shen, P. U. Freda, and S. B. Heymsfield. 2004. Intermuscular adipose tissue-free skeletal muscle mass: Estimation by dual-energy X-ray absorptiometry in adults. *Journal of Applied Physiology* 97: 655–660.

Kirchengast, S. 2000. Differential reproductive success and body size in !Kung San people from northern Namibia. *Collegium Antropologicum* 24: 121–132.

Knauft, B. M. 1987. Reconsidering violence in simple human societies: Homicide among the Gebusi of Papua New Guinea. *Current Anthropology* 28: 457–500.

Knauft, B. M. 1991. Violence and sociality in human evolution. *Current Anthropology* 32: 391–428.

Kummer, H. 1968. *Social Organization of Hamadryas Baboons*. Chicago: University of Chicago Press.

Lancaster, J. B., and H. Kaplan. 2009. The endocrinology of the human adaptive complex. In P. T. Ellison and P. B. Gray, eds., *Endocrinology of Social Relationships*. Cambridge, MA: Harvard University Press.

Langergraber, K. E., J. C. Mitani, D. P. Watts, and L. Vigilant. 2013. Male-female sociospatial relationships and reproduction in wild chimpanzees. *Behavioral Ecology and Sociobiology* 67: 861–873.

Lassek, W. D., and S. J. C. Gaulin. 2009. Costs and benefits of fat-free muscle mass in men: Relationship to mating success, dietary requirements, and native immunity. *Evolution and Human Behavior* 30: 322–328.

Lawler, R. R. 2009. Monomorphism, male-male competition, and mechanisms of sexual dimorphism. *Journal of Human Evolution* 57: 321–325.

Lawler, R. R., A. F. Richard, and M. A. Riley. 2005. Intrasexual selection in Verreaux's sifaka *(Propithecus verreauxi verreauxi)*. *Journal of Human Evolution* 48: 259–277.

Layton, R., S. O'Hara, and A. Bilsborough. 2012. Antiquity and social functions of multilevel social organization among human hunter-gatherers. *International Journal of Primatology* 33: 1215–1245.

Lebatard, A.-E., D. L. Boulès, P. Duringer, M. Jolivet, R. Braucher, J. Carcaillet, M. Schuster, N. Arnaud, P. Monié, F. Lihoreau, A. Likius, H. T. Mackaye, P. Vignaud, and M. Brunet. 2008. Cosmogenic nuclide dating of *Sahelanthropus tchadensis* and *Australopithecus bahrelghazali*: Mio-Pliocene hominids from Chad. *Proceedings of the National Academy of Sciences* 105: 3226–3231.

Lee, R. B. 1979. *The !Kung San: Men, Women, and Work in a Foraging Society*. Cambridge: Cambridge University Press.

Lehmann, J., G. Fickenscher, and C. Boesch. 2006. Kin biased investment in wild chimpanzees. *Behaviour* 143: 931–955.

Leigh, S. R., J. M. Setchell, M. Charpentier, L. A. Knapp, and E. J. Wickings. 2008. Canine tooth size and fitness in male mandrills *(Mandrillus sphinx)*. *Journal of Human Evolution* 55: 75–85.

Lockwood, C. A. 1999. Sexual dimorphism in the face of *Australopithecus africanus*. *American Journal of Physical Anthropology* 108: 97–127.

Lockwood, C. A., C. G. Menter, J. Moggi-Cecchi, and A. W. Keyser. 2007. Extended male growth in a fossil hominin species. *Science* 318: 1443–1445.

Lovejoy, C. O. 1981. The origin of man. *Science* 211: 341–350.

Machanda, Z. P., I. C. Gilby, and R. W. Wrangham. 2013. Male-female association patterns among free-ranging chimpanzees *(Pan troglodytes schweinfurthii)*. *International Journal of Primatology* 34: 917–938.

Marlowe, F. W. 2000. The patriarch hypothesis: An alternative explanation of menopause. *Human Nature* 11: 27–42.

Marlowe, F. W. 2003. The mating system of foragers in the standard cross-cultural sample. *Cross-Cultural Research* 37: 282–306.

Marlowe, F. W. 2004. Mate preferences among Hadza hunter-gatherers. *Human Nature* 15: 365–376.

Marlowe, F. W. 2005. Hunter-gatherers and human evolution. *Evolutionary Anthropology* 14: 54–67.

Marlowe, F. W. 2010. *The Hadza: Hunter-Gatherers of Tanzania*. Berkeley: University of California Press.

Marlowe, F. W., and J. C. Berbesque. 2012. The human operational sex ratio: Effects of marriage, concealed ovulation, and menopause on mate competition. *Journal of Human Evolution* 63: 834–842.

Matsumoto-Oda, A., M. Hamai, H. Hayaki, K. Hosaka, K. Hunt, E. Kasuya, K. Kawanaka, J. C. Mitani, H. Takasaki, and Y. Takahata. 2007. Estrus cycle asynchrony in wild female chimpanzees, *Pan troglodytes schweinfurthii*. *Behavioral Ecology and Sociobiology* 61: 661–668.

McFarland, R. 1997. Female primates: Fat or fit? In M. E. Morbeck, A. Galloway, and A. L. Zihlman, eds., *The Evolving Female*. Princeton, NJ: Princeton University Press.

Mitani, J. C., J. GrosLouis, and J. H. Manson. 1996. Number of males in primate groups: Comparative tests of competing hypotheses. *American Journal of Primatology* 38: 315–332.

Muller, M. N. 2002. Agonistic relations among Kanyawara chimpanzees. In C. Boesch, G. Hohmann, and L. Marchant, eds., *Behavioral Diversity in Chimpanzees and Bonobos*, 112–124. Cambridge: Cambridge University Press.

Muller, M. N., and M. Emery Thompson. 2012. Mating, parenting and male reproductive strategies. In J. C. Mitani, J. Call, P. M. Kappeler, R. A. Palombit, and J. B. Silk, eds., *Evolution of Primate Societies*, 387–411. Chicago: University of Chicago Press.

Muller, M. N., M. Emery Thompson, S. Kahlenberg, and R. W. Wrangham. 2011. Sexual coercion by male chimpanzees shows that female choice may be more apparent than real. *Behavioral Ecology and Sociobiology* 65: 921–933.

Muller, M. N., M. Emery Thompson, and R. W. Wrangham. 2006. Male chimpanzees prefer mating with old females. *Current Biology* 16: 2234–2238.

Muller, M. N., and J. C. Mitani. 2005. Conflict and cooperation in wild chimpanzees. *Advances in the Study of Behavior* 35: 275–331.

Muller, M. N., and R. W. Wrangham. 2004. Dominance, aggression and testosterone in wild chimpanzees: A test of the "Challenge Hypothesis." *Animal Behaviour* 67: 113–123.

Muñoz-Reyes, J. A., C. Gil-Burmann, B. Fink, and E. Turiegano. 2012. Physical strength, fighting ability, and aggressiveness in adolescents. *American Journal of Human Biology* 24: 611–617.

Murdock, G. P., and C. Provost. 1973. Factors in the division of labor by sex: A cross-cultural analysis. *Ethnology* 12: 203–225.

Murray, C. M., M. A. Stanton, E. V. Lonsdorf, E. E. Wroblewski, and A. E. Pusey. 2016. Chimpanzee fathers bias their behaviour towards their offspring. *Royal Society Open Science* 3: 160441.

Muscarella, F., and M. R. Cunningham. 1996. The evolutionary significance and social perception of male pattern baldness and facial hair. *Ethology and Sociobiology* 17: 99–117.

Nascimento, J. M., L. Z. Shi, S. Meyers, P. Gagneux, N. M. Loskutoff, E. L. Botvinick, and M. W. Berns. 2008. The use of optical tweezers to study sperm competition and motility in primates. *Journal of the Royal Society Interface* 5: 297–302.

Neave, N., and K. Shields. 2008. The effects of facial hair manipulation on female perceptions of attractiveness, masculinity, and dominance in male faces. *Personality and Individual Differences* 45: 373–377.

Nishida, T. 2012. *Chimpanzees of the Lakeshore: Natural History and Culture at Mahale.* Cambridge: Cambridge University Press.

Okada, D., and P. M. Bingham. 2008. Human uniqueness: Self-interest and social cooperation. *Journal of Theoretical Biology* 253: 261–270.

Opie, C., Q. D. Atkinson, R. I. M. Dunbar, and S. Shultz. 2013. Male infanticide leads to social monogamy in primates. *Proceedings of the National Academy of Sciences* 110: 13328–13332.

Palombit, R. A. 2012. Infanticide: Male strategies and female counterstrategies. In J. C. Mitani, J. Call, P. M. Kappeler, R. A. Palombit, and J. B. Silk, eds., *Evolution of Primate Societies,* 432–468. Chicago: University of Chicago Press.

Phillips, T., J. Li, and G. Kendall. 2014. The effects of extra-somatic weapons on the evolution of human cooperation towards non-kin. *PLoS ONE* 9: e95742.

Pisanski, K., P. J. Fraccaro, C. C. Tigue, J. J. M. O'Connor, S. Röder, P. W. Andrews, B. Fink, L. M. DeBruine, B. C. Jones, and D. R. Feinberg. 2014. Vocal indicators of body size in men and women: A meta-analysis. *Animal Behaviour* 95: 89–99.

Plavcan, M. J. 2001. Sexual dimorphism in primate evolution. *Yearbook of Physical Anthropology* 44: 25–53.

Plavcan, M. J. 2003. Scaling relationships between craniofacial sexual dimorphism and body mass dimorphism in primates: Implications for the fossil record. *American Journal of Physical Anthropology* 120: 38–60.

Plavcan, J. M. 2004. Sexual selection, measures of sexual selection, and sexual dimorphism in primates. In P. M. Kappeler and C. P. van Schaik, eds., *Sexual Selection in Primates: New and Comparative Perspectives*, 230–252. Cambridge: Cambridge University Press.

Plavcan, J. M. 2011. Understanding dimorphism as a function of changes in male and female traits. *Evolutionary Anthropology* 20: 143–155.

Plavcan, J. M. 2012a. Body size, size variation, and sexual size dimorphism in early *Homo*. *Current Anthropology* 53: S409–S423.

Plavcan, J. M. 2012b. Sexual size dimorphism, canine dimorphism, and male-male competition in primates: Where do humans fit in? *Human Nature* 23: 45–67.

Plavcan, J. M., C. A. Lockwood, W. H. Kimbel, M. R. Lague, and E. H. Harmon. 2005. Sexual dimorphism in *Australopithecus afarensis* revisited: How strong is the case for a human-like pattern of dimorphism? *Journal of Human Evolution* 48: 313–320.

Plavcan, J. M., and C. P. van Schaik. 1997a. Interpreting hominid behavior on the basis of sexual dimorphism. *Journal of Human Evolution* 32: 345–374.

Plavcan, J. M., and C. P. van Schaik. 1997b. Intrasexual competition and body size dimorphism in anthropoid primates. *American Journal of Physical Anthropology* 103: 37–68.

Plavcan, J. M., C. P. van Schaik, and P. M. Kappeler. 1995. Competition, coalitions and canine size in primates. *Journal of Human Evolution* 28: 245–276.

Pontzer, H., D. A. Raichlen, B. M. Wood, S. B. Racette, A. Z. P. Mabulla, and F. W. Marlowe. 2012. Hunter-gatherer energetics and modern human obesity. *PLoS ONE* 7: e40503.

Pusey, A. E., G. W. Oehlert, J. M. Williams, and J. Goodall. 2005. Influence of ecological and social factors on body mass of wild chimpanzees. *International Journal of Primatology* 26: 3–31.

Pusey, A. E., and K. Schroepfer-Walker. 2013. Female competition in chimpanzees. *Philosophical Transactions of the Royal Society B* 368: 20130077.

Puts, D. A. 2005. Mating context and menstrual phase affect women's preferences for male voice pitch. *Evolution and Human Behavior* 26: 388–397.

Puts, D. A. 2010. Beauty and the beast: Mechanisms of sexual selection in humans. *Evolution and Human Behavior* 31: 157–175.

Puts, D. A., C. L. Apicella, and R. A. Cardenas. 2012. Masculine voices are honest signals of men's threat potential in foraging and industrial societies. *Proceedings of the Royal Society B: Biological Sciences* 279: 601–609.

Puts, D. A., L. M. Doll, and A. K. Hill. 2014. Sexual selection on human voices. In V. Shackelford and T. K. Shackelford, eds., *Evolutionary Perspectives on Human Sexual Psychology and Behavior*, 69–86. New York: Springer.

Puts, D. A., S. J. C. Gaulin, and K. Verdolini. 2006. Dominance and the evolution of sexual dimorphism in human voice pitch. *Evolution and Human Behavior* 27: 283–296.

Puts, D. A., C. R. Hodges, R. A. Cárdenas, and S. J. C. Gaulin. 2007. Men's voices as dominance signals: Vocal fundamental and formant frequencies influence dominance attributions among men. *Evolution and Human Behavior* 28: 340–344.

Quinlan, R. J. 2008. Human pair-bonds: Evolutionary functions, ecological variation, and adaptive development. *Evolutionary Anthropology* 17: 227–238.

Quinlan, R. J., and M. B. Quinlan. 2007. Evolutionary ecology of human pair-bonds: Cross-cultural tests of alternative hypotheses. *Cross-Cultural Research* 41: 149–169.

Quinlan, R. J., and M. B. Quinlan. 2008. Human lactation, pair-bonds, and alloparents. *Human Nature* 19: 87–102.

Reno, P. L., R. S. Meindl, M. A. McCollum, and C. O. Lovejoy. 2003. Sexual dimorphism in *Australopithecus afarensis* was similar to that of modern humans. *Proceedings of the National Academy of Sciences* 100: 9404–9409.

Rodseth, L., R. W. Wrangham, A. M. Harrigan, and B. B. Smuts. 1991. The human community as a primate society. *Current Anthropology* 32: 221–254.

Rosenbaum, S., J. Silk, and T. S. Stoinski. 2011. Male-immature relationships in multimale groups of mountain gorillas *(Gorilla beringei beringei)*. *American Journal of Primatology* 73: 356–365.

Rosenbaum, S., J. P. Hirwa, J. B. Silk, L. Vigilant, and T. S. Stoinski. 2015. Male rank, not paternity, predicts male–immature relationships in mountain gorillas, *Gorilla beringei beringei*. *Animal Behaviour* 104: 13–24.

Salas-Wright, C. P., and M. G. Vaughn. 2016. Size matters: Are physically large people more likely to be violent? *Journal of Interpersonal Violence* 31: 1274–1292.

Saxton, T. K., P. G. Caryl, and S. C. Roberts. 2006. Vocal and facial attractiveness judgments of children, adolescents and adults: The ontogeny of mate choice issue. *Ethology* 112: 1179–1185.

Sayers, K., M. A. Raghanti, and C. O. Lovejoy. 2012. Human evolution and the chimpanzee referential doctrine. *Annual Review of Anthropology* 41: 119–138.

Sell, A., G. Bryant, L. Cosmides, J. Tooby, D. Sznycer, C. von Rueden, A. Krauss, and M. Gurven. 2010. Adaptations in humans for assessing physical strength and fighting ability from the voice. *Proceedings of the Royal Society B* 277: 3509–18.

Sell, A., L. Hone, and N. Pound. 2012. The importance of physical strength to human males. *Human Nature* 23: 30–44.

Sell, A., J. Tooby, and L. Cosmides. 2009. Formidability and the logic of human anger. *Proceedings of the National Academy of Sciences* 106: 15073–15078.

Senut, B., M. Pickford, D. Gommery, P. Mein, K. Cheboi, and Y. Coppens. 2001. First hominid from the Miocene (Lukeino formation, Kenya). *Comptes Rendus des Academies Sciences Paris, Sciences de la Terre et des planètes* 1: 1–9.

Seuanez, H., A. Carothers, D. Martin, and R. Short. 1977. Morphological abnormalities in spermatozoa of man and great apes. *Nature* 270: 345–347.

Shackelford, T. K., and A. T. Goetz. 2007. Adaptation to sperm competition in humans. *Current Directions in Psychological Science* 16: 47–50.

Simpson, S. C., L. Kleinsasser, J. Quade, N. E. Levin, W. C. McIntosh, N. Dunbar, S. Semaw, and M. J. Rogers. 2015. Late Miocene hominin teeth from the Gona Paleoanthropological Research Project area, Afar, Ethiopia. *Journal of Human Evolution* 81: 68–82.

Smith, F. 1980. Sexual differences in European Neanderthal crania with special reference to the Krapina remains. *Journal of Human Evolution* 9: 359–375.

Smith, R. L. 1984. Human sperm competition. In R. L. Smith, ed. *Sperm Competition and the Evolution of Animal Mating Systems.* New York: Academic Press.

Smith, R. J., and W. L. Jungers. 1997. Body mass in comparative primatology. *Journal of Human Evolution* 32: 523–559.

Snyder, J. K., D. M. T. Fessler, L. Tiokhin, D. A. Frederick, S. W. Lee, and C. D. Navarrete. 2011. Trade-offs in a dangerous world: Women's fear of crime predicts preferences for aggressive and formidable mates. *Evolution & Human Behavior* 32: 127–137.

Spoor, F., M. G. Leakey, P. N. Gathogo, F. H. Brown, S. C. Anton, I. McDougall, C. Kiarie, F. K. Manthi, and L. N. Leakey. 2007. Implications of new early *Homo* fossils from Ileret, east of Lake Turkana, Kenya. *Nature* 448: 688–691.

Stanford, C. B. 1996. The hunting ecology of wild chimpanzees: Implications for the behavioral ecology of Pliocene hominids. *American Anthropologist* 98: 96–113.

Stewart-Williams, S., and A. G. Thomas. 2013. The ape that thought it was a peacock: Does evolutionary psychology exaggerate human sex differences? *Psychological Inquiry* 24: 137–168.

Stumpf, R., and C. Boesch. 2005. Does promiscuous mating preclude female choice? Female sexual strategies in chimpanzees *(Pan troglodytes verus)* of the Taï National Park, Côte d'Ivoire. *Behavioral Ecology and Sociobiology* 57: 511–524.

Suwa, G., R. T. Kono, S. W. Simpson, B. Asfaw, C. O. Lovejoy, and T. D. White. 2009. Paleobiological implications of the *Ardipithecus ramidus* dentition. *Science* 326: 94–99.

Symons, D. 1995. Beauty is in the adaptations of the beholder: The evolutionary psychology of human female sexual attractiveness. In P. R. Abramson and S. D. Pinkerton, eds., *Sexual Nature, Sexual Culture.* Chicago: University of Chicago Press.

Thornhill, N. W., and R. Thornhill. 1990. An evolutionary analysis of psychological pain following rape, I: The effects of victim's age and marital status. *Evolution & Human Behavior* 11: 155–176.

Truswell, E. S., and J. D. L. Hansen. 1976. Medical research among the Kung. In R. B. Lee and I. DeVore, eds., *Kalahari Hunter-Gatherers,* 166–194. Cambridge, MA: Harvard University Press.

Tutin, C. E. G. 1979. Mating patterns and reproductive strategies in a community of wild chimpanzees. *Behavioral Ecology and Sociobiology* 6: 39–48.

Uehara, S., and T. Nishida. 1987. Body weights of wild chimpanzees *(Pan troglodytes schweinfurthii)* of the Mahale Mountains National Park, Tanzania. *American Journal of Physical Anthropology* 72: 315–321.

Walker, R. S., and K. R. Hill. 2003. Modeling growth and senescence in physical performance among the Ache of eastern Paraguay. *American Journal of Human Biology* 15: 196–208.

Walker, R. S., K. R. Hill, H. Kaplan, and G. MacMillan. 2002. Age-dependency in hunting ability among the Ache of Eastern Paraguay. *Journal of Human Evolution* 42: 639–657.

Watts, D. P. 2007. Effects of male group size, parity, and cycle stage on female chimpanzee copulation rates at Ngogo, Kibale National Park, Uganda. *Primates* 48: 222–231.

Wild, K. 2009. A failed attempt to force a consortship at Kanyawara. In M. N. Muller and R. W. Wrangham, eds., *Sexual Coercion in Primates and Humans*, 202–203. Cambridge, MA: Harvard University Press.

Wilson, M. L., C. Boesch, T. Furuichi, I. C. Gilby, C. Hashimoto, C. L. Hobaiter, G. Hohmann, N. Itoh, K. Koops, J. N. Lloyd, T. Matsuzawa, J. C. Mitani, D. C. Mjungu, D. Morgan, R. Mundry, M. N. Muller, M. Nakamura, J. Pruetz, A. E. Pusey, J. Riedel, C. Sanz, A. M. Schel, N. Simmons, M. Waller, D. P. Watts, F. White, R. Wittig, K. Zuberbühler, and R. W. Wrangham. 2014. Lethal aggression in *Pan* is best explained by adaptive strategies, rather than human impacts. *Nature* 513: 414–417.

Winking, J., H. Kaplan, M. Gurven, and S. Rucas. 2007. Why do men marry, and why do they stray? *Proceedings of the Royal Society B: Biological Sciences* 274: 1643–1649.

Wolff, S. E., and D. A. Puts. 2010. Vocal masculinity is a robust dominance signal in men. *Behavioral Ecology and Sociobiology* 64: 1673–1683.

Wolpoff, M. H. 1976. Some aspects of the evolution of hominid sexual dimorphism. *Current Anthropology* 17: 579–606.

Woodburn, J. 1979. Minimal politics: The political organization of the Hadza of north Tanzania. In W. A. Shack and P. S. Cohen, eds., *Politics in Leadership*. Oxford: Clarendon.

Woodburn, J. 1982. Egalitarian societies. *Man* 17: 431–451.

Wrangham, R. W. 2002. The cost of sexual attraction: Is there a trade-off in female *Pan* between sex appeal and received coercion? In C. Boesch, G. Hohmann, and L. Marchant, eds., *Behavioral Diversity in Chimpanzees and Bonobos*, 204–215. Cambridge: Cambridge University Press.

Wrangham, R. W., and L. Glowacki. 2012. Intergroup aggression in chimpanzees and war in hunter-gatherers. *Human Nature* 23: 5–24.

Wrangham, R. W., M. L. Wilson, and M. N. Muller. 2006. Comparative rates of violence in chimpanzees and humans. *Primates* 47: 14–26.

Wroblewski, E. E. 2010. Paternity and father-offspring relationships in wild chimpanzees, *Pan troglodytes schweinfurthii*. Ph.D. dissertation, University of Minnesota.

Wroblewski, E. E., C. M. Murray, B. F. Keele, J. C. Schumacher-Stankey, B. H. Hahn, and A. E. Pusey. 2009. Male dominance rank and reproductive success in chimpanzees, *Pan troglodytes schweinfurthii*. *Animal Behaviour* 77: 873–885.

Young, K. A., K. L. Gobrogge, Y. Liu, and Z. Wang. 2011. The neurobiology of pair bonding: Insights from a socially monogamous rodent. *Frontiers in Neuroendocrinology* 32: 53–69.

Zihlman, A. L., and D. Bolter. 2015. Body composition in *Pan paniscus* compared with *Homo sapiens* has implications for changes during human evolution. *Proceedings of the National Academy of Sciences* 112: 7466–7471.

Zihlman, A. L., and R. K. McFarland. 2000. Body mass in lowland gorillas: A quantitative analysis. *American Journal of Physical Anthropology* 113: 61–78.

12

From Chimpanzee Society to Human Society

Bridging the Kinship Gap

BERNARD CHAPAIS

In this chapter I examine the hypothesis that human society evolved from a chimpanzee-like society characterizing the last common ancestor (LCA) of the *Pan* and *Homo* lineages. Although I am concerned with the evolution of human social organization as a whole, the focus is on kinship throughout the paper. The reason for this is that kinship lies at the heart of human social organization, a statement that may not be immediately obvious considering the extent of the diversity of human societies and the fact that kinship plays a relatively minor role in many of them, especially large ones. But the expression "social organization" refers here to the set of traits that characterize human societies collectively and set them apart from all other animal societies. In other words, it refers to the cross-culturally *deep structure* of human society (Chapais 2008, 2010), the biologically grounded template upon which, presumably, cultural evolution built all past and present human societies. Kinship thus pervades the deep structure of human society.

The importance of kinship in human social organization stems from the very nature of the parameters defining primate social organizations—group size, age-sex composition, degree of spatial cohesion, mating system, dispersal patterns, between-group relations. In humans those parameters converge in generating the most extensive set of differentiated kin types—or domains of kin recognition—in the animal world. Humans form large mixed-sex groups including up to four generations of relatives; residence patterns produce extensive patrilines and matrilines; kin recognition encompasses both the maternal and paternal lines, it is independent of whether or not kin reside in the same local group, and it includes affines (in-laws). Not surprisingly, the unparalleled extent of kin recognition in humans correlates with an unprecedented impact of kinship on social life. Table 12.1 lists kinship-related traits characterizing the human species as a whole (for general discussions, see Fox 1967, 1980; Rodseth et al. 1991; Holy 1996; Chapais 2008, 2013, 2014, 2016b; Shenk and Mattison 2011; Stone 2013). I refer to this set of traits as the *human kinship configuration*. Most components are cross-culturally universal. A minority, such as matrimonial exchange and marital arrangements, are not, but they are the rule in small-scale societies (discussed below), and when they are absent the underlying parental motivation takes different forms (Apostolou 2013, 2014; Chapais 2014).

Given the central importance of kinship in human affairs, the idea that the deep structure of human society evolved from a chimpanzee-like social organization is not immediately obvious. There is indeed a pronounced discrepancy between chimpanzees and humans in the importance of kinship as a factor regulating social life (Table 12.1). The domain of kin recognition in chimpanzees appears to be extremely limited compared to the situation in humans. If only as a consequence of this, the role of kinship in the patterning of social relationships is comparatively minor. This dual discrepancy might seem to imply that the transition from a chimpanzee-like ancestor to the human kinship configuration involved a large number of social and cognitive changes. But, as will be argued here, the discrepancy may be parsimoniously accounted for. My aim in this paper is to show that two major features of human society, stable breeding bonds and peaceful contact between interbreeding groups, help bridge the gap to a considerable extent. After describing the respective domains of kin recognition in chimpanzees and humans, I show how the integration of the two features to a *Pan* social structure produces a primitive version of the human kinship configuration. I then address the origins of stable breeding bonds and peaceful between-group contact by presenting a general, five-stage evolutionary sequence, which begins with an ancestral *Pan* social structure—multimale-multifemale group,

TABLE 12.1. Main characteristics of the human kinship configuration compared with the situation in chimpanzees. Traits are explained in the text.

	Present in Chimpanzees
Features of human society affecting kinship	
Multimale-multifemale groups	Yes
Multigenerational groups	Yes
Stable breeding bonds and multifamily groups	No
Dual-phase residence (prebreeding/breeding)	Yes
Sex-biased philopatry/dispersal	Yes
Multigroup, multilevel social organization	No
Domain of kin recognition (differentiated kin types)	
Mother-offspring recognition	Yes
Father-offspring recognition	Fragmentary
Recognition of maternal siblings	Yes
Recognition of paternal siblings	No
Recognition of other matrilineal kin types	Unusual
Recognition of other patrilineal kin types	Unlikely
Intergroup kin recognition (i.e., independent of co-residence)	No
Recognition of kin's spouses (affines)	No
Recognition of spouses' kin (affines)	No
Recognition of affines independent of co-residence	No
Symmetrical recognition of affines (husband's and wife's sides)	No
Kinship phenomena	
Biparental family	No
Kin biases in cooperation	Yes
Incest avoidance between matrilineal kin	Yes
Incest avoidance between patrilineal kin	Unlikely
Lineal transmission of status	No
Kinship-based nested structure of rivalries and alliances	No
Bonds between kin and affines living in distinct groups	No
Kin biases in group fissions	Unlikely
Bisexual (ambisexual) dispersal	No
Bisexual (ambisexual) residence	Unusual
Lifetime bonds between cross-sex kin	No
Brother-sister kinship structure	No
Bilateral grandparenting	No
Preferential bonds between affines	No
Residential variability across societies	Limited
Kinship-regulated marital arrangements	No
Kinship terminologies	No
Recognition of lineages and descent groups	No

fission-fusion pattern of association, sexual promiscuity, male philopatry, female transfer, territoriality—and ends with modern human social structure. That sequence is compatible whether the LCA resembled chimpanzees (Wrangham and Pilbeam 2001), bonobos (Zihlman 1996), or both (Prüfer et al. 2012). I argue that monogamy in humans did not evolve directly from chimpanzee-like sexual promiscuity, as is commonly assumed, but from groups composed of several polygynous units, as in multilevel primate societies (e.g., hamadryas baboons, *Papio hamadryas*), and that monogamy was a prerequisite for the evolution of the first between-group alliances involving interpersonal bonds. I then turn to the far-reaching consequences of stable breeding bonds and peaceful between-group contact on human kinship by describing how they generated the majority of the features listed in Table 12.1. This paper constitutes an updated and original synthesis of many ideas presented elsewhere (Chapais 2008, 2010, 2011, 2013, 2014, 2015, 2016a, 2016b).

The Kin Recognition Gap

Figure 12.1a illustrates the minimal domain of kin recognition in humans. Ego differentiates all twenty-six depicted kin types in addition to many others such as Ego's great-grandparents, grandchildren, and great-grandchildren. Kin recognition in humans is *bilineal:* individuals recognize their kin through both the female line and the male line, which includes their ascending lineal kin (e.g., their maternal and paternal grandparents), descending lineal kin (e.g., their daughters' and sons' offspring), and collateral kin (e.g., their sisters' and brothers' offspring). Kin recognition in humans is also independent of co-residence, a phenomenon I refer to as *intergroup kin recognition:* Ego recognizes his kin whether he lives with them in the same local group or not. Considering the two points together, kin recognition in humans may be said to be *fully bilineal.*

Humans also recognize their affines, or in-laws (Fox 1980; Chapais 2008), a fact that considerably complexifies human kinship networks. A husband' affines include the spouses of his own kin: those of his children (daughters- and sons-in-law), siblings (brothers- and sisters-in-law), siblings' children (nieces- and nephews-in-law), and parents' siblings (aunts- and uncles-in-law), among others. The husband's affines also include his wife's relatives: her parents, siblings, and nieces and nephews, among others. The same applies to

FROM CHIMPANZEE SOCIETY TO HUMAN SOCIETY 431

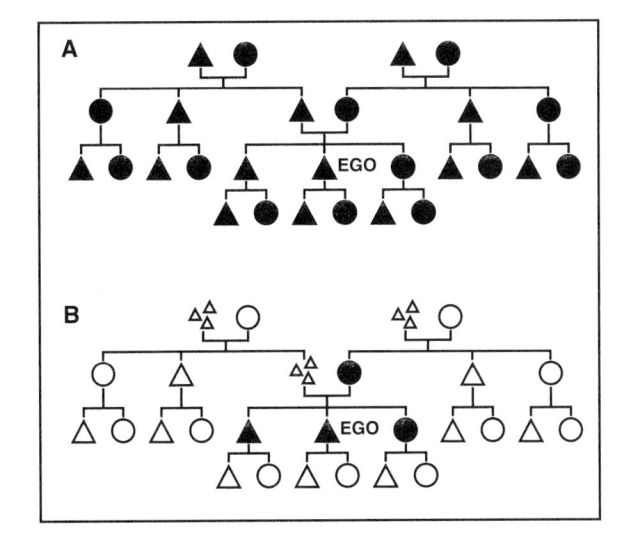

FIGURE 12.1. The domain of kin recognition in (a) humans and (b) chimpanzees from Ego's viewpoint, assuming complete male philopatry and complete female dispersal in chimpanzees. Triangles: males. Circles: females. Black symbols: kin types recognized by Ego. Sets of three triangles symbolize sexual promiscuity in chimpanzees.

the wife, who recognizes both the spouses of her kin and her husband's relatives. The recognition of affines is thus symmetrical, or bilateral. Exogamy being the rule in humans, many of Ego's affines reside in groups other than Ego's group, so that the symmetrical character of affinal kinship implies that it is independent of co-residence. Moreover, in addition to what may be called single-link (or "first-degree") affines, Ego also recognizes two-link (second-degree) affines (for example, his wife's brother's wife), and even third-degree affines (for example, the kin of the latter).

The contrast with the situation in chimpanzees and bonobos is striking. Both species display male philopatry and female dispersal (Doran et al. 2002; Stumpf 2011). This residence pattern creates genealogical structures that usually include small, two-generation *matrilines,* each composed of an immigrated mother and her offspring, but extensive *patrilines* composed of several kin types and spanning two to three generations. Male philopatry thus creates male kin groups, in which matrilineally related adult females form a very small proportion of all female dyads, as empirically documented by Langergraber et al. (2009). Figure 12.1b illustrates the domain of kin recognition in chimpanzees from male Ego's viewpoint. The figure assumes complete male philopatry and complete female dispersal. It also depicts four generations of individuals in order to include Ego's parents, offspring, siblings, nieces and nephews, uncles and aunts, and grandparents (chimpanzee groups normally include no more than three co-resident generations).

Matrilineal Kinship

The recognition of matrilineal kin in a chimpanzee group under complete male philopatry and female dispersal is limited to the small fraction of those that are co-resident, namely mother-offspring dyads and maternal sibling dyads. Empirical data show that mothers and daughters, mothers and sons, and maternal siblings maintain preferential relationships, that females avoid incest with their sons and brothers, and that mothers maintain lifetime bonds with their sons, the resident sex (Goodall 1986; Pusey 2001; Lehmann et al. 2006; Langergraber et al. 2007; Mitani 2009). Other matrilineal kin types, such as grandmothers and their daughters' offspring, or uncles and their maternal nieces live in different groups and do not meet. Significantly, however, anecdotal evidence indicates that when females fail to emigrate, and breed in their natal group, males may recognize their maternal grandmothers and maternal aunts, minimally (Goodall 1990: 112, 169–170). This shows that the actual domain of matrilineal kin recognition in chimpanzees reflects group composition constraints rather than limits in kin discrimination capabilities.

Matrilineal kin recognition is more extensive in female philopatric primate species such as macaques and baboons, and appears to involve principally familiarity-based processes (Berman 2004, 2011; Cheney and Seyfarth 2004; Rendall 2004; Chapais 2008: 40). In those species, young females come in contact with their sisters, grandmothers, aunts, and other kin types on a regular basis, thanks to the enduring bonds their mothers maintain with those kin. The mother's mediation sets in motion at least two distinct processes likely involved in kin recognition. First, it generates different levels of familiarity between the young female and her maternal kin types as the mother herself spends different amounts of time in proximity with them (Chapais 2001). Second, the young female's mother has different types of bonds with her own mother, daughters, and sisters, so that the young female may learn to differentiate kin types on the basis of how her mother associates and interacts with them (e.g., Cheney and Seyfarth 2004).

Importantly, both processes require mothers and daughters to maintain lifetime bonds (Chapais 2008). Those principles are general and, as will be discussed later, they were probably co-opted in the recognition of patrilineal and affinal kin types in the course of human evolution.

Patrilineal Kinship

Based on the available evidence, the domain of patrilineal kinship in chimpanzees appears to be even more limited than that of matrilineal kinship. Lehmann et al. (2006) compared the frequencies of grooming and play between father-offspring dyads and father-nonoffspring dyads for three different age categories of offspring in the Taï population of chimpanzees. Out of eight comparisons, only two showed significant, but weak, father-offspring biases. A study carried out in the Gombe Kasekala chimpanzee community similarly reported inconsistent results. Wroblewski (2010) found that juvenile and adolescent males did not exhibit significantly higher rates of association and affiliative behavior with their fathers than with unrelated males, nor lower rates of agonistic interactions. She also found that rates of total sexual activity were higher between daughters and fathers compared to unrelated males, in contravention of the principle of incest avoidance based on kin recognition. However, there was a nonsignificant trend for lower rates of successful copulations between daughters and fathers due to daughters resisting adult males in general, including their fathers. No such trends were found for other measures relating to mate guarding by males and to rates of association and grooming between males and maximally tumescent females. Recently, Murray et al. (2016) reported stronger evidence for paternity recognition in the same community. Fathers were found to spend more time with their offspring than with same-age offspring of unrelated mothers, but those differences were significant only for offspring aged less than six months. Further evidence for the inconsistent character of paternity recognition in chimpanzees comes from the observation that paternal half-brothers do not associate preferentially with each other (Lehmann et al. 2006; Langergraber et al. 2007), which they would be expected to do to some extent if they recognized their father through familiarity with him.

Taken together, these results suggest that paternity recognition in chimpanzees is highly fragmentary. Even positing a consistent pattern of paternity recognition, the observation that father-offspring associations "waned during infancy" and that by the time their offspring had reached eighteen months, fathers associated with the mother at levels "indistinguishable from [their] association with other lactating females" (Murray et al. 2016: 8) makes it unlikely that chimpanzees recognize their father's kin, such as their paternal grandfather, uncles, and cousins. Indeed, for Ego to recognize, for instance, his father's brother, Ego must maintain a long-term bond with his father so as to be in a position to recognize the preferential bond between his father and his paternal uncle. Moreover, Ego's capacity to recognize a patrilineal kin is expected to decrease steeply with the number of father-offspring links separating the two kin, because the inconsistent character of father-offspring recognition would likely have an additive effect across father-offspring links. I conclude that although chimpanzees live with various patrilineal kin types, they probably do not differentiate them, a conclusion that remains to be substantiated empirically.

Bridging the Kin Recognition Gap

The discrepancy between chimpanzees and humans as far as kin recognition is concerned can hardly be more pronounced, but, as will now be argued, much of it boils down, basically, to two major characteristics of human society: the existence of stable breeding bonds (which include monogamous, polygynous, and polyandrous unions) and the existence of peaceful contact between interbreeding groups manifested in intergroup visiting and meetings.

The effect of each factor on a chimpanzee-like society is examined separately. Figure 12.2a depicts the kin types Ego would differentiate in a hypothetical chimpanzee-like group in which stable breeding bonds substituted for sexual promiscuity. In this context, enduring associations between mothers and fathers make it possible for fathers and offspring to recognize each other systematically and to maintain an association in the long run. This in turn sets the stage for the recognition of patrilineal kin through familiarity-based processes similar to those involved in the recognition of matrilineal kin. Ego is now in a position to recognize his own offspring, his brother's offspring, and his father's relatives, including his paternal grand-

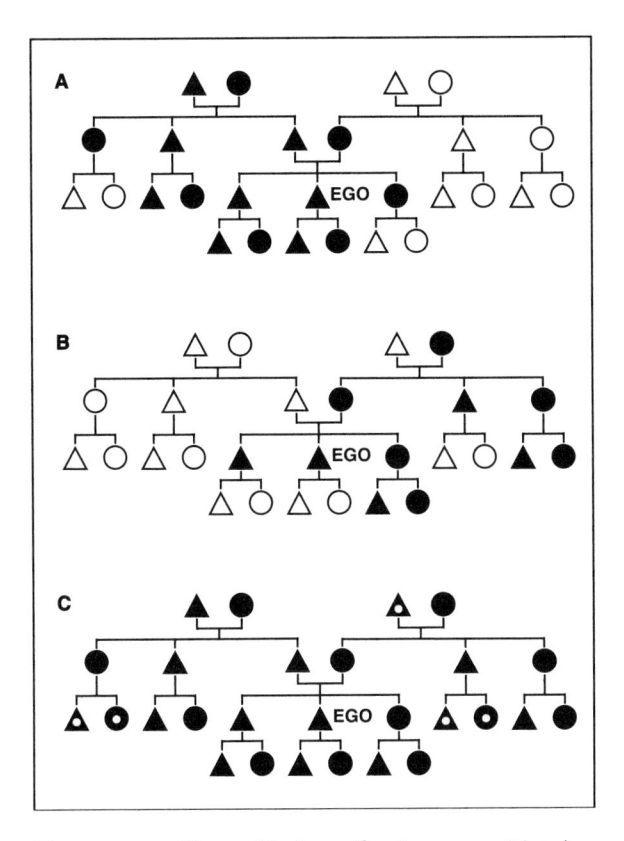

FIGURE 12.2. The separate effects of father-offspring recognition (as a correlate of pair-bonding) and peaceful contact between groups on kin recognition in a chimpanzee-like social organization. (a) Kin types differentiated by Ego (black symbols) assuming that pair-bonding substituted for sexual promiscuity. (b) Kin types differentiated by Ego assuming that peaceful contact between interbreeding groups substituted for avoidance and hostility. (c) The combined effects of both factors (recognition of kin types with a white dot requires both factors simultaneously). Diagram (c) corresponds to the human domain of kin recognition pictured in Figure 12.1a.

parents, uncles, aunts, and cousins. Those processes of kin recognition are not possible in chimpanzees, owing to their high levels of sexual promiscuity (Wrangham 2002) and the fact that they do not form pair-bonds (for a discussion of the latter point, see Muller and Pilbeam, this volume). Familiarity-based mechanisms of kin recognition would have paved the way for cognitively more sophisticated processes involving, for example, the conceptualization of the criteria used by primates to differentiate kin (sex, age, generation, lineal distance, collateral distance; Chapais 2016a) and language-based classifications of kin types (e.g., Fox 1979; Jones 2010).

Figure 12.2b illustrates the domain of kin recognition in a hypothetical population of chimpanzees practicing sexual promiscuity, but in which peaceful contact substituted for hostility between groups and dispersed kin maintain contact with each other through intergroup visiting. Compared to the situation in chimpanzees (Figure 12.1b), Ego may now recognize several of his mother's relatives in addition to his sister's offspring born in other groups.

Figure 12.2c shows the combined effects of both factors (stable breeding bonds and peaceful between-group contact) and identifies the few kin types whose recognition by Ego requires both factors simultaneously (mother's father, maternal uncle's offspring, and paternal aunt's offspring). Adding up the three figures produces the human domain of kin recognition pictured in Figure 12.1a.

The evolution of stable breeding bonds and peaceful contacts between interbreeding groups also explains the advent of affinal kinship. Pair-bonding is, by definition, an intrinsic feature of the recognition of affines. With regard to peaceful contact between groups, in their absence stable breeding bonds would generate a fragmentary domain of affinal kinship. In a patrilocal society for example, a man would recognize the wives of his co-resident kinsmen, but not the husbands of his kinswomen nor his wife's relatives, because they live in other groups. In all likelihood, the processes involved in the early recognition of affinal kin were similar to those governing the recognition of biological kin. The reason that Ego would have been able to recognize his sister's husband, for instance, was that he maintained an enduring bond with that sister, who herself maintained an enduring bond with her husband: the sister mediated proximity, familiarity, and affiliation between the two males (Chapais 2008, 2014).

A General Evolutionary Sequence

The idea that stable breeding bonds and peaceful contact between groups explain many of the profound discrepancies between chimpanzee and human kinship raises the question of the evolutionary history of those traits. From a comparative interspecific perspective, human societies may be defined as *federated communities of multifamily groups* (Chapais 2013), an expression purposefully emphasizing the importance of the two traits. Human groups are associations of breeding units, the vast majority of which are monogamous—they are communities of conjugal families (Rodseth et al. 1991), and more specifically, communities of predominantly monogamous families. Indeed, even though polygyny is practiced in the vast majority (80 percent) of human societies (Murdock and White 1969; Low 2003), only a minority of men have more than one wife (Marlowe 2003; Apostolou 2007; Layton et al. 2012).

The expression "federated communities" refers to multifamily groups being strongly bonded to each other and combining to form multileveled, nested social structures; for example, in hunter-gatherer societies, families assemble in bands that belong to multiband regional entities, which are themselves part of more inclusive (e.g., linguistic) entities such as tribes (Dunbar 1993; Hamilton et al. 2007). *Strong* bonds between groups refer to the existence of actual interpersonal relationships between non-co-resident individuals—notably between kin and between affines—those bonds being maintained through intergroup visiting and occasional but regular meetings between groups. Strong bonds are the essential ingredient of *between-group alliances*. Other important aspects of federated communities are that the number of organization levels is in theory unlimited, and that two or more groups may coordinate their actions in relation to other groups: there is multigroup coordination.

Other primate species, such as hamadryas baboons, gelada baboons *(Theropithecus gelada)*, Guinea baboons *(Papio papio)*, and Asian colobines *(Rhinopithecus* spp.), exhibit a multilevel social structure (Grueter et al. 2012; Patzelt et al. 2014; Fisher et al. 2017), but differ from human society in two important respects. In those species, groups are composed of several polygynous, or one-male units (OMUs), so that a number of males have no females at any point in time (Swedell 2011; Swedell and Plummer 2012)—polygyny is the rule. Human societies are the only multilevel primate societies in which reproductive units are predominantly monogamous. The other difference is that associations between multi-OMU groups amount to loose aggregations around common resources and appear to be exempt from actual interpersonal bonds (Grueter et al. 2012). For this reason, multi-OMU groups may be said to be *weakly* bonded (Chapais 2013). Between-group alliances involving dyadic bonds would thus be a uniquely human phenomenon. So is the unlimited number of nested levels of organization, which is fixed in other multilevel primate societies, and multigroup coordination, which is embryonic in those societies.

How did multifamily groups and strong alliances between multifamily groups evolve from a chimpanzee-like social structure? On strictly logical grounds, alliances between multimale-multifemale groups might have evolved first, followed by a transition from sexual promiscuity to monogamy, with this producing the first communities of multifamily groups. This sequence is unlikely, however, because relationships between patrilineally related kin and between affines are among the most powerful bonds connecting groups together in foraging societies (e.g., Hill et al. 2011), and both types of bonds could only come into being after stable breeding bonds had evolved.

Alternatively, monogamy might have substituted for sexual promiscuity in multimale-multifemale groups, followed by the evolution of alliances between the resulting mulitifamily groups. This sequence is also problematic, for it implies a direct transition from sexual promiscuity to monogamy. Such a transition does not appear to be documented in the evolutionary history of nonhuman primates. What is documented is a transition from sexual promiscuity to independent, territorial monogamous families (Schultz et al. 2011), a sequence not relevant for understanding the origin of cohesive groups of monogamous units. For ecological constraints to account for the fragmentation of large mixed-sex groups into monogamous units dispersing during the day and reuniting later, one must assume that those foraging units cannot support more than one adult female. This restrictive condition is not observed in other primates, in which foraging units are polygynous as a rule. Accordingly, models propounding a direct transition from sexual promiscuity to monogamy explain the very origin of the enduring bond between the father and the mother not in terms of ecologically driven mating strategies, but as a *parental* strategy. They invoke the benefits of paternal provisioning of mothers and children in response to an increase in the biological costs of raising children (Kaplan et al. 2000, 2009; Fisher 2006; Flinn et al. 2007; Lovejoy 2009; Gavrilets 2012). I refer to this principle as the *parental cooperation hypothesis*. There are two major problems with it.

The first problem is that this hypothesis implies a chronologically compressed view of the evolutionary history of the human family. All major components of the modern human family—the lengthened period of juvenile dependency, the enduring sexual bond, continuous female sexual receptivity, maternal provisioning of children, paternal provisioning of mothers, and sexual specialization in subsistence activities—are seen as evolving concomitantly. This implicitly assumes that because all components are intimately and causally related in the modern human family, they were necessary for the initial system to work and therefore evolved as a functional whole, or adaptive suite (Chapais 2011). Notwithstanding their present-day interconnectedness, the various components of the family may have evolved sequentially and contingently, in which case some alleged causal connections are spurious. In particular, phylogenetic analyses of the distribution of pair-bonding and paternal care in mammals point to a sequential and independent evolution of the two components (Brotherton and Komers 2003; Lukas and Clutton-Brock 2013), with monogamy evolving as a mating strategy and paternal care evolving subsequently, if need be. This accords with the observation that paternal care is absent in many monogamous and polygynous primates (van Schaik and Kappeler 2003). In this view, parental cooperation is a possible consequence rather than the cause of stable breeding bonds (Dunbar 1995), and the increase in the costs of maternity did not select for pair-bonding per se, but for paternal investment at a time when mothers and fathers were already bonded. (See also Hawkes 2004 for a functional argument on the evolution of human monogamy in terms of mating rather than parental effort.)

The second, related problem is that modeling the transition from sexual promiscuity to monogamy in terms of the benefits of paternal investment (Lovejoy 2009; Gavrilets 2012) requires an unusually large number of assumptions. One must indeed posit that several, if not all of the following conditions were met at about the same time: (1) the costs of rising children had begun to increase and did so relatively early in the human lineage; (2) individuals had the capacity to transport food; (3) the provisioning of mothers had a significant impact on the alleviation of the costs of raising children; (4) provisioning performed by the mothers' kin was insufficient, and (5) one possible reason for this was that males already contributed different food types (meat), which implies that hunting or scavenging was practiced on a regular basis; (6) males collected enough food to be in a position to share some of it with others; (7) females evolved continuous sexual receptivity as a means to engage in enduring food-for-sex exchanges; (8) females underwent profound motivational changes that translated into a dramatic lowering of their drive for mating with multiple males and led them to seek enduring cooperative relationships with particular males; (9) females mated with males conditionally upon those males provisioning them on a regular basis; (10) males also underwent profound motivational changes that resulted in a dramatic diminution of their drive for sexual promiscuity and led them to form enduring cooperative relationships with particular females; (11) females were faithful to males who provisioned them; and (12) males stayed with the same female long after she had conceived and paternity was ensured.

The point here is not that such changes could not have occurred in the course of human evolution—several of them did take place—but that it is extremely unlikely that they occurred concomitantly as part of the same adaptive suite. It is certainly possible to assume that this was indeed the case, build models in which values are assigned to the corresponding variables, and conclude on that basis that the transition is mathematically possible. The problem then lies in the validity of the primary assumption.

As a way of eliminating the problems raised by the promiscuity-to-monogamy transition, I proposed an alternative evolutionary sequence integrating an intermediate stage of polygyny between sexual promiscuity and monogamy. According to this model, stable breeding bonds—polygynous and monogamous—evolved as mating strategies as in other mammals, not as parental ones, and the various components of the family came into being progressively and cumulatively in the course of human evolution (Chapais 2008, 2011, 2013). This sequence was inferred on the basis of various structural and phylogenetic constraints, independently of paleoanthroplogical data, and for that reason it generates hypotheses testable against those data. The overall sequence is seen as the most parsimonious evolutionary pathway between a *Pan* social structure and human social structure; that is to say, there were certainly a larger number of intermediate stages but, arguably, not a smaller number, and the intermediate stages would be necessary steps in the evolution of human society. Figure 12.3 depicts two closely related versions of the proposed sequence. In both versions, sexual promiscuity in chimpanzee-like social groups (phase 1) gives way to groups composed of several polygynous units, or multi-OMU groups, with their associated peripheral males or all-male groups (phase 2). Next, the advent of various constraints on polygyny produces groups in which the reproductive units are mostly monogamous—multifamily groups (phase 3). This social structure in turn paves the way for the evolution of the first between-groups alliances involving interpersonal bonds, a stage I refer to as the primitive tribe (phase 4). Between the primi-

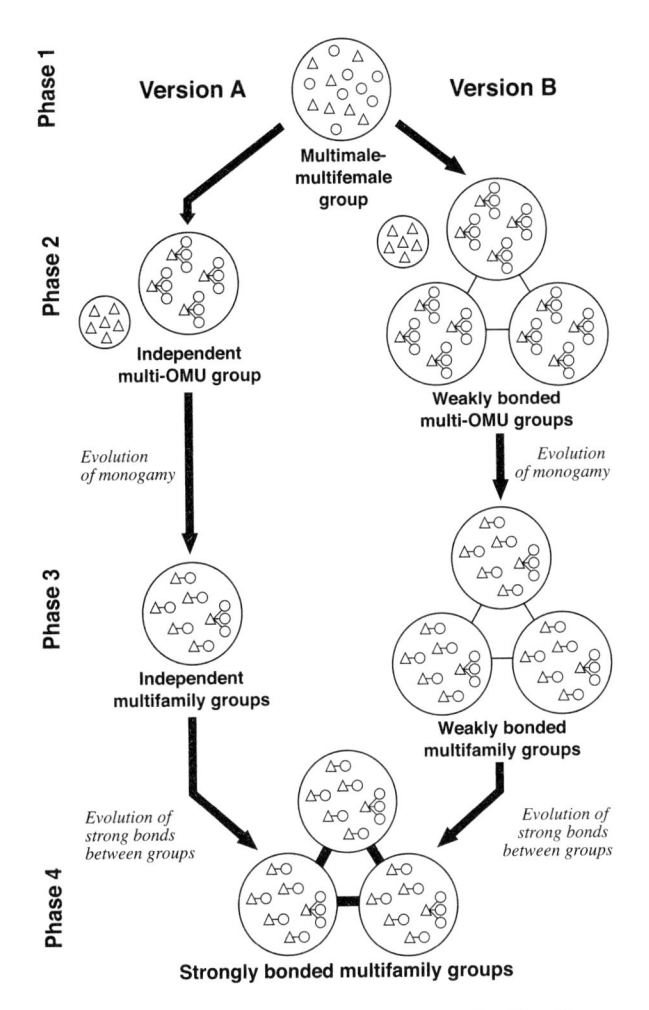

FIGURE 12.3. Two closely related evolutionary sequences leading from a chimpanzee-like social organization to human social organization. The number of phases shown is the minimum required. Phase 5 (Chapais 2013) has been omitted for the sake of clarity. In phase 2, multi-OMU groups are shown with a peripheral all-male group. Thin lines between groups represent loose aggregations in which groups are weakly bonded; thick lines represent between-group alliances based on interpersonal bonds between kin and affines.

tive tribe and modern human social structure—the federated community of multifamily groups (phase 5, not pictured in Figure 12.3)—lay a number of important characteristics, including matrimonial exchange and multigroup coordination. In the present paper, I concentrate on the first four phases, because stable breeding bonds and between-group alliances presumably evolved during those phases, but I nonetheless briefly discuss the origin of matrimonial exchange and marital arrangements. The two versions of the sequence presented in Figure 12.3 differ in that in phase 2, multi-OMU groups are pictured either as independent groups (version A), or as weakly bonded (version B), in line with the situation in some other living multilevel primate species such as hamadryas and Guinea baboons.

The Evolution of Stable Breeding Bonds

Several arguments support the feasibility of the transition from sexual promiscuity (phase 1) to multi-OMU groups (phase 2). Polygyny, rather than monogamy, is the rule in all multilevel primate species, presumably because the ecologically constrained spatial distribution of females in those species, compared to a situation of territorial monogamy, allows males to herd or defend more than one female simultaneously. Second, the transition from the sexually promiscuous multimale-multifemale group to the multi-OMU structure receives support from a phylogenetic analysis of the evolution of multi-OMU groups in hamadryas and gelada baboons from a multimale-multifemale ancestor (Barton 1999; Grueter et al. 2012). Third, various theoretical models accounting for the transition from multimale-multifemale groups to groups composed of several breeding units—ecological model, time constraint model, social model, and social brain model (reviewed in Grueter et al. 2012)—predict the formation of polygynous, not monogamous, units. Finally, paleoanthropological evidence supports the view that the evolutionary sequence leading to modern human social structure included one or more phases of polygyny. Data on body mass indicate that *Australopithecus afarensis* and *Au. africanus* were more sexually dimorphic than *Homo erectus*, humans, and chimpanzees, but less so than gorillas (reviewed in Muller and Pilbeam, this volume; Grabowski et al. 2015). Based on various socioecological arguments, a primate multilevel social structure has been invoked to characterize specific hominin species: *Australopithecus* spp. (Wilson and Glowacki, this volume), *H. erectus* (Swedell and Plummer 2012), and *H. heidelbergensis* (Layton et al. 2012).

If, then, monogamy (phase 3) evolved from a prior state of polygyny (phase 2) rather than from sexual promiscuity, this implies that models on its origins need not account for the origin of the stable breeding bond itself, which is already present, but for the reduction of the number of females monopolized by males. One must therefore look into the factors that might have increased the costs of polygyny, a line of inquiry that has been rather neglected in prior studies. I consider below four different types of costs and corresponding hypotheses about the origin of monogamy (Chapais 2008, 2011, 2013): the paternal load hypothesis, the leveling (or weapon) hypothesis, the trade-off (or cooperation) hypothesis, and the ecological (or herding) hypothesis. Considering that the processes described under each hypothesis are expected to have had a cumulative impact on the transition from polygyny to monogamy, my objective is not to argue in favor of a particular hypothesis, but to illustrate the range of possibilities raised by the polygyny-to-monogamy transition.

Like the parental cooperation hypothesis, the *paternal load hypothesis* posits a rise in the costs of raising children, for example, as a result of increased body size and/or brain size, slower maturation rates, and increased altriciality, except that it posits a prior state of polygyny. In that context, the leaders of polygynous units would have been selected to provision their females. The provisioning of several females, however, would have been costly—if possible in the first place—and rather inefficient from the females' viewpoint. Females belonging to smaller polygynous units, or those monogamously mated, would have received larger shares of the paternal contribution, with this creating selective pressures for the evolution of smaller polygynous units up to the unit size that maximizes paternal investment from the female's viewpoint—the monogamous pair. Contrary to the parental cooperation hypothesis, the paternal load hypothesis explains the reduction in the size of breeding units, not the origin of pair-bonding.

The *leveling* (or *weapon*) *hypothesis* invokes a dramatic rise in the costs of physical aggression between males following the invention and systematic use of weapons against conspecifics. In mammals in general, clear discrepancies in dominance between males translate into significant differences in their capacity to defend females, from which it follows that any factor reducing discrepancies in physical strength among males should reduce the variance in the size of polygynous units. Because weapons considerably increase the capacity of aggressors to inflict lethal wounds while minimizing the time required to do so, they help compensate for a deficit in physical strength. The use of lethal weapons by hominins would have substantially reduced discrepancies in actual power between larger and smaller individuals. In conflicts between equally armed opponents, stronger individuals would still have been able to overpower weaker ones, but the risks of fatal wounds to dominants would have increased substantially, and, perhaps more importantly, subordinates acting on their own would have been in position to easily wound or kill stronger opponents when the latter were vulnerable (e.g., inattentive or sleeping). Before the advent of lethal weapons, killing an inattentive but stronger individual would have been extremely risky, as suggested by the observation that when chimpanzees attack strangers, they outnumber the victim by a median 8:1 ratio (Wilson et al. 2014), they do not use weapons, and the killing involves a lot of fighting (Wrangham 1999; Wilson and Wrangham 2003; Wilson and Glowacki, this volume). Moreover, the efficiency of projectile weapons such as bows and arrows or hand-thrown spears and darts (Shea 2006, Churchill and Rhodes 2009) depends less on sheer physical power than on skill. When male competition among hominins came to rely on the systematic use of weapons, the monopolization of several females by a single male would have become extremely risky, creating selective pressures for a reduction in the size of polygynous units.

The *trade-off* (or *cooperation*) *hypothesis* rests on the social costs of polygyny. In a typical dominance relation, the subordinate has virtually no leverage to prevent the dominant from monopolizing defensible resources. The situation is expected to be different if the two individuals cooperate on a regular basis in relation to various fitness-related objectives, and if cooperation is essential for attaining those objectives. Because the dominant partner is dependent on the lower-ranking one to satisfy primary needs, subordinates now have leverage: they may cooperate conditionally upon the dominants sharing their resources. The substantial increase in the range and importance of cooperative activities in the course of human evolution (Tomasello et al. 2009) may be expected to have substantially increased the social costs of male sexual competition. Prominent among these activities are hunting, with its high levels of cooperation, especially in terms of meat sharing (Gurven 2004), and the formation of within-group male coalitions as between-group conflicts intensified (e.g., Flinn et al. 2005). The resulting increase in mutual dependence among males would have favored the evolution of some trade-offs between sexual competition and cooperation, and the likely outcome would have been a reduced variance in the size of polygynous units, a drift toward monogamy.

TABLE 12.2. Four hypotheses about the processes involved in the evolution of monogamy (multifamily groups) from a prior state of polygyny (multi-OMU groups) in the course of human evolution.

Hypothesis	Changes Initiating the Transition	Nature of Costs of Polygyny	Consequences Selecting for Reduction in Unit Size
Paternal load hypothesis	Increased costs of raising children	Energetic costs of provisioning several females	Males unable to provision several females efficiently
Leveling hypothesis	Use of lethal weapons against conspecifics	Rise in biological costs of aggression	Dominant males unable to defend several females against other males
Trade-off hypothesis	Expansion of essential cooperative activities among males	Dominant, polygynous males losing cooperative partners or opportunities	Dominant males benefit from sharing females with competent partners
Ecological hypothesis	Ecological changes favoring small foraging units	Energetic costs of herding spatially dispersed females	Males unable to guard several females efficiently

The *ecological hypothesis* rests on the energetic costs of guarding several spatially dispersed females. Those costs would become prohibitive to polygynous males if, as a result of a change in ecological niche, individuals had to disperse widely over large home ranges to obtain their food and were thereby forced into monogamy. As mentioned earlier, in nonhuman primates, ecologically induced monogamy is associated with territorial monogamy, not with group living as in the present case, but the observation that nuclear families in hunter-gatherer societies may forage independently and coalesce on a regular basis (Low 2003) points to the adaptiveness of this pattern in certain circumstances.

Table 12.2 summarizes the four hypotheses. Positing a prior state of polygyny, and considering that an increase in the costs of maternity, the use of lethal weapons against conspecifics, a greater dependence on cooperative partners, and the spatial distribution of foragers over large territories are among many distinct phenomena that did occur during human evolution, it is difficult to ignore their negative impact on the profitability of polygyny and their role in the origin of monogamy. The polygyny-to-monogamy transition also has important implications for other models on the origin of monogamy. For example, some data support the view that monogamy in primates evolved to prevent infanticide, as, for instance, in gibbons (Borries et al. 2011). But for infanticide to explain the evolution of human monogamy *from a prior state of polygyny,* one must posit that monogamously mated females incurred lower risks of infanticide than polygynously mated females. This is not immediately obvious, given that in both situations females benefit from the protection of a male (Chapais 2013). From that perspective, if infanticide did play a role in the evolution of the human mating system, it would likely be in the transition from sexual promiscuity to polygyny.

The Evolution of Patrilineal Kinship

The evolution of stable breeding bonds deeply enriched the domain of kin recognition and, in consequence, the impact of kinship on social relations. It accounts for the existence of several features of the human kinship configuration (Table 12.1). As mentioned earlier, stable breeding bonds provide a reliable, familiarity-based means of father-offspring recognition, which in turn allows individuals to recognize their father's kin and brothers' kin. In humans, the recognition of patrilineal kin translates into group-wide *patrilineal kinship structures.* In their simplest forms, these may be defined as recurrent patterns of social interactions and relationships mapping a group's patrilineal structure. For example, patrilineal descent groups include all individuals descended from a common ancestor through the male line and sharing, as a result, a number of prerogatives and obligations (Fortes 1953; Fox 1967; Stone 2013).

Although patrilineal kinship structures have no counterparts in nonhuman primates, matrilineal kinship structures are observed in female philopatric species such as macaques and baboons (for reviews of the influence of matrilineal kinship on behavior, see Silk 2009; Berman 2011; Langergraber 2012), and their analysis provides insights into the origins of human patrilineal structures. Primate matrilineal kinship structures may be broken down into a number of more elementary features having their own biological underpinnings and evolutionary history (Chapais 2014, 2016a), namely: (1) a multimale-multifemale group composition maximizing the domain of matrilineal kin recognition; (2) a pattern of female philopatry and male dispersal producing spatially localized matrilines, with this resulting in (3) the female kin group being the outbreeding unit; (4) lifetime bonds between mothers and daughters, a necessary condition for (5) the operation of familiarity and association-based mechanisms of kin recognition centered on the mother as a point of reference and allowing the recognition of several matrilineal kin types (lineal and collateral); (6) the occurrence of various types of cooperative activities among females, such as grooming and alliance formation; (7) kinship biases in the patterning of those cooperative activities; (8) status differentials between females in the form of stable dominance relationships; (9) the transmission of status from mothers to daughters, based on consistent maternal favoritism in conflicts between daughters and lower-ranking females (Chapais 1992, 2004); and (10) a pattern of group fission along maternal kinship lines, resulting in the formation of two independent female kin groups (Okamoto 2004; Di Fiore 2012). Primate matrilineal kinship structures may be seen as emergent products of the co-occurrence of those features.

This analysis provides what may be called a *phylogenetic model* of hominin kinship structures, according to which some sort of matrilineal kinship structure was bound to emerge in any species meeting those conditions. The rationale underlying the construction of phylogenetic models may be summarized as follows: if a human social phenomenon has formal counterparts in other primate species, if the nonhuman forms may be shown to be emergent products of the combination of elementary components, and if several of those components are homologous or homoplasious between humans and other primates, then the nonhuman forms provide a model of the evolutionary origins of the human forms (Chapais 2014). Note that it is not the whole phenomenon itself (for example, the kinship structure) that is homologous or homoplasious between humans and other primates, but its evolutionary building blocks.

Applying this analysis to the origins of patrilineal kinship structures in hominins, one expects the latter to have emerged whenever the following ten elements co-occurred in the same species: (1) a multimale-multifemale group composition maximizing the domain of patrilineal kin recognition; (2) a pattern of male philopatry and female dispersal producing spatially localized patrilines, with this resulting in (3) the male kin group being the outbreeding unit; (4) lifetime bonds between fathers and sons, a necessary condition for (5) the operation of familiarity- and association-based mechanisms of kin recognition centered on the father as a point of reference and allowing the recognition of several patrilineal kin types; (6) the occurrence of various types of cooperative activities among males, such as grooming and alliance formation; (7) kinship biases in the patterning of those cooperative activities; (8) status differentials between males in the form of stable dominance relationships; (9) the transmission of status from fathers to sons based on consistent paternal favoritism in conflicts between sons and lower-ranking males; and (10) a pattern of group fission along patrilineal kinship lines producing independent male kin groups.

The ten components co-occur only in humans, but, remarkably, chimpanzees exhibit six of them, with the four missing ones stemming from the same source: a mating system featuring sexual promiscuity rather than stable breeding bonds. In the absence of enduring relationships between mothers and fathers, fathers and children cannot recognize each other on the basis of their common association with the same female, and hence cannot form enduring bonds (component 4). Such bonds in turn are a necessary condition for the patrilineal transmission of status (component 9), a well-documented and uniquely human phenomenon (e.g., Fox 1967; Stone 2013). Stable breeding bonds are also a prerequisite for the recognition of patrilineal kin (component 5), without which group fissions cannot proceed along paternal kinship lines (component 10). Considering that group fissions have been observed in chimpanzees (Goodall 1986; Wilson and Wrangham 2003), one may speculate that the recognition of patrilineal kin would translate into those fissions proceeding along paternal kinship line—for the same reasons they take place along maternal kinship lines in female philopatric species. Group fissions along paternal kinship lines are well documented in humans (Evans-Pritchard 1940; Chagnon 1979; Hunley et al. 2008; Walker and Hill 2014).

The Evolution of Between-Group Alliances

At this stage (phase 3) in the evolution of human social structure, multifamily groups were independent entities (version A in Figure 12.3), or weakly bonded ones (version B); there were no interpersonal bonds between kin or between affines living in distinct groups, no true between-group alliances. However, the recognition of patrilineal kinship would have created conditions favorable to the evolution of social bonds between interbreeding groups (Chapais 2008, 2010, 2013). After having developed intimate bonds with their close patrilineal kin over several years prior to emigrating—female chimpanzees emigrate at eleven years on average—females moved to a new group in which they formed stable breeding bonds with particular males. When the two groups came into contact, several adult males in both groups would have recognized many of their female relatives living in the other group—for example, their daughters, granddaughters, sisters, nieces—and refrained from attacking them and their newborn offspring. This *consanguinity route* to the pacification of between-group relations would have created a partial state of mutual tolerance between interbreeding groups, based on the continuation of kinship bonds between non-co-resident kin.

At the same time, between-group meetings would have allowed in-laws to recognize each other and instigated the *affinity route* to pacification. "Husbands" would meet their wives' close relatives and, reciprocally, the latter would meet the husbands of their close female kin. From an evolutionary perspective, a number of considerations suggest that affines (for example, brothers-in-law) were biased for developing preferential bonds. First, brothers-in-law would have become disproportionately familiar and predictable to each other thanks to their connection to the same female. Second, brothers-in-law shared a vested interest in the same female *and* her children, one as a brother and uncle, the other as a husband and father, and hence would have been de facto partners as far as protecting those vulnerable individuals was concerned. Third, the female herself had a vested interest in the well-being of both her husband and her brother, and therefore would be expected to interfere in conflicts between them. Fourth, owing to incest avoidance between siblings, the husband and the brother were not sexual competitors for that female. Finally, because they belonged to the same generation, brothers-in-law shared similar age-related interests, levels of experience, and social objectives. Together, those factors would have created biases for the formation of peaceful, if not friendly, relationships between brothers-in-law (compared to other random assortments of males), which would eventually have translated into cooperative activities. The affinity route to between-group pacification is particularly significant, as it involves relationships between adult males, the initiators of intergroup aggression in chimpanzees (Wrangham 1999; Wilson and Wrangham 2003; Wilson and Glowacki, this volume).

The impact of the consanguinity and affinity routes to between-group pacification is expected to have been substantially affected by the exact pattern of female transfer between groups, notably by the proportion of females in a particular group who transferred into the same group and by the extent to which transfer was bilateral between groups (Chapais 2008). Whatever the exact pace of pacification, however, the end result would have been a variable state of mutual tolerance between interbreeding groups, based on the existence of actual intergroup social bonds among non-co-resident kin and affines, a state I refer to as the *primitive tribe* (phase 4).

The Kinship Correlates of the Primitive Tribe

Following the evolution of the primitive tribe, interbreeding groups were pacified in relation to each other, whereas groups belonging to different tribes were not. This likely brought about a number of major changes with regard to human kinship and social organization as a whole. In nonhuman primates, philopatry and dispersal patterns are often strongly sex-biased (Pusey 2004), and as a result adult cross-sex kin (e.g., brothers and sisters) rarely live in the same local group. Philopatry and dispersal patterns are also relatively invariant across populations of a given species, and even across species of the same genus—for example, *Pan* (Doran et al. 2002; Stumpf 2011) and *Macaca* (Thierry et al. 2000). The advent of the primitive tribe among hominins set the stage for bisexual (or ambisexual) dispersal, in which both males and females moved freely between groups of the same tribe but, importantly, not between different tribes. Bisexual dispersal was a necessary condition for the diversification of postmarital residence patterns from ancestral patrilocality. It paved the way for matrilocality, bilocality, neolocality, and avunculocality. The case of matrilocality is particularly significant in this respect. In humans, matrilocality is associated with matrilineal descent (Schneider 1961), in which the line of authority and the inheritance of property run from men to their sisters' sons (it is avuncular), not from women to their daughters, as one would expect if matrilineal descent was the mirror image of patrilineal descent. The association of matrilocality with the political control of men implies that brothers and sisters living in different groups nonetheless maintain lifetime bonds. This in turn suggests that human-like matrilocality evolved only after the evolution of the primitive tribe and hence did not characterize early human society—*contra* Knight (2008) and Opi and Power (2008).

From the time both sexes could move between groups (bisexual dispersal), both sexes could stay and marry in their natal group (bisexual philopatry) without contravening the exogamy rule. For example, a brother and his sister could remain in their birth group and marry spouses coming from other groups. Under a rule of exogamy, bisexual dispersal is a necessary condition for bisexual philopatry. This reasoning accords with empirical data on actual residence patterns in hunter-gatherers. Based on a worldwide sample of thirty-two foraging societies, Hill et al. (2011) reported a pattern of bisexual dispersal and residence in which parents were often equally likely to live with their adult sons or daughters, and adult brothers and sisters were often co-resident (see also Marlowe 2004; Walker et al. 2012).

The capacity of cross-sex kin to maintain lifetime bonds whether they were co-resident or not had profound consequences for human kinship. Lifetime bonds between brothers and sisters enabled brothers to recognize their sororal nieces and nephews, sisters to recognize their fraternal nieces and nephews, and cross-cousins to recognize each other (cross-cousins are the offspring of a brother and a sister, as opposed to parallel cousins, the offspring of two brothers or two sisters). This was a necessary condition for the evolution of what I refer to as the *brother-sister kinship structure,* a set of preferential relationships, matrimonial privileges, matrimonial proscriptions, and terminological classifications stemming from the brother-sister bond (Chapais 2008: figure 7.1), which Lévi-Strauss (1969: 124) described as "being second only to the incest prohibition" in its universal character. For example, preferential bonds between maternal uncles and their sororal nephews— avuncular relationships—are cross-culturally widespread, occupy a central place in anthropological theory (Lévi-Strauss 1963, 1969; Bloch and Sperber 2003), and constitute a particularly original aspect of human society (Fox 1993; Chapais 2008, 2010). Such bonds could evolve only after non-co-resident brothers and sisters were in a position to maintain lifetime bonds (phase 4).

The advent of intergroup kinship networks would also have created several opportunities for cooperative activities between individuals living in different groups of the same primitive tribe, including alloparenting, reciprocal access to territories and resources, coordination in subsistence activities, and alliance formation in the context of conflicts between groups of different tribes. To take an example, in removing co-residence as a necessary condition for alloparenting, the primitive tribe allowed *bilateral grandparenting*— performed by all four categories of grandparents. In other primate species forming multimale-multifemale groups, relationships between grandparents and grandoffspring are limited to maternal grandmothers (e.g., Fairbanks

1988; Nozaki 2009) because only they, among all four grandparents, simultaneously live with and recognize their grandchildren. From the time fathers could recognize their offspring (phase 2) and parents could maintain lifetime bonds with their sons and daughters (phase 4), any child was in a position to recognize his two sets of grandparents. Empirical studies indicate that grandparenting in humans may be performed by all four grandparents, but that the relative importance of each type of bond varies considerably across societies (Sear and Mace 2008; Hrdy 2009; Strassmann and Garrard 2011).

Human Kinship beyond the Primitive Tribe

At that stage in the evolution of human society (phase 4), some important aspects of between-group alliances were still lacking. Intergroup relations in contemporary human societies go well beyond mutual tolerance, interactional biases, and even preferential cooperative bonds. They notably involve agreements between families about the formation of marital unions. Cross-cultural studies indicate that marital arrangements are the rule in hunter-gatherer societies (Apostolou 2007) and in horticultural, agricultural, and pastoral societies (Apostolou 2010). A phylogenetic analysis of marriage practices in hunter-gatherer societies using mitochondrial DNA sequences suggests a deep history of arranged marriages going back at least to the first human migrations out of Africa (Walker et al. 2011). How did marital arrangements evolve? Their high degree of variability across societies, coupled with the fact that they are not cross-culturally universal, suggests that they are cultural creations devoid of any evolutionary foundation. An evolutionary perspective, however, suggests otherwise, as will now be argued (see Chapais 2014 and 2015 for more detail).

Although marital arrangements take various cultural forms—sister (daughter) exchange, cross-cousin marriage, levirate, sororate, bride-price, bride service—all forms share a small number of fundamental properties: (1) they are planned and regulated by parents, (2) they are governed by the principle of contingent reciprocity, (3) children have a high value as commodities of exchange, and (4) both parents gain from children exchange, which explains why in many societies women participate actively in the organization of marital unions (Stone 2013). The high value of children reflects both the importance of securing a spouse in a species in which reproduction is based on enduring breeding bonds and involves high levels of parental care, and the fact that because parents maintain lifetime bonds with their children, exchanging them generates long-lasting alliances between affines.

A phylogenetic perspective suggests that matrimonial exchange evolved when the components of human kinship characterizing the primitive tribe later combined with two uniquely human abilities: contingent reciprocity (reviewed in Silk 2007; see also Brosnan et al. 2009) and the *instructional control of others,* which I define as the capacity to orient the behavior of others by spelling out directives—in nonhuman primates the control of others is noninstructional (Chapais 2014, 2015). Instructional control in humans involves language, thanks to which specific directives may be spelled out, and a full-blown theory of mind, which makes it possible to know, and hence attempt to affect, the thoughts and beliefs of others; chimpanzees and other apes exhibit several, but not all, of the skills involved in human-like theory of mind (Call and Tomasello 2008; Krupenye et al. 2016). After these (and probably other) cognitive abilities had evolved in the human lineage (phase 5), individuals were in a position to orient the thoughts and behaviors of dependent and less powerful individuals in ways that satisfied their self-interest. The parent-child relationship would have lent itself particularly well to instructional control, owing to the large age difference between parents and children and the high degree of dependence of children on parents.

In the proposed model, the dual capacity to control children and to engage in contingent reciprocity, in combination with the high value of children as commodities of exchange, generates a universal parental propensity for controlling mate choice. That propensity would be similar to many others, such as interrupting children's disputes, inducing children to take care of their younger siblings, contribute to subsistence activities, and adopt specific beliefs. All such propensities would be internally (intuitively) generated, that is, individually learned by all parents; they would be neither Darwinian adaptations nor cultural creations. They would be universal because their emotional and cognitive underpinnings and the socioecological contexts eliciting them (children forming pair-bonds) are themselves cross-culturally universal (Chapais 2014).

Even though the parental propensity for controlling mate choice is assumed to be universal, it is expected to translate into matrimonial exchange only when two conditions are met (Chapais 2014). First, parents must benefit from exchanging children, which is the case in a vast number of societies in which children exchange is a primary means of obtaining spouses, resources, services, and alliances (Lévi-Strauss 1969, 1985; Apostolou 2007, 2010); it is not expected in societies where families do not depend on marital exchange to obtain such benefits, as is often the case in Western societies. Second, parents must be in a position to impose their preferences on their children. If this condition is often met in preindustrial societies, where children are economically dependent on their parents, it is often not satisfied in Western societies, where children may be financially independent of their parents at an earlier age (Apostolou 2013, 2014). When the socioecological contexts are favorable, parents would take control of mate selection. Whenever they do so, families become dependent on each other to obtain spouses; they have no choice but to exchange children.

A further aspect of the present model is that individually learned social propensities are seen as operating as prime movers of cultural evolution. Specifically, universal social propensities would impose strong constraints on creativity and social learning, so that the higher the degree of harmony of a cultural innovation in relation to a social propensity, the higher the degree of appealingness of that innovation and the greater the probability that it will be adopted by others (Chapais 2017). In the present case, sister exchange, cross-cousin marriage, levirate, sororate, bride-price, and bride service would be among the many distinct cultural innovations that were highly congruent with the parental propensity for controlling mate choice and had widespread distributions as a result (Chapais 2014).

Conclusion

Chimpanzees and bonobos are our closest relatives, but, somewhat paradoxically, the role of kinship in their social life is much less important than the role of kinship in species much more distantly related to humans, such as macaques and baboons. This is largely a matter of social structure, as we saw. In female philopatric species, genetic relatedness translates into preferential social bonds and relatively complex kinship structures, because females maintain lifetime bonds with their mothers and recognize their matrilineal kin through her mediation. In male philopatric species such as chimpanzees, genetic relatedness cannot translate into kinship structures, because males do not maintain enduring bonds with their fathers. Were chimpanzees and bonobos to form pair-bonds, they would likely exhibit complex patrilineal kinship structures, which might be even more sophisticated than what is observed in macaques and baboons. Thus, the gap is certainly wide between chimpanzee kinship and human kinship, but at the same time it is relatively shallow. To a large extent, it appears to reflect structural constraints on the capacity to recognize kin, rather than cognitive limitations. The evolution of stable breeding bonds and peaceful between-group contact in the human lineage removed those structural constraints and resulted in a dramatic expansion of the domain of kin recognition, thereby launching the deployment of the human kinship configuration.

The proposition that major cognitive changes were not required for the evolution of primitive versions of most of the traits listed in Table 12.1 does not mean that human cognition played a minor role in the evolution of human kinship. The discussion of marital arrangements illustrates the importance of new cognitive abilities in producing uniquely human kinship traits, and fundamental ones at that. As another example, the conjunction of nonhuman primate kinship features with uniquely human cognitive (e.g., linguistic) abilities generated the universal set of principles governing the highly diversified array of kin classifications and kinship terminologies characterizing human societies (Fox 1967, 1979; Barnard 2008; Jones 2010). This enabled individuals to trace kinship through several generations of deceased ancestors, and on this basis to categorize themselves as members of the same lineage and descent group (Fortes 1953, 1983; Stone 2013; Chapais 2016a). Human cognition thus tremendously increased the impact of kinship on human social life.

Summary

There is a pronounced discrepancy between chimpanzees and humans in the importance of kinship as a factor regulating social life. The domain of kin recognition (differentiated kin types) in our closest relatives is extremely limited compared to the situation in humans, and the role of kinship in the patterning of social relationships is comparatively minor. This might lead one to infer that the transition from chimpanzee-like society to human society involved a large number of social and cognitive changes. As argued here, however, two major features of human society—stable breeding bonds and peaceful contact between groups—generate many fundamental features of human kinship, including patrilineal kinship structures, affinal kinship, postmarital residential diversity and, in combination with new cognitive abilities, marital arrangements. The integration of stable breeding bonds and between-group alliances to a chimpanzee-like social structure produces a simplified version of the domain of kin recognition in humans. The comparative analysis of human and nonhuman primate societies provides what I call *phylogenetic models* of human behavior, and such models, somewhat paradoxically, help explain not only the origins of social phenomena that we share with other primates, but of many others that are exclusively human.

Acknowledgments

I thank the editors and anonymous reviewers for helpful comments on the manuscript.

References

Apostolou, M. 2007. Sexual selection under parental choice: The role of parents in the evolution of human mating. *Evolution and Human Behavior* 28: 403–409.

Apostolou, M. 2010. Sexual selection under parental choice in agropastoral societies. *Evolution and Human Behavior* 31: 39–47.

Apostolou, M. 2013. Do as we wish: Parental tactics of mate choice manipulation. *Evolutionary Psychology* 11: 795–813.

Apostolou, M. 2014. The context-dependent expression of the predisposition to control mating. *Current Anthropology* 55: 765–766.

Barnard, A. 2008. The co-evolution of language and kinship. In N. J. Allen, H. Callan, R. Dunbar, and W. James, eds., *Early Human Kinship: From Sex to Social Reproduction*, 232–243. Oxford: Blackwell.

Barton, R. A. 1999. Socioecology of baboons: The interaction of male and female strategies. In P. M. Kappeler, ed., *Primate Males: Causes and Consequences of Variation and Group Composition*, 97–107. Cambridge: Cambridge University Press.

Berman, C. M. 2004. Developmental aspects of kin bias in behavior. In B. Chapais and C. M. Berman, eds., *Kinship and Behavior in Primates*, 317–346. New York: Oxford University Press.

Berman, C. M. 2011. Kinship: Family ties and social behavior. In C. J. Campbell, A. Fuentes, K. C. MacKinnon, S. K. Bearder, and R. M. Stumpf, eds., *Primates in Perspective,* 576–587. New York: Oxford University Press.

Bloch, M., and D. Sperber. 2003. Kinship and evolved psychological dispositions: The mother's brother controversy reconsidered. *Current Anthropology* 43: 723–748.

Borries, C., T. Savini, and A. Koenig. 2011. Social monogamy and the threat of infanticide in larger mammals. *Behavioral Ecology and Sociobiology* 65: 685–693.

Brosnan, S. F., J. B. Silk, J. Heinrich, M. C. Mareno, S. P. Lamberth, and S. J. Schapiro. 2009. Chimpanzees *(Pan troglodytes)* do not develop contingent reciprocity in an experimental task. *Animal Cognition* 12: 587–597.

Brotherton, P. M. N., and P. E. Komers. 2003. Mate guarding and the evolution of social monogamy in mammals. In U. H. Reichard and C. Boesch, eds., *Monogamy: Mating Strategies and Parternships in Birds, Humans and Other Mammals,* 42–58. Cambridge: Cambridge University Press.

Call, J., and M. Tomasello. 2008. Does the chimpanzee have a theory of mind? 30 years later. *Trends in Cognitive Sciences* 12: 187–192.

Chagnon, N. A. 1979. Mate competition, favoring close kin, and village fissioning among the Yanomamö Indians. In N. A. Chagnon and W. A. Irons, eds., *Evolutionary Biology and Human Behavior,* 86–131. North Sciutate, MA: Duxbury.

Chapais, B. 1992. Role of alliances in the social inheritance of rank among female primates. In A. H. Harcourt and F. B. M. de Waal, eds., *Coalitions and Alliances in Humans and Other Animals,* 29–60. New York: Oxford University Press.

Chapais, B. 2001. Primate nepotism: What is the explanatory value of kin selection? *International Journal of Primatology* 22: 223–229.

Chapais, B. 2004. How kinship generates dominance structures in macaques: A comparative perspective. In B. Thierry, M. Singh, and W. Kaumanns, eds., *Macaque Societies: A Model for the Study of Social Organization,* 186–208. Cambridge: Cambridge University Press.

Chapais, B. 2008. *Primeval Kinship: How Pair-Bonding Gave Birth to Human Society.* Cambridge, MA: Harvard University Press.

Chapais, B. 2010. The deep structure of human society: Primate origins and evolution. In P. M. Kappeler and J. B. Silk, eds., *Mind the Gap: Tracing the Origins of Human Universals,* 19–51. Berlin: Springer.

Chapais, B. 2011. The evolutionary history of pair-bonding and parental collaboration. In C. Salmon and T. Shackelford, eds., *The Oxford Handbook of Evolutionary Family Psychology,* 33–50. Oxford: Oxford University Press.

Chapais, B. 2013. Monogamy, strongly-bonded groups, and the evolution of human social structure. *Evolutionary Anthropology* 22: 52–65.

Chapais, B. 2014. Complex kinship patterns as evolutionary constructions, and the origins of sociocultural universals. *Current Anthropology* 55: 751–783.

Chapais, B. 2015. *Liens de sang: Aux origines biologiques de la société humaine.* Montréal: Boréal.

Chapais, B. 2016a. The evolutionary origins of kinship structures. *Structure and Dynamics: eJournal of Anthropological and Related Sciences* 9: 33–51.

Chapais, B. 2016b. Universal aspects of kinship. In T. K. Shackelford and V. A. Weekes-Shackelford, eds., *Encyclopedia of Evolutionary Psychological Science*. New York: Springer.

Chapais, B. 2017. Psychological adaptations and the production of culturally polymorphic social universals. *Evolutionary Behavioral Sciences* 11: 63–82.

Cheney, D. L., and R. M. Seyfarth. 2004. The recognition of other individuals' kinship relationships. In B. Chapais and C. M. Berman, eds., *Kinship and Behavior in Primates*, 347–364. New York: Oxford University Press.

Churchill, S. E., and J. A. Rhodes. 2009. The evolution of the human capacity for killing at a distance: The human fossil evidence for the evolution of projectile weaponry. In J. J. Hublin and M. P. Richards, eds., *The Evolution of Hominin Diets: Integrating Approaches to the Study of Paleolithic Subsistence*, 201–210. Berlin: Springer.

Di Fiore, A. 2012. Genetic consequences of primate social organization. In J. Mitani, J. Call, P. M. Kappeler, R. A. Palombit, and J. B. Silk, eds.. *The Evolution of Primate Societies*, 269–292. Chicago: University of Chicago Press.

Doran, D. M., W. L. Jungers, Y. Sugiyama, J. G. Fleagle, and C. P. Heesy. 2002. Multivariate and phylogenetic approaches to understanding chimpanzee and bonobo behavioral diversity. In C. Boesch, G. Hohmann, and L. F. Marchant, eds., *Behavioural Diversity in Chimpanzees and Bonobos*, 14–34. Cambridge: Cambridge University Press.

Dunbar, R. I. M. 1993. The coevolution of neocortical size, group size and language in humans. *Behavioral and Brain Sciences* 16: 681–735.

Dunbar, R. I. M. 1995. The mating system of callitrichid primates, I: Conditions for the coevolution of pair bonding and twining. *Animal Behavior* 50: 1057–1070.

Evans-Pritchard, E. E. 1940. *The Nuer: A Description of the Modes of Livelihood and Political Institutions of a Nilotic People*. London: Oxford University Press.

Fairbanks, L. 1988. Vervet monkey grandmothers: Interactions with infant grandoffspring *International Journal of Primatology* 9: 425–441.

Fisher, H. 2006. *Why We Love: The Nature and Chemistry of Romantic Love*. New York: Henry Holt.

Fisher, J., G. H. Kopp, F. Dal Pesco, A. Goffe, K. Hammerschmidt, U. Kalbitzer, M. Klapproth, P. Marciej, I. Ndao, A. Patzel, and D. Zinner. 2017. Charting the neglected West: The social system of Guinea baboons. *Yearbook of Physical Anthropology* 162 Suppl 63: 15–31.

Flinn, M. V., D. C. Geary, and C. V. Wards. 2005. Ecological dominance, social competition, and coalitionary arms races: Why humans evolved extraordinary intelligence. *Evolution and Human Behavior* 26: 10–46.

Flinn, M. V., R. J. Quinlan, K. Coe, and C. V. Ward. 2007. Evolution of the human family: Cooperative males, long social childhoods, smart mothers, and extended kin networks. In C. A. Salmon and T. K. Shackelford, eds., *Family Relationships: An Evolutionary Perspective*, 16–38. Oxford: Oxford University Press.

Fortes, M. 1953. The structure of unilineal descent groups. *American Anthropologist* 55: 17–41.

Fortes, M. 1983. *Rules and the Emergence of Society,* London: Royal Anthropological Institute of Great Britain and Ireland.

Fox, R. 1967. *Kinship and Marriage: An Anthropological Perspective.* London: Penguin.

Fox, R. 1979. Kinship categories as natural categories. In N. A. Chagnon and W. Irons, eds., *Evolutionary Biology and Human Behavior,* 132–144. North Scituate, MA: Duxbury.

Fox, R. 1980. *The Red Lamp of Incest.* New York: Dutton.

Fox, R. 1993. *Reproduction and Succession: Studies in Anthropology, Law and Society.* New Brunswick, NJ: Transaction.

Gavrilets, S. 2012. Human origins and the transition from promiscuity to pair-bonding. *Proceedings of the National Academy of Sciences* 109: 9923–9928.

Goodall, J. 1986. *The Chimpanzees of Gombe: Patterns of Behavior.* Cambridge, MA: Harvard University Press.

Goodall, J. 1990. *Through a Window: My Thirty Years with the Chimpanzees of Gombe.* Boston: Houghton Mifflin.

Grabowski, M., K. G. Hatala, W. L. Jungers, and B. G. Richmond. 2015. Body mass estimates of hominin fossils and the evolution of human body size. *Journal of Human Evolution.* 85: 75–93.

Grueter, C. C., B. Chapais, and D. Zinner. 2012. Evolution of multilevel social systems in nonhuman primates and humans. *International Journal of Primatology* 33: 1002–1037.

Gurven, M. 2004. To give and to give not: The behavioral ecology of human transfers. *Behavioral and Brain Sciences* 27: 543–583.

Hamilton, M. J., B. T. Milne, R. S. Walker, O. Burger, and J. H. Brown. 2007. The complex structure of hunter-gatherer social networks. *Proceedings of the Royal Society B* 274: 2195–2202.

Hawkes, K. 2004. Mating, parenting, and the evolution of human pair bonds. In B. Chapais and C. B. Berman, eds., *Kinship and Behavior in Primates,* 443–473. New York: Oxford University Press.

Hill, K. R., R. S. Walker, M. Božičević, J. Eder, T. Headland, B. Hewlett, M. A. Hurtado, F. Marlowe, P. Wiessner, and B. Wood. 2011. Co-residence patterns in hunter-gatherer societies show unique human social structure. *Science* 33: 1286–1289.

Holy, L. 1996. *Anthropological Perspectives on Kinship.* London: Pluto.

Hrdy, S. B. 2009. *Mothers and Others: The Evolutionary Origins of Mutual Understanding.* Cambridge, MA: Harvard University Press.

Hunley, K. L., J. E. Spence, and D. A. Merriwether. 2008. The impact of group fissions on genetic structure in Native South America and implications for human evolution. *American Journal of Physical Anthropology* 135: 195–205.

Jones, D. 2010. Human kinship: From conceptual structure to grammar. *Behavioral and Brain Sciences* 33: 367–416.

Kaplan, H. S., K. R. Hill, J. B. Lancaster, and A. M. Hurtado. 2000. A theory of human life history evolution: Diet, intelligence and longevity. *Evolutionary Anthropology* 9: 156–185.

Kaplan, H. S., P. L. Hooper, and M. Gurven. 2009. The evolutionary and ecological roots of human social organization. *Philosophical Transactions of the Royal Society B* 364: 3289–3299.

Knight, C. 2008. Early human kinship was matrilineal. In N. J. Allen, H. Callan, R. Dunbar, and W. James, eds., *Early Human Kinship: From Sex to Social Reproduction*, 61–82. Oxford: Blackwell.

Krupenye, C., F. Kano, H. Satoshi, J. Call, and M. Tomasello. 2016. Great apes anticipate that other individuals will act according to false beliefs. *Science* 354: 110–114.

Langergraber, K. E. 2012. Cooperation among kin. In J. C. Mitani, J. Call, P. M. Kappeler, R. A. Palombit, and J. B. Silk, eds., *The Evolution of Primate Societies*, 491–513. Chicago: University of Chicago Press.

Langergraber, K. E., J. C. Mitani, and L. Vigilant. 2007. The limited impact of kinship on cooperation in wild chimpanzees. *Proceedings of the National Academy of Sciences* 104: 7786–7790.

Langergraber, K. E., J. C. Mitani, and L. Vigilant. 2009. Kinship and social bonds in female chimpanzees *(Pan troglodytes)*. *American Journal of Primatology* 71: 840–851.

Layton, R., S. O'Hara, and A. Bilsborough. 2012. Antiquity and social functions of multilevel social organization among human hunter-gatherers. *International Journal of Primatology* 33: 1215–1245.

Lehmann, J., G. Fickenscher, and C. Boesch. 2006. Kin biased investment in wild chimpanzees. *Behaviour* 143: 931–955.

Lévi-Strauss, C. 1963. *Structural Anthropology.* New York: Basic Books.

Lévi-Strauss, C. 1969. *The Elementary Structures of Kinship.* Boston: Beacon.

Lévi-Strauss, C. 1985. *The View from Afar.* Oxford: Blackwell.

Lovejoy, C. O. 2009. Reexamining human origins in light of *Ardipithecus ramidus*. *Science* 326: 108–115.

Low, B. S. 2003. Ecological and social complexities in human monogamy. In U. H. Reichard and C. Boesch, eds., *Monogamy: Mating Strategies and Partnerships in Birds, Humans and Other Mammals*, 161–176. Cambridge: Cambridge University Press.

Lukas, D., and T. Clutton-Brock. 2013. The evolution of social monogamy in mammals. *Science* 34: 526–530.

Marlowe, F. W. 2003. The mating system of foragers in the cross-cultural Standard Cross-Cultural Sample. *Cross-Cultural Research.* 37: 282–306.

Marlowe, F. W. 2004. Marital residence among foragers. *Current Anthropology* 45: 277–284.

Mitani, J. C. 2009. Male chimpanzees form enduring and equitable social bonds. *Animal Behavior* 77: 633–640.

Murdock, G. P., and D. R. White. 1969. Standard cross-cultural sample. *Ethnology* 8: 329–369.
Murray, C. M., M. A. Stanton, E. V. Lonsdorf, E. E. Wroblewski, and A. E. Pusey. 2016. Chimpanzee fathers bias their behavior towards their offspring. *Royal Society Open Science,* http://dx.doi.org/10.1098/rsos.160441.
Nozaki, M. 2009. Grandmothers care for orphans in a provisioned troop of Japanese macaques *(Macaca fuscata). Primates* 50: 85–88
Okamoto, K. 2004. Patterns of group fission. In B. Thierry, M. Singh, and W. Kaumanns, eds., *Macaque Societies: A Model for the Study of Social Organization,* 112–116. New York: Cambridge University Press.
Opi, K., and C. Power. 2008. Grandmothering and female coalitions: A basis for matrilineal priority? In N. J. Allen, H. Callan, R. Dunbar, and W. James, eds., *Early Human Kinship: From Sex to Social Reproduction,* 168–186. Oxford: Blackwell.
Patzelt, A., G. H. Kopp, I. Ndao, U. Kalbitzer, D. Zinner, and J. Fisher. 2014. Male tolerance and male-male bonds in a multilevel primate society. *Proceedings of the National Academy of Sciences* 111: 14740–14745.
Prüfer, K., K. Munch, I. Hellman, et al. 2012. The bonobo genome compared to the chimpanzee and human genomes. *Nature* 486: 527–531.
Pusey, A. E. 2001. Of genes and apes: Chimpanzee social organization and reproduction. In F. B. M. de Waal, ed., *Tree of Origin: What Primate Behavior Can Tell Us about Human Social Evolution,* 9–37. Cambridge, MA: Harvard University Press.
Pusey, A. E. 2004. Inbreeding avoidance in primates. In A. P. Wolf and W. H. Durham, eds., *Inbreeding, Incest, and the Incest Taboo: The State of Knowledge at the Turn of the Century,* 61–75. Stanford, CA: Stanford University Press.
Rendall, D. 2004. "Recognizing" kin: Mechanisms, media, minds, modules, and muddles. In B. Chapais and C. M. Berman, eds., *Kinship and Behavior in Primates,* 295–316. New York: Oxford University Press.
Rodseth, L., R. W. Wrangham, A. M. Harrigan, and B. B. Smuts. 1991. The human community as a primate society. *Current Anthropology* 32: 221–254.
Schneider, D. M. 1961. Introduction: The distinctive features of matrilineal descent groups. In D. M. Schneider and K. Gough, eds., *Matrilineal Kinship,* 1–29. Berkeley: University of California Press.
Shultz, S., C. Opie, and Q. D. Atkinson. 2011. Stepwise evolution of stable sociality in primates. *Nature* 479: 219–222.
Sear, R., and R. Mace. 2008. Who keeps children alive? A review of the effects of kin on child survival. *Evolution and Human Behavior* 29: 1–18.
Shea, J. J. 2006. The origins of lithic projectile point technology: Evidence from Africa, the Levant, and Europe. *Journal of Archaeological Sciences* 33: 823–846.
Shenk, M. K., and S. M. Mattison. 2011. The rebirth of kinship: Evolutionary and quantitative approaches in the revitalization of a dying field. *Human Nature* 22: 1–15.
Silk, J. B. 2007. The strategic dynamics of cooperation in primate groups. *Advances in the Study of Behavior* 37: 1–40.

Silk, J. B. 2009. Nepotistic cooperation in nonhuman primate groups. *Philosophical Transactions of the Royal Society B* 364: 3243–3254.

Strassman, B. I., and W. M. Garrard. 2011. Alternatives to the grandmother hypothesis. *Human Nature* 22: 201–222.

Stone, L. 2013. *Kinship and Gender*. Boulder, CO: Westview.

Stumpf, R. 2011. Chimpanzees and bonobos: Inter- and intraspecies diversity. In C. J. Campbell, A. Fuentes, K. C. MacKinnon, S. K. Bearder, and R. M. Stumpf, eds., *Primates in Perspective*, 340–356. New York: Oxford University Press.

Swedell, L. 2011. African papionins: Diversity of social organization and ecological flexibility. In C. J. Campbell, A. Fuentes, K. C. MacKinnon, S. K. Bearder, and R. M. Stumpf, eds., *Primates in Perspective*, 241–277. New York: Oxford University Press.

Swedell, L, and T. Plummer. 2012. A papionin multilevel society as a model for hominin social evolution. *International Journal of Primatology* 33: 1165–1193.

Thierry, B., A. N. Iwaniuk, and S. M. Pellis, S. M. 2000. The influence of phylogeny on the social behaviour of macaques (Primates: Cercopithecidae, genus *Macaca*). *Ethology* 106: 713–728.

Tomasello, M., A. P. Melis, C. Tennie, E. Wyman, and E. Herrmann. 2012. Two key steps in the evolution of human cooperation: The interdependence hypothesis. *Current Anthropology* 53: 673–692.

van Schaik, C. P., and P. M. Kappeler. 2003. The evolution of social monogamy in primates. In U. H. Reichard and C. Boesch, eds., *Monogamy: Mating Strategies and Partnerships in Birds, Humans and Other Mammals*, 59–80. Cambridge: Cambridge University Press.

Walker, R. S., S. Beckerman, M. V. Flinn, M. Gurven, C. R. von Rueden, K. L. Kramer, R. D. Greaves, L. Córdoba, D. Villar, and E. H. Hagen. 2012. Living with kin in lowland horticultural societies. *Current Anthropology* 54: 96–103.

Walker, R. S., and K. R. Hill. 2014. Causes, consequences, and kin biases of human group fissions. *Human Nature* 25: 465–475.

Walker, R. S., K. R. Hill, M. V. Flinn, and R. M. Ellsworth. 2011. Evolutionary history of hunter-gatherer marriage practices. *PLoS ONE* 6: e19066.

Wilson, M. L., C. Boesch, B. Fruth, T. Furuichi, I. C. Gilby, C. Hashimoto, C. Hobaiter, G. Hohmann, N. Itoh, K. Koops, J. Lloyd, T. Matsuzawa, J. C. Mitani, D. C. Mjungu, D. Morgan, R. Mundry, M. N. Muller, M. Nakamura, J. D. Pruetz, A. E. Pusey, J. Riedel, C. Sanz, A. M. Schel, N. Simmons, M. Waller, D. P. Watts, F. J. White, R. M. Wittig, K. Zuberbühler, and R. W. Wrangham. 2014. Lethal aggression in *Pan* is better explained by adaptive strategies than human impacts. *Nature* 513: 414–417.

Wilson, M. L., and R. W. Wrangham. 2003. Intergroup relations in chimpanzees. *Annual Review of Anthropology* 32: 363–392.

Wrangham, R. W. 1999. Evolution of coalitionary killing. *Yearbook of Physical Anthropology* 42: 1–30.

Wrangham, R. W. 2002. The cost of sexual attraction: Is there a trade-off in female *Pan* between sex appeal and received coercion? In C. Boesch, G. Hohmann, and

L. Marchant, eds., *Behavioral Diversity in Chimpanzees and Bonobos*, 204–215. Cambridge: Cambridge University Press.

Wrangham, R. W., and D. Pilbeam. 2001. African apes as time machines. In B. Galdikas, N. E. Briggs, L. K. Sheeran, G. L. Shapiro, and J. Goodall, eds., *All Apes Great and Small*, vol. 1: *African Apes*, 5–17. New York: Kluwer Academic/Plenum.

Wroblewski, E. E. 2010. Paternity and father-offspring relationships in wild chimpanzees, *Pan troglodytes schweinfurthii*. Ph.D. dissertation, University of Minnesota, Minneapolis.

Zihlman, A. 1996. Reconstructions reconsidered: Chimpanzee models and human evolution. In W. C. McGrew, L. F. Marchant, and T. Nishida, eds., *Great Ape Societies*, 293–304. Cambridge: Cambridge University Press.

13

Violent Cousins

Chimpanzees, Humans, and the Roots of War

MICHAEL L. WILSON *and* LUKE GLOWACKI

In 1916, during the Battle of the Somme, millions of soldiers from the German, French, and British Empires spent months using machine guns, artillery, aircraft, and the first battle tanks to injure or kill roughly a million people; during the course of which the French and British forces advanced the line some ten kilometers into German territory (Philpott 2009). The scale of such battles far exceeds anything observed in other vertebrates. Only in some social insects, such as supercolonies of Argentine ants, do battles involve similar numbers of individuals (Moffett 2012). Like other examples of modern warfare, the Battle of the Somme depended on many uniquely human traits: advanced technology, including not only weapons but also ships, trains, and horses transporting millions of soldiers to the battlefront; complex hierarchical societies; division of labor, with agriculture and industry providing food and *matériel* for full-time soldiers; motivations for fighting rooted in views about morality and honor; and many historical contingencies, including cultural glorification of war and particular tangles of alliances.

Such industrial-scale warfare emerged only recently in history, and following the devastation of the world wars, great powers have fought mainly through proxy wars rather than direct confrontation (Pinker 2011).

The empires that fought the Battle of the Somme have long since dissolved, and the German, French, and British nations now peacefully coexist within Europe. Even with Brexit looming, war between France, Britain, and Germany now seems unthinkable. The observation that such patterns of warfare vary greatly over time leads many scholars to view warfare as entirely the result of historical contingency and cultural inventions, such as weapons (Lorenz 1966), agriculture and population growth (Haas 2001), or segmentary societies (Kelly 2000). Nonetheless, despite recognition that particular instances of warfare depend on many specifics of history and technology, the fact that warfare occurs so widely across human societies, and throughout recorded history, suggests that war has deeper evolutionary roots, and can be explained within the same evolutionary framework used to explain intergroup aggression in other species (Ghiglieri 1989; Manson and Wrangham 1991; Gat 2006; Crofoot and Wrangham 2010; LeBlanc 2014).

Observations of intergroup violence in chimpanzees have provided important contributions to the debate about warfare's origins. Male chimpanzees defend group territories and sometimes kill members of other groups (Goodall 1986). While chimpanzee "warfare" occurs at a vastly smaller scale than modern industrial warfare, the resulting rates of mortality from intergroup aggression are similar in scale to those of subsistence-level human societies (Wrangham et al. 2006; Gurven et al. 2013).

Controversy continues over whether the roots of warfare are deep, shallow, or somewhere in between. One possibility is that the roots of warfare extend at least as deep as the last common ancestor (LCA) of humans, chimpanzees, and bonobos (Wrangham and Glowacki 2012; Gat 2015). Alternatively, coalitionary killing may have evolved separately in chimpanzees and humans, but for similar reasons (Manson and Wrangham 1991; Wilson and Wrangham 2003). Some have argued that warfare has deep or moderately deep roots, but has varied in frequency over time. For example, Knauft (1991) proposed that intergroup killing followed a U-shaped trajectory, occurring frequently in the LCA but infrequently for most of the evolutionary history of *Homo sapiens,* then becoming more common in more complex prestate societies. Otterbein (2004) proposed that warfare emerged with *Homo,* occurred frequently among heavily armed Paleolithic big game hunters, became rare following the depletion of big game, and then reemerged with the development of agricultural societies. An extreme version of the shallow roots hypothesis posits that neither humans nor chimpanzees have a biological propensity toward coalitionary killing. Instead, such killing has emerged only recently as a consequence of cultural inventions (in humans) and human disturbance (in chimpanzees; Ferguson 2011; Fry 2011). We consider the most extreme version of the shallow roots hypothesis unlikely, given widespread archaeological evidence for warfare among hunter-gatherers (Allen and Jones 2014), and evidence that rates of lethal aggression in chimpanzees are unrelated to measures of human disturbance (Wilson et al. 2014). In this chapter we will examine similarities and differences in patterns of coalitionary killing in chimpanzees and humans, and attempt to infer whether the roots of war in humans are deep, shallow, or somewhere in between.

In considering the time depth of warfare's roots, it is worth noting that no serious discussion of the evolution of warfare assumes it to be the result of a single psychological mechanism or genetic pathway. Instead, warfare is a complex phenomenon that emerges from a combination of psychological adaptations, under the influence of demographic, ecological, and cultural factors. Many psychological mechanisms involved in warfare surely have roots extending far deeper than the LCA. Aggression occurs almost universally among animals and involves highly conserved neuroendocrine mechanisms (Waltes et al. 2016). The capacity for assessing the relative costs and benefits of fighting is likely also widespread (Parker 1974). A preference for avoiding fights if outnumbered may be nearly universal among animals capable of fighting, because even animals that never fight conspecifics in coalitions may nonetheless face risks posed by group-hunting animals such as lions and wolves. Recent observations of lethal coalitionary aggression by mountain gorillas (Rosenbaum et al. 2016) suggest that psychological mechanisms shared broadly among primates can be sufficient to support coalitionary killing when socioecological circumstances make this an effective strategy.

In the following, we begin with an overview of patterns of intergroup violence in chimpanzees and hunter-gatherers. We then review evidence for whether earlier members of our lineage likely had similar patterns of intergroup aggression. In doing so, we focus on key similarities and differences between chimpanzees and humans, discussing their potential effects on the evolution of intergroup aggression in the human lineage. Humans and chimpanzees resemble one another in having (1) fission-fusion societies, (2) intergroup hostility, (3) male coalitions, (4) territorial behavior, and (5) coalitionary killings. At the same time, humans differ strikingly from chimpanzees in many ways, including: (1) weapons, (2) the kinds of benefits gained by aggressors, (3) multilevel societies, and (4) language.

Intergroup Violence in Chimpanzees and Hunter-Gatherers

Chimpanzees

Since 1960, long-term field studies have logged a total of more than 426 years of observing chimpanzees (Wilson et al. 2014), providing a detailed picture of chimpanzee social structure and intergroup relations. Considerable diversity exists across sites, but many aspects of chimpanzee social behavior are nonetheless similar across a broad range of habitats (Boesch et al. 2002; Mitani 2009; Nishida 2012). Chimpanzees live in fission-fusion communities in which members travel in parties of varying size (Figure 13.1a). Communities change in size slowly, through births, deaths, and migration, whereas parties change in size and composition throughout the day. Parties may consist of lone individuals, small groups of females with young offspring, or mixed-sex groups of various sizes. Males almost always spend their entire lives in their natal community, whereas females generally transfer to a new community at adolescence, presumably to avoid the risk of breeding with closely related males (Pusey 1980). Male chimpanzees show higher levels of affiliation than female chimpanzees (Machanda et al. 2013), and engage in various group activities, including hunting, mutual grooming, coalitions, boundary patrols, and intergroup aggression (Goodall 1986; Mitani 2009).

Each chimpanzee community occupies a territory that males defend from other communities during intergroup encounters (Figure 13.1a; Goodall et al. 1979; Nishida 1979). Substantial overlap may exist between the ranges of neighboring communities (Herbinger et al. 2001). Nonetheless, intercommunity interactions occur infrequently (e.g., 0.67 per month at Kanyawara [Wilson et al. 2012]; 0.94 per month at Kasekela [Gilby et al. 2013]). This is partly because chimpanzees live in large ranges at low population densities (median home range = 12.2 km^2, range: 4.1–86; median density = 3.3 individuals / km^2, range: 0.37–9.2; Wilson et al. 2014; Figure 13.2a). Even given low densities, intergroup interactions occur less frequently than expected based on Waser's (1976) gas law models (Kanyawara: 2.4 / month predicted,

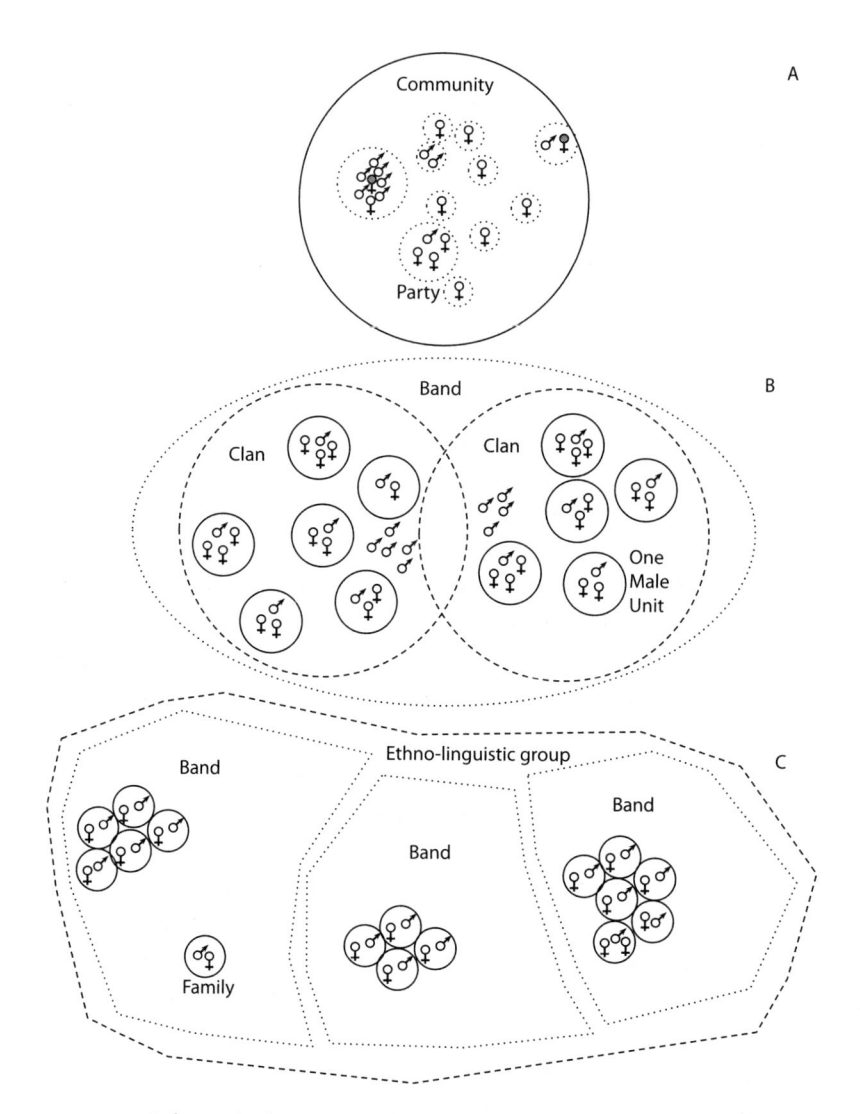

FIGURE 13.1. Schematic depictions of social organization. Panels illustrate societies of (a) chimpanzees, (b) multilevel papionins, and (c) hunter-gatherers. Closed circles enclose individuals with stable, regular associations and social bonds. Dotted and dashed lines enclose temporary or fluid groupings.

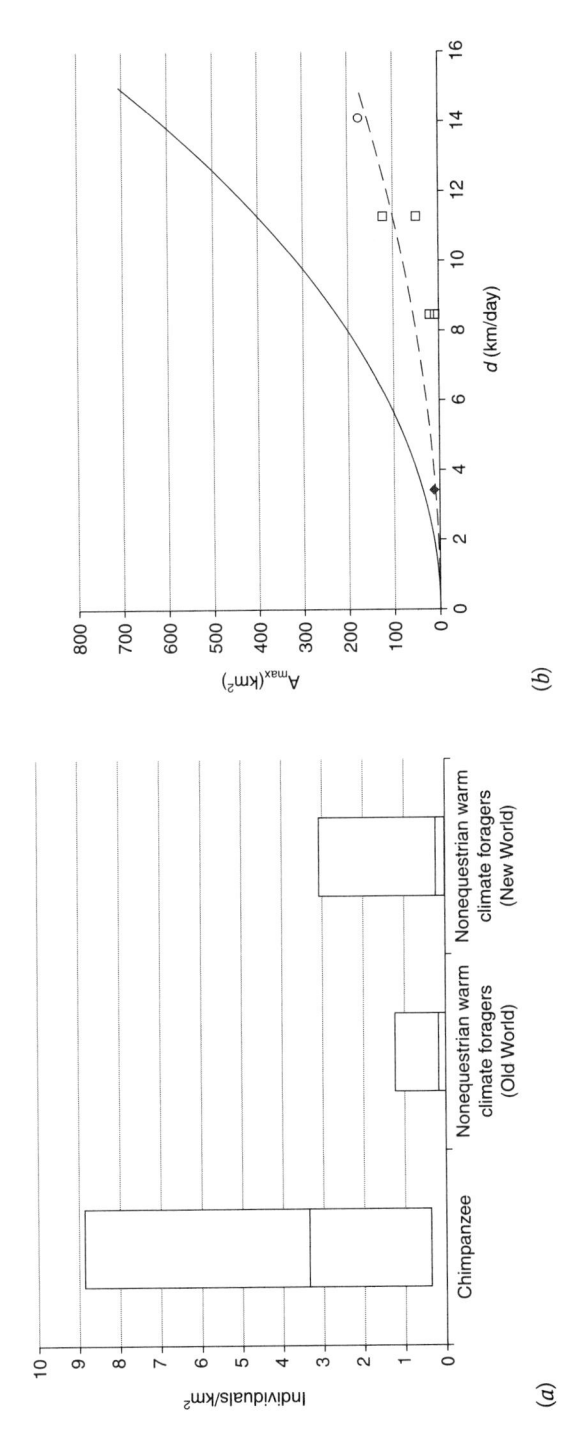

FIGURE 13.2. Population density and territory defendability for chimpanzees and humans. (a) Range of population densities reported for chimpanzees and hunter-gatherers. (b) Predicted economic defendability of territories for chimpanzees and humans. Curves indicate the predicted maximum economically defendable territory size as a function of daily travel distance, with societies with one (dashed line) or two (solid line) parties capable of defending boundaries at any given time. Estimated mean daily travel distances and home range sizes are given for chimpanzees (closed diamond), hunter-gatherer bands (open circle), and a range of four estimates for australopiths (open squares). The average home range sizes for both chimpanzees and hunter-gatherers are close to or just above the limit for predicted economically defendable range if only a single party is involved in defense, but well within the defendable range if more than one party is involved in defense. Estimates for australopith home ranges are highly uncertain but fall within the range of economic defendability.

0.67 / month observed; Kasekela: 3.2 / month predicted, 0.94 / month observed; predicted rates calculated with the following assumptions: velocity $v = 3.4$ km / day (Williams 1999); party spread $s = 100$ m; approach distance $d = 100$ m). The rarity of intergroup encounters thus likely reflects active avoidance of other groups (found also by Wrangham et al. 2007).

Intergroup interactions have been described in detail for Mahale (Nishida and Kawanaka 1972; Nishida 1979; Nishida et al. 1985; Kutsukake and Matsusaka 2002); Gombe (Goodall et al. 1979; Goodall 1986; Williams et al. 2004; Wilson et al. 2004), Taï (Boesch and Boesch-Achermann 2000; Boesch et al. 2008), Ngogo (Watts and Mitani 2001; Mitani and Watts 2005; Mitani et al. 2010), and Kanyawara (Wilson et al. 2012). The following description draws on these studies, focusing on features that appear to be common across these sites. For chimpanzees, the community boundary marks the frontier of friendly social relations. Males may visit the periphery of their range for various reasons, including searching for food, females, and boundary patrols, which appear to involve intentional searches for signs of strangers. When visiting border areas, chimpanzees frequently stop to look and listen for neighbors, sometimes standing bipedally. Any sign of strangers is treated with apparent fear and hostility. On hearing calls of strangers in the distance, males may erect their hair, embrace, touch each other's genitalia in reassurance, and either give loud calls in response, or silently move toward or away from the calls. Whether they approach or move away from the strangers, and whether they give a vocal response, depends mainly on the number of males present. Parties with many males are more likely to give a loud vocal response and approach distant calls, both in experimentally simulated (Wilson et al. 2001) and naturally occurring encounters (Wilson et al. 2012). The great majority of intercommunity encounters in chimpanzees involve "shouting matches" between parties that are far out of visible range of one another, and may be hundreds of meters apart (Wilson et al. 2012). Chimpanzees give pant-hoots and other loud calls, and also "drum," producing resonant booming sounds by kicking or hitting tree buttresses.

Chimpanzees come within visual range of their neighbors infrequently, but when they do, the outcome almost invariably involves displays of mutual hostility. Upon seeing strangers, chimpanzees frequently raise their hair (presumably to appear larger and more intimidating) and may perform charge displays, rapidly moving toward their opponents, sometimes dragging or throwing vegetation, sticks, and/or stones.

The outcome of intergroup encounters depends mainly on the number of males present on each side. If many males are present on both sides, a "battle" may ensue, with individuals displaying at, charging, and chasing one another. Fatalities rarely occur in such battles, unless a lone male or mother becomes isolated from the rest of the party. Males often behave peacefully toward females with sexual swellings and without dependent infants, as these females are potential immigrants (Williams et al. 2004). Males behave aggressively toward individuals in all other age-sex classes.

Most fatalities occur when many males encounter a lone stranger of either sex (Wilson et al. 2014). In such cases, the isolated individual generally attempts to flee, while those in the more numerous party seek to capture and attack the victim en masse. Sixty-three percent of killings (62/99 cases) involved intercommunity violence. Attackers outnumbered victims by a median 8:1 ratio. Just over half of all victims were infants, which were usually forcibly removed from the mother (32/62 = 52 percent of intergroup killings). Adult males were the next most common category of victims (20/62 = 32 percent of intercommunity killings) Adult females were often beaten severely but were less commonly killed (6/62 = 10 percent of intergroup killings; all figures from Wilson et al. 2014).

In some ways it is surprising that adult females are ever killed, because they are potential future mates for males. Males may gain fitness benefits from such attacks by reducing feeding competition, either by directly eliminating competitors or (if the females survive) inducing them to avoid border regions (Williams et al. 2004; Pradhan et al. 2014). However, given that in many cases of intercommunity infanticide, the many males attacking the infant could presumably have killed the mother as well, males generally appear to show restraint toward adult females.

Infants taken from their mother are often killed quickly by bites to the head and/or abdomen, and are frequently consumed. Adults are typically pinned to the ground by attackers and are bitten, hit, and kicked until immobile. Victims typically suffer numerous canine puncture wounds. Attackers commonly bite fingers and toes and other soft tissue, including the face, throat, and genitalia. In at least nine cases attackers removed or otherwise damaged the genitalia of adult male victims (Wilson et al. 2015), and in at least two cases attackers tore out the victim's trachea (Figure 13.3; Watts et al. 2006; Wilson et al. 2015).

Rates of intercommunity killing vary greatly among study communities. For example, the percentage of deaths attributed to intercommunity killing was only 5/130 = 3.8 percent for M-group at Mahale (Nishida et al. 2003), whereas at Gombe, intercommunity killing accounted for 8/86 = 9.3 percent for Kasekela (Williams et al. 2008) and 2/15 = 13 percent for Kalande (Rudicell et al. 2010). While comparisons of chimpanzee and human aggression tend to focus on intercommunity killing, intracommunity killing also occurs, and can cause an even greater proportion of mortality (e.g., 13/130 = 10 percent of deaths for M-group [Nishida et al. 2003]; 9/86 = 11 percent of deaths for the Kasekela community [Williams et al. 2008]). As with intercommunity killing, intracommunity killing typically involves coalitions of males, who usually target grown males or infants (Wilson et al. 2014). Females sometimes participate in coalitionary killing, but at a much lower rate than males (Wilson et al. 2014).

FIGURE 13.3. Adult male victim of intergroup killing by M-group males in Kalinzu Forest, Uganda (photo courtesy Kathelijne Koops).

Male chimpanzees have been proposed to gain two major benefits from intercommunity aggression: food and females. Considerable evidence now supports the view that male chimpanzees defend a group feeding territory for themselves, their mates, and their offspring. At Kanyawara, the great majority of intercommunity interactions occurred when chimpanzees were drawn to groves of an abundant food source (fruits of *Uvariopsis congensis* trees) located in a border region. The Ngogo community expanded their range by 22 percent by occupying a region where they killed thirteen strangers in ten years (Mitani et al. 2010). In contrast, the Kanyawara chimpanzees contracted their range away from an area where they frequently encountered powerful neighbors (Wilson et al. 2012). At Gombe, when territory size was larger, chimpanzees traveled in larger parties (Williams et al. 2004), females had shorter interbirth intervals (Williams et al. 2004), and, controlling for age, sex, and reproductive condition, individuals had heavier body mass (Pusey et al. 2005), indicating that food was more abundant in larger territories (Figure 13.4).

While males sometimes increase the number of females in their communities through intercommunity killing, it remains unclear whether males attack other groups mainly to acquire new females or to gain other benefits, such as acquiring key food resources. Boesch and colleagues (2008) describe males taking stranger females "prisoner," but it is not clear whether those females subsequently transferred to the attackers' community. Williams and colleagues (2004) found that even peripheral females appear to affiliate with a single community, contracting their ranges when the community's range contracted, indicating that males are unlikely to acquire new females merely by expanding their range. Parous females rarely transfer into a new community, doing so only when the number of males in their resident community is dramatically reduced (Nishida et al. 1985; Rudicell et al. 2010). Nulliparous females generally transfer from their natal communities in any case, though perhaps their decision of where to settle is affected by male displays of numerical strength. Presumably, females seek to transfer to a community with a large, well-defended territory and many potential mates. In cases of community extinction, males may be able to acquire females from vanquished rivals. Of eighteen chimpanzee communities in long-term studies (Wilson et al. 2014), at least five (28 percent) have suffered precipitous declines. One or more parous females transferred following the declines of Kahama (Williams et al. 2004), K-group (Nishida et al. 1985), and Kalande (Rudicell

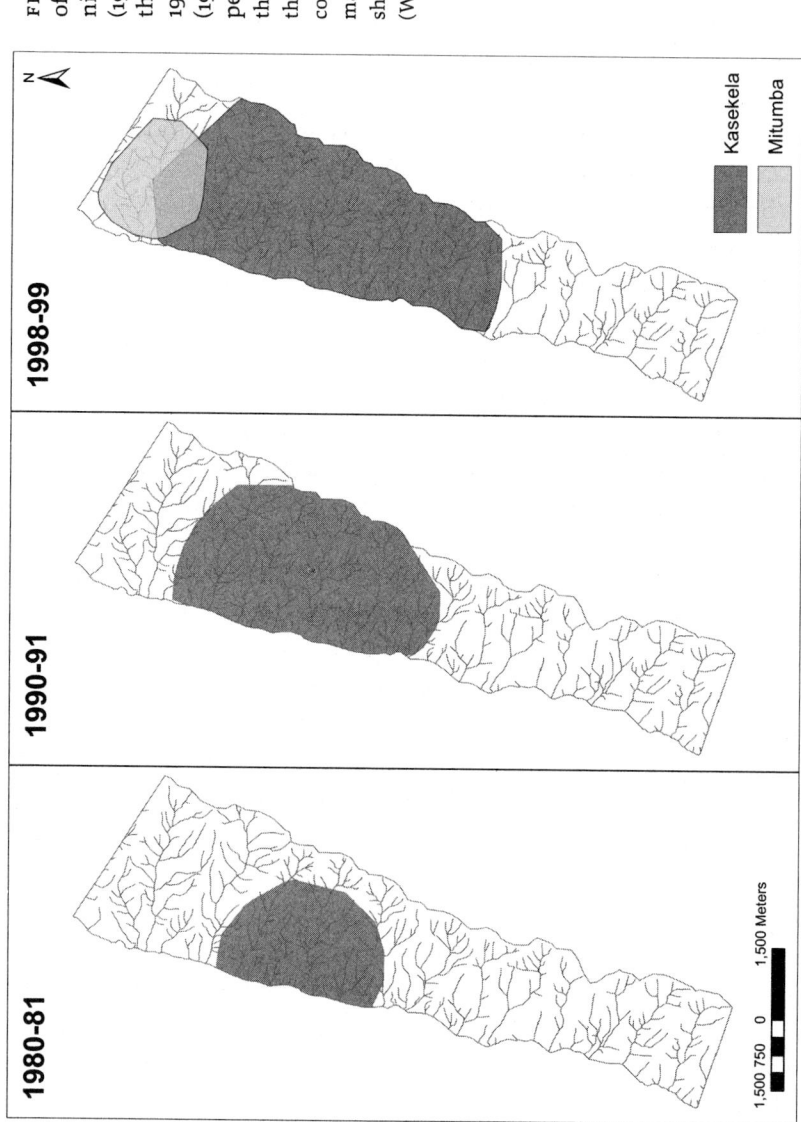

FIGURE 13.4. Home range of the Kasekela community for selected years (1973–2000), including their minimum (1980–1981) and maximum (1998–1999) ranges for this period. Previous studies of this community have found that larger home range was correlated with heavier body mass (Pusey et al. 2005) and shorter interbirth interval (Williams et al. 2004).

et al. 2010), and at least one of two declining groups at Taï (Roman Wittig, personal communication).

While the evidence for acquiring females as a result of intergroup aggression remains ambiguous, multiple lines of evidence support the view that by increasing territory size, victors in intergroup aggression increase their food supply. Controlling for age, sex, and reproductive state, Kasekela chimpanzees weighed more when territory size was larger (Pusey et al. 2005), indicating a general benefit of increased food supply. Females particularly benefit by having shorter interbirth intervals (Williams et al. 2004). Males that are successful at mating will benefit disproportionately from faster reproduction by females. Not surprisingly, then, participation in boundary patrols is correlated with mating success (Watts and Mitani 2001).

Group-living species commonly compete over food and females. Intergroup killing, however, is less common. The best supported reason that chimpanzees kill their opponents, rather than simply chasing them off, is the imbalance of power hypothesis (Manson and Wrangham 1991). This hypothesis begins with the observation that in some species, including chimpanzees, intense scramble competition for food leads to variation in party size. When parties with many males encounter isolated individuals from rival communities, they can kill at low risk of being injured in return. In support of this hypothesis, male chimpanzees avoid the periphery of their territory except when in parties with many males (Wilson et al. 2007); are more likely to conduct boundary patrols when in parties with many males (Mitani and Watts 2005); and are more likely to approach and call in response to simulated (Wilson et al. 2001) and real (Wilson et al. 2012) intruders, the more males they have in their party. Moreover, the great majority of inter-community killings occur when attackers have an overwhelming numerical advantage (median ratio of attackers to victim = 8:1; Wilson et al. 2014). By killing adult males, attackers reduce the current coalition size of rival groups, increasing their odds of winning future intergroup interactions (Wrangham 1999), and thereby improving their ability to acquire territory, food, and/or females.

Hunter-Gatherers

Hunter-gatherers live in societies ranging from egalitarian, mobile bands such as those of the Ju/'hoansi (Lee 1979), Aché (Hill and Hurtado 1996), and Hadza

(Marlowe 2010) to hierarchical, sedentary chiefdoms such as the Nootka (Drucker 1951). Because the development of stratified societies is thought to be relatively recent, dependent upon newer technologies such as fishing weirs and food storage, anthropologists have focused on mobile foragers as the best living models of Paleolithic societies. The possibility exists that this focus on mobile foragers underestimates the importance of more densely settled societies that may have existed in areas such as lakeshores and river valleys that are now densely settled by farmers (LeBlanc 2014). In any case, even within the more limited subset of mobile hunter-gatherers, considerable diversity exists across varied habitats and cultural regions (Kelly 1995; Binford 2001).

While hunter-gatherers are diverse, compared to the range of possible societies exhibited by other primate species, the variation documented in human societies is actually rather limited (Rodseth et al. 1991). In no human society do people regularly live as solitary foragers (as in orangutans), as isolated monogamous families (like gibbons), as isolated polygynous families (like gorillas), or in promiscuously mating communities (like chimpanzees and bonobos). Instead, humans everywhere live in multilevel societies that include diverse and overlapping social networks (Rodseth et al. 1991; Rodseth and Wrangham 2004; Hill et al. 2009; Chapais 2010). Unlike chimpanzees, in which females sever their social ties on transferring to a new community, both men and women maintain lifelong relations with kin of both sexes (Rodseth et al. 1991). Humans differ strikingly from the promiscuous mating of chimpanzees (Tutin 1979), in that men and women commonly form long-lasting and relatively exclusive sexual relationships (Chapais 2010; Walker et al. 2011).

In contrast to chimpanzees, in which friendly relations are limited to members of a single community, forager societies include at least three distinct levels of social organization, known by various names in the literature, which we call here *family, band,* and *ethnolinguistic group*. Families, usually consisting of a husband, a wife, their children, and sometimes one or more close kin, such as a parent or son-in-law, constitute the basic unit of hunter-gatherer societies. In some societies, such as Western Shoshoni (Dyson-Hudson and Smith 1978), families forage independently during times when food resources are scarce. Usually, however, several families stay together in a group variously called a *band, camp,* or *the local group*, staying at a temporary campsite within a larger home range. Estimates of median band size range from 18.3 (range: 5.80–81.6; $n = 32$) from a detailed study of census records (Hill et al. 2011) to 26 people (range: 13–250; $n = 130$) for a sample containing more societies but less detailed records (Marlowe 2005). Each family usually builds its own simple shelter and maintains its own fire for cooking. Bands move to different campsites periodically (median = 7 times per year; Marlowe 2005) when local food supplies run short. Band members may regard themselves as part of a larger ethnolinguistic group that numbers a median 565 individuals (range: 23–11,800; Marlowe 2005; Figure 13.1c). Some or all bands of an ethnolinguistic group may come together periodically during times of food abundance to socialize and exchange goods and members (Grubb 1911; Meggitt 1962; Clastres 2000; Burch 2005). Kinship relations provide an additional, overlapping set of memberships that cut across band and even ethnolinguistic boundaries (Rodseth and Wrangham 2004; Chapais 2010).

In contrast to chimpanzees, in which males and females spend most of their time searching for the same sorts of foods, a striking sexual division of labor exists in humans. During the day, men and women usually forage separately. Women gather food such as fruit, seeds, and underground storage organs (USOs) of plants, usually in small groups with other women and their children. Men travel separately, hunting alone or in small groups. Some individuals of both sexes may stay in camp throughout the day, looking after small children, making tools or clothing, resting, or socializing. Both sexes bring food back to camp for further preparation and sharing with other camp members. Men share meat widely with other camp members, whereas women share foods they have gathered mainly with their own families (Kelly 1995). Women do the great majority of the cooking of both meat and plant foods (Kelly 1995). Bands and/or individual members of bands may claim ownership of specific territories (Heinz 1972; Peterson 1975; Dyson-Hudson and Smith 1978; Cashdan 1983).

Anthropologists continue to debate the extent to which hunter-gatherers engage in intergroup violence, and whether it merits the term "warfare." Some reviews find that warfare is widespread and deadly in hunter-gatherers (Ember 1978; Gat 1999a; LeBlanc and Register 2004; Otterbein 2004; Wrangham and Glowacki 2012; Allen 2014), whereas others argue that intergroup killing is relatively unimportant among most mobile foragers (Kelly 2000; Fry and Söderberg 2013).

Fatal fighting, both within and between groups, has been documented in most hunter-gatherer societies. For example, Fry and Söderberg (2013) reported that in a sample of twenty-one hunter-gatherer societies, a total of 148 killings occurred. A median of four killings occurred per society (range: 0–69), and killings occurred in all but three societies. It is impossible to compare rates of killings for this dataset, since only the numerator (number of killings) was reported, not the denominator (such as observation time and number of people observed per society, or number of observations of mortality from other causes). Nonetheless, even though Fry and Söderberg argued that these data indicated that intergroup killing was rare in hunter-gatherers, intergroup killings made up 34 percent of the total. Additionally, Fry and Söderberg excluded thirteen cases that would normally be considered examples of aggression between social groups, such as executing outsiders and interfamilial vendettas. Including these cases, intergroup killings made up 43 percent of all killings. Furthermore, the sample chosen for this study included many groups that have been pacified by more powerful neighbors or were living in a state society (Ju/'hoansi, Hadza, Mbuti, Semang, Kaska, Paiute, Vedda, and Yukaghir) or who live in such sparsely populated regions that they rarely encounter neighbors, such as the Copper Inuit (Wrangham and Glowacki 2012). Thus, even in a sample presented as evidence that lethal intergroup is rare in hunter-gatherers, such killings were widespread and represented a substantial proportion of all killings.

Wrangham and colleagues (2006) found that hunter-gatherers and chimpanzees have similar rates of mortality from intergroup aggression. Hunter-gatherers had a median mortality rate from intergroup aggression of 164.5 per 100,000 people per year (range: 0–1,000; $n = 12$), which was within the upper and lower estimates for chimpanzees (median = 69–287 individuals killed per 100,000 chimpanzees per year; $n = 5$ populations). An independent estimate based on different datasets also found that violence caused a similar proportion of deaths for chimpanzees and people living as hunter-gatherers and hunter-horticulturalists (Gurven 2013). Similarly, a recent review found that ethnographically documented "bands" (mostly hunter-gatherers) had a median percentage of death from violence that was higher than any other category of society, including tribes, chiefdoms, and states (Gómez et al. 2016). The median percentage of deaths attributed to violence for prehistoric societies was lower, but highly variable (Gómez et al. 2016), and must be treated with caution, given the many uncertainties involved with prehistoric data. For modern hunter-gatherers, where ethnographers can be confident they have obtained data on all causes of death, the likelihood of dying from violence is high even compared to modern state societies, including the horrific world wars of the twentieth century (Pinker 2011).

Despite considerable variation among groups, recent reviews have found that several aspects of warfare were remarkably similar among a broad range of societies in different ecological and subsistence contexts, from the deserts of Australia to the rainforests of South America (Gat 1999a; Wrangham and Glowacki 2012). While relationships between groups can often be peaceful, fatal fighting occurs at every level of social organization, including among individuals, bands, and ethnolinguistic groups. The most common method of attack is the ambush, which results in the majority of deaths from warfare (Gat 1999a). Most ambushes are by a group of male warriors who seek out an enemy settlement. While women rarely participate in offensive attacks, they sometimes provide support by accompanying raiding parties to cook and act as lookouts (Funk 2010). Raiders usually hide outside an enemy village at night or in the early morning with the hope of attacking the first person or two to exit the village. While raiders might have a specific victim in mind, in practice any individual from the enemy group will usually suffice as a victim. Immediately after the attack, raiders generally flee lest they be discovered by the enemy group or overtaken on their retreat. Opportunities for plunder are limited by the concerns about being overtaken by the enemy group, but sometimes include material goods, or more often captives, especially women and children, who are incorporated into the capturing group (Bridges 1948; Meggitt 1962; Clastres 1972; Burch 2005; Mendoza 2007). Deaths and injuries to raiders are rare but do occur, usually due to being detected by the enemy group while en route to the ambush (Wrangham and Glowacki 2012).

While ambushes were the most common type of intergroup violence, many groups were distrustful of strangers and would attack any they encountered (Burch 2005; Mendoza 2007; Wrangham and Glowacki 2012). Clastres (1972: 163) succinctly summarizes the ethos when two strange parties encountered each other: "If two strange bands meet by chance in the forest they either try to massacre each other or flee in opposite directions."

Some hunter-gatherers occasionally have what are described as "battles," in which dozens or even hundreds of warriors gather (Bridges 1948; Meggitt 1962; Burch 2007). Conflict in battles usually appears to involve small skirmishes along a fluid front rather than pitched conflict with a concerted effort. Formal leadership and chains of command are rare to nonexistent, and conflicts usually cease after one or two deaths or serious injuries. Battles could occasionally result in mass casualties if one side is routed (Burch 2007), but documented instances of this occurring are rare.

Whereas in chimpanzees the benefits from intergroup aggression are primarily through increased territory, humans may benefit through multiple pathways including access to territory, the acquisition of captives, transportable goods, and within-group cultural rewards. While in some cases, successful groups take over territory from defeated groups (e.g., Meggitt 1977), warfare can also occur without any obvious resource competition or territorial gain on the part of the victors (Glowacki and Wrangham 2013). Women and children are commonly captured in warfare, but there is significant intercultural variation, and in some societies, taking captives is rare. Commonly, warriors in small-scale societies gain status and prestige and may have increased reproductive success (Chagnon 1988; Glowacki and Wrangham 2015). The crucial difference between chimpanzees and humans is that human culture allows for multiple pathways by which warriors can benefit from their participation, some of which involve intangible rewards such as status or prestige. These pathways may create the opportunity for reproductive benefits for participants in intergroup aggression. These may increase the incentives for warriors as compared to chimpanzees, and account for the increased risk-taking humans encounter in warfare through conflict with armed individuals.

Coalitionary Killing in the Human Lineage

Chimpanzees and humans thus share many similarities in their patterns of intergroup violence. However, whether these patterns have been inherited from a common ancestor, or evolved convergently, or followed some more complicated path (such as being present in the LCA but rare or absent in some subsequent species), remains unclear. Little direct evidence exists for coalitionary killing for any early hominin, but this is not surprising, given the scarcity of fossils and the difficulty of unambiguously determining the cause of skeletal trauma. In the absence of direct evidence of killing, we can make inferences based on theoretical predictions and evidence related to key socioecological traits. According to the imbalance of power hypothesis (Manson and Wrangham 1991; Wrangham 1999), coalitionary killing should be most likely to occur in species with fission-fusion grouping patterns, intergroup hostility, and coalitionary bonds, which together make killing more likely by lowering costs to the attackers. In the following, we consider whether earlier hominins were likely to have these traits. Given the importance of territorial competition in intergroup relations of both chimpanzees and modern humans, we also consider whether earlier hominins were likely to defend territories. Territorial competition appears neither necessary nor sufficient for coalitionary killing to occur, given that many primates are territorial but few kill. Nonetheless, territorial behavior provides a useful marker of the intensity of intergroup competition, as the territory contains the key resources over which males compete: food, water, shelter, and females.

Grouping Patterns

Did australopiths live in large, stable groups like olive baboons? In small, stable groups like gorillas? Were they solitary, like orangutans, or did they live in large fission-fusion communities like chimpanzees? Or in complex multilevel societies like some papionins living in dry habitats: hamadryas baboons, Guinea baboons, and gelada monkeys? Such grouping patterns leave little in the way of fossil evidence, and what evidence exists is difficult to interpret. For example, if we possessed only skeletons of hamadryas and olive baboons, would we ever guess that these two species live in such strikingly different societies? Speculations are prone to wishful thinking, confirmation bias, and the limitations of our imaginations. Adding to the challenge of the task, the feeding adaptations of australopiths differed from any living apes, including humans (Nelson and Hamilton, this volume). Aspects of feeding ecology unique to australopiths surely affected their patterns of grouping and foraging, but in ways we can only dimly infer. Moreover, considerable variation likely existed both among and within species distributed over broad and varied landscapes, just as we observe today.

Recognizing the limitations of what we can know, we can nonetheless make some inferences based on what we know about living species. Modern humans and chimpanzees both live in communities with fission-fusion dynamics, because they rely on high-quality food resources that occur in patches of variable size and abundance (Wrangham 1987; Aureli et al. 2008). At some times of the year, abundant food supplies support many individuals foraging together, while at other times, individuals must spend much of their time alone or in small parties. As Wrangham (2001) argues, it seems unlikely that any hominins could have lived in large, stable groups in dry open country as several species of baboons do, because they presumably lacked the feeding and digestive adaptations peculiar to baboons.

Chapais (this volume) has argued that a multilevel society consisting of multiple polygynous family groups is a likely intermediate between chimpanzee-like promiscuity and the human pattern of pair-bonds with extensive paternal provisioning. Chapais does not attempt to infer the social systems exhibited by particular hominin species, but available evidence suggests that australopiths are reasonable candidates for having multilevel societies. They appear to have had higher sexual dimorphism in body size and facial structure than chimpanzees (Muller and Pilbeam, this volume), consistent with a more polygynous social structure. Living in drier, more open, and seasonal habitat likely would prevent individuals from foraging in large, stable troops throughout the year, forcing them to adopt fission-fusion dynamics. Nonetheless, suitable sleeping sites, such as cliffs and groves of trees, may have been sufficiently scarce to promote nighttime aggregations even when food scarcity prevented large foraging parties.

In contrast to australopiths, the body size, shape, and dentition of earlier *Homo* were broadly similar to those of modern humans. It therefore seems likely that diet and social organization were also similar to modern humans, and included fission-fusion dynamics. An increasing reliance on meat eating likely increased the extent to which males and females foraged separately, if males obtained most of the meat (as they do in chimpanzees and modern humans).

Intergroup Hostility

Primate groups commonly have hostile and competitive intergroup relations (Crofoot and Wrangham 2010). Even bonobos, well known for having less violent intergroup relations than chimpanzees, nonetheless commonly exhibit hostility toward other groups: 87 percent of encounters observed at Lomako involved aggressive displays, 35 percent involved physical aggression, and females unaccompanied by males fled from strangers (Hohmann and Fruth 2002).

One of the primary drivers proposed for the evolution of social behavior is improved effectiveness in competing for resources, since larger groups tend to defeat smaller groups (Wrangham 1980). Intergroup hostility is expected to be absent only if groups rely on resources that are too dispersed to be defensible. For example, gelada monkeys, like many grass-eating mammals, form herds of variable size (Dunbar 1992). Male geladas defend their females against other males, and individuals may defend valuable food patches, but no effort appears to be made to prevent new subgroups from joining the herd, presumably because the costs of doing so outweigh the benefits. Thus, australopiths, like the majority of other group-living primates, would be expected to have generally hostile intergroup relations, unless they relied on an undefendable resource such as grass. Additionally, whatever australopiths ate, they likely required some resources that were scarce and defensible, including water and sleeping sites. Baboon groups living in environments similar to those reconstructed for australopiths compete over access to food, water, and sleeping sites (Markham et al. 2012). Moreover, as with other primates, it is likely that males would have been hostile to extragroup males as potential competitors for mates. We therefore consider intergroup hostility a likely aspect of australopith social life.

As with australopiths, groups of earlier *Homo* probably competed for key resources, including plant foods, water, shelter, and mates. Additionally, the increasing proportion of meat in the diet likely increased intergroup hostility (Schaller and Lowther 1969; Smith et al. 2012), because of the importance of territory for hunting. Social carnivores such as lions, spotted hyenas, and wolves defend group territories and are intolerant of other groups, presumably to prevent depletion of the prey population in their hunting territories (Smith et al. 2012).

Male Coalitions

In both chimpanzees and hunter-gatherers, intergroup fighting is generally undertaken by coalitions of males. Whether various hominins also had male coalitions likely depended on two factors: dispersal patterns and reproductive skew.

Dispersal patterns affect coalition patterns by determining whether members of either sex are likely to have access same-sex coalition partners. In many species, members of one or both sexes disperse at sexual maturity, presumably to avoid inbreeding (Pusey and Packer 1987). Due to kin selection, close kin are generally preferred as coalition partners (Hamilton 1964).

Accordingly, sex-biased dispersal appears to favor the evolution of cooperation among members of the nondispersing sex. In many cercopithecine monkeys, males disperse and females participate in matrilineal coalitions (Wrangham 1980). In chimpanzees, females disperse and rarely form coalitions, whereas males stay in their natal group and engage in many group-level behaviors, including hunting and territory defense (Pusey 2001). Determining dispersal patterns of extinct species is challenging, but recent studies using stable isotope signatures and ancient DNA have begun to provide intriguing hints of female-biased dispersal in australopiths (Copeland et al. 2011) and Neanderthals (Lalueza-Fox et al. 2011). If these species did in fact have female-biased dispersal, that would favor the evolution of male coalitions. Nonetheless, dispersal patterns appear to be only part of the story, given that modern human foragers have strong coalitions among males despite only moderate female bias in dispersal patterns (Hill et al. 2011).

High levels of reproductive skew may be associated with low levels of cooperation among males. For example, in gorillas, a single male usually monopolizes reproduction, and male coalitions are rare (Robbins and Sawyer 2007). Conversely, in chimpanzees, reproductive skew is lower and male coalitions are common (Mitani 2009). Sexual dimorphism in body size and canine length is generally higher when reproductive skew is high (Plavcan 2012). Given the consistent correlation between high sexual dimorphism and high reproductive skew, high sexual dimorphism may also indicate a low level of cooperation among males. This relationship is not precise, however. For example, if reproductive competition is mainly between male coalitions, rather than within male coalitions, significant body size dimorphism might be favored. This appears to be the case in lions, where males are both much larger than females and form strong coalitions with other males in order to control access to female prides and associated territories (Packer et al. 1988). Likewise, in hamadryas baboons, males are substantially larger than females, but within clans, males appear to respect the mating bonds of other males. Instead, most male-male competition appears to take place during encounters between bands, when males seek to abduct females from other bands (Kummer 1968; Swedell and Plummer 2012).

If australopith males were substantially larger than females (Muller and Pilbeam, this volume), australopiths likely had a relatively high level of reproductive skew. In this case, males would likely have had highly competitive relationships, which would reduce incentives for males to cooperate closely with one another. Coalitions among males therefore may have been weaker among australopiths than in chimpanzees, unless (as in lions and hamadryas baboons) males competed mainly against coalitions of other males. Starting with earlier *Homo,* however, sexual dimorphism appears to decrease compared to australopiths (Muller and Pilbeam, this volume), which may indicate reduced reproductive skew, and thus stronger coalitions among males.

Territorial Behavior

In both chimpanzees and human foragers, competition over territory plays a central role in intergroup relations. As in many other species, chimpanzees and humans depend for their existence on limited resources associated with specific locations: food, water, and shelter. By defending territories, territory owners prevent outsiders from depleting the resources they need to survive and reproduce (Maher and Lott 2000). Whether animals defend territories depends on whether it is economical for them to do so: the costs of territory defense must be outweighed by the benefits (Brown 1964). Whether earlier members of the human lineage defended territories likewise would depend upon whether it was economically feasible to do so.

Among primates, territorial behavior varies considerably among species, and among populations within species. Mitani and Rodman (1979) proposed a simple index to assess the economic defensibility of territories. They found that in primates, species that defended territories generally had average daily travel distances that were at least as long as the diameter of their home range. More recently, Lowen and Dunbar (1994), argued that in species with fission-fusion dynamics, the maximum defendable area should take into account the number of independent foraging parties, or at least those parties that contain members of the territorial sex: Lowen and Dunbar (1994) estimated that chimpanzees have on average 5.5 parties containing males. This estimate seems high, however, given that parties with few males are unlikely to visit boundaries (Wilson et al. 2007) or approach strangers (Wilson et al. 2001, 2012), and only parties containing a substantial portion of the community's males are likely to have sufficient males to repel parties from other communities. It seems reasonable to assume, though, that chimpanzee communities commonly contain more than one party large enough to visit boundaries.

Similarly, in human foragers, men travel a median 14.1 km each day (Marlowe 2005), yielding a maximum defendable home range of 156 km², which is smaller than the median home range size for bands of nonequestrian, warm-climate foragers (median = 175 km², range: 22–4,500 km², $n = 125$; Marlowe 2005). However, if bands on average contain at least two independently foraging parties with multiple men, the maximum defendable home range increases to 625 km². Thus, for both chimpanzees and human foragers, home ranges of typical size should be economically defendable, though extremely large home ranges should be too large to defend. This is consistent with reports that many foragers defend territories, but that in areas of extremely low density, foragers are nonterritorial (Dyson-Hudson and Smith 1978).

Australopiths appear to have lived in drier, more seasonal habitats than those favored by chimpanzees (Cerling et al. 2011; Nelson and Hamilton, this volume). They therefore would likely have required larger home ranges, and would have defended territories only if their daily travel distances also increased (Mitani and Rodman 1979). Evidence reviewed by Pontzer (this volume) indicates that australopiths had a humanlike efficiency of locomotion, and thus longer daily travel distances than chimpanzees. Even if, as Kramer and Eck (2000) argue, *Australopithecus afarensis* traveled 20–40 percent less each day than modern humans, reducing the median 14.1 km/day traveled by male hunter-gatherers (Marlowe 2005) by this amount yields a daily travel distance of 8.5–11.3 km/day for male *A. afarensis*, which is still much further than the average of 3.4 km/day traveled by male chimpanzees at Gombe (Williams 1999). Anton and colleagues (2002) estimated per capita home range requirements for australopiths based on whether they had apelike or human-like ranging requirements. We don't know the likely group size of australopiths, nor whether their home range requirements scaled linearly with group size. Using two estimates of group size (twenty or fifty individuals) to bracket the median local group size for humans ($n = 26$; Marlowe 2005) and community size for chimpanzees ($n = 42.3$; Wilson et al. 2014), we estimate that australopith home ranges commonly would have been economically defendable as territories (Figure 13.2b).

Given that earlier *Homo* had body size and shape similar to modern humans, it seems probable that they would have had economically defendable ranges in typical habitats. Additionally, in earlier *Homo*, the intensity of territorial competition was likely increased by a combination of four factors. First, as the proportion of meat in the diet increased, *Homo* entered a higher trophic level, and thus required larger territories (Anton et al. 2002). Second, earlier *Homo* shows evidence of adaptations for improved endurance running and/or distance walking (Bramble and Lieberman 2004). Whether these adaptations evolved mainly for running or walking, they indicate an increased daily travel distance for earlier *Homo*, which would permit economical defense of larger territories. Third, at some point during the evolution of *Homo*, our ancestors stopped sleeping in cliffs and trees and began sleeping on the ground instead. This is presumably due to more effective defense against predators (either through improved weapons, use of fire, or both; Wrangham 2010). This independence from sleeping cliffs and trees would permit more efficient range use, enabling *Homo* to set up camp wherever they liked, rather than being limited to sites with safe sleeping locations. Prior to this, scarcity of sleeping sites may have limited populations and provided natural spacing between them. Fourth, compared to *Pan*, modern humans have shorter interbirth intervals (Schwartz 2012). The ability to produce infants more quickly likely depends on increases in available energy due to practices already present in earlier *Homo*: increased meat in the diet, more efficient extraction of energy from plant and animal foods, and the availability of softer, more easily chewed weaning foods, due to food processing and possibly cooking (Wrangham 2010). All else being equal, shorter interbirth intervals would lead to faster population growth, which in turn would increase competition for available resources.

Coalitionary Killing

Australopiths thus seem likely to have had fission-fusion grouping patterns, hostile intergroup relations, and economically defendable ranges, but at the same time seem likely to have had weaker coalitions among males than exist in chimpanzees and humans. Additionally, even if australopiths had fission-fusion grouping patterns, individuals may have rarely foraged alone. If females regularly foraged in small parties, as modern humans do when utilizing similar resources, and if males regularly foraged with those females, as in the one-male units (OMUs) of hamadryas baboons and geladas, then australopiths may have rarely foraged alone, and thus had few opportunities for coalitionary killing. In this case, australopiths would have had lower rates of coalitionary killing than either chimpanzees or modern humans.

Like australopiths, early members of *Homo* likely had fission-fusion dynamics, hostile intergroup relations, and defendable territories. In contrast to australopiths, reduced sexual dimorphism in body size suggests reduced reproductive skew, and thus higher levels of within-group cooperation among males, which in turn would promote coalitionary killing. If, as in chimpanzees and modern humans, males in earlier *Homo* were more engaged than females in obtaining meat, it is possible that males and females foraged separately. This would increase the vulnerability of both sexes to intergroup aggression. Thus, several lines of evidence suggest that circumstances favoring coalitionary killing occurred in earlier *Homo*. Consistent with this, by the time of later *Homo*, evidence consistent with coalitionary killing exists, including cut-marks on human bones (suggesting processing of carcasses for cannibalism) and injuries to bones (Churchill et al. 2009; Trinkaus 2012).

Key Differences in Intergroup Aggression between *Pan* and Hominins

While chimpanzees and humans exhibit some striking similarities in their patterns of intergroup aggression, they also differ in important ways. Here we review some of the key differences, including weapons, benefits gained by aggressors, multilevel societies, and language.

Weapons

One of the striking differences between chimpanzees and humans is that in humans, combat is usually armed. Chimpanzees mainly fight with their hands and teeth, though they sometimes throw stones (Goodall 1986) and use sticks to hit other chimpanzees with sticks while fighting, and probe for and sometimes kill prey items (Pruetz and Bertolani 2007). In contrast, in hunter-gatherer societies, men routinely spend much of their time foraging far from camp, armed with weapons for hunting big game. While many hunter-gatherers made weapons specifically for fighting people, such as the Aleuts (who used the atlatl when hunting sea mammals but the bow and arrow against human enemies; Maschner and Mason 2013), weapons that are effective at killing prey generally can be used to kill people as well. Hunter-gatherer men are thus nearly always armed and dangerous when away from camp, capable of killing at a distance if provoked or threatened, and disposed to strike preemptively.

As Darwin (1871) noted, the evolution of bipedality freed the hands from locomotion to making and using tools, carrying items, and communicating. Australopith hands show clear evidence of evolution toward greater manual dexterity, including more powerful and more precise grips (Young 2003), suggesting the use of tools such as digging sticks. If australopiths regularly used sticks and/or bones for obtaining USOs, insects, and other foods, these same tools also may have been useful as weapons. Marlowe (2005) notes that among the Hadza, "even women armed with only their digging sticks occasionally scare off a leopard and take its kill." Hadza digging sticks are substantially larger and sturdier than those used by savanna-woodland chimpanzees (Herndandez-Aguilar et al. 2007), but if australopiths regularly dug for USOs they would presumably have used sturdier sticks than those used by chimpanzees. In addition to sticks, australopiths would have been able to throw stones more effectively than chimpanzees, and probably would have done so more often (Young 2003).

Tools for fighting would have complicated and potentially contrasting effects on coalitionary killing: reducing the costs of attacking unarmed or outnumbered opponents, while raising the costs against armed opponents.

By the time of early *Homo*, tool use had expanded to include various shaped stone implements. If stones were used as weapons, they would have had effects similar to those described above for australopiths, but to a greater extent, given that the sharp cutting edges of stones would have made for more dangerous weapons than the simple sticks and stones of earlier hominins.

It seems likely that regular use of even simple weapons such as sticks and stones would have affected the frequency of coalitionary killing, though scholars have drawn different conclusions about the likely effects of weapons. For example, it is commonly argued that effective hunting weapons during the evolution of *Homo* made the killing of people more feasible and common (Otterbein 2004). In contrast, Kelly (2005) argues that thrown spears made fighting so dangerous that a long period of Paleolithic warlessness ensued. This argument seems implausible to us, given the widespread prevalence of intergroup killing among documented peoples with projectile weapons. But whatever the overall effects on rates of killing, it seems clear that weapons increase the advantage of armed attackers against unarmed victims.

Individuals who used weapons would thus obtain strong advantages against those who didn't, promoting the cultural and biological evolution of more effective weapon use (Young 2003). Two further contrasting effects of weapons can be predicted. First, weapons, especially projectile weapons, can increase the advantage of numerical superiority, because weapons enable attackers to concentrate force at a single target (e.g., Bingham 2000). Chimpanzees manage to concentrate force in gang attacks by piling up on isolated victims, with some attackers pinning the victim's hands and feet to the ground while others bite and beat the victim to death. Projectile weapons presumably make such concentration of force even more effective, as many attackers can simultaneously throw rocks, spears, or other projectiles at a victim from a distance. Second, while weapons can amplify numerical advantages in gang attacks, in dyadic conflict weapons can have an opposite effect, serving to equalize competitive ability among individuals of different sizes, and thus reducing the advantage of large body size. In contrast with hand fighting sports such as boxing and wrestling, in weapon fighting sports such as fencing, no weight classes are used, because success depends more on skill than body size (Faurie and Raymond 2003). Fights between individuals armed with weapons (even simple weapons such as stones for throwing and sticks for clubbing) can thus become costly for both participants, with outcomes less easily predicted by variables such as body size. One of several possible explanations for the apparent reduction of body mass dimorphism in *Homo* is that weapons increased the costs of within-group fighting, reducing the selection pressure for large body mass in males (Chapais 2013). However, it is worth noting that men have much greater upper body musculature than women (Puts 2010; Muller and Pilbeam, this volume), suggesting that upper body strength has continued to provide men with important competitive advantages in hunting and/or fighting.

With the advent of modern humans, rapid advances in weapons technology appear to have occurred, particularly a proliferation of weapons made from multiple parts (composite weapons). Many of these involve improvements in projectile weapons, promoting the ability to kill at a distance: boomerangs, hafted weapons, spear throwers, bow and arrow. Improved weapons technology may have been important in the replacement of Neanderthals by modern humans (Shea and Sisk 2010).

We suggest that rather than causing qualitative changes in patterns of coalitionary killing, changes in weapons technology would have provided continued pressure for trends that likely began with the use of simple weapons by australopiths. Within-group fighting would become increasingly dangerous. In a world where an angered group member could kill with a single poisoned arrow, maintaining good intragroup relations would be of paramount importance. Between-group fighting would also become increasingly dangerous, perhaps requiring greater rewards to make fighting worthwhile for participants (Glowacki and Wrangham 2013). Additionally, as Gat (1999a) argues, the use of manufactured weapons created a "first strike capability" in humans. In contrast to animals armed with horns, claws, or teeth, humans can be disarmed, for example, when sleeping. This unusual feature of human weapons may destabilize intergroup relations by favoring preemptive surprise attacks (Gat 1999a).

Weapons, whether used for hunting, fighting, or both, enabled humans to attain an unprecedented level of ecological dominance (Alexander 1990). Circumstantial evidence suggests that as early humans expanded into the carnivore niche, they profoundly impacted the populations of competitors and prey species alike, with species of large carnivores (Werdelin and Lewis 2013) and terrestrial primates (Klein 1988) becoming extinct. Anthropologists have long speculated that violent competition with modern humans contributed to the extinction of Neanderthals, supposing that modern humans were more effective because of either larger social groups (Gat 1999b) or more sophisticated weapons (Shea 2003). Hortolà and Martínez-Navarro (2013) argue that "Neanderthal extinction should be seen as a mere branch of the Quaternary Megafaunal extinction," in which killing and predation by modern humans caused the extinction of more than 178 large mammal species worldwide. Due to sparse evidence, such explanations remain speculative (Villa and Roebroeks 2014). Nonetheless, such scenarios are consistent with observations from history that asymmetries in weapons technology have led repeatedly to wars of conquest and sometimes extermination (Diamond 1997; Morris 2014).

Benefits Gained by Aggressors

To the extent that australopiths engaged in intergroup aggression, aggressors likely gained benefits similar to those gained by chimpanzees: increased territory size, improved access to key resources, and repelling competitors for mates. Similarly, intergroup conflict in *Homo* likely involved direct benefits similar to patterns found in chimpanzees and foragers. Successful groups would have gained increased access to territory that included resources such as water sources and hunting areas.

Among small-scale societies with warfare, including band-level societies, warriors are almost universally accorded cultural benefits such as status (Glowacki and Wrangham 2013). It is impossible to directly infer when uniquely human cultural benefits such as status and prestige developed. However, status and prestige likely originated with the development of group living with advanced social learning capacities (Henrich and Gil-White 2001). Several lines of evidence indicate that such capacities were probably present at least from *Homo erectus*. Evidence for increased meat consumption around two million years ago has been attributed to the emergence of cooperative foraging behaviors (Anton et al. 2014), including taking meat back to camp to share with others rather than consuming it directly (Potts 2012). The development of the Acheulean stone tool tradition around 1.76 million year ago (Lepre et al. 2011; Beyene et al. 2013) may have required teaching and significant social learning for these sophisticated flint-knapping techniques to spread. Some faunal assemblages from large game dating to approximately 1.7 million years ago have been interpreted as evidence of male displays and status-seeking behaviors (O'Connell et al. 2002). By the emergence of *Homo*, the sophistication of cultural behavior certainly far exceeded that exhibited by chimpanzees. It seems likely that by the time modern *Homo sapiens* emerged, within-group status mechanisms would have been present. However, these were unlikely to have been particularly well developed until the emergence of complex cultural systems found in more recent groups.

Multilevel Societies and Prospects for Peaceful Intergroup Relations

One of the striking differences between chimpanzees and humans is that intergroup relations in chimpanzees are always hostile, whereas in multilevel human societies, relationships among various kinds of groups are sometimes peaceful. The origin of multilevel societies, either in australopiths or in later hominins, would have affected the prospects for peaceful intergroup relations in two major ways: (1) by increasing the range of group identities and (2) by increasing potential for affiliative interactions between kin of both sexes (Chapais 2010, this volume; Swedell and Plummer 2012).

First, in a closed or single-level society, group identity is essentially binary: ingroup (friends) or outgroup (foes). Immigrants must negotiate the transition from foe to friend when they join a new group, and they do not always manage this successfully (Pusey et al. 2008). In contrast, in a multilevel society, individuals have a hierarchy of group identities, such as one-male unit, band, and troop (Swedell and Plummer 2012). This raises the possibility that members of other social groups are evaluated on a sliding scale, from friendly to hostile. For example, within a band, males in one-male units may compete with other band males for access to females or feeding sites, but may join together with those same males to exclude other troops from occupying their sleeping cliffs or abducting females. Thus, if australopiths lived in multilevel groups, they may have had a similar range of group identities, with intergroup relationships ranging from more friendly to more hostile depending on the level.

FIGURE 13.5. Members of a multilevel primate society: gelada baboons (*Theropithecus gelada*). Here a former leader male (Tony) carries an infant (presumably his) on his back while facing off against the current leader male (Ptolomey, to the left, not pictured), while a male coalition partner (Cthulu, behind Tony, to the left) provides support. (Photographed at Guassa, Ethiopia, by M. L. Wilson.)

Second, multilevel societies can make it possible for both sexes to maintain affiliative relationships with kin (Chapais 2010, this volume). In chimpanzees, males stay with male kin, but when females transfer to a new community, they are cut off from female kin for the rest of their lives (unless multiple females immigrate to the same community). In contrast, in multilevel societies, both males and females may have opportunities to interact with kin across some social boundaries. For example, in hamadryas baboons, both males and females commonly stay within their natal band (Swedell et al. 2011; Städele et al. 2016). Females are usually separated from close kin when forcibly transferred from their natal unit by males; nonetheless, because females generally stay close to other female kin, the possibility exists for affiliative relations among female kin, though this may rarely be expressed (Swedell et al. 2011; Städele et al. 2016). Such a scenario might eventually favor the evolution of both sexes maintaining kinship ties, as seen in modern humans (Rodseth et al. 1991; Chapais, this volume).

Language Evolution in Earlier *Homo*

Language would have caused many changes that are relevant for intergroup relations. Different languages divide the social world into two distinct groups: Us and Them. Those who speak a language or dialect that we can understand are real people: we can communicate with them, reason with them, and negotiate peace. When language boundaries are sharp, relations may be more hostile (Marlowe 2005). For example, in the Andaman Islands, speakers of mutually intelligible dialects also shared cultural practices for negotiating peace, whereas across the linguistic divide relations were implacably hostile (Kelly 2000). However, fighting does occur among speakers of the same language; because people who share a language commonly share borders, they may also be more likely to have conflicts.

Language can facilitate relationships across group boundaries, even at the tribal level. It helps men and women maintain lifelong ties with both kin and nonkin members of other groups. For example, because female chimpanzees sometimes disperse to neighboring communities, males sometimes encounter their female relatives during intergroup encounters. In the few such encounters that have been observed between known relatives, there is little evidence of kin recognition. However, even if individuals do recognize each other as kin, lacking language, they cannot easily communicate this to one another. Language enables people to say to apparent strangers, "I am the friend of your uncle," or "I am your half-sister, born after your father left your group and joined our group," or "I am the cousin you have never met, but I knew your mother."

With language, group membership does not depend on physical proximity. Even people who have never previously met can use language or other symbolic communication (such as tattoos or other ornaments) to communicate that they are members of the same social group. Language thus enables social groups to expand beyond the number of people who can afford to forage together at a given time. Language also facilitates conflict resolution by enabling peaceful communication, such as negotiation of peaceful relations after periods of strife. Language can produce mutually beneficial relationships such as the promotion of trade networks.

Language facilitates planning and coordination, enabling advanced tactics and strategies found only in humans. Chimpanzees going on border patrol communicate with one another through body posture, facial expression, and vocalizations (in the case of border patrols, by staying unusually quiet). But they cannot plan their course of action in detail. In contrast, people can make elaborate plans for raids, setting up ambushes, inviting rivals to treacherous feasts, and so forth.

Kinship, Marriage, and Trade in *Homo*

In hunter-gatherers, three factors are particularly important for promoting peaceful intergroup interactions: (1) long-term recognition of bilateral kinship, (2) marital exchange, and (3) exchange of goods through gifts and trade. Language permeates all of these interactions, and perhaps none of them were possible before the evolution of language.

Humans are unique among primates in that both males and females maintain lifelong relationships with kin of both sexes (Rodseth et al. 1991). Bilateral recognition of kinship, combined with dispersal from the natal group by one or both sexes, results in a complex network of kinship patterns among social groups (Chapais, this volume). Recognizing kin across group boundaries may promote peaceful interactions. Even during battles, men may position themselves to avoid injuring or killing kinsmen among the opposition (Meggitt 1977). Because language enables people to maintain relations with kin despite long separations, and to discover kinship relations that would be impossible to ascertain without language, such extensive recognition of bilateral kinship may not have existed prior to the evolution of language. Nonetheless, a multilevel social structure may have enabled both sexes to maintain kin relations prior to the evolution of language.

Second, people commonly use marriage as a way to build kinship ties across groups (Chapais 2010, this volume). Exchanging spouses connects groups through kinship, and thus promotes kin-based affiliation. Spousal exchange also creates reciprocal obligations, closely related to those involved in the exchange of goods through gifts and trade.

Third, the exchange of goods has an enormous impact on intergroup relations. In chimpanzees, intergroup encounters are always zero-sum events: any territory gained by group A represents a loss for group B. With trade, however, intergroup interactions can result in mutual benefits. Trade can provide access to otherwise unobtainable materials, such as obsidian, needed to make weapons. Evidence of trade begins in the late Stone Age in Africa, with the long-distance exchange of stone for tools (McBrearty and Brooks 2000), and likely depended both on language (to negotiate trade) and on the development of a sufficiently sophisticated material culture to make trade worthwhile.

Conclusions

In summary, chimpanzees and modern humans share many intriguing similarities in their patterns of intergroup aggression, but also striking differences (Table 13.1). In both species, attacks are conducted mainly by coalitions of males, who defend territories and attack rivals when favorable numerical advantages reduce the risks to attackers. If the LCA was very much like a chimpanzee, then the roots of intergroup violence in humans may be deep indeed. However, our understanding of the LCA remains sketchy, and several unusual features of australopiths suggest a more complicated history.

One great challenge for answering these questions is determining the likely social behavior of australopiths. Recent advances, particularly in stable isotope studies, have provided new windows into the past (Nelson and Hamilton, this volume), and we can hope that future research will continue to provide both new fossils and new means of gaining information from them.

TABLE 13.1. Summary of hypothesized traits for LCA (if chimpanzee-like) and hominins.

	LCA	Australopiths	Earlier Homo	Later Homo
Fission-fusion dynamics	Yes	Yes	Yes	Yes
Intergroup hostility	Yes	Yes	Yes	Yes
Male coalitions	Yes	Maybe	Yes	Yes
Territorial (when economical)	Yes	Yes	Yes	Yes
Coalitionary killing	Yes	Rare	Yes	Yes
Weapons	Simple (sticks, stones; rarely used)	Simple (sticks, stones; commonly used)	More complex (stone, shaped; routinely used)	Complex (stone, composite; routinely used)
Rewards to aggressors	Direct (territory, mating)	Direct (territory, acquiring females)	Direct and indirect (cultural?)	Direct and indirect (cultural)
Multilevel societies	No	Maybe	Yes	Yes
Language	No	No	Proto-language?	Yes

Because australopiths likely experienced seasonal food scarcity, and lived where nighttime refuges from predators were scarce, one possibility is that australopiths lived in multilevel societies, much like those of dry country papionins. Given that coalitionary killing has not been observed among multilevel papionins, coalitionary killing may have occurred rarely, if at all, among australopiths living in similar societies. Whatever form australopith societies took, coalitionary killing seems likely to have varied among populations and species, being more frequent at some times and places than at others.

By the origin of *Homo*, however, several factors appear likely to have increased rates of coalitionary killing: more efficient walking and running, leading to longer daily travel distances; more reliance on meat, leading to increased trophic level and thus a requirement for larger territories; increased energy from meat and food-processing technology, providing more energy, leading to more rapidly growing populations, further increasing territorial competition; increasingly effective weapons, providing greater advantages to numerically superior attackers; and stronger alliances among males due to reduced reproductive skew.

By the origin of *Homo sapiens,* if not earlier, the evolution of language multiplied the complexity of intergroup relations, creating ethnolinguistic groups, lifelong relationships among kin and nonkin, leading to larger alliances; prospects for peaceful interactions and mutual benefits of trade; and the ability to communicate long memories of past wrongs, leading to blood revenge and feuds. Intergroup competition fueled ongoing arms races, leading to improvements in projectile technology, body armor, and other tools of war. Archaeology, ethnology, and history document more recent elaborations of warfare within particular human groups: age-graded warrior sets in pastoralist tribes, hierarchically organized armies with professional soldiers in states; industrialized world wars; and, more recently, cyber warfare and "chair-forces" of remote-controlled drones (Morris 2014). At the same time, and despite the seemingly endless parade of dreadful news on our televisions and social media feeds, our species has demonstrated a remarkable ability to forge peaceful intergroup relations across increasingly large geographic and political scales (Pinker 2011). We may have inherited demonic tendencies from our ancestors, but we can learn to listen to our angels.

Summary

Controversy continues over whether the roots of warfare are deep, shallow, or somewhere in between. One possibility is that the roots of warfare extend at least as deep as the last common ancestor (LCA) of humans, chimpanzees, and bonobos. Alternatively, coalitionary killing may have evolved separately in chimpanzees and humans, but for similar reasons. We reviewed patterns of intergroup violence in chimpanzees and hunter-gatherers, and evidence for whether earlier members of our lineage likely had similar patterns of intergroup aggression. In doing so, we focused on key similarities and differences between chimpanzees and humans and their potential effects on the evolution of intergroup aggression. Humans and chimpanzees resemble one another in having (1) fission-fusion societies, (2) intergroup hostility, (3) male coalitions, (4) territorial behavior, and (5) coalitionary killings. At the same time, humans differ strikingly from chimpanzees in having (1) weapons, (2) different benefits gained by aggressors, (3) multilevel societies, and (4) language. If the LCA closely resembled chimpanzees, then the roots of intergroup violence in humans may be deep indeed. However, our understanding of the LCA remains sketchy, and several unusual features of australopiths suggest a more complicated history.

Acknowledgments

MLW thanks the Erasmus Mundus Master Programme in Evolutionary Biology and the Institut des Sciences de l'Evolution-Montpellier for support during the writing of this chapter, and Michel Raymond, Elise Huchard, and Timo Brockmeyer for discussion. We thank Carrie Miller, Gilliane Monnier, Martin Muller, David Pilbeam, Richard Wrangham, and anonymous reviewers for helpful comments on the manuscript.

References

Alexander, R. D. 1990. How did humans evolve? Reflections on the uniquely unique species. Ann Arbor: Museum of Zoology, University of Michigan.

Allen, M. W. 2014. Hunter-gatherer conflict: The last bastion of the pacified past? In M. W. Allen and T. L. Jones, eds., *Violence and Warfare among Hunter-Gatherers*. Walnut Creek, CA: Left Coast Press.

Allen, M. W., and T. L. Jones. 2014. *Violence and Warfare among Hunter-Gatherers*. Walnut Creek, CA: Left Coast Press.

Anton, S. C., W. R. Leonard, and M. L. Robertson. 2002. An ecomorphological model of the initial hominid dispersal from Africa. *Journal of Human Evolution* 43: 773–785.

Anton, S. C., R. Potts, and L. C. Aiello. 2014. Evolution of early *Homo*: An integrated biological perspective. *Science* 345: 1236828.

Aureli, F., C. M. Schaffner, C. Boesch, S. K. Bearder, J. Call, C. A. Chapman, R. Connor, A. Di Fiore, R. I. M. Dunbar, S. P. Henzi, K. Holekamp, A. H. Korstjens, R. Layton, P. Lee, J. Lehmann, J. H. Manson, G. Ramos-Fernandez, K. B. Strier, and C. P. van Schaik. 2008. Fission-fusion dynamics: New research frameworks. *Current Anthropology* 49: 627–654.

Beyene, Y., S. Katoh, G. WoldeGabriel, W. K. Hart, K. Uto, M. Sudo, M. Kondo, M. Hyodo, P. R. Renne, G. Suwa, and B. Asfaw. 2013. The characteristics and chronology of the earliest Acheulean at Konso, Ethiopia. *Proceedings of the National Academy of Sciences* 110: 1584–1591.

Binford, L. R. 2001. *Constructing Frames of Reference*. Berkeley: University of California Press.

Bingham, P. M. 2000. Human evolution and human history: A complete theory. *Evolutionary Anthropology* 9: 248–257.

Boesch, C., and H. Boesch-Achermann. 2000. *The Chimpanzees of the Taï Forest: Behavioral Ecology and Evolution*. Oxford: Oxford University Press.

Boesch, C., C. Crockford, I. Herbinger, R. M. Wittig, Y. Moebius, and E. Normand. 2008. Intergroup conflicts among chimpanzees in Taï National Park: Lethal violence and the female perspective. *American Journal of Primatology* 70: 519–532.

Boesch, C., G. Hohmann, and L. F. Marchant, eds. 2002. *Behavioural Diversity in Chimpanzees and Bonobos*. Cambridge: Cambridge University Press.

Bramble, D. M., and D. E. Lieberman. 2004. Endurance running and the evolution of Homo. *Nature* 432: 345–352.

Bridges, E. 1948. *Uttermost Part of the Earth*. London: Hodder & Stoughton.

Brown, J. L. 1964. The evolution of diversity in avian social systems. *Wilson Bulletin* 76: 160–169.

Burch, E. S., Jr. 2005. *Alliance and Conflict: The World System of the Inupiaq Eskimos*. Lincoln: University of Nebraska Press.

Burch, E. S., Jr. 2007. Traditional native warfare in western Alaska. In R. J. Chacon and R. G. Mendoza, eds., *North American Indigenous Warfare and Ritual Violence*, 11–29. Tucson: University of Arizona Press.

Cashdan, E. 1983. Territoriality among human foragers: Ecological models and an application to four Bushman groups. *Current Anthropology* 24: 47–66.

Cerling, T. E., J. G. Wynn, S. A. Andanje, M. I. Bird, D. K. Korir, N. E. Levin, W. Mace, A. N. Macharia, J. Quade, and C. H. Remien. 2011. Woody cover and hominin environments in the past 6 million years. *Nature* 476: 51–56.

Chagnon, N. A. 1988. Life histories, blood revenge, and warfare in a tribal population. *Science* 239: 985–992.

Chapais, B. 2010. *Primeval Kinship: How Pair-Bonding Gave Birth to Human Society*. Cambridge, MA: Harvard University Press.

Chapais, B. 2013. Monogamy, strongly bonded groups, and the evolution of human social structure. *Evolutionary Anthropology* 22: 52–65.

Churchill, S. E., R. G. Franciscus, H. A. McKean-Peraza, J. A. Daniel, and B. R. Warren. 2009. Shanidar 3 Neandertal rib puncture wound and paleolithic weaponry. *Journal of Human Evolution* 57: 163–178.

Clastres, P. 1972. The Guyaki. In M. G. Bicchieri, ed., *Hunters and Gatherers Today: A Socioeconomic Study of Eleven Such Cultures in the Twentieth Century*. New York: Holt, Reinhart and Winston.

Clastres, P. 2000. *Chronicle of the Guayaki Indians*. New York: Zone.

Copeland, S. R., M. Sponheimer, D. J. de Ruiter, J. A. Lee-Thorp, D. Codron, P. J. le Roux, V. Grimes, and M. P. Richards. 2011. Strontium isotope evidence for landscape use by early hominins. *Nature* 474: 76–U100.

Crofoot, M., and R. W. Wrangham. 2010. Intergroup aggression in primates and humans: The case for a unified theory. In P. M. Kappeler and J. B. Silk, eds., *Mind the Gap*. Berlin: Springer.

Darwin, C. 1871. *The Descent of Man and Selection in Relation to Sex*. New York: Modern Library.

Diamond, J. 1997. *Guns, Germs, and Steel: The Fates of Human Societies*. New York: W. W. Norton.

Drucker, P. 1951. *The Northern and Central Nootkan Tribes*. Washington, DC: Smithsonian Institution.

Dunbar, R. I. M. 1992. A model of the gelada socio-ecological system. *Primates* 33: 69–83.

Dyson-Hudson, R., and E. A. Smith. 1978. Human territoriality: An ecological assessment. *American Anthropologist* 80: 21–42.

Ember, C. R. 1978. Myths about hunter-gatherers. *Ethnology* 17: 439–448.

Faurie, C., and M. Raymond. 2003. Handedness: Neutral or adaptive? *Behavioral and Brain Sciences* 26: 220.

Ferguson, R. B. 2011. Born to live: Challenging killer myths, In R. W. Sussman and C. R. Cloninger, eds., *Origins of Altruism and Cooperation*, 249–270. New York: Springer.

Fiske, A. P., and T. S. Rai. 2015. *Virtuous Violence: Hurting and Killing to Create, Sustain, End, and Honor Social Relationships*. Cambridge: Cambridge University Press.

Fitch, W. T. 2010. *The Evolution of Language*. Cambridge: Cambridge University Press.

Fry, D. P. 2011. Human nature: The nomadic forager model. In R. W. Sussman and C. R. Cloninger, eds., *Origins of Altruism and Cooperation*, 227–247. New York: Springer.

Fry, D. P., and P. Söderberg. 2013. Lethal aggression in mobile forager bands and implications for the origins of war. *Science* 341: 270–273.

Funk, C. 2010. The Bow and Arrow War days on the Yukon-Kuskokwim Delta of Alaska. *Ethnohistory* 57: 523–569.

Gat, A. 1999a. The pattern of fighting in simple, small-scale, prestate societies. *Journal of Anthropological Research*. 55: 563–583.

Gat, A. 1999b. Social organization, group conflict and the demise of Neanderthals. *Mankind Quarterly* 39: 437–454.

Gat, A. 2006. *War in Human Civilization*. Oxford: Oxford University Press.

Gat, A. 2015. Proving communal warfare among hunter-gatherers: The quasi-Rousseauan error. *Evolutionary Anthropology* 24: 111–126.

Ghiglieri, M. P. 1989. Hominid sociobiology and hominid social evolution. In P. G. Heltne and L. A. Marquardt, eds., *Understanding Chimpanzees*, 370–379. Cambridge, MA: Harvard University Press.

Gilby, I. C., M. L. Wilson, and A. E. Pusey. 2013. Ecology rather than psychology explains co-occurrence of predation and border patrols in male chimpanzees. *Animal Behaviour* 86: 61–74.

Glowacki, L., and R. W. Wrangham. 2013. The role of rewards in motivating participation in simple warfare. *Human Nature* 24: 444–460.

Glowacki, L., and R. W. Wrangham. 2015. Warfare and reproductive success in a tribal population. *Proceedings of the National Academy of Sciences* 112: 348–353.

Gómez, J. M., M. Verdú, A. González-Megías, and M. Méndez. 2016. The phylogenetic roots of human lethal violence. *Nature* 538: 233–237.

Goodall, J. 1986. *The Chimpanzees of Gombe: Patterns of Behavior.* Cambridge, MA: Harvard University Press.

Goodall, J., A. Bandora, E. Bergmann, C. Busse, H. Matama, E. Mpongo, A. Pierce, and D. Riss. 1979. Intercommunity interactions in the chimpanzee population of the Gombe National Park. In D. A. Hamburg and E. R. McCown, eds., *The Great Apes,* 13–53. Menlo Park, CA: Benjamin/Cummings.

Grove, M., E. Pearce, and R. I. M. Dunbar. 2012. Fission-fusion and the evolution of hominin social systems. *Journal of Human Evolution* 62: 191–200.

Grubb, W. B. 1911. *An Unknown People in an Unknown Land.* London: Seeley, Service.

Gurven, M. 2013. Human survival and life history in evolutionary perspective. In J. C. Mitani, J., Call, P. M. Kappeler, R. A. Palombit, and J. B. Silk, eds., *The Evolution of Primate Societies.* Chicago: University of Chicago Press.

Haas, J. 2001. Warfare and the evolution of culture. In T. D. Price and G. Feinman, eds., *Archaeology at the Millennium: A Sourcebook.* New York: Kluwer Academic/Plenum.

Hamilton, W. D. 1964. The genetic evolution of social behavior, I, II. *Journal of Theoretical Biology* 7: 1–52.

Heinz, H. J. 1972. Territoriality among Bushmen in general and Ko in particular. *Anthropos* 67: 405–416.

Henrich, J., and F. J. Gil-White. 2001. The evolution of prestige: Freely conferred deference as a mechanism for enhancing the benefits of cultural transmission. *Evolution and Human Behavior* 22: 165–196.

Herbinger, I., C. Boesch, and H. Rothe. 2001. Territory characteristics among three neighboring chimpanzee communities in the Taï National Park, Ivory Coast. *International Journal of Primatology* 22: 143–167.

Hernandez-Aguilar, R. A., J. Moore, and R. P. Travis. 2007. Savanna chimpanzees use tools to harvest the underground storage organs of plants. *Proceedings of the National Academy of Sciences* 104: 19210–19213.

Hill, K., M. Barton, and A. M. Hurtado. 2009. The emergence of human uniqueness: Characters underlying behavioral modernity. *Evolutionary Anthropology* 18: 187–200.

Hill, K., and M. A. Hurtado. 1996. *Ache Life History: The Ecology and Demography of a Foraging People.* New York: Aldine de Gruyter.

Hill, K. R., R. S. Walker, M. Bozicevic, J. Eder, T. Headland, B. Hewlett, A. M. Hurtado, F. W. Marlowe, P. Wiessner, and B. Wood. 2011. Co-residence patterns in hunter-gatherer societies show unique human social structure. *Science* 331: 1286–1289.

Hohmann, G., and B. Fruth. 2002. Dynamics in social organization of bonobos (*Pan paniscus*). In C. Boesch, G. Hohmann, and L. F. Marchant, eds., *Behavioural Diversity in Chimpanzees and Bonobos.* Cambridge: Cambridge University Press.

Hortolà, P., and B. Martínez-Navarro. 2013. The Quaternary megafaunal extinction and the fate of Neanderthals: An integrative working hypothesis. *Quaternary International* 295: 69–72.

Kelly, R. C. 2000. *Warless Societies and the Origin of War.* Ann Arbor: University of Michigan Press.

Kelly, R. C. 2005. The evolution of lethal intergroup violence. *Proceedings of the National Academy of Sciences* 102: 15294–15298.
Kelly, R. L. 1995. *The Foraging Spectrum: Diversity in Hunter-Gatherer Lifeways.* Washington, DC: Smithsonian Institution.
Klein, R. 1988. The causes of "robust" australopithecine extinction. In F. F. Grine, ed., *The Evolutionary History of the Robust Australopithecines,* 499–505. New York: Aldine de Gruyter.
Knauft, B. 1991. Violence and sociality in human evolution. *Current Anthropology* 32: 391–428.
Kramer, P. A., and G. G. Eck. 2000. Locomotor energetics and leg length in hominid bipedality. *Journal of Human Evolution* 38: 651–666.
Kummer, H. 1968. *Social Organization of Hamadryas Baboons.* Chicago: University of Chicago Press.
Kutsukake, N., and T. Matsusaka. 2002. Incident of intense aggression by chimpanzees against an infant from another group in Mahale Mountains National Park, Tanzania. *American Journal of Primatology* 58: 175–180.
Lalueza-Fox, C., A. Rosas, A. Estalrrich, E. Gigli, P. F. Campos, A. Garcia-Tabernero, S. Garcia-Vargas, F. Sanchez-Quinto, O. Ramirez, S. Civit, M. Bastir, R. Huguet, D. Santamaria, M. T. P. Gilbert, E. Willerslev, and M. de la Rasilla. 2011. Genetic evidence for patrilocal mating behavior among Neandertal groups. *Proceedings of the National Academy of Sciences* 108: 250–253.
LeBlanc, S. A. 2014. Forager warfare and our evolutionary past. In M. W. Allen and T. L. Jones, eds., *Violence and Warfare among Hunter-Gatherers.* Walnut Creek, CA: Left Coast Press.
LeBlanc, S. A., Register, K. E. 2004. *Constant Battles: Why We Fight.* New York: St. Martin's Griffin.
Lee, R. B. 1979. *The !Kung San: Men, Women and Work in a Foraging Society.* New York: Cambridge University Press.
Lepre, C. J., H. Roche, D. V. Kent, S. Harmand, R. L. Quinn, J. P. Brugal, P. J. Texier, A. Lenoble, and C. S. Feibel. 2011. An earlier origin for the Acheulian. *Nature* 477: 82–85.
Lorenz, K. 1966. *On Aggression.* New York: Harcourt Brace.
Lowen, C., and R. I. M. Dunbar. 1994. Territory size and defendability in primates. *Behavioral Ecology and Sociobiology* 35: 347–354.
Machanda, Z. P., I. C. Gilby, and R. W. Wrangham. 2013. Male-female association patterns among free-ranging chimpanzees *(Pan troglodytes schweinfurthii). International Journal of Primatology* 34: 917–938.
Maher, C. R., and D. F. Lott. 2000. A review of ecological determinants of territoriality within vertebrate species. *American Midland Naturalist* 143: 1–29.
Manson, J. H., and R. W. Wrangham. 1991. Intergroup aggression in chimpanzees and humans. *Current Anthropology* 32: 369–390.
Markham, A. C., S. C. Alberts, and J. Altmann. 2012. Intergroup conflict: Ecological predictors of winning and consequences of defeat in a wild primate population. *Animal Behaviour* 84: 399–403.

Marlowe, F. W. 2005. Hunter-gatherers and human evolution. *Evolutionary Anthropology* 14: 54–67.

Marlowe, F. W. 2010. *The Hadza: Hunter-Gatherers of Tanzania.* Berkeley: University of California Press.

Maschner, H., and O. K. Mason. 2013. The bow and arrow in northern North America. *Evolutionary Anthropology* 22: 133–138.

McBrearty, S., and A. S. Brooks. 2000. The revolution that wasn't: A new interpretation of the origin of modern human behavior. *Journal of Human Evolution* 39: 453–563.

McDonald, M. M., C. D. Navarrete, and M. van Vugt. 2012. Evolution and the psychology of intergroup conflict: The male warrior hypothesis. *Philosophical Transactions of the Royal Society B: Biological Sciences* 367: 670–679.

Meggit, M. 1962. *Desert People: A study of the Walbiri Aborigines of Central Australia.* Sydney: Angus and Robertson.

Meggitt, M. 1977. *Blood Is Their Argument: Warfare among the Mae Enga Tribesmen of the New Guinea Highlands.* Palo Alto, CA: Mayfield.

Mendoza, M. 2007. Hunter-gatherers' aboriginal warfare in Western Chaco. In R. J. Chacon and R. G. Mendoza, eds., *Latin American Indigenous Warfare and Ritual Violence,* 198–211. Phoenix: University of Arizona Press.

Mitani, J. C. 2009. Cooperation and competition in chimpanzees: Current understanding and future challenges. *Evolutionary Anthropology* 18: 215–227.

Mitani, J. C., and P. S. Rodman. 1979. Territoriality: The relation of ranging pattern and home range size to defendability, with an analysis of territoriality among primate species. *Behavioral Ecology and Sociobiology* 5: 241–251.

Mitani, J. C., and D. P. Watts. 2005. Correlates of territorial boundary patrol behaviour in wild chimpanzees. *Animal Behaviour* 70: 1079–1086.

Mitani, J. C., D. P. Watts, and S. J. Amsler. 2010. Lethal intergroup aggression leads to territorial expansion in wild chimpanzees. *Current Biology* 20: R507–R508.

Moffett, M. W. 2012. Supercolonies of billions in an invasive ant: What is a society? *Behavioral Ecology* 23: 925–933.

Morris, I. 2014. *War! What Is It Good For? Conflict and the Progress of Civilization from Primates to Robots.* New York: Farrar, Straus and Giroux.

Nishida, T. 1979. The social structure of chimpanzees of the Mahale Mountains. In D. A. Hamburg and E. R. McCown, eds., *The Great Apes,* 73–121. Menlo Park, CA: Benjamin/Cummings.

Nishida, T. 2012. *Chimpanzees of the Lakeshore: Natural History and Culture at Mahale.* New York: Cambridge University Press.

Nishida, T., N. Corp, M. Hamai, T. Hasegawa, M. Hiraiwa-Hasegawa, K. Hosaka, K. D. Hunt, N. Itoh, K. Kawanaka, A. Matsumoto-Oda, J. C. Mitani, M. Nakamura, K. Norikoshi, T. Sakamaki, L. Turner, S. Uehara, and K. Zamma. 2003. Demography, female life history, and reproductive profiles among the chimpanzees of Mahale. *American Journal of Primatology* 59: 99–121.

Nishida, T., M. Hiraiwa-Hasegawa, T. Hasegawa, and Y. Takahata. 1985. Group extinction and female transfer in wild chimpanzees in the Mahale National Park, Tanzania. *Zeitschrift für Tierpsychologie* 67: 284–301.

Nishida, T., and K. Kawanaka. 1972. Inter-unit-group relationships among wild chimpanzees of the Mahali Mountains. *Kyoto University African Studies* 7: 131–169.

O'Connell, J. F., K. Hawkes, K. D. Lupo, and N. G. B. Jones. 2002. Male strategies and Plio-Pleistocene archaeology. *Journal of Human Evolution* 43: 831–872.

Otterbein, K. F. 2004. *How War Began*. College Station: Texas A&M University Press.

Packer, C., L. Herbst, A. E. Pusey, J. D. Bygott, J. P. Hanby, S. J. Cairns, and M. Borgerhoff Mulder. 1988. Reproductive success of lions. In T. H. Clutton-Brock, ed., *Reproductive Success,* 363–383. Chicago: University of Chicago Press.

Parker, G. A. 1974. Assessment strategy and the evolution of fighting behavior. *Journal of Theoretical Biology* 47: 223–243.

Peterson, N. 1975. Hunter-Gatherer Territoriality: The Perspective from Australia. *American Anthropologist* 77: 53–68.

Philpott, W. 2009. *Bloody Victory: The Sacrifice on the Somme and the Making of the Twentieth Century*. New York: Little, Brown.

Pinker, S. 2011. *The Better Angels of Our Nature: Why Violence Has Declined*. New York: Viking.

Plavcan, J. M. 2012. Sexual size dimorphism, canine dimorphism, and male-male competition in primates: Where do humans fit in? *Human Nature* 23: 45–67.

Potts, R. 2012. Evolution and environmental change in early human prehistory. In D. Brenneis and P. T. Ellison, eds., *Annual Review of Anthropology,* vol. 41, 151–167. Palo Alto, CA: Annual Reviews.

Pradhan, G. R., R. K. Pandit, and C. P. van Schaik. 2014. Why do chimpanzee males attack the females of neighboring communities? *American Journal of Physical Anthropology* 155: 430–435.

Pruetz, J. D., and P. Bertolani. 2007. Savanna chimpanzees, *Pan troglodytes verus,* hunt with tools. *Current Biology* 17: 412–417.

Pusey, A. E. 1980. Inbreeding avoidance in chimpanzees. *Animal Behaviour* 28: 543–582.

Pusey, A. E. 2001. Of genes and apes: Chimpanzee social organization and reproduction. In F. B. M. de Waal, ed., *Tree of Origin,* 9–38. Cambridge, MA: Harvard University Press.

Pusey, A. E., C. Murray, W. R. Wallauer, M. L. Wilson, E. Wroblewski, and J. Goodall. 2008. Severe aggression among female chimpanzees at Gombe National Park, Tanzania. *International Journal of Primatology* 29: 949–973.

Pusey, A. E., G. W. Oehlert, J. M. Williams, and J. Goodall. 2005. The influence of ecological and social factors on body mass of wild chimpanzees. *International Journal of Primatology* 26: 3–31.

Pusey, A. E., and C. Packer. 1987. Dispersal and philopatry. In B. B. Smuts, D. L. Cheney, R. M. Seyfarth, T. T. Struhsaker, and R. W. Wrangham, eds., *Primate Societies*, 250–266. Chicago: University of Chicago Press.

Puts, D. A. 2010. Beauty and the beast: Mechanisms of sexual selection in humans. *Evolution and Human Behavior* 31: 157–175.

Robbins, M. M., and S. C. Sawyer. 2007. Intergroup encounters in mountain gorillas of Bwindi Impenetrable National Park, Uganda. *Behaviour* 144: 1497–1519.

Rodseth, L., and R. W. Wrangham. 2004. Human kinship: A continuation of politics by other means? In B. Chapais and C. M. Berman, eds., *Kinship and Behavior in Primates*, 389–419. New York: Oxford University Press.

Rodseth, L., R. W. Wrangham, A. M. Harrigan, and B. B. Smuts. 1991. The human community as a primate society. *Current Anthropology* 32: 221–254.

Rosenbaum, S., V. Vecellio, and T. Stoinski. 2016. Observations of severe and lethal coalitionary attacks in wild mountain gorillas. *Scientific Reports* 6: 8.

Rudicell, R. S., J. H. Jones, E. E. Wroblewski, G. H. Learn, Y. Li, J. Robertson, E. Greengrass, F. Grossmann, S. Kamenya, L. Pintea, D. C. Mjungu, E. V. Lonsdorf, A. Mosser, C. Lehman, D. A. Collins, B. F. Keele, J. Goodall, B. H. Hahn, A. E. Pusey, and M. L. Wilson. 2010. Impact of simian immunodeficiency virus infection on chimpanzee population dynamics. *PLoS Pathogens* 6: e1001116.

Schaller, G. B., and G. R. Lowther. 1969. The relevance of carnivore behavior to the study of early hominids. *Southwestern Journal of Anthropology* 25: 307–341.

Schwartz, G. T. 2012. Growth, development, and life history throughout the evolution of *Homo*. *Current Anthropology* 53: S395–S408.

Shea, J. J. 2003. Neandertals, competition, and the origin of modern human behavior in the Levant. *Evolutionary Anthropology* 12: 173–187.

Shea, J. J., and M. L. Sisk. 2010. Complex projectile technology and *Homo sapiens* dispersal into western Eurasia. *PaleoAnthropology* 100–122.

Smith, J. E., E. M. Swanson, D. Reed, and K. E. Holekamp. 2012. Evolution of cooperation among mammalian carnivores and its relevance to hominin evolution. *Current Anthropology* 53: S436–S452.

Städele, V., M. Pines, L. Swedell, and L. Vigilant. 2016. The ties that bind: Maternal kin bias in a multilevel primate society despite natal dispersal by both sexes. *American Journal of Primatology* 78: 731–744.

Swedell, L., and T. Plummer. 2012. A papionin multilevel society as a model for hominin social evolution. *International Journal of Primatology* 33: 1165–1193.

Swedell, L., J. Saunders, A. Schreier, B. Davis, T. Tesfaye, and M. Pines. 2011. Female "dispersal" in hamadryas baboons: Transfer among social units in a multilevel society. *American Journal of Physical Anthropology* 145: 360–370.

Trinkaus, E. 2012. Neandertals, early modern humans, and rodeo riders. *Journal of Archaeological Science* 39: 3691–3693.

Tutin, C. E. G. 1979. Mating patterns and reproductive strategies in a community of wild chimpanzees *(Pan troglodytes schweinfurthii)*. *Behavioral Ecology and Sociobiology* 6: 29–38.

Villa, P., and W. Roebroeks. 2014. Neandertal demise: An archaeological analysis of the modern human superiority complex. *PLoS ONE* 9: e96424.

Walker, R. S., K. R. Hill, M. V. Flinn, and R. M. Ellsworth. 2011. Evolutionary history of hunter-gatherer marriage practices. *PLoS ONE* 6: e19066.

Waltes, R., A. G. Chiocchetti, and C. M. Freitag. 2016. The neurobiological basis of human aggression: A review on genetic and epigenetic mechanisms. *American Journal of Medical Genetics Part B: Neuropsychiatric Genetics* 171: 650–675.

Waser, P. M. 1976. *Cercocebus albigena:* Site attachment, avoidance, and intergroup spacing. *American Naturalist* 110: 911–935.

Watts, D. P., and J. C. Mitani. 2001. Boundary patrols and intergroup encounters in wild chimpanzees. *Behaviour* 138: 299–327.

Watts, D. P., M. N. Muller, S. J. Amsler, G. Mbabazi, and J. C. Mitani. 2006. Lethal intergroup aggression by chimpanzees in Kibale National Park, Uganda. *American Journal of Primatology* 68: 161–180.

Werdelin, L., and M. E. Lewis. 2013. Temporal change in functional richness and evenness in the eastern African Plio-Pleistocene carnivoran guild. *PLoS ONE* 8: e57944.

Williams, J. M. 1999. Female strategies and the reasons for territoriality in chimpanzees: Lessons from three decades of research at Gombe. Ph.D. dissertation, University of Minnesota.

Williams, J. M., E. V. Lonsdorf, M. L. Wilson, J. Schumacher-Stankey, J. Goodall, and A. E. Pusey. 2008. Causes of death in the Kasekela chimpanzees of Gombe National Park, Tanzania. *American Journal of Primatology* 70: 766–777.

Williams, J. M., G. Oehlert, J. Carlis, and A. E. Pusey. 2004. Why do male chimpanzees defend a group range? Reassessing male territoriality. *Animal Behaviour* 68: 523–532.

Wilson, M. L., C. Boesch, B. Fruth, T. Furuichi, I. C. Gilby, C. Hashimoto, C. Hobaiter, G. Hohmann, N. Itoh, K. Koops, J. Lloyd, T. Matsuzawa, J. C. Mitani, D. C. Mjungu, D. Morgan, R. Mundry, M. N. Muller, M. Nakamura, J. D. Pruetz, A. E. Pusey, J. Riedel, C. Sanz, A. M. Schel, N. Simmons, M. Waller, D. P. Watts, F. J. White, R. M. Wittig, K. Zuberbühler, and W. R. Wrangham. 2014. Lethal aggression in *Pan* is better explained by adaptive strategies than human impacts. *Nature* 513: 414–417.

Wilson, M. L., J. Cossette, K. Koops, I. Lipende, E. V. Lonsdorf, J. C. Mitani, J. D. Pruetz, N. Simmons, D. Travis, and D. P. Watts. 2015. The most unkindest cut: Genital wounding by chimpanzees. *American Journal of Physical Anthropology* 156(S60): 326.

Wilson, M. L., M. D. Hauser, and R. W. Wrangham. 2001. Does participation in intergroup conflict depend on numerical assessment, range location, or rank for wild chimpanzees? *Animal Behaviour* 61: 1203–1216.

Wilson, M. L., M. D. Hauser, and R. W. Wrangham. 2007. Chimpanzees *(Pan troglodytes)* modify grouping and vocal behaviour in response to location-specific risk. *Behaviour* 144: 1621–1653.

Wilson, M. L., S. M. Kahlenberg, M. T. Wells, and R. W. Wrangham. 2012. Ecological and social factors affect the occurrence and outcomes of intergroup encounters in chimpanzees. *Animal Behaviour* 83: 277–291.

Wilson, M. L., W. Wallauer, and A. E. Pusey. 2004. New cases of intergroup violence among chimpanzees in Gombe National Park, Tanzania. *International Journal of Primatology* 25: 523–549.

Wilson, M. L., and R. W. Wrangham. 2003. Intergroup relations in chimpanzees. *Annual Review of Anthropology* 32: 363–392.

Wrangham, R. W. 1980. An ecological model of female-bonded primate groups. *Behaviour* 75: 262–300.

Wrangham, R. W. 1987. The significance of African apes for reconstructing human social evolution. In W. G. Kinzey, ed., *The Evolution of Human Behavior: Primate Models*, 51–71. Albany: State University of New York Press.

Wrangham, R. W. 1999. The evolution of coalitionary killing. *Yearbook of Physical Anthropology* 42: 1–30.

Wrangham, R. W. 2001. Out of the *Pan*, into the fire: From ape to human. In F. B. M. de Waal, ed., *Tree of Origin*, 119–143. Cambridge: Harvard University Press.

Wrangham, R. 2010. *Catching Fire: How Cooking Made Us Human*. New York: Basic Books.

Wrangham, R. W., M. Crofoot, R. Lundy, and I. C. Gilby. 2007. Use of overlap zones among group-living primates: A test of the risk hypothesis. *Behaviour* 144: 1599–1619.

Wrangham, R. W., and L. Glowacki. 2012. Intergroup aggression in chimpanzees and war in nomadic hunter-gatherers: Evaluating the chimpanzee model. *Human Nature* 23: 5–29.

Wrangham, R. W., M. L. Wilson, and M. N. Muller. 2006. Comparative rates of violence in chimpanzees and humans. *Primates* 47: 14–26.

Young, R. W. 2003. Evolution of the human hand: The role of throwing and clubbing. *Journal of Anatomy* 202: 165–174.

14

Cooperative and Competitive Relationships within Sexes

RICHARD W. WRANGHAM *and* JOYCE BENENSON

In this chapter, we compare within-sex social relationships among humans, chimpanzees, bonobos, and gorillas. A combination of genetic and comparative morphological evidence suggests that the last common ancestor (LCA) of humans, chimpanzees, and bonobos was morphologically more similar to a chimpanzee than to a bonobo, gorilla, australopithecine, or human (Pilbeam and Lieberman, this volume). Accordingly, some likely ecological similarities between the LCA and living chimpanzees include habitat (rainforest), diet (ripe fruits backed up by leaves and piths), locomotor style (terrestrial knuckle-walking, arboreal climbing), and body size (~25–40 kg for females). Our question is whether there is evidence of the LCA and chimpanzees sharing similarities in social behavior as well, specifically in interactions among adults of the same sex and community.

Behavioral similarities among humans and any of the African apes may represent phylogenetic continuities, with implications for understanding the evolution of both human and ape behavior. However, human-chimpanzee putative synapomorphies (derived traits found in both humans and chimpanzees, but not in the other apes) are of special interest because they raise questions of evolutionary persistence and adaptive significance. In line with this idea, researchers have for decades drawn attention to similarities in the social relationships of humans and chimpanzees, particularly those among adult males (de Waal 1982, 1984; Wrangham and Peterson 1996). Here we review recent studies so as to consider the behavioral similarities between chimpanzees and humans in greater detail than before, and also to present equivalent comparisons from data on bonobos and gorillas.

Our review focuses on sex differences within populations. Sex differences provide a useful perspective for cross-species comparisons, partly because they tend to be patterned more consistently than the behaviors themselves. For example, Whiting and Whiting (1975) studied children's behavior in six human cultures. They found that the average scores of girls' and boys' aggressiveness (or intimacy) overlapped extensively between populations: girls in some cultures were scored as being more aggressive (or less intimate) than boys in other cultures. The variation meant that if comparisons were made without regard to population, average aggressiveness (or intimacy) between girls and boys was not clearly different. Yet within each of the same six societies, on average boys were always more aggressive (and less intimate) than girls (Whiting and Whiting 1975). Sex differences within populations were thus more consistent than the behavioral scores were. We suggest that, in a similar way, sex differences within ape populations allow useful comparisons across species, even when differences in ecology, demography, or culture cause substantial behavioral variation among populations.

In addition to the merit of their consistency, we regard sex differences in cooperation and competition as being especially interesting because of their potential contribution to intersexual dominance. Most anthropologists consider humans to be a patriarchal species, a fact of great importance for society generally (Ortner and Whitehead 1981). To quote Rosaldo (1974: 19–21):

> Male, as opposed to female, activities are always recognized as predominantly important ... and cultural systems give authority and value to the roles and activities of men.... Everywhere, from those societies we might want to call most egalitarian to those in which sexual stratification is most marked, men are the locus of cultural value. Some area of activity is always seen as exclusively or predominantly male, and therefore overwhelmingly and morally important.... This observation has its corollary in the fact that everywhere men have authority over women, that they have a culturally legitimated right to her subordination and compliance.

Why is culturally approved patriarchy so pervasive in humans? Various causes have been proposed that are not directly concerned with within-sex relationships, such as greater male physical strength, freedom of males from child-rearing commitments, male interest in controlling female sexuality, language-based ideologies, and male control of resources (Smuts 1995). In addition, however, the struggle for power between men and women is often seen as being influenced by the degree to which individuals are effective in forming alliances with other members of the same sex (Smuts 1995). A question that arises, therefore, is whether male success in using alliances results partly from the evolution of psychological adaptations for cooperation. We consider this question here by comparing the dynamics of cooperation and competition within sexes of humans and the African great apes.

To anticipate our conclusion, we find that the human social relationships considered here tend to resemble those of chimpanzees more than other African apes. This raises the question whether the similarities between humans and chimpanzees are homologous, which leads us to consider how certain aspects of human social relationships might have evolved from a chimpanzee-like system. Current evidence indicates that male success in using alliances is a human-chimpanzee synapomorphy. We suggest that male humans and chimpanzees were selected for effective use of male-male alliances as a result of their vulnerability to lethal intergroup violence. The bonobo system, in which male alliances are less important, would accordingly have been derived.

We begin our species reviews with a brief account of social structure.

Social Structure of the Community: Grouping and Philopatry

Humans

Humans have lived for several hundred thousand years as nomadic hunter-gatherers, who currently form relatively discrete ethnolinguistic societies. Warm-climate, nonequestrian nomadic groups of hunter-gatherers average 991 (Old World) to 1,451 individuals (New World) (median 528 and 875, respectively) and control access to a shared territory (Marlowe 2005; Wilson and Glowacki, this volume). Within such societies, the smallest level of consistent grouping is the household, a domestic group that incorporates "women and infant children, aspects of child care, commensality, and the preparation of food" (Rosaldo 1980: 398). These two levels, the society and the household, are regarded as the most universal of human social units (Rodseth and Novak 2000).

Various kinds of social groups occur at intervening levels. The most prominent in warm-climate, nonequestrian nomadic hunter-gatherers is the semistable band (or camp). Bands are composed of a median of twenty-five to thirty-two individuals (Old World and New World, respectively). Based on data assembled by Marlowe (2005), this means that the median number of bands per ethnolinguistic group averages twenty-one (Old World) to twenty-seven (New World).

Within bands the average number of co-resident adults ranges from twelve to twenty-two, that is, about half the band, with more females than males (Hill et al. 2011; Dyble et al. 2015). Friendly relationships tend to predominate among bands, permitting easy movement between them (Chapais 2009). This leads to substantial variation in kin composition. For example, the widespread obligations of bride service mean that after a man marries he often leaves his parents to join a new band and live with his wife's family for a few years, before returning home with them (Alvarez 2004). On the other hand, a divorced woman might reside with her natal family until she remarries (an important dynamic given that divorce can be common; Blurton Jones et al. 2000). Or married individuals can live in a different band from either of their parents. Such variety means that there is considerable variation in the degree of philopatry by each sex, and that bands are not kin groups: less than 10 percent of individuals in thirty-two well-studied recent societies were generally primary kin (descendants or siblings) (Hill et al. 2011). Nevertheless, brothers have been recorded to be in the same band more often than sisters, because at the band level, male philopatry tends to be more common than female philopatry (Rodseth et al. 1991; Hill et al. 2011). We also note that although many recent studies have stressed that there is only a weak tendency for patrilocality in contemporary nomadic hunter-gatherers, these studies have been conducted in societies where war has largely ceased (Hill et al. 2011). Since war tends to be associated with male fraternal kin groups (Otterbein 1994), it is likely that the degree of male philopatry was higher when hunter-gatherers lived in a world of hunter-gatherers and therefore had more frequent war (Wrangham and Glowacki 2012).

Like humans everywhere, grouping in hunter-gatherers is fission-fusion, in the sense that during the day individuals can move alone or in subgroups

FIGURE 14.1. Foraging by hunter-gatherers is commonly segregated by sex. Here a group of Hadza women process baobab fruits. Photo by Martin N. Muller.

of changeable size and composition. Every night band members reunite in camp, so sleeping can be said to occur in a single camp-group, albeit broken into hut-groups that are largely familial. Mixed-sex groups also occur in camp by day, depending on how many people have left camp on foraging excursions or other activities. Foraging, in contrast, is generally segregated by sex. Women forage mainly for plants, burrowing animals, or shellfish, while men tend to hunt large animals, either alone or in small parties (Figure 14.1). Outside camp, therefore, mixed-sex foraging parties tend to be uncommon, but they may be formed for cooperative activities such as poisoning fish or driving prey (Kelly 1995).

In farming societies and among complex hunter-gatherer groups, the core human patterns of households belonging to a society that controls a shared territory, and of daily activities involving fission-fusion working subgroups, are universal. Farming societies vary from being occasionally smaller than those of hunter-gatherers (e.g., some Amazonian horticultural groups) to millions or more (in a nation-state). Within them there are extensive variations on the intermediate grouping levels between household and society (e.g., men's houses, bachelor camps, villages, cities) (Rodseth et al. 1991).

In nomadic hunter-gatherers, relationships among bands (like those among farming villages) are mostly peaceful, and when tensions mount to violence, explicit peacemaking mechanisms tend to come into play (Kelly 2000). By contrast, relations between neighboring ethnolinguistic groups (societies) are consistently tense and hostile, such that there is a nearly constant potential for violence (Wilson and Glowacki, this volume). This implies a steady selection pressure on males to form relationships that are effective in intersociety warfare.

Chimpanzees and Bonobos

Chimpanzees and bonobos resemble each other in their patterns of community organization and grouping. Both species live in communities that defend stable home ranges as territories (Mitani 2009; Furuichi 2011). Table 14.1 shows that eastern chimpanzee communities average forty-eight individuals, 41 percent more than western chimpanzee communities (thirty-four), and 54 percent more than bonobo communities (thirty-one). This suggests that eastern chimpanzees have relatively large communities, but more data are needed to be sure if the differences are representative. One community of western chimpanzees at Taï had seventy-nine individuals before it went into a long-term decline (Boesch and Boesch-Achermann 2000).

In both species, adult females normally outnumber adult males, but the adult sex ratio varies over time and among communities. Importantly, considering that females are relatively powerful in bonobos, bonobo communities average no more adult females per male than those of chimpanzees (Table 14.1).

In both species, a very strong tendency for males to be philopatric means that adult males are more closely related to each other within communities than are adult females (Langergraber et al. 2007; Furuichi 2011). In eastern chimpanzees, the number of adult or adolescent males within a community averages 11.5, rising to more than thirty (Mitani et al. 2010). Though any two males may be maternally or paternally related, nevertheless the cohort is large enough that many dyads are essentially unrelated (Langergraber et al. 2007). Smaller communities in western chimpanzees and bonobos may mean that male-male kin relationships are more prevalent.

In both species, females typically transfer to a new community at adolescence, but in at least two chimpanzee communities, some females (up to

TABLE 14.1. Community size and sex ratio in chimpanzees and bonobos.

Clade	No. Populations	No. Communities	Years of Study	Community Size	Adult Sex Ratio (females per male)
Bonobo	4	5	21	31.2 (21–44)	1.8 (1.4–2.7)
Western chimpanzee	3	6	24	34.1 (7–43)	2.9 (0.6–3.5)
Eastern chimpanzee	7	12	18.5	48.1 (16–144)	1.5 (0.7–2.7)

Note: Data are summarized from Wilson et al. (2014: extended data figure 1), with the addition of Georgiev et al. (2011) for the Kokolopori bonobo site. Numbers for "Years of Study," "Community Size," and "Adult Sex Ratio" are medians across communities; ranges are in parentheses. Community size includes all ages. Populations (with number of communities) are as follows. Bonobos: Lomako (2), LuiKotale (1), Wamba (1), Kokolopori (1); western chimpanzees: Bossou (1), Fongoli (1), Taï (1); eastern chimpanzees: Budongo (1), Gombe (3), Goualougo (1), Kalinzu (1), Kibale (3), Kyambura (1), Mahale (2).

50 percent in Gombe) breed in their natal community (Stumpf et al. 2009). Therefore, as in humans, some brother-sister co-residence can occur in chimpanzees. Mother-son co-residence also occurs in both species, but is more important in bonobos because mothers provide critical coalitionary support to their adult sons competing for dominance rank (Surbeck and Hohmann 2013). A comparison of five chimpanzee and three bonobo communities indicates that adult male bonobos are more likely to have mothers alive in the same community, apparently because female bonobos survive better (Schubert et al. 2013).

Like humans, both species have fission-fusion grouping, with modal subgroup (party) size in the range of five to fifteen, and sometimes more. But unlike humans, in both species subgroups are mostly mixed-sex (Boesch and Boesch-Achermann 2000). During food-rich periods and when adult females are sexually attractive, subgroups tend to be larger, up to a majority of the entire community, but at any level of food abundance adults can sometimes be found traveling alone (males) or (in the case of mothers) only with their offspring (Chapman et al. 1994; Hashimoto et al. 2001). Relative party size (subgroup size as a percentage of the community size) tends to be higher in bonobos (27–51 percent) than in chimpanzees (9–30 percent), indicating that bonobos are in general somewhat more gregarious (Furuichi 2009, 2011).

Despite the many similarities in community structure and fission-fusion, the two species differ strikingly in the relative sociability of each sex. In chimpanzees, males are the more gregarious, affiliative, and dominant sex;

in bonobos, females take those roles. Thus, within mixed-sex subgroups males tend to outnumber females in chimpanzees, whereas in bonobos females outnumber males (Furuichi 2009). In chimpanzees, within-sex grooming is more common among males than females, whereas in bonobos it is more common among females (Furuichi 2009). In chimpanzees, all males are almost always dominant to all females (Boesch and Boesch-Achermann 2000), whereas in bonobos females tend to be higher-ranked than males, including occupying the highest positions in a mixed-sex hierarchy (Furuichi 2011; Surbeck and Hohmann 2013; Hare and Wrangham, this volume). The species differences in sociability include both cooperation and competition, which we consider below.

Gorillas

The grouping system of gorillas differs markedly from those in humans, chimpanzees, or bonobos, because gorillas live in stable groups (troops) that forage, travel, and sleep as a single unit (Harcourt and Stewart 2007). Fission-fusion is limited to relatively brief separations of two subgroups spending a few hours apart before reuniting. Groups of mountain gorillas, the best-studied clade, last on average ten years, and most often consist of one top-ranking (silverback) adult male and three to five unrelated adult females with their growing offspring, for an average of eight total individuals (excluding infants) across all populations (Robbins et al. 2004). Additional solitary males or bachelor groups of adolescent and adult males roam looking for adult females with whom they could begin a breeding group.

Groups do not defend a territory; there is much overlap of home ranges, and their locations can vary across years. Breeding groups, solitary males, and bachelor groups of males generally meet each other without any violence, because resident silverbacks tend to be successful in intimidating outside males, and resident females prefer to remain with the father of their infants. However, occasionally adult females without infants leave and join another male, an interaction normally accompanied by aggression between winning and losing males (Harcourt and Stewart 2007). When conflicts happen they are often intense. Infanticide can occur, and in recent years at least three alpha males have been killed in fights with solitary males (Rosenbaum et al. 2016). The risk of being severely attacked might explain why breeding groups occasionally initiate violent, potentially lethal, attacks on solitary males, possibly as a preemptive move (Rosenbaum et al. 2016).

Around 40 percent of groups in mountain gorillas, but only 5 percent in western gorillas, are multimale, typically including an alpha male, his male kin and nonkin, and more adult females than are found in a one-male group (Bradley et al. 2005; Robbins et al. 2016). Groups can number as many as sixty-five individuals, including nine adult males (Rosenbaum et al. 2016). Mothers leave multimale groups at lower rates than they leave one-male groups (Robbins et al. 2013). Earlier findings also suggested that multimale groups attract and retain nulliparous females at higher rates, that females in multimale groups experience less infanticide, and that females in multimale groups have higher reproductive success, but long-term data have not confirmed these trends (Robbins et al. 2013). The advantages to females of multimale groups are therefore not fully clear, but reduced infant mortality as a result of longer male tenure is a leading candidate (Rosenbaum et al. 2016).

Although male philopatry is found in mountain gorillas, it is virtually absent in western gorillas, given that they are all one-male groups. Thus, almost all subordinate males emigrate from their natal groups, even though in 68 percent of cases the alpha male at the time is their father (Robbins et al. 2016). Nevertheless, an important similarity to male philopatry has been proposed for western gorillas based on genetic data, namely that relationships between breeding males from different groups are structured by a tendency for male kin to reside close to each other (Bradley et al. 2004). Bradley et al. (2004) suggest that this occurs as a result of adult males tending to disperse only short distances from their natal groups compared to females. The resulting kin relationships could in theory contribute to explaining why intergroup interactions among western gorillas are often peaceful.

More importantly for the purposes of this chapter, the idea suggests that in all African apes and humans there is a consistent tendency for patrilocality (male philopatry) going back to their common ancestor (Bradley et al. 2004). Against this, however, adult males in neighboring groups of western gorillas have not been observed associating with each other (Robbins et al. 2004). Furthermore, even mountain gorillas, despite their frequent multimale groups, do not exhibit typical features of primates living in multimale groups, such as morphological or physiological adaptations for sperm competition. These points suggest that multimale groups are a recent adaptation

(Robbins et al. 2016), and raise the possibility that relationships among male kin in western gorillas are not functionally philopatric, that is, that western gorillas have experienced a long evolutionary history without any meaningful tendency for male kin to cooperate. Since mountain gorillas appear to represent a recent evolutionary lineage compared to the more widely distributed western gorillas, Robbins et al. (2016) suggested that male philopatry in mountain gorillas has converged with male philopatry in the human-chimpanzee-bonobo lineage. In the absence of any clear functional benefits, the evolutionary significance of male gorilla philopatry for the African hominoids therefore remains unresolved.

Sex Differences in Cooperation and Competition

In humans and chimpanzees, males both cooperate and compete more than females do. By contrast, in bonobos, females cooperate more than males, and competition is relatively reduced in both sexes; in gorillas, competition is more prevalent than cooperation in both sexes. The difference between chimpanzees and bonobos is particularly provocative, and its evolution is not well understood. However, the difference has been suggested to stem ultimately from differences in food supply that lead bonobos to have more stable subgroups, more gregarious females, and longer durations of sexual attractiveness. These factors supposedly increase the social power of females relative to males, partly by facilitating the formation of female-female alliances and partly by increasing sexual leverage (Furuichi 2011; Hare et al. 2012). The large differences in feeding ecology and sexual relationships between humans and the two *Pan* species mean that such comparisons cannot easily be extended to humans. Here we consider a complementary explanation for the behavioral similarities and differences among the three species at a level of social dynamics that can be compared more easily than feeding ecology, namely the intensity of intergroup aggression.

A common explanation for understanding variation in patterns of social affiliation is that more intense competition between groups leads to stronger bonding within groups, including more fraternal interest groups (Otterbein 1994; McDonald et al. 2012; Radford et al. 2016). In line with this idea, we suggest that a major reason for the disparity between humans and chimpanzees on the one hand and bonobos on the other is that there is a difference in the value of using aggression in intergroup competition. In humans and chimpanzees, cooperative male aggression against members of neighboring groups enhances fitness (Wilson and Glowacki, this volume). Aggression pays because, as a result of each species' unique ecology, members of neighboring groups are often vulnerable to being brutally attacked at very low cost to the aggressors (Wrangham and Glowacki 2012; Wilson et al. 2014). In bonobos, by contrast, cooperative male aggression is expected not to pay, because vulnerable victims are not likely to be found. The relative safety of bonobos is putatively explained by features of their feeding ecology (namely high fruit density and/or reliance on extensive "meadows" of herbaceous food) that allow subgroups to be relatively stable; and in stable subgroups, both sexes can form alliances that grant defenses against aggressors. According to this logic, in bonobos natural selection has favored males who do not search for opportunities to express cooperative aggression toward members of neighboring communities. Their reduced aggressive tendency saves them from unnecessary energetic expenses and risks.

Given that violence against neighboring communities is beneficial, male humans and chimpanzees must balance their desires to attain high status through within-group competition against their simultaneous need to cooperate with those same community members to attack or defend against outside groups. However, where violence against neighboring communities is not favored, as in bonobos, the pressure for males to cooperate is greatly reduced. We therefore suggest that variation in the importance of between-group aggression contributes a critical factor explaining the differences in cooperation and competition between humans and chimpanzees, on the one hand, and bonobos, on the other.

Humans: Same-Sex Cooperation

Humans exhibit a striking sex difference in the patterning of cooperative relationships. Female investment in other females tends to be focused on kin and a few close friends. By contrast, males tend to invest more in relationships with same-sex peers, especially where individuals of different status are concerned (Baumeister and Sommer 1997).

Starting in childhood, girls and boys voluntarily segregate into groups that tend to be single-sex. In the United States, sex segregation starts by the age of three to four years (Mehta and Strough 2009). The separation is asymmetrical: boys are less willing to interact with girls than vice versa, and they tend to enforce the separation more forcefully (Maccoby 1998). Cross-culturally, the childhood separation of groups and activities develops into the sexual division of labor, namely an economic complementarity between the sexes that has its roots in the production and/or preparation of food (Rosaldo et al. 1974; Kaplan et al. 2009). In societies with extensive leisure activities, which are mostly larger and more complex societies, even leisure time is sexually divided (Craig 2002).

Beyond the extended family, women do not generally form exclusively all-female social groups but they are certainly capable of doing so. In some West African cultures, all-female groups up to one hundred strong are formed by economically independent wives, most of them married polygynously (Leis 1974). Groups include secret societies, tax rebels, and market traders. Women cooperate in work groups such as cleaning up communal areas. They also police behavior, such as by adjudicating disputes among women, controlling promiscuity, excluding outsiders from selling in the market, and forcing debtors (largely men) to meet their obligations. The West African groups tend to fall apart when they are too large to allow decisions to be unanimous (Leis 1974). The fact that such all-female groups occur raises the question of why they are relatively rare.

One answer is that men unite to stop women from getting political power. Bamberger (1974) showed that small-scale societies often claim that in the distant past women were in charge, which caused life to be chaotic and humiliating for men. So men took over and restored order. The myth of primitive matriarchy provides a justification for men to keep women under control so as to avoid a return to societal ineptitude. In such societies, women are liable to be gang raped or even executed for challenging the patriarchy. Punishable offenses could include entering the "men's country" (Walbiri, Australia: Meggitt 1987), disobeying men's rules (Yamana, Tierra del Fuego: Bamberger 1974), defying sexual demands (Mangaia, Cook Islands: Marshall 1971), being unfaithful (Cheyenne, North America: Hoebel 1954), seeing the men's house or sacred objects (Mehinaku and Mundurucu, Brazil: Murphy and Murphy 1974; Gregor 1985), or touching men's weapons, heads, or ritual ornaments (New Guinea: Herdt 1987). The equivalent political domination of women that is found in all societies is predicated on legal systems that justify male control of females, albeit normally with lighter sanctions than gang rape or execution (Hoebel 1954).

But a power struggle is not the only potential source of sex differences in cooperation. A complementary kind of hypothesis for why women tend to express same-sex cooperation in smaller groups and in different ways from men is that the sex differences in social relationships result from sex differences in evolutionary psychology. Differences include the identity and number of cooperating partners, and the nature of the cooperation.

First is the tendency at all ages for human females to focus their cooperative activity on a smaller set of same-sex individuals than males do (Lever 1978; Baumeister and Sommer 1997; Maccoby 1998). Where children can exert their own wishes, girls interact more with close female kin, while boys spend relatively more time with unrelated same-sex peers. They also differentiate less than girls do between friends and acquaintances (Douvan and Adelson 1966). These generalizations apply even when the two sexes are equally gregarious, which is how women and men are perceived across twenty-five cultures (Williams and Best 1990). Among adults, analysis of social networking sites from nine world regions shows that men are much more likely than women to display their allegiance to large single-sex groups (David-Barrett et al. 2015). The loose networks formed by affiliated males are easily institutionalized into larger groupings, including immense groups such as those found in military organizations (Baines and Blatchford 2009; Rodseth 2012). By contrast, all-female networks are rare and relatively small. Jealousy over who is friends with whom often appears to interfere with the integration of female friendship pairs into larger groupings (Parker et al. 2005). Further reducing the size of female groupings is the fact that females tend to set higher standards for friendships, as indicated by their lower tolerance for transgressions by same-sex peers (Benenson et al. 2009; MacEvoy and Asher 2012).

The sexes differ not only in who they choose to affiliate with, but also in the nature of their cooperation. In small same-sex groups in Western societies, males more often engage in collaborative actions, while females are more likely to engage with each other while conducting their own activities side by side (Winstead and Griffin 2001). The sex difference starts by the age of five (Barbu et al. 2011). Girls are more likely to engage in parallel play than united activity, but they nevertheless cooperate by talking to each other at the same time (Barbu et al. 2011).

A comparable sex difference is seen in the foraging of hunter-gatherers. In a rare example of a quantitative study of cooperation during foraging, Hill

TABLE 14.2. Patterns of cooperative foraging reported for hunter-gatherer societies listed by Kaplan et al. (2000).

	Female Cooperation in Foraging	Male Cooperation in Foraging	More Parallel Activity within Group	More Shared Actions
Onge, Andaman Islands, India	Bose, p. 301: "go collecting in a group."	Bose, p. 301: "hunting is collective."	Not clear	Men
Anbarra, Australia	Meehan, p. 86: A group of women goes to collect, ending with each choosing her spot and digging individually.	Similar to rest of Arnhem Land, Thomson.	Women	Men
Arnhem Land, Australia	Thomson, p. 30: "[W]omen generally go out in a body. But at such time, although they work closely together, there is no pooling of resources; each collects food independently, working for her own family or family group."	Communal activities include burning, drives, and fishing	Women	Men
Aché, Paraguay	Thirty-three minutes cooperation per day, mostly with husband (55 percent of time).	Forty-one minutes cooperation per day, mostly not with spouse (73 percent of time).	Not clear	Men
Nukak, Colombia	Women gather foods that can be collected without assistance.	Communal hunts for peccary, communal tree-cutting for honey. Women and men cooperate in fishing.	Women	Men
Hiwi, Venezuela	Women often assist spouses.	Hill, p. 122: "All men hunt in pairs or larger groups."	Not clear	Men
Ju/'hoansi, Botswana	Independent collection of fruits and tubers.	Share arrows, collaboration in hunting and carrying kill	Women	Men
G/wi, Botswana	Silberbauer, p. 200: "The women and girls forage in groups. They cooperate by pooling information on what is to be found in particular spots, but each gathers only for her own household, and, normally, sharing only occurs between mothers and young daughters."	Usually hunt in pairs, with much communication, planning of hunt, butchery, and transport. Silberbauer, p. 212: "The commonest approach is a pincers movement.... Ideally each puts an arrow into the target."	Women	Men
Hadza, Tanzania	Forage in groups of 3–8. Mostly collect individually but often cooperate in digging tubers (taking turns, or collaborating).	Hunt alone by day, in pairs by night. Cooperate in tracking, butchery, transport.	Women	Men

Sources: Onge: Radcliffe-Brown (1922), Bose (1964); Anbarra: Jones (1980), Meehan (1982); Arnhem Land: Thomson (1949); Aché: Hill (2002); Nukak: Politis (2009); Hiwi: Hill (2002); Ju/'hoansi: Lee (1979); G/wi: Silberbauer (1981); Hadza: Marlowe (2010).

(2002) found that Aché men cooperated with each other more than Aché women did. Table 14.2 illustrates this finding more broadly by assembling information from the nine societies for which Kaplan et al. (2000) found quantitative data on hunter-gatherer food production. The populations come from Australia, India, Africa, and South America, and include extensive variation in foraging patterns depending on habitat, season, and the species being gathered or hunted. Yet there is considerable consistency in the nature of cooperation. Although women typically forage in larger groups than men, the process of obtaining the food is most commonly an individual (or side-by-side) activity among women, whereas it is more collaborative among men (Table 14.2). This sex difference may be imposed by the nature of the resources being sought, but it is a reminder that mere gregariousness does not directly predict cooperation.

A sex difference is also seen in food sharing. Among hunter-gatherers, men's foods are shared more widely than women's foods (Kelly 1995). Women give primarily to their husbands and close kin, while men share their food extensively with heads of other households, many of whom are nonkin. Men's wider sharing is subject to strong social norms, so it does not necessarily mean that individuals are highly motivated to be generous to peers (Gurven 2004). However, it does show that the networks of social obligation in which men operate are larger than those among women.

We suggest that the evidence that males cooperate with each other more intensely and across a wider set of individuals contributes critically to explaining why in general it is men who form the public sphere of most human communities, and hence are able to use their political power to dominate women (Rosaldo et al. 1974). Men are more likely than women to cooperate in both formal institutions, including the military, government, business, and religious organizations (Tiger 2005; Baumeister 2010), and informal ones such as sports teams (Craig 2002) and gangs (Thrasher and Shor 1963). Accordingly, from middle childhood onward, males more than females care about one another's skills, including how athletic, financially astute, intelligent, creative, and socially astute each is (Goodwin 1990; Vigil 2007). Girls and women, in contrast, tend to value their relationships more for social and emotional similarities and overlapping vulnerabilities (Buhrmester and Prager 1995).

Humans: Same-Sex Competition

For humans, cooperation occurs alongside competition. As in most species, beginning early in life, human males compete more frequently, overtly, and intensely than females with same-sex peers (Archer 2004). Competition can lead to serious verbal and physical aggression, and the more intense the aggression, the greater the sex difference (Moffitt et al. 2001). By twenty-four months, compared to females, males in diverse cultures use more physical force in competition over resources (Hay et al. 2011), and by three years more competitive speech forms (Maltz et al. 1982). The sex difference in frequency and intensity of overt physical and verbal aggression continues into adolescence (Maccoby and Jacklin 1974; Archer 2004) and adulthood (Archer 2009). The most extreme form of competition, homicide, universally occurs ten times more frequently between men than between women, with the peak incidence between eighteen and twenty-five years (Daly and Wilson 1988). The primary causes of within-community aggression between men are competition for status and access to women (Muller and Pilbeam, this volume). The outcomes of continual competition lead boys and men across diverse cultures to produce hierarchical organizations based first on physical prowess and later on skills that a society values (Sapolsky 2004), even in the simplest and most egalitarian hunter-gatherer societies (Apicella et al. 2012).

Given the importance of overt competition and aggression to males' lives, it makes sense that from childhood onward physical strength and coordination are critical for achieving high status (Tuddenham 1951; Cillessen and Mayeux 2004). As boys grow older, more complex skills, including self-confidence, intelligence, social astuteness, and access to resources become increasingly important (Vigil 2007). Intriguingly, males with high status are much admired by their lower-ranked peers, and partnerships exist between high- and low-ranked males (Benenson et al. 2014). We suggest that this unusual combination of continual contests accompanied by cooperation across ranks is adaptive because it tends to facilitate success in intergroup conflicts.

Whether women organize themselves hierarchically as well has not been extensively investigated. Most cross-cultural evaluations of women conclude that women are less hierarchical than men (Bem 1974), and studies show that with unrelated same-sex peers, from early childhood onward females more than males privilege equality (Maltz et al. 1982; Benenson et al. 2014). However, in contrast to males, females generally eschew overt competition and aggression, so outcomes of conflicts are more difficult to observe. Instead, girls and women engage in more subtle forms of competition and aggression, including utilization of discreet nonverbal gestures directed toward the target, verbal denigration of absent targets, and coalitionary exclusion of lone targets (Western societies: Björkqvist 1994; Archer and Coyne 2005; foragers: Hess et al. 2010), all of which help reduce potentially harmful retaliation. Despite low levels of direct aggression, by middle childhood and increasingly in adolescence, at least in Western societies, human females are ranked according to status (Merten 2012). Whereas the mechanism for achieving high status remains unclear, attractiveness to males is believed to play a critical role, along with familial support and resources (Merten 2012). In adulthood, across varied marital systems, status for females appears to be based on obtaining and retaining a mate who can provide resources and social support. This is because, across diverse cultures, a high-quality mate is the most important asset a woman can provide to her children in terms of enhancing their survival in addition to her preexisting familial ties and resources (Burbank 1987; Rucas et al. 2012). Thus, whereas males first compete with one another to attain status, then later attract females, attractiveness to males seems to precede, and perhaps determine, status with other females. The strong bond that a female forms with a male has been suggested to contribute a protective screen that allows females to minimize the risk of physical injury to themselves or their infants, while using indirect competitive strategies such as gossip to harm the reputations of rivals (Opie et al. 2013; Pusey and Schroepfer-Walker 2013). Male-female bonding may thus contribute to explaining why physical aggression is relatively uncommon among human females compared to chimpanzees and bonobos. It can also account for the apparent contradiction that across cultures females behave in more egalitarian ways than males with their same-sex peers, yet also occupy different ranks in a community's hierarchy (Maltz et al. 1982; Aries 1996).

Regardless of how high status is attained, for both sexes it tends to increase survival and reproductive success (Sapolsky 2004; Marmot and Wilkinson 2006; von Rueden et al. 2011). The mechanism appears to vary between the sexes. Females who are able to exclude others may benefit because of competition for scarce resources, including investment from husbands. In line with this idea, newcomer females, females who are attractive to males, and unprotected, lone females all threaten the diversion of valuable resources from current females' mates, and are common targets of social exclusion (Ein-Dor et al. 2015). Consequently, females are expected to prefer relationships with others whose situation means they are not competitors. This can explain why, beginning in early childhood, females prefer same-sex friends of similar status and dislike higher-ranked females (Maltz et al. 1982; Goodwin 1990; Rose et al. 2011).

Somewhat paradoxically, the outcome of males' greater competitiveness is a hierarchical social structure in which individuals cooperate across differing ranks, whereas females form primarily egalitarian relationships with one or two same-sex peers at a time (Maltz et al. 1982; Aries 1996). In line with this conclusion, in a predictive test males were found to cooperate more often across ranks than females did (Benenson et al. 2014).

Greater integration of cooperation and competition in male than female activities is illustrated by Lever's (1978) year-long naturalistic observations of children's interactions. Boys repeatedly engaged in cooperative activities, while simultaneously competing both to outdo one another individually and to defeat other male groups. By contrast, girls were less involved in both competition and cooperation. Boys' cooperative activities endured for longer intervals, because when a conflict occurred, they successfully renegotiated social rules as well as each other's status, whereas girls who experienced a conflict were more likely to abruptly stop cooperating. Thus, despite higher levels of competitiveness, in order to remain part of a group of cooperators, males are more willing than females to accept an individual loss (Goodwin 1990). Similar sex differences are found in several behavioral contexts, including affiliation following a sports match (Benenson and Wrangham 2016), tensions among roommates in college (Benenson et al. 2009), acceptance of ex-spouses' new partners (Hetherington and Kelly 2003), and role-playing scenarios (Benenson et al. 2014). In every case, men make up with each other faster and/or more effectively than women do. Presumably, the male pattern creates advantages during fluid and complex conflicts by facilitating effective coalitions.

Chimpanzees: Same-Sex Cooperation and Competition

The sex differences in same-sex relationships of chimpanzees resemble those of humans, although they are even more pronounced in some respects. Compared to female chimpanzees, male chimpanzees spend more time in close proximity to same-sex adults. For example, in each of three two-year periods in Kanyawara, even the males who had the weakest association indices of any male dyads had substantially stronger association indices than the mode for female dyads (Gilby and Wrangham 2008). Males also groom one another more than females do (Lehmann and Boesch 2008; Wakefield 2013), and they greet each other through specialized calls, kisses, and embraces up to twenty times more often than females do (Nishida and Hiraiwa-Hasegawa 1987). Males hunt together and share meat with each other more often than females do, especially male coalition partners (Mitani 2009; Gilby et al. 2015). Females are relatively solitary, and even when they are more gregarious they typically prefer male over female company, probably partly because males intervene to reduce conflicts between females (Boesch and Boesch-Achermann 2000; Kahlenberg et al. 2008).

Long-term bonds can be important for both sexes, but they are used in different ways. Those formed among females tend to be relatively stable (Langergraber et al. 2009; Lehmann and Boesch 2009). For example, in Gombe, where a high proportion of females breed in their natal groups, female-female bonds are commonly between mother and daughter, and between similar-aged individuals, both of which can last until one member of the pair dies (Pusey and Schroepfer-Walker 2013). Since female ranks are stable and generally age-dependent, this means that bonds among females are normally with those of similar dominance status. Females use their same-sex bonds in aggression in defending core feeding areas against immigrant females, in infanticidal attacks, and in maintaining dominance rank (Pusey and Schroepfer-Walker 2013). However, compared to males, females rarely engage in coalitionary attacks. Males, by contrast, regularly form coalitions with other males. They also cooperate with many males from the community to patrol the periphery of their territory and to launch raids on neighbors (Wilson et al. 2014). In addition, they form coalitions when competing for mates or attempting to improve their position in the dominance hierarchy (Watts 1998; Gilby et al. 2013; Foerster et al. 2016). The contexts in which males use bonds to compete with their peers require flexible partner choices, meaning that males at times cooperate with all other adult males in their community, including those of widely different ranks (Wilson et al. 2014; Wilson and Glowacki, this volume).

Despite their more extensive network of cooperation, within the community males compete more frequently, overtly, and intensely than females.

For example, in Kanyawara, same-sex aggression was approximately fourteen times more common between males than between females (Muller 2002). Overt competition between males is predominantly a consequence of lower-ranked males challenging higher-ranked males for dominance status. Competition is intense after an alpha male is displaced until the hierarchy becomes settled. A male's individual competitive success depends on his strength, size, intelligence, motivation, and ability to form coalitions. Alpha males and other high-ranking males then excel as coalition partners, in part due to their physical prowess but also due to their social acumen, along with their motivation and intelligence. They are therefore sought out as coalition partners (Goodall 1986; Mitani 2009).

In contrast to males' aggression, female competition and aggression are less frequent. Dyadic contests occur most often over plant food and protection of infants, rather than in competition for status (Goodall 1986; Foerster et al. 2016). Female survival and reproductive success increase with high status and quality of the core area occupied by each female (Pusey et al. 1997; Thompson et al. 2007), but female ranks are so stable that rank reversals rarely occur (Foerster et al. 2016). Why ranks are so stable is unclear, but females appear to show less motivation than males to challenge their superiors (Pusey and Schroepfer-Walker 2013), preferring to queue so that with time, long-term residents eventually become high ranked (Foerster et al. 2016). Unrelated females who differ in rank are therefore less likely to be bonded with each other than an equivalent pair of males. Accordingly, they engage less often in coalitional behavior, with the exception of mother-daughter pairs (Goodall 1986). Since female-female contests are frequently over food, low-ranked females tend to avoid high-ranked ones particularly in times of food scarcity (Murray et al. 2006).

Social exclusion serves as a valuable strategy for females. Because immigrant females inevitably intensify feeding competition (Boesch and Boesch-Achermann 2000), residents benefit by excluding them from their core areas (Murray et al. 2007; Kahlenberg et al. 2008). Resident females, particularly those of high status who inhabit core areas richest in foods, attack immigrant newcomers and their infants, either individually or as part of a coalition with another high-ranked female (Townsend et al. 2007; Miller et al. 2014). An attack that drives away a female represents successful competition for food. Thus, in contrast to males, whose victory in intercommunity contests depends on numerical strength (Wrangham 1999), females have little reason not to restrict each other's numbers.

As expected from the importance to males of being able to cooperate opportunistically with other males in the community, their high levels of aggression are often followed by reconciliation (i.e., post-conflict affiliation). Among wild males, reconciliation tends to occur at high rates (19 percent at Ngogo: Watts 2006; 23 percent at Kanyawara: Hartel 2015; 15 percent at Taï: Wittig and Boesch 2003; 14 percent at Mahale: Kutsukake and Castles 2004; 33 percent at Budongo: Arnold and Whiten 2001; numbers show corrected conciliatory tendency as defined by Veenema et al. 1994). Rates of reconciliation among females vary more. In three studies, females reconciled much less than males (3 percent at Kanyawara: Hartel 2015; 6 percent at Taï: Wittig and Boesch 2003; 0 percent at Budongo: Arnold and Whiten 2001), whereas in one study females reconciled more (21 percent at Mahale: Kutsukake and Castles 2004). The variation is not understood, but small sample sizes may be partly responsible. In the two studies with large sample sizes (>630 conflicts each, i.e., three times as many as in other studies), males reconciled much more often than females (Taï: Wittig and Boesch 2003; Kanyawara: Hartel 2015). Current data therefore indicate that males have a greater conciliatory tendency than females, but more needs to be known about sources of variation.

In sum, sex differences in same-sex competition and cooperation among chimpanzees are strong and consistent, and resemble sex differences found in humans. Males are more gregarious with same-sex individuals, more aggressive in pursuit of high rank, and more flexible in their choice of coalition partners. They cooperate with all members of their community, as opposed to a few, and in the most complete studies they reconcile with peers at higher rates. Since these conclusions apply to adults, they raise the question of how the sex differences develop. Intriguingly, Lonsdorf et al. (2014) found that among twenty infant chimpanzees (2.5–3 years old), males interacted affiliatively with more individuals than females did, including more adult males. Their study controlled for maternal sociability, so it suggested that male infants were more motivated than female infants to develop a wide social network. In support of that idea, older male juveniles (7–9 years) were more likely than female juveniles to lead their mothers to social groups (Pusey 1983). In addition to innate sex differences, maternal influences may also be important. Murray et al. (2014) showed that mothers of very young infants (up to six months old) tended to spend more time in parties with adult males present if their infants were male than if they were female. Mothers may therefore have been preferentially giving their male infants opportunities to interact with a wide social network of males.

Bonobos: Same-Sex Cooperation and Competition

Bonobos collectively defend community territories, but their interactions with neighbors are much less dangerous than chimpanzees'. Killing has never been recorded among bonobos, and sufficient observations have accumulated to show that bonobos are in this respect less aggressive than chimpanzees (Wilson et al. 2014; Wilson and Glowacki, this volume). Indeed, unlike chimpanzees, intercommunity interactions among bonobos are often affiliative (Furuichi 2011). Positive interactions are due to females, who appear responsible for leading their companions toward a group from a neighboring community. Females merge into the neighbors' space, where they readily groom and have sex with females of the other community. While two such groups relax together, which they may do for several hours, females even copulate more frequently with males of the neighboring community than with their own males. Males from the different communities do not interact positively with each other and apparently try to initiate departures, but despite these signs of unease they do not show aggression toward the other males (Furuichi 2011). Compared to chimpanzees, the level of tension between males of neighboring bonobo communities is thus enormously reduced, suggesting that male-male cooperation is much less important in bonobos. This change in the quality of intergroup relationships appears capable of explaining several ways in which the grouping patterns of bonobos differ from those of humans and chimpanzees.

First, within communities, unrelated females are more likely than unrelated males to associate with each other among bonobos than among chimpanzees (Furuichi 2011; Tokuyama and Furuichi 2016). Hohmann et al. (1999) analyzed the frequency of nonrandom associations at Lomako and found that female-female dyads were about three times as common as male-male dyads, indicating that females formed more affiliative bonds (Hohmann et al. 1999). Likewise, unrelated female bonobos groom one another about three times as often as unrelated males do (Kanō 1992). Homosexual interactions ("genito-genital rubbing," or "GG-rubbing"), which bonobos use as a form of affiliation or tension reduction, are also substantially more common among females than among males (Kanō 1992; Parish 1994). Females show more cooperation than males do by sharing food more, including fruits and meat (Kanō 1992; Parish 1994; Hohmann and Fruth 2002; Furuichi 2011). Finally, females form coalitions with each other more frequently than males do (Tokuyama and Furuichi 2016).

Differentiated long-term female-female bonds are rare, however, except among the oldest and highest-ranking females in a community. In Lomako, most female-female associations lasted only one year (Hohmann et al. 1999). Not only are bonds rare, they do not structure the pattern of coalitions; that is, strong affiliation between two females in the form of high rates of spatial association, grooming, and homosexual interaction does not predict coalition frequency (Surbeck and Hohmann 2013; Tokuyama and Furuichi 2016). Instead, the frequency of forming coalitions depends merely on who is in the subgroup at the time. This pattern of a generalized willingness to support peers differs from most group-living primates, in which same-sex coalitions within groups are associated with social bonds. It is more similar to the way in which males or females support each other in intergroup aggression, except that in the case of female bonobos, Tokuyama and Furuichi (2016) found that 100 percent of fifty-eight female-female coalitions formed in Wamba, were formed to fight against males. Thus, the "enemies" for female-female coalitions are males of the same group.

In contrast to these important relationships among females, male bonobos show little evidence of forming close bonds with each other. They rarely give agonistic support or share foods, and their relationships are so obscure that it is not even clear if males who spend more time together groom each other more often (Surbeck and Hohmann 2014).

Competition and aggression among female bonobos occur primarily over food, leading to a dominance hierarchy in which older, resident females are dominant to younger immigrants. High-ranked bonobo females obtain more food than low-ranked females, suggesting that they may experience greater rates of survival and reproductive success. Strikingly, unlike human and chimpanzee females, bonobo females also cooperate across ranks, such that an immigrant female has several times been seen to assimilate into her new community by forming a relationship with a high-ranked female (Furuichi 2011). Younger females respond positively to older ones:

genital rubbing, food sharing, and coalitionary alliances often occur between females differing in status (Hohmann and Fruth 2000; Tokuyama and Furuichi 2016).

Similar to humans and chimpanzees, male bonobos engage in much higher rates of competition and aggression than females, with the ratio as high as 18:1 (White and Chapman 1994). This produces a male dominance hierarchy as steep as in despotic species of primates, with higher-ranking males achieving higher mating success than lower-ranking males (Surbeck et al. 2011). Unlike humans and chimpanzees, a male's mother normally plays an active role in determining his rank, such that males tend to lose rank after their mothers die (Furuichi 2011; Surbeck et al. 2011). Compared to the top-ranking chimpanzee males, therefore, the highest-ranking bonobo males are often young, in late adolescence or early adulthood (Furuichi 2011). However, they do not predictably outrank females (Surbeck and Hohmann 2013).

In contrast to humans and chimpanzees, the only study to date of sex differences in the conciliatory tendency of bonobos found that after a same-sex conflict, females reconciled at much higher rates than males (Palagi et al. 2004). Palagi et al.'s (2004) study was in captivity, so caution is necessary, but it conformed to the expectation that individuals in more valuable relationships are more likely to reconcile (de Waal 1986). In the wild, high rates of post-conflict affiliation have been described among female bonobos (Hohmann and Fruth 2000), but data for males have not been reported.

As with chimpanzees, our questions with bonobos are whether they are similar to humans in terms of males being more likely to segregate themselves from females, exhibiting greater gregariousness, and organizing their cooperative activities more hierarchically. Though less is known about bonobos than about chimpanzees, the answer is clear. The striking differences that bonobos show from chimpanzees are all in the direction of making them less similar to humans than chimpanzees are. Thus, male bonobos stay with their mothers when possible, do not show strong affiliative tendencies with other males, rarely cooperate, and achieve high rank more as a result of their mother's help than through their own abilities. By contrast, females are strongly cohesive with each other and cooperate regularly, including in joint defense against any threats directed by males toward their young or in feeding competition (Surbeck and Hohmann 2013; Tokuyama and Furuichi 2016).

Gorillas: Same-Sex Cooperation and Competition

In theory, gorillas offer a chance to test the polarity of evolutionary changes in social relationships. Unfortunately, however, gorillas do not have strong similarities in intrasexual relationships to either chimpanzees or bonobos. Neither male nor female gorillas show much evidence of close intrasexual association or bonding other than that between females and their close kin (siblings or parent-offspring). Nevertheless, there are some useful points of comparison.

First, among males a major point of interest derives from multimale groups being common in mountain gorillas, since this raises the question of whether males then form relationships more like chimpanzees or bonobos. Bradley et al. (2005) noted that multimale groups of gorillas are chimpanzee-like because they are long-term assemblages of kin and nonkin, but they did not compare with bonobos. They noted that not only does the alpha male tolerate the presence of adult male kin, but he also has been found to recruit nonkin as juveniles and allow them to mature and compete alongside him (Harcourt and Stewart 2007; Stoinski et al. 2009). As immatures, such males often groom the alpha, but as they become blackbacks they spend less time near him and rarely interact. After eleven to twelve years of age, when adulthood is reached, males living within the same group tend to avoid one another, and subordinate males obtain some copulations (which can cumulatively total more than 50 percent in groups with multiple males) and a low proportion of paternity (Nsubuga et al. 2008; Stoinski et al. 2009). Often the two top-ranked males are unrelated to each other and compete for copulations (Bradley et al. 2005).

While affiliative relationships among males thus appear relatively unimportant as influences on within-group social interactions, the presence of multiple males does appear to be beneficial for the alpha, perhaps because of cooperation among males against other groups or extra-group males (Rosenbaum et al. 2016). The potential advantage to the alpha of accommodating multiple males suggested to Stoinski et al. (2009) that part of the reason why subordinate males are sometimes able to mate is that the alpha allows them to do so: he refrains from intense competition by conceding matings to them as a way to induce them not to leave. Reduced competition between individuals of different rank thus seems to be important in multimale groups because of the advantages of cooperation.

In sum, multimale groups of gorillas echo a pattern of male-male tolerance found throughout humans, chimpanzees, and bonobos. However, given that even in bonobos there are benefits to males of cooperating in contests against neighboring groups, it is unclear whether male relationships in multimale groups of gorillas are more like those of chimpanzees rather than those of bonobos.

Second, affiliative relationships among female gorilla nonkin are rarely important, as shown by each female tending to form stronger affiliative relationships with the dominant silverback than with each other. For example, female mountain gorillas groom the silverback more than they do any female. By contrast when close female relatives breed in the same group, they form bonds with each other, shown in grooming and supporting each other in conflicts (Watts 2001).

Competition occurs more frequently among female mountain gorillas than cooperation, and is most often expressed as mild threats (Watts 1994; Harcourt and Stewart 2007). Relationships between females tend to be more egalitarian than hierarchical, in the sense that aggression can occur reciprocally (Watts 2001). As with humans and chimpanzees, newcomer females are frequent targets of aggression from residents, and food is the most common resource over which competition occurs overall. Similar to human and chimpanzee males who intervene to protect their mates, the silverback male provides protection to a female who is losing a conflict, which promotes the welfare of subordinate females.

Although these patterns seem to make sense in terms of competition for group membership and food, based on a sample of 214 births Robbins et al. (2007) found no evidence that group size influenced female reproductive success. Furthermore, although higher-ranking females did have relatively higher reproductive success, even this effect was very small. The value of competition among female mountain gorillas seems to be reduced by the fact that their principal food (herbaceous leaves and stems) is abundant and evenly distributed (Robbins et al. 2007).

In keeping with the low value of same-sex social relationships, Watts (1995) found that although reconciliation occurred in mountain gorillas following aggression between males and females, it did not occur following same-sex aggression. His two study groups included two adult males each, so there was some opportunity for males to affiliate.

In sum, the stable bisexual groups and low importance of social bonds within sexes mean that for gorillas, meaningful sex differences in degree of investment in same-sex relationships are not found. However, the presence of some male philopatry and male-male cooperation with kin and nonkin echo the more elaborate social relationships of humans, chimpanzees, and bonobos, while persistent female competition within groups makes gorillas more similar to humans and chimpanzees than to bonobos.

The Evolution of Same-Sex Relationships in African Hominoids

Our review shows that in several important respects the same-sex relationships of humans resemble those of chimpanzees more than those of bonobos or gorillas. In both humans and chimpanzees, females tend to focus their cooperation on relatively few partners compared to males. In both species, males are more gregarious than females, and they are more liable than females to cooperate with peers with whom their relationship is less intimate, and with peers who have a different status. In both cases, there is also evidence that males are more likely to reconcile with each other following a conflict. In humans, the sex differences in gregariousness and cooperative tendency start in early childhood, and there is some evidence that the same is true of chimpanzees.

Bonobos are not as well studied as chimpanzees, but they nevertheless provide a stark contrast because female bonobos are more gregarious and cooperative than males, and preliminary data suggest they reconcile at higher rates. In gorillas, on the other hand, except among female kin, same-sex relationships are so lacking in affiliation and cooperation that sex differences are undetectable.

These results conform to the hypothesis that the last common ancestor (LCA) of humans, chimpanzees, and bonobos was chimpanzee-like (Pilbeam and Lieberman, this volume). They suggest that humans and chimpanzees share a legacy of males cooperating over a wider network than females do, and they are consistent with the hypothesis that bonobos represent a unique system of social behavior that evolved following their split from chimpanzees (Hare and Wrangham, this volume).

The hypothesis of a chimpanzee-like LCA implies that human social relationships have evolved from a chimpanzee-like system, and raises the question of why same-sex relationships should show the similarities documented here. Like humans, chimpanzees engage in lethal conflicts between communities. Outcomes are more successful when the number of males cooperating during boundary patrols or offensive raids is higher (Wilson and Glowacki, this volume). In both species, the need for males to cooperate against external communities can therefore explain why males affiliate with a large number of same-sex peers (cf. McDonald et al. 2012). As discussed in this chapter, the fact that intercommunity interactions among bonobos are much less dangerous than among chimpanzees or humans readily accounts for male bonobos being less motivated to form strong coalitionary relationships with their peers.

If similarities in same-sex relationships between humans and chimpanzees indeed represent a shared legacy from the LCA, the intermediate species of australopithecines and early *Homo* must have had the same patterns as well. This makes sense only if the species leading to *Homo sapiens* lived in groups that included multiple males. Recent discussions of early human or australopithecine social structure have considered that such groups were indeed likely, and were probably multilevel systems (Swedell and Plummer 2012; Chapais 2013, this volume; Gintis et al. 2015).

Even accepting the likelihood of a multilevel social organization, the breeding systems of fossil hominins cannot yet be reconstructed with any confidence: polygynous and multimale mating systems are both possible (Muller and Pilbeam, this volume). However, this difficulty does not constrain the evolutionary reconstruction of same-sex relationships, because as the comparison of humans and chimpanzees shows, essential features of same-sex relationships can be found in widely different breeding systems. The critical constraints that we suggest are responsible for the characteristic same-sex relationships of humans and chimpanzees lie elsewhere. For females, a high value to excluding rivals of different competitive ability favors cooperation among a small set of intimate individuals, often kin. For males, a high value to attacking members of neighboring groups with a large number of cooperating partners favors investment in a wide social network.

According to this analysis, therefore, the ability of male chimpanzees and humans to use their relatively large coalitions to dominate females is an incidental consequence of their same-sex relationships.

Summary

We addressed the hypothesis that with regard to same-sex relationships, humans show more similarities with chimpanzees than with bonobos and gorillas. Using ethnographies, field studies, and social science data, including developmental studies, we focused on competition and cooperation, with special attention paid to sex differences within populations. We found that in two main respects humans and chimpanzees are more similar to each other than they are to bonobos or gorillas. First, in both species cooperative behavior tends to be spread among a larger number of individuals in males than in females. Second, intense competition among same-sex peers interferes less with male-male than with female-female cooperation. Bonobos and gorillas each show unique patterns of competition and cooperation that differ from humans and chimpanzees. A likely functional explanation contributing to the similarities in the behavior of male humans and chimpanzees is that intergroup aggression is particularly dangerous for males in these species. This favors behavioral mechanisms promoting male-male cooperation against hostile neighbors, despite competing for status within communities. Similarities in the behavior of females, who cooperate more intensely with a small number of allies, especially kin, may be related to the advantages of excluding competitors. Whether the LCA exhibited similar traits cannot be determined with certainty, but the behaviors found in common between humans and chimpanzees are those most likely to have been characteristic of the LCA.

Acknowledgments

We thank Rose McDermott, Martin Muller, David Pilbeam, and an anonymous reviewer for helpful comments.

References

Alvarez, H. 2004. Residence groups among hunter-gatherers: A view of the claims and evidence for patrilocal bands. In B. Chapais and C. Berman, eds., *Kinship and Behavior in Primates*, 400–442. Oxford: Oxford University Press.

Apicella, C. L., F. W. Marlowe, J. H. Fowler, and N. A. Christakis. 2012. Social networks and cooperation in hunter-gatherers. *Nature* 481: 497–501.

Archer, J. 2004. Sex differences in aggression in real-world settings: A meta-analytic review. *Review of General Psychology* 8: 291–322.

Archer, J. 2009. Does sexual selection explain human sex differences in aggression? *Behavioral and Brain Sciences* 32: 249–266.

Archer, J., and S. M. Coyne. 2005. An integrated review of indirect, relational, and social aggression. *Personality and Social Psychology Review* 9: 212–230.

Aries, E. 1996. Men and women in interaction: Reconsidering the differences. New York: Oxford University Press.

Arnold, K., and A. Whiten. 2001. Post-conflict behaviour of wild chimpanzees *(Pan troglodytes schweinfurthii)* in the Budongo forest, Uganda. *Behaviour* 138: 649–690.

Baines, E., and P. Blatchford. 2009. Sex differences in the structure and stability of children's playground social networks and their overlap with friendship relations. *British Journal of Developmental Psychology* 27: 743–760.

Bamberger, J. 1974. The myth of matriarchy: Why men rule in primitive society. In M. Z. Rosaldo, and L. Lamphere, eds., *Women, Culture and Society*, 263–280. Stanford, CA: Stanford University Press.

Barbu, S., G. Cabanes, and G. Le Maner-Idrissi. 2011. Boys and girls on the playground: Sex differences in social development are not stable across early childhood. *PLoS ONE* 6: e16407.

Baumeister, R. F. 2010. *Is There Anything Good about Men? How Cultures Flourish by Exploiting Men.* New York: Oxford University Press.

Baumeister, R. F., and K. L. Sommer. 1997. What do men want? Gender differences and two spheres of belongingness: Comment on Cross and Madson (1997). *Psychological Bulletin* 122: 38–44.

Bem, S. L. 1974. The sex role inventory. *Journal of Personality and Social Psychology* 42: 122–162.

Benenson, J. F., H. Markovits, C. Fitzgerald, D. Geoffroy, J. Flemming, S. M. Kahlenberg, and R. W. Wrangham. 2009. Males' greater tolerance of same-sex peers. *Psychological Science* 20: 184–190.

Benenson, J. F., H. Markovits, and R. W. Wrangham. 2014. Rank influences human sex differences in dyadic cooperation. *Current Biology* 24: R190–R191.

Benenson, J. F., and R. W. Wrangham. 2016. Cross-cultural sex differences in post-conflict affiliation following sports matches. *Current Biology* 26: 2208–2212.

Björkqvist, K. 1994. Sex differences in physical, verbal, and indirect aggression: A review of recent research. *Sex Roles* 30: 177–188.

Blurton Jones, N. G., F. W. Marlowe, K. Hawkes, and J. F. O'Connell. 2000. Paternal investment and hunter-gatherer divorce rates. In L. Cronk, N. Chagnon, and W. Irons, eds., *Adaptation and Human Behavior: An Anthropological Perspective*, 69–90. New York: Aldine de Gruyter.

Boesch, C., and H. Boesch-Achermann. 2000. *The Chimpanzees of the Taï Forest: Behavioral Ecology and Evolution.* Oxford: Oxford University Press.

Bose, S. 1964. Economy of the Onge of Little Andaman. *Man in India* 44: 298–310.

Bradley, B. J., D. M. Doran-Sheehy, D. Lukas, C. Boesch, and L. Vigilant. 2004. Dispersed male networks in western gorillas. *Current Biology* 14: 510–513.

Bradley, B. J., M. M. Robbins, E. A. Williamson, H. D. Steklis, N. G. Steklis, N. Eckhardt, C. Boesch, and L. Vigilant. 2005. Mountain gorilla tug-of-war: Silverbacks have limited control over reproduction in multimale groups. *Proceedings of the National Academy of Sciences* 102: 9418–9423.

Buhrmester, D., and K. Prager. 1995. Patterns and functions of self-disclosure during childhood and adolescence. In K. Rotenberg, ed., *Disclosure Processes in Children and Adolescents*, 10–56. Cambridge: Cambridge University Press.

Burbank, V. K. 1987. Female aggression in cross-cultural perspective. *Behavior Science Research* 21: 70–100.

Chapais, B. 2009. *Primeval Kinship: How Pair-Bonding Gave Birth to Human Society.* Cambridge, MA: Harvard University Press.

Chapais, B. 2013. Monogamy, strongly bonded groups, and the evolution of human social structure. *Evolutionary Anthropology* 22: 52–65.

Chapman, C. A., F. J. White, and R. W. Wrangham. 1994. Party size in chimpanzees and bonobos. In R. W. Wrangham, W. C. McGrew, F. B. M. de Waal, and P. G. Heltne, eds., *Chimpanzee Cultures*, 41–58. Cambridge, MA: Harvard University Press.

Cillessen, A. H. N., and L. Mayeux. 2004. From censure to reinforcement: Developmental changes in the association between aggression and social status. *Child Development* 75: 147–163.

Craig, S. 2002. *Sports and Games of the Ancients.* Westport, CT: Greenwood Publishing Group.

Daly, M., and M. Wilson. 1988. *Homicide.* Hawthorne, NY: Aldine de Gruyter.

David-Barrett, T., A. Rotkirch, J. Carney, I. Behncke Izquierdo, J. A. Krems, D. Townley, E. McDaniell, A. Byrne-Smith, and R. I. M. Dunbar. 2015. Women favour dyadic relationships, but men prefer clubs: Cross-cultural evidence from social networking. *PLoS ONE* 10: e0118329.

de Waal, F. B. M. 1982. *Chimpanzee Politics: Power and Sex among Apes.* New York: Harper and Row.

de Waal, F. B. M. 1984. Sex differences in the formation of coalitions among chimpanzees. *Ethology and Sociobiology* 5: 239–255.

de Waal, F. B. M. 1986. The integration of dominance and social bonding in primates. *Quarterly Review of Biology* 61: 459–479.

Douvan, E. A. M., and J. Adelson. 1966. *The Adolescent Experience.* New York: Wiley.

Dyble, M., G. D. Salali, N. Chaudhary, A. Page, D. Smith, J. Thompson, L. Vinicius, R. Mace, and A. B. Migliano. 2015. Sex equality can explain the unique social structure of hunter-gatherer bands. *Science* 348: 796–798.

Ein-Dor, T., A. Perry-Paldi, G. Hirschberger, G. E. Birnbaum, and D. Deutsch. 2015. Coping with mate poaching: Gender differences in detection of infidelity-related threats. *Evolution and Human Behavior* 36: 17–24.

Foerster, S., M. Franz, C. M. Murray, I. C. Gilby, J. T. Feldblum, K. K. Walker, and A. E. Pusey. 2016. Chimpanzee females queue but males compete for social status. *Scientific Reports* 6: 35404.

Furuichi, T. 2009. Factors underlying party size differences between chimpanzees and bonobos: A review and hypotheses for future study. *Primates* 50: 197–209.

Furuichi, T. 2011. Female contributions to the peaceful nature of bonobo society. *Evolutionary Anthropology* 20: 131–142.

Georgiev, A. V., M. E. Thompson, A. L. Lokasola, and R. W. Wrangham. 2011. Seed-predation by bonobos *(Pan paniscus)* at Kokolopori, Democratic Republic of the Congo. *Primates* 52: 309–314.

Gilby, I. C., L. J. Brent, E. E. Wroblewski, R. S. Rudicell, B. H. Hahn, J. Goodall, and A. E. Pusey. 2013. Fitness benefits of coalitionary aggression in male chimpanzees. *Behavioral Ecology and Sociobiology* 67: 373–381.

Gilby, I. C., Z. Machanda, D. C. Mjungu, J. Rosen, M. N. Muller, A. Pusey, and R. W. Wrangham. 2015. Impact hunters catalyze cooperative hunting in two wild chimpanzee communities. *Philosophical Transactions of the Royal Society* 370: 1–12.

Gilby, I. C., and R. W. Wrangham. 2008. Association patterns among wild chimpanzees *(Pan troglodytes schweinfurthii)* reflect sex differences in cooperation. *Behavioral Ecology and Sociobiology* 62: 1831–1842.

Gintis, H., C. van Schaik, and C. Boehm. 2015. The evolutionary origins of human political systems. *Current Anthropology* 56: 327–353.

Goodall, J. 1986. *The Chimpanzees of Gombe: Patterns of Behavior.* Cambridge, MA: Harvard University Press.

Goodwin, M. H. 1990. *He-Said-She-Said: Talk as Social Organization Among Black Children.* Indiana University Press, Bloomington, IN.

Gregor, T. 1985. *Anxious Pleasures: The Sexual Lives of an Amazonian People.* Chicago: University of Chicago Press.

Gurven, M. 2004. To give and to give not: The behavioral ecology of human food transfers. *Behavioral and Brain Sciences* 27: 543–583.

Harcourt, A. H., and K. J. Stewart. 2007. *Gorilla Society: Conflict, Compromise, and Cooperation between the Sexes.* Hawthorne, NY: Aldine de Gruyter.

Hare, B., V. Wobber, and R. W. Wrangham. 2012. The self-domestication hypothesis: Bonobos evolved due to selection against male aggression. *Animal Behavior* 83: 573–585.

Hartel, J. A. 2015. *Social Dynamics of Intragroup Aggression and Conflict Resolution in Wild Chimpanzees* (Pan troglodytes) *at Kanyawara, Kibale National Park, Uganda.* Los Angeles: Integrative and Evolutionary Biology Program, University of Southern California.

Hashimoto, C., T. Furuichi, and Y. Tashiro. 2001. What factors affect the size of chimpanzee parties in the Kalinzu Forest, Uganda? Examination of fruit abundance and number of estrous females. *International Journal of Primatology* 22: 947–959.

Hay, D. F., A. Nash, M. Caplan, J. Swartzentruber, F. Ishikawa, and J. E. Vespo. 2011. The emergence of gender differences in physical aggression in the context of conflict between young peers. *British Journal of Developmental Psychology* 29: 158–175.

Herdt, G. 1987. *The Sambia: Ritual and Gender in New Guinea.* Fort Worth, TX: Harcourt Brace Jovanovich.

Hess, N., C. Helfrecht, E. Hagen, A. Sell, and B. Hewlett. 2010. Interpersonal aggression among Aka hunter-gatherers of the Central African Republic: assessing the effects of sex, strength, and anger. *Human Nature* 21: 330–354.

Hetherington, E. M., and J. Kelly. 2003. *For Better or for Worse: Divorce Reconsidered.* New York: W. W. Norton.

Hill, K. 2002. Altruistic cooperation during foraging by the Ache, and the evolved human predisposition to cooperate. *Human Nature* 13: 105–128.

Hill, K. R., R. S. Walker, M. Božičević, J. Eder, T. Headland, B. Hewlett, A. M. Hurtado, F. Marlowe, P. Wiessner, and B. Wood. 2011. Co-residence patterns in hunter-gatherer societies show unique human social structure. *Science* 331: 1286–1289.

Hoebel, E. A. 1954. *The Law of Primitive Man: A Study in Comparative Legal Dynamics.* Cambridge, MA: Harvard University Press.

Hohmann, G., and B. Fruth. 2000. Use and function of genital contacts among female bonobos. *Animal Behaviour* 60: 107–120.

Hohmann, G., and B. Fruth. 2002. Dynamics in social organization of bonobos *(Pan paniscus)*. In C. Boesch, G. Hohmann, and L. Marchant, eds., *Behavioural Diversity in Chimpanzees and Bonobos,* 138–150. Cambridge: Cambridge University Press.

Hohmann, G., U. Gerloff, D. Tautz, and B. Fruth. 1999. Social bonds and genetic ties: Kinship, association and affiliation in a community of bonobos *(Pan paniscus)*. *Behaviour* 136: 1219–1235.

Jones, R. 1980. Hunters in the Australian coastal savanna. In D. Harris, ed., *Human Ecology in Savanna Environments,* 107–147. New York: Academic Press.

Kahlenberg, S. M., M. Emery Thompson, M. N. Muller, and R. W. Wrangham. 2008. Immigration costs for female chimpanzees and male protection as an immigrant counterstrategy to intrasexual aggression. *Animal Behaviour* 76: 1497–1509.

Kanō, T. 1992. *The Last Ape: Pygmy Chimpanzee Behavior and Ecology.* Stanford, CA: Stanford University Press.

Kaplan, H., K. Hill, J. Lancaster, and A. M. Hurtado. 2000. A theory of human life history evolution: Diet, intelligence, and longevity. *Evolutionary Anthropology* 9: 156–185.

Kaplan, H. S., P. L. Hooper, and M. Gurven. 2009. The evolutionary and ecological roots of human social organization. *Philosophical Transactions of the Royal Society B: Biological Sciences* 364: 3289–3299.

Kelly, R. C. 2000. *Warless Societies and the Origins of War.* Ann Arbor: University of Michigan Press.

Kelly, R. L. 1995. *The Foraging Spectrum: Diversity in Hunter-Gatherer Lifeways.* Washington, DC: Smithsonian Institution.

Kutsukake, N., and D. L. Castles. 2004. Reconciliation and post-conflict third-party affiliation among wild chimpanzees in the Mahale Mountains, Tanzania. *Primates* 45: 157–165.

Langergraber, K. E., J. C. Mitani, and L. Vigilant. 2007. The limited impact of kinship on cooperation in wild chimpanzees. *Proceedings of the National Academy of Sciences* 104: 7786–7790.

Langergraber, K., J. C. Mitani, and L. Vigilant. 2009. Kinship and social bonds in female chimpanzees *(Pan troglodytes)*. *American Journal of Primatology* 71: 840–851.

Lee, R. B. 1979. *The !Kung San: Men, Women and Work in a Foraging Society.* Cambridge: Cambridge University Press.

Lehmann, J., and C. Boesch. 2008. Sexual differences in chimpanzee sociality. *International Journal of Primatology* 29: 65–81.

Lehmann, J., and C. Boesch. 2009. Sociality of the dispersing sex: the nature of social bonds in West African female chimpanzees, *Pan troglodytes*. *Animal Behaviour* 77: 377–387.

Leis, N. B. 1974. Women in groups: Ijaw women's associations. In M. Rosaldo and L. Lamphere, eds., *Women, Culture and Society,* 223–242. Stanford, CA: Stanford University Press.

Lever, J. 1978. Sex differences in the complexity of children's play and games. *American Sociological Review* 43: 471–483.

Lonsdorf, E. V., K. E. Anderson, M. A. Stanton, M. Shender, M. R. Heintz, J. Goodall, and C. M. Murray. 2014. Boys will be boys: Sex differences in wild infant chimpanzee social interactions. *Animal Behaviour* 88: 79–83.

Maccoby, E. E. 1998. *The Two Sexes: Growing Up Apart, Coming Together.* Cambridge, MA: Belknap Press/Harvard University Press.

Maccoby, E. E., and C. N. Jacklin. 1974. *The Psychology of Sex Differences.* Stanford, CA: Stanford University Press.

MacEvoy, J. P., and S. R. Asher. 2012. When friends disappoint: Boys' and girls' responses to transgressions of friendship expectations. *Child Development* 83: 104–119.

Maltz, D. N., R. A. Borker, and J. A. Gumperz. 1982. A cultural approach to male-female mis-communication. In L. Monaghan, J. E. Goodman, and J. M. Robinson, eds., *Language and Social Identity,* 195–216. Marblehead, MA: John Wiley & Sons.

Marlowe, F. W. 2005. Hunter-gatherers and human evolution. *Evolutionary Anthropology* 14: 54–67.

Marlowe, F. W. 2010. *The Hadza: Hunter-Gatherers of Tanzania.* Berkeley: University of California Press.

Marmot, M. G., and R. G. Wilkinson. 2006. *Social Determinants of Health.* Oxford: Oxford University Press.

Marshall, D. S. 1971. Sexual behavior on Mangaia. In D. S. Marshall and R. C. Suggs, eds., *Human Sexual Behavior: Variations in the Ethnographic Spectrum,* 103–162. New York: Basic Books.

McDonald, M. M., C. D. Navarrete, and M. Van Vugt. 2012. Evolution and the psychology of intergroup conflict: The male warrior hypothesis. *Philosophical Transactions of the Royal Society* 367: 670–679.

Meehan, B. 1982. *Shell Bed to Shell Midden.* Canberra: Australian Institute of Aboriginal Studies.

Meggitt, M. J. 1987. Understanding Australian Aboriginal society: Kinship systems or cultural categories? In W. H. Edwards, ed., *Traditional Aboriginal Society: A Reader,* 113–137. Melbourne: Macmillan.

Mehta, C. M., and J. N. Strough. 2009. Sex segregation in friendships and normative contexts across the life span. *Developmental Review* 29: 201–220.

Merten, D. E. 2012. Being there awhile: An ethnographic perspective on popularity. In A. H. N. Cillessen, D. Schwartz, and L. Mayeux, eds., *Popularity in the Peer System,* 57–76. New York: Guilford.

Miller, J. A., A. E. Pusey, I. C. Gilby, K. Schroepfer-Walker, A. C. Markham, and C. M. Murray. 2014. Competing for space: female chimpanzees are more aggressive inside than outside their core areas. *Animal Behaviour* 87: 147–152.

Mitani, J. C. 2009. Cooperation and competition in chimpanzees: Current understanding and future challenges. *Evolutionary Anthropology* 18: 215–227.

Mitani, J. C., D. P. Watts, and S. J. Amsler. 2010. Lethal intergroup aggression leads to territorial expansion in wild chimpanzees. *Current Biology* 20: R507–R508.

Moffitt, T. E., A. Caspi, M. Rutter, and P. A. Silva. 2001. *Sex Differences in Antisocial Behaviour: Conduct Disorder, Delinquency, and Violence in the Dunedin Longitudinal Study.* New York: Cambridge University Press.

Muller, M. N. 2002. Agonistic relations among Kanyawara chimpanzees. In C. Boesch, G. Hohmann, and L. Marchant, eds., *Behavioural Diversity in Chimpanzees and Bonobos,* 112–124. Cambridge: Cambridge University Press.

Murphy, Y., and R. F. Murphy. 1974. *Women of the Forest.* New York: Columbia University Press.

Murray, C. M., L. E. Eberly, and A. E. Pusey. 2006. Foraging strategies as a function of season and rank among wild female chimpanzees *(Pan troglodytes). Behavioral Ecology* 17: 1020–1028.

Murray, C. M., E. V. Lonsdorf, M. A. Stanton, K. R. Wellens, J. A. Miller, J. Goodall, and A. E. Pusey. 2014. Early social exposure in wild chimpanzees: Mothers with sons are more gregarious than mothers with daughters. *Proceedings of the National Academy of Sciences* 111: 18189–18194.

Murray, C. M., S. V. Mane, and A. E. Pusey. 2007. Dominance rank influences female space use in wild chimpanzees, *Pan troglodytes:* Towards an ideal despotic distribution. *Animal Behaviour* 74: 1795–1804.

Nishida, T., and M. Hiraiwa-Hasegawa. 1987. Chimpanzees and bonobos: Cooperative relationships among males. In B. B. Smuts, D. L. Cheney, R. M. Seyfarth, R. W. Wrangham, and T. T. Struhsaker, eds., *Primate Societies,* 165–177. Chicago: University of Chicago Press.

Nsubuga, A. M., M. M. Robbins, C. Boesch, and L. Vigilant. 2008. Patterns of paternity and group fission in wild multimale mountain gorilla groups. *American Journal of Physical Anthropology* 135: 263–274.

Opie, C., Q. D. Atkinson, R. I. Dunbar, and S. Shultz. 2013. Male infanticide leads to social monogamy in primates. *Proceedings of the National Academy of Sciences* 110: 13328–13332.

Ortner, S. B., and H. Whitehead. 1981. *Sexual Meanings: The Cultural Construction of Gender and Sexuality*. Cambridge: Cambridge University Press.

Otterbein, K. F. 1994. *Feuding and Warfare: Selected Works of Keith F. Otterbein*. Langhorne, PA: Gordon & Breach.

Palagi, E., T. Paoli, and S. B. Tarli. 2004. Reconciliation and consolation in captive bonobos *(Pan paniscus)*. *American Journal of Primatology* 62: 15–30.

Parish, A. R. 1994. Sex and food control in the "uncommon chimpanzee": How bonobo females overcome a phylogenetic legacy of male dominance. *Ethology and Sociobiology* 15: 157–179.

Parker, J. G., C. M. Low, A. R. Walker, and B. K. Gamm. 2005. Friendship jealousy in young adolescents: Individual differences and links to sex, self-esteem, aggression, and social adjustment. *Developmental Psychology* 41: 235–250.

Politis, G. 2009. *Nukak: Ethnoarchaeology of an Amazonian People*. Walnut Creek, CA: Left Coast Press.

Pusey, A. E. 1983. Mother-offspring relationships in chimpanzees after weaning. *Animal Behaviour* 31: 363–377.

Pusey, A. E., and K. Schroepfer-Walker. 2013. Female competition in chimpanzees. *Philosophical Transactions of the Royal Society B* 368: 20130077.

Pusey, A. E., J. Williams, and J. Goodall. 1997. The influence of dominance rank on the reproductive success of female chimpanzees. *Science* 277: 828–831.

Radcliffe-Brown, A. 1922. *The Andaman Islanders: A Study in Social Anthropology*. Cambridge: Cambridge University Press.

Radford, A. N., B. Majolo, and F. Aureli. 2016. Within-group behavioural consequences of between-group conflict: a prospective review. *Proceedings of the Royal Society B* 283: 20161567.

Robbins, A. M., M. Gray, A. Basabose, P. Uwingeli, I. Mburanumwe, E. Kagoda, and M. M. Robbins. 2013. Impact of male infanticide on the social structure of mountain gorillas. *PLoS ONE* 8: e78256.

Robbins, A. M., M. Gray, T. Breuer, M. Manguette, E. J. Stokes, P. Uwingeli, I. Mburanumwe, E. Kagoda, and M. M. Robbins. 2016. Mothers may shape the variations in social organization among gorillas. *Royal Society Open Science* 3: 160533.

Robbins, M. M., M. Bermejo, C. Cipolletta, F. Magliocca, R. J. Parnell, and E. Stokes. 2004. Social structure and life-history patterns in western gorillas *(Gorilla gorilla)*. *American Journal of Primatology* 64: 145–159.

Robbins, M. M., A. M. Robbins, N. Gerald-Steklis, and H. D. Steklis. 2007. Socioecological influences on the reproductive success of female mountain gorillas *(Gorilla beringei beringei)*. *Behavioral Ecology and Sociobiology* 61: 919–931.

Rodseth, L. 2012. From bachelor threat to fraternal security: Male associations and modular organization in human societies. *International Journal of Primatology* 33: 1194–1214.

Rodseth, L., and S. A. Novak. 2000. The social modes of men. *Human Nature* 11: 335–366.

Rodseth, L., R. W. Wrangham, A. M. Harrigan, and B. B. Smuts. 1991. The human community as a primate society. *Current Anthropology* 32: 221–254.

Rosaldo, M. Z. 1974. Women, culture and society: A theoretical overview. In M. Z. Rosaldo and L. Lamphere, eds., *Woman, Culture and Society,* 17–42. Stanford, CA: Stanford University Press.

Rosaldo, M. Z. 1980. The use and abuse of anthropology: Reflections on feminism and cross-cultural understanding. *Signs* 5: 389–417.

Rosaldo, M. Z., L. Lamphere, and J. Bamberger. 1974. *Woman, Culture, and Society.* Stanford, CA: Stanford University Press.

Rose, A. J., G. C. Click, and R. L. Smith. 2011. Popularity and gender. In A. H. N. Cillessen, D. Schwartz, and L. Mayeux, eds., *Popularity in the Peer System,* 103–122. New York: Guilford.

Rosenbaum, S., V. Vecellio, and T. Stoinski. 2016. Observations of severe and lethal coalitionary attacks in wild mountain gorillas. *Scientific Reports* 6: 37018.

Rucas, S. L., M. Gurven, J. Winking, and H. Kaplan. 2012. Social aggression and resource conflict across the female life-course in the Bolivian Amazon. *Aggressive Behavior* 38: 194–207.

Sapolsky, R. M. 2004. Social status and health in humans and other animals. *Annual Review of Anthropology* 33: 393–418.

Schubert, G., L. Vigilant, C. Boesch, R. Klenke, K. E. Langergraber, R. Mundry, M. Surbeck, and G. Hohmann. 2013. Co-residence between males and their mothers and grandmothers is more frequent in bonobos than chimpanzees. *PLoS ONE* 8: e83870.

Silberbauer, G. B. 1981. *Hunter and Habitat in the Central Kalahari Desert.* Cambridge: Cambridge University Press.

Smuts, B. 1995. The evolutionary origins of patriarchy. *Human Nature* 6: 1–32.

Stoinski, T. S., S. Rosenbaum, T. Ngaboyamahina, V. Vecellio, F. Ndagijimana, and K. Fawcett. 2009. Patterns of male reproductive behaviour in multi-male groups of mountain gorillas: Examining theories of reproductive skew. *Behaviour* 146: 1193–1215.

Stumpf, R. M., M. Emery Thompson, M. N. Muller, and R. W. Wrangham. 2009. The context of female dispersal in Kanyawara chimpanzees. *Behaviour* 146: 629–656.

Surbeck, M., and G. Hohmann. 2013. Intersexual dominance relationships and the influence of leverage on the outcome of conflicts in wild bonobos *(Pan paniscus). Behavioral Ecology and Sociobiology* 67: 1767–1780.

Surbeck, M., and G. Hohmann. 2014. Social preferences influence the short-term exchange of social grooming among male bonobos. *Animal Cognition* 18: 573–579.

Surbeck, M., R. Mundry, and G. Hohmann. 2011. Mothers matter! Maternal support, dominance status and mating success in male bonobos *(Pan paniscus). Proceedings of the Royal Society B: Biological Sciences* 278: 590–598.

Swedell, L., and T. Plummer. 2012. A papionin multilevel society as a model for hominin social evolution. *International Journal of Primatology* 33: 1165–1193.

Thompson, M. E., S. M. Kahlenberg, I. C. Gilby, and R. W. Wrangham. 2007. Core area quality is associated with variance in reproductive success among female chimpanzees at Kanyawara, Kibale National Park. *Animal Behaviour* 73: 501–512.

Thomson, D. F. 1949. *Economic Structure and the Ceremonial Exchange Cycle*. Melbourne: Macmillan.

Thrasher, F. M., and J. F. Shor. 1963. *The Gang*. Chicago: University of Chicago Press.

Tiger, L. 2005. *Men in Groups*. Piscataway, NJ: Transaction.

Tokuyama, N., and T. Furuichi. 2016. Do friends help each other? Patterns of female coalition formation in wild bonobos at Wamba. *Animal Behaviour* 119: 27–35.

Townsend, S. W., K. E. Slocombe, M. Emery Thompson, and K. Zuberbühler. 2007. Female-led infanticide in wild chimpanzees. *Current Biology* 17: R355–R356.

Tuddenham, R. D. 1951. Studies in reputation, I: Sex and grade differences in school children's evaluation of their peers; II: The diagnosis of social adjustment. *Psychological Monographs* 333: 1–39.

Veenema, H. C., M. Das, and F. Aureli. 1994. Methodological improvements for the study of reconciliation. *Behavioral Processes* 31: 29–38.

Vigil, J. M. 2007. Asymmetries in the friendship preferences and social styles of men and women. *Human Nature* 18: 143–161.

von Rueden, C., M. Gurven, and H. Kaplan. 2011 Why do men seek status? Fitness payoffs to dominance and prestige. *Proceedings of the Royal Society B* 278: 2223–2232.

Wakefield, M. L. 2013. Social dynamics among females and their influence on social structure in an East African chimpanzee community. *Animal Behaviour* 85: 1303–1313.

Watts, D. P. 1994. Agonistic relationships between female mountain gorillas *(Gorilla gorilla beringei)*. *Behavioral Ecology and Sociobiology* 34: 347–358.

Watts, D. P. 1995. Post-conflict social events in wild mountain gorillas (Mammalia, Hominoidea), I: Social interactions between opponents. *Ethology* 100: 139–157.

Watts, D. P. 1998. Coalitionary mate guarding by male chimpanzees at Ngogo, Kibale National Park, Uganda. *Behavioral Ecology and Sociobiology* 44: 43–55.

Watts, D. P. 2001. Social relationships of female mountain gorillas. In M. Robbins, P. Sicotte, and K. Stewart, eds., *Mountain Gorillas: Three Decades of Research at Karisoke*, 215–240. Cambridge: Cambridge University Press.

Watts, D. P. 2006. Conflict resolution in chimpanzees and the valuable-relationships hypothesis. *International Journal of Primatology* 27: 1337–1364.

White, F. J., and C. A. Chapman. 1994. Contrasting chimpanzees and bonobos: Nearest neighbor distances and choices. *Folia Primatologica* 63: 181–191.

Whiting, J. W. M., and B. B. Whiting. 1975. Children of six cultures: A psycho-cultural analysis. Cambridge, MA: Harvard University Press.

Williams, J. E., and D. L. Best. 1990. *Measuring Sex Stereotypes: A Multination Study*. Newbury Park, CA: Sage Publications.

Wilson, M. L., C. Boesch, B. Fruth, T. Furuichi, I. C. Gilby, C. Hashimoto, C. Hobaiter, G. Hohmann, N. Itoh, K. Koops, J. N. Lloyd, T. Matsuzawa, J. C. Mitani, D. C. Mjungu, D. Morgan, M. N. Muller, R. Mundry, M. Nakamura, J. Pruetz, A. Pusey, J. Riedel, C. Sanz, A. M. Schel, N. Simmons, M. Waller, D. P. Watts, F. White, R. M. Wittig, K. Zuberbühler, and R. W. Wrangham. 2014. Lethal aggression in *Pan* is better explained by adaptive strategies than human impacts. *Nature* 513: 414–417.

Winstead, B. A., and J. L. Griffin. 2001. Friendship styles. In J. Worell, ed., *Encyclopedia of Women and Gender*, 481–492. Boston: Academic Press.

Wittig, R. M., and C. Boesch. 2003. The choice of post-conflict interactions in wild chimpanzees *(Pan troglodytes)*. *Behaviour* 140: 1527–1559.

Wrangham, R. W. 1999. Evolution of coalitionary killing. *Yearbook of Physical Anthropology* 29: 1–30.

Wrangham, R. W., and L. Glowacki. 2012. War in chimpanzees and nomadic hunter-gatherers: Evaluating the chimpanzee model. *Human Nature* 23: 5–29.

Wrangham, R. W., and D. Peterson. 1996. *Demonic Males: Apes and the Origins of Human Violence*. Boston: Houghton Mifflin.

Wrangham, R. W., and D. Pilbeam. 2001. African apes as time machines. In B. M. F. Galdikas, N. Briggs, L. K. Sheeran, G. L. Shapiro, and J. Goodall, eds., *All Apes Great and Small*, vol. 1: *Chimpanzees, Bonobos, and Gorillas*, 5–18. New York: Kluwer Academic/Plenum.

15

Cooperation between the Sexes

ADRIAN V. JAEGGI, PAUL L. HOOPER, ANN E. CALDWELL,
MICHAEL D. GURVEN, JANE B. LANCASTER,
and HILLARD S. KAPLAN

Despite a shared evolutionary history and many common characteristics—intelligence, tool use, slow life histories, and other traits discussed throughout this volume—patterns of mating and parental investment between humans and chimpanzees are remarkably dissimilar. These dissimilarities have a profound impact on the nature of social relationships between the sexes. In this chapter we synthesize recent research on free-ranging chimpanzees, human foragers, and forager-horticulturalists,[1] documenting these distinct patterns of interaction between females and males.

We begin by developing an evolutionary and ecological framework to understand the divergence in patterns of cooperation between the sexes in *Pan* and *Homo*. We review the state of evidence bearing on this topic from the study of wild chimpanzees and human foragers, respectively. We then address the implications of these patterns for the life history of family formation and reproduction in each species. We conclude by discussing prospects for future research.

Theory: Mating, Parental Investment, and Sex Roles

The evolutionary theories of parental investment and sexual selection provide a framework for understanding the divergent mating systems of chimpanzees and humans. This section describes a theory that integrates classic and recent models of parental investment and sexual selection (Trivers 1972; Emlen and Oring 1977; Maynard Smith 1977; Grafen and Sibly 1978; Clutton-Brock 1991; Kokko and Johnstone 2002; Kokko and Jennions 2008) applied to the case of these two species. This theory—synthesized in Hooper et al. (2014)—suggests that the evolution of sex roles in mating and parenting depend on (1) the returns to providing different forms of investment in offspring, (2) the shape of the trade-offs between these investments and mating effort, and (3) the dynamics of best response to the behavior of the opposite sex.

According to the theory, reproductive-age animals face a decision between investing in conceived offspring versus pursuing new opportunities for fertilization and conception. Investments in offspring come principally in the form of energy (i.e., metabolic resources, food) or care (protection from harm, direct attention to needs). Selection acts on the amount of energy and care delivered by parents, given that increased investment in care will tend to reduce investments in energy (and vice versa), and that both of these investments will tend to deplete resources available for seeking future reproductive opportunities.

When providing care trades off harshly with producing energy—when carrying infants while foraging, for example, increases risks of mortality or greatly reduces returns—or when there are increasing returns to specialization in the production of energy or care, a single parent cannot efficiently provide the mix of both energy and care necessary for offspring success. Under these conditions, two parents that specialize and combine their inputs may be able to achieve a level of offspring fitness more than double that of a single parent balancing both roles (Hooper et al. 2014). This multiplicative advantage of joint provisioning can provide an important motive for males to remain with mates and offspring, rather than deserting them in favor of the next mating opportunity (Maynard Smith 1977; Kokko and Johnstone 2002). It is likely that these advantages, combined with extraordinarily high need of offspring concomitant with the slow human life history, are fundamental factors motivating cooperative biparental investment in human foragers, as discussed below.

There are a number of conditions under which the fitness generated by two contributing parents is unlikely to be greater than twice that generated by one parent serving both roles. When optimal diets are more permissive of simultaneous infant care and foraging, when young can be safely hidden away during foraging, or when there are diminishing returns to specialization in energy or care, one sex (usually the male) is likely to face greater gains from a resumption of mating effort than continued parental investment. Generally, the greater the substitutability (rather than complementarity) of female and male inputs, the more likely is the outcome of desertion by one sex and uniparental care by the other (Kokko and Johnstone 2002; Hooper et al. 2014).

Among mammals, maternal commitments to lactation, typically low paternity certainty, and few gains from specialization make females most often responsible for the full cost of investment in offspring (Clutton-Brock 1989). Given that the vast majority (>95 percent) of the chimpanzee diet consists of foods that can be collected or extracted with a clinging infant or accompanying juvenile (Kaplan et al. 2000; Watts et al. 2012; Carmody, this volume), trade-offs between the efficiency of foraging and care are relatively weak. These conditions are likely to underlie the observed *Pan* equilibrium of female-only investment in offspring, and relatively undifferentiated patterns of foraging behavior between the sexes.

In contrast to other primates, hunted foods play a critical role in human diets, accounting for an average of about 60 percent of calories consumed (Kaplan et al. 2000). Hunting has two important characteristics that are critical for understanding human pair-bonds and male parental investment. First, it is largely incompatible with childcare (except under special circumstances; Gurven and Hill 2009). Second, it provides protein and lipids, and thus may provide more valuable calories than plant foods (Hill 1988). For both reasons, hunting increases the marginal value of male inputs to both women and children, which we propose is a prime driver in shifting the human mating system away from the standard mammalian pattern.

In this theory, the returns to renewed mating effort bind the equilibrium level and duration of parental investment for each sex. All else being equal, paternal investment will tend to decline as opportunities for further mating increase (Blurton Jones et al. 2000; Schacht and Borgerhoff Mulder 2015). In general, male parental investment tends to decrease when the costs of mate search are low (e.g., when females are spatially clustered), or when the ratio of females to males in the mating pool is high (Clutton-Brock 1991; Kokko and Jennions 2008). In the human case, where pair-bonds form and males invest, the extent of monogamy versus polygyny is affected by the degree of inequality in male resources, and the extent to which the value of male inputs diminishes with division between multiple wives and offspring (Borgerhoff Mulder 1992). Among foragers, both food sharing and the inability to accumulate durable resources tend to reduce inequality in resources among males, leading to more monogamous mating systems (Marlowe 2005; Kaplan et al. 2009; Kelly 2013; Hooper et al. 2014; Jaeggi et al. 2016).

Paternal investment is predicted to decrease as females place relatively greater value on a male's genetic quality than on offered parental investment during mate choice. Factors affecting the marginal benefits derived from additional male parental investment (such as the compatibility of offspring care with food production) are likely to affect both the direct fitness benefits of male investment, as well as female choice criteria regarding the relative weight of investment offers versus genetic quality. These two effects should mutually reinforce each other. When male parental investment is relatively less important, on the other hand, the result may be promiscuity (as in *Pan*) or polygynous mating without significant male parental investment (as in most nonhuman primates, with the exception of callitrichids).

It is relevant to note that male coercion may limit the scope of female choice, particularly among chimpanzees (Muller et al. 2011; Muller, this volume). This phenomenon probably reinforces the effect of a care-compatible optimal diet on the outcome of uncooperative sex roles in chimpanzees. The scope for female choice may be greater where coalitions can be formed against aggressive males, as in bonobos (Furuichi 2011; Hare et al. 2012; Jaeggi et al. 2016). In the case of humans, the scope for female choice is highly variable. A lack of direct female control over partner choice and marriage is clearly manifest in some human societies, particularly those relying on defensible, durable, and inherited resources (Boone 1986; Borgerhoff Mulder 1992). Among contemporary foragers, the evidence discussed below shows that females exert considerable choice in mating decisions for their own benefit (Draper 1975; Marlowe 2004; Pillsworth 2008; Gurven et al. 2009), with the important exception of rape, which occurs at apparently low levels ubiquitously (Burbank 1992; Marlowe 2010; Muller, this volume).

Two additional considerations have important implications for the theory of mating effort and parental investment in chimpanzees and humans. First, there is evidence in both species that male support for females or their offspring can be motivated in part by mating effort (Anderson et al. 1999; Kaburu and Newton-Fisher 2015). Second, there has been considerable debate in the literature on humans over whether the production and sharing of food by men are motivated by payoffs to parental investment per se, or increased access to fertility through signaling or trade (Hawkes and Bliege Bird 2002; Gurven and Hill 2009, 2010; Hawkes et al. 2010, 2014; Wood and Marlowe 2013, 2014). The empirical data on contemporary chimpanzees and human foragers are discussed in light of these theoretical considerations in the following two sections.

Chimpanzees

Despite highly dimorphic reproductive behavior, there are typically small differences between chimpanzee females and males in diet and foraging behavior. As in most species in which female reproduction is limited by access to food and males are larger than females, female chimpanzees are expected to consume relatively more high-quality foods—ripe fruit, insects, and nuts, rather than lower-quality leaves and unripe fruit—compared to males (Gaulin 1979; Sailer et al. 1985). Because chimpanzee males more frequently participate in hunts and have greater access to and control over carcasses, however, male chimpanzees tend to eat greater quantities of meat (0–3 percent of total diet) (Teleki 1973; Boesch and Boesch-Achermann 1989; Tennie et al. 2008; Wood and Gilby, this volume). Dimorphism in diet could also be affected by tendencies for females to occupy and defend core areas, and for males to range more widely across the breadth of territories (Williams et al. 2002; Emery Thompson et al. 2007; Murray et al. 2007).

The reproductive roles of female and male chimpanzees are strikingly different. Females bear the full cost of reproduction and parental investment after fertilization, including gestation, lactation, carrying, care, food transfers (mostly passive), and maintaining access to core feeding areas (Emery Thompson et al. 2007, 2012; Murray et al. 2007; Jaeggi and van Schaik 2011). Male chimpanzees cooperate to defend territories capable of supporting the reproduction of multiple females (Wilson et al. 2014) and compete for access to estrous females within linear dominance hierarchies (Wroblewski et al. 2009).

The promiscuous mating system of chimpanzees yields diffuse expectations of paternity certainty across males in a group (Chapais, this volume). Females actively seek matings with multiple males and advertise their fertility through sexual swellings (Stumpf and Boesch 2006). While this may have a protective effect against infanticide, it also produces minimal inclusive fitness motivation for parental investment by males. Thus, there is little evidence of direct paternal investment in terms of direct care or food provisioning, with the potential exception of males adopting orphaned offspring (Boesch et al. 2010).

The cooperation of chimpanzee males in the defense of territory against other groups can be considered a form of blanket protection to females and their offspring, particularly if territorial takeovers entail a threat of direct attacks or infanticide (Watts et al. 2002; Mitani et al. 2010; Wilson et al. 2014). Individual males may also sometimes protect females from others' aggression within communities. Recently immigrated females may preferentially associate with adult males in order to be protected against the aggression of resident females (Kahlenberg et al. 2008), which can also lead to infanticide (Townsend et al. 2007).

Within chimpanzee communities, rates of association and affiliative interactions (e.g., grooming) are greatest within male-male dyads—who rely on alliances with each other for rank and territorial defense—followed by female-male dyads, then female-female dyads (Watts 1998, 2002; Stumpf and Boesch 2006; Machanda et al. 2013). There is evidence that this pattern changes when female choice is more pronounced—for example, when female-female coalitions protect against male coercion, as in bonobos (White and Wood 2007; Tokuyama and Furuichi 2016).

Adult female-male associations in chimpanzees sometimes reflect kinship, principally between mothers and sons (Pusey 1983). Mother-son bonds in adulthood are especially important in bonobos, where male-male bonds are weak or absent; here, males rely on maternal support for acceptance by dominant females, and for access to high rank, resources, and mating opportunities (Surbeck et al. 2011). Given dominant patterns of male philopatry and female dispersal, however, there are relatively few opportunities for female and male relatives (other than mother and offspring) to interact or even fully recognize each other as kin (Wroblewski 2010; Chapais this volume).

Female-male interactions are shaped in large part by the dynamics of the mating system. There is some evidence that preferred—older, parous, higher-ranking, and/or estrous—females receive higher rates of affiliative behavior from males (e.g., association and spatial proximity: Muller et al. 2006; Machanda et al. 2013; grooming: Kaburu and Newton-Fisher 2015; meat sharing: Gomes and Boesch 2009; Wood and Gilby this volume). Ironically, preferred females also experience higher rates of aggression from males as a form of mate guarding, particularly during estrus (Muller et al. 2007, 2011).

There is evidence for pair-specific heterogeneity in frequencies of copulation, paternity, and interaction within groups, which may be motivated by preferences of one or both individuals (Newton-Fisher et al. 2010; Gomes and Boesch 2011), or which may simply arise as a by-product of other factors, such as overlapping spatial patterns (Langergraber et al. 2013; Machanda et al. 2013). Male chimpanzees sometimes attempt to draw or coerce estrous females away from other males for periods of "consortship" (Watts 1998; Wroblewski et al. 2009), yet there is no evidence that particular dyads form consortships more than expected by chance. Several males may cooperate to consort with females and exclude other males (Watts 1998), a pattern also observed in dolphins (Connor et al. 2001).

There are important differences across sites and across time in the form and intensity of male-female relationships in chimpanzees and bonobos (Langergraber et al. 2013). For instance, levels of female sociality (i.e., time associating with others, mating activity, and frequency and duration of sexual swellings) increase with food availability and decrease with feeding competition (van Schaik 1989; Chapman et al. 1994; Hohmann and Fruth 2002; Mitani et al. 2002; Stumpf 2007). With greater food availability, females can afford to be more sexually active—increasing the ratio of receptive females to males—and more sociable, which broadens the scope for forming long-term bonds and alliances with both females and males (Stumpf 2007; Jaeggi et al. 2016). Thus, the ratio of males to females with maximal sexual swellings ranges from two to three among bonobos and Taï chimpanzees, to twelve at Gombe (Stumpf 2007). The percentage of time that females spend away from other adults averages 30 percent across chimpanzees, but varies considerably across site: females at Taï, for example, average only 4 percent, while bonobos average 2–3 percent (Stumpf 2007). By reducing male-male competition over estrous females and increasing female social capital, greater food availability thus increases the potential for female choice and may even lead to dominance over males, as in bonobos (Stumpf and Boesch 2006; White and Wood 2007; Hare et al. 2012; Jaeggi et al. 2016).

Humans

Patterns of cooperation between the sexes in human societies stand in stark contrast to those of *Pan*. The sexual division of labor is a ubiquitous and well-documented feature of all traditional human societies (Murdock and Provost 1973). There is an extensive division of labor by sex—illustrated in Figure 15.1 for the Tsimané—including not only direct childcare and energy production, but also food processing, collecting firewood and cooking, construction of housing, and manufacture of tools (Marlowe 2007). The behavior of females and males reflects complementary adjustments to the investments of the opposite sex and the needs of offspring. This allows each sex to intensively learn the skills necessary for one role, and disregard the skills of the other (Kaplan et al. 2001).

Compatibility of labor with simultaneous childcare is an important predictor of women's activities (Minge-Klevana 1980; Hurtado et al. 1985, 1992).

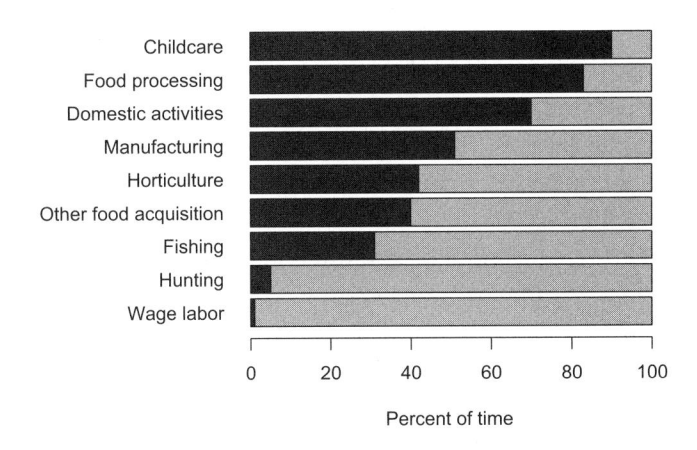

FIGURE 15.1. The sexual division of productive labor among Tsimané women (dark gray) and men (light gray). The horizontal axis denotes percent of time in each activity devoted by women versus men. Values are derived from 11,971 spot observations (adapted from Gurven et al. 2009).

Among the Aché who did not maintain permanent camps with safe places for children, women spent more than 90 percent of daytime hours in tactile contact with children under age three (Hurtado et al. 1985). Even in safer environments where cleared spaces are maintained, as among the Tsimané, children are actively cared for by their mothers 30 percent of their waking hours (Winking et al. 2009). While mothers are the predominant caregivers in apparently all forager societies, fathers also engage in some direct care, though there is significant variation in direct paternal care across groups: Aka fathers provide around 22 percent of the care received by young children as measured by time investment, while Ju/'hoansi, Hadza, Efe, Agta, and Tsimané fathers provide 1.9–7 percent (Griffin and Griffin 1992; Hewlett 1992; Marlowe 1999; Hewlett and Macfarlan 2002; Winking et al. 2009).

The weight of evidence suggests substantial economic contributions to familial and offspring well-being from both mothers and fathers. Across a sample of ten foraging societies, females produce a mean of 32 percent of all calories and 12 percent of all protein, while males produce 68 percent of all calories and 88 percent of all protein (Kaplan et al. 2000). There are no foraging societies reported in which men are not important providers of animal protein and lipids (Kaplan et al. 2000; Marlowe 2005; Kelly 2013). While there is a consistent specialization of females and males in complementary forms of production, there is also considerable variability across foragers depending on conditions. Hunting by women, for example, has been well described for the Aka and Agta, where hunting methods (nets among the Aka; bows, arrows, and dogs among the Agta) were less incompatible with pregnancy and childcare, small and medium game were relatively abundant near camps, and (among the Agta) there were considerable gains from trading meat for carbohydrates (Goodman et al. 1985; Noss and Hewlett 2001; Gurven and Hill 2009).

Among the Tsimané, mothers, grandmothers, fathers, and grandfathers all contribute substantial net transfers of food to dependent young. Fathers' rate of net provisioning to children is roughly twice that of mothers, while grandfathers give around 25 percent more to grandchildren than do grandmothers (Hooper et al. 2015). The importance of offspring need in motivating Tsimané women and men's work effort is illustrated in Table 15.1: women produce 310 additional calories through horticulture per day for each child over three, while men produce 240 additional calories through hunting for each additional child. As shown in Figure 15.2, 65–75 percent of Tsimané par-

TABLE 15.1. Tsimané economic production as a function of dependent offspring (multilevel regression with community-level random effects).

	Hours per Day			Calories per Day		
Predictor	B	β	p	B	β	P
A. Women's horticultural effort and productivity (n=253)							
(Intercept)	−0.116	−0.061	0.746	−171.1	−0.134	0.965
Age	0.070	1.030	0.010	117.4	0.326	0.446
Age²	−0.001	−0.665	0.078	−0.2	−0.037	0.935
Offspring 0–2	−0.180	−0.175	0.004	−828.3	−0.153	0.017
Offspring 3–19	0.043	0.130	0.087	313.7	0.182	0.033
B. Men's hunting effort and productivity (n=281)							
(Intercept)	−0.060	0.228	0.798	−1,205.8	0.196	0.206
Age	0.054	0.657	0.020	162.3	0.770	0.019
Age²	−0.001	−0.407	0.148	−2.4	−0.664	0.039
Offspring 0–19	0.065	0.204	0.002	240.5	0.295	<0.001

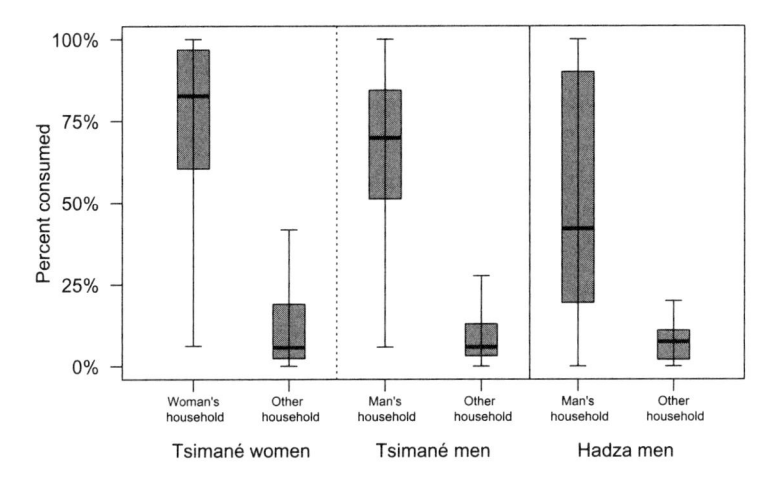

FIGURE 15.2. Estimated percent of food consumed by members of an adult producer's household versus members of other households, for Tsimané women and men and Hadza men. The values received by other households represent the mean received by households that ever received food from the producer (for the Tsimané) or that resided in the same camp (for the Hadza). Tsimané values reflect sums over 93 (±40 SD) days from 371 women and 490 men (Hooper et al. 2015), while Hadza values reflect 98 distribution events from 44 men (adapted from figure 5 in Wood and Marlowe 2013).

ents' production is consumed within the nuclear family. Among the Hadza, who share food more widely, men's families consume around 40 percent of the food they produce (Wood and Marlowe 2013).

Consistent with an emphasis on providing direct care with the economic support of husbands and other kin, female foragers significantly decrease time spent foraging with the presence of young children. Among the Aché and Hiwi, nursing women show significant reductions in production rates and time allocated to production (Hurtado et al. 1985, 1992). Hadza mothers with nursing infants likewise show reduced foraging effort and productivity, while married men and fathers are significantly more productive than unmarried and childless men (Marlowe 2003, 2010; Wood and Marlowe 2013). Table 15.1 shows that Tsimané women produce roughly 830 fewer calories per day for each child under three. Shuar women who are pregnant or lactating show reduced physical activity, while their husbands are relatively more active (Madimenos et al. 2011). The ability of female foragers to decrease their burden of work during lactation pays off in survival and fertility; among other primates, in contrast, lactation is often the most vulnerable part of the adult life course, because females must work to support the nutritional needs of both themselves and their nursling (Lancaster and Kaplan 2009).

A number of accounts have emphasized the role of male-male competition, mate guarding, and infanticide avoidance in the origins of human pair-bonds (Blurton Jones et al. 2000; Hawkes and Bliege Bird 2002; Coxworth et al. 2015); these factors have also been associated with variation in pair-bonding in other primates and mammals (van Schaik and Kappeler 1997; Palombit 1999; Lukas and Clutton-Brock 2013; Opie et al. 2013). Chapais has proposed that polygynous—but basically uncooperative—pair-bonds may have been present as a result of these factors in hominins preceding *Homo*, with implications for paternity certainty and alliance formation (Chapais 2008, this volume). Muller and Pilbeam (this volume) discuss the evidence bearing on whether polygynous pair-bonds (as in gorillas) or promiscuity (as in chimpanzees) were characteristic of the common ancestor of *Pan* and *Homo*.

Despite generally high levels of paternal investment in human foragers compared to other mammals, cases of desertion, infidelity, and disinvestment by men are also clearly common (Winking et al. 2007; Marlowe 2010; Stieglitz et al. 2012, 2014). Disinvestment becomes more likely where there are relatively low costs and high potential benefits to extrapair mating effort, and mothers and other kin can more easily compensate for the loss of men's contributions. The data show that dissolution of pair-bonds and spousal abuse are most frequent early in marriages, and that marriages tend to be quite stable once more than two children are born to the same parents (Early and Headland 1998; Kaplan et al. 2001, 2010; Marlowe 2010; Stieglitz et al. 2011).

While there is some evidence of motivations for men to produce and provision women or their offspring to secure mating opportunities, either within or outside current pair-bonds (Anderson et al. 1999; Smith et al. 2003), mating effort alone appears insufficient to account for the magnitude of fathers' contributions to their families as described for contemporary foragers and forager-horticulturalists (Hewlett and Macfarlan 2002; Hill and Hurtado 2009; Howell 2010; Wood and Marlowe 2013; Hooper et al. 2015). As such, the pattern of cooperative pair-bonding and biparental care observed in humans is probably best understood in light of returns to biparental investment. Ultimately, the origins of cooperative pair-bonds in humans may have more in common with birds than with other primates and mammals.

Further Implications for the Biodemography of *Homo* and *Pan*

Complementarities between female and male parental roles structure patterns of pair-bonding, marriage, and reproduction in human foraging societies. In a cross-cultural sample of 145 foraging groups, the modal percentage of monogamous marriages is 96–100 percent, and in the majority of societies, fewer than 10 percent of marriages are polygynous (Binford 2001). As a result, female and male reproductive schedules tend to be closely linked: age-specific fertility and expected future fertility for women and men are similar in shape, with the male curves shifted three to five years to the right (i.e., higher ages) and having a slightly longer tail (Tuljapurkar et al. 2007). The demographic linkage between female and male foragers is also manifest in the tendency for males to cease reproducing when their wives reach menopause. Among the Tsimané and Aché, only 10 percent and 17 percent of men reproduce again after their wife (or first wife) reaches menopause, respectively (Hill and Hurtado 1996; Kaplan et al. 2010).

Because chimpanzees and bonobos lack cooperative pair-bonds, the life histories of reproduction are likely to be substantially more dimorphic compared to human foragers. Chimpanzees—like most mammals with vanishing levels of paternal investment and high male reproductive skew—perpetuate and endure high levels of male-male competition, which can lead to later onset and earlier termination of reproduction and higher mortality among males compared to females—males live faster and die younger (Clutton-Brock 1991; van Noordwijk and van Schaik 2004; Clutton-Brock and Isvaran 2007). In bonobos, too, males have substantially lower life expectancy than females, at least in captivity (Jeroen Stevens, personal communication), which seems to point to a similarly dimorphic life history.

The importance of alliances for achieving rank in chimpanzees may somewhat reduce this effect compared to species with more solitary forms of competition, as older, politically connected males can continue to enjoy reproductive success (de Waal 1982; Duffy et al. 2007; Wroblewski et al. 2009). While published data on age-specific fertility and mortality rates in wild chimpanzees are still sparse, these predictions can soon be tested given a growing number of studies on genetic paternity (Boesch et al. 2006; Wroblewski et al. 2009; Newton-Fisher et al. 2010).

In Chapter 10, Chapais describes the role of human pair-bonding in expanding the scope of cooperative alliances through both affinal and consanguineous kinship. As Chapais indicates, while the chimpanzee pattern of migration at puberty halves the potential extent of one's kin network, human marriage doubles it. Indeed, data from contemporary foragers and forager-horticulturalists indicate that cooperative pair-bonds provide a nucleus for networks of support across extended families. These families show patterns of investment in offspring by nonparents across three generations—grandparents, uncles, aunts, and siblings (Gurven et al. 2000; Hooper et al. 2015)—which have been described as a system of cooperative breeding (Kramer 2005; Sear and Mace 2008; Hill and Hurtado 2009; Hrdy 2009). The support that mothers and offspring receive from spouses, parents, and other members of the extended family described above likely underlies the relatively shorter interbirth intervals and higher fertility of humans compared to other great apes (Hrdy 2009; Isler and van Schaik 2012a, 2012b).

Conclusion

We have an increasingly clear picture of the modal equilibrium pattern of behavioral and biodemographic characteristics in human foragers and wild chimpanzees. The coevolutionary sequence that gave rise to these modal patterns, on the other hand, remains more difficult to illuminate (Foley and Gamble 2009).

We suggest that knowledge of shifting natural environments can provide an anchor for theorizing social and behavioral change among the great apes from the Miocene to the present. Given the time scales involved, and the expected transience of unstable constellations of traits, models will improve by specifying the conditions that support the evolutionary stability of the equilibrium distribution of traits and behaviors in each step of the historical sequence. The ecological factors emphasized here are the returns from extracted, scavenged, and hunted food, and the returns to economic and reproductive specialization between the sexes.

The paleoanthropological and archaeological signatures of the variables highlighted in this chapter include evidence of subsistence technology and practices (Wrangham 2009; Ferraro et al. 2013; Estalrrich and Rosas 2015; Roach and Richmond 2015), diet (Balter et al. 2012; Wood and Gilby, this volume; Carmody et al. this volume), sexual dimorphism (Plavcan 2012), brain size (Holloway et al. 2004), and developmental rates (Bermúdez de Castro et al. 2010; Zollikofer and Ponce de León 2010). Current evidence is generally consistent with (in that it does not refute) the mainstream view that cooperative pair-bonding and the sexual division of labor arose with the genus *Homo* ca. 2–3 Ma (Lovejoy 2009). Evidence of meat eating and uncertainty surrounding the extent of sexual dimorphism in *Australopithecus*, however, may push these possibilities farther back in time (McPherron et al. 2010); contradictory developmental evidence, however, could push them later, toward the origin of modern *Homo sapiens* (Ramirez Rozzi and Bermudez de Castro 2004; Dean 2006; Ruff and Burgess 2015). In this context, it might be useful to consider a gradual evolution of meat-eating in hominins (Thompson et al. in revision), starting with chimpanzee-like opportunistic hunting of small game, the consumption of bone marrow using percussive technology (at which females excel: Boesch and Boesch 1981) once chimpanzee-like cognition increasingly encountered animal bones left by carnivores in more open habitats, to habitual scavenging, cooperative hunting with spears, and eventually solitary hunting with projectile weapons; the first stages of this sequence would already have provided females and their infants with better nutrition, allowing the evolution of larger brains, but only the latter stages would have required a sexual division of labor. New empirical methods combined with the development of explicit, testable theoretical models will provide greater certainty regarding the natural histories of the mating systems and linked life histories of *Pan* and *Homo* in coming years.

Endnotes

1. The term forager is used to refer to "pure" foragers (i.e., people subsisting almost entirely from hunting and gathering, such as the Aché, Hadza, or Ju/'hoansi). Forager-horticulturalists (such as the Shuar or Tsimané) may derive the bulk of their calories from cultivated foods, but share with contemporary foragers a reliance on hunting or fishing for animal protein and fats, communal property rights, and relatively egalitarian, small-scale sociopolitical organization.

Acknowledgments

Thanks to Jessica Thompson, Dietrich Stout, Kim Hill, Brian Wood, Samuel Bowles, Monique Borgerhoff Mulder, Martin Muller, Paul Seabright, Jonathan Stieglitz, and Jeffrey Winking for helpful input and feedback. Tsimané research was supported by NSF BCS-0422690, NIH/NIA R01AG024119-01, and NIH/NIA 2P01AG022500-06A1.

References

Anderson, K. G., H. Kaplan, D. Lam, and J. Lancaster. 1999. Paternal care by genetic fathers and stepfathers, II: Reports by Xhosa high school students. *Evolution and Human Behavior* 20: 433–451.

Balter, V., J. Braga, P. Télouk, and J. F. Thackeray. 2012. Evidence for dietary change but not landscape use in South African early hominins. *Nature* 489: 558–560.

Bermúdez de Castro, J. M., M. Martinón-Torres, L. Prado, A. Gómez-Robles, J. Rosell, L. López-Polín, J. L. Arsuaga, and E. Carbonell. 2010. New immature hominin fossil from European Lower Pleistocene shows the earliest evidence of a modern human dental development pattern. *Proceedings of the National Academy of Sciences* 107: 11739–11744.

Binford, L. F. 2001. *Constructing Frames of Reference: An Analytical Method for Archaeological Theory Building Using Ethnographic and Environmental Data Sets*. Berkeley: University of California Press.

Blurton Jones, N. G., F. W. Marlowe, K. Hawkes, and J. F. O'Connell. 2000. Paternal investment and hunter-gatherer divorce rates. In L. Cronk, N. A. Chagnon, and

W. Irons, eds., *Adaptation and Human Behavior: An Anthropological Perspective*, 31–35. New York: Aldine de Gruyter.

Boesch, C., and H. Boesch. 1981. Sex differences in the use of natural hammers by wild chimpanzees: A preliminary report. *Journal of Human Evolution* 10: 585–593.

Boesch, C., and H. Boesch-Achermann. 1989. Hunting behavior of wild chimpanzees in the Taï National Park. *American Journal of Physical Anthropology* 78: 547–573.

Boesch, C., C. Bolé, N. Eckhardt, and H. Boesch. 2010. Altruism in forest chimpanzees: The case of adoption. *PLoS ONE* 5: e8901.

Boesch, C., G. Kohou, H. Néné, and L. Vigilant. 2006. Male competition and paternity in wild chimpanzees of the Taï forest. *American Journal of Physical Anthropology* 130: 103–15.

Boone, J. L. 1986. Parental investment and elite family structure in preindustrial states: A case study of late medieval–early modern Portuguese genealogies. *American Anthropologist* 88: 859–878.

Borgerhoff Mulder, M. 1992. Women's strategies in polygynous marriage. *Human Nature* 3: 45–70.

Burbank, V. K. 1992. Sex, gender, and difference. *Human Nature* 3: 251–277.

Chapais, B. 2008. *Primeval Kinship: How Pair-Bonding Gave Birth to Human Society.* Cambridge, MA: Harvard Univ Press.

Chapman, C., F. White, and R. W. Wrangham. 1994. Party size in chimpanzees and bonobos. In R. W. Wrangham, W. C. McGrew, F. B. M. de Waal, and P. G. Heltne, eds., *Chimpanzee Cultures*, 41–57. Cambridge, MA: Harvard University Press.

Clutton-Brock, T. H. 1989. Review lecture: Mammalian mating systems. *Proceedings of the Royal Society of London B: Biological Sciences* 236: 339–372.

Clutton-Brock, T. H. 1991. *The Evolution of Parental Care.* Princeton, NJ: Princeton University Press.

Clutton-Brock, T. H., and K. Isvaran. 2007. Sex differences in ageing in natural populations of vertebrates. *Proceedings of the Royal Society of London B: Biological Sciences* 274: 3097–3104.

Connor, R. C., M. R. Heithaus, and L. M. Barre. 2001. Complex social structure, alliance stability and mating access in a bottlenose dolphin "super-alliance." *Proceedings of the Royal Society of London B: Biological Sciences* 268: 263–267.

Coxworth, J. E., P. S. Kim, J. S. McQueen, and K. Hawkes. 2015. Grandmothering life histories and human pair bonding. *Proceedings of the National Academy of Sciences* 112: 11806–11811.

Dean, M. C. 2006. Tooth microstructure tracks the pace of human life-history evolution. *Proceedings of the Royal Society of London B: Biological Sciences* 273: 2799–2808.

de Waal, F. B. M. 1982. *Chimpanzee Politics: Power and Sex Among Apes.* London: Jonathan Cape.

Draper, P. 1975. !Kung women: Contrasts in sexual egalitarianism in foraging and sedentary contexts. In R. R. Reiter, ed., *Toward an Anthropology of Women*, 77–109. New York: Monthly Review Press.

Duffy, K. G., W. Richard, and J. B. Silk. 2007. Male chimpanzees exchange political support for mating opportunities. *Current Biology* 17: R586–R587.

Early, J. D., and T. N. Headland. 1998. *Population Dynamics of a Philippine Rain Forest People: The San Ildefonso Agta.* Gainesville: University of Florida Press.

Emery Thompson, M., S. M. Kahlenberg, I. C. Gilby, and R. W. Wrangham. 2007. Core area quality is associated with variance in reproductive success among female chimpanzees at Kibale National Park. *Animal Behaviour* 73: 501–512.

Emery Thompson, M., M. N. Muller, and R. W. Wrangham. 2012. The energetics of lactation and the return to fecundity in wild chimpanzees. *Behavioral Ecology* 23: 1234–1241.

Emlen, S., and L. Oring. 1977. Ecology, sexual selection, and the evolution of mating systems. *Science* 197: 215–223.

Estalrrich, A., and A. Rosas. 2015. Division of labor by sex and age in Neandertals: An approach through the study of activity-related dental wear. *Journal of Human Evolution* 80: 51–63.

Ferraro, J. V., T. W. Plummer, B. L. Pobiner, J. S. Oliver, L. C. Bishop, D. R. Braun, P. W. Ditchfield, J. W. Seaman, K. M. Binetti, J. W. Seaman, F. Hertel, and R. Potts. 2013. Earliest archaeological evidence of persistent hominin carnivory. *PLoS ONE* 8: e62174.

Foley, R., and C. Gamble. 2009. The ecology of social transitions in human evolution. *Philosophical Transactions of the Royal Society B* 364: 3267–3279.

Furuichi, T. 2011. Female contributions to the peaceful nature of bonobo society. *Evolutionary Anthropology* 20: 131–142.

Gaulin, S. 1979. A Jarman/Bell model of primate feeding niches. *Human Ecology* 7: 1–20.

Gomes, C. M., and C. Boesch. 2009. Wild chimpanzees exchange meat for sex on a long-term basis. *PLoS ONE* 4: e5116.

Gomes, C. M., and C. Boesch. 2011. Reciprocity and trades in wild West African chimpanzees. *Behavioral Ecology and Sociobiology* 65: 2183–2196.

Goodman, M. J., P. B. Griffin, A. A. Estioko-Griffin, and J. S. Grove. 1985. The compatibility of hunting and mothering among the Agta hunter-gatherers of the Philippines. *Sex Roles* 12: 1199–1209.

Grafen, A., and R. Sibly. 1978. A model of mate desertion. *Animal Behaviour* 26: 645–652.

Griffin, P. B., and M. B. Griffin. 1992. Fathers and childcare among the Cagayan Agta. In B. S. Hewlett, ed., *Father-Child Relations: Cultural and Biosocial Contexts,* 297–320. New York: Aldine de Gruyter.

Gurven, M., and K. Hill. 2009. Why do men hunt? A reevaluation of "Man the Hunter" and the sexual division of labor. *Current Anthropology* 50: 51–74.

Gurven, M., and K. Hill. 2010. Moving beyond stereotypes of men's foraging goals: Reply to Hawkes, O'Connell, and Coxworth. *Current Anthropology* 51: 265–267.

Gurven, M., K. Hill, H. Kaplan, A. Hurtado, and R. Lyles. 2000. Food transfers among Hiwi foragers of Venezuela: Tests of reciprocity. *Human Ecology* 28: 171–218.

Gurven, M., J. Winking, H. Kaplan, C. Rueden, and L. McAllister. 2009. A bioeconomic approach to marriage and the sexual division of labor. *Human Nature* 20: 151–183.

Hare, B., V. Wobber, and R. Wrangham. 2012. The self-domestication hypothesis: Evolution of bonobo psychology is due to selection against aggression. *Animal Behaviour* 83: 573–585.

Hawkes, K., and R. Bliege Bird. 2002. Showing off, handicap signaling, and the evolution of men's work. *Evolutionary Anthropology* 11: 58–67.

Hawkes, K., J. F. O'Connell, and N. G. Blurton Jones. 2014. More lessons from the Hadza about men's work. *Human Nature* 25: 596–619.

Hawkes, K., J. F. O'Connell, and J. E. Coxworth. 2010. Family provisioning is not the only reason men hunt: A comment on Gurven and Hill. *Current Anthropology* 51: 259–264.

Hewlett, B. S. 1992. Husband-wife reciprocity and the father-infant relationship among Aka pygmies. In B. S. Hewlett, ed., *Father-Child Relations: Cultural and Biosocial Contexts,* 153–176. New York: Aldine de Gruyter.

Hewlett, B. S., and S. J. Macfarlan. 2002. Fathers' roles in hunter-gatherer and other small-scale cultures. In M. E. Lamb, ed., *The Role of the Father in Child Development*, 413–434. Hoboken, NJ: Wiley.

Hill, K. 1988. Macronutrient modifications of optimal foraging theory: An approach using indifference curves applied to some modern foragers. *Human Ecology* 16: 157–197.

Hill, K., and A. M. Hurtado. 1996. *Ache Life History: The Ecology and Demography of a Foraging People.* New York: Aldine de Gruyter.

Hill, K., and A. M. Hurtado. 2009. Cooperative breeding in South American hunter-gatherers. *Proceedings of the Royal Society of London B: Biological Sciences* 276: 3863–3870.

Hohmann, G., and B. Fruth. 2002. Dynamics in social organization of bonobos *(Pan paniscus)*. In C. Boesch, G. Hohmann, and L. F. Marchant, eds., *Behavioural Diversity in Chimpanzees and Bonobos,* 138–150. Cambridge: Cambridge University Press.

Holloway, R. L., D. C. Broadfield, and M. S. Yuan. 2004. *The Human Fossil Record*, vol. 3: *Brain Endocasts: The Paleoneurological Evidence*. Hoboken, NJ: Wiley & Sons.

Hooper, P. L., M. Gurven, and H. S. Kaplan. 2014. Social and economic underpinnings of human biodemography. In M. Weinstein and M. A. Lane, eds., *Sociality, Hierarchy, Health: Comparative Biodemography,* 169–195. Washington, DC: National Academies Press.

Hooper, P. L., M. Gurven, J. Winking, and H. S. Kaplan. 2015. Inclusive fitness and differential productivity across the life course determine intergenerational transfers in a small-scale human society. *Proceedings of the Royal Society B: Biological Sciences* 282: 20142808.

Howell, N. 2010. *Life Histories of the Dobe !Kung: Food, Fatness, and Well-Being over the Life-Span.* Berkeley: University of California Press.

Hrdy, S. 2009. *Mothers and Others: The Evolutionary Origins of Mutual Understanding.* Cambridge, MA: Harvard University Press.

Hurtado, A. M., K. Hawkes, K. Hill, and H. Kaplan. 1985. Female subsistence strategies among Ache hunter-gatherers of eastern Paraguay. *Human Ecology* 13: 1–28.

Hurtado, A. M., K. Hill, I. Hurtado, and H. Kaplan. 1992. Trade-offs between female food acquisition and child care among Hiwi and Ache foragers. *Human Nature* 3: 185–216.

Isler, K., and C. P. van Schaik. 2012a. Allomaternal care, life history and brain size evolution in mammals. *Journal of Human Evolution* 63: 52–63.

Isler, K., and C. P. van Schaik. 2012b. How our ancestors broke through the gray ceiling: Comparative evidence for cooperative breeding in early *Homo*. *Current Anthropology* 53: S453–S465.

Jaeggi, A. V., K. F. Boose, F. J. White, and M. Gurven. 2016. Obstacles and catalysts of cooperation in humans, bonobos, and chimpanzees: Behavioural reaction norms can help explain variation in sex roles, inequality, war and peace. *Behaviour* 153: 1015–1051.

Jaeggi, A. V, and C. P. van Schaik. 2011. The evolution of food sharing in primates. *Behavioral Ecology and Sociobiology* 65: 2125–2140.

Kaburu, S. S. K., and N. E. Newton-Fisher. 2015. Trading or coercion? Variation in male mating strategies between two communities of East African chimpanzees. *Behavioral Ecology and Sociobiology* 69: 1039–1052.

Kahlenberg, S. M., M. E. Thompson, M. N. Muller, and R. W. Wrangham. 2008. Immigration costs for female chimpanzees and male protection as an immigrant counterstrategy to intrasexual aggression. *Animal Behaviour* 76: 1497–1509.

Kaplan, H., M. Gurven, J. Winking, P. L. Hooper, and J. Stieglitz. 2010. Learning, menopause, and the human adaptive complex. *Annals of the New York Academy of Sciences* 1204: 30–42.

Kaplan, H. S., K. Hill, A. M. Hurtado, and J. Lancaster. 2001. The embodied capital theory of human evolution. In P. T. Ellison, ed., *Reproductive Ecology and Human Evolution*, 293–317. Hawthorne, NY: Aldine de Gruyter.

Kaplan, H., K. Hill, J. Lancaster, and A. M. Hurtado. 2000. A theory of human life history evolution: Diet, intelligence, and longevity. *Evolutionary Anthropology* 9: 156–185.

Kaplan, H. S., P. Hooper, and M. Gurven. 2009. The evolutionary and ecological roots of human social organization. *Philosophical Transactions of the Royal Society B: Biological Sciences* 364: 3289–3299.

Kelly, R. L. 2013. *The Lifeways of Hunter-Gatherers: The Foraging Spectrum*. Cambridge: Cambridge University Press.

Kokko, H., and M. D. Jennions. 2008. Parental investment, sexual selection and sex ratios. *Journal of Evolutionary Biology* 21: 919–48.

Kokko, H., and R. A. Johnstone. 2002. Why is mutual mate choice not the norm? Operational sex ratios, sex roles and the evolution of sexually dimorphic and monomorphic signalling. *Philosophical Transactions of the Royal Society of London B: Biological Sciences* 357: 319–330.

Kramer, K. L. 2005. Children's help and the pace of reproduction: Cooperative breeding in humans. *Evolutionary Anthropology* 14: 224–237.

Lancaster, J., and H. Kaplan. 2009. The endocrinology of the human adaptive complex. In P. T. Ellison and P. G. Gray, eds., *Endocrinology of Social Relationships*, 95–119. Cambridge, MA: Harvard University Press.

Langergraber, K. E., J. C. Mitani, D. P. Watts, and L. Vigilant. 2013. Male–female sociospatial relationships and reproduction in wild chimpanzees. *Behavioral Ecology and Sociobiology* 67: 861–873.

Lovejoy, C. O. 2009. Reexamining human origins in light of *Ardipithecus ramidus*. *Science* 326: 74–74e8.

Lukas, D., and T. H. Clutton-Brock. 2013. The evolution of social monogamy in mammals. *Science* 341: 526–530.

Machanda, Z. P., I. C. Gilby, and R. W. Wrangham. 2013. Male-female association patterns among free-ranging chimpanzees *(Pan troglodytes schweinfurthii)*. *International Journal of Primatology* 34: 917–938.

Madimenos, F. C., J. J. Snodgrass, A. D. Blackwell, M. A. Liebert, and L. S. Sugiyama. 2011. Physical activity in an indigenous Ecuadorian forager-horticulturalist population as measured using accelerometry. *American Journal of Human Biology* 23: 488–497.

Marlowe, F. W. 1999. Showoffs or providers? The parenting effort of Hadza men. *Evolution and Human Behavior* 20: 391–404.

Marlowe, F. W. 2003. A critical period for provisioning by Hadza men: Implications for pair bonding. *Evolution and Human Behavior* 24: 217–229.

Marlowe, F. W. 2004. Mate preferences among Hadza hunter-gatherers. *Human Nature* 15: 365–376.

Marlowe, F. W. 2005. Hunter-gatherers and human evolution. *Evolutionary Anthropology* 14: 54–67.

Marlowe, F. W. 2007. Hunting and gathering: The human sexual division of foraging labor. *Cross-Cultural Research* 41, 170–195.

Marlowe, F. W. 2010. *The Hadza: Hunter-gatherers of Tanzania*. Berkeley: University of California Press.

Maynard Smith, J. 1977. Parental investment: A prospective analysis. *Animal Behaviour* 25: 1–9.

McPherron, S. P., Z. Alemseged, C. W. Marean, J. G. Wynn, D. Reed, D. Geraads, R. Bobe, and H. A. Béarat. 2010. Evidence for stone-tool-assisted consumption of animal tissues before 3.39 million years ago at Dikika, Ethiopia. *Nature* 466: 857–860.

Minge-Klevana, W. 1980. Does labor time decrease with industrialization? A survey of time-allocation studies. *Current Anthropology* 21: 279–298.

Mitani, J. C., D. P. Watts, and S. J. Amsler. 2010. Lethal intergroup aggression leads to territorial expansion in wild chimpanzees. *Current Biology* 20: R507–R508.

Mitani, J. C., D. P. Watts, and J. S. Lwanga. 2002. Ecological and social correlates of chimpanzee party size and composition. In C. Boesch, G. Hohmann, and L. F.

Marchant, eds., *Behavioural Diversity in Chimpanzees and Bonobos*, 102–111. Cambridge: Cambridge University Press.

Muller, M. N., S. M. Kahlenberg, M. Emery Thompson, and R. W. Wrangham. 2007. Male coercion and the costs of promiscuous mating for female chimpanzees. *Proceedings of the Royal Society B: Biological Sciences* 274: 1009–1014.

Muller, M. N., M. E. Thompson, S. M. Kahlenberg, and R. W. Wrangham. 2011. Sexual coercion by male chimpanzees shows that female choice may be more apparent than real. *Behavioral Ecology and Sociobiology* 65: 921–933.

Muller, M. N., M. E. Thompson, and R. W. Wrangham. 2006. Male chimpanzees prefer mating with old females. *Current Biology* 16: 2234–2238.

Murdock, G. P., and C. Provost. 1973. Factors in the division of labor by sex: A cross-cultural analysis. *Ethnology* 12: 203–225.

Murray, C. M., S. V. Mane, and A. E. Pusey. 2007. Dominance rank influences female space use in wild chimpanzees, *Pan troglodytes:* Towards an ideal despotic distribution. *Animal Behaviour* 74: 1795–1804.

Newton-Fisher, N. E., M. E. Thompson, V. Reynolds, C. Boesch, and L. Vigilant. 2010. Paternity and social rank in wild chimpanzees *(Pan troglodytes)* from the Budongo Forest, Uganda. *American Journal of Physical Anthropology* 142: 417–428.

Noss, A. J., and B. S. Hewlett. 2001. The contexts of female hunting in central Africa. *American Anthropologist* 103: 1024–1040.

Opie, C., Q. D. Atkinson, R. I. M. Dunbar, and S. Shultz. 2013. Male infanticide leads to social monogamy in primates. *Proceedings of the National Academy of Sciences* 110: 13328–13332.

Palombit, R. A. 1999. Infanticide and the evolution of pair bonds in nonhuman primates. *Evolutionary Anthropology* 7: 117–129.

Pillsworth, E. G. 2008. Mate preferences among the Shuar of Ecuador: Trait rankings and peer evaluations. *Evolution and Human Behavior* 29: 256–267.

Plavcan, J. M. 2012. Sexual size dimorphism, canine dimorphism, and male-male competition in primates: Where do humans fit in? *Human Nature* 23: 45–67.

Pusey, A. E. 1983. Mother-offspring relationships in chimpanzees after weaning. *Animal Behaviour* 31: 363–377.

Ramirez Rozzi, F. V., and J. M. Bermudez de Castro. 2004. Surprisingly rapid growth in Neanderthals. *Nature* 428: 936–939.

Roach, N. T., and B. G. Richmond. 2015. Clavicle length, throwing performance and the reconstruction of the *Homo erectus* shoulder. *Journal of Human Evolution* 80: 107–113.

Ruff, C. B., and M. L. Burgess. 2015. How much more would KNM-WT 15000 have grown? *Journal of Human Evolution* 80: 74–82.

Sailer, L. D., S. J. C. Gaulin, J. S. Boster, and J. A. Kurland. 1985. Measuring the relationship between dietary quality and body size in primates. *Primates* 26: 14–27.

Schacht, R., and M. Borgerhoff Mulder. 2015. Sex ratio effects on reproductive strategies in humans. *Royal Society Open Science* 2: 140402.

Sear, R., and R. Mace. 2008. Who keeps children alive? A review of the effects of kin on child survival. *Evolution And Human Behavior* 29: 1–18.

Smith, E. A., R. B. Bird, and D. W. Bird. 2003. The benefits of costly signaling: Meriam turtle hunters. *Behavioral Ecology* 14: 116–126.

Stieglitz, J., M. Gurven, H. Kaplan, and J. Winking. 2012. Infidelity, jealousy, and wife abuse among Tsimane forager-farmers: Testing evolutionary hypotheses of marital conflict. *Evolution and Human Behavior* 33: 438–448.

Stieglitz, J., A. V. Jaeggi, A. D. Blackwell, B. C. Trumble, M. Gurven, and H. S. Kaplan. 2014. Work to live and live to work: Productivity and psychological well-being in adulthood and old age. In M. Weinstein and M. A. Lane, eds., *Sociality, Hierarchy, Health: Comparative Biodemography*, 195–220. Washington, DC: National Academies Press.

Stieglitz, J., H. Kaplan, M. Gurven, J. Winking, and B. V. Tayo. 2011. Spousal violence and paternal disinvestment among Tsimané forager-horticulturalists. *American Journal of Human Biology* 23: 445–457.

Stumpf, R. 2007. Chimpanzees and bonobos: diversity within and between species. In C. J. Campbell, A. Fuentes, K. C. MacKinnon, M. Panger, and S. K. Bearder, eds., *Primates in Perspective*, 321–344. New York: Oxford University Press.

Stumpf, R. M., and C. Boesch. 2006. The efficacy of female choice in chimpanzees of the Taï forest, Côte d'Ivoire. *Behavioral Ecology and Sociobiology* 60: 749–765.

Surbeck, M., R. Mundry, and G. Hohmann. 2011. Mothers matter! Maternal support, dominance status and mating success in male bonobos *(Pan paniscus)*. *Proceedings of the Royal Society B: Biological Sciences* 278: 590–598.

Teleki, G. 1973. *The Predatory Behavior of Wild Chimpanzees*. Lewisburg, PA: Bucknell University Press.

Tennie, C., I. C. Gilby, and R. Mundry. 2008. The meat-scrap hypothesis: Small quantities of meat may promote cooperative hunting in wild chimpanzees *(Pan troglodytes)*. *Behavioral Ecology and Sociobiology* 63: 421–431.

Thompson, J. C., S. Carvalho, C. W. Marean, and Z. Alemseged. In revision. The origins of the human predatory pattern: The transitions to large animal exploitation by early hominins. *Current Anthropology*.

Tokuyama, N., and T. Furuichi. 2016. Do friends help each other? Patterns of female coalition formation in wild bonobos at Wamba. *Animal Behaviour* 119: 27–35.

Townsend, S. W., K. E. Slocombe, M. Emery Thompson, and K. Zuberbühler. 2007. Female-led infanticide in wild chimpanzees. *Current Biology* 17: 355–356.

Trivers, R. 1972. Parental investment and sexual selection. In B. Campbell, ed., *Sexual Selection and the Descent of Man*, 136–179. Chicago: Aldine.

Tuljapurkar, S. D., C. O. Puleston, and M. D. Gurven. 2007. Why men matter: Mating patterns drive evolution of human lifespan. *PLoS ONE* 2: e785.

van Noordwijk, M. A., and C. P. van Schaik. 2004. Sexual selection and the careers of primate males: Paternity concentration, dominance-acquisition tactics and transfer decisions. In P. M. Kappeler and C. P. van Schaik, eds., *Sexual Selection in*

Primates: New and Comparative Perspectives. Cambridge: Cambridge University Press. 208–229.

van Schaik, C. P. 1989. The ecology of social relationships amongst female primates. In V. Standen and R. A. Foley, eds., *Comparative Socioecology. The Behavioural Ecology of Humans and Other Mammals,* 195–218. Oxford: Blackwell Scientific.

van Schaik, C. P., and P. M. Kappeler. 1997. Infanticide risk and the evolution of male-female association in primates. *Proceedings of the Royal Society B: Biological Sciences* 264: 1687–1694.

Watts, D. P. 1998. Coalitionary mate guarding by male chimpanzees at Ngogo, Kibale National Park, Uganda. *Behavioral Ecology and Sociobiology* 44: 43–55.

Watts, D. P. 2002. Reciprocity and interchange in the social relationships of wild male chimpanzees. *Behaviour* 139: 343–370.

Watts, D. P., K. B. Potts, J. S. Lwanga, and J. C. Mitani. 2012. Diet of chimpanzees *(Pan troglodytes schweinfurthii)* at Ngogo, Kibale National Park, Uganda, I: Diet composition and diversity. *American Journal of Primatology* 74: 114–129.

Watts, D. P., H. M. Sherrow, and J. C. Mitani. 2002. New cases of inter-community infanticide by male chimpanzees at Ngogo, Kibale National Park, Uganda. *Primates.* 43: 263–270.

White, F. J., and K. D. Wood. 2007. Female feeding priority in Bonobos, *Pan paniscus,* and the question of female dominance. *American Journal of Primatology* 69: 1–14.

Williams, J. M., H.-Y. Liu, and A. E. Pusey. 2002. Costs and benefits of grouping for female chimpanzees at Gombe. In C. Boesch, G. Hohmann, and L. F. Marchant, eds., *Behavioural Diversity in Chimpanzees and Bonobos,* 192–203. Cambridge: Cambridge University Press.

Wilson, M. L., C. Boesch, B. Fruth, T. Furuichi, I. C. Gilby, C. Hashimoto, C. L. Hobaiter, G. Hohmann, N. Itoh, K. Koops, J. N. Lloyd, T. Matsuzawa, J. C. Mitani, D. C. Mjungu, D. Morgan, M. N. Muller, R. Mundry, M. Nakamura, J. Pruetz, A. E. Pusey, J. Riedel, C. Sanz, A. M. Schel, N. Simmons, M. Waller, D. P. Watts, F. White, R. M. Wittig, K. Zuberbühler, and R. W. Wrangham. 2014. Lethal aggression in *Pan* is better explained by adaptive strategies than human impacts. *Nature* 513: 414–417.

Winking, J., M. Gurven, H. Kaplan, and J. Stieglitz. 2009. The goals of direct paternal care among a South Amerindian population. *American Journal of Physical Anthropology* 139: 295–304.

Winking, J., H. Kaplan, M. Gurven, and S. Rucas. 2007. Why do men marry and why do they stray? *Proceedings of the Royal Society of London B: Biological Sciences.* 274: 1643–1649.

Wood, B. M., and F. W. Marlowe. 2013. Household and kin provisioning by Hadza men. *Human Nature* 24: 280–317.

Wood, B. M., and F. W. Marlowe. 2014. Toward a reality-based understanding of Hadza men's work. *Human Nature* 25: 620–630.

Wrangham, R. W. 2009. *Catching Fire: How Cooking Made Us Human.* New York: Basic Books.

Wroblewski, E. E. 2010. *Paternity and Father-Offspring Relationships in Wild Chimpanzees,* Pan troglodytes schweinfurthii. Ph.D. dissertation, University of Minnesota.

Wroblewski, E. E., C. M. Murray, B. F. Keele, J. C. Schumacher-Stankey, B. H. Hahn, and A. E. Pusey. 2009. Male dominance rank and reproductive success in chimpanzees, *Pan troglodytes schweinfurthii. Animal Behaviour* 77: 873–885.

Zollikofer, C. P. E., and M. S. Ponce de León. 2010. The evolution of hominin ontogenies. *Seminars in Cell and Developmental Biology* 21: 441–452.

16

Sexual Coercion in Chimpanzees and Humans

MARTIN N. MULLER

Sexual conflict occurs when the reproductive strategies employed by one sex impose reproductive costs on members of the opposite sex. Such conflict is pervasive in sexually reproducing species (Chapman et al. 2003; Arnqvist and Rowe 2005). In primates, a wide range of traits related to reproduction has likely been shaped by sexual conflict, including both anatomical and physiological features (e.g., penile spines, copulatory plugs) and behavioral strategies (e.g., infanticide by males; reviewed in Stumpf et al. 2011). Sexual coercion represents a specific, behavioral manifestation of sexual conflict, in which males use aggression to overcome female mating preferences (Watson-Capps 2009).

The essential definition of sexual coercion was provided by Smuts and Smuts (1993): "use by a male of force, or threat of force, that functions to increase the chances that a female will mate with him at a time when she is likely to be fertile, and to decrease the chances that she will mate with other males, at some cost to the female." This definition includes two key elements. First, it draws a distinction between *direct* coercion, which is aggression used by a male to overcome female mating reluctance, and *indirect* coercion (or *coercive mate guarding*), which is aggression used by a male to prevent a female from mating with other males (Muller et al. 2009b). Although both involve male aggression against females, these represent distinct mechanisms of coercion, each with its own logic and tactics. Second, the Smuts and Smuts definition highlights the necessity of considering fitness costs to females. Although receiving aggression might appear to be inherently detrimental, a female might benefit from it reproductively if aggression signals male quality (Szykman et al. 2003; Arnqvist and Rowe 2005). If the fitness benefits of screening males in this way outweighed the costs of the aggression, then this would not qualify as coercion, or sexual conflict.

Male aggression against females shows considerable variation among primates. In ring-tailed lemurs, at one extreme, females are the dominant sex, and clear observations of adult males victimizing adult females are notably absent (e.g., Pereira and Kappeler 1997). In chimpanzees, at the other extreme, females routinely experience male violence, including extended beatings that can result in serious injury (Goodall 1986; Muller et al. 2009a). Old World monkeys and apes show comparatively high rates of sexual aggression, probably because many species are diurnal, terrestrial, and sexually dimorphic, and males most easily target females by day, on the ground, and when females are smaller (Clarke et al. 2009). Although male aggression against females serves a variety of functions, current evidence suggests that it often represents sexual coercion (Smuts and Smuts 1993; Muller et al. 2009b; Muller and Wrangham 2009).

In this chapter I compare patterns of male aggression against females in chimpanzees and humans. Despite substantial differences in social structure and mating systems (see Muller and Pilbeam, this volume), these two species show remarkable similarities in sexual coercion that are not all shared with other great apes. I examine the selection pressures that likely drove these similarities, and evaluate the hypothesis that the similarities reflect homology rather than homoplasy. Finally, I consider whether unique aspects of human social life have generated new forms of gendered violence that fall outside the Smuts and Smuts definition of sexual coercion.

Male Aggression against Females in Chimpanzees

Although male aggression varies in frequency and intensity across study sites, female chimpanzees everywhere are victimized more often than most primate females, and at rates similar to those experienced by male chimpanzees[1] (Muller et al. 2009a). Aggression can be mild, such as an arm raised in

FIGURE 16.1. Male-female aggression in Gombe National Park. In chimpanzees, male aggression against females encompasses everything from threats, such as this charging display (directed by adult male Goblin to swollen female Gigi), to more serious violence, including hitting, kicking, biting, and dragging. Photo by Jane Goodall, © The Jane Goodall Institute.

perfunctory threat. However, males also launch vicious attacks on females that can last for many minutes, include hitting, biting, stomping, and dragging, and result in wounding (e.g., Kahlenberg 2009). The brutality of these episodes is indicated by significant fractions of chimpanzee skeletons showing signs of healed cranial trauma, and at equivalent frequency in the two sexes (Novak and Hatch 2009). The placement of wounds and large sex differences in violence suggest that males are the perpetrators in most cases.

Male aggression against females is not a unitary phenomenon, and some forms do not involve sexual coercion. For example, adolescent male chimpanzees frequently torment females, but primarily in nonmating contexts, and with the apparent goal of establishing dominance, not sexual access (Goodall 1986; Pusey 1990; Nishida 2003; Muller et al. 2009a). Adult males occasionally fight with females over food—usually meat—but this accounts for only 1–12 percent of male-female aggression across sites (Muller et al. 2009a). Finally, males sometimes intervene aggressively to stop fights among females. This is often done to protect and retain new female immigrants, who are regular targets of aggression from female residents (Kahlenberg et al. 2008). In one year at Kanyawara (Kibale National Park, Uganda), such policing accounted for approximately 7 percent of male aggression against females (Muller et al. 2009a).

Evidence from multiple study sites, however, suggests that chimpanzee males often use aggression as a coercive mating tactic, making females more likely to mate with them, or less likely to mate with other males. First, males are routinely most aggressive toward the females who are most likely to conceive. In a study from Kanyawara, mothers received more aggression than nulliparas (females who have not yet given birth), who experience a long period of adolescent subfecundity. Cycling mothers received more aggression when they were maximally swollen (i.e., near ovulation) than during periods of lactational amenorrhea (Muller et al. 2007). In Kasekela (Gombe National Park, Tanzania), cycling mothers received more aggression when they were maximally swollen than when nonswollen (Feldblum et al. 2014). And in both M-group (Mahale Mountains National Park, Tanzania) and Sonso (Budongo Forest Reserve, Uganda), cycling females received more aggression than lactating females (Kaburu and Newton-Fisher 2015). In Sonso, but not in M-group, maximally swollen females received more aggression than females who were cycling but nonswollen (Kaburu and Newton-Fisher 2015).

Second, aggression directed toward individual females is frequently associated with increased rates of mating or conception with those females. In Kanyawara, males not only showed increased rates of copulation with the females that they were most aggressive toward, but were actively solicited for copulation at the highest rates by those females during the periovulatory period, when conception was most likely (Muller et al. 2007, 2011). Sonso males similarly achieved a higher proportion of copulations with the females they were most aggressive toward (Kaburu and Newton-Fisher 2015). In Kasekela, male aggression toward individual cycling females was predictive of actual paternity (Feldblum et al. 2014, currently the only test of coercion to employ genetic data). In M-group, by contrast, there was no relationship between male aggression against females and mating success, and such aggression was comparatively rare (Kaburu and Newton-Fisher 2015).

Kanyawara is the only site where the costs of male aggression against females have been quantified systematically, as long-term hormonal monitoring has documented patterns of cortisol production by females (Muller et al. 2007; Emery Thompson et al. 2010). Although acute cortisol secretion represents an adaptive response to stress, it also incurs physiological costs, redirecting energy from long-term processes such as growth and reproduction to meet the demands of the stressor (Sapolsky 2002). Chronic activation of the stress response potentially incurs additional costs, as sustained exposure to cortisol can result in muscle wasting, immunosuppression, gastric ulcers, atherosclerosis, and other pathologies (Sapolsky 2002).

Confirming the physiological costs of victimization, cortisol secretion by Kanyawara females was correlated with fecundity in the same manner as male aggression. Specifically, cycling mothers exhibited higher cortisol levels than did cycling nulliparas, and mothers maintained elevated cortisol levels when they were maximally swollen, compared to periods of lactational amenorrhea (Muller et al. 2007). A six-year study at Kanyawara showed that aggression received from males was the strongest predictor of urinary cortisol levels in maximally swollen mothers (Emery Thompson et al. 2010). Although long-term patterns of wounding have not been analyzed at Kanyawara, it is clear that females are often physically injured by male aggression, or in their attempts to escape it (e.g., falling from trees while fleeing: Muller et al. 2009a).

Do the fitness costs of male aggression outweigh the potential benefits that a female could receive from males signaling their quality as sires? Aspects of chimpanzee cognition and social organization suggest the answer is yes. The idea that females might resist male mating advances to test persistence and stamina comes primarily from studies of invertebrates that lack individual recognition (reviewed in Arnqvist and Rowe 2005). Chimpanzee females, however, live in stable communities and form long-term relationships with individual males. They presumably have years to assess male quality, and should be motivated to avoid costly aggression by considering prior social interactions when making mating decisions. Consequently, the pervasive aggression that characterizes chimpanzee mating and premating is more likely to result from a true conflict of interest than from female assessment strategies (Muller et al. 2009a).

Because male aggression in chimpanzees both constrains female mate choice and imposes costs on females, it appears to meet the Smuts and Smuts definition of sexual coercion. What is the actual mechanism at work, though—direct coercion or coercive mate guarding?

Direct coercion "functions to increase the chances that a female will mate" with a particular male, and involves the use of force to overcome female resistance (Smuts and Smuts 1993). This is the simplest form of coercion, and the primary one recognized by Clutton-Brock and Parker (1995), who suggested three distinct categories: *forced copulation, harassment,* and *intimidation*. Forced copulation involves violent restraint, resulting in immediate mating. Harassment involves repeated attempts to mate that impose costs on females, inducing eventual submission. Intimidation involves physical retribution against female refusals to mate, conditioning submission to future advances. All of these are expected to primarily involve nonpreferred males, whom females are more likely to reject as suitors.

Indirect coercion, or coercive mate guarding, refers to the use of force "to decrease the chances that [a female] will mate with other males." Thus, it functions to constrain female promiscuity rather than overcome female resistance (Smuts and Smuts 1993; van Schaik et al. 2004). Consequently, both preferred and nonpreferred males might be expected to practice it. In many species, however, coercive mate guarding is primarily a strategy of high-ranking males (Clarke et al. 2009). Muller and colleagues (2009b) proposed three basic categories: *herding, punishment,* and *sequestration*. Herding involves short-term aggression directed toward females to induce immediate separation from rival males, and restore proximity to the guarding male. Punishment involves physical retribution against females mating with other males, decreasing the likelihood of such behavior in the future. Sequestration involves forceful separation of females from a social group, particularly during periods of maximal fecundity, to prevent copulations with rivals.

Forced copulation, the most extreme form of direct coercion, has been observed in chimpanzees, but it is rare. Goodall (1986) describes a few cases, all of which involved males forcibly mating with their mothers or maternal sisters. Harassment and intimidation are more common than forced copulation, but are still infrequent, probably because chimpanzee females are generally eager to mate with multiple partners, and rarely rebuff male advances. At Gombe, for example, Goodall (1986) reported that over a five-year period, females responded positively to 95.9 percent of adult male copulatory advances within one minute ($n = 1,475$). Instances where they did not almost all involved a single sterile female, or females refusing male relatives. Stumpf and Boesch (2006) reported lower—though still high—rates of female receptivity from Taï National Park (Côte d'Ivoire), where 80 percent of 938 male advances led to successful copulation. This sample included adolescent males,[2] however, who are known at other sites to have lower success rates in soliciting females (e.g., Gombe: Goodall 1986; Mahale: Matsumoto-Oda 1999). Thus, the figure for adult males at Taï is probably higher.

When female chimpanzees do resist male advances, it often appears to result from a fear of other, higher-ranked males in the party, rather than an aversion to the suitor himself. For example, in M-group, male courtship was more likely to succeed when higher-ranked males were absent (Matsumoto-Oda 1999). And in Kanyawara, maximally swollen mothers solicited all males at increased rates when the alpha male was absent (Muller et al. 2011). Thus, when harassment and intimidation do occur, they often involve low-ranking males attempting to overcome female reluctance to mate in the presence of higher-ranked, mate-guarding males (Muller et al. 2009a).

Coercive mate guarding, then, is probably the prevailing form of coercion in chimpanzee mating, with males employing aggression to constrain female promiscuity. Consistent with this hypothesis, data from Kanyawara show that high-ranking males direct more aggression toward maximally swollen mothers than do low-ranking males (Muller et al. 2009a). In Kasekela, males of all ranks showed a correlation between aggression directed at specific females and the probability of conception with those females, but the effect was stronger for high-ranking males (Feldblum et al. 2014). Assuming that, all else being equal, females prefer high-status partners, these patterns support a strategy of coercive mate guarding by high-ranking males.

The most extreme form of coercive mate guarding is sequestration, which in chimpanzees can occur when a male forces a female to accompany him to a peripheral part of their territory for several days to more than a month. During this "consortship," he enjoys exclusive mating access (Tutin 1979; Goodall 1986; Smuts and Smuts 1993). Consortships appear designed to avoid competition from rival males, rather than to overcome female mating reluctance, because the males who initiate them are normally accepted by females as mating partners in multimale settings. Although some consortships are clearly initiated by male aggression, others are not, and the degree to which female cooperation is necessary to ensure a successful consortship is not known (see also Muller and Pilbeam, this volume). At some sites, forced consortships are rare or absent (e.g., Taï: Boesch and Boesch-Achermann 2000). The reproductive significance of consortships, whether coerced or voluntary, is not well studied, and remains an important area for future research.

Much of the aggression that females receive during periods of maximal swelling qualifies as herding or punishment, as males attempt to prevent them from mating with rival suitors. However, cycling females also experience significant aggression during periods when they are not ovulating, or have no sexual swelling, and males have no immediate interest in mating. In Kanyawara, for example, males copulated at the highest rates with the females that they were most aggressive toward throughout periods of cycling, not just during periods of maximal swelling (Muller et al. 2007, 2009a). Similarly, in Kasekela, aggression directed at nonswollen females was positively associated with likelihood of paternity (Feldblum et al. 2014). And in Sonso, male aggression directed at cycling females—both swollen and nonswollen—was positively associated with mating success at the dyadic level (Kaburu and Newton-Fisher 2015). These patterns support the hypothesis that male aggression against females often represents a long-term strategy of *conditioning aggression* (Wrangham and Muller 2009). Males intimidate cycling females, who subsequently become more likely to respond to their advances, less likely to respond to the advances of other males in their presence, or both (Clutton-Brock and Parker 1995; Muller et al. 2009a).

Infanticide by males represents an additional form of long-term coercion under the Smuts and Smuts (1993) definition (Wrangham and Muller 2009). By killing unweaned infants, males accelerate a female's return to reproductive cycling, and thus increase their chances of mating with her sooner (reviewed in Palombit 2012). Male chimpanzees sometimes kill infants from both their own and neighboring groups, and, consistent with this hypothesis, most within-group cases involve infants that are not likely to have been sired by the killer (e.g., because a female had a prolonged absence from the community range: Palombit 2012). The risk of male infanticide places pressure on females to mate broadly, giving all males the perception of paternity, and thus protecting the resulting offspring (Hrdy 1979; van Schaik and Janson 2000; Wolff and Macdonald 2004). This counterstrategy generates a new conflict of interest between the sexes, however, leading to coercive mate guarding by males to constrain female promiscuity (van Schaik et al. 2004; Clarke et al. 2009).

Male Aggression against Females in Humans

Although rates of male aggression against females vary widely across human societies, women are universally more likely to be victimized by husbands or intimate partners than by strangers (Counts et al. 1999; Krug et al. 2002; World Health Organization 2005). The World Health Organization reviewed forty-eight surveys from around the world in which 10 percent to 69 percent of women reported being assaulted by an intimate partner at some point in their lives. The percentage of women reporting assault by an intimate partner within the previous twelve months varied from 3 percent or less in countries such as Canada and the United States, to more than 30 percent in parts of the Middle East (Krug et al. 2002).

Anthropologists have generally concluded that women in hunting and gathering societies experience lower rates of abuse, or less grievous abuse, than women in agricultural, pastoralist, or industrial societies (e.g., Draper 1999; Hess et al. 2010; Marlowe 2010). Forager women are also more likely to fight back against their partners when assaulted (Lee 1979; Tonkinson 1991; Hewlett and Hewlett 2008; Butovskaya 2013). A potential explanation for both of these patterns is that women are more likely to live among kin in foragers than in nonforagers, particularly during the early years of a marriage (Marlowe 2004). The presence of kin, combined with the open, public nature of forager camps, often leads to third parties intervening when disputes between spouses turn violent (Tonkinson 1991; Draper 1999; Marlowe 2010). Pacifying interventions are less common when wives live away from kin, or in private compounds where they can be abused behind locked doors (Smuts 1992; Rodseth and Novak 2009).

Nevertheless, even in foraging societies men physically abuse their intimate partners. Published rates are rare, but in one Hadza camp 9.1 percent of wives reported having been the objects of violence in disputes with their spouses in the previous year (Butovskaya 2013). (For husbands the figure was 5 percent.) In a camp with easy access to alcohol, owing to non-Hadza neighbors, wives reported much higher rates of abuse (37.5 percent in the previous year). Over three years of fieldwork with the Ju/'hoansi, Lee (1979) recorded fourteen instances of men attacking women, ten of which were husbands attacking their wives. Among the Aché, Hill and Hurtado (1996: 230) noted that extramarital affairs were common, and that men "often beat their wives under such circumstances." Among the Mardu of Western Australia, women "sometimes, are severely beaten by their husbands" (Tonkinson 1991: 153). Hewlett and Hewlett (2008: 53) report that Aka men do hit their wives, but rarely ("few times, if ever during their relationship"), and that fights are as likely to be initiated by wives.

Many of the foragers that have been studied recently maintain strong norms condemning interpersonal violence (e.g., Draper 1999; Hess et al. 2010). In part, this attitude probably reflects the threat of intervention by the state when violence does occur, together with the political subordination of foragers by their more powerful farming or pastoralist neighbors (Hill and Hurtado 1996; Hill et al. 2007). In the past, when foragers were surrounded by other foragers, aggression between groups, in the form of lethal raiding, was more common (Wrangham and Glowacki 2012). Such conditions would have favored increased rates of male-female aggression, for two reasons. First, the threat of intergroup aggression promotes strong alliances among males, and these produce antagonism between the sexes (Smuts 1992). Second, intergroup raiding undermines cultural norms against aggression, as children are socialized to be warriors (Ember and Ember 1994). Consequently, groups like the Ju/'hoansi or the Hadza probably exhibit less male-female aggression today than was usual in the past. Archaeological evidence of cranial trauma is consistent with domestic abuse having been common in some foraging populations (e.g., Webb 1995; Lambert 1997; Schwitalla et al. 2014). It is impossible, however, to definitively attribute trauma in such samples to a specific source.

The evidence that men's aggression frequently functions as sexual coercion is similar to that previously discussed for chimpanzees. Specifically, the likelihood of a woman being victimized is positively associated with her fecundity (i.e., potential for conception). Wilson and Daly (2009) reviewed data from the United States, Canada, and Australia showing that both lethal and nonlethal assaults by male partners decreased with female age. Similarly, a study by the World Health Organization presenting data from more than 24,000 women in ten countries (Bangladesh, Brazil, Ethiopia, Japan, Namibia, Peru, Samoa, Serbia and Montenegro, Thailand, and Tanzania) showed that younger women were more likely than older women to have been physically abused by a partner within the preceding twelve months (World Health Organization 2005). In many cases the difference in risk was substantial. In Bangladesh, for example, 48 percent of fifteen- to nineteen-year-old women reported experiencing intimate partner violence within the previous twelve months, versus 10 percent of (perimenopausal) forty-five- to forty-nine-year-olds. In Peru, this difference was 41 percent versus 8 percent.

Although young women tend to be married to young men, who are more violent than old men, this does not fully explain the decline in abuse with age. Wilson and colleagues (1997), for example, found that young women married to much older husbands were substantially more likely to be the victims of lethal abuse than those married to men of their own age. They argue that men are more strongly motivated to defend women of high reproductive value (i.e., women with longer future reproductive careers), and that older men paired with young women are especially apprehensive about potential infidelity (see also Wilson and Daly 2009).

Evidence also suggests that women are at greater risk of abuse when cycling compared to periods of infecundity. Flinn (1988), for example, collected detailed data in a Caribbean village, where partly open houses facilitated direct observation. He found that men had more agonistic interactions with their partners when the latter were cycling versus pregnant or lactating. Unfortunately, most studies of domestic abuse do not systematically investigate the influence of female fecundity. A notable exception is the literature on abuse during pregnancy. Large studies in the United States, Mexico, Belgium, and Canada have tracked women's experiences of physical abuse from twelve to twenty-four months before pregnancy through parturition. In support of the coercion hypothesis, these studies show a clear pattern in which women ($n > 140,000$) report lower rates of intimate partner violence during pregnancy, compared to the months immediately preceding it (Castro et al. 2003; Saltzman et al. 2003; Martin et al. 2004; Daoud et al. 2012; van Parys et al. 2014). Relatedly, several psychological studies report that men are more vigilant and self-assertive in their relationships when their primary partners are most fecund, just prior to ovulation (Gangestad et al. 2002, 2005, 2014; Pillsworth and Haselton 2006).

Additional support for the idea that men's intimate partner violence often functions as sexual coercion comes from the "remarkably consistent" list of triggers for such violence reported by both men and women cross-culturally (Krug et al. 2002: 95). According to the World Health Organization's global survey, these include (1) not obeying the man, (2) arguing back, (3) not having food ready on time, (4) not caring adequately for the children and home, (5) questioning the man about money and girlfriends, (6) going somewhere without the man's permission, (7) refusing the man sex, and (8) the man suspecting the woman of infidelity (Krug et al. 2002). Although several of these triggers involve general assertions of men's dominance, or disagreements over household work, items 6 and 8 clearly involve indirect coercion (i.e., coercive mate guarding), and item 7 provides an example of direct coercion.

The most commonly cited triggers for domestic violence among foragers resemble those documented globally by the World Health Organization. Most violence between Hadza spouses results either from sexual jealousy or "disregard of household duties and child neglect" (Butovskaya 2013: 281). Similarly, most Mardu men "beat their wives only if they wantonly neglect their domestic responsibilities or are excessively active extramaritally" (Tonkinson 1991: 100). When Aché women have affairs, men "are not angry with the man who cuckolded them, but they are upset with their spouse. Aché men often beat their wives under such circumstances" (Hill and Hurtado 1996: 230). For both men and women, "Hitting in response to the fear of loosing [sic] a mate was a common theme among the Aka" (Hewlett and Hewlett 2008: 54). Lee (1979: 377) notes that the causes of fights among the Ju/'hoansi were "not always easy to discern," but that "[a]dultery was the most common single factor."

As with chimpanzees, then, much male aggression against females in humans appears designed to constrain female sexuality. This similarity is surprising, because although female chimpanzees are famously promiscuous, human females show relatively low levels of infidelity by the standards of other socially monogamous species (Emery Thompson and Muller 2016). Rates of extrapair paternity are not known from most pair-living primates, but are reported to be high in two socially monogamous lemurs (47 percent *Cheirogaleus medius,* 30 percent *Phaner furcifer*) and one tarsier (20 percent *Tarsius lariang*), intermediate in one gibbon (7 percent *Hylobates lar*) and one sportive lemur (7 percent *Lepilemur ruficaudatus*), and low in one platyrrhine (none detected in infants from seventeen pairs, *Aotus azarae*) (reviewed in Huck et al. 2005). Another pair-living gibbon, *Nomascus gabriellae,* has an extrapair paternity rate around 10 percent (Kenyon et al. 2011), close to the average for socially monogamous birds (11 percent: Griffith et al. 2002). In humans, by contrast, genetic analyses suggest that nonpaternity rates of 1–3 percent are the norm for a broad range of Western and non-Western agricultural and industrial populations (Anderson 2006; Voracek et al. 2008; Strassman et al. 2012; Greeff and Erasmus 2015; Larmuseau et al. 2016).

Although the rate of cuckoldry is generally low for a range of human populations, high levels of paternal investment may necessitate enhanced vigilance and sexual jealousy within pair-bonds. Humans are unique in maintaining a post-weaning period of juvenile dependence on food provided by adults, with men in foraging groups contributing resources to offspring for fifteen to twenty-two years (Kaplan and Lancaster 2003). Given this pattern of investment, even low levels of cuckoldry could impose significant costs on men, and selection should favor mechanisms to reduce its likelihood (Daly et al. 1982). This may explain why patterns of indirect coercion are broadly similar in humans and chimpanzees, despite chimpanzee females being substantially more promiscuous.

Direct coercion, by contrast, appears to be more common in humans than in chimpanzees, at least in its most extreme manifestation of forced copulation. Despite some claims to the contrary (e.g., Sanday 1981), rape appears to be present to some degree in all human societies (Palmer 1989). Assessing prevalence is difficult, however, because both the definition of rape and the proportion of rapes reported to authorities vary cross-culturally (Emery Thompson 2009). Accordingly, estimates differ widely by country and by survey. For example, the percentage of women reporting that their first experience of sexual intercourse was forced varied from less than 1 percent in Japan, to 30 percent in rural Bangladesh (World Health Organization 2005). The percentage of women reporting attempted or completed forced sex by an intimate partner in the previous year varied from less than 1 percent in the United States to more than 22 percent in a Peruvian sample (Krug et al. 2002).

As with domestic violence, rape is apparently less common among foragers than in other societies, at least within groups (e.g., Lee 1979, Marlowe 2010). This conclusion might partly reflect a tendency by ethnographers to emphasize stranger rape over acquaintance or marital rape. For example, Turnbull (1965) claimed to "know of no cases of rape" among the Mbuti, while simultaneously reporting that "The men say that once they lie down with a girl ... if they want her they take her by surprise, when petting her, and force her to their will." Primarily, though, the low incidence of rape within forager groups seems to reflect the fact that such behavior is difficult to conceal in small-scale, intimate societies. Knowledge of a rape in such groups is likely to become widespread, potentially leading to retaliation by the victim, retribution by the victim's kin, and loss of status to the perpetrator (Smith et al. 2001).

The risk of rape is higher between groups. Hadza women, for example, avoid leaving camp alone, for fear of rape or capture by non-Hadza men (Marlowe 2010). Similarly, Efe men normally escort Efe women between camps to prevent their harassment by men from other clans (Bailey and Aunger 1989). As with other forms of violence, this type of rape was probably more prevalent in the past, before state-level intervention in local conflicts was possible. This would have been especially true during episodes of intergroup conflict.

Why is rape more common in humans than in chimpanzees? One might have predicted the reverse pattern, as chimpanzees show slightly more sexual dimorphism in body mass than do humans, giving chimpanzee males a greater strength advantage. Female chimpanzees are also more likely than forager women to travel alone, presumably making them more vulnerable to assault. The critical difference appears to lie in the strength of female preferences. Because female chimpanzees gain large benefits from paternity confusion, intense resistance to male sexual advances is rare, unless close relatives are involved (Muller et al. 2009a). Consequently, forced copulation must also be rare. Women, by contrast, are highly selective about their mating partners, and thus more likely to resist male solicitation (Muller and Emery Thompson 2012).

Human rape has sometimes been characterized as an alternative reproductive tactic employed by men with poor sexual prospects (Thornhill and Palmer 2000). Alternatively, rape has been proposed as part of a longer-term strategy of male coercion and control, similar to the conditioning aggression discussed previously for male chimpanzees (Emery Thompson 2009). Three lines of evidence support the conditioning hypothesis (reviewed in Emery Thompson 2009; Emery Thompson and Alvarado 2012). First, most rape appears to be marital or acquaintance rape, and not stranger rape. Consistent with conditioning aggression, women often stay in relationships with men who have raped them (e.g., Ellis et al. 2009), and intimate partner rape is more likely to occur when a relationship is unstable. Second, rapists are not necessarily low-status or "nonpreferred" males. Instead, they are often men with extensive sexual experience who are not accustomed to female rejection. Third, the risk of rape is more closely associated with reproductive value (i.e., youth) than it is with fecundity (i.e., the probability of immediate successful conception). This pattern makes little sense if immediate conception, rather than a longer-term relationship, is the proposed benefit.

Homology or Homoplasy?

Do similar patterns of indirect coercion in humans and chimpanzees represent homology or homoplasy? This question is difficult to answer, because relatively few data are available on male aggression against females in the other African apes. Patterns are starting to emerge, however, and these are consistent with a common ancestor that showed chimpanzee-like patterns of coercion.

Bonobos are as closely related to humans as are chimpanzees, but they show little evidence of any kind of male sexual coercion, direct or indirect (Kano 1992; Paoli 2009; Furuichi 2011). In one study from LuiKotale (Salonga National Park, Democratic Republic of the Congo), bonobo males showed increased rates of aggression in the presence of fecund females, but the aggression was not directed at those females (Surbeck et al. 2012). An earlier study from Lomako (Democratic Republic of the Congo) reported that male aggression against females was generally rare, and not obviously associated with mating (Hohmann and Fruth 2003). The lack of male coercion in bonobos is the logical result of a status hierarchy in which males are not universally dominant to females, and in fact most of the highest-ranked individuals are female, and the lowest-ranked male (Furuichi 2011). The ultimate reasons for the high status of bonobo females are the subject of debate, but the relevant point here is that the lack of sexual coercion in bonobos is part of a suite of behavioral traits that are likely derived, as described by Hare and Wrangham (this volume).

There is more evidence for sexual coercion in gorillas. As with chimpanzees, direct coercion in the form of forced copulation is rare. Among mountain gorillas, several decades of research have produced only four published descriptions (Watts 1990; Sicotte 2001). Gorilla females are not promiscuous in the chimpanzee manner, and similar to human females, they might be expected to have strong mate preferences. However, gorilla females also have significant latitude in choosing a silverback with whom to associate (Sicotte 2001). Consequently, females might rarely face situations in which nonpreferred males attempt to mate with them.

Across Africa, it is estimated that 97 percent of gorilla breeding groups contain only one adult male (Harcourt and Stewart 2007). Consequently, indirect coercion in the mating context is not likely to be a concern for most silverbacks—when females are fecund, there are no other males to mate with. Consistent with this idea, among mountain gorillas, males in single-male groups are less aggressive toward estrous females than toward anestrous females (Harcourt 1979).

In mountain gorillas, more than 40 percent of silverbacks live in multimale groups, however, and in these groups there is evidence for coercive mate guarding. In Karisoke (Volcanoes National Park, Rwanda), males in multimale groups directed more aggression at estrous females than anestrous females (Robbins 2003). And when Karisoke males interrupted other individuals' mating attempts, they often attacked the female, rather than the male (Robbins 1999). In Bwindi Impenetrable National Park (Uganda), cycling females received more aggression from silverbacks than lactating females, and males were more aggressive toward females on the days that they mated with them (Robbins 2009). There was, however, no difference in the amount of aggression given by a silverback when he was the only male in the group, compared to when the group had multiple males. Further, most aggression was mild, leading Robbins (2009) to conclude that it may sometimes represent courtship rather than coercion.

A clearer example of coercion in gorillas comes from the herding behavior exhibited by males during intergroup encounters. It is during such encounters that females are most likely to exercise mate choice, leaving one male to join another. Consequently, males often charge at females who are moving toward another group, sometimes biting them (Sicotte 1993, 2001; Robbins and Sawyer 2007). In Karisoke, females without dependent offspring were the most frequent victims of male charging displays during intergroup encounters (Sicotte 2001). This is consistent with male coercion, as females with unweaned infants are unlikely to transfer, owing to the risk of infanticide from the new male.

Similar patterns of herding behavior have been observed among male western gorillas in Mbeli Bai (Nouabalé-Ndoki National Park, Republic of Congo). Females there received increased rates of male aggression during intergroup encounters (Breuer et al. 2016). As in Karisoke, potential migrants (females without unweaned infants) received higher rates of aggression from males, particularly during such encounters.

Indirect coercion, then, is practiced by male gorillas, and at higher rates in groups that look more like those of humans and chimpanzees, by including multiple males. Consequently, a parsimonious scenario is that indirect coercion was present in the last common ancestor of humans, chimpanzees, and gorillas, and represents a behavioral homology in these close relatives, with a loss in the bonobo lineage after the split from chimpanzees. The alternative, that indirect coercion evolved independently, is possible, but is it consistent with the likely mating systems of early hominins? Our knowledge of these systems is severely limited, but, as discussed by Muller and Pilbeam (this volume), body mass dimorphism in several australopithecines and paranthropines is estimated to be higher than that of chimpanzees, with males 50 to 60 percent heavier than females (Grabowski et al. 2015). Facial dimorphism in many of these species is also substantial (e.g., Lockwood 1999; Lockwood et al. 2007). These patterns likely reflect a mating system with some degree of polygyny, if not as extreme as that exhibited by gorillas (Dixson 2009). Grueter and colleagues (2012) describe plausible scenarios, based on modern primates, for how such a social system could have evolved from a promiscuous, chimpanzee-like mating system (see also Chapais 2013, this volume). If such a system took the form of independent, one-male groups, as in gorillas, one would expect to see indirect coercion, particularly in groups with multiple males. If the mating system took the form of polygynous units clustered in a broader multimale, multifemale society (Grueter et al. 2012), then the risk of extrapair copulation would have been higher, and indirect coercion even more intense, as in hamadryas baboons (Swedell and Schreier 2009). In either case, some degree of polygyny in australopithecines would support the idea that similar patterns of indirect coercion in gorillas, chimpanzees, and humans represent homology. By contrast, the extent of forced copulation in humans appears to represent a novel trait, probably related to the evolution of strong mate preferences in women.

Unique Aspects of Human Sexual Conflict

A large proportion of men's aggression against women probably represents sexual coercion by the Smuts and Smuts definition, and is conspicuously similar to what is seen in chimpanzees. Other aspects of human sexual conflict, however, appear to be derived, and perhaps unique among primates. These include coercive sexual behavior that is not directly related to reproduction, and conflict between spouses over the management of labor and resources within the family.

Coercive Sexual Behavior That Is Not Sexual Coercion

The Smuts and Smuts definition of sexual coercion addresses men's use of force to control women's sexuality for the purposes of mating and conception. However, the ethnographic record suggests that men are often motivated to control women's sexuality in order to further political goals that are not directly related to reproduction (Wrangham and Muller 2009). Wife lending and wife exchange provide prominent examples. In the Standard Cross-Cultural Sample, a subset of 110 societies has information on wife exchange or lending (Broude and Greene 1976). Some form of wife lending occurs in one-third of these. For foraging societies in the sample, wife lending occurs in just over half of those with relevant information (eleven of twenty-one groups: Table 16.1). In the most common context cross-culturally, wives are offered sexually, on a one-time basis, for hospitality or ritual purposes that function to establish or confirm men's alliances. Institutional arrangements are also observed in which men are expected to loan their wives to specific individuals (for example, a brother-in-law) or groups (such as men from their age grade).

In some cases, wife lending appears to be a form of spousal protection and paternal investment. For example, among the north Alaskan Eskimo, Spencer (1959: 83–84) reported that if a man left the community on a prolonged trading or hunting expedition, he might loan his wife to a friend or neighbor during the absence. The friend would have sexual rights over the wife during this interval, but with the expectation that he would protect her and render assistance to her children. Such relationships between men were often reciprocal. In other cases, wives appear to be exchanged simply for sexual novelty. One Ju/'hoansi informant told Lorna Marshall that "If you want to sleep with someone's wife, you get him to sleep with yours, then neither of you goes after the other with poisoned arrows" (1959: 360).

In many cases, however, the sharing of wives is explicitly designed to promote bonds among men. Among Australian foragers, for example, Elkin (1938: 128) noted that "Just before a revenge expedition sets out on its dangerous enterprise, its members temporarily exchange wives, thus expressing their unity and friendship to one another." Wives could also be loaned to visitors, or offered to men with whom the husband had quarreled, in order to erase a debt or reconcile (Elkin 1938; Berndt and Berndt 1964). Finally, wives were sometimes exchanged temporarily to make peace between two groups

TABLE 16.1. Wife sharing among foragers in the standard cross-cultural sample (SCCS).

SCCS ID	Society	Region	Score
2	!Kung	Africa	4
9	Hadza	Africa	7
13	Mbuti	Africa	7
77	Semang	East Eurasia	7
79	Andamanese	East Eurasia	—
80	Vedda	East Eurasia	7
86	Badjau	North America	—
90	Tiwi	North America	—
91	Aranda	North America	6
118	Ainu	North America	—
119	Gilyak	North America	2
120	Yukaghir	North America	—
122	Ingalik	North America	7
123	Aleut	North America	—
124	C. Eskimo	North America	4
125	Montagnais	North America	—
126	Micmac	North America	7
127	Saulteaux	North America	—
128	Slave	North America	7
129	Kaska	North America	4
130	Eyak	North America	4
131	Haida	North America	2
132	Bellacoola	North America	—
133	Twana	North America	7
134	Yurok	North America	6
135	Pomo	North America	—
136	Yokuts	North America	—
137	Paiute	North America	—
138	Klamath	North America	—
139	Kutenai	North America	7
162	Warrau	South America	—
173	Siriono	South America	2
178	Botocudo	South America	—
179	Shavante	South America	3
180	Aweikoma	South America	1
186	Yahgan	South America	7

Note: Scores are from Broude and Greene (1976): 1 = extramarital sex of any kind allowed for wives; 2 = wife lending and/or exchange institutionalized vis-à-vis a woman and a group of men (e.g., any man in husband's age grade, husband's clansmen); 3 = wife lending and/or exchange institutionalized vis-à-vis some specific man other than husband (e.g., brother-in-law); 4 = wife lending and/or exchange only on occasion and specifically for sexual satisfaction; 5 = wife lending and/or exchange occurs for reason that benefits husband (e.g., wife exchanged in return for labor, money); 6 = wife lending and/or exchange occurs on a one-time basis for a specific purpose over and above sexual satisfaction (e.g., hospitality, alliance, ceremonial); 7 = no wife lending or exchange allowed. A dash means that information on wife sharing was not available for that group.

(Elkin 1938). Even in societies where this is not the declared purpose of wife sharing, the practice strengthens male bonds.

Although much of the literature in evolutionary psychology and behavioral ecology focuses on men's coercive sexual behavior as a means to ensure paternity, the widespread practice of wife exchange suggests that male-male bonds are often more important to men's fitness than is partner fidelity. A similar dynamic has been documented in wild chimpanzees. In an initial study from Kanyawara, the alpha male was observed to trade mating opportunities for social support. Male allies who supported this alpha in conflicts were permitted to mate with estrous females undisturbed, whereas other males were harassed during their mating attempts (Duffy et al. 2007). A more recent study from Gombe looked at thirty-six years of data encompassing the tenures of eight alpha males (Bray et al. 2016). Similar to Kanyawara, Gombe males who spent more time grooming with the alpha were more likely to mate with females when the alpha was present.[3] The primary difference between this kind of reciprocity in chimpanzees and human patterns of wife exchange is that coercion is not generally necessary to induce female chimpanzees to mate with multiple males.

Wife exchange, to be clear, does not always involve coercion. Among the Ju/'hoansi, for example, Marshall observed that "two men may agree to exchange wives temporarily, provided the wives consent" (1959: 359–360). In many societies, however, this decision is made by the husband, and the wife is expected to comply. Among the Eskimo, for example, Spencer wrote that "The two men involved agreed to exchange their wives and did not consult the women involved. A woman was told, 'Go over and stay with so and so'" (1956: 84). Similar practices were documented among Australian foragers. According to Berndt and Berndt, "a husband can lend his wife to another man without her consent, although usually the matter of consent is incidental where a wife has been brought up to regard this as one of her duties" (1964: 159–160). Elkin noted that although aboriginal women "may often not object" to being shared, they "sometimes live in terror of the use which is made of them at some ceremonial times" (1938: 130). Coercion is thus often involved in wife sharing and exchange, though not in a form consistent with the Smuts and Smuts definition.

Sexual Conflict over Men's Infidelity and Family Disinvestment

Because human pair-bonds represent economic as well as sexual unions, conflict between spouses often involves the distribution of family labor and resources, something that is not a factor for most primates. As noted by the World Health Organization, one of the common triggers for spousal abuse worldwide is a wife questioning her husband about girlfriends and money (Krug et al. 2002). This dynamic has been documented in the Tsimané, a Bolivian group that combines farming with foraging and some wage labor (Stieglitz et al. 2011, 2012).

As in many groups, spousal abuse by Tsimané men is often associated with women's infidelity. More often, however, such abuse results from infidelity by the husband (Stieglitz et al. 2012). This appears to represent a clear case of sexual conflict, as wives are objecting to men directing household resources toward prostitutes or mistresses, and husbands respond with violence, defending their prerogative to pursue extrapair matings. It is not clear whether this represents sexual coercion according to the Smuts and Smuts definition. If men's aggression ultimately functions to keep women from leaving the marriage, despite men's infidelity, then this would represent a form of direct coercion. Stieglitz and colleagues (2011), however, suggest that even when there is no threat of a woman leaving, aggression might function to keep wives from retaliating against male infidelity by neglecting their duties within the household. In either case, it is clear that the sexual division of labor in human pair-bonds creates opportunities for sexual conflict that are not present in other primates. These are likely exacerbated when men have access to cash through wage labor.

Summary

Male aggression against females occurs more frequently in both chimpanzees and humans than in many other primates. Although women in foraging societies probably experience less abuse than those in agricultural, pastoralist, or industrial societies, violence still occurs, with husbands or intimate partners the most common perpetrators. In both humans and chimpanzees, much male-female aggression functions as sexual coercion, and specifically coercive mate guarding, making females less likely to mate with males other than the aggressor. The evidence for coercion is similar in both species: females at the greatest risk of conception, whether because of their age or cycling status, are the most likely to receive aggression, with sexual jealousy the primary contributing factor. Females in both species are more likely to mate with their aggressors than with other males. These similarities are ostensibly surprising, because human women are substantially less promiscuous than chimpanzee females, which should render coercive mate guarding less necessary. However, human men are expected to be more sensitive to infidelity than chimpanzee males because of their greater investment in offspring. Patterns of direct coercion, by contrast, are divergent in humans and chimpanzees, with forced copulation occurring more frequently in humans, particularly in the context of intergroup conflict. This difference is probably driven by stronger mate preferences in human women than in chimpanzees.

Patterns of indirect coercion across the African apes are consistent with homology and a last common ancestor that practiced coercive mate guarding. Humans are unique, however, in that men practice derived forms of coercive sexual behavior that are not directly related to reproduction. In many foraging societies, for example, husbands share or exchange wives to strengthen their bonds with other men, and thus to further their own political goals. These practices support the hypothesis that male bonding is sometimes more important to men's long-term fitness than is spousal fidelity. Additionally, human men are often aggressive in conflicts with their spouses over the distribution of family labor and resources. This derived behavior is a consequence of the sexual division of labor, which establishes an economic aspect to human pair-bonding that is absent in the relationships of other primates.

Endnotes

1. Aggression here is defined as violence or the threat of violence. Chimpanzee threats include displays directed at individuals, together with agonistic chases. Violence generally takes the form of kicking, hitting, dragging, and biting.
2. Ten- to fifteen-year-olds. This classification was justified by the fact that males produce viable sperm at these ages, and occasionally reproduce. However, males at these ages are clearly not physically or socially mature, and in human terms would be considered adolescents.
3. Although time spent grooming is one important indicator of a social bond, it does not reflect political support as directly as actual assistance during an agonistic encounter.

Acknowledgments

I thank Christopher Boehm, Kristen Hawkes, Richard Wrangham, and an anonymous reviewer for helpful comments on this chapter, and the Jane Goodall Institute for the use of the photo in Figure 16.1.

References

Anderson, K. 2006. How well does paternity confidence match actual paternity? *Current Anthropology* 47: 513–520.

Arnqvist, G., and L. Rowe. 2005. *Sexual Conflict*. Princeton, NJ: Princeton University Press.

Bailey, R. C., and R. Aunger. 1989. Significance of the social relationships of Efe pygmy men in the Ituri Forest, Zaire. *American Journal of Physical Anthropology* 78: 495–507.

Berndt, R. M., and C. H. Berndt. 1964. *The World of the First Australians*. London: Angus and Robertson.

Boesch, C., and H. Boesch-Achermann. 2000. *The Chimpanzees of the Taï Forest: Behavioural Ecology and Evolution*. New York: Oxford University Press.

Bray, J., A. E. Pusey, and I. C. Gilby. 2016. Incomplete control and concessions explain mating skew in male chimpanzees. *Proceedings of the Royal Society of London B: Biological Sciences* 283: 20162071.

Breuer, T., A. M. Robbins, and M. M. Robbins. 2016. Sexual coercion and courtship by male western gorillas. *Primates* 57: 29–38.

Broude, G. J., and S. J. Greene. 1976. Cross-cultural codes on twenty sexual attitudes and practices. *Ethnology* 15: 409–429.

Butovskaya, M. L. 2013. Aggression and conflict resolution among the nomadic Hadza of Tanzania as compared with their pastoralist neighbors. In D. P. Fry, ed., *War, Peace, and Human Nature: The Convergence of Evolutionary and Cultural Views*, 278–296. New York: Oxford University Press.

Castro, R., C. Peek-Asa, and A. Ruiz. 2003. Violence against women in Mexico: A study of abuse before and during pregnancy. *American Journal of Public Health* 93: 1110–1116.

Chapais, B. 2013. Monogamy, strongly bonded groups, and the evolution of human social structure. *Evolutionary Anthropology* 22: 52–65.

Chapman, T., G. Arnqvist, J. Bangham, and L. Rowe. 2003. Sexual conflict. *Trends in Ecology and Evolution* 18: 41–47.

Clarke, P., G. Pradhan, C. van Schaik, M. N. Muller, and R. W. Wrangham. 2009. Intersexual conflict in primates: Infanticide, paternity allocation, and the role of coercion. In M. N. Muller and R. W. Wrangham, eds., *Sexual Coercion in Primates and Humans: An Evolutionary Perspective on Male Aggression against Females*, 42–77. Cambridge, MA: Harvard University Press.

Clutton-Brock, T. H., and G. A. Parker. 1995. Sexual coercion in animal societies. *Animal Behaviour* 49: 1345–1365.

Counts, D. A., J. K. Brown, and J. Campbell, eds. 1999. *To Have and to Hit: Cultural Perspectives on Wife Beating.* Urbana: University of Illinois Press.

Daly, M., M. Wilson, and S. J. Weghorst. 1982. Male sexual jealousy. *Ethology and Sociobiology* 3: 11–27.

Daoud, N., M. L. Urquia, P. O'Campo, M. Heaman, P. A. Janssen, J. Smylie, and K. Thiessen. 2012. Prevalence of abuse and violence before, during, and after pregnancy in a national sample of Canadian women. *American Journal of Public Health* 102: 1893–1901.

Dixson, A. F. 2009. *Sexual Selection and the Evolution of Human Mating Systems.* New York: Oxford University Press.

Draper, P. 1999. Room to maneuver: !Kung women cope with men. In D. A. Counts, J. K. Brown, and J. Campbell, eds., *To Have and to Hit: Cultural Perspectives on Wife Beating,* 53–72. Urbana: University of Illinois Press.

Duffy, K. G., R. W. Wrangham, and J. B. Silk. 2007. Male chimpanzees exchange political support for mating opportunities. *Current Biology* 17: 586–587.

Elkin, A. P. 1938. *The Australian Aborigines: How to Understand Them.* Sydney: Angus and Robertson.

Ellis, L., A. Widmayer, and C. T. Palmer. 2009. Perpetrators of sexual assault continuing to have sex with their victims following initial assault. *International Journal of Offender* 53: 454–463.

Ember, C. R., and M. Ember. 1994. War, socialization, and interpersonal violence: A cross-cultural study. *Journal of Conflict Resolution* 38: 620–646.

Emery Thompson, M. 2009. Human rape: Revising evolutionary perspectives. In M. N. Muller and R. W. Wrangham, eds., *Sexual Coercion in Primates and Humans: An Evolutionary Perspective on Male Aggression against Females,* 346–374. Cambridge, MA: Harvard University Press.

Emery Thompson, M., and L. Alvarado. 2012. Human sexual conflict and sexual coercion in comparative evolutionary perspective. In T. K. Shackelford and A. T. Goetz, eds., *Oxford Handbook of Sexual Conflict in Humans,* 100–121. New York: Oxford University Press.

Emery Thompson, M., and M. N. Muller. 2016. Comparative perspectives on human reproductive behavior. *Current Opinion in Psychology* 7: 61–66.

Emery Thompson, M., M. N. Muller, S. M. Kahlenberg, and R. W. Wrangham. 2010. Dynamics of social and energetic stress in wild female chimpanzees. *Hormones and Behavior* 58: 440–449.

Feldblum, J. T., E. E. Wroblewski, R. S. Rudicell, B. H. Hahn, T. Paiva, M. Cetinkaya-Rundel, A. E. Pusey, and I. C. Gilby. 2014. Sexually coercive male chimpanzees sire more offspring. *Current Biology* 24: 2855–2860.

Flinn, M. V. 1988. Mate guarding in a Caribbean village. *Ethology and Sociobiology* 9: 1–28.

Furuichi, T. 2011. Female contributions to the peaceful nature of bonobo society. *Evolutionary Anthropology* 20: 131–142.

Gangestad, S. W., Garver-Apgar, C. E., Cousins, A. J., Thornhill, R. 2014. Intersexual conflict across women's ovulatory cycle. *Evolution and Human Behavior* 35, 302–308.

Gangestad, S. W., R. Thornhill, and C. E. Garver. 2002. Changes in women's sexual interests and their partner's mate-retention tactics across the menstrual cycle: Evidence for shifting conflicts of interest. *Proceedings of the Royal Society of London B: Biological Sciences* 269: 975–982.

Gangestad, S. W., R. Thornhill, and C. E. Garver-Apgar. 2005. Adaptations to ovulation: Implications for sexual and social behavior. *Current Directions in Psychological Science* 14: 312–316.

García-Moreno, C., H. A. F. M. Jansen, M. Ellsberg, L. Heise, and C. Watts. 2005. *WHO Multi-Country Study on Women's Health and Domestic Violence against Women: Initial Results on Prevalence, Health Outcomes and Women's Responses*. Geneva: World Health Organization.

Goodall, J. 1986. *The Chimpanzees of Gombe: Patterns of Behavior*. Cambridge, MA: Harvard University Press.

Grabowski, M., K. G. Hatala, W. L. Jungers, and B. G. Richmond. 2015. Body mass estimates of hominin fossils and the evolution of human body size. *Journal of Human Evolution*. 85: 75–93.

Greeff, J. M., and J. C. Erasmus. 2015. Three hundred years of low non-paternity in a human population. *Heredity* 115: 396–404.

Griffith, S. C., I. P. F. Owens, and K. A. Thuman. 2002. Extra pair paternity in birds: A review of interspecific variation and adaptive function. *Molecular Ecology* 11: 2195–2212.

Grueter, C. C., B. Chapais, and D. Zinner. 2012. Evolution of multilevel social systems in nonhuman primates and humans. *International Journal of Primatology* 33: 1002–1037.

Harcourt, A. H. 1979. Contrasts between male relationships in wild gorilla groups. *Behavioral Ecology and Sociobiology* 5: 39–49.

Harcourt, A. H., and K. J. Stewart. 2007. *Gorilla Society: Conflict, Compromise, and Cooperation between the Sexes*. Chicago: University of Chicago Press.

Hess, N., C. Helfrecht, E. Hagen, A. Sell, and B. Hewlett. 2010. Interpersonal aggression among Aka hunter-gatherers of the Central African Republic. *Human Nature* 21: 330–354.

Hewlett, B. L., and B. S. Hewlett. 2008. A biocultural approach to sex, love, and intimacy in Central African foragers and farmers. In W. R. Jankowiak, ed., *Intimacies: Love and Sex across Cultures*. New York: Columbia University Press.

Hill, K. R., and A. M. Hurtado. 1996. *Ache Life History: The Ecology and Demography of a Foraging People*. New York: Aldine de Gruyter.

Hill, K., A. M. Hurtado, and R. S. Walker. 2007. High adult mortality among Hiwi hunter-gatherers: Implications for human evolution. *Journal of Human Evolution* 52: 443–454.

Hohmann, G., and B. Fruth. 2003. Intra- and inter-sexual aggression by bonobos in the context of mating. *Behaviour* 140: 1389–1413.

Hrdy, S. B. 1979. Infanticide among animals: A review, classification, and examination of the implications for the reproductive strategies of females. *Ethology and Sociobiology* 1: 13–40.

Huck, M., P. Lottker, U. R. Bohle, and E. W. Heymann. 2005. Paternity and kinship patterns in polyandrous moustached tamarins *(Saguinus mystax)*. *American Journal of Physical Anthropology* 127: 449–464.

Kaburu, S. S. K., and N. E. Newton-Fisher. 2015. Trading or coercion? Variation in male mating strategies between two communities of East African chimpanzees. *Behavioral Ecology and Sociobiology* 69: 1039–1052.

Kahlenberg, S. M. 2009. Males attack an anestrous female at Kanyawara: June 22, 2002. In M. N. Muller and R. W. Wrangham, eds., *Sexual Coercion in Primates and Humans: An Evolutionary Perspective on Male Aggression against Females.* Cambridge, MA: Harvard University Press. 188.

Kahlenberg, S. M., M. E. Thompson, M. N. Muller, and R. W. Wrangham. 2008. Immigration costs for female chimpanzees and male protection as an immigrant counterstrategy to intrasexual aggression. *Animal Behaviour* 76: 1497–1509.

Kano, T. K. 1992. *The Last Ape: Pygmy Chimpanzee Behavior and Ecology.* Stanford, CA: Stanford University Press.

Kaplan, H. S., and J. B. Lancaster. 2003. An evolutionary and ecological analysis of human fertility, mating patterns, and parental investment. In K. W. Wachter and R. A. Bulatao, eds., *Offspring: Human Fertility Behavior in Biodemographic Perspective,* 170–223. Washington, DC: National Academies Press.

Kenyon, M., C. Roos, V. T. Binh, and D. Chivers. 2011. Extrapair paternity in golden-cheeked gibbons *(Nomascus gabriellae)* in the secondary lowland forest of Cat Tien National Park, Vietnam. *Folia Primatologica* 82: 154–164.

Krug, E. G., J. A. Mercy, L. L. Dahlberg, and A. B. Zwi. 2002. *The World Report on Violence and Health.* Geneva: World Health Organization.

Lambert, P. M. 1997. Patterns of violence in prehistoric hunter-gatherer societies of coastal Southern California. In D. L. Martin and D. W. Frayer, eds., *Troubled Times: Violence and Warfare in the Past,* 77–109. New York: Routledge.

Larmuseau, M. H. D., K. Matthijs, and T. Wenseleers. 2016. Cuckolded fathers rare in human populations. *Trends in Ecology and Evolution* 31: 327–329.

Lee, R. B. 1979. *The !Kung San: Men, Women, and Work in a Foraging Society.* Cambridge: Cambridge University Press.

Lockwood, C. A. 1999. Sexual dimorphism in the face of *Australopithecus africanus*. *American Journal of Physical Anthropology* 108: 97–127.

Lockwood, C. A., C. G. Menter, J. Moggi-Cecchi, and A. W. Keyser. 2007. Extended male growth in a fossil hominin species. *Science* 318: 1443–1445.

Marlowe, F. W. 2004. Marital residence among foragers. *Current Anthropology* 45: 277–284.

Marlowe, F. W. 2010. *The Hadza: Hunter-Gatherers of Tanzania.* Berkeley: University of California Press.

Marshall, L. 1959. Marriage among !Kung Bushmen. *Africa: Journal of the International African Institute* 29: 335–365.

Martin, S. L., A. Harris-Britt, Y. Li, K. E. Moracco, L. L. Kupper, and J. C. Campbell. 2004. Changes in intimate partner violence during pregnancy. *Journal of Family Violence* 19: 201–210.

Matsumoto-Oda, A. 1999. Female choice in the opportunistic mating of wild chimpanzees *(Pan troglodytes schweinfurthii)* at Mahale. *Behavioral Ecology and Sociobiology* 46: 258–266.

Muller, M. N., S. M. Kahlenberg, M. E. Thompson, and R. W. Wrangham. 2007. Male coercion and the costs of promiscuous mating for female chimpanzees. *Proceedings of the Royal Society of London B: Biological Sciences* 274: 1009–1014.

Muller, M. N., and R. W. Wrangham, eds. 2009. *Sexual Coercion in Primates and Humans.* Cambridge, MA: Harvard University Press.

Muller, M. N., S. M. Kahlenberg, and R. W. Wrangham. 2009a. Male aggression against females and sexual coercion in chimpanzees. In M. N. Muller and R. W. Wrangham, eds., *Sexual Coercion in Primates and Humans: An Evolutionary Perspective on Male Aggression against Females,* 184–217. Cambridge, MA: Harvard University Press.

Muller, M. N., S. M. Kahlenberg, and R. W. Wrangham. 2009b. Male aggression and sexual coercion of females in primates. In M. N. Muller and R. W. Wrangham, eds., *Sexual Coercion in Primates and Humans: An Evolutionary Perspective on Male Aggression against Females,* 3–22. Cambridge, MA: Harvard University Press.

Muller, M. N., M. E. Thompson, S. M. Kahlenberg, and R. W. Wrangham. 2011. Sexual coercion by male chimpanzees shows that female choice may be more apparent than real. *Behavioral Ecology and Sociobiology* 65: 921–933.

Muller, M. N., and M. Emery Thompson. 2012. Mating, parenting and male reproductive strategies. In J. C. Mitani, J. Call, P. M. Kappeler, R. A. Palombit, and J. B. Silk, eds., *Evolution of Primate Societies.* Chicago: University of Chicago Press.

Nishida, T. 2003. Harassment of mature female chimpanzees by young males in the Mahale Mountains. *International Journal of Primatology* 24: 503–514.

Novak, S., and M. Hatch. 2009. Intimate wounds: Craniofacial trauma in women and female chimpanzees. In M. N. Muller and R. W. Wrangham, eds., *Sexual Coercion in Primates and Humans: An Evolutionary Perspective on Male Aggression against Females,* 322–345. Cambridge, MA: Harvard University Press.

Palmer, C. 1989. Is rape a cultural universal? A re-examination of the ethnographic data. *Ethnology* 28: 1–16.

Palombit, R. A. 2012. Infanticide: Male strategies and female counterstrategies. In J. C. Mitani, J. Call, P. M. Kappeler, R. A. Palombit, and J. B. Silk, eds., *The Evolution of Primate Societies,* 432–468. Chicago: University of Chicago Press.

Paoli, T. 2009. The absence of sexual coercion in bonobos. In M. N. Muller and R. W. Wrangham, eds., *Sexual Coercion in Primates and Humans: An Evolutionary Perspec-*

tive on Male Aggression against Females, 410–423. Cambridge, MA: Harvard University Press.

Pereira, M. E., and P. M. Kappeler. 1997. Divergent systems of agonistic behaviour in lemurid primates. *Behaviour* 134: 225–274.

Pillsworth, E. G., and M. G. Haselton. 2006. Male sexual attractiveness predicts differential ovulatory shifts in female extra-pair attraction and male mate retention. *Evolution and Human Behavior* 27: 247–258.

Pusey, A. E. 1990. Behavioural changes at adolescence in chimpanzees. *Behaviour* 115: 203–246.

Robbins, M. M. 1999. Male mating patterns in wild multimale mountain gorilla groups. *Animal Behaviour* 57: 1013–1020.

Robbins, M. M. 2003. Behavioral aspects of sexual selection in mountain gorillas. In C. B. Jones, ed., *Sexual Selection and Reproductive Competition in Primates: New Perspectives and Directions,* 477–501. Norman, OK: American Society of Primatologists.

Robbins, M. M. 2009. Male aggression against females in mountain gorillas: Courtship or coercion? In M. N. Muller and R. W. Wrangham, eds., *Sexual Coercion in Primates and Humans: An Evolutionary Perspective on Male Aggression against Females,* 112–127. Cambridge, MA: Harvard University Press.

Robbins, M. M., and S. C. Sawyer. 2007. Intergroup encounters in mountain gorillas of Bwindi Impenetrable National Park, Uganda. *Behaviour* 144: 1497–1519.

Rodseth, L., and S. A. Novak. 2009. The political significance of gender violence. In M. N. Muller and R. W. Wrangham, eds., *Sexual Coercion in Primates and Humans: An Evolutionary Perspective on Male Aggression against Females,* 292–321. Cambridge, MA: Harvard University Press.

Saltzman, L. E., C. H. Johnson, B. C. Gilbert, and M. M. Goodwin. 2003. Physical abuse around the time of pregnancy: An examination of prevalence and risk factors in 16 states. *Maternal and Child Health Journal* 7: 31–43.

Sanday, P. R. 1981. The socio-cultural context of rape: A cross-cultural study. *Journal of Social Issues* 37: 5–27.

Sapolsky, R. M. 2002. Endocrinology of the stress-response. In J. B. Becker, S. M. Breedlove, D. Crews, and M. M. McCarthy, eds., *Behavioral Endocrinology,* 409–450. Cambridge, MA: MIT Press.

Schwitalla, A. W., T. L. Jones, M. A. Pilloud, B. F. Codding, and R. S. Wiberg. 2014. Violence among foragers: The bioarchaeological record from central California. *Journal of Anthropological Archaeology* 33: 66–83.

Sicotte, P. 1993. Inter-group encounters and female transfer in mountain gorillas: Influence of group composition on male behavior. *American Journal of Primatology* 30: 21–36.

Sicotte, P. 2001. Female mate choice in mountain gorillas. In M. M. Robbins, P. Sicotte, and K. J. Stewart, eds., *Mountain Gorillas: Three Decades of Research at Karisoke,* 59–88. Cambridge: Cambridge University Press.

Smith, E. A., M. B. Mulder, and K. Hill. 2001. Controversies in the evolutionary social sciences: A guide for the perplexed. *Trends in Ecology and Evolution* 16: 128–135.

Smuts, B. 1992. Male aggression against women: An evolutionary perspective. *Human Nature* 3: 1–44.

Smuts, B. B., and R. W. Smuts. 1993. Male aggression and sexual coercion of females in nonhuman primates and other mammals: Evidence and theoretical implications. *Advances in the Study of Behavior* 22: 1–63.

Spencer, R. F. 1959. *The North Alaskan Eskimo: A Study in Ecology and Society.* Smithsonian Institution Bureau of American Ethnology, Bulletin No. 171. Washington, DC: U.S. Government Printing Office.

Stieglitz, J., M. Gurven, H. Kaplan, and J. Winking. 2012. Infidelity, jealousy, and wife abuse among Tsimane forager-farmers: Testing evolutionary hypotheses of marital conflict. *Evolution and Human Behavior* 33: 438–448.

Stieglitz, J., H. Kaplan, M. Gurven, J. Winking, and B. V. Tayo. 2011. Spousal violence and paternal disinvestment among Tsimane forager-horticulturalists. *American Journal of Human Biology* 23: 445–457.

Strassmann, B. I., N. T. Kurapati, B. F. Hug, E. E. Burke, B. W. Gillespie, T. M. Karafet, and M. F. Hammer. 2012. Religion as a means to assure paternity. *Proceedings of the National Academy of Sciences* 109: 9781–9785.

Stumpf, R. M., and C. Boesch. 2006. The efficacy of female choice in chimpanzees of the Taï Forest, Côte d'Ivoire. *Behavioral Ecology and Sociobiology* 60: 749–765.

Stumpf, R. M., R. Martinez-Mota, K. M. Milich, N. Righini, and M. R. Shattuck. 2011. Sexual conflict in primates. *Evolutionary Anthropology* 20: 62–75.

Surbeck, M., T. Deschner, G. Schubert, A. Weltring, and G. Hohmann. 2012. Mate competition, testosterone and intersexual relationships in bonobos, *Pan paniscus*. *Animal Behaviour* 83: 659–669.

Swedell, L., and A. Schreier. 2009. Male aggression toward females in hamadryas baboons: Conditioning, coercion, and control. In M. N. Muller and R. W. Wrangham, eds., *Sexual Coercion in Primates and Humans: An Evolutionary Perspective on Male Aggression against Females,* 244–268. Cambridge, MA: Harvard University Press.

Szykman, M., A. L. Engh, R. C. Van Horn, E. E. Boydston, K. T. Scribner, and K. E. Holekamp. 2003. Rare male aggression directed toward females in a female-dominated society: Baiting behavior in the spotted hyena. *Aggressive Behavior* 29: 457–474.

Thornhill, R., and C. T. Palmer. 2000. *A Natural History of Rape.* Cambridge, MA: MIT Press.

Tonkinson, R. 1991. *The Mardu Aborigines: Living the Dream in Australia's Desert.* Fort Worth, TX: Holt, Rinehart and Winston.

Turnbull, C. M. 1965. *The Mbuti Pygmies: An Ethnographic Survey.* New York: American Museum of Natural History.

Tutin, C. E. G. 1979. Mating patterns and reproductive strategies in a community of wild chimpanzees *(Pan troglodytes schweinfurthii)*. *Behavioral Ecology and Sociobiology* 6: 29–38.

van Parys, A.-S., E. Deschepper, K. Michielsen, M. Temmerman, and H. Verstraelen. 2014. Prevalence and evolution of intimate partner violence before and during pregnancy: A cross-sectional study. *BMC Pregnancy and Childbirth* 14: 294.

van Schaik, C. P., and C. H. Janson. 2000. *Infanticide by Males and Its Implications*. Cambridge: Cambridge University Press.

van Schaik, C. P., G. R. Pradhan, and M. A. van Noordwijk. 2004. Mating conflict in primates: Infanticide, sexual harassment and female sexuality. In P. Kappeler and C. van Schaik, eds., *Sexual Selection in Primates: New and Comparative Perspectives*, 131–150. Cambridge: Cambridge University Press.

Voracek, M., T. Haubner, and M. L. Fisher. 2008. Recent decline in nonpaternity rates: A cross-temporal meta-analysis. *Psychological Reports* 103: 799–811.

Watson-Capps, J. J. 2009. Evolution of sexual coercion with respect to sexual selection and sexual conflict theory. In M. N. Muller and R. W. Wrangham, eds., *Sexual Coercion in Primates and Humans: An Evolutionary Perspective on Male Aggression against Females*, 23–41. Cambridge, MA: Harvard University Press.

Watts, D. P. 1990. Mountain gorilla life histories, reproductive competition, and sociosexual behavior and some implications for captive husbandry. *Zoo Biology* 9: 185–200.

Webb, S. 1995. *Palaeopathology of Aboriginal Australians: Health and Disease across a Hunter-Gatherer Continent*. Cambridge: Cambridge University Press.

Wilson, M. I., M. Daly, and J. Scheib. 1997. Femicide: An evolutionary psychological perspective. In P. A. Gowaty, ed., *Feminism and Evolutionary Biology*. New York: Chapman Hall.

Wilson, M., and M. Daly. 2009. Coercive violence by human males against their female partners. In M. N. Muller and R. W. Wrangham, eds., *Sexual Coercion in Primates and Humans: An Evolutionary Perspective on Male Aggression against Females*, 271–291. Cambridge, MA: Harvard University Press.

Wolff, J. O., and D. W. Macdonald. 2004. Promiscuous females protect their offspring. *Trends in Ecology and Evolution* 19: 127–134.

World Health Organization. 2005. *WHO Multi-Country Study on Women's Health and Domestic Violence against Women: Summary Report of Initial Results on Prevalence, Health Outcomes and Women's Responses*. Geneva: WHO Press.

Wrangham, R. W., and L. Glowacki. 2012. Intergroup aggression in chimpanzees and war in nomadic hunter-gatherers. *Human Nature* 23: 5–29.

Wrangham, R. W., and M. N. Muller. 2009. Sexual coercion in humans and other primates: The road ahead. In M. N. Muller and R. W. Wrangham, eds., *Sexual Coercion in Primates and Humans: An Evolutionary Perspective on Male Aggression against Females*, 451–468. Cambridge, MA: Harvard University Press.

17

Tool Use and Manufacture in the Last Common Ancestor of *Pan* and *Homo*

CAMPBELL ROLIAN *and* SUSANA CARVALHO

Cumulative culture, including complex forms of tool use, is often seen as a defining feature of hominins, but the manufacture and use of tools are by no means restricted to humans and their ancestors. Although tool use is a rare behavioral strategy in the animal kingdom, the use of tools for both subsistence purposes (i.e., extractive foraging tasks) and aggressive display is well documented in other primates, and in other vertebrates (Shumaker et al. 2011). Among nonhuman primates, chimpanzees have been the focus of numerous studies reporting on the regional variation, sometimes labeled cultural, in the behavioral repertoires of different groups (Whiten et al. 1999; Carvalho and McGrew 2012). After humans, chimpanzees *(Pan troglodytes)* have the largest repertoires of tool use and manufacture known among primates (McGrew 2017). Bonobos *(Pan paniscus)* have been the focus of much less research, and appear to be infrequent tool users in natural contexts (Whiten et al. 1999; McGrew 2004; Haslam 2014; for in-depth comparisons between *Pan troglodytes* and *Pan paniscus* and their tool use, see Koops et al. 2015; Gruber and Clay 2016). Because chimpanzees and bonobos are both our sister species and, notwithstanding their different degrees of reliance on technology, both are known to use tools, a comparative study of tool behavioral variation, but also of the morphological and cognitive correlates of tool use (e.g., hand morphology: Marzke 1997) between *Homo* and *Pan* may be useful in making inferences on the tools and tool-making behaviors of our last common ancestor, the LCA.

In this chapter, we will address two questions pertaining to tool use and manufacture in the LCA. Was the LCA able to use and/or manufacture tools? And if so, what type of tools likely composed the tool kit of this Miocene ape, and in which ecological contexts was it more likely to use tools to solve problems? To answer these questions, we will explore four complementary lines of evidence: (1) comparative tool use in extant apes and other primates, (2) comparative anatomy and biomechanics of primate hands, (3) comparative cognition in hominoids, and (4) evidence from the fossil and archaeological records of early hominins. It is not our intent to provide exhaustive reviews of each line of evidence, but rather to leverage current knowledge in each, especially with regards to the genus *Pan*, to paint the most likely picture of what tool use and manufacture in the LCA would have looked like.

We will limit our discussion primarily to tool use in the wild. Tool manufacture and use in captivity are well documented for a number of primates that have not been observed to do so in the wild (reviewed in Shumaker et al. 2011). However, captive settings lack ecological validity, making it difficult to evaluate the roles of group size, kinship, food availability, and other factors in understanding the emergence and duration of novel technological behavior.

What Is a Tool?

Before answering the questions outlined above, it is important to define what we consider to be tools in the context of primate behavioral ecology. We define a tool as a noncorporeal, unattached object that an animal makes and/or uses to modify itself, another individual, or its proximate environment (see also Shumaker et al. 2011). Tools may be deliberately modified objects (tool manufacture), or they may be objects used without prior modification—and modified only by use (tool use). Tools may be used in flexible, complex ways, sequentially, as part of tool composites or serving multiple functions. They are employed in multiple tasks or behaviors in the context of a primate's natural habitat and ecology. The most commonly reported uses of tools in wild primates fall into three categories: (1) tools used for extractive foraging and/or food processing, (2) tools used in grooming or increasing physical comfort/hygiene, and (3) tools used for communication.

Tool Use and Manufacture in Nonhuman Primates

Chimpanzee *(Pan troglodytes)*

Apart from humans, tool use and manufacture is best documented in chimpanzees, who show variation in their technological repertoires—for example, sequential tool use and tool composite use that have yet to be observed in any other extant nonhuman primate (Carvalho et al. 2009; Carvalho and McGrew 2012). Over fifty-five years of observations in the wild, from multiple long-term sites in East, Central, and West Africa, have painted a detailed picture of the diversity and complexity of their tool kits and behaviors (Whiten et al. 1999; see McGrew 2017 for an update on long-term sites). Unlike all other tool-using nonhuman primates, there is no well-studied chimpanzee population for which tool use has not been reported. As with humans, technology is present independently from habitat type, group size, predation pressure, and availability of resources. Every few years, a previously unknown type of tool or tool behavior is reported for specific chimpanzee populations (see, e.g., Sanz et al. 2004; Pruetz and Bertolani 2007; Koops et al. 2010, Lapuente et al. 2016; Musgrave et al. 2016 for recent examples). A thorough discussion of each tool, its use(s), and contexts is beyond the scope of this chapter, and several volumes have been written on the subject (see, e.g., McGrew 1992, 2004; Shumaker et al. 2011). Here we will limit our discussion to a few key features of chimpanzee tool behavior that make it unique among nonhuman primates.

Chimpanzees habitually use many different tools for communication with conspecifics, and/or as deterrents against perceived predators (or human observers) (Matsuzawa 1999). Tools used in this context include branches shaken, waved, or dropped from trees, objects (e.g., stones) thrown as part of conspecific displays (e.g., male-male aggression), and leaves that are ripped or torn to produce sounds that may be involved in courtship (Whiten et al. 1999). Tools are also used for grooming or personal comfort. These include leaves used as napkins or as sponges (for drinking), and branches used to swat away insects, or to scratch an itch (Goodall 1986; Sanz and Morgan 2007).

The most complex and diverse tools in chimpanzees are used for extractive foraging, for example, the use of stone tools for nut cracking (Carvalho et al. 2008) or the sequential use of five different tools to obtain honey (Boesch et al. 2009). Chimpanzees are omnivores, and while the fruits that make up large portions of their diets do not typically require tools to consume, many other consumed foods require tool-assisted extraction or processing. Sticks and twigs are among the most common foraging tools used by chimpanzees. Sticks of varying diameters are used to probe holes in insect mounds, to pry or dig them open, and, most famously, to dip the sticks in these mounds and extract termites and other insects (aka "fishing") (e.g., Goodall 1963; McGrew 1974; Sanz et al. 2004). Using sticks to extract honey has also been documented from multiple long-term study sites (e.g., Boesch et al. 2009; Sanz and Morgan 2009). Savanna chimpanzees at Fongoli, Senegal, have also been observed to use wooden spears to hunt for bushbabies (Pruetz and Bertolani 2007).

Habitual lithic technology has been observed only in West African chimpanzees, especially at Taï, Ivory Coast (Boesch and Boesch-Achermann 2000), and Bossou, Guinea (Fushimi et al. 1991; Carvalho et al. 2008, 2009), and constitutes a tiny proportion of the tool-use repertoire. The most common use of stone tools is as hammers, which are used to crack open the nuts of several tree species (Figure 17.1). On occasion, and only at Bossou, another large yet movable stone is used as an anvil upon which the nut is placed, and smaller stones have also been used by chimpanzees at that site to prop up and stabilize the anvil. Experiments in a "natural laboratory" setting at Bossou, in Guinea, show that the chimpanzees discriminate features of stone tools when selecting hammers or anvils, and often transport and reuse them for nut cracking at later times or places, before ultimately discarding them (i.e., "future planning": see Rosati, this volume; Carvalho et al. 2008; Carvalho and McGrew 2012). Moreover, there is regional variation between the sites in these operational sequences *(chaînes opératoires)* that maps onto local environmental factors (e.g., the availability of raw material), leading to differences in the stone assemblages left behind (Carvalho et al. 2008; Koops et al. 2013). This sort of cognitive flexibility related to lithic tools has not been documented in any other primate except for humans and their hominin ancestors.

FIGURE 17.1. Adult male in the outdoor laboratory of Bossou, Guinea, using a pair of stones—anvil and hammer—to crack open nuts (*Elaeis guineensis* and *Coula edulis* sp.). Photo: Susana Carvalho.

In addition to their frequent tool use, wild chimpanzees are also accomplished tool makers. Many of the organic tools described above require one or more steps to make. The simplest form of manufacture involves removing a branch from a tree, in a single step, to use as a signaling device (see below). Manufacturing more complex tools, especially for extractive foraging, typically requires greater modification of the raw materials before use. For example, dipping sticks may be stripped of leaves or bark before they are used to fish for insects. In a further elaboration of this activity, Sanz et al. (2009) reported that chimpanzees at Goualougo, Republic of Congo, deliberately fray the tips of the stick, turning it into a brush that is more efficient for extracting ants (Whiten 2011).

In contrast to these organic tools, the stone tools used by West African chimpanzees are not deliberately manufactured. Although they are used together (hence are a composite tool; see below), the hammers and anvils are unmodified stones that are modified only by the use (Figure 17.2). On two occasions, Carvalho et al. (2008) observed that stones were unintentionally fractured as part of the nut cracking process. In one case, a large flake was

FIGURE 17.2. Pounding tools after being used by the subject in Figure 1 during one experimental nut-cracking session. Note the damage on the edges of the anvil as a result of missed strikes—depending on the type of raw material used, this missed blow may produce objects that are very similar to human-produced flakes. Photo: Susana Carvalho.

detached from the anvil, and the user subsequently attempted to use the detached flake as a hammer, while the fractured tool was discarded, making this one of the few reported cases of the reuse of a by-product of a percussive activity by nonhuman primates.

Tool use and manufacture have been documented in every chimpanzee population studied to date, and hence is considered to be universal (and likely plesiomorphic) among chimpanzees (McGrew 1992, 2004). Still, the wealth of data accumulated over the last fifty years from multiple long-term study sites shows that there is substantial local variation in tool repertoires and behaviors among populations (reviewed in Whiten et al. 1999). Some tools and behaviors are universally shared among all documented populations, such as dragging, shaking, or waving branches in conspecific communication, probing with sticks, or using leaves as sponges (Whiten et al. 1999). Others may be absent from particular sites for ecological reasons—for example, the absence of a particular food source or a low probability of encountering the resource in the home range (Koops et al. 2013).

Many tools, however, are unique to specific populations, though ecological factors do not explain their absence elsewhere. This is exemplified by the stone tools, which are only found in West Africa, despite the availability of suitable stones and nuts at sites in Central and East Africa (Whiten et al. 1999; but see Koops et al. 2014 for an opposing view on the "method of exclusion," which identifies culture in wild animals by excluding ecological and genetic factors to explain geographical variation in behavior). This suggests that tool traditions have evolved locally among chimpanzees at the subspecies and population levels. More broadly, it also indicates that tools and tool behaviors have the potential to evolve rapidly, and in parallel, among populations and even species. This has important implications for reconstructing tool behavior in the LCA, as it suggests that convergent evolution (homoplasy) of technological behavior is likely, or even frequent, and thus, although not mutually exclusive with homoplasy, parsimony-based arguments in favor of the presence of tool behaviors in the LCA must be examined critically (see below).

The rate of evolution of these behavioral traditions may relate to the pathways of transmission of information in chimpanzees. There is little doubt that social influences are important in the process of acquisition and diffusion of tool use in primates (Biro et al. 2003; Lonsdorf et al. 2004; Liu et al. 2011; Hobaiter et al. 2014). The process of learning a tool behavior has been described as "Education by Master-Apprenticeship" (Matsuzawa et al. 2001), in which learning is facilitated by tolerant conspecifics allowing juveniles to closely observe their behavior during extended periods of time (see also Henrich and Tennie, this volume). In the absence of active teaching (but see Musgrave et al. 2016 for a potential exception), a combination of trial and error, with the recycling by youngsters of tools made or used by adults (Biro et al. submitted), combined with high levels of tolerance, would likely favor the emergence and endurance across generations of more complex tool behaviors.

Bonobos *(Pan paniscus)*

Tool use and manufacture has also been documented in wild bonobos, but it has been described as very rare (Gruber and Clay 2016). Although this may reflect the paucity of bonobo field studies compared with chimpanzees (McGrew 2010), the factors that best explain the major differences in tool use between chimpanzees and bonobos have been identified as one of the most urgent questions to address in future *Pan* behavioral studies (Gruber and Clay 2016). Bonobos do not use tools for extractive foraging, but have been seen using tools for communication and personal grooming. Much like chimpanzees, bonobos have been observed to use modified branches for displays, to swat away flies with leafy twigs, and to use moss sponges to absorb water (Hohmann and Fruth 2003). Probing tools and behaviors, and more specifically insect fishing, have never been observed directly in wild bonobos, though Badrian (1981) reports indirect evidence of this behavior from Lomako Forest (Democratic Republic of Congo) in the form of broken termite mounds with nearby harvesting sticks (see also McGrew et al. 2007).

Until very recently the use of stone tools had not been observed for bonobos outside of captivity. Neufuss et al. (2017) recently reported on the nut-cracking behavior of wild-born, rehabilitated bonobos in the Lola ya Bonobo sanctuary in the Democratic Republic of the Congo. Although not free-ranging, the bonobos live in a thirty-hectare natural environment, with access to both high canopy forest areas with palm oil trees, as well as more open areas. The bonobos engaged in palm nut cracking behaviors frequently, using a variety of hand grips (see below) that matched the varying sizes and weights of the hammerstones at their disposal. The authors also noted that individuals selected the most appropriate stones and hand grips to improve the efficiency of the nut cracking (i.e., the number of hits required to crack a nut and the number of nuts per minute), matching the efficiencies previously observed in chimpanzees (Biro et al. 2006; Neufuss et al. 2017). It remains to be seen whether the behavior exhibited by these semi-wild bonobos is also found in the wild, or whether their use of stones for nut cracking is a technological innovation acquired more recently in the sanctuary. Nonetheless, this shared ability to perform stone tool cracking behaviors in chimpanzees and bonobos indicates that this is likely a plesiomorphic behavior for the genus *Pan*, with implications for its presence in the LCA (see below).

Other Hominoids

Breuer et al. (2005) reported the only two, anecdotal, cases of western gorillas *(Gorilla gorilla)* using tools in the wild. One adult female was observed using a stick to gauge the depth of water in an elephant pool. In the second case, another adult female detached the trunk of a small shrub and used it to support and stabilize herself as she dredged food from a swamp. She then used this trunk as a bridge upon which she walked bipedally to cross an area of the swamp. It is noteworthy that in these individuals, the tools were not used directly for extractive foraging, but rather to assist them in adapting to an atypical environment. As the authors note, the fact that tool use has rarely been documented in wild gorillas may be precisely because the exploitation of their primarily food resources (e.g., piths, leaves, fruit) involves extracting or processing that can be accomplished by hands and/or teeth alone (Breuer et al. 2005). However, a juvenile female mountain gorilla in Volcanoes National Park (Rwanda) was also recently observed using a stick as a probe to extract ants from their nest (Kinani and Zimmerman 2015).

Wild orangutans are frequent tool makers and users (Meulman and van Schaik 2013). As in chimpanzees, tools are used in different contexts, including extractive foraging (e.g., using sticks as chisels to pierce the shell of durian fruit), communication with conspecifics (e.g., branch dragging displays), and as a means of increasing physical comfort (e.g., use of leaves as napkins, shelter against sun/rain). In contrast to chimpanzees, the tool kits of orangutans do not include inorganic materials, or at least their use has not yet been documented in the wild—which may be linked with the degree of arboreality of this ape, less exposed to the opportunity to encounter and manipulate lithic objects. Cultural variation in tool use across orangutan subspecies, and across different sites, is well documented (van Schaik et al. 2003). In fact, very few tools and tool behaviors are universally shared among different populations of Bornean and Sumatran orangutans, and none of these relate to extractive foraging (van Schaik et al. 2003). This suggests that, as in chimpanzees, local ecological conditions, especially the availability of food resources that require extractive processing, play an important role in the nature, diversity, and frequency of tools and tool use in these omnivorous primates.

Gibbons have been observed on occasion to use branches as tools, in the context of displays and communication with conspecifics (Carpenter 1940; Deng and Zhou 2016). To date, however, they have not been observed to use tools for extractive foraging purposes in the wild (van Schaik et al. 1999). The absence of tool use in these primates may be attributed at least in part to their diet, which consists mainly of easily accessible fruits (e.g., figs) and leaves (Palombit 1997). Although insects may make up a substantial portion of some species' diets as well (e.g., Fan et al. 2009), this appears to be opportunistic or seasonal, and gibbons do not feed on insect species that require extractive techniques, such as termites.

Other Primates

1. Old World Monkeys: Tool use among wild cercopithecoid monkeys has been documented in baboons and macaques. In *The Descent of Man* (1871), Darwin reported secondhand on observations of chacma baboons using stones as defensive tools. Specifically, the baboons would drop or roll stones down the rocky walls of the canyons they inhabit onto predators below. These anecdotal reports were later confirmed by Hamilton and colleagues (1975), who observed three separate baboon troops engaging in this behavior—aimed at them—in a desert floor canyon in Namibia. Tool use in extractive foraging contexts in the wild has not been documented for papionins. Among macaques, semi-wild pig-tailed macaques *(Macaca nemestrina)* have been observed using leaves to wash dirt/mud off seeds and fruit before eating them (Chiang 1967). In addition, long-tailed macaques in Thailand *(Macaca fascicularis aurea)* have been observed to use unmodified stones to pound and crack open the shells of oysters (Malaivijitnond et al. 2007; Gumert et al. 2009; Haslam et al. 2016a). These authors report that the handheld stone tools were occasionally transported from one activity site to another. Thus, long-tailed macaques are one of only two nonhominoid species known to engage in percussive tool use in the wild, albeit this behavior seems to be restricted to a few groups living in the coastal areas of Thailand.

2. New World Monkeys: The most commonly reported use of tools is in wild capuchin monkeys (especially bearded capuchins, *Sapajus libidinosus*) (reviewed in Ottoni and Izar 2008). Specifically, capuchins are proficient nut crackers, habitually using pounding stones and anvils to crack open palm nuts (Visalberghi et al. 2007), and have seemingly been doing so for at least seven hundred years (Haslam et al. 2016b). Capuchins are selective in the type and size of rock they use as a hammer, and have been observed to transport these to different anvils around their home ranges (Visalberghi and Fragaszy 2013). Recent evidence also suggests that capuchins deliberately break stones using stone-on-stone percussion, a behavior that accidentally produces flakes with attributes (e.g., conchoidal fracture, bulb of force, platform, weight, sharp edges) that are indistinguishable from intentionally produced flakes from early hominin stone tool assemblages (Proffitt et al. 2016)

FIGURE 17.3. Stone-on-stone percussive behavior in capuchins (*Sapajus libidinosus*). In the left panel, a juvenile observes as the adult prepares to strike the rocky conglomerate with the stone hammer. In the middle panel, the hammer shatters into several pieces. Some of the flakes that resulted from the percussive event are shown in the right panel. Adapted by permission from Macmillan Publishers Ltd.: *Nature* (Proffitt et al. 2016. Wild monkeys flake stone tools. *Nature* 539: 85–88), Copyright 2016.

(Figure 17.3). The functionality of this behavior is not well understood, but the fact that the capuchins lick the rock each time it fractures raises the possibility that they are getting trace minerals from the rocks. This is an excellent example of a process by which flakes are produced (albeit apparently not used, hence they are not "tools" *sensu stricto*) that has not been considered in previous analyses of stone tool assemblages. There is also marked geographical variation and limitation in the habitual use of percussive technology in bearded capuchins, with mostly those inhabiting savanna environments exhibiting these behaviors. It has been suggested that increased terrestriality may be partly responsible for this variation (see below; Meulman et al. 2012; Visalberghi and Fragaszy 2013).

3. Prosimians: Strepsirrhine primates do not use or make tools, either in the wild or in captivity (Santos et al. 2005).

Hand Anatomy, Dexterity, and Tool Use in Modern Primates

Tool use and manufacture among primates is predicated on their ability to manipulate and/or modify objects for specific tasks with their hands. In other words, hands are the "hardware" that make primate tool use possible. Hand anatomy and morphology vary substantially among primates (Napier 1980). Hence, there has been a long-standing interest in correlating variation in hand morphology and dexterity among primates with interspecific variation in the ability to make and use tools (van Schaik et al. 1999). Our survey of tool use among nonhuman primates clearly shows that a specific hand morphology, by itself, is not a prerequisite for tool use. However, hand morphology likely influences the *nature* of the tools that can be made or how they are used. Hence, comparing hand anatomy among chimpanzees and humans, especially as it relates to dexterity, can yield insights into the hand anatomy of the LCA and, by extrapolation, into its ability to make and use tools.

Primate Hand Grips

John Napier (Napier 1956) first proposed a classification system for hand grips in primates based on a dichotomy between power grips and precision grips. Power grips are those in which the object, be it a tool or an arboreal support, is pinched by the fingers and stabilized across the palm. The thumb is more or less involved in power grips. As the name implies, power grips are used when large muscle forces are necessary to hold or manipulate objects. In contrast, precision grips involve mainly the digits, as opposed to the palm. Precision grips involve opposition of part(s) of the thumb to the other digits. Napier argued that humans performed better at precision grips, while other primates relied more on power grips for manipulation and locomotor behaviors. There is likely more of a continuum between these grips, and all primates are capable of using both to variable extents (Marzke 1997).

Additional analyses of manipulative behavior in primates, especially in modern humans replicating stone tool use, have refined this grip typology. Marzke and colleagues' studies of manipulative behavior in humans and nonhuman primates (e.g., Marzke and Shackley 1986; Marzke et al. 1992) found that when archaeologists experimentally replicate stone tools, they mainly use three types of precision grip. In the "three-jaw chuck," the palmar aspect of the thumb opposes a variably flexed third digit, and the index rests atop and stabilizes the object, which is partly buttressed against the second metacarpal. This grip is primarily used for holding hammerstones during hard-hammer percussion. The "two-jaw chuck" opposes the distal thumb pad to the side of the index, that is, as one would hold a key. This grip is often used in handling stone flakes for more delicate tasks. Finally, cores were often manipulated using "cradle grips," in which the thumb and four fingers firmly pinch and maneuver large stone cores, exposing surfaces for flake removal by the hammerstone (Marzke 1997) (Figure 17.4).

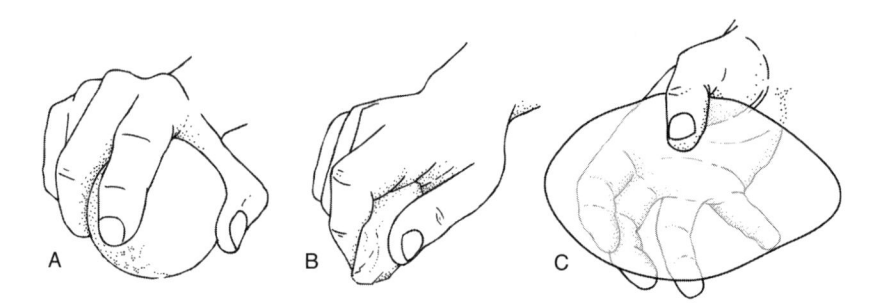

FIGURE 17.4. Precision grips used to wield Oldowan tools. (a) Three-jaw chuck grip used to hold a hammerstone. (b) Key pinch grip used to hold a flake. (c) Cradle five-jaw grip used for large cores. Modified from Rolian et al. (2011), with permission from Elsevier.

One feature these three grips have in common is that they require forceful and sustained opposition of the thumb to the radial digits, either to make a stone flake or to use it (Marzke 2013). By extension, the ability to wield and make stone tools may thus be directly linked to the user's ability to forcefully oppose the thumb to the lateral digits. If it also has correlates in hand musculoskeletal anatomy, and these can be traced in the fossil record or inferred in the LCA, then it may provide additional information on the types of tools the LCA would have used.

Chimpanzee Hand Anatomy and Grips

Unlike humans, the hands of chimpanzees are used in locomotion, as well as to manipulate objects. The chimpanzee hand may thus reflect an evolutionary compromise between selective pressures related to manipulation, if any, and those related to locomotion. One of the most salient features of the chimpanzee hand skeleton is its digit proportions, specifically the ratio of thumb length to finger length (also known as intrinsic hand proportions). After orangutans, chimpanzees have the lowest ratio among hominoids, and one of the lowest among all primates (Marzke 1997; Almécija et al. 2015; Pilbeam and Lieberman, this volume). The thumb is also relatively gracile (Rolian et al. 2011). In addition, its proximal phalanges are highly curved in the dorsopalmar plane. This feature is also prominent in gibbons and orangutans, as well as a number of fossil hominoids (see below). The degree of phalangeal curvature likely relates to the biomechanics of below-branch, orthograde suspension and/or brachiation, although it may be largely epigenetic (Richmond 2007).

As in all hominoids, the thumb is opposable, though the articular facet between the thumb metacarpal and the trapezium is strongly curved (saddle-shaped), limiting its mobility but increasing its stability (Tocheri and Marzke 2007; Tocheri et al. 2008). The distal phalanges have narrow apical tufts (Mittra et al. 2007). In terms of pertinent arm and hand muscles, chimpanzees have powerful elbow, wrist, and digital flexors, the latter reflected in strong flexor sheath ridges on the phalanges of the fingers (Thorpe et al. 1999). In contrast, the intrinsic muscles of the hand, and in particular those that support the thumb (thenar muscles), are poorly developed (Marzke 1992).

The sum of these morphological features indicates that the chimpanzee hand would not perform well in manipulative tasks that require the forceful opposition of the thumb to the other digits. In fact, such forceful, single-handed precision grips have not been observed in the wild (Marzke 2013). This further suggests that chimpanzee hands are constrained in their ability to make and use stone tools, which requires firm grips. The most significant limitation on this ability is not its short thumb relative to long fingers. Most primates, including chimpanzees and Miocene apes (Lovejoy et al. 2009b; Almécija et al. 2012), are able to oppose the digit tips in a *fine* precision grip, but such grips are not useful for manipulating stone tools (Marzke and Shackley 1986).

A more significant constraint on the ability of chimpanzees to produce the forceful precision grips required for stone tool use stems from their relatively weak intrinsic hand muscles and gracile thumbs. Using an experimental biomechanics approach, Rolian et al. (2011) determined that the total muscle force potential necessary to stabilize the fingers and thumb joints during simulated Oldowan stone tool use exceeds the available force potential of the chimpanzee intrinsic muscles of the hand. Moreover, even if chimpanzees were physically able to wield hammerstones and flakes in the way modern humans can, the relatively gracile thumbs and overall smaller joint surfaces in the carpus and thumb metacarpal would produce large joint stresses under the kinds of loads required to wield stone tools.

Given these considerations, it is perhaps not surprising that the tool kit of chimpanzees and bonobos is limited largely to tools that do not require the forceful opposition of the thumb to lateral digits to produce or use, such as twigs and leaves, or conversely large-diameter branches and other objects (e.g., stone pestles) that are held using power grips, that is, with the objects buttressed against the palm. As discussed above, chimpanzees do use unmodified stone tools, albeit not universally. Importantly, however, the grips that chimpanzees employ to use hammers and anvils are not forceful precision grips. They are full-fledged power grips, in which the bulk of the force produced by smashing nuts is distributed across the palm(s). These grips do not involve the thumbs in forceful opposition to the other digits, but rather in the same plane as the fingers (Boesch and Boesch 1993). Similarly, the vast majority of grips used by wild bonobos are power grips, although forceful precision grips involving the thumb and lateral digits without the palm were occasionally used by two individuals in nut cracking (Neufuss et al. 2017).

Human Hand Anatomy and Grips

Several important features of the human hand are correlated with its enhanced ability to manipulate objects, and in particular to generate forceful precision grips. The thumb to digit ratio is the highest of all primates, reflecting not only an absolutely longer thumb, but also relatively shorter lateral digits. These proportions facilitate the opposition of the thumb to the tips of the other digits. The thumb bones are also robust, with larger articular facets, allowing joint reaction forces to be distributed over a greater surface area when loaded (Susman 1998; Rolian et al. 2011). In addition, the distal phalanges of the digits bear prominent apical tufts that underlie the division of the broad fleshy pulp of the fingertips into movable proximal and more firm distal compartments (Shrewsbury et al. 2003). A number of derived carpal shapes relate to increased range of motion of the wrist, increased opposability of the thumb and fingers, and increased internal stability between the carpal bones (reviewed in Marzke and Marzke 2000).

The human hand also has derived musculotendinous features that correlate with the ability to generate forceful precision grips. It has a fully separate and well-developed flexor pollicis longus, whereas in other apes it is absent or is still a part of the deep digital flexors of the digits (Linscheid et al. 1991; Marzke et al. 1999). Its intrinsic hand musculature is particularly well developed, both on the thumb (thenar) and fifth digit (hypothenar) sides. For example, the total physiological cross-sectional area of the thenar muscles in humans, which reflects the total force these thumb muscles can produce, is over twice as large as it is in chimpanzees (Marzke et al. 1999).

This combination of short digits, a long robust thumb, and well-developed hand muscles is at least in part responsible for the ability of humans to generate and withstand large forces within the hand and fingers. Inasmuch as these forces are associated with lithic technology, and with stone tool *manufacture* in particular, they suggest that a specific hand anatomy is required to wield stone tools in the way that humans and their ancestors did.

Hand Anatomy, Grips, and Tool Kits in the LCA

The comparative morphological evidence suggests that *Pan*-like hand anatomy constrains the types of tools chimpanzees can manipulate, and the extent to which they can deliberately modify such tools. Thus, if we could directly examine the morphology of the LCA's hand, we might gain insight into the type of tools it used. Tocheri et al. (2008) used the principles of parsimony and the fossil record of hominin hand elements (see below) to reconstruct the most likely musculoskeletal anatomy of the LCA hand. They conclude that the hand of the LCA would have looked like that of an African ape. Specifically, they argue that skeletal features present among extant apes were also present in the LCA (i.e., they are homologies in the African apes), such as long fingers relative to the thumb, dorsopalmar curvature of the phalanges, marked ridges for the insertion of the digital flexor sheaths on the sides of the proximal phalanges. They also argue from parsimony that the thenar muscles would have had a relatively small cross-sectional area, and, as in African apes, there would have been no or a poorly developed flexor pollicis longus, but strongly developed digital flexor muscles.

Almécija et al. (2015) used a phylogenetically broad morphometric approach coupled with Ornstein-Uhlenbeck (OU) evolutionary modeling to reconstruct the ancestral states of digit lengths and proportions among hominoids. Using shape variables based on the ratio of metacarpal and phalangeal lengths to the cube root of body mass, they arrived at radically different conclusions from Tocheri et al. (2008): (1) the digital elongation of chimpanzees and orangutans is convergently derived, (2) human and gorilla proportions are plesiomorphic for great apes, and (3) the LCA's proportions are closer to this "plesiomorphic" condition exemplified by humans and gorillas, with "moderate" proportions between those of chimpanzees and modern humans. In other words, while its hands may not have looked like those of modern *Pan*, the LCA had intrinsic hand proportions capable of producing the precision grips associated with stone tool use, provided it also had well-developed hand muscles.

Almécija et al.'s (2015) conclusions, however, may have been influenced by their choice of characters used in the OU models, namely shape ratios of thumb and fourth digit element lengths to the cube root of body mass. Body mass is a common proxy for size; however, as Almécija et al.'s (2015) supplementary figure 4 suggests, hominoids have different linear relationships between body mass and digit length (with respect to their intercepts). For example, at comparable body sizes, *Pan* and *Pongo* fourth digits are absolutely longer than humans, while the reverse is true for gorillas. As a result, the reconstruction of ancestral hand extrinsic proportions in the LCA and in the last common ancestor of African apes and humans with short lateral digits and relatively long thumbs may have been driven at least in part by the low length-to-mass ratio of gorillas and humans. Put differently, even slight differences in the ratios obtained using different size proxies, such as skeletal measurements from the proximal limb skeleton, could lead to different LCA reconstructions, and hence to different interpretations of the relative evolutionary changes in each lineage (see, e.g., Pilbeam and Lieberman, this volume).

Based on this anatomical evidence, what sort of grips could the LCA produce? Its inferred hand anatomy is indicative of a hand that was capable of strong power grips that involve the fingers exclusive of the thumb. Similarly, the inferred proportions and muscle distribution suggest it would have performed as well as a modern chimpanzee (Tocheri et al. 2008), if not better (Almécija et al. 2015), in producing the forceful precision grips required to use and manufacture stone tools. Put simply, the inferred anatomy of the LCA hand suggests, if anything, a tool repertoire similar to modern chimpanzees, with mostly organic material that can be used with relatively little finger/thumb muscle force, and perhaps unmodified hammerstones and anvils.

Cognition and Tool Use

If hands are the hardware that facilitate tool use, then the brain is the operating "software" that directs the act of using and making stone tools. Though we cannot observe behavior or cognitive ability in the LCA directly, we can use comparative evidence and studies from modern primates to make inferences on the range of cognitive abilities required to use, and make, tools. Such studies reveal the existence, at least superficially, of a hierarchy of cognitive abilities associated with ever more complex tool kits and tool behaviors (e.g., Matsuzawa 1996; but see Shumaker et al. 2011 for a discussion of the myth that all tool behavior is "intelligent").

Perhaps the most basic cognitive requirement to make and use tools is an understanding of, and mental ability to correlate, cause and effect (i.e., causality, Urbani and Garber 2002). Many experiments in captivity indicate that most primates have a basic understanding of causality in the context of tool use and manufacture. These experiments usually take the form of a test that, if successfully completed, will lead to a food reward. Some of the experiments are specifically designed so that the reward cannot be obtained without using an object to modify the immediate environment of the individual, for example, using a stick to probe for food rewards that are otherwise inaccessible. By our definition, these are instances of tool use. These experiments are contrived, and do not necessarily reflect the ability of individuals (or species) to develop these skills in the wild. Still, they are informative for two reasons. First, they show that anthropoid primates are capable of correlating cause and effect, regardless of whether an understanding of causality has an impact on their evolutionary fitness in the wild. Second, they show that in novel contexts, primates are able to learn through unguided experimentation or emulation (see Henrich and Tennie, this volume; Rosati, this volume).

A higher cognitive requirement for using tools is a technical understanding of the objects and interactions involved in the task to be performed and, based on this understanding, selecting an appropriate tool for the task (Mulcahy and Call 2006). In most instances of primate tool use described above, appropriate tools were selected for given tasks, presumably based on the individual's ability to discriminate good from poor tools. Oftentimes, this entails an understanding of the physics of the tools and tasks. For example, the primates that habitually crack open nuts or mollusk shells must know, perhaps from experience or learning from others, that the tools they use must have certain physical properties (e.g., hardness, weight) that make them suitable for cracking open hard objects (Visalberghi et al. 2007). Thus, they select stones or large-diameter sticks rather than twigs and leaves for this task. Similarly, the use of thin probes or wands for insect fishing in chimpanzees, and at least in one instance in gorillas (Kinani and Zimmerman 2015), indicates that they understand that only tools of a certain diameter will fit in the termite nest's holes, and/or that using alternative tools such as clubs to gain access to the termite mound will have undesirable consequences.

Similarly, following the selection of an appropriate tool, primates must have some understanding of the physics of the interactions between tool and substrate, especially in the case of kinematics of percussive tool use. Adjusting the force required to break open a nut, for example, may not only improve the efficiency of the extractive task (i.e., optimize it), but can also minimize the risk of injury. A recent kinematic study in wild capuchins suggests that they adjust the magnitude and frequency of strikes necessary to open nuts with shells of varying hardness, even though they may use the same stone in each case (Mangalam et al. 2016). Similar observations have been made in bonobos (Neufuss et al. 2017) and chimpanzees (Boesch and Boesch 1983).

Associative tool use is cognitively more complex. Shumaker et al. (2011: 19) define associative tool use as "tools used in any combination to achieve an outcome." This includes tools that are used one after the other (sequential tools), tools that are used to acquire other tools, or combinations of tools, usually with different functions, that are used simultaneously to achieve a single outcome (composite tools; Hayashi 2015). Wild capuchins have been observed to use tools sequentially, such as stones used to pound on cavities in tree trunks, followed by the use of sticks as probes to access insect nests on the inside (Mannu and Ottoni 2009). One of the best examples of composite tool use comes from observations of nut cracking by Bossou chimpanzees (Sakura and Matsuzawa 1991; Carvalho et al. 2008, 2009). The composite tools here are the hammerstone and a movable anvil, in contrast to other sites such as Taï, where the anvil is typically an unmovable rocky outcrop. This composite tool can be even more complex, as some chimpanzees at Bossou have been observed to prop and stabilize the anvil with additional stones, leading to a final composite tool made up of four stones.

Tool manufacture also requires more advanced cognitive skills. Whereas many forms of tool behavior in primates rely on the immediate use of unmodified objects for specific tasks, tool manufacture involves a sequence of events qualitatively different from sequential or composite tool use. First, because the tool does not yet exist, a mental representation must be made of it *and* the use to which it will be put in some more or less distant time and place, that is, an understanding of causality in the absence of its immediate manifestation ("future planning": Rosati, this volume). A second mental image must also be formed that defines how the original object must be modified or manufactured into the final tool such that it will actually perform the required task.

The simplest and most frequent form of tool manufacture occurs in the context of conspecific or interspecific communication. The most common manifestation of this behavior is the use of "detached" objects, typically branches, that are then dropped from trees, thrown, or waved at others (Shumaker et al. 2011). This behavior qualifies as tool manufacture since the branch, prior to being a missile or a display, is not yet an unattached object. This behavior has been observed in the wild in many anthropoid primates, including New World monkeys (e.g., capuchins: Oppenheimer 1977), cercopithecoids (e.g., red colobus monkeys: Starin 1990), and all hominoids (gibbons: Carpenter 1940; orangutans: Galdikas 1982; gorillas: Wittiger and Sunderland-Groves 2007; bonobos: Hohmann and Fruth 2003; chimpanzees: Goodall 1986).

More complex forms of tool manufacture are associated with extractive foraging. Tools used as displays or deterrents are likely manufactured without regard to their physical properties. In other words, whatever is at hand will do to deter a potential predator or to attract the attention of a potential mate. In contrast, tools manufactured for extractive foraging must meet specific criteria. Thus, these tools require both a technical understanding *and* a mental template to produce and use. For example, probing an insect nest requires a twig of the right diameter, which must also be stripped of its leaves. Similarly, sticks or stones can't be used as sponges to absorb water, and hence chimpanzees deliberately strip and use leaves to soak up water from concavities (McGrew 1992). Additionally, it has been shown that chimpanzees select particular species of vegetation to make sponges for drinking water, perhaps because those raw materials are more efficient at absorbing liquid than other vegetation readily available in the same location (Tonooka 2001; Sousa 2011)

Only a few nonhuman primate taxa have been observed to manufacture tools of this kind in the wild, namely capuchin monkeys, orangutans, and chimpanzees (Goodall 1986; van Schaik et al. 1996; Mannu and Ottoni 2009). Interestingly, two manufactured tool types are common to these three species: modified sticks or twigs used for probing and collecting arthropods, and branches modified to pound, chisel, or hammer into insect nests or to open tough-skinned fruits and nuts. Additionally, wild chimpanzees have been observed on occasion to use stone tools that were modified in the sense that they broke during use, and one of the fragments was then used to continue the extractive foraging task, although the breakage was likely accidental (Carvalho et al. 2008).

Tool use and manufacture in the wild also involves a type of cognitive skill indirectly linked with the tool and its use, namely the ability to learn, either by trial and error or by emulation (social learning; van Schaik et al. 1999). It is unclear how much tools are spontaneously discovered or invented in the wild (but see Hobaiter et al. 2014 for a case of innovation and transmission), although simpler tools such as waving displays may arise out of a basic understanding of causality. In the context of associative tool use and tool manufacture, primatologists have observed that tool users are often in the presence of other individuals. This social influence, which also requires tolerance and gregariousness on the part of the tool user, is most frequently observed in orangutans and especially chimpanzees (Tomasello et al. 1987; Nagell et al. 1993; Call and Tomasello 1994; see also Henrich and Tennie, this volume, for a detailed discussion of teaching and social learning in chimpanzees).

The comparative evidence thus indicates that chimpanzees, and to a lesser extent orangutans and capuchins, have all the cognitive skills associated with complex tool use (causality, technical understanding, associative tool use), and also with tool manufacture (mental representation or projection). These skills are acquired in humans at a very young age (Nagell et al. 1993). Chimpanzees also have the social context that facilitates the learning and cultural transmission of tool use/manufacture skills, as do humans. Captive studies indicate that most anthropoid primates have at least the basic cognitive skills to use tools as well, suggesting that the capacity to use tools is primitive for anthropoids. Unless the more advanced cognitive skills associated with tool manufacture found in *Pan* and *Homo* evolved convergently, as likely happened between hominids and capuchins *(Cebus/Sapajus)*, then the most parsimonious explanation is that the LCA should have had advanced technical-cognitive skills as well.

Fossil and Archaeological Evidence

Fossil primates, especially early hominins, can provide important details on the evolution of morphological correlates of tool use/manufacture, especially hand morphology. In addition, paleoenvironmental reconstructions for fossil taxa provide important clues as to the habitats they lived in, and whether they favored tool innovation. Finally, the archaeological record itself provides indisputable proof of tool use, and especially manufacture. How complex the tools are, and when they first appear, can provide insights into the likelihood that the LCA used tools, and what kind these were.

Miocene Apes

Numerous species of Miocene ape have been described (reviewed in Begun 2010). Among these, a few taxa from different geological ages stand out because hand remains have been recovered, allowing for reconstructions of features relevant to the grips involved in locomotion versus dexterous manipulation. In a review of available hand remains from Miocene apes, Almécija and colleagues (2012) conclude that overall, these primates had hand proportions with long, highly curved finger bones and a relative thumb length intermediate between extant apes and modern humans (Almécija et al. 2012; see also Almécija et al. 2007). The structure of the thumb skeleton shows that its proximal phalanx was also highly curved, while its distal phalanx was elongate and possessed strong flexor attachments, but no apical tufts that would have supported a distal pulp (Almécija et al. 2012). An analysis of the distal thumb phalanx of the late Miocene putative hominin *Orrorin tugenensis* (~6 million years ago [Ma]) by the same authors suggests a morphology that is more similar to modern humans than to extant and other Miocene apes, with distinct apical tufts, and an ungual fossa and elevated ridge marking the insertion of the flexor pollicis longus (Gommery and Senut 2006; Almécija et al. 2010).

From this fossil evidence, Almécija and colleagues infer that the primitive condition for Middle and Late Miocene apes was distinct from extant apes, with long curved fingers but a relatively long thumb, which moreover in *Orrorin* possessed all the markings of a thumb capable of producing strong precision grips. They conclude that the skeletal morphology of the hand in the LCA would have been more primitive than is generally assumed, with only subtle changes in proportions and thumb anatomy necessary to produce the human condition, and more marked and convergent elongation of the non-pollical digits in the extant apes (Almécija et al. 2015) (but see above, and Pilbeam and Lieberman, this volume, for an alternative interpretation of hand evolution in gorillas based on the scaling of skeletal dimensions to body size in the context of terrestrial locomotion). If true, this suggests that the evolution of hand proportions in the hominins was not necessarily the result of selection for increased manual dexterity (Alba et al. 2003; Rolian et al. 2010).

Paleoenvironmental reconstructions for the Middle Miocene apes most often point to seasonal, subtropical forest habitats (reviewed in Elton 2008). The inference made from these reconstructions, supported to some extent by craniodental remains, is that the Middle Miocene apes relied on the exploitation of arboreal resources as a significant portion of their diets. In contrast, paleoecological reconstructions for Late Miocene putative stem hominins such as *Sahelanthropus tchadensis* and *Orrorin tugenensis* reflect a mix of habitats including closed-canopy forests but also open woodlands and open water environments (Elton 2008). Such mixed habitats may have spurred technological innovation as new food sources became increasingly available. There is no archaeological record from this time period.

Ardipithecus ramidus

Ar. ramidus is a remarkably complete fossil dated to 4.4 Ma recovered from the Afar region of Ethiopia (White et al. 2009). The fossil remains of a female individual include the skull, a virtually complete forelimb, pelvis, and many elements of the hind limb (Suwa et al. 2009a, 2009b; Lovejoy et al. 2009a, 2009b, 2009d). These remains have enabled a detailed reconstruction of its paleobiology, although some details are still a matter of debate (Lovejoy 2009a, 2009b, 2009c; Stanford 2012; Young et al. 2015). *Ar. ramidus*'s hand remains indicate digit proportions more similar to gorillas than to either modern *Pan* or *Homo*. It also possessed relatively large joint surfaces in the wrist, a relatively long and robust thumb with a marked gable on its distal phalanx that reflects the insertion of a flexor pollicis longus (Lovejoy et al. 2009b). Lovejoy and colleagues have suggested that the hand of *Ar. ramidus* was capable of strong palmar grips with good opposability afforded by a more flexible wrist. Similar to Almécija et al. (2015), they conclude that the primitive condition of the LCA hand was more similar to humans than to modern African apes, although, as discussed above, and in Pilbeam and Lieberman (this volume), these conclusions may depend on the choice of body size proxies for interspecific comparisons.

Paleoenvironmental reconstructions suggest that *Ar. ramidus* lived in a mixed woodland habitat with small forest patches and access to water resources, rather than in open grasslands (Louchart et al. 2009; Suwa et al. 2009b; White et al. 2009). Based on its dentition, Suwa and colleagues suggest that *Ardipithecus* was an omnivore, feeding in both arboreal and terrestrial environments, but showing neither the dental features associated with a C_3 diet, based on ripe fruit frugivory (chimpanzees), nor those associated with a more C_4 diet consisting of more abrasive foods found in open grassland / savanna environments.

Australopithecines

The australopithecines are a morphologically diverse group of hominins that lived between 4.2 and ~2 Ma in both East and South Africa. Australopithecines were committed bipeds, though postcranial remains are also indicative of climbing and arboreal behaviors (e.g., climbing for food resources, escape from predators, or nesting) (Berger et al. 2011; Ward et al. 2011). Hand fossils of one of the earliest members of the genus, *Au. afarensis,* indicate digit proportions that are at the lower end of the human range, overlapping with gorillas, although this is based on reconstructions from unassociated hand remains from multiple sites in the Afar region of Ethiopia (Marzke 1983; Alba et al. 2003; Rolian and Gordon 2013). The thumb metacarpal is relatively gracile and has weakly developed apical tufts, but shows a clear insertion for the flexor pollicis longus (Bush et al. 1982; Ward et al. 2012). Additional hand fossils from South African members of the genus also evince hands that were modern in appearance, possibly as early as 3.7 Ma (*Au. africanus:* Clarke 1999), and certainly by 1.9 Ma (*Au. sediba:* Kivell et al. 2011).

Reconstructions of the paleoenvironmental context of australopithecines point to habitats that were cooler and more similar to modern-day woodlands and grasslands than to closed-canopy forests associated with earlier hominins (Bobe et al. 2002). The environmental shift to more open grasslands with more fragmented forest patches is often seen as being partly responsible for increased terrestriality and hence the emergence of committed bipedalism in these hominins (Dominguez-Rodrigo 2014). It is also thought to have opened up opportunities for extractive foraging of new food resources, including tubers and perhaps even scavenged meat. In other words, it may have provided an impetus for technological innovation in order to gain access to these new resources.

Two recent archaeological discoveries suggest *Au. afarensis* did in fact use and make stone tools. In Dikika, a 3.39 Ma site in Ethiopia, bones bearing cutmarks as well as percussive marks have been interpreted by their discoverers as indicating the use of sharp-edged stone tools for removing flesh from the bones and smashing them to access the marrow, respectively (McPherron et al. 2010; but see Dominguez-Rodrigo et al. 2010). No stone tools have been recovered from that site, though a nearby site in the same formation has yielded hominin fossils, including a juvenile attributed to *Au. afarensis*, suggesting this species may have been the maker/user of these tools (Alemseged et al. 2006). In 2015, Harmand and colleagues described over one hundred lithic artifacts recovered from Lomekwi, Kenya, dated to 3.3 Ma. These artifacts were recovered *in situ*, and in the same spatiotemporal context as hominins (as yet undescribed). The assemblage is distinctive from any Oldowan collection, and includes a diversity of large artifacts, such as sharp-edged flakes, hammerstones, and potential anvils, all pointing to the significance of percussive activities in these Pliocene hominins (Harmand et al. 2015; Lewis and Harmand 2016) (Figure 17.5).

Early *Homo*

Our own genus first appears around 2.8 Ma (Villmoare et al. 2015). Early putative members of the genus, such as *Homo habilis*, had postcranial proportions more similar to *Australopithecus* than to later *Homo*, a slightly larger cranial capacity, and smaller teeth than geologically earlier australopithecines. A partial hand, OH7, recovered from Olduvai Gorge in Tanzania, was initially attributed to *Homo habilis*, although it may have belonged to a robust australopithecine (Moyà-Solà et al. 2008). The OH7 hand does not permit reconstructions of digit proportions, but possesses a distal thumb phalanx with none of the derived features of *Homo* hands (Almécija et al. 2010). The types of grips *H. habilis* could have generated are thus hard to reconstruct.

Regardless of the paucity of its hand remains, the first evidence of stone tool use fitting the Oldowan techno-complex is usually associated with *H. habilis*, though the 2.6 Ma assemblage from Gona, Ethiopia (Semaw et al. 1997), has been suggested to be a product of *Australopithecus garhi*, and multiple species living at the time were likely using stone tools, in-

FIGURE 17.5. The Lomekwi stone tool assemblage, dated to 3.3 MYA. (a) *In situ* core (1.85 kg) and refitting surface flake (650 g), both displaying dispersed percussion marks on cortex. (b) *In situ* unifacial core. (c) Flakes showing scars of previous removals. Adapted by permission from Macmillan Publishers Ltd.: *Nature* (Harmand et al. 2015. 3.3-million-year-old stone tools from Lomekwi 3, West Turkana, Kenya. *Nature* 521: 310–315), Copyright 2015.

cluding robust australopithecines (e.g., Susman 1994). The *H. habilis* taxon name ("handy man") comes from its discovery in Olduvai Gorge in association with modified stone artifacts (Leakey et al. 1964). This discovery led to the description of the first lithic industry, known as the Oldowan. The Oldowan stone tool kit contains deliberately modified stone artifacts such as choppers and sharp-edged flakes, along with hammerstones involved in the manufacture of the former, most likely through one-handed percussion. The tools are assumed to have been involved in butchering activities.

All later members of the genus *Homo*, beginning around 2 Ma and up to the present day, are systematically associated with lithic technology, with the exception of the recently described *Homo naledi* from South Africa (Berger et al. 2015). The lithic tool kits of hominins expands and becomes increasingly sophisticated throughout the Pleistocene, shifting away from simple but highly variable Oldowan flakes, choppers, and hammerstones, to tools that required more systematic and lengthier reduction processes (Toth and Schick 2009). This systematic processing leads to highly reproducible and uniform tools, such as the Acheulean bifaces from ~1.6 Ma (Whiten et al. 2009). Such tools presumably required, at a minimum, the cognitive skill to produce a mental template of the final product and proceed through the complex series of steps that would produce it (but see Moore and Perston 2016 for an alternative view).

Synthesizing the Evidence

Did the last common ancestor of *Homo* and *Pan* use and/or make tools? The sum of the evidence described above suggests that it could. The comparative evidence from extant primates suggests that tool use has been reported in all great apes, albeit at very different frequencies, and that tool manufacture is well developed in chimpanzees and orangutans. Furthermore, anatomical evidence from the hand in extant primates suggests that a derived human-like hand is not a prerequisite for the efficient use of tools. The fossil evidence from Miocene apes suggests that the LCA hand may actually have been more human-like than previously thought, meaning that it may have been able to generate the forceful precision grips associated with stone tool use and manufacture. Finally, the evidence indicates that most primates possess the cognitive requirements to use tools. Anthropoid primates show the cognitive skills necessary to make simple tools used for displays. Wild chimpanzees show the greatest cognitive flexibility to not only use and make tools, but to make composite tools that are task-specific and require multiple sequential steps to produce. The tolerance toward conspecifics, allowing youngsters to observe closely technological behaviors, combined with long periods of trial and error, normally within a social context, could also help to explain the emergence of more complex tool behaviors and contribute to its endurance across generations.

Thus, chimpanzees share with humans many of the anatomical, cognitive, behavioral, and even social attributes related to tool use and manufacture (Henrich and Tennie, this volume; Rosati, this volume). Based purely on principles of parsimony, then, one can conclude that the LCA had the capacity to use and make tools (Panger et al. 2002; Duda and Zrzavy 2013). However, one must also consider the possibility that tool behavior in *Pan* and *Homo* evolved convergently. After all, tool use in the wild is relatively rare among primates, and the three nonhuman primates that are thought to use and make tools for extractive foraging purposes more frequently—*Pan, Pongo,* and *Sapajus (Cebus)*—are not sister taxa.

The striking similarity in the tool behaviors of wild chimpanzees and capuchins further indicates that a more elaborate tool kit, including the use of stone tools, and the manufacture of dipping sticks and pounding sticks to extract invertebrate food resources, has evolved in parallel at least once among anthropoids (Visalberghi et al. 2015). It is important to note, however, that whereas complex and habitual tool use/manufacture appears to be universal among wild chimpanzees, it is not present among all species in the genus *Sapajus,* nor even in the one that has been observed most frequently to engage in this behavior (*Sapajus libidinosus:* Ottoni and Izar 2008). Its evolution in *Sapajus* groups may be an isolated event spurred on by local ecological conditions in specific populations (see below).

Whether one believes that the LCA had the capacity to use and make tools ultimately boils down to how much homoplasy one is ready to accept among sister taxa. Complex tool behavior, including variability in cultural traditions across groups and habitual tool manufacture, is universal among chimpanzees and humans, and has now been documented in a semi-wild context for bonobos (Neufuss et al. 2017). As such, we find it unlikely that such an event would have evolved in parallel between sister taxa, to the extent that it became fixed as an autapomorphy among all populations in both lineages. Simply put, we believe the last common ancestor of *Pan* and *Homo* could use tools.

Yet arguing that the LCA could use tools is not the same as stating that it did. In other words, the primitive condition in the LCA was not necessarily the *existence* of complex tool culture and behavior prior to the divergence of *Pan* and hominins, but rather the *presence of the necessary features* to develop it under the right circumstances. In fact, looking across anthropoid primates, the comparative behavioral and cognitive evidence discussed in this chapter, in captivity and in the wild, suggests that the *potential* for tool use/manufacture may be the primitive condition for all anthropoid primates. If true, then this suggests that the absence of tool behaviors among most wild primates is not due to a cognitive or manipulative inability to use, or even make, tools, but rather because the circumstances that favor technological behaviors and innovation are absent for most species. If the *propensity* for technological behavior is a primitive condition in higher primates, then this may also explain the convergent nature of tool behavior among distantly related primates: all higher primates have the potential use tools, but the impetus to do so is context-dependent.

So what exactly are the contexts that are conducive to realizing tool using/making potential in primates? One of the most important conditions for the presence or development of tool use in primates may be the availability and likelihood of finding food resources that require extractive processing. These foods, such as shelled nuts, nest-dwelling arthropods, tubers, roots, and mollusks, have high nutritional payoffs, but require extraction beyond the capabilities of hands or teeth. As in captive experiments, they present a problem that must be solved, and only the most innovative individuals capable of leveraging their immediate environment will gain access to the food. The opportunity for innovation is thus a key prerequisite, and a predictor, of tool use and manufacture (van Schaik et al. 1999; Koops et al. 2015).

In this context, it is not a coincidence that primate species that use tools most frequently are exploiting resources with high nutritional payoffs, but high investments in terms of extractive processing. Conversely, many primates that show no evidence of tool use/manufacture in the wild may simply lack the opportunity or motivation to do so. Lack of opportunity may stem from the absence of such high-reward food items in their proximate environment. More likely, the lack of motivation may be because these primates, especially frugivores and folivores, can "get by" on diets consisting of foods with lower nutritional value but equally low investment, and there is little reproductive advantage in accessing other, higher-payoff foods.

Some habitats may be more conducive to tool use/manufacture than others, for example, because of the availability of food sources that create opportunities for innovation and social learning. Meulman and colleagues (Meulman et al. 2012) have argued that partially terrestrial environments play an important role in the development of complex tool technology in primates, especially lithic technology. They cite four lines of evidence to support this view. First, the capuchin monkeys most often observed using stone tools in the wild live in savanna-like environments, while more arboreal populations of the same taxon do not exhibit this behavior (Visalberghi and Fragaszy 2013). Second, chimpanzees, which are more terrestrial than orangutans, show more complex tool variants than the latter (the authors define complex tools as those that are acquired in part through social learning, are used flexibly, and can be accumulated). Third, among chimpanzees, tools used in terrestrial contexts are more complex (e.g., composite or sequential) than those used in exclusively arboreal environments. The authors ascribe this difference to availability of extractive foraging opportunities, to the opportunity to use their hands without the constraints of an arboreal setting, and to the greater opportunities for social learning in terrestrial settings. Finally, they suggest that the greater frequency of tool behaviors observed in captivity may be due to a "terrestriality effect," in which even the most arboreal primates spend more time on the ground than they would in the wild.

Thus, knowing whether the LCA actually used and made tools may greatly depend on reconstructions of its paleoenvironment, and especially whether this environment facilitated terrestrial behaviors (Elton 2008). As there are no fossils from the LCA, it is challenging to reconstruct its exact paleoenvironment. Nonetheless, if one accepts that it lived in Africa between 6 and 9 Ma (Moorjani et al. 2016), then indirect evidence of its habitats may be gleaned from Late Miocene fossil/geological sites in Africa. The Late Miocene record in Africa is spotty at best (reviewed in Elton 2008), and in fact many of our insights into the LCA's habitat come from paleoenvironmental reconstructions of the earliest putative hominins, *Sahelanthropus tchadensis* (~7 Ma: Vignaud et al. 2002), *Orrorin tugenensis* (~6 Ma: Pickford and Senut 2001) and *Ardipithecus kadabba* (~5.2–5.8 Ma: WoldeGabriel et al. 2001).

Habitat reconstructions for these putative stem hominins are fraught with difficulties, but the general pattern that emerges is one of mosaic environments, the product of increasing climatic variability. Mosaic habitats, as the name implies, include "a range of different habitat types, scattered across and interspersed within a given area" (Elton 2008: 381). While this may not be a satisfactory way of reconstructing the Late Miocene environment of the LCA, it does suggest that its habitat was neither closed tropical forest, nor open grassland/savanna. It was more likely woodland with interrupted canopy forests, greater seasonality, and perhaps access to more open spaces, which may have required some degree of terrestriality, all the while preserving an important arboreal component in the LCA behavioral repertoire. Some degree of bipedalism has also been inferred for these Late Miocene stem hominins, and for *Ardipithecus ramidus,* reinforcing the notion that they and the LCA may have been semi-terrestrial.

If, as the evidence suggests, the LCA lived in a mosaic environment with increasing opportunities for terrestriality and access to food resources associated with terrestrial environments (e.g., roots, tubers, termite mounds), and filled all the prerequisites for tool use/manufacture, then it is highly likely that it did use and make tools. Specifically, its tool kit was probably most similar to the tool kit of chimpanzee populations that live in mosaic habitats, subject to higher ecological variability, with more seasonal environments in which availability of arboreal versus terrestrial foods varies due to fluctuations in temperature and/or rainfall, such as Bossou in West Africa (Yamakoshi 1998; Hockings et al. 2015).

In concrete terms, its tool kit likely included organic tools that were used in extractive foraging, grooming, and display activities, such as leaf sponges, dipping sticks, and modified branches. Some of these tools need to be deliberately modified, implying tool manufacture in the LCA as well. Whether it used stone tools is an open question (Whiten 2011; Haslam 2014). Whereas many organic tools and associated behaviors are shared across African chimpanzee populations and bonobos (e.g., branch shaking and leaves as tools), the use of stone tools for pounding and opening nuts, and nut cracking itself, is best known in West African chimpanzees (Whiten et al. 1999; Carvalho et al. 2008), although recent evidence indicates that semi-wild (i.e., rehabilitated) bonobos also partake in these activities (Neufuss et al. 2017). Outside of hominins, stone tool use for pounding appears to have evolved independently in *Pan troglodytes, Sapajus libidinosus,* and *Macaca fascicularis aurea.* Because it is not customary among all chimpanzee subspecies or populations, and because it is exceedingly rare in bonobos, Haslam (2014) argues that stone tool use was unlikely in the ancestor of bonobos and chimpanzees. If this is true, then the most parsimonious scenario is that the *Pan-Homo* last common ancestor also did not use stone tools. However, the absence of stone tool use in chimpanzees in East Africa, and in bonobos, could equally be due to lack of ecological opportunity to innovate, develop, and maintain stone tool behaviors, and the most recent evidence (Harmand et al. 2015) reinforces the idea that stone tool use is shared by a larger number of hominin genera *(Australopithecus, Kenyanthropus, Homo),* as well as by modern *Pan* and *Homo.* If the LCA did use stone tools, however, we expect based on comparative evidence from extant primates that these assemblages would have been scarce, with relatively low numbers of artifacts per site, and largely composed of unmodified hammers, and possibly anvils, used in percussive activities.

Conclusion

We may never recover fossil or archaeological evidence of the LCA, and thus know with certainty that it used or made tools. However, given the shared complexity of tool use and manufacture in wild chimpanzees and hominins, which is both qualitatively and quantitatively different from other primates, the most parsimonious scenario, at least from a phylogenetic perspective, is that the LCA did use and make tools. Even without invoking parsimony, the comparative anatomical, cognitive, fossil, and archaeological evidence discussed in this chapter strongly suggests that the LCA had the capacity to and did use and manufacture tools. These tools were probably mostly organic, used for displays, grooming, and especially extractive foraging, but also likely included a small lithic component. In the absence of "smoking gun" fossil and archaeological evidence, future comparative studies on the habitat and behavioral ecology of extant primates in the wild (e.g., terrestriality), and on paleoenvironmental reconstructions of the LCA's habitat during the Late Miocene of Africa, will continue to refine our understanding of the tool types and behaviors of the LCA.

Summary

Although often still considered a hallmark of humans and their ancestors, the use and manufacture of tools for hygiene, communication, and extractive foraging is now well documented in other primates, including capuchins, macaques, and most hominoids. A so-called tool culture is particularly well documented in chimpanzees, and present in bonobos. Consequently, the comparative study of the behavioral variation in tool repertoires, but also of the morphological and cognitive prerequisites for tool use in extant *Homo* and *Pan*, can be used to make inferences about the tools and tool-making capabilities of the LCA. Comparative tool use in nonhuman primates, comparative anatomy and biomechanics of primate hands, comparative cognition in hominoids, and evidence from the fossil and archaeological records of early hominins, all support the idea that the LCA was capable of using and manufacturing stone tools for percussive activities (e.g., cracking nuts), and likely also organic tools (e.g., modified sticks for extractive foraging or displays). Whether the LCA actually used or made tools was likely contingent on whether its environment provided enough opportunities to encounter and exploit key resources. Specifically, the distribution of food sources and raw material availability and quality would have been important factors determining opportunities for innovation and for transmission and diffusion of technological behaviors.

Acknowledgments

We thank Martin Muller, David Pilbeam, and Richard Wrangham for inviting us to contribute to this volume, as well as for providing helpful feedback on earlier versions of the manuscript.

References

Alba, D. M., S. Moyà-Solà, and M. Kohler. 2003. Morphological affinities of the *Australopithecus afarensis* hand on the basis of manual proportions and relative thumb length. *Journal of Human Evolution* 44: 225–254.

Alemseged, Z., F. Spoor, W. H. Kimbel, R. Bobe, D. Geraads, D. Reed, and J. G. Wynn. 2006. A juvenile early hominin skeleton from Dikika, Ethiopia. *Nature* 443: 296–301.

Almécija, S., D. M. Alba, and S. Moyà-Solà. 2012. The thumb of Miocene apes: New insights from Castell de Barbera (Catalonia, Spain). *American Journal of Physical Anthropology* 148; 436–450.

Almécija, S., D. M. Alba, S. Moyà-Solà, and M. Kohler. 2007. Orang-like manual adaptations in the fossil hominoid *Hispanopithecus laietanus:* First steps towards great ape suspensory behaviours. *Proceedings of the Royal Society B: Biological Sciences* 274: 2375–2384.

Almécija, S., S. Moyà-Solà, and D. M. Alba. 2010. Early origin for human-like precision grasping: A comparative study of pollical distal phalanges in fossil hominins. *PLoS ONE* 5: e11727.

Almécija, S., J. B. Smaers, and W. L. Jungers. 2015. The evolution of human and ape hand proportions. *Nature Communications* 6: 7717.

Badrian, N. 1981. Preliminary observations on the feeding behavior of *Pan paniscus* in the Lomako forest of central Zaïre. *Primates* 22: 173–181.

Begun, D. R. 2010. Miocene hominids and the origins of the African apes and humans. *Annual Review of Anthropology* 39: 67–84.

Berger, L. R., J. Hawks, D. J. de Ruiter, S. E. Churchill, P. Schmid, L. K. Delezene, T. L. Kivell, H. M. Garvin, S. A. Williams, J. M. DeSilva, M. M. Skinner, C. M. Musiba, N. Cameron, T. W. Holliday, W. Harcourt-Smith, R. R. Ackermann, M. Bastir, B. Bogin, D. Bolter, J. Brophy, Z. D. Cofran, K. A. Congdon, A. S. Deane, M. Dembo, M. Drapeau, M. C. Elliott, E. M. Feuerriegel, D. Garcia-Martinez, D. J. Green, A. Gurtov, J. D. Irish, A. Kruger, M. F. Laird, D. Marchi, M. R. Meyer, S. Nalla, E. W. Negash, C. M. Orr, D. Radovcic, L. Schroeder, J. E. Scott, Z. Throckmorton, M. W. Tocheri, C. VanSickle, C. S. Walker, P. P. Wei, and B. Zipfel. 2015. *Homo naledi*, a new species of the genus *Homo* from the Dinaledi Chamber, South Africa. *Elife* 4: e09560.

Berger, L. R., J. Kibii, S. Churchill, P. Schmid, K. Carlson, B. de Klerk, D. de Ruiter, T. Holliday, T. Kivell, J. Gurche, B. Zipfel, J. DeSilva, and R. Kidd. 2011. New remains of *Australopithecus sediba* from the Malapa site, South Africa. *American Journal of Physical Anthropology* 144: 88–88.

Biro, D., N. Inoue-Nakamura, R. Tonooka, G. Yamakoshi, C. Sousa, and T. Matsuzawa. 2003. Cultural innovation and transmission of tool use in wild chimpanzees: evidence from field experiments. *Animal Cognition* 6: 213–223.

Biro, D., C. Sousa, and T. Matsuzawa. 2006. Ontogeny and cultural propagation of tool use by wild chimpanzees at Bossou, Guinea: Case studies in nut cracking and leaf folding. In T. Matsuzawa, M. Tomonaga, and M. Tanaka, eds., *Cognitive Development in Chimpanzees*, 476–508. Tokyo: Springer Tokyo.

Bobe, R., A. K. Behrensmeyer, and R. E. Chapman. 2002. Faunal change, environmental variability and late Pliocene hominin evolution. *Journal of Human Evolution* 42: 475–497.

Boesch, C., and H. Boesch. 1983. Optimization of nut-cracking with natural hammers by wild chimpanzees. *Behaviour* 83: 265–286.

Boesch, C., and H. Boesch. 1993. Different hand postures for pounding nuts with natural hammers by wild chimpanzees. In H. Preuschoft and D. J. Chivers, eds.), *Hands of Primates*, 31–43. Vienna: Springer.

Boesch, C., and H. Boesch-Achermann. 2000. *The Chimpanzees of the Taï Forest: Behavioural Ecology and Evolution*. Oxford: Oxford University Press.

Boesch, C., J. Head, and M. M. Robbins. 2009. Complex tool sets for honey extraction among chimpanzees in Loango National Park, Gabon. *Journal of Human Evolution* 56: 560–569.

Breuer, T., M. Ndoundou-Hockemba, and V. Fishlock. 2005. First observation of tool use in wild gorillas. *PLoS Biology* 3: 2041–2043.

Bush, M. E., C. O. Lovejoy, D. C. Johanson, and Y. Coppens. 1982. Hominid carpal, metacarpal, and phalangeal bones recovered from the Hadar formation: 1974–1977 collections. *American Journal of Physical Anthropology* 57: 651–677.

Call, J., and M. Tomasello. 1994. Production and comprehension of referential pointing by orangutans *(Pongo pygmaeus)*. *Journal of Comparative Psychology* 108: 307–317.

Carpenter, C. R. 1940. *A Field Study in Siam of the Behavior and Social Relations of the Gibbon* (Hylobates lar). Baltimore: Johns Hopkins Press.

Carvalho, S., D. Biro, W. C. McGrew, and T. Matsuzawa. 2009. Tool-composite reuse in wild chimpanzees *(Pan troglodytes)*: Archaeologically invisible steps in the technological evolution of early hominins? *Animal Cognition* 12: S103–S114.

Carvalho, S., E. Cunha, C. Sousa, and T. Matsuzawa. 2008. Chaînes opératoires and resource-exploitation strategies in chimpanzee *(Pan troglodytes)* nut cracking. *Journal of Human Evolution* 55: 148–163.

Carvalho, S., and W. McGrew. 2012. The origins of the Oldowan: Why chimpanzees *(Pan troglodytes)* still are good models for technological evolution in Africa. In M. Domínguez-Rodrigo, ed., *Stone Tools and Fossil Bones: Debates in the Archaeology of Human Origins*, 201–221. Cambridge: Cambridge University Press.

Chiang, M. 1967. Use of tools by wild macaque monkeys in Singapore. *Nature* 214: 1258–1259.

Clarke, R. J. 1999. Discovery of complete arm and hand of the 3.3 million-year-old *Australopithecus* skeleton from Sterkfontein. *South African Journal of Science* 95: 477–480.

Darwin, C. 1871. *The Descent of Man, and Selection in Relation to Sex*. New York: D. Appleton.

Deng, H., and J. Zhou. 2016. "Juggling" behavior in wild Hainan gibbons, a new finding in nonhuman primates. *Scientific Reports* 6: 23566.

Dominguez-Rodrigo, M. 2014. Is the "Savanna Hypothesis" a dead concept for explaining the emergence of the earliest hominins? *Current Anthropology* 55: 59–81.

Dominguez-Rodrigo, M., A. Z. P. Mabulla, H. T. Bunn, F. Diez-Martin, E. Baquedano, D. Barboni, R. Barba, S. Dominguez-Solera, P. Sanchez, G. M. Ashley, and J. Yravedra. 2010. Disentangling hominin and carnivore activities near a spring at FLK North (Olduvai Gorge, Tanzania). *Quaternary Research* 74: 363–375.

Duda, P., and J. Zrzavy. 2013. Evolution of life history and behavior in Hominidae: Towards phylogenetic reconstruction of the chimpanzee-human last common ancestor. *Journal of Human Evolution* 65: 424–446.

Elton, S. 2008. The environmental context of human evolutionary history in Eurasia and Africa. *Journal of Anatomy* 212: 377–393.

Fan, P. F., Q. Y. Ni, G. Z. Sun, B. Huang, and X. L. Jiang. 2009. Gibbons under seasonal stress: The diet of the black crested gibbon *(Nomascus concolor)* on Mt. Wuliang, central Yunnan, China. *Primates* 50: 37–44.

Fushimi, T., O. Sakura, T. Matsuzawa, H. Ohno, and Y. Sugiyama. 1991. Nut-cracking behavior of wild chimpanzees *(Pan troglodytes)* in Bossou, Guinea (West Africa). In A. Ehara, T. Kimura, O. Takenaka, and M. Iwamoto, eds., *Primatology Today*, 695–696. Amsterdam: Elsevier.

Galdikas, B. M. F. 1982. Orangutan tool-use at Tanjung-Puting-Reserve, central Indonesian Borneo (Kalimantan-Tengah). *Journal of Human Evolution* 11: 19–33.

Gommery, D., and B. Senut. 2006. The terminal thumb phalanx of *Orrorin tugenensis* (Upper Miocene of Kenya). *Geobios-Lyon* 39: 372–384.

Goodall, J. 1963. My life among wild chimpanzees. *National Geographic* 124: 273–308.

Goodall, J. 1986. *The Chimpanzees of Gombe: Patterns of Behavior.* Cambridge, MA: Harvard University Press.

Gruber, T., and Z. Clay. 2016. A comparison between bonobos and chimpanzees: A review and update. *Evolutionary Anthropology* 25: 239–252.

Gumert, M. D., M. Kluck, and S. Malaivijitnond. 2009. The physical characteristics and usage patterns of stone axe and pounding hammers used by long-tailed macaques in the Andaman Sea region of Thailand. *American Journal of Primatology* 71: 594–608.

Hamilton, W. J., R. E. Buskirk, and W. H. Buskirk. 1975. Chacma baboon tactics during intertroop encounters. *Journal of Mammalogy* 56: 857–870.

Harmand, S., J. E. Lewis, C. S. Feibel, C. J. Lepre, S. Prat, A. Lenoble, X. Boes, R. L. Quinn, M. Brenet, A. Arroyo, N. Taylor, S. Clement, G. Daver, J. P. Brugal, L. Leakey, R. A. Mortlock, J. D. Wright, S. Lokorodi, C. Kirwa, D. V. Kent, and H. Roche. 2015. 3.3-million-year-old stone tools from Lomekwi 3, West Turkana, Kenya. *Nature* 521: 310–315.

Haslam, M. 2014. On the tool use behavior of the bonobo-chimpanzee last common ancestor, and the origins of hominine stone tool use. *American Journal of Primatology* 76: 910–918.

Haslam, M., L. Luncz, A. Pascual-Garrido, T. Falótico, S. Malaivijitnond, and M. Gumert. 2016a. Archaeological excavation of wild macaque stone tools. *Journal of Human Evolution* 96: 134–138.

Haslam, M., L. V. Luncz, R. A. Staff, F. Bradshaw, E. B. Ottoni, and T. Falótico. 2016b. Pre-Columbian monkey tools. *Current Biology* 26: R521–R522.

Hayashi, M. 2015. Perspectives on object manipulation and action grammar for percussive actions in primates. *Philosophical Transactions of the Royal Society B: Biological Sciences* 370: 20140350.

Hobaiter, C., T. Poisot, K. Zuberbuhler, W. Hoppitt, and T. Gruber. 2014. Social network analysis shows direct evidence for social transmission of tool use in wild chimpanzees. *PLoS Biology* 12: e1001960.

Hockings, K. J., M. R. McLennan, S. Carvalho, M. Ancrenaz, R. Bobe, R. W. Byrne, R. I. M. Dunbar, T. Matsuzawa, W. C. McGrew, E. A. Williamson, M. L. Wilson, B. Wood, R. W. Wrangham, and C. M. Hill. 2015. Apes in the Anthropocene: Flexibility and survival. *Trends in Ecology and Evolution* 30: 215–222.

Hohmann, G., and B. Fruth. 2003. Culture in bonobos? Between-species and within-species variation in behavior. *Current Anthropology* 44: 563–571.

Kinani, J.-F., and D. Zimmerman. 2015. Tool use for food acquisition in a wild mountain gorilla *(Gorilla beringei beringei)*. *American Journal of Primatology* 77: 353–357.

Kivell, T. L., J. M. Kibii, S. E. Churchill, P. Schmid, and L. R. Berger. 2011. *Australopithecus sediba* hand demonstrates mosaic evolution of locomotor and manipulative abilities. *Science* 333: 1411–1417.

Koops, K., T. Furuichi, and C. Hashimoto. 2015. Chimpanzees and bonobos differ in intrinsic motivation for tool use. *Scientific Reports* 5: 11356.

Koops, K., W. C. McGrew, and T. Matsuzawa. 2010. Do chimpanzees *(Pan troglodytes)* use cleavers and anvils to fracture Treculia africana fruits? Preliminary data on a new form of percussive technology. *Primates* 51: 175–178.

Koops, K., W. C. McGrew, and T. Matsuzawa. 2013. Ecology of culture: Do environmental factors influence foraging tool use in wild chimpanzees, *Pan troglodytes verus*? *Animal Behaviour* 85: 175–185.

Koops, K., E. Visalberghi, and C. P. van Schaik. 2014. The ecology of primate material culture. *Biology Letters* 10: 20140508.

Lapuente, J., T. C. Hicks, and K. E. Linsenmair. 2016. Fluid dipping technology of chimpanzees in Comoé National Park, Ivory Coast. *American Journal of Primatology* 79: e22628.

Leakey, L. S. B., J. R. Napier, and P. V. Tobias. 1964. New species of genus *Homo* from Olduvai Gorge. *Nature* 202: 7–9.

Lewis, J. E., and S. Harmand. 2016. An earlier origin for stone tool making: Implications for cognitive evolution and the transition to *Homo*. *Philosophical Transactions of the Royal Society B: Biological Sciences* 371: 20150233.

Linscheid, R., K. An, and M. Gross. 1991. Quantitative analysis of the intrinsic muscles of the hand. *Clinical Anatomy* 4: 265–284.

Liu, Q., D. Fragaszy, B. Wright, K. Wright, P. Izar, and E. Visalberghi. 2011. Wild bearded capuchin monkeys *(Cebus libidinosus)* place nuts in anvils selectively. *Animal Behaviour* 81: 297–305.

Lonsdorf, E. V., L. E. Eberly, and A. E. Pusey. 2004. Sex differences in learning in chimpanzees. *Nature* 428: 715–716.

Louchart, A., H. Wesselman, R. J. Blumenschine, L. J. Hlusko, J. K. Njau, M. T. Black, M. Asnake, and T. D. White. 2009. Taphonomic, avian, and small-vertebrate indicators of *Ardipithecus ramidus* habitat. *Science* 326: 66e1–4.

Lovejoy, C. O. 2009. Reexamining human origins in light of *Ardipithecus ramidus*. *Science* 326: 74e1–8.

Lovejoy, C. O., B. Latimer, G. Suwa, B. Asfaw, and T. D. White. 2009a. Combining prehension and propulsion: The foot of *Ardipithecus ramidus*. *Science* 326: 72e71–78.

Lovejoy, C. O., S. W. Simpson, T. D. White, B. Asfaw, and G. Suwa. 2009b. Careful climbing in the Miocene: The forelimbs of *Ardipithecus ramidus* and humans are primitive. *Science* 326: 70e1–8.

Lovejoy, C. O., G. Suwa, S. W. Simpson, J. H. Matternes, and T. D. White, T. D. 2009c. The great divides: *Ardipithecus ramidus* reveals the postcrania of our last common ancestors with African apes. *Science* 326: 100–106.

Lovejoy, C. O., G. Suwa, L. Spurlock, B. Asfaw, and T. D. White. 2009d. The pelvis and femur of *Ardipithecus ramidus*: The emergence of upright walking. *Science* 326: 71e1–6.

Malaivijitnond, S., C. Lekprayoon, N. Tandavanittj, S. Panha, C. Cheewatham, and Y. Hamada. 2007. Stone-tool usage by Thai long-tailed macaques *(Macaca fascicularis)*. *American Journal of Primatology* 69: 227–233.

Mangalam, M., P. Izar, E. Visalberghi, and D. M. Fragaszy. 2016. Task-specific temporal organization of percussive movements in wild bearded capuchin monkeys. *Animal Behaviour* 114: 129–137.

Mannu, M., and E. B. Ottoni. 2009. The enhanced tool-kit of two groups of wild bearded capuchin monkeys in the Caatinga: Tool making, associative use, and secondary tools. *American Journal of Primatology* 71: 242–251.

Marzke, M. W. 1983. Joint functions and grips of the *Australopithecus afarensis* hand, with special reference to the region of the capitate. *Journal of Human Evolution* 12: 197–211.

Marzke, M. W. 1992. Evolutionary development of the human thumb. *Hand Clinics* 8: 1–8.

Marzke, M. W. 1997. Precision grips, hand morphology, and tools. *American Journal of Physical Anthropology* 102: 91–110.

Marzke, M. W. 2013. Tool making, hand morphology and fossil hominins. *Philosophical transactions of the Royal Society of London B* 368: 20120414.

Marzke, M. W., and R. F. Marzke. 2000. Evolution of the human hand: Approaches to acquiring, analysing and interpreting the anatomical evidence. *Journal of Anatomy* 197: 121–140.

Marzke, M. W., R. F. Marzke, R. L. Linscheid, P. Smutz, B. Steinberg, S. Reece, and K. N. An. 1999. Chimpanzee thumb muscle cross sections, moment arms and potential torques, and comparisons with humans. *American Journal of Physical Anthropology* 110: 163–178.

Marzke, M. W., and M. Shackley. 1986. Hominid hand use in the Pliocene and Pleistocene: Evidence from experimental archaeology and comparative morphology. *Journal of Human Evolution* 15: 439–460.

Marzke, M. W., K. L. Wullstein, and S. F. Viegas. 1992. Evolution of the power (squeeze) grip and its morphological correlates in hominids. *American Journal of Physical Anthropology* 89: 283–298.

Matsuzawa, T. 1996. Chimpanzee intelligence in nature and in captivity: Isomorphism of symbol use and tool use. In W. McGrew, L. F. Marchant, and J. Nishida, eds., *Great Ape Societies*. Cambridge: Cambridge University Press.

Matsuzawa, T. 1999. Communication and tool use in chimpanzees: Cultural and social contexts. In M. Hauser and M. Konishi, eds., *The Design of Animal Communication,* 645–671. Cambridge, MA: MIT Press.

Matsuzawa, T., D. Biro, T. Humle, N. Inoue-Nakamura, R. Tonooka, and G. Yamakoshi. 2001. Emergence of culture in wild chimpanzees: Education by masterapprenticeship. In T. Matsuzawa, ed., *Primate Origins of Human Cognition and Behavior,* 557–574. Tokyo: Springer Japan.

McGrew, W. C. 1974. Tool use by wild chimpanzees in feeding upon driver ants. *Journal of Human Evolution* 3: 501–504.

McGrew, W. C. 1992. *Chimpanzee Material Culture: Implications for Human Evolution.* Cambridge: Cambridge University Press.

McGrew, W. C. 2004. *The Cultured Chimpanzee: Reflections on Cultural Primatology.* Cambridge: Cambridge University Press.

McGrew, W. C. 2010. In search of the last common ancestor: New findings on wild chimpanzees. *Philosophical Transactions of the Royal Society of London B* 365: 3267–3276.

McGrew, W. C. 2017. Field studies of *Pan troglodytes* reviewed and comprehensively mapped, focusing on Japan's contribution to cultural primatology. *Primates* 58: 237–258.

McGrew, W. C., L. F. Marchant, M. M. Beuerlein, D. Vrancken, B. Fruth, and G. Hohmann. 2007. Prospects for bonobo insectivory: Lui Kotal, Democratic Republic of Congo. *International Journal of Primatology* 28: 1237–1252.

McPherron, S. P., Z. Alemseged, C. W. Marean, J. G. Wynn, D. Reed, D. Geraads, R. Bobe, and H. A. Bearat. 2010. Evidence for stone-tool-assisted consumption of animal tissues before 3.39 million years ago at Dikika, Ethiopia. *Nature* 466: 857–860.

Meulman, E. J. M., C. M. Sanz, E. Visalberghi, and C. P. van Schaik. 2012. The role of terrestriality in promoting primate technology. *Evolutionary Anthropology* 21: 58–68.

Meulman, E. J. M., and C. P. van Schaik. 2013. Orangutan tool use and the evolution of technology. In C. Sanz, J. Call, and C. Boesch, eds., *Tool Use in Animals: Cognition and Ecology,* 176–202. Cambridge: Cambridge University Press.

Mittra, E. S., H. F. Smith, P. Lemelin, and W. L. Jungers. 2007. Comparative morphometrics of the primate apical tuft. *American Journal of Physical Anthropology* 134: 449–459.

Moore, M. W., and Y. Perston. 2016. Experimental insights into the cognitive significance of early stone tools. *PLoS ONE* 11: e0158803.

Moorjani, P., C. E. G. Amorim, P. F. Arndt, and M. Przeworski. 2016. Variation in the molecular clock of primates. *Proceedings of the National Academy of Sciences* 113: 10607–10612.

Moyà-Solà, S., M. Kohler, D. M. Alba, and S. Almécija. 2008. Taxonomic attribution of the Olduvai Hominid 7 manual remains and the functional interpretation of hand morphology in robust australopithecines. *Folia Primatologica* 79: 215–250.

Mulcahy, N. J., and J. Call. 2006. Apes save tools for future use. *Science* 312: 1038–1040.

Musgrave, S., D. Morgan, E. Lonsdorf, R. Mundry, and C. Sanz. 2016. Tool transfers are a form of teaching among chimpanzees. *Scientific Reports* 6: 34783.

Nagell, K., R. S. Olguin, and M. Tomasello. 1993. Processes of social-learning in the tool use of chimpanzees *(Pan troglodytes)* and human children *(Homo sapiens). Journal of Comparative Psychology* 107: 174–186.

Napier, J. R. 1956. The prehensile movements of the human hand. *Journal of Bone and Joint Surgery* 38: 902–913.

Napier, J. R. 1980. *Hands.* New York: Pantheon.

Neufuss, J., T. Humle, A. Cremaschi, and T. L. Kivell. 2017. Nut-cracking behaviour in wild-born, rehabilitated bonobos *(Pan paniscus):* A comprehensive study of hand-preference, hand grips and efficiency. *American Journal of Primatology* 79: 1–16.

Oppenheimer, J. 1977. Forest structure and its relation to activity of the capuchin monkey *(Cebus)*. In T. C. A. Kumar and M. R. N. Prasad, eds., *Use of Non-Human Primates in Biomedical Research,* 74–84. New Delhi: Indian National Science Academy.

Ottoni, E. B., and P. Izar. 2008. Capuchin monkey tool use: Overview and implications. *Evolutionary Anthropology* 17: 171–178.

Palombit, R. A. 1997. Inter- and intraspecific variation in the diets of sympatric siamang *(Hylobates syndactylus)* and lar gibbons *(Hylobates lar)*. *Folia Primatologica* 68: 321–337.

Panger, M. A., A. S. Brooks, B. G. Richmond, and B. Wood. 2002. Older than the Oldowan? Rethinking the emergence of hominin tool use. *Evolutionary Anthropology* 11: 235–245.

Pickford, M., and B. Senut. 2001. The geological and faunal context of Late Miocene hominid remains from Lukeino, Kenya. *Comptes Rendus de l'Academie des Sciences, Serie II, Fascicule A: Sciences de la Terre et des Planetes* 332 : 145–152.

Proffitt, T., L. V. Luncz, T. Falótico, E. B. Ottoni, I. de la Torre, and M. Haslam. 2016. Wild monkeys flake stone tools. *Nature* 539: 85–88.

Pruetz, J. D., and P. Bertolani. 2007. Savanna chimpanzees, *Pan troglodytes verus,* hunt with tools. *Current Biology* 17: 412–417.

Richmond, B. G. 2007. Biomechanics of phalangeal curvature. *Journal of Human Evolution* 53: 678–690.

Rolian, C., and A. D. Gordon. 2013. Reassessing manual proportions in *Australopithecus afarensis*. *American Journal of Physical Anthropology* 152: 393–406.

Rolian, C., D. E. Lieberman, and B. Hallgrimsson. 2010. The coevolution of human hands and feet. *Evolution* 64: 1558–1568.

Rolian, C., D. E. Lieberman, and J. P. Zermeno. 2011. Hand biomechanics during simulated stone tool use. *Journal of Human Evolution* 61: 26–41.

Sakura, O., and T. Matsuzawa. 1991. Flexibility of wild chimpanzee nut-cracking behavior using stone hammers and anvils: An experimental analysis. *Ethology* 87: 237–248.

Santos, L. R., N. Mahajan, and J. L. Barnes. 2005. How prosimian primates represent tools: Experiments with two lemur species (*Eulemur fulvus* and *Lemur catta*). *Journal of Comparative Psychology* 119: 394–403.

Sanz, C., D. Morgan, and S. Gulick. 2004. New insights into chimpanzees, tools, and termites from the Congo basin. *American Naturalist* 164: 567–581.

Sanz, C. M., and D. B. Morgan. 2007. Chimpanzee tool technology in the Goualougo Triangle, Republic of Congo. *Journal of Human Evolution* 52: 420–433.

Sanz, C. M., and D. B. Morgan. 2009. Flexible and persistent tool-using strategies in honey-gathering by wild chimpanzees. *International Journal of Primatology* 30: 411–427.

Semaw, S. P. Renne, J. W. K. Harris, C. S. Feibel, R. L. Bernor, N. Fesseha, and K. Mowbray. 1997. 2.5-million-year-old stone tools from Gona, Ethiopia. *Nature* 385: 333–336.

Shrewsbury, M. M., M. W. Marzke, R. L. Linscheid, and S. P. Reece. 2003. Comparative morphology of the pollical distal phalanx. *American Journal of Physical Anthropology* 121: 30–47.

Shumaker, R. W., K. R. Walkup, and B. B. Beck. 2011. *Animal Tool Behavior: The Use and Manufacture of Tools by Animals*. Baltimore: Johns Hopkins University Press.

Sousa, C. 2011. Use of leaves for drinking water. In T. Matsuzawa, T. Humle, and Y. Sugiyama, eds., *The Chimpanzees of Bossou and Nimba*, 85–96. Tokyo: Springer Japan.

Stanford, C. B. 2012. Chimpanzees and the behavior of *Ardipithecus ramidus*. *Annual Review of Anthropology* 41: 139–149.

Starin, E. D. 1990. Object manipulation by wild red colobus monkeys living in the Abuko Nature-Reserve, the Gambia. *Primates* 31: 385–391.

Susman, R. L. 1994. Fossil evidence for early hominid tool use. *Science* 265: 1570–1573.

Susman, R. L. 1998. Hand function and tool behavior in early hominids. *Journal of Human Evolution* 35: 23–46.

Suwa, G., B. Asfaw, R. T. Kono, D. Kubo, C. O. Lovejoy, and T. D. White. 2009a. The *Ardipithecus ramidus* skull and its implications for hominid origins. *Science* 326: 68e1–7.

Suwa, G., R. T. Kono, S. W. Simpson, B. Asfaw, C. O. Lovejoy, and T. D. White. 2009b. Paleobiological implications of the *Ardipithecus ramidus* dentition. *Science* 326: 94–99.

Thorpe, S. K. S., R. H. Crompton, M. M. Gunther, R. F. Ker, and R. M. Alexander. 1999. Dimensions and moment arms of the hind- and forelimb muscles of common chimpanzees *(Pan troglodytes)*. *American Journal of Physical Anthropology* 110: 179–199.

Tocheri, M. W., and M. W. Marzke. 2007. The recent evolutionary history and adaptive significance of the hominin hand. *Journal of Anatomy* 210: 771.

Tocheri, M. W., C. M. Orr, M. C. Jacofsky, and M. W. Marzke. 2008. The evolutionary history of the hominin hand since the last common ancestor of *Pan* and *Homo*. *Journal of Anatomy* 212: 544–562.

Tomasello, M., M. Davis-Dasilva, L. Camak, and K. A. Bard. 1987. Observational learning of tool-use by young chimpanzees. *Human Evolution* 2: 175–183.

Tonooka, R. 2001. Leaf-folding behavior for drinking water by wild chimpanzees *(Pan troglodytes verus)* at Bossou, Guinea. *Animal Cognition* 4: 325–334.

Toth, N., and K. Schick. 2009. The Oldowan: The tool making of early hominins and chimpanzees compared. *Annual Review of Anthropology* 38: 289–305.

Urbani, B., and P. A. Garber. 2002. A stone in their hands ... are monkeys tool users? *Anthropologie* 40: 183–191.

van Schaik, C. P., A. Ancrenaz, G. Borgen, B. Galdikas, C. D. Knott, I. Singleton, A. Suzuki, S. S. Utami, and M. Merrill. 2003. Orangutan cultures and the evolution of material culture. *Science* 299: 102–105.

van Schaik, C. P., R. O. Deaner, and M. Y. Merrill. 1999. The conditions for tool use in primates: Implications for the evolution of material culture. *Journal of Human Evolution* 36: 719–741.

van Schaik, C. P., E. A. Fox, and A. F. Sitompul. 1996. Manufacture and use of tools in wild Sumatran orangutans: Implications for human evolution. *Naturwissenschaften* 83: 186–188.

Vignaud, P., P. Duringer, H. T. Mackaye, A. Likius, C. Blondel, J. R. Boisserie, L. de Bonis, V. Eisenmann, M. E. Etienne, D. Geraads, F. Guy, T. Lehmann, F. Lihoreau, N. Lopez-Martinez, C. Mourer-Chauvire, O. Otero, J. C. Rage, M. Schuster, L. Viriot, A. Zazzo, and M. Brunet. 2002. Geology and palaeontology of the Upper Miocene Toros-Menalla hominid locality, Chad. *Nature* 418: 152–155.

Villmoare, B., W. H. Kimbel, C. Seyoum, C. J. Campisano, E. N. DiMaggio, J. Rowan, D. R. Braun, J. R. Arrowsmith, and K. E. Reed. 2015. Early *Homo* at 2.8 Ma from Ledi-Geraru, Afar, Ethiopia. *Science* 347: 1352–1355.

Visalberghi, E., and D. Fragaszy. 2013. The Etho-Cebus Project: Stone-tool use by wild capuchin monkeys. In C. Sanz, J. Call, and C. Boesch, eds., *Tool Use in Animals: Cognition and Ecology,* 203–222. Cambridge: Cambridge University Press.

Visalberghi, E., D. Fragaszy, E. Ottoni, P. Izar, M. G. de Oliveira, and F. R. D. Andrade. 2007. Characteristics of hammer stones and anvils used by wild bearded capuchin monkeys *(Cebus libidinosus)* to crack open palm nuts. *American Journal of Physical Anthropology* 132: 426–444.

Visalberghi, E., G. Sirianni, D. Fragaszy, and C. Boesch. 2015. Percussive tool use by Taï western chimpanzees and Fazenda Boa Vista bearded capuchin monkeys: A comparison. *Philosophical Transactions of the Royal Society B: Biological Sciences* 370: 20140351.

Ward, C. V., W. H. Kimbel, E. H. Harmon, and D. C. Johanson. 2012. New postcranial fossils of *Australopithecus afarensis* from Hadar, Ethiopia (1990–2007). *Journal of Human Evolution* 63: 1–51.

Ward, C. V., W. H. Kimbel, and D. C. Johanson. 2011. Complete fourth metatarsal and arches in the foot of *Australopithecus afarensis*. *Science* 331: 750–753.

White, T. D., B. Asfaw, Y. Beyene, Y. Haile-Selassie, C. O. Lovejoy, G. Suwa, and G. WoldeGabriel. 2009. *Ardipithecus ramidus* and the paleobiology of early hominids. *Science* 326: 75–86.

Whiten, A. 2011. The scope of culture in chimpanzees, humans and ancestral apes. *Philosophical Transactions of the Royal Society B: Biological Sciences* 366: 997–1007.

Whiten, A., J. Goodall, W. C. McGrew, T. Nishida, V. Reynolds, Y. Sugiyama, C. E. G. Tutin, R. W. Wrangham, and C. Boesch. 1999. Cultures in chimpanzees. *Nature* 399: 682–685.

Whiten, A., K. Schick, and N. Toth. 2009. The evolution and cultural transmission of percussive technology: Integrating evidence from palaeoanthropology and primatology. *Journal of Human Evolution* 57: 420–435.

Wittiger, L., and J. L. Sunderland-Groves. 2007. Tool use during display behavior in wild Cross River gorillas. *American Journal of Primatology* 69: 1307–1311.

WoldeGabriel, G., Y. Haile-Selassie, P. R. Renne, W. K. Hart, S. H. Ambrose, B. Asfaw, G. Heiken, and T. White. 2001. Geology and palaeontology of the Late Miocene Middle Awash valley, Afar rift, Ethiopia. *Nature* 412: 175–178.

Yamakoshi, G. 1998. Dietary responses to fruit scarcity of wild chimpanzees at Bossou, Guinea: Possible implications for ecological importance of tool use. *American Journal of Physical Anthropology* 106: 283–295.

Young, N. M., T. D. Capellini, N. T. Roach, and Z. Alemseged. 2015. Fossil hominin shoulders support an African ape-like last common ancestor of humans and chimpanzees. *Proceedings of the National Academy of Sciences* 112: 11829–11834.

18

Cultural Evolution in Chimpanzees and Humans

JOSEPH HENRICH *and* CLAUDIO TENNIE

Over the last few decades, researchers from diverse disciplines have developed cultural evolutionary and gene-culture coevolutionary theory (Cavalli-Sforza and Feldman 1981; Boyd and Richerson 1985; Laland et al. 2011). Applied to humans, this approach has yielded new insights into our species' evolution, behavior, and cognition, and broadened into research programs in biology, anthropology, psychology, archaeology, and economics (Shennan 2003; Richerson and Boyd 2005; Hoppitt and Laland 2013; Henrich 2016). Here, we apply these theoretical developments to chimpanzees by reviewing the now large body of evidence on chimpanzee social learning, culture, and traditions. Along the way, we provide comparative evidence for humans to assess the similarities and differences between these two species. By asking theoretically driven questions about the nature of culture in each species, we aim to assess their shared phylogenetic heritage, and to isolate the selective forces that have distinguished these lineages over the last six to nine million years.

Our review begins with a brief introduction to gene-culture coevolutionary theory, though specific elements of the theory will be rolled out as we go along. We emphasize that our goal here is to apply theoretically derived insights to the available evidence from chimpanzees, and not to provide a general review of all work on this topic (for this, see Whiten 2011). The literature on chimpanzee culture is, perhaps unavoidably, loaded with ad hoc and often vague concepts that are frequently used flexibly to argue for the presence or absence of qualitatively distinct, human psychological capacities, abilities, or motivations in chimpanzees, often with little or no emphasis on quantifying these differences or exploring the implications of such quantitative differences for cultural evolution and gene-culture interactions. By sticking close to the theory, we hope to avoid these traps.

Theorizing Culture

Dual inheritance or gene-culture coevolutionary theory arose from the recognition that humans, unlike most other species, are heavily reliant on learning from others, and that social learning created a second system of inheritance that evolves and interacts with genetic inheritance in a coevolutionary duet (Campbell 1965; Pulliam and Dunford 1980; Cavalli-Sforza and Feldman 1981; Lumsden and Wilson 1981; Boyd and Richerson 1985; Durham 1991). "Culture" in this view is the emergent product of individuals of various generations or ages interacting with and learning from each other over the course of their lives. This conceptualization focuses our attention on (1) the abilities of individuals to learn from each other (or *as a consequence* of each other) and (2) the importance of interaction and sociality. This means that "culture" is now anchored in brains, and traceable to individual cognitive abilities or learning strategies (Tomasello 1999b; Laland 2004; Henrich and McElreath 2007). But this also means that culture is not reducible to these abilities, since culture is what arises from a combination of learning and social interaction. At any given time, it is the statistical distribution of ideas, beliefs, values, or practices stored in the minds of individuals in a population. Isolated individuals can learn, but they can't have culture.

Spreading out from this conceptualization of culture, several possibilities open up. First, culture can evolve over time, as individuals learn from each other and across generations. Cavalli-Sforza and Feldman (1973, 1981) showed how such *population processes* can be formally modeled using mathematical tools drawn from population genetics and epidemiology. This permits researchers to connect individual-level psychological or cognitive abilities through social interaction and social structure to ask what the outcome is for cultural evolution—the change in the distribution of practices over time. Second, there is no reason to take these social learning abilities as given. Instead, Boyd and Richerson (1985) began approaching them as genetic adaptations that have evolved to allow individuals to more effectively extract useful information from patterns of behavior exhibited by those around them.[1] This insight opened the door for full-blown models of culture-gene coevolution, in which culture and genes mutually influence each other (Aoki 1986; Laland 1994; Laland et al. 1995a, 1995b; Feldman and Laland 1996; Henrich and Boyd 1998; McElreath et al. 2003).

Since the turn of the century, this approach has been fruitfully applied to humans to understand individual-level psychological abilities and population-level patterns, structures, and processes (Boyd et al. 2011; Mesoudi 2011; Whiten 2012; Henrich 2016). This research has:

1. Tested various hypotheses regarding "what," "when," and "from whom" people will apply their adaptive social learning abilities (e.g., Rendell et al. 2011; Morgan et al. 2012; Chudek et al. 2013; Wood et al. 2013; Muthukrishna et al. 2016), as well as "when" and "to whom" they will transmit (teach) cultural information (Kline et al. 2013; Kline 2015).
2. Shown how social learning mechanisms can respond to local ecological variation to generate adaptive population patterns of cultural variation (Henrich and Henrich 2010; Henrich 2016).
3. Established empirical relationships that link the size and interconnectedness of human societies to the complexity of their tool kits and technologies (e.g., Henrich 2004b; Kline and Boyd 2010; Derex et al. 2013; Muthukrishna et al. 2014).
4. Linked the spread of specific genes in response to specific cultural practices (e.g., Holden and Mace 1997; Dediu and Ladd 2007; Chiao and Blizinsky 2010; Laland et al. 2010; Richerson et al. 2010).
5. Explored the degree to which culture, like genes, is a process of descent with modification that builds tree-like patterns of descent (Lipo et al. 2006; Tehrani et al. 2010; Walker et al. 2010).

This blossoming now permits us to readily view chimpanzees and other primates through the same evolutionary lens (van Schaik and Burkart 2011; Whiten 2011). Though our comparative focus is on chimpanzees, we will at times bring in evidence from other primates as well as other species.

Do Chimpanzees Learn Socially?

Culture and cultural evolution are consequences of social learning (Boyd and Richerson 1995; Hoppitt and Laland 2013). If a species does not engage in social learning, in some form or fashion, it cannot have culture and will not experience cultural evolution. An immense amount of evidence shows that humans are automatic, unconscious, and frequent social learners (Bandura 1977; Tomasello 1999b; Csibra and Gergely 2006). As we discuss later, human attention and social learning abilities appear functionally honed to adaptively extract information from the minds and behaviors of other members of our groups; this information ranges from the meaning of words and the proper use of artifacts to the existence of invisible agents like germs or angels (Corriveau and Harris 2009a, 2009b; Harris and Corriveau 2011; Chudek et al. 2012, 2013; Herrmann et al. 2013). Across diverse societies, children, adolescents, and young adults socially learn vast repertoires of practices and bodies of know-how that are crucial for survival, such as how to find food, detoxify plants, build shelters, organize social groups, track animals, and make fire (Boyd et al. 2011; Henrich and Broesch 2011; Kline et al. 2013; Henrich 2016). Social learning is so powerful in humans that we readily copy actions, motivations, and beliefs that contradict our innate intuitions, tastes, and direct experiences (Rozin and Schiller 1980; Rozin et al. 1981; Billing and Sherman 1998; Henrich 2009a). Social learning even influences our opioid and cannabinoid systems to alter how much pain we experience for the same stimuli (Craig and Prkachin 1978; Craig 1986; Benedetti et al. 2013; Henrich 2016).

So, do chimpanzees socially learn? Yes, though as we will see in later sections the character, frequency, and life history of their social learning is different from humans in crucial ways. The most decisive evidence on this question comes from studies of captive chimpanzees. Typically, these studies take the following form: a trained demonstrator (a human or chimpanzee[2]) shows an observer how to open a baited puzzle box in one of two ways (the "two-target method"). Half of the observers see one way to open the box, and the other half see an alternative method. For example, half the observers might see the demonstrator push a bolt to open a door, while the other half see the bolt pulled out. If observers tend to match the method of their demonstrators, then some form of social learning is taking place. Typically, observer chimpanzees do indeed match these demonstrations to some detectable degree (Horner and Whiten 2005; Hopper et al. 2008; Kendal et al. 2015; Tennie et al. 2010b), though the degree of matching is often not substantial and sometimes is indistinguishable from zero (Tennie et al. 2006). Such results are often interpreted as revealing human-like social learning (e.g., Whiten et al. 2005; Bonnie et al. 2007; Hopper et al. 2007, 2008), even though the learning may involve a range of psychological mechanisms that cannot be pinpointed by the experimental methods deployed (Tennie et al. 2009, 2010a). Nevertheless, from the point of view of existing evolutionary models, what matters most is transmission fidelity (Henrich 2004b; Lewis and Laland 2012), and not the specific psychological details; thus, we focus on transmission fidelity.

One large study allows us to directly assess transmission fidelities across a battery of eight two-target tasks deployed within a single study involving human toddlers (aged two years), chimpanzees, gorillas, orangutans, and bonobos using conspecific (same species) demonstrators (Tennie et al. 2010b). Three different conditions in this study tested for different social learning mechanisms: (1) Full Demonstration, where a conspecific demonstrated the target methods, (2) Intention, where a conspecific demonstrated failed attempts, and (3) Endstate, where subjects observed only the physical endstates of the apparatuses. The tasks were presented in two sets, each with four tasks. Due to limited sample size ($n = 36$), the apes were tested repeatedly while the children were tested in one trial per condition.

Across all five species, only the toddlers showed evidence for copying across all conditions. The other great apes showed evidence for copying only in the Full Demonstration condition. However, even in the Full Demonstration condition, only the toddlers revealed any evidence of copying in the first trial data. For the other apes, only when all of their trials were analyzed together—in this one condition—did they reveal any evidence for copying at all, though this only occurred in one set of tasks. In sum, the nonhuman ape species showed rather weak and inflexible copying abilities compared to children, who copied reliably and robustly across contexts and informational conditions. So, regardless of the psychological mechanisms, chimpanzees and other apes reveal only very low transmission fidelities relative to humans (except sometimes in particularly simple tasks: Kendal et al. 2015). Notably, older children and adults show even higher transmission fidelities than toddlers (McGuigan et al. 2011), rendering the discrepancy even more extreme. We consider the implications of such results below.

While captive studies have proven invaluable for assessing chimpanzee social learning, there are important concerns regarding how transferable such findings are to the field. In other words, how ecologically valid are these findings? Chimpanzees in typical captive studies may be showing more or less social learning than they would in the wild. It may be tempting to simply choose the best performers, but only the most ecologically valid populations should be used (1) to explain wild behavior patterns and (2) to directly inform evolutionary scenarios, though understanding latent capacities, unexpressed in the wild, may still be important. High-performing populations may show *potentials* that are never expressed in wild populations. For example, chimpanzees trained in sign language may use a gestural sign for the color blue, but this never happens in the wild (Terrace 1979).

To assess ecological validity, captive apes can be heuristically divided into four categories: (1) those that have received extensive human training and enculturation (hereafter *highly enculturated;* e.g., Kanzi, the bonobo); (2) apes who have received extensive human interaction, training, and some enculturation, such as some of the apes housed in zoos (hereafter *semi-enculturated* chimpanzees; e.g., human-reared chimpanzees who then continue to live among conspecifics); (3) apes living in conspecific groups under nondeprived conditions—but without having received intensive human interaction (hereafter *enriched captive* apes; e.g., many zoos and sanctuaries); and (4) those who have been traumatized and/or have experienced prolonged socially and physically deprived conditions (e.g., isolation; hereafter *deprived;* some Hollywood/circus-trained apes or those isolated in medical laboratories). Again, we should not base our choice of study population on performance, but note that a choice is necessary, since these populations can differ in their levels of skills. For example, in terms of social learning, and in related abilities such as pointing (Leavens et al. 2010), highly enculturated chimpanzees are generally superior (Tomasello et al. 1993; Bjorklund et al. 2002; van Schaik and Burkart 2011), followed by semi-enculturated ones (Furlong et al. 2008), then enriched captives, and finally the deprived apes (Menzel et al. 1970).

Consequently, though far from ideal, we believe that enriched captive apes provide the best available population from which to draw the most valid conclusions about wild populations (apart from studying wild populations directly, of course).[3] Clearly, deprived apes are not acceptable—though short-term deprivations during certain developmental periods may not create enduring cognitive or motivational problems (Ferdowsian et al. 2011; though see

Wobber and Hare 2011). Similarly, highly and semi-enculturated apes are unlikely to be the best model for wild apes, since extensive and intimate contact with humans does not occur in the wild, and this does seem to alter cognitive skills and motivations in significant ways. Enriched captive populations live in social groups, are well fed (better than wild apes), and experience (somewhat) enriched physical environments (wild chimpanzees also live in a range of environments). Note that the social learning studies above all involved enriched captive apes.

There is little doubt that chimpanzees have at least some forms of social learning. Psychologists, aiming to distinguish qualitatively different types of social learning in humans and other species, have worked extensively to distinguish cognitive mechanisms such as imitation from others such as emulation and local enhancement. Cultural evolutionary theory, however, suggests that while these psychological categories provide useful proximate distinctions, our focus here should remain on thinking quantitatively (not qualitatively) about the frequency, fidelity, and durability of social learning. This is especially important since high-fidelity transmission might be achieved by using a combination of different psychological mechanisms, such as by copying some motor patterns (imitation), inferring some goals (goal emulation), and noting some mechanical affordances—often helped by various sorts of socially enhanced individual learning. Showing that some chimpanzees can sometimes imitate, for example, doesn't tell us whether this imitation is likely to give rise to *any* cultural diffusion or evolution. If not enough chimpanzees can copy and spread the behavior further, and/or if the imitation is too crude or rare, there will not be any resulting cultural evolution. A little imitation is the same as no imitation in many situations. However, this does not mean that other types of social learning could not provide the basis for some cultural evolution (unless, of course, the trait in question is purely action based and imitation fueled). Thus, we refer readers interested in various psychological categories of social learning mechanisms to the many excellent reviews (see Zentall 2006; Tennie et al. 2009; Whiten 2011; Hoppitt and Laland 2013).

Next, we consider whether chimpanzee social learning has been shaped by natural selection, as it appears to have been in humans, to expand and hone the behavioral repertoires of individuals by facilitating the acquisition of adaptive practices from others. Alternatively, it is plausible that chimpanzees possess some degree of social learning as a by-product of having brains selected for individual (asocial) learning—that is, for the ability to figure things out on their own. Individual and social learning involve many of the same cognitive skills and neurological resources (Reader and Laland 2002; Heyes 2012), so it is possible that selection for one delivers some amount of the other.

Do Chimpanzees Show the Predicted Social Learning Mechanisms or Biases?

Theorists have explored how natural selection might have shaped the cognition of learners to allow them to most effectively extract information from both the environment and their social milieu. Here we review the evidence for these hypotheses in chimpanzees, and provide a comparative perspective with humans.

Uncertainty and Conformity

Much theoretical work has focused on how learners should respond to uncertainty or task difficulty (Boyd and Richerson 1988, 1995; Henrich and Boyd 1998; Laland 2004; Wakano et al. 2004; Nakahashi et al. 2012; Aoki and Feldman 2014). Under many conditions, learners should respond to greater uncertainty or task difficulty (including poorer individual information or ambiguous environmental cues) by increasing their reliance on social learning—thus prioritizing social information over their own perceptions and inferences. Psychologists have termed this response *informational conformity*. The predicted shifts have been observed in humans (Baron et al. 1996; Efferson et al. 2008b; McElreath et al. 2008; Morgan et al. 2012; Muthukrishna et al. 2016) and in non-primate taxa such as rats and fish (Kendal et al. 2005; Galef et al. 2008; Galef 2009b; Laland et al. 2011).

Testing this copy-when-uncertain bias, Kendal and her collaborators (2015) studied the open diffusion of the practice of sliding a door to the right or left to access a grape (i.e., a simple two-target task). They found that the more experience an individual had with sliding the door, the less they relied on the observations of others in deciding which way to slide the door. The idea is that if you aren't sure which way the door might slide, you might as well try sliding it in the direction you've previously observed, either most recently or most frequently.

Beyond this, we know of no other tests of these predictions in chimpanzees, though claims of "conformity" are common (Whiten et al. 2005; Whiten and van Schaik 2007; Hopper et al. 2011a, 2011b; van de Waal et al. 2013). In our view, however, a combination of methodological problems and conceptual ambiguities deflate such interpretations (Galef and Whiskin 2008; van Leeuwen and Haun 2013). For example, evidence for conformity has been claimed from "reversion designs" in which individuals first acquire and master one technique as it spreads to become common in their group. Then, later, if some individuals perform a different technique, these individuals may drop their new techniques, and instead revert ("conform," is the claim) to the technique they first learned. For example, in the diffusion experiments described above, after the initial spread of either the "poke" or "lift" techniques in different groups, researchers have argued that the fact that some individuals subsequently tried a different technique but then switched back to their initial technique is evidence of "conformity" (Whiten et al. 2005).

Conceptually, these studies fail to distinguish *informational conformity* from either *conservatism* or *normative conformity*.[4] Conservatism is a tendency to "stick with" or revert to old habits—previously acquired and more deeply ingrained practices or preferences. Normative conformity is a tendency to "go along with the group" to avoid appearing deviant, which could result in sanctions or ostracism (it is not a form of social learning in the sense currently used by theorists). The observations of reversions in chimpanzees could be informational conformity, conservatism, or even normative conformity. Since most studies show that chimpanzees are conservative (though see Manrique et al. 2013; Yamamoto et al. 2013), this is a likely alternative explanation (Whiten 1998). While we think—on theoretical grounds—that normative conformity is unlikely to be found in chimpanzees (Henrich 2016), it is not ruled out in these experimental designs (see van Leeuwen and Haun 2013).[5]

In one well-designed study focused on these issues, van Leeuwen et al. (2013) tested the strength of chimpanzee conservatism. In contrast to most such research, chimpanzees in this study first individually learned their own ways to solve a task (either to place one of two tokens into the same container or the same token into one of two containers—upon which food rewards were handed out). The main question was whether chimpanzees would ever abandon their first-learned behavior in favor of another one shown by the majority of subjects in their social group, which would have been evidence for informational conformity. Chimpanzees did not show such conformity. Instead, they stuck to their initially learned (asocial) solution. This occurred despite the fact that chimpanzees who performed the minority strategy paid greater attention to what majority chimpanzees were doing. Thus, conservatism appears to be a potent tendency in chimpanzees.

Nevertheless, this conservatism can be overridden by social-payoff factors. In another condition, chimpanzees did abandon their first-learned strategy in favor of the demonstrated alternative. This was not due to conformity, however, but because the new method yielded a fivefold increase in food rewards.

Oblique Transmission Using Age, Success, Knowledge, and Prestige Biases

A great deal of theoretical work has examined the conditions under which natural selection will favor social learners who strategically target their attention toward those individuals most likely to possess fitness-enhancing behaviors, beliefs, motivations, or practices (Boyd and Richerson 1985; McElreath et al. 2003; Laland 2004; McElreath and Strimling 2008; Rendell et al. 2010). Theorists have argued that learners should use "model-based" cues such as skill, competence, success, age, experience (or perceived knowledge), prestige, and self-similarity cues like sex or ethnicity (based on cues related to language or dialect). Combinations of these cues help learners rapidly identify those individuals most likely to have adaptive information, which could be useful to the learner in the roles they will assume, and problems they will encounter, later in life. An immense amount of empirical work, much of it within the last fifteen years, has substantiated these predictions in adults, children, and even infants (Koenig and Harris 2005; Jaswal and Neely 2006; Efferson et al. 2008a; McElreath et al. 2008; Corriveau and Harris 2009a, 2009b; Corriveau et al. 2009, 2013; Rendell et al. 2011; Buttelmann et al. 2012; Chudek et al. 2012, 2013; Morgan et al. 2012; Wood et al. 2013), as well as providing some evidence in other non-primate species (Galef 2009b; Laland et al. 2011; Rendell et al. 2011).

Building on these insights, researchers have proposed that learners should take into account the costs of accessing their preferred models (those who they deem skilled, successful, and prestigious). Placing this within a life history framework, infants and children are expected to first learn all they can from their parents, siblings, and other accessible models, and then subsequently pay access costs to update cultural traits from their preferred models (Henrich 2004b; Henrich and Broesch 2011; Kline et al. 2013). The idea here is that children have easy access to their family and household members, who themselves have kinship incentives for transmitting useful cultural information to the learner. However, potentially more valuable models, with greater skill, success, and prestige, will often be available outside the household. Accessing these preferred models will require learners to pay costs in the form of spending time with these individuals, and in paying them deference in the form of gifts and services in exchange for access and potentially instruction (Henrich and Gil-White 2001). This can be characterized as a switch from primarily vertical cultural transmission to various forms of biased oblique transmission over the life course. Broadly, fieldwork in small-scale human societies provides evidence consistent with these predictions (Tehrani and Collard 2009; Henrich and Broesch 2011; Hewlett et al. 2011; Henrich et al. 2015).

Chimpanzees also show some of these patterns, although the evidence is limited. Among wild populations, detailed observational studies focused on three different practices—termite fishing, ant dipping, and nut cracking—indicate a clear shift from primarily watching the mother to increasingly watching others engaged in the practice. This is the expected vertical to oblique shift in attention. Moreover, the data make it clear that chimpanzee learners are preferentially attending to older and more experience practitioners, and largely ignoring their younger and less experienced conspecifics (Biro et al. 2003; Melber et al. 2007; Humle et al. 2009; Lonsdorf 2013). This is consistent with some form of age, experience, or skill bias in attention. Attention, however, is merely a necessary precursor to social learning (Corp and Byrne 2002). It needs to be shown that this extra attention results in social learning, and is not simply part of a scrounging strategy, in which skilled or experienced individuals are scrutinized because they are more productive targets for scrounging (Stammbach 1988; Henrich and Gil-White 2001).

In one study, observational data indicate not only the transmission of a specific part of a behavior (in this case, the length of their termite dips), but also a sex bias in transmission, with female offspring preferentially learning from their mothers, relative to males from their mothers (Lonsdorf et al. 2004). To our knowledge, other studies have not revealed similar patterns among chimpanzees (Lonsdorf 2013). However, natural selection may adjust sex biases in attention to adapt to different forms of social organization. For example, since vervet monkeys are female philopatric (females stay home), we would expect females to be the most locally knowledgeable. In accordance with this hypothesis, females are indeed the most attentively observed by others (van de Waal et al. 2010; Renevey et al. 2013), including males.

Two experiments address the possibility of adaptive biases in captive chimpanzees. Both reveal some selective tendencies in either social learning or attention, but it is not clear as to whether these confirm a priori theoretical predictions. In the more recent paper (mentioned above), Kendal et al. (2015) studied the diffusion of door-sliding practices (left versus right) as chimpanzees repeatedly operated a slide-box to access grapes. In some groups, all individuals were initially naïve to the apparatus, while in others one middle-ranking female was trained to operate it by always going to one side. The data show three patterns: (1) low- and middle-ranking individuals tended to *copy* their side choices more than dominant individuals, (2) dominant individuals were *watched* more by inexperienced lower rankers, and (3) trained females were *watched* more by inexperienced individuals of the same or lower rank. Notably, the data do not show that dominant individuals or the trained females were *copied* more, only watched more. The authors argue that the variation in the choice data was insufficient to reveal biased copying, but that the visual attention biases were likely "for learning" (as opposed to, say, "for scrounging"), because this attention was limited to inexperienced individuals. But, as noted above, attention differences have failed to translate into actual learning in another recent study (van Leeuwen et al. 2013).

Theoretically, we don't see how the patterns of "copy when not dominant" and "watch the dominant" arise from the logic of natural selection applied to social learning. Alternative explanations are available for the selectivity observed. Dominants, for reasons related to status competition, may tend to garner attention when they are engaged in novel activities. And lower-ranking individuals may look around more, monitoring more dominant individuals for threats. Consequently, these patterns might represent nonadaptive biases arising from status competition.

In the other study, Horner et al. (2010) show that chimpanzees possess some ability to distinguish among potential models during social learning. Each of two social groups was exposed to two different female models from their own group, one "experienced model" and one "inexperienced model." The experienced model was roughly two decades older and more dominant than the inexperienced model, who was just barely out of her juvenile period. Moreover, the experienced model had previously introduced successful innovations in a series of other experiments, so the experimenters knew she was a good transmitter. Of the twenty-two chimpanzees exposed to these two models, fourteen decided to participate (which meant effectively copying one of the two models). Of these fourteen participants, eight revealed no significant preference for either model. The remaining six tended to copy the experienced model by making deposits in the same location.[6]

We think both studies are interesting and should spur further research. Together they support some tendency to copy experienced or knowledgeable mid-ranking females. Perhaps these individuals are successful enough to be worth attending to, but not so dominant that watching them is dangerous or uncomfortable. It is problematic, however, that demonstrators were all carefully selected by the researchers for training. Horner et al. used previously successful transmitters, and Kendall et al. selected their models because they were "comfortable being briefly separated from their group for training" (extroverted) and "fast learners" (ideal individuals to scrounge from). Thus, more research will be needed to figure out why mid- and low-ranking chimpanzees tended to watch or copy these individuals.

Majority and Conformist Transmission Biases

Theorists have examined the conditions under which learners should rely on conformist transmission over other strategies for social and individual learning (Boyd and Richerson 1985; Kendal et al. 2009; Nakahashi et al. 2012; Perreault et al. 2012). Conformist transmission is the tendency to disproportionately "copy the plurality." For example, suppose there are three behavioral variants, A, B, and C, at frequencies of 40 percent, 30 percent, and 30 percent, respectively, in a population. If the new generation of learners picks a model at random, the next generation would—on average—have the same frequencies of A, B, and C. However, if individuals are using conformist transmission, the frequencies will shift to favor the plurality, changing to, say, 60 percent, 20 percent, and 20 percent, in the next generation. All else being equal, variant A will eventually spread to fixation. Largely consistent with predictions derived from formal models, research shows that humans use conformist transmission under some conditions (Efferson et al. 2008b; Morgan and Laland 2012; Morgan et al. 2012; Muthukrishna et al. 2016). Other than in primates, conformist transmission has been shown, perhaps most decisively, in fish (Pike and Laland 2010).[7]

No study has isolated conformist transmission by showing the requisite disproportionate tendency to copy the plurality or majority in chimpanzees or any other primates (van Leeuwen and Haun 2013; Acerbi et al. 2016).[8] To the contrary, neither Kendal et al. (2015) nor van Leeuwen et al. (2013) found support for conformist transmission in their diffusion experiments.

However, while conformist transmission has not emerged, chimpanzees may sometimes still use the frequency with which a trait is demonstrated by different individuals as a cue about whether to adopt it. Revealing what they termed "majoritarian bias," Haun and colleagues (2012) used a carefully designed experiment that controlled for both the frequency of times learners observed the use of a particular location for deposit (for dropping an object into an apparatus), and the number of different models observed using each location. Their evidence suggests that chimpanzees—but not orangutans—use the prevalence of a particular location among their models as a learning cue (though they could also have copied with a bias toward any of the demonstrator chimpanzees, which would have led to the same overall effect).

Teaching in Chimpanzees

We now shift our focus away from the learner toward the model, who can facilitate the acquisition of useful practices by the learner. From an evolutionary perspective, teaching involves paying at least small costs to help another individual—so it is a type of altruism or cooperation. Evolutionary models suggest that teaching and social learning can coevolve, but because of the costs to self and benefits to others, the conditions favoring teaching are narrower than those favoring social learning (Castro and Toro 2002; Fogarty et al. 2011). It is primarily expected to emerge between parents and their offspring.

Teaching is any costly behavior by the model that facilitates learning in conspecifics (Caro and Hauser 1992). Variously, teachers may (1) structure the environment to enhance the learners' chances of individually figuring things out (e.g., by leaving the right tools around); (2) approve or disapprove of their pupil's activities, guiding learners via reinforcement (Castro and Toro 2002);

(3) actively draw the learners' attention to key elements of a demonstration with cues such as pointing or eye contact ("pedagogical cues"; see Csibra and Gergely 2006, 2009); (4) mold the learners' hands, position their feet, or orient their bodies; (5) slow demonstrations down or exaggerate key aspects in order to make it easier for the learner to take in; and (5) scaffold the learner by providing challenges just above their current skill level (Hoppitt et al. 2008; Boesch 2012). This behavioral definition permits us to cast a wide net and to compare teaching in humans with other species.[9]

Characterizing teaching in humans is problematic, because most studies come from developmental psychologists studying children in Western societies (e.g., Tomasello 1999a; Csibra and Gergely 2009). Middle- and upper-class Westerners place immense emphasis on active and often verbal forms of teaching, molding their students' hands, and providing explicit feedback. In the smallest-scale human societies, including foragers, teaching exists—but is much less common and largely passive (Lancy 1996, 2009; Fiske 1998; Gaskins and Paradise 2010; Hewlett et al. 2011; Strauss and Ziv 2012). Moreover, some of the teaching observed by ethnographers in small-scale societies may have been culturally introduced by so called WEIRD societies (Henrich et al. 2010). In our view, many in this debate about teaching across human societies may have missed key questions by focusing on a "presence" versus "absence" debate, which frequently come down to arguments about definitions. Recent quantitative studies in Fijian villages show patterns of teaching or pedagogy quite unlike those common among Westerners, but largely consistent with the predictions from evolutionary reasoning (Kline 2015). Thus, the real puzzle for evolutionary researchers may be why WEIRD people teach as much as they do and in the ways they do.[10]

In nature, as expected from theory, teaching is much rarer than social learning. However, several studies furnish solid evidence of teaching (Hoppitt et al. 2008; Thornton and Raihani 2008). For example, tutor meerkats provide live—but previously disarmed—scorpions to inexperienced meerkats, who in turn learn to handle scorpions earlier than untutored meerkats (Thornton and McAuliffe 2006). In chimpanzees, three long-term and detailed studies have focused on understanding the factors that influence the acquisition of the skills for termite fishing, ant dipping, and nut cracking. In both termite fishing and ant dipping, chimpanzees make a probing tool out of immediately available materials, and then dip the tool into the habitat of the insects. To open nuts, chimpanzees use stone or wooden "hammers" to smash the shells on "anvils." This set of skills is an ideal place to look for teaching, since, as we argue below, social learning likely plays at least a facilitating role in their acquisition in the wild (Tennie et al. 2009). Researchers have studied how chimpanzees between the ages of about one and six years of age acquire these skills, observing, coding, and analyzing the behaviors of both mothers and other nearby adults and juveniles for any hint of teaching (Inoue-Nakamura and Matsuzawa 1997; Biro et al. 2003; Lonsdorf et al. 2004; Lonsdorf 2005, 2006, 2013; Humle et al. 2009).

The results are consistent across different researchers and field sites. Adults, particularly mothers, were highly tolerant of the activities of young chimpanzees (<5 yrs), permitting them to play with tools and "steal" or "scrounge" the tool and/or harvest, but they did not *actively facilitate* learning. Mothers generally reacted neutrally to their infants' efforts, provided no feedback of any kind, and never molded learners' hands, pointed, made eye contact, or provided other pedagogical cues. Eye contact was rarely made at all, as mothers were focused on their own foraging activities. Mothers never handed their offspring a tool or some of the harvest.[11] In short, no teaching was found (also see Moore and Tennie 2015).

The one potential exception occurs in dipping for army ants. This foraging task can be done at more dangerous nests or on less dangerous trails. Mothers with infant learners showed a bias to ant-dip at the less productive trails (paying a cost), thereby providing a safer environment for their offspring to learn in (Humle et al. 2009). An important question is whether this is merely a by-product of mothers' concerns about their offspring being attacked by army ants (or even about themselves, as mothers are less mobile, being handicapped by offspring), or if it was selected (by mothers or natural selection) because it facilitates social learning.

In the laboratory, there has been one detailed comparative study of teaching in children (under five years of age), chimpanzees, and capuchin monkeys. Dean et al. (2012) presented participants with a three-step task in which solving each step supplied the learner with a reward and opened the opportunity to complete the next step to obtain an even larger reward. While teaching in the children was common, and increased with task difficulty (as predicted by theory), neither monkeys nor chimpanzees engaged in any teaching. In the children, teaching correlated with greater success at the task, and may help explain why so many children reached the final stage but so few nonhumans advanced. We return to the presence and importance of teaching in humans when we discuss cumulative cultural evolution.

Population-Level Patterns of Behavior

Cultural evolutionary models show that social learning abilities can, under some conditions, give rise to stable behavioral variation between groups. Practices, beliefs, and ideas—cultural variants—clearly spread via cultural transmission among humans within groups and from group to group (Rogers 1995; Henrich 2001; Bell et al. 2009; Jordan 2015). Alternatively, sometimes groups expand, fission, and spread geographically, taking their cultural variants with them. Both kinds of processes can create spatially structured networks of cultural similarity, and in some cases tree-like patterns of descent with modification (Shennan 2009; Walker et al. 2010; Watts et al. 2015).

These cultural patterns are often adaptive, and are systematically associated with ecological variables for several reasons (Billing and Sherman 1998; Jordan and Shennan 2003; Shennan 2003; Henrich and Henrich 2010; Hruschka and Henrich 2013). First, as discussed above, human social learning has likely been honed by natural selection to use a wide range of adaptive cues, such as success, age, and prestige, to more effectively target attention and learning. This means that cultural evolution will respond to local environments and spread locally adaptive practices through populations. Second, since natural selection also influences cultural inheritance, those with locally less well-adapted repertoires will tend to be less available to transmit their cultural variants (Richerson and Boyd 2005). Third, human groups compete, and those with better-adapted cultural repertoires, including norms and forms of social organization, spread at the expense of those with less well-adapted cultural packages (Diamond 1997; Henrich 2004a; Currie and Mace 2009; Richerson et al. 2016).

To illustrate this, consider that the practice of constructing and inhabiting snow houses—as seen among Inuit foragers—is closely correlated with climatic temperature or latitude. The practice itself requires substantial culturally learned know-how, and cannot be figured out by, for example, lost Arctic explorers even when their survival depends on it (Boyd et al. 2011; Henrich 2016). However, cultural evolution only assembles the relevant know-how when the environmental conditions favor the practice. Thus, we should expect cultural evolution to create correlations between ecology and behavior.

Of course, evolutionary approaches to cultural transmission also predict, at least under some conditions, that cultural transmission can spread and stabilize neutral or even maladaptive variants. This can occur through a variety of mechanisms that need not concern us here, but whatever the mechanism, much empirical evidence supports the existence and persistence of neutral or maladaptive cultural variation among groups (Boyd and Richerson 1985; Durham 1991; Edgerton 1992; Henrich and Henrich 2010).

In light of the available theory, the evidence from humans, and the presence of some degree of social learning in chimpanzees, we can ask two questions:

1. Does chimpanzee social learning contribute to the spread of certain behaviors that remain locally stable and vary among groups?
2. Are these patterns of variation broadly adaptive, showing predictable and patterned ecological variation?

Field evidence gleaned from nine different chimpanzee populations scattered across tropical Africa does indeed reveal substantial *behavioral* variation across populations (Whiten et al. 1999, 2001). This research team isolated and categorized sixty-nine different behavioral variants across their sites. These variants included using (1) probes (e.g., sticks) to obtain ants, termites, or honey (or to clear the nose); (2) leaves as sponges, wipes, or brushes; (3) stones as hammers and anvils for nuts; and (4) sticks as levers to open and access the nests of birds or insects. Some categories include several variants. For example, nut hammering accounts for five variants, with some variants merely swapping the materials used for the hammers and anvils (stone versus wood). Each of the sixty-nine variants was classified according to its local frequency as (1) "customary" (most adults do it, or most of some subclass do it—e.g., all females), (2) "habitual" (commonly observed but not customary), (3) "present," (4) "absent," or (5) "status not established."

The tricky part turns out to be showing that this substantial and important *behavioral* variation is in fact *cultural* variation, as opposed to (1) genetic variation (Galef 2009a; Laland et al. 2009; Tennie et al. 2009; Langergraber et al. 2011) or (2) locally adaptive responses to ecological variation that depend only on individual learning or other ontogenetic responses to environmental cues (Galef 1992; Tomasello 1994; Laland et al. 2009; Tennie et al. 2009). The authors recognized these problems, and attempted to mitigate them by removing variants that were (1) universal, (2) very rare, or (3) readily explained by ecological variables, to arrive at a list of thirty-nine putative cultural variants. This catalogue of behaviors is impressive, and analyses of it have led researchers to argue that chimpanzee cultures are special (Whiten and van Schaik 2007) and more sophisticated than those of crows (McGrew 2013).

We are sympathetic to this effort, but the approach has several interpretive limitations. First, cultural evolution is adaptive, at least in humans, so removing things that are universal or explained by ecology potentially removes important cultural variants (Laland and Janik 2006; Byrne 2007). Second, in removing the "rarities," the authors suggest that social learning will cause traits to be common in groups—implying that rare traits are not socially learned. Theoretically, this is not true. How common a cultural trait becomes within a group depends on many factors, including how hard it is to learn, how easy it is to forget, how adaptively important it is, how the social network of the group interconnects, and what other variants it might be competing with. These first two limitations suggest that the number of cultural traits may in fact be underestimated by Whiten et al. Third, however, the local forces that shape individual learning or other noncultural ontogenetic responses may arise from nonobvious or even subtle ecological differences (e.g., differences in the number of available nuts to crack). Indeed, a recent review concluded that ecological opportunities were one of the main drivers of tool use patterns in chimpanzees, as well as in orangutans and capuchin monkeys (Koops et al. 2014). And, fourth, these nine groups span a vast range, and can be classified into three subspecies, with much internal genetic structure. Thus, genes are a potentially important competing explanation for the behavioral variation. These last two limitations imply a tendency toward overestimating the number of cultural traits.

Subsequent analyses of these putative cultural variants have shown that it is difficult to exclude genetic variation as a potential cause. Langergraber et al. (2011) assembled mitochondrial DNA data on the nine populations, and correlated measures of both cultural and genetic dissimilarity for all possible pairs of groups. The correlations range from 0.36 to 0.52, suggesting that genes might be important. However, when the data are analyzed at the level of particular variants, the authors establish that genetic variation is unlikely to explain five, and possibly as many as twenty, of the variants. It is still possible that most or all of these variants are cultural, but we cannot tell for many or even most of the traits.

In this study, the correlation between the geographical distance between communities and the genetic distance (mtDNA) was 0.96. Some argue that this means that genes and culture were merely moving together as populations expanded, a common pattern in human migration. This would mean that correlations between genes and behavior revealed by Langergraber et al. were noncausal associations created by a spreading population. However, this view overlooks two key differences between humans and chimpanzees: (1) all human migrations that have been studied in this fashion are relatively recent, and therefore shallow compared to the spread of chimpanzees across Africa (and the emergence of different subspecies), and (2) the fidelity of human cultural transmission is substantially higher than in chimpanzees (and both are much lower in fidelity than genetic transmission). This implies, given the temporal depths involved in the spread of chimpanzee populations, that there should be no remaining correlation between behavioral dissimilarity and geographic distance due to shared cultural inheritance. Given enough time, cultural drift, losses, inventions, and transmission noise will eventually wipe out the correlation between geography and culture created by migration. In humans, correlations between culture and genes exist only because the temporal depths of human expansions are recent, and the fidelity of cultural transmission is high. A firmer answer to this question awaits proper modeling, but on a first pass, the correlation between genes and culture in chimpanzees is unlikely to be due to persistence created by high-fidelity inheritance.[12]

In light of this evolutionary logic, we are concerned about recent efforts to apply phylogenetic techniques to broad patterns of chimpanzee behavioral variation. Lycett et al. (2010, 2011) analyzed Whiten et al.'s thirty-nine traits using the tools of cladistic analysis, which were developed to infer *genetic* phylogenies from extant variation. They argue that their analysis reveals a "phylogenetic" signal, which they use to construct a phylocultural tree for chimpanzees. Given the low fidelity of chimpanzee cultural transmission, high rates of both loss and reinvention, and the deep time scales associated with the expansion of chimpanzees across Africa, we find it unlikely that the signal revealed by Lycett et al. represents cultural descent with modification at the group level from an ancestral population of chimpanzees. To illustrate this, consider that the deepest human cultural phylogeny, which was constructed based on "ultraconserved words," goes back only 15,000 years (Pagel et al. 2013). By contrast, the trans-African geographic spread that eventually led to chimpanzee subspeciation occurred over a million years ago, and again about 500,000 years ago (Bjork et al. 2011). So cultural signals in humans don't last more than 15,000 years, but chimpanzee cultural signals endure for half a million years? For the reasons given above, this seems unlikely. To be clear, this is not to argue that the observed differences are not cultural. Theoretically, it is perfectly plausible that these variants are all 100 percent cultural, yet virtually no phylocultural signal remains, given the time scales involved. Social learning need not produce either group-level heritability or tree-like patterns of descent.

Because of these issues, we prefer regional or local studies of specific variants over continental-level analyses, because they reduce or eliminate concerns with genetic variation, narrow the potential sources of ecologically induced variation, and provide direct observation of the potential learning processes involved for different aspects of behavior (see Byrne 2007). Here, we focus again on three practices: (1) termite fishing, (2) ant dipping, and (3) nut cracking. At the local and regional level, all three (a) appear adaptively responsive to ecological or environmental changes, including seasonal variation, (b) are learned by young chimpanzees in a manner that is likely facilitated by mothers engaging in these skills (and others to a much lesser degree), and yet (c) show some patterned variation among communities that cannot be readily traced to obvious ecological differences.

Termite fishing is a good place to start since it was one of the five behavioral variants that Langergraber et al. (2011) evaluated as unlikely to be due to genetic differences among chimpanzee groups, and it is widespread across Africa (Whiten et al. 1999), but not found in some populations where termite mounds do exist. In termite fishing, chimpanzees fashion simple tools out of vegetation found near the mounds, and insert these tools to extract the termites. Within chimpanzee groups, the frequency of termite fishing varies seasonally with rainfall and temperature, and constitutes an important food source in some populations (Bogart and Pruetz 2009), particularly among populations living in savanna-woodlands (Bogart and Pruetz 2011). Meanwhile, in locales with more limited opportunities for exploiting termites relative to other resources, the practice is nonexistent (Koops et al. 2013; Sanz and Morgan 2013). Overall, termite fishing is indeed responsive to ecology and environment.

It is clear that termite fishing involves substantial individual learning, through practice and trial and error. The key question is how this learning is enhanced by social factors. Detailed studies of the acquisition of termite fishing skills in East Africa have helped illuminate the process. As mentioned above, Lonsdorf (2006) studied termite fishing by following eleven infants (unweaned, typically less than age five) and juveniles, along with their five mothers, for sixty-five hours. The rate at which these young wild chimpanzees increased their fishing skills depended on what the mother did, for how long, and with whom. Being exposed to a small group of fishers helped early on, that is, at a time when learners were mostly watching. Then, later, being alone with mother helped more, perhaps by reducing competition for access to the mound and tools. The correlation between the proficiency of the mother and her older offspring (over age six) was 0.63. This correlation may be due, entirely or in part, to genetic similarities between mothers and their offspring. However, this seems unlikely since mother-offspring correlations on other such tasks are generally small or zero. This is consistent with other work suggesting that the complexity of termite fishing rods depends on direct experience and learning opportunities (Sanz and Morgan 2011).

All of this is consistent with at least "exposure" learning (Thorndike 1911), meaning that youngsters were aided in learning to fish because hanging around their mothers provided access to termite mounds, tools, and opportunities to practice (similar to local enhancement). However, as noted above, a comparison of male and female learners revealed that females watched their mothers (and other females) more, achieved proficiency faster than their brothers, and ended up more skilled. Instead of watching, the males engaged in more individual experimentation (play). Moreover, these analyses reveal that daughters tended to match their mother's dipping strategy (in terms of stick length alone—not necessarily a sign of high-fidelity copying—see Moore 2013), while their sons did not (Lonsdorf et al. 2004; Lonsdorf 2005). In a manner consistent with the theoretical expectations discussed above, this suggests that more may be afoot than mere exposure or local enhancement.

The practice of ant dipping shows patterns that parallel termite fishing. Like termite fishing, ant dipping is widespread across Africa, often sea-

FIGURE 18.1. Termite fishing by an eight-year-old female chimpanzee (Gaia) in the Kasekela community in Gombe National Park, Tanzania (2001). Photo by Ian Gilby.

sonal, and responsive to ecological variation (Mobius et al. 2008; Schöning et al. 2008). Nevertheless, patterns of variation remain that are not readily accounted for as direct adaptive responses to ecological variation, unmediated by social interaction. In a study similar to that just described, Humle et al. (2009) studied ant dipping among thirteen mother-offspring pairs at Bossou in East Africa. Young chimpanzees tended to watch their mothers dipping, and then increasingly engaged in dipping as they got older. The time spent ant dipping by juveniles (weaned offspring) correlated highly with the time spent dipping by their mothers. Dipping proficiency, as measured by failed dips (or errors), increased with age (error rates declined). And juveniles with mothers who dipped a lot made fewer errors. Dipping proficiency, as measured by dip duration, was correlated at 0.87 between mothers and their juvenile offspring. Mom provides access to ants and tools as well as time and tolerance. This permits offspring to learn through direct experience. The more time mom provides, the better both she and her offspring get.

Humle et al. looked for correlations between mother-offspring (1) dipper stick lengths and (2) techniques used, but did not find any—unlike in the termite fishing study above. This is not altogether surprising, given that in an earlier study, Humle and Matsuzawa (2002) had already found that differences in ant characteristics (species and current behavior and location of the ants) were the major drivers of dipper stick length and perhaps also—in turn—of dipping technique (Humle et al. 2009).

Nut cracking, the use of wooden or stone "hammers" to crack nuts of various kinds, was once thought to be found exclusively among West African chimpanzees (see recent findings below), which highlighted the possibility of genetic influences (Langergraber et al. 2011). Many other wild chimpanzees inhabit environments with the requisite nuts, stones, and wood, but do not crack nuts. Nut cracking—including the choices of particular nut species and the tool materials used—also appears to be influenced by ecological factors in adaptive ways (Yamakoshi 1998; Biro et al. 2003), but not solely determined by ecology. For example, Luncz et al. (2012) studied the nut cracking behavior of three neighboring communities of chimpanzees in the Taï National Park, Côte d'Ivoire. Though they found few or no differences in ecology between these three communities, they did observe some differences in the nut cracking behavior. These differences were relatively subtle, being related to the selection of hammer material and size, rather than to the technique of nut cracking itself. Nevertheless, such differences are unlikely to be related to genetic variation among these neighbors, since they are known to interbreed.

As with both ant dipping and termite fishing, observational studies reveal that it is the exposure to, and possibly the observation of, nut crackers, their tools, and the fruits of their labors that stimulates the trial-and-error process necessary for chimpanzees to acquire nut cracking skills. This work also identified a sensitive window for the acquisition of nut cracking, between about age three and five years (Inoue-Nakamura and Matsuzawa 1997; Biro et al. 2003; Marshall-Pescini and Whiten 2008), though if the ability to crack one kind of nut is acquired during the window, this ability can be extended to different kinds of nuts later in life. As noted, younger individuals tend to watch older nutcrackers (especially the mother), though they do not copy the mother's specific use of her right or left hand for hammering.

So far, we have reviewed evidence showing the existence of experimentally induced "traditions" and field evidence of patterns of behavioral differences among captive chimpanzee populations. This evidence seems sufficient to establish that social learning can facilitate the spread of novel prac-

FIGURE 18.2. Nut cracking by an adult male chimpanzee (Jeje) at Bossou, Guinea, West Africa (2012). Photo by Kathelijne Koops.

tices. And, without it, novel inventions disappear. But, as noted above, what the studies of captive chimpanzees also show is that the fidelity of chimpanzee social learning is not sufficient to explain the sustained persistence of arbitrarily different, maladaptive, or otherwise costly practices (Claidière and Sperber 2009), as it does in humans. Most of the patterns we have reviewed are consistent with social learning facilitating the spread of practices, but with individual learning in response to the economics of the local ecology maintaining the practices, and accounting for why ecology seems so important to their distribution (Koops et al. 2014).

Chimpanzees and Cumulative Cultural Evolution

The survival of humans, including hunter-gatherers, depends critically on socially learned skills, know-how, motivations, tastes, and practices. Stripped of this culturally acquired information, humans cannot survive as foragers.

This fact has been repeatedly demonstrated as lost or stranded European explorers struggled to survive in "hostile" environments where local populations of hunter-gatherers had been living for centuries or millennia (Boyd et al. 2011; Henrich 2016). Thus, the massive ecological success and global expansion of our species into an immense diversity of environments, from the frozen Arctic to the arid deserts of Australia, was made possible by the ability of human populations to gradually, over generations, accrete large bodies of skills and know-how that no individual could ever figure out in one lifetime. Tomasello refers to this process as the "ratchet effect," capturing the idea that each generation can "ratchet up" in know-how from where the last generation left off (Tomasello 1999b). Unfortunately, the ratchet metaphor occludes the fact that groups can lose cultural traits, practices, and know-how in a variety of ways.

Our species' addiction to cultural information has led culture-gene coevolutionary theorists to propose that many aspects of human psychology, anatomy, and physiology are products of an ongoing interaction between culture and genes (Aoki 1986; Laland et al. 2010; Richerson et al. 2010; Henrich 2016). We are a "cultural species," meaning that cultural evolution has driven much of our genetic evolution. For example, the know-how and skills surrounding cooking and fire making are clearly culturally transmitted, at least in part. Yet the length of our colons and the size of our stomachs, teeth, and gape only make sense in a species that has genetically adapted to eating cooked food (Henrich and McElreath 2007; Wrangham and Carmody 2010; Henrich 2016). More broadly, researchers have suggested that this process of cumulative cultural evolution created genetic selection pressures for our long-distance running abilities (e.g., foot anatomy), folkbiological and artifact cognition, "overimitative" tendencies, status psychology (prestige), and verbal mimicry, among other aspects of our species (Henrich 2016).

Applying this theoretical work, we next ask how much cumulative cultural evolution exists in chimpanzees. Since our answer is that little or no cumulative cultural evolution has emerged, we then ask: why so little?

Cumulative cultural evolution creates practices, bodies of manufacturing know-how, and whole behavioral repertoires that no single individual could invent in their lifetime (Tomasello 1999a; Tennie et al. 2009). The question is, then, how much of chimpanzee repertoires could a group of naïve chimpanzees reinvent without any social input? The question is not, can everyone reinvent everything, but can anyone in the group reinvent it?

In the field, the existence of variation in tool-using skills, as discussed with nut cracking, termite fishing, and ant dipping, among populations suggests that practices aren't easily reinvented. This may be the case. However, since these practices have already been shown to be susceptible to ecological pressures, including variation in the relative frequency of certain resources (not just the existence of the resource), it is hard to exclude subtle influences rooted in the economics of various resource distributions and the availability of learning opportunities (Sanz and Morgan 2013). Moreover, few of these variants among local communities have been shown to be stable for long periods. In some cases, we may be looking at ephemeral fluctuations, as practices are repeatedly lost and reinvented over years or decades.

In captivity, this issue has recently been put to the test by giving naïve chimpanzees opportunities to independently invent practices that have been observed in the field. "Leaf swallowing" behavior, a proposed tool against internal parasites, develops fully in naïve chimpanzees (Huffman and Hirata 2004; Huffman et al. 2010; Menzel et al. 2013). Similarly, food washing and food mining behavior—that is, the classic cases of primate culture—also spontaneously reemerge in naïve chimpanzees (Allritz et al. 2013). The same is true for other behaviors and/or for other primates. For example, mountain gorilla nettle feeding behavior—a complex non-tool-use behavior—reappeared in naïve captive gorillas (Tennie et al. 2008; Masi 2011).[13] Similarly, leaf swallowing also occurs in naïve bonobos (Menzel et al. 2013), and both food washing and food mining behaviors reemerge in naïve orangutans (Allritz et al. 2013).

However, tests in captivity on naïve individuals still remain outstanding for most cases of purported cultural traits in wild chimpanzees (e.g., ant dipping). In the case of termite fishing, Lonsdorf and colleagues (2009) ran a test in both naïve enriched captive chimpanzees and gorillas. Both species became proficient at the task, with many chimpanzees and some gorillas engaging with the task and succeeding on Day 1 (on average, chimpanzees learned more quickly than gorillas). Though this study presented the task to the whole group, rather than individually, it nevertheless shows that at the very least one of the apes in each species developed the technique spontaneously, that is, without the need to observe others engage in it. For other behavioral traits, such as nut cracking, research has often not taken seriously the possibility that subjects would develop the target behavior on their own, and thus baseline conditions in which naïve individuals are given time to learn individually have been largely neglected (Sumita et al. 1985; Hayashi et al. 2005). For example, in a paper on the emergence of stone tools, Hayashi et al. (2005) did not include any asocial baseline, instead providing demonstrations even before the first trial.[14] Given that capuchin monkeys develop nut cracking *without* social cues (Visalberghi 1987), it would be surprising if chimpanzees could not also figure it out by themselves.

Recent field evidence converges with this captive work, indicating that chimpanzee behavioral traits can be individually invented and reinvented. Nut cracking, once thought to be locally restricted, has now also been found in chimpanzees living 1,700 km to the east of its originally described occurrence (Morgan and Abwe 2006). Similarly, the most interesting "two-handed" ant dipping techniques likewise appear in several populations (Bossou, Guinea, and Gombe, Tanzania), thousands of kilometers apart (e.g., Yamakoshi and Myowa-Yamakoshi 2004). Finally, termite fishing also occurs in widely disconnected populations of chimpanzees, such as in both Fongoli, Senegal, as well as in Gombe, Tanzania—again, thousands of kilometers apart (e.g., Bogart and Pruetz 2011).

To be clear, we are not arguing that social learning plays no role in these practices. On the contrary, it likely plays a big role in spreading behaviors that are only occasionally reinvented by some individuals (Tennie et al. 2009); for example, nut cracking can spread socially once one individual invents it (Marshall-Pescini and Whiten 2008). Deploying social learning in these cases is adaptive, since these skills are easier to learn using a combination of individual and social learning. For example, after observing subjects who showed a leaf swallowing behavior, others who beforehand resisted reinvention expressed the same behavior themselves (Huffman and Hirata 2004; Huffman et al. 2010; Menzel et al. 2013). Nevertheless, the fact that these behaviors appear in some naïve individuals without any social input means that it is not so complicated or nonintuitive that no single individual can reinvent it in their lifetime. Thus, it is not cumulative culture.[15]

Currently, the best candidate for a cumulative cultural evolutionary product is a particular ant dipping rod used in the Goualougo Triangle, in the Republic of Congo (Sanz et al. 2009; Sanz and Morgan 2011). Using camera traps, Sanz et al. found that chimpanzees in one particular location use several tools in succession to access army ants, with the last one being a stick whose tip has been "brushed" using the chimpanzees' teeth. The brushed tip is more efficient at gathering the target prey than a non-brushed tip. While broadly similar behaviors have been inferred elsewhere (Boesch 2012), Sanz et al. argue that theirs is a cumulative cultural case because the videos show that the chimpanzees brush the tip of the stick even before this tool is used.[16] We concur that, currently, this type of brush tool is the best candidate example for cumulative culture in chimpanzees, though we would not be surprised if it, too, would reappear in naïve subjects.

Finally, recent analyses by Kamilar and Atkinson (2013) of Whiten et al.'s thirty-nine traits, while not showing evidence of cumulative cultural evolution, do reveal a kind of nested structuring of traits, which the authors argue presents a precursor to cumulative cultural evolution. We agree that this could be consistent with a reliance on social learning, but the same patterns could arise from purely individual learning, if learning one trait tends to bias the acquisition of other traits. Since, as we have seen, Whiten et al.'s thirty-nine traits include several versions of different variants (five forms of nut cracking, six types of dipping, and three types of food pounding), it is not hard to see why this might be. Work by Gruber et al. (2011, 2009), for example, shows that prior knowledge of how to "fluid dip" increases an ape's chances of individually figuring out how to "honey dip" in a somewhat novel context. Finally, we also worry that the tendency of different researchers to either split or lump variants into subvarieties may actually account for part of the apparent nested structure.

Research on whether naïve chimpanzees can individually reinvent the various practices found among wild chimpanzees has just begun, so it remains to be seen which practices (if any) prove too difficult. So far, albeit with only a handful of cases, captive apes have readily reinvented the behaviors seen in the field.[17] Moreover, bonobos and gorillas invented traits that their wild brethren don't perform. Whatever the final score turns out to be on cumulative cultural evolution in chimpanzees, the important theoretical point already seems clear: chimpanzees have little (or no) cumulative cultural evolution compared to humans. Thus, at this point, there is no reason to suspect that they have gone down the same (or even a somewhat similar) culture-gene coevolutionary pathway.

Factors Influencing Cumulative Cultural Evolution, or Lack Thereof

What might account for the relative lack of cumulative cultural evolution in chimpanzees? Theoretical work has isolated four areas that influence the emergence and rate of cumulative cultural evolution: (1) individual inventiveness, trial-and-error exploration, or general cognitive abilities; (2) high transmission fidelity via social learning due to cognitive abilities or motivations; (3) sociality (including teaching), network size, and social structures; and (4) a life history with extended periods of brain plasticity and learning. We briefly discuss each of these in turn.[18]

Cognitive abilities for, or motivations to, individually figure out novel practices foster greater cumulative cultural evolution (van Schaik and Pradhan 2003; Henrich 2004b, 2009b; Kobayashi and Aoki 2012). Of all four factors that influence cumulative cultural evolution, we suspect that this one creates the least hindrance for chimpanzees, as well as other apes. Chimpanzees are excellent individual learners and keen explorers—the latter more in captivity (though see Forss et al. 2015). Captive studies show that their cognitive skills related to number, space, and quantities are equivalent to human toddlers (Herrmann et al. 2007), and their working memories are competitive with undergraduates (Inoue and Matsuzawa 2007; Silberberg and Kearns 2009). Chimpanzees can even outcompete children in figuring out the most efficient way to accomplish a task, in part because children slavishly rely on imitation (Nagell et al. 1993; Horner and Whiten 2005). Moreover, field studies have repeatedly shown that while wild chimpanzees often invent novel behaviors, these novelties are not picked up by others, and eventually peter out (Biro et al. 2003; Nishida et al. 2009; O'Malley et al. 2012). This is not to say that chimpanzees' cognitive skills and motivations are sufficient for humanlike cumulative cultural evolution, merely that it is not a show stopper for getting the process started. This is underlined by theoretical work showing that individual smarts are often relatively less important for generating cumulative cultural evolution than sociality and transmission fidelity (Henrich 2009b; Kobayashi and Aoki 2012; Lewis and Laland 2012; Pradhan et al. 2012).

The *fidelity* of social learning is a different story. While arguments about the details, categories, and classifications of various forms of chimpanzee social learning are not settled, a vast body of experimental work shows that chimpanzee social learning is generally of lower fidelity than human social learning (Tennie et al. 2009; Whiten et al. 2009). Notably, theoretical work shows that transmission fidelity is crucial for cumulative cultural evolution (Henrich 2004b, 2009b; Kobayashi and Aoki 2012; Lewis and Laland 2012; Pradhan et al. 2012; Kolodny et al. 2015). In many direct comparisons of humans and chimpanzees, the children are near ceiling and the apes near floor in performance—at least with regard to action copying (Nagell et al. 1993; Whiten et al. 1996; Call et al. 2005; Herrmann et al. 2007; Tennie et al. 2010b). Thus, compared with humans, chimpanzees are worse at copying motor patterns (Tomasello and Call 1997; Tennie et al. 2012), but also at inferring underlying goals, strategies, and motivations (Tennie et al. 2010a; Dean et al. 2012), and especially poor at actively transmitting them (teaching). Chimpanzees copy less frequently and usually require clear incentives to do any copying—and even then, their copying is very restricted (Tennie et al. 2012). Meanwhile, children are "imitation machines" (Tomasello 1999a), copying automatically, unconsciously, and persistently (Bandura 1977; Nielsen and Tomaselli 2010). Consistent with this, recent neuroimaging studies have found major deficits in brain structures enabling detailed action copying in chimpanzees[19] relative to humans (Hecht et al. 2013). These action-copying deficits can result in an effective blocking of certain types of cultural evolution, namely those that depend on the transmission of action styles (e.g., dance, sign language)—but they can also have detrimental effects on the overall copying fidelity of tasks that additionally involve other types of information (Acerbi and Tennie 2016).

Another relevant element may be the degree to which learners rely on their own intuitions and experience over information gleaned from social learning—the *informational conformity* mentioned earlier. In humans, various forms of "overimitation," which involve copying apparently unnecessary steps, are a persistent and potent feature of social learning (Lyons et al. 2007; Nielsen and Tomaselli 2010; Herrmann et al. 2013). By contrast, chimpanzees readily drop any unnecessary steps once they perceive that specific steps are superfluous (Horner and Whiten 2005)—including steps consisting of specific actions (Tennie et al. 2012). Overall, most analyses of chimpanzee data strive to detect a transmission fidelity above zero. But only in a few cases involving trivially easy tasks does the data support a fidelity comparable to humans (Hopper et al. 2008).

Cultural evolutionary models also show how *sociality* influences the emergence and rate of cumulative cultural evolution (Henrich 2016; Muthukrishna and Henrich 2016). In short, the larger and more interconnected populations are, the more likely the emergence of cumulative cultural evolution is, and the faster the rate if it does emerge. In humans, these predictions have been tested using a combination of laboratory experiments (Derex et al. 2013; Muthukrishna et al. 2014), field studies (Kline and Boyd 2010;

Collard et al. 2013), and ethno-historical cases (Henrich 2004b; Boyd et al. 2011). Strikingly, when populations suddenly shrink or become disconnected from larger social networks, they begin to lose complex technologies over generations. Overall, growing up in a larger, more interconnected network gives people access to more models to select among and learn from.

From this perspective, chimpanzees and other apes have several strikes against them. First, the fission-fusion social structure of chimpanzees and their overall group size means that young chimpanzees have access to only a very limited range of potential models. For the most part, they can access only their mothers, and essentially never get to access individuals from other residential groups. To help them acquire nut cracking skills, for example, the percentage of time that infants have access to models beyond the mother increases from 0 percent at six months of age to a mere 10 percent at age 3.5 years. When given the opportunity, young chimpanzees do attend to others besides their mother, but they just do not get many opportunities (Lonsdorf 2013). By contrast, human foragers live enmeshed in vast social webs that network together hundreds or thousands of people across many residential groups (Wiessner 2002; Henrich and Broesch 2011; Hill et al. 2011, 2014; Apicella et al. 2012; Salali et al. 2016).

Second, from a life history perspective, the intersection of broadening opportunities for social learning and developmental timing of learning windows may be crucial. Infant chimpanzees wean at about age four to five years, after which time they begin interacting in a wider social circle (though still sticking relatively close to their mother for several more years). But the developmental window on learning to nut crack, ant dip, and termite fish seems to narrow around age five to six (Inoue-Nakamura and Matsuzawa 1997; Biro et al. 2003; Marshall-Pescini and Whiten 2008; Lonsdorf 2013). This means that there may only be a short time when young chimpanzees are developmentally ready and able to learn these (and presumably other) skills *and* able to access a broad range of models. Note that, as in humans, we don't expect these windows to entirely shut, but merely to narrow: some flexibility is retained into adulthood in, for example, what type of hammer (wood or stone) to use for cracking nuts (Luncz and Boesch 2014).

This suggests that part of the secret of human cumulative cultural evolution may lie in creating a situation in which learners can access a broad range of models while their brains remain highly plastic (Henrich 2008, 2016; Muthukrishna and Henrich n.d.), and the developmental window for many skills remains open. This implies a different form of social organization and a different life history, one that adds middle childhood and adolescence (Bogin 2009). Moreover, humans retain much greater brain plasticity into adulthood compared to chimpanzees (Miller et al. 2012), and they have longer lives, which gives them more time to meet and learn from a broader range of individuals. This has an effect similar to increasing group size or social interconnectedness.

The importance of group size in the creation of large behavioral repertoires, including tools, may help explain why bonobos and gorillas do not show the repertoires seen in chimpanzees and orangutans (Tennie et al. 2009; Henrich 2016) despite showing substantial individual-level cognitive abilities in captivity. Realize first that, limited as they are, current analyses of the available data from both chimpanzees and orangutans reveal that larger or more socially connected populations have more extensive behavioral repertoires (van Schaik et al. 2003; Lind and Lindenfors 2010), though in chimpanzees it is the number of females that matters. Gorillas tend to live in small groups with only a silverback, his mates, and one or two other males. The particulars of their social life also seem less suited for social transmission than those of chimpanzees (Lonsdorf et al. 2009; Robbins et al. 2016). Meanwhile, though bonobos live in larger groups (mean size twenty-three), their average group size is half that of chimpanzees (mean size forty-six).[20] By expanding the size of the cultural repertoires of chimpanzees, group size differences may have generated more tools and techniques, thereby precipitating a genetic response that led to greater object-focused *individual learning* and exploration in chimpanzees (Koops et al. 2015).

The prosociality of potential models is the final important element—for example, teaching is a form of altruism that greatly facilitates the evolution of culture (Tennie et al. 2009; Dean et al. 2012). There is now a substantial literature comparing the sociality of chimpanzees to that of humans, including both children and adults. No matter how you look at the comparative data, humans are much more prosocial across a wide range of circumstances than chimpanzees. As with imitation studies, the issue is never whether chimpanzees are as prosocial as humans (or more prosocial), but only whether non-zero levels of prosociality can be detected, and what lengths researchers go to in order to pry any prosociality out of these apes (Henrich 2004c; Silk et al. 2005; Jensen et al. 2006, 2007a, 2007b; Warneken et al. 2006; Warneken and Tomasello 2006; Vonk et al. 2008; Brosnan et al. 2009; Silk and House 2011; House et al. 2012, 2013; Henrich and Silk 2013; Tennie et al. 2016).

Experimental work described above highlights how important sociality is for cumulative cultural evolution. Dean et al. (2012) not only show that children teach and act prosocially toward each other and chimpanzees do not, but that children's success in acquiring a multistep procedure was associated with their willingness to actively assist and reward each other. The lack of any chimpanzee teaching, assisting, or rewarding in this experiment is consistent with most field observations.

The culture-gene coevolutionary approach predicts that forms of social organization, life history, and prosociality (including teaching) may be as much a consequence of cumulative cultural evolution as its cause (Burkart et al. 2009; Chudek and Henrich 2010; van Schaik and Burkart 2011; van Schaik et al. 2012; Henrich 2016). However, any ape species that for unrelated reasons had a form of social organization, prosocial motivations, or life history more conducive to cumulative cultural evolution would have had an advantage in crossing the threshold into a regime of culture-driven genetic evolution.

Jump-Starting Cumulative Cultural Evolution

Cultural evolutionary theorists have identified what they call the "start-up problem," which aims to explain why something as seemingly valuable to survival and reproduction as cumulative cultural evolution is so rare in nature (Boyd and Richerson 1996; Henrich 2016). The core of the idea is that cumulative cultural evolution drove human brain expansion, selecting for bigger brains and longer juvenile periods to facilitate acquiring, storing, and organizing vast amounts of cultural know-how (Boyd et al. 2011; Muthukrishna and Henrich n.d.). The more cultural know-how accumulates in the form of adaptive practices, the stronger the selective pressures are for brains capable of acquiring all that know-how from the minds of others. To see the challenge in starting this process, first realize that big, powerful (and energetically expensive) brains capable of sophisticated high-fidelity social learning can only pay for themselves if there is highly adaptive information already out there in the minds of others, perhaps in the form of numerous valuable practices related to tool making and food processing, waiting to be learned. Once there is a lot of complex cultural information in the world, natural selection has no choice but to favor brains that are better at acquiring, organizing, and storing this information. However, in the beginning, before cumulative cultural evolution got going, there would not have been very much out there, in terms of valuable practices, in the minds and behavior of others. What there was could have been discovered on one's own, using individual learning (Tennie et al. 2009). One might think that a little culture can accumulate, and natural selection will incrementally favor bigger brains that are better at cultural learning. However, the problem is that natural selection faces a choice between investing either in brains that are better at individual learning *or* social learning. Either you are spending your time engaged in trial-and-error experimentation or you are watching and hanging around others. Early on, individual learning will often be favored by natural selection because not only does improved individual learning increase one's chances of figuring stuff out on one's own, it also improves some simple forms of social learning (e.g., if you hang around nut crackers, you tend to be around nuts and anvils more, so improved individual learning focused on objects increases your chances of figuring out how those nuts and anvils go together). However, when natural selection invests in individual learning, it inhibits cumulative cultural evolution.

To bypass the start-up problem, Henrich (2016) has recently suggested that human ancestors may have experienced ecological conditions favorable to creating cumulative cultural evolution without an initial change in social learning abilities. In the Late Pliocene, fluctuating environmental conditions could have favored greater social learning (as predicted by theory; see Richerson and Boyd 2000), and a much larger predator guild in Africa would have forced terrestrial primates into larger social groups. As just noted, much theory suggests that larger and more interconnected groups will experience greater cultural accumulations, as will groups more reliant on social learning. These two factors would have enlarged the sizes of learned repertoires of these primates, potentially shifting the balance of costs and benefits in favor of investing specifically in social learning abilities over individual learning. Henrich also argues that larger groups, induced by territoriality and predation, would have favored greater pair-bonding (Langergraber et al. 2009), which could have expanded the circle of identifiable kin and the potential for alloparenting. Greater alloparenting by fathers, aunts, and grandmothers would have permitted longer juvenile periods and more opportunities for teaching—which would have further fueled cumulative cultural evolution.

Early cultural evolution would not have involved the continuous improvements in technical know-how and skills that many paleoanthropologists seem to expect (Henrich 2016). Instead, it would have had many fits and starts, with some groups occasionally surging ahead and other groups losing tools and know-how.[21] This is because both the size of tool kits and their complexity is heavily influenced by the size and sociality of groups. Environmental shocks, climatic fluctuations, and migrations would have consistently set groups back in cultural complexity. In light of such theoretical insights, Henrich argues that the oldest tool assemblages (before about 2 Ma) were likely not static, but instead stood on the precipice of cumulative cultural evolution, and reveal a diversity consistent with repeated gains and losses. After about two million years ago, a pattern of cumulative cultural evolution does begin to emerge, at least in some populations. By 750,000 years ago, based on findings at Gesher Benot Ya'aqov, Henrich argues that some populations were clearly reliant on diverse behavior repertoires that, taken together, no single individual could reinvent in their lifetime. These repertoires may also have included individual tools or techniques (e.g., perhaps hand axes) that represented true cumulative cultural evolution.[22]

Inferences to the Last Common Ancestor

What can we say about cultural evolution in our last common ancestor with chimpanzees and bonobos? Cultural evolutionary theory predicts that the selection pressures for more sophisticated forms of social learning will increase with environmental variability (Boyd and Richerson 1985). Empirically, the available data from lake and ice cores suggest that after about three million years ago, paleoclimates increased in variability, plausibly on time scales favoring social learning (Richerson et al. 2005). High levels of variability continued until about 10,000 years ago. This combination of theory and evidence suggests that *after* humans and chimpanzees split from the LCA, climatic changes may have increased selection pressures for social learning in both lineages, as well as in other taxa.

This view is consistent with the argument that selection for social learning or behavioral flexibility drove the expansion of brains in several taxa, including in both primates and birds (Reader and Laland 2002; Reader et al. 2011). Various measures of brain size are correlated with both social learning and innovation, and brains appear to have expanded across many taxa over several million years. Indeed, culture-gene coevolutionary simulations can reproduce the extant empirical relations observed across species, between group size and brain size, between brain size and juvenile period, and between social learning, innovation, and brain size (Muthukrishna and Henrich 2016, n.d.).

The upshot of this is that chimpanzees likely set an upper boundary for the social learning abilities, traditions, and culturally evolved patterns that we might expect in the LCA. The LCA likely had some social learning abilities, as we observe in other apes, and these abilities probably help foster community differences in local traditions. But, unlike in humans, these traditions were probably only ephemeral—subject to periodic disappearances and reinventions over time. There was likely little scope for culture-driven genetic evolution.

Summary

We have applied a culture-gene coevolutionary approach to compare humans to one of our closest living relatives. In doing so, we pointed out a few potential theoretical pitfalls—which we aimed to avoid—and focused instead on what we argue to be the most ecologically valid type of data: namely that derived from either wild chimpanzees, or from what we call "enriched captive apes." Focusing in this way illuminated some clear differences between the two species: humans, in contrast to chimpanzees, show higher fidelity in their social learning, clear evidence for teaching, and a wider range of adaptive social learning biases (e.g., conformist transmission). In addition, human social structures, effective population sizes, and life history traits are all much better suited for cumulative cultural evolution than their equivalents in chimpanzees. Perhaps resulting from these differences, there is currently no clear evidence for cumulative cultural evolution in chimpanzee populations. Instead, these behaviors seem to be (re)inventable at the individual level. This line of research suggests that small initial differences between proto-humans and proto-chimpanzees after the LCA led our lineage into a regime of culture-driven genetic evolution that has not been experienced by any other ape, including chimpanzees. Since cultural evolution created a series of autocatalytic selective pressures on the genes in our lineage (e.g., due to cultural products such as fire, cooking, projectile weapons, cutting edges, and plant knowledge), humans have diverged much further from the LCA than chimpanzees. The human species alone has become "addicted to culture."

Endnotes

1. For the earliest model of the genetic evolution of a cultural learning ability, see Feldman and Cavalli-Sforza (1976).
2. For chimpanzee observers, the species of the demonstrator does not seem to have a significant impact on the outcome (Boesch 2007; Marshall-Pescini and Whiten 2008; Dean et al. 2012). For humans, the impact of other-species demonstrators has not been explored, though efforts to co-rear humans and chimpanzees suggest that infants and young children will readily copy older, more physically skilled apes (Kellogg and Kellogg 1933; Henrich and McElreath 2003).
3. Not everyone agrees: Boesch (2012) argues that human training and high human exposure are substitutes for the rich environments found in the wild. Perhaps. But there is no data to support this. In contrast, there are clear differences between these two populations. For example, finger pointing is quite common in enculturated chimpanzees but extremely rare in wild chimpanzees (Leavens et al. 2010).
4. Drawing terminology from psychology, dual inheritance theorists have long made and explored the distinction between *informational* and *normative* conformity (Boyd and Richerson 1985: 224; Henrich and Boyd 2001: 81; Henrich and McElreath 2007; Henrich and Henrich 2007: 22–27; Chudek and Henrich 2010), though see Claidière and Whiten (2012) for an incorrect claim to the contrary.
5. Nevertheless, for evidence against normative conformity in chimpanzees, see Haun et al. (2014).
6. The labels "experienced" and "inexperienced" are ours. The authors of this study interpret their findings as showing the effects of "prestige" cues on social learning, testing the Dominance-Prestige Theory (Henrich and Gil-White 2001). Unfortunately, this experiment cannot test this idea since their potential models are distinguished by many cues, including age, experience, competence, dominance, and past success. What the authors do show is that chimpanzees will continue selectively copying those they have copied in the past. By contrast, young children track others' visual attention (a carefully manipulated "prestige cue") and preferentially attend to and learn from those who are watched more by others (Chudek et al. 2012). Similarly, adults copy those who have been imitated more in the past by others (Atkisson et al. 2012), independent of other factors.
7. Data on wild great tits is consistent with conformist transmission (Aplin et al. 2015), though other important alternative interpretations exist for these findings (van Leeuwen et al. 2015; Acerbi et al. 2016).

8. Whiten et al. (2005) imply they found conformist transmission by using the term "conformity bias" and citing Richerson and Boyd (2005), who only discuss conformist transmission biases.
9. We avoid mentalistic approaches to teaching in order to facilitate comparisons across species (Kline 2015).
10. Arguably, from the perspective of small-scale societies, Westerners have to teach so much because they begin transmitting before learners are ready—in terms of maturation—to learn things on their own.
11. These otherwise consistent patterns, showing no teaching, contradict earlier work on nut cracking at Taï forest (Boesch 1991, 2012). Aside from two anecdotes (see Maestripieri 1995 for a critique), much of the seeming discrepancy comes from whether the young chimpanzees were "stealing" hammers and nuts from their mother, with her tolerating it, or whether she was actively "giving" the hammers and nuts. What all three of the other research team coded as "stealing" and "scrounging" appears to have been coded as the mother "giving" by Boesch (Lonsdorf 2013). Similarly, a recent study of wild chimpanzees argues for "teaching" via the transfer of a needed tool from mother to juveniles in the context of termite fishing (Musgrave et al. 2015, 2016)—but here, too, the youngsters and tool recipients were the active force in the transfer of these tools. Most importantly, even though the recipients showed increased rates of termite fishing following tool transfers, it is unclear whether the development of the underlying technique required, or even benefited, from these transfers.
12. For more on the debate see Langergraber et al. (2011) and Langergraber and Vigilant (2011).
13. But see also Byrne et al. (2011).
14. In an earlier study (Sumita et al. 1985), five chimpanzees were tested, but four of the five subjects only received a single baseline session (of about one hour each).
15. Tennie et al. (2009) labeled such behaviors "latent solutions."
16. Since some reports only relied on the tool descriptions without having actually seen the behavior being performed (Boesch et al. 2009), there is a possibility that in other places brush/fray tools are also produced prior to usage. Indeed, Boesch (2012: 132) claims that most other tool modifications in Taï chimpanzees are made prior to use. Thus, modifications prior to use may actually be common in chimpanzees.
17. Incidentally, this also fits great ape gestures and vocalizations; see Slocombe and Scott-Philips (this volume) for an overview. Thus, great ape communication does not seem to require high-fidelity social transmission, either.
18. Some argue that language explains cumulative culture. While language certainly increases the fidelity and volume of cultural transmission in some domains, languages are themselves clearly the product of cumulative cultural evolution. So pointing to language to explain culture would be like pointing to archery technology to explain hunting success. Language likely first developed as a consequence of cumulative cultural evolution before it began increasing

transmission fidelity and fostering further cumulative cultural evolution (Henrich 2016: chapter 13).
19. Importantly (as far as we know), these data stem from enriched captive chimpanzees.
20. These averages were generated using the data in Wilson et al. (2014).
21. In any case, recent findings suggest that anatomy, especially hand anatomy, was not the bottleneck for early stone tool production (e.g., see Rolian and Carvalho, this volume)
22. The pathway to full-blown cumulative cultural evolution, in which individuals acquire specific skills, techniques, or bodies of know-how that no single individual could figure out in their lifetime, has a number of less sophisticated way stations along the way (Dean et al. 2014). For example, individuals may acquire a diverse repertoire of different skills, techniques, and bits of knowledge that they would be unlikely to assemble on their own despite the fact that they could individually figure out each trait in the set. More specifically, suppose a total repertoire consisted of ten different tool-using and tool-making skills, and each individual has a 30 percent chance of figuring out each skill on his or her own. This, nevertheless, makes the chances of figuring out all ten skills by themselves vanishingly small, at 0.00006 percent. Thus, social learning skills can play a big role in building up the cultural aggregations before tipping into full-blown cumulative cultural evolution; see Henrich (2016) for further discussion.

References

Acerbi, A., and C. Tennie. 2016. The role of redundant information in cultural transmission and cultural stabilization. *Journal of Comparative Psychology* 130: 62–70.

Acerbi, A., E. J. C. van Leeuwen, D. B. M. Haun, and C. Tennie. 2016. Conformity cannot be identified based on population-level signatures. *Scientific Reports* 6: 36068.

Allritz, M., C. Tennie, and J. Call. 2013. Food washing and placer mining in captive great apes. *Primates* 54: 361–370.

Aoki, K. 1986. A stochastic-model of gene culture coevolution suggested by the culture historical hypothesis for the evolution of adult lactose absorption in humans. *Proceedings of the National Academy of Sciences* 83: 2929–2933.

Aoki, K., and M. W. Feldman. 2014. Evolution of learning strategies in temporally and spatially variable environments: A review of theory. *Theoretical Population Biology* 91: 3–19.

Apicella, C. L., F. W. Marlowe, J. H. Fowler, and N. A. Christakis. 2012. Social networks and cooperation in hunter-gatherers. *Nature* 481: 497–501.

Aplin, L. M., D. R. Farine, J. Morand-Ferron, A. Cockburn, A. Thornton, and B. C. Sheldon. 2015. Experimentally induced innovations lead to persistent culture via conformity in wild birds. *Nature* 518: 538–541.

Atkisson, C., M. J. O'Brien, and A. Mesoudi. 2012. Adult learners in a novel environment use prestige-biased social learning. *Evolutionary Psychology* 10: 519–537.
Bandura, A. 1977. *Social Learning Theory*. Oxford: Prentice-Hall.
Baron, R., J. Vandello, and B. Brunsman. 1996. The forgotten variable in conformity research: Impact of task importance on social influence. *Journal of Personality and Social Psychology* 71: 915–927.
Bell, A. V., P. J. Richerson, and R. McElreath. 2009. Culture rather than genes provides greater scope for the evolution of large-scale human prosociality. *Proceedings of the National Academy of Sciences* 106: 17671–17674.
Benedetti, F., W. Thoen, C. Blanchard, S. Vighetti, and C. Arduino. 2013. Pain as a reward: Changing the meaning of pain from negative to positive co-activates opioid and cannabinoid systems. *Pain* 154: 361–367.
Billing, J., and P. W. Sherman. 1998. Antimicrobial functions of spices: Why some like it hot. *Quarterly Review of Biology* 73: 3–49.
Biro, D., N. Inoue-Nakamura, R. Tonooka, G. Yamakoshi, C. Sousa, and T. Matsuzawa. 2003. Cultural innovation and transmission of tool use in wild chimpanzees: Evidence from field experiments. *Animal Cognition* 6: 213–223.
Bjork, A., W. Liu, J. O. Wertheim, B. H. Hahn, and M. Worobey. 2011. Evolutionary history of chimpanzees inferred from complete mitochondrial genomes. *Molecular Biology and Evolution* 28: 615–623.
Bjorklund, D. F., J. L. Yunger, J. M. Bering, and P. Ragan. 2002. The generalization of deferred imitation in enculturated chimpanzees *(Pan troglodytes)*. *Animal Cognition* 5: 49–58.
Boesch, C. 1991. Teaching among wild chimpanzees. *Animal Behaviour* 41: 530–532.
Boesch, C. 2007. What makes us human *(Homo sapiens)*? The challenge of cognitive cross-species comparison. *Journal of Comparative Psychology* 121: 227–240.
Boesch, C. 2012. *Wild Cultures: A Comparison between Chimpanzee and Human Cultures*. New York: Cambridge University Press.
Boesch, C., J. Head, and M. M. Robbins. 2009. Complex tool sets for honey extraction among chimpanzees in Loango National Park, Gabon. *Journal of Human Evolution* 56: 560–569.
Bogart, S. L., and J. D. Pruetz. 2008. Ecological context of savanna chimpanzee *(Pan troglodytes verus)* termite fishing at Fongoli, Senegal. *American Journal of Primatology* 70: 605–612.
Bogart, S. L., and J. D. Pruetz. 2011. Insectivory of savanna chimpanzees *(Pan troglodytes verus)* at Fongoli, Senegal. *American Journal of Physical Anthropology* 145: 11–20.
Bogin, B. 2009. Childhood, adolescence, and longevity: a multilevel model of the evolution of reserve capacity in human life history. *American Journal of Human Biology* 21: 567–577.
Bonnie, K. E., V. Horner, A. Whiten, and F. B. M. de Waal. 2007. Spread of arbitrary conventions among chimpanzees: A controlled experiment. *Proceedings of the Royal Society B: Biological Sciences* 274: 367–372.

Boyd, R., and P. J. Richerson. 1985. *Culture and the Evolutionary Process*. Chicago: University of Chicago Press.

Boyd, R., and P. J. Richerson. 1988. An evolutionary model of social learning: The effects of spatial and temporal variation. In T. R. Zentall and B. G. Galef, eds., *Social Learning: Psychological and Biological Perspectives*, 29–48. Hillsdale, NJ: Lawrence Erlbaum Associates.

Boyd, R., and P. J. Richerson. 1995. Why does culture increase human adaptability. *Ethology and Sociobiology* 16: 125–143.

Boyd, R., and P. J. Richerson. 1996. Why culture is common, but cultural evolution is rare. *Proceedings of the British Academy* 88: 77–93.

Boyd, R., P. J. Richerson, and J. Henrich. 2011. The cultural niche: Why social learning is essential for human adaptation. *Proceedings of the National Academy of Sciences* 108: 10918–10925.

Brosnan, S. F., J. B. Silk, J. Henrich, M. C. Mareno, S. P. Lambeth, and S. J. Schapiro. 2009. Chimpanzees *(Pan troglodytes)* do not develop contingent reciprocity in an experimental task. *Animal Cognition* 12: 587–597.

Burkart, J. M., S. B. Hrdy, and C. P. van Schaik. 2009. Cooperative breeding and human cognitive evolution. *Evolutionary Anthropology* 18: 175–186.

Buttelmann, D., N. Zmyj, M. M. Daum, and M. Carpenter. 2012. Selective imitation of in-group over out-group members in 14-month-old infants. *Child Development.* 84: 422–428.

Byrne, R. W. 2007. Culture in great apes: Using intricate complexity in feeding skills to trace the evolutionary origin of human technical prowess. In N. Emery, N. Clayton, and C. Frith, eds., *Social Intelligence: From Brain to Culture*. Oxford: Oxford University Press.

Byrne, R. W., C. Hobaiter, and M. Klailova. 2011. Local traditions in gorilla manual skill: Evidence for observational learning of behavioral organization. *Animal Cognition* 14: 683–693.

Call, J., M. Carpenter, and M. Tomasello. 2005. Copying results and copying actions in the process of social learning: Chimpanzees *(Pan troglodytes)* and human children *(Homo sapiens)*. *Animal Cognition* 8: 151–163.

Campbell, D. T. 1965. Variation and selective retention in socio-cultural evolution. In H. R. Barringer, G. I. Glanksten, and R. W. Mack, eds., *Social Change in Developing Areas: A Reinterpretation of Evolutionary Theory*, 19–49. Cambridge, MA: Schenkman.

Caro, T. M., and M. D. Hauser. 1992. Is there teaching in nonhuman animals? *Quarterly Review of Biology* 67: 151–174.

Castro, L., and M. A. Toro. 2002. Cultural transmission and the capacity to approve or disapprove of offspring's behavior. *Journal of Memetics* 6.

Cavalli-Sforza, L. L., and M. W. Feldman. 1973. Models for cultural inheritance, 1: Group mean and within group variation. *Theoretical Population Biology* 4: 42–55.

Cavalli-Sforza, L. L., and M. W. Feldman. 1981. *Cultural Transmission and Evolution*. Princeton, NJ: Princeton University Press.

Chiao, J. Y., and K. D. Blizinsky. 2010. Culture-gene coevolution of individualism-collectivism and the serotonin transporter gene. *Proceedings of the Royal Society B: Biological Sciences* 277: 529–537.

Chudek, M., P. Brosseau, S. Birch, and J. Henrich. 2013. Culture-gene coevolutionary theory and children's selective social learning. In M. Banaji and S. Gelman, eds., *The Development of Social Cognition*. New York: Oxford University Press.

Chudek, M., S. Heller, S. Birch, and J. Henrich. 2012. Prestige-biased cultural learning: Bystander's differential attention to potential models influences children's learning. *Evolution and Human Behavior* 33: 46–56.

Chudek, M., and J. Henrich. 2010. Culture-gene coevolution, norm-psychology, and the emergence of human prosociality. *Trends in Cognitive Sciences* 15: 218–226.

Claidière, N., and D. Sperber. 2009. Imitation explains the propagation, not the stability of animal culture. *Proceedings of the Royal Society B: Biological Sciences* 277: 651–659.

Claidière, N., and A. Whiten. 2012. Integrating the study of conformity and culture in humans and nonhuman animals. *Psychological Bulletin* 138: 126–145.

Collard, M., A. Ruttle, B. Buchanan, and M. J. O'Brien. 2013. Population size and cultural evolution in nonindustrial food-producing societies. *PLoS ONE* 8: e72628.

Corp, N., and R. Byrne. 2002. The ontogeny of manual skill in wild chimpanzees: Evidence from feeding on the fruit of *Saba florida*. *Behaviour* 139: 137–168.

Corriveau, K., and P. L. Harris. 2009a. Choosing your informant: Weighing familiarity and recent accuracy. *Developmental Science* 12: 426–437.

Corriveau, K., and P. L. Harris. 2009b. Preschoolers continue to trust a more accurate informant 1 week after exposure to accuracy information. *Developmental Science* 12: 188–193.

Corriveau, K. H., K. D. Kinzler, and P. L. Harris. 2013. Accuracy trumps accent in children's endorsement of object labels. *Developmental Psychology* 49: 470–479.

Corriveau, K. H., K. Meints, and P. L. Harris. 2009. Early tracking of informant accuracy and inaccuracy. *British Journal of Developmental Psychology* 27: 331–342.

Craig, K. D. 1986. Social modeling influences: Pain in context. In R. A. Sternbach, ed., *The Psychology of Pain*. New York: Raven Press.

Craig, K. D., and K. M. Prkachin. 1978. Social modeling influences on sensory decision-theory and psychophysiological indexes of pain. *Journal of Personality and Social Psychology* 36: 805–815.

Csibra, G., and G. Gergely. 2006. Social learning and social cognition: The case of pedagogy. In M. H. Johnson and Y. Munakata, eds., *Processes of Change in Brain and Cognitive Development. Attention and Performance XXI,* 249–274. Oxford: Oxford University Press.

Csibra, G., and G. Gergely. 2009. Natural pedagogy. *Trends in Cognitive Sciences* 13: 148–153.

Currie, T. E., and R. Mace. 2009. Political complexity predicts the spread of ethnolinguistic groups. *Proceedings of the National Academy of Sciences* 106: 7339–7344.

Dean, L. G., R. L. Kendal, S. J. Schapiro, B. Thierry, and K. N. Laland. 2012. Identification of the social and cognitive processes underlying human cumulative culture. *Science* 335: 1114–1118.

Dean, L. G., G. L. Vale, K. N. Laland, E. Flynn, and R. L. Kendal. 2014. Human cumulative culture: A comparative perspective. *Biological Reviews* 89: 284–301.

Dediu, D., and D. R. Ladd. 2007. Linguistic tone is related to the population frequency of the adaptive haplogroups of two brain size genes, ASPM and Microcephalin. *Proceedings of the National Academy of Sciences* 104: 10944–10949.

Derex, M., M.-P. Beugin, B. Godelle, and M. Raymond. 2013. Experimental evidence for the influence of group size on cultural complexity. *Nature* 503: 389–391.

Diamond, J. M. 1997. *Guns, Germs, and Steel: The Fates of Human Societies.* New York: W. W. Norton.

Durham, W. 1991. *Coevolution: Genes, Culture, and Human Diversity.* Stanford, CA: Stanford University Press.

Edgerton, R. B. 1992. *Sick Societies: Challenging the Myth of Primitive Harmony.* New York: Free Press.

Efferson, C., R. Lalive, and E. Fehr. 2008a. The coevolution of cultural groups and in-group favoritism. *Science* 321: 1844–1849.

Efferson, C., R. Lalive, P. J. Richerson, R. McElreath, and M. Lubell. 2008b. Conformists and mavericks: The empirics of frequency-dependent cultural transmission. *Evolution and Human Behavior* 29: 56–64.

Feldman, M. W., and L. L. Cavalli-Sforza. 1976. Cultural and biological evolutionary processes: Selection for a trait under complex transmission. *Theoretical Population Biology* 9: 238–259.

Feldman, M. W., and K. N. Laland. 1996. Gene-culture coevolutionary theory. *Trends in Ecology & Evolution* 11: 453–457.

Ferdowsian, H. R., D. L. Durham, C. Kimwele, G. Kranendonk, E. Otali, T. Akugizibwe, J. B. Mulcahy, L. Ajarova, and C. M. Johnson. 2011. Signs of mood and anxiety disorders in chimpanzees. *PLoS ONE* 6: e19855.

Fiske, A. P. 1998. *Learning a Culture the Way Informants Do: Observing, Imitating, and Participating.* Washington, DC: American Anthropological Association.

Fogarty, L., P. Strimling, and K. N. Laland. 2011. The evolution of teaching. *Evolution* 65: 2760–2770.

Forss, S. I. F., C. Schuppli, D. Haiden, N. Zweifel, and C. P. van Schaik. 2015. Contrasting responses to novelty by wild and captive orangutans. *American Journal of Primatology* 77: 1109–1121.

Furlong, E. E., K. J. Boose, and S. T. Boysen. 2008. Raking it in: The impact of enculturation on chimpanzee tool use. *Animal Cognition* 11: 83–97.

Galef, B. G. 1992. The question of animal culture. *Human Nature* 3: 157–178.

Galef, B. G. 2009a. Culture in animals? In K. N. Laland and B. G. Galef, eds., *The Question of Animal Culture,* Cambridge, MA: Harvard University Press.

Galef, B. G. 2009b. Strategies for social learning: Testing predictions from formal theory. *Advances in the Study of Behavior* 39: 117–151.

Galef, B. G., K. E. Dudley, and E. E. Whiskin. 2008. Social learning of food preferences in "dissatisfied" and "uncertain" Norway rats. *Animal Behaviour* 75: 631–637.

Galef, B. G., and E. E. Whiskin. 2008. "Conformity" in Norway rats? *Animal Behaviour* 75: 2035–2039.

Gaskins, S., and R. Paradise. 2010. Learning through observation in daily life. In D. F. Lancy, J. Bock, and S. Gaskins, eds., *The Anthropology of Learning in Childhood,* 85–117. Lanham, MD: Rowman & Littlefield.

Gruber, T., M. N. Muller, V. Reynolds, R. Wrangham, and K. Zuberbuhler. 2011. Community-specific evaluation of tool affordances in wild chimpanzees. *Scientific Reports* 1: 128.

Gruber, T., M. N. Muller, P. Strimling, R. Wrangham, and Z. Zuberbuhler. 2009. Wild chimpanzees rely on cultural knowledge to solve an experimental honey acquisition task. *Current Biology* 19: 1806–1810.

Harris, P. L., and K. H. Corriveau. 2011. Young children's selective trust in informants. *Philosophical Transactions of the Royal Society B: Biological Sciences* 366: 1179–1187.

Haun, D. B. M., Y. Rekers, and M. Tomasello. 2012. Majority-biased transmission in chimpanzees and human children, but not orangutans. *Current Biology* 22: 727–731.

Haun, D. B. M., Y. Rekers, and M. Tomasello. 2014. Children conform to the behavior of peers: Other great apes stick with what they know. *Psychological Science* 25: 2160–2167.

Hayashi, M., Y. Mizuno, and T. Matsuzawa. 2005. How does stone-tool use emerge? Introduction of stones and nuts to naive chimpanzees in captivity. *Primates* 46: 91–102.

Hecht, E. E., D. A. Gutman, T. M. Preuss, M. M. Sanchez, L. A. Parr, and J. K. Rilling. 2013. Process versus product in social learning: Comparative diffusion tensor imaging of neural systems for action execution-observation matching in macaques, chimpanzees, and humans. *Cerebral Cortex* 23: 1014–1024.

Henrich, J. 2001. Cultural transmission and the diffusion of innovations: Adoption dynamics indicate that biased cultural transmission is the predominate force in behavioral change and much of sociocultural evolution. *American Anthropologist* 103: 992–1013.

Henrich, J. 2004a. Cultural group selection, coevolutionary processes and large-scale cooperation. *Journal of Economic Behavior & Organization* 53: 3–35.

Henrich, J. 2004b. Demography and cultural evolution: Why adaptive cultural processes produced maladaptive losses in Tasmania. *American Antiquity* 69: 197–214.

Henrich, J. 2004c. Inequity aversion in capuchins? *Nature* 428: 139.

Henrich, J. 2008. A cultural species. In M. Brown, ed., *Explaining Culture Scientifically,* 184–210. Seattle: University of Washington Press.

Henrich, J. 2009a. The evolution of costly displays, cooperation, and religion: Credibility enhancing displays and their implications for cultural evolution. *Evolution and Human Behavior* 30: 244–260.

Henrich, J. 2009b. The evolution of innovation-enhancing institutions. In S. J. Shennan and M. J. O'Brien, eds., *Innovation in Cultural Systems: Contributions in Evolution Anthropology.* Cambridge, MA: MIT Press.

Henrich, J. 2016. *The Secret of Our Success: How Culture Is Driving Human Evolution, Domesticating Our Species, and Making Us Smart.* Princeton, NJ: Princeton University Press.

Henrich, J., and R. Boyd. 1998. The evolution of conformist transmission and the emergence of between-group differences. *Evolution and Human Behavior* 19: 215–242.

Henrich, J., and R. Boyd. 2001. Why people punish defectors: Weak conformist transmission can stabilize costly enforcement of norms in cooperative dilemmas. *Journal of Theoretical Biology* 208: 79–89.

Henrich, J., R. Boyd, M. Derex, M. A. Kline, A. Mesoudi, M. Muthukrishna, A. Powell, S. Shennan, and M. G. Thomas. 2015. Understanding cumulative cultural evolution: A reply to Vaesen, Collard, et al. *Proceedings of the National Academy of Sciences* 113: 201610005.

Henrich, J., and J. Broesch. 2011. On the nature of cultural transmission networks: Evidence from Fijian villages for adaptive learning biases. *Philosophical Transactions of the Royal Society B: Biological Sciences* 366: 1139–1148.

Henrich, J., and F. Gil-White. 2001. The evolution of prestige: Freely conferred deference as a mechanism for enhancing the benefits of cultural transmission. *Evolution and Human Behavior* 22: 165–196.

Henrich, J., S. J. Heine, and A. Norenzayan. 2010. The weirdest people in the world? *Behavior and Brain Sciences* 33: 1–23.

Henrich, J., and N. Henrich. 2010. The evolution of cultural adaptations: Fijian taboos during pregnancy and lactation protect against marine toxins. *Proceedings of the Royal Society B: Biological Sciences* 366: 1139–1148.

Henrich, J., and R. McElreath. 2003. The evolution of cultural evolution. *Evolutionary Anthropology* 12: 123–135.

Henrich, J., and R. McElreath. 2007. Dual inheritance theory: The evolution of human cultural capacities and cultural evolution. In R. Dunbar and L. Barrett, eds., *Oxford Handbook of Evolutionary Psychology,* 555–570. Oxford: Oxford University Press.

Henrich, J., and J. B. Silk. 2013. Interpretative problems with chimpanzee ultimatum games. *Proceedings of the National Academy of Sciences* 110: E3049.

Henrich, N., and J. Henrich. 2007. *Why Humans Cooperate: A Cultural and Evolutionary Explanation.* Oxford: Oxford University Press.

Herrmann, E., J. Call, M. V. Hernández-Lloreda, B. Hare, and M. Tomasello. 2007. Humans have evolved specialized skills of social cognition: The cultural intelligence hypothesis. *Science* 317: 1360–1366.

Herrmann, P. A., C. H. Legare, P. L. Harris, and H. Whitehouse. 2013. Stick to the script: The effect of witnessing multiple actors on children's imitation. *Cognition* 129: 536–543.

Hewlett, B. S., H. N. Fouts, A. H. Boyette, and B. L. Hewlett. 2011. Social learning among Congo Basin hunter-gatherers. *Philosophical Transactions of the Royal Society B: Biological Sciences* 366: 1168–1178.

Heyes, C. 2012. What's social about social learning? *Journal of Comparative Psychology* 126: 193–202.

Hill, K. R., R. S. Walker, M. Božičević, J. Eder, T. Headland, B. Hewlett, A. M. Hurtado, F. Marlowe, P. Wiessner, and B. Wood. 2011. Co-residence patterns in hunter-gatherer societies show unique human social structure. *Science* 331: 1286–1289.

Hill, K. R., B. M. Wood, J. Baggio, A. M. Hurtado, and R. T. Boyd. 2014. Hunter-gatherer inter-band interaction rates: Implications for cumulative culture. *PLoS ONE* 9: e102806.

Holden, C., and R. Mace. 1997. Phylogenetic analysis of the evolution of lactose digestion in adults. *Human Biology* 69: 605–628.

Hopper, L. M., S. J. Schapiro, S. P. Lambeth, and S. F. Brosnan. 2011. Chimpanzees' socially maintained food preferences indicate both conservatism and conformity. *Animal Behaviour* 81 (6): 1195–1202.

Hopper, L. M., S. P. Lambeth, S. J. Schapiro, and A. Whiten. 2008. Observational learning in chimpanzees and children studied through "ghost" conditions. *Proceedings of the Royal Society B: Biological Sciences* 275: 835–840.

Hopper, L. M., S. J. Schapiro, S. P. Lambeth, and S. F. Brosnan. 2011b. Chimpanzees' socially maintained food preferences indicate both conservatism and conformity. *Animal Behaviour* 81: 1195–1202.

Hopper, L. M., A. Spiteri, S. P. Lambeth, S. J. Schapiro, V. Horner, and A. Whiten. 2007. Experimental studies of traditions and underlying transmission processes in chimpanzees. *Animal Behaviour* 73: 1021–1032.

Hoppitt, W., and K. N. Laland. 2013. *Social Learning: An Introduction to Mechanisms, Methods, and Models*. Princeton, NJ: Princeton University Press.

Hoppitt, W. J. E., G. R. Brown, R. Kendal, L. Rendell, A. Thornton, M. M. Webster, and K. N. Laland. 2008. Lessons from animal teaching. *Trends in Ecology & Evolution* 23: 486–493.

Horner, V., D. Proctor, K. E. Bonnie, A. Whiten, and F. B. M. de Waal. 2010. Prestige affects cultural learning in chimpanzees. *PLoS ONE* 5: e10625.

Horner, V., and A. Whiten. 2005. Causal knowledge and imitation/emulation switching in chimpanzees *(Pan troglodytes)* and children *(Homo sapiens)*. *Animal Cognition* 8: 164–181.

House, B. R., J. Henrich, S. F. Brosnan, and J. B. Silk. 2012. The ontogeny of human prosociality: Behavioral experiments with children aged 3 to 8. *Evolution and Human Behavior* 33: 291–308.

House, B. R., J. B. Silk, J. Henrich, H. C. Barrett, B. A. Scelza, A. H. Boyette, B. S. Hewlett, R. McElreath, and S. Laurence. 2013. Ontogeny of prosocial behavior across diverse societies. *Proceedings of the National Academy of Sciences* 110: 14586–14591.

Hruschka, D. J., and J. Henrich. 2013. Institutions, parasites and the persistence of ingroup preferences. *PLoS ONE* 8: e63642.

Huffman, M. A., and S. Hirata. 2004. An experimental study of leaf swallowing in captive chimpanzees: Insights into the origin of a self-medicative behavior and the role of social learning. *Primates* 45: 113–118.

Huffman, M. A., C. Spiezio, A. Sgaravatti, and J.-B. Leca. 2010. Leaf swallowing behavior in chimpanzees *(Pan troglodytes):* Biased learning and the emergence of group level cultural differences. *Animal Cognition* 13: 871–880.

Humle, T., and T. Matsuzawa. 2002. Ant-dipping among the chimpanzees of Bossou, Guinea, and some comparisons with other sites. *American Journal of Primatology* 58: 133–148.

Humle, T., C. T. Snowdon, and T. Matsuzawa. 2009. Social influences on ant-dipping acquisition in the wild chimpanzees *(Pan troglodytes verus)* of Bossou, Guinea, West Africa. *Animal Cognition* 12: S37–S48.

Inoue, S., and T. Matsuzawa. 2007. Working memory of numerals in chimpanzees. *Current Biology* 17: R1004–R1005.

Inoue-Nakamura, N., and T. Matsuzawa. 1997. Development of stone tool use by wild chimpanzees *(Pan troglodytes). Journal of Comparative Psychology* 111: 159–173.

Jaswal, V. K., and L. A. Neely. 2006. Adults don't always know best: Preschoolers use past reliability over age when learning new words. *Psychological Science* 17: 757–758.

Jensen, K., J. Call, and M. Tomasello. 2007a. Chimpanzees are rational maximizers in an ultimatum game. *Science* 318: 107–109.

Jensen, K., J. Call, and M. Tomasello. 2007b. Chimpanzees are vengeful but not spiteful. *Proceedings of the National Academy of Sciences* 104: 13046–13050.

Jensen, K., B. Hare, J. Call, and M. Tomasello. 2006. What's in it for me? Self-regard precludes altruism and spite in chimpanzees. *Proceedings of the Royal Society B: Biological Sciences* 273: 1013–1021.

Jordan, P. 2015. *Technology as Human Social Tradition: Cultural Transmission among Hunter-Gatherers.* Oakland: University of California Press.

Jordan, P., and S. Shennan. 2003. Cultural transmission, language, and basketry traditions amongst the California Indians. *Journal of Anthropological Archaeology* 22: 42–74.

Kamilar, J. M., and Q. D. Atkinson. 2013. Cultural assemblages show nested structure in humans and chimpanzees but not orangutans. *Proceedings of the National Academy of Sciences* 111: 111–115.

Kellogg, W., and L. Kellogg. 1933. *The Ape and the Child: A Study of Environmental Influence upon Early Behavior.* New York: McGraw-Hill.

Kendal, J., L. A. Giraldeau, and K. Laland. 2009. The evolution of social learning rules: Payoff-biased and frequency-dependent biased transmission. *Journal of Theoretical Biology* 260: 210–219.

Kendal, R., L. M. Hopper, A. Whiten, S. F. Brosnan, S. P. Lambeth, S. J. Schapiro, and W. Hoppitt. 2015. Chimpanzees copy dominant and knowledgeable individuals: Implications for cultural diversity. *Evolution and Human Behavior* 36: 65–72.

Kendal, R. L., I. Coolen, Y. van Bergen, and K. N. Laland. 2005. Trade-offs in the adaptive use of social and asocial learning. *Advances in the Study of Behavior* 35: 333–379.

Kline, M. A. 2015. How to learn about teaching: An evolutionary framework for the study of teaching behavior in humans and other animals. *Behavioral and Brain Sciences* 38: e31.

Kline, M. A., and R. Boyd. 2010. Population size predicts technological complexity in Oceania. *Proceedings of the Royal Society B: Biological Sciences* 277: 2559–2564.

Kline, M. A., R. Boyd, and J. Henrich. 2013. Teaching and the life history of cultural transmission in Fijian villages. *Human Nature* 24: 351–374.

Kobayashi, Y., and K. Aoki. 2012. Innovativeness, population size and cumulative cultural evolution. *Theoretical Population Biology* 82: 38–47.

Koenig, M. A., and P. L. Harris. 2005. Preschoolers mistrust ignorant and inaccurate speakers. *Child Development* 76: 1261–1277.

Kolodny, O., N. Creanza, and M. W. Feldman. 2015. Evolution in leaps: The punctuated accumulation and loss of cultural innovations. *Proceedings of the National Academy of Sciences* 112: E6762–E6769.

Koops, K., T. Furuichi, and C. Hashimoto. 2015. Chimpanzees and bonobos differ in intrinsic motivation for tool use. *Scientific Reports* 5: 11356.

Koops, K., W. C. McGrew, and T. Matsuzawa. 2013. Ecology of culture: Do environmental factors influence foraging tool use in wild chimpanzees, *Pan troglodytes verus*? *Animal Behaviour* 85: 175–185.

Koops, K., E. Visalberghi, and C. P. van Schaik. 2014. The ecology of primate material culture. *Biology Letters* 10: 20140508.

Laland, K. 1994. Sexual selection with a culturally-transmitted mating preference. *Theoretical Population Biology* 45: 1–15.

Laland, K. N. 2004. Social learning strategies. *Learning & Behavior* 32: 4–14.

Laland, K. N., N. Atton, and M. M. Webster. 2011. From fish to fashion: Experimental and theoretical insights into the evolution of culture. *Philosophical Transactions of the Royal Society B: Biological Sciences* 366: 958–968.

Laland, K. N., and V. M. Janik. 2006. The animal cultures debate. *Trends in Ecology & Evolution* 21: 542–547.

Laland, K. N., J. R. Kendal, and R. L. Kendal. 2009. Animal culture: Problems and solutions. In K. N. Laland and B. G. Galef, eds., *The Question of Animal Culture*, 174–197. Cambridge, MA: Harvard University Press.

Laland, K. N., J. Kumm, and M. W. Feldman. 1995a. Gene-culture coevolutionary theory: A test-case. *Current Anthropology* 36: 131–156.

Laland, K. N., J. Kumm, J. D. Vanhorn, and M. W. Feldman. 1995b. A gene-culture model of human handedness. *Behavior Genetics* 25: 433–445.

Laland, K. N., J. Odling-Smee, and S. Myles. 2010. How culture shaped the human genome: Bringing genetics and the human sciences together. *Nature Reviews Genetics* 11: 137–148.

Lancy, D. 1996. *Playing on Mother Ground: Cultural Routines for Children's Development*. London: Guilford.

Lancy, D. 2009. *The Anthropology of Childhood: Cherubs, Chattel and Changelings*. Cambridge: Cambridge University Press.

Langergraber, K., J. Mitani, and L. Vigilant. 2009. Kinship and social bonds in female chimpanzees *(Pan troglodytes)*. *American Journal of Primatology* 71: 840–851.

Langergraber, K. E., C. Boesch, E. Inoue, M. Inoue-Murayama, J. C. Mitani, T. Nishida, A. Pusey, V. Reynolds, G. Schubert, R. W. Wrangham, E. Wroblewski, and L. Vigilant. 2011. Genetic and "cultural" similarity in wild chimpanzees. *Proceedings of the Royal Society B: Biological Sciences* 278: 408–416.

Langergraber, K. E., and L. Vigilant. 2011. Genetic differences cannot be excluded from generating behavioural differences among chimpanzee groups. *Proceedings of the Royal Society B: Biological Sciences* 278: 2094–2095.

Leavens, D. A., K. A. Bard, and W. D. Hopkins. 2010. BIZARRE chimpanzees do not represent "the chimpanzee." *Behavioral and Brain Sciences* 33: 100–101.

Lewis, H. M., and K. N. Laland. 2012. Transmission fidelity is the key to the build-up of cumulative culture. *Philosophical Transactions of the Royal Society B: Biological Sciences* 367: 2171–2180.

Lind, J., and P. Lindenfors. 2010. The number of cultural traits is correlated with female group size but not with male group size in chimpanzee communities. *PLoS ONE* 5: e9241.

Lipo, C. P., M. J. O'Brien, M. Collard, and S. Shennan. 2006. *Mapping Our Ancestors*. Piscataway, NJ: Aldine Transactions.

Lonsdorf, E. V. 2005. Sex differences in the development of termite-fishing skills in the wild chimpanzees, *Pan troglodytes schweinfurthii*, of Gombe National Park, Tanzania. *Animal Behaviour* 7: 673–683.

Lonsdorf, E. V. 2006. What is the role of mothers in the acquisition of termite-fishing behaviors in wild chimpanzees *(Pan troglodytes schweinfurthii)*? *Animal Cognition* 9: 36–46.

Lonsdorf, E. V. 2013. The role of mothers in the development of complex skills in chimpanzees. In K. B. H. Clancy, K. Hinde, and J. N. Rutherford, eds., *Building Babies: Primate Development in Proximate and Ultimate Perspective*. New York: Springer.

Lonsdorf, E. V., L. E. Eberly, and A. E. Pusey. 2004. Sex differences in learning in chimpanzees. *Nature* 428: 715–716.

Lonsdorf, E. V., S. R. Ross, S. A. Linick, M. S. Milstein, and T. N. Melber. 2009. An experimental, comparative investigation of tool use in chimpanzees and gorillas. *Animal Behaviour* 77: 1119–1126.

Lumsden, C., and E. O. Wilson. 1981. *Genes, Mind, and Culture: The Coevolutionary Process*, 25th Anniversary ed. Cambridge, MA: Harvard University Press.

Luncz, L. V., and C. Boesch. 2014. Tradition over trend: Neighboring chimpanzee communities maintain differences in cultural behavior despite frequent immigration of adult females. *American Journal of Primatology* 76: 649–657.

Luncz, L. V., R. Mundry, and C. Boesch. 2012. Evidence for cultural differences between neighboring chimpanzee communities. *Current Biology* 22: 922–926.

Lycett, S. J., M. Collard, and W. C. McGrew. 2010. Are behavioral differences among wild chimpanzee communities genetic or cultural? An assessment using tool-use data and phylogenetic methods. *American Journal of Physical Anthropology* 142: 461–467.

Lycett, S. J., M. Collard, and W. C. McGrew. 2011. Correlations between genetic and behavioural dissimilarities in wild chimpanzees *(Pan troglodytes)* do not undermine the case for culture. *Proceedings of the Royal Society B: Biological Sciences* 278: 2091–2093.

Lyons, D. E., A. G. Young, and F. C. Keil. 2007. The hidden structure of overimitation. *Proceedings of the National Academy of Sciences* 104: 19751–19756.

Maestripieri, D. 1995. First steps in the macaque world: Do rhesus mothers encourage their infants' independent locomotion? *Animal Behaviour* 49: 1541–1549.

Manrique, H. M., C. J. Völter, and J. Call. 2013. Repeated innovation in great apes. *Animal Behaviour* 85: 195–202.

Marshall-Pescini, S., and A. Whiten. 2008. Social learning of nut-cracking behavior in east African sanctuary-living chimpanzees *(Pan troglodytes schweinfurthii)*. *Journal of Comparative Psychology* 122: 186–194.

Masi, S. 2011. Differences in gorilla nettle-feeding between captivity and the wild: Local traditions, species typical behaviors or merely the result of nutritional deficiencies? *Animal Cognition* 14: 921–925.

McElreath, R., A. V. Bell, C. Efferson, M. Lubell, P. J. Richerson, and T. Waring. 2008. Beyond existence and aiming outside the laboratory: Estimating frequency-dependent and pay-off-biased social learning strategies. *Philosophical Transactions of the Royal Society B: Biological Sciences* 363, 3515–3528.

McElreath, R., R. Boyd, and P. J. Richerson. 2003. Shared norms and the evolution of ethnic markers. *Current Anthropology* 44: 122–129.

McElreath, R., and P. Strimling. 2008. When natural selection favors imitation of parents. *Current Anthropology* 49: 307–316.

McGrew, W. C. 2013. Is primate tool use special? Chimpanzee and New Caledonian crow compared. *Philosophical Transactions of the Royal Society B: Biological Sciences* 368: 20120422.

McGuigan, N., J. Makinson, and A. Whiten. 2011. From over-imitation to super-copying: Adults imitate causally irrelevant aspects of tool use with higher fidelity than young children. *British Journal of Psychology* 102: 1–18.

Melber, T. N., E. V. Lonsdorf, and S. R. Ross. 2007. Social learning of tool-use skills in captive chimpanzees *(Pan troglodytes)* and gorillas *(Gorilla gorilla gorilla)*. *American Journal of Primatology* 69: 100–101.

Menzel, C., A. Fowler, C. Tennie, and C. Josep. 2013. Leaf surface roughness elicits leaf swallowing behavior in captive chimpanzees *(Pan troglodytes)* and bonobos *(P. paniscus)*, but not in gorillas *(Gorilla gorilla)* or orangutans *(Pongo abelii)*. *International Journal of Primatology* 34: 533–553.

Menzel, E. W., Jr., R. K. Davenport, and C. M. Rogers. 1970. The development of tool using in wild-born and restriction-reared chimpanzees. *Folia Primatologica* 12: 273–283.

Mesoudi, A. 2011. *Cultural Evolution: How Darwinian Theory Can Explain Human Culture and Synthesize the Social Sciences*. Chicago: University of Chicago Press.

Miller, D. J., T. Duka, C. D. Stimpson, S. J. Schapiro, W. B. Baze, M. J. McArthur, A. J. Fobbs, A. M. M. Sousa, N. Sestan, D. E. Wildman, L. Lipovich, C. W. Kuzawa, P. R. Hof, and C. C. Sherwood. 2012. Prolonged myelination in human neocortical evolution. *Proceedings of the National Academy of Sciences* 109: 16480–16485.

Mobius, Y., C. Boesch, K. Koops, T. Matsuzawa, and T. Humle. 2008. Cultural differences in army ant predation by West African chimpanzees? A comparative study of microecological variables. *Animal Behaviour* 76: 37–45.

Moore, R. 2013. Social learning and teaching in chimpanzees. *Biology & Philosophy* 28: 879–901.

Moore, R., and C. Tennie. 2015. Cognitive mechanisms matter—but they do not explain the absence of teaching in chimpanzees. *Behavioral and Brain Sciences* 38: 32–33.

Morgan, B. J., and E. E. Abwe. 2006. Chimpanzees use stone hammers in Cameroon. *Current Biology* 16: R632–R633.

Morgan, T. J. H., and K. Laland. 2012. The biological bases of conformity. *Frontiers in Neuroscience* 6: 1–7.

Morgan, T. J. H., L. E. Rendell, M. Ehn, W. Hoppitt, and K. N. Laland. 2012. The evolutionary basis of human social learning. *Proceedings of the Royal Society B: Biological Sciences* 279: 653–662.

Musgrave, S., E. Bell, D. Morgan, E. Lonsdorf, and C. Sanz. 2015. Preliminary report on the acquisition of tool-using elements during termite gathering among chimpanzees of the Goualougo Triangle, Republic of Congo. *American Journal of Physical Anthropology* 156: 232–232.

Musgrave, S., D. Morgan, E. Lonsdorf, R. Mundry, and C. Sanz. 2016. Tool transfers are a form of teaching among chimpanzees. *Scientific Reports* 6: 34783.

Muthukrishna, M., and J. Henrich. 2016. Innovation in the collective brain. *Philosophical Transactions of the Royal Society B: Biological Sciences* 371: 20150192.

Muthukrishna, M., and J. Henrich. N.d. *The Cultural Brain Hypothesis: How Culture Can Drive Brain Expansion and Alter Life Histories.*

Muthukrishna, M., T. J. H. Morgan, and J. Henrich. 2016. The when and who of social learning and conformist transmission. *Evolution and Human Behavior* 37: 10–20.

Muthukrishna, M., B. W. Shulman, V. Vasilescu, and J. Henrich. 2014. Sociality influences cultural complexity. *Proceedings of the Royal Society B: Biological Sciences* 281: 20132511.

Nagell, K., R. S. Olguin, and M. Tomasello. 1993. Processes of social learning in the tool use of chimpanzees *(Pan troglodytes)* and human children *(Homo sapiens)*. *Journal of Comparative Psychology* 107: 174–186.

Nakahashi, W., J. Y. Wakano, and J. Henrich. 2012. Adaptive social learning strategies in temporally and spatially varying environments. *Human Nature* 23: 386–418.

Nielsen, M., and K. Tomasello. 2010. Overimitation in Kalahari bushman children and the origins of human cultural cognition. *Psychological Science* 21: 729–736.

Nishida, T., T. Matsusaka, and W. C. McGrew. 2009. Emergence, propagation or disappearance of novel behavioral patterns in the habituated chimpanzees of Mahale: A review. *Primates* 50: 23–36.

O'Malley, R. C., W. Wallauer, C. M. Murray, and J. Goodall. 2012. The appearance and spread of ant fishing among the Kasekela chimpanzees of Gombe: A possible case of intercommunity cultural transmission. *Current Anthropology* 53: 650–663.

Pagel, M., Q. D. Atkinson, A. S. Calude, and A. Meade. 2013. Ultraconserved words point to deep language ancestry across Eurasia. *Proceedings of the National Academy of Sciences* 110: 8471–8476.

Perreault, C., C. Moya, and R. Boyd. 2012. A Bayesian approach to the evolution of social learning. *Evolution and Human Behavior* 33: 449–459.

Pike, T. W., and K. N. Laland. 2010. Conformist learning in nine-spined sticklebacks' foraging decisions. *Biology Letters* 6: 466–468.

Pradhan, G. R., C. Tennie, and C. P. van Schaik. 2012. Social organization and the evolution of cumulative technology in apes and hominins. *Journal of Human Evolution* 63: 180–190.

Pulliam, H. R., and C. Dunford. 1980. *Programmed to Learn: An Essay on the Evolution of Culture.* New York: Columbia University Press.

Reader, S. M., Y. Hager, and K. N. Laland. 2011. The evolution of primate general and cultural intelligence. *Philosophical Transactions of the Royal Society B: Biological Sciences* 366: 1017–1027.

Reader, S. M., and K. N. Laland. 2002. Social intelligence, innovation, and enhanced brain size in primates. *Proceedings of the National Academy of Sciences* 99: 4436–4441.

Rendell, L., R. Boyd, D. Cownden, M. Enquist, K. Eriksson, M. W. Feldman, L. Fogarty, S. Ghirlanda, T. Lillicrap, and K. N. Laland. 2010. Why copy others? Insights from the social learning strategies tournament. *Science* 328: 208–213.

Rendell, L., L. Fogarty, W. J. E. Hoppitt, T. J. H. Morgan, M. M. Webster, and K. N. Laland. 2011. Cognitive culture: Theoretical and empirical insights into social learning strategies. *Trends in Cognitive Sciences* 15: 68–76.

Renevey, N., R. Bshary, and E. van de Waal. 2013. Philopatric vervet monkey females are the focus of social attention rather independently of rank. *Behaviour* 150: 599–615.

Richerson, P. J., R. Baldini, A. Bell, K. Demps, K. Frost, V. Hillis, S. Mathew, E. Newton, N. Narr, L. Newson, C. Ross, P. Smaldino, T. Waring, and M. Zefferman. 2016. Cultural group selection plays an essential role in explaining human cooperation: A sketch of the evidence. *Behavioral and Brain Sciences* 39: e30.

Richerson, P. J., R. L. Bettinger, and R. Boyd. 2005. Evolution on a restless planet: Were environmental variability and environmental change major drivers of human evolution? In F. M. Wuketits and F. J. Ayala, eds., *Handbook of Evolution: Evolution of Living Systems (Including Hominids),* 223–242. Weinheim: Wiley-VCH.

Richerson, P. J., and R. Boyd. 2000. Climate, culture and the evolution of cognition. In C. M. Heyes, ed., *The Evolution of Cognition,* 329–345. Cambridge, MA: MIT Press.

Richerson, P. J., and R. Boyd. 2005. *Not by Genes Alone: How Culture Transformed Human Evolution.* Chicago: University of Chicago Press.

Richerson, P. J., R. Boyd, and J. Henrich. 2010. Gene-culture coevolution in the age of genomics. *Proceedings of the National Academy of Sciences* 107: 8985–8992.

Robbins, M. M., C. Ando, K. A. Fawcett, C. C. Grueter, D. Hedwig, Y. Iwata, J. L. Lodwick, S. Masi, R. Salmi, T. S. Stoinski, A. Todd, V. Vercellio, and J. Yamagiwa. 2016. Behavioral variation in gorillas: Evidence of potential cultural traits. *PLoS ONE* 11: e0160483.

Rogers, E. M. 1995. *Diffusion of Innovations.* New York: Free Press.

Rozin, P., M. Mark, and D. Schiller. 1981. The role of desensitization to capsaicin in chili pepper ingestion and preference. *Chemical Senses* 6: 23–31.

Rozin, P., and D. Schiller. 1980. The nature and acquisition of a preference for chili pepper by humans. *Motivation and Emotion* 4: 77–101.

Salali, G. D., N. Chaudhary, J. Thompson, O. M. Grace, X. M. van der Burgt, M. Dyble, A. E. Page, D. Smith, J. Lewis, R. Mace, L. Vinicius, and A. B. Migliano. 2016. Knowledge-sharing networks in hunter-gatherers and the evolution of cumulative culture. *Current Biology* 26: 2516–2521.

Sanz, C., J. Call, and D. Morgan. 2009. Design complexity in termite-fishing tools of chimpanzees *(Pan troglodytes). Biology Letters* 5: 293–296.

Sanz, C. M., and D. B. Morgan. 2011. Elemental variation in the termite fishing of wild chimpanzees *(Pan troglodytes). Biology Letters* 7: 634–637.

Sanz, C. M., and D. B. Morgan. 2013. Ecological and social correlates of chimpanzee tool use. *Philosophical Transactions of the Royal Society B: Biological Sciences* 368: 20120416.

Schöning, C., T. Humle, Y. Möbius, and W. C. McGrew. 2008. The nature of culture: Technological variation in chimpanzee predation on army ants revisited. *Journal of Human Evolution* 55: 48–59.

Shennan, S. 2003. *Genes, Memes, and Human History: Darwinian Archaeology and Cultural Evolution.* London: Thames & Hudson.

Shennan, S. 2009. *Pattern and Process in Cultural Evolution.* Berkeley: University of California Press.

Silberberg, A., and D. Kearns. 2009. Memory for the order of briefly presented numerals in humans as a function of practice. *Animal Cognition* 12: 405–407.

Silk, J. B., S. F. Brosnan, J. Vonk, J. Henrich, D. J. Povinelli, A. S. Richardson, S. P. Lambeth, J. Mascaro, and S. J. Shapiro. 2005. Chimpanzees are indifferent to the welfare of unrelated group members. *Nature* 437: 1357–1359.

Silk, J. B., and B. R. House. 2011. Evolutionary foundations of human prosocial sentiments. *Proceedings of the National Academy of Sciences* 108: 10910–10917.

Stammbach, E. 1988. Group responses to specially skilled individuals in a *Macaca fascicularis* group. *Behavior* 107: 241–266.

Strauss, S., and M. Ziv. 2012. Teaching is a natural cognitive ability for humans. *Mind, Brain, and Education* 6: 186–196.

Sumita, K., J. Kitahara-Frisch, and K. Norikoshi. 1985. The acquisition of stone-tool use in captive chimpanzees. *Primates* 26: 168–181.

Tehrani, J. J., and M. Collard. 2009. On the relationship between interindividual cultural transmission and population-level cultural diversity: A case study of weaving in Iranian tribal populations. *Evolution and Human Behavior* 30: 286–300.

Tehrani, J. J., M. Collard, and S. J. Shennan. 2010. The cophylogeny of populations and cultures: Reconstructing the evolution of Iranian tribal craft traditions using trees and jungles. *Philosophical Transactions of the Royal Society B: Biological Sciences* 365: 3865–3874.

Tennie, C., J. Call, and M. Tomasello. 2006. Push or pull: Imitation vs. emulation in great apes and human children. *Ethology* 112: 1159–1169.

Tennie, C., J. Call, and M. Tomasello. 2009. Ratcheting up the ratchet: On the evolution of cumulative culture. *Philosophical Transactions of the Royal Society B: Biological Sciences* 364: 2405–2415.

Tennie, C., J. Call, and M. Tomasello. 2010a. Evidence for emulation in chimpanzees in social settings using the floating peanut task. *PLoS ONE* 5: e10544.

Tennie, C., J. Call, and M. Tomasello. 2012. Untrained chimpanzees *(Pan troglodytes schweinfurthii)* fail to imitate novel actions. *PLoS ONE* 7: e41548.

Tennie, C., K. Greve, H. Gretscher, and J. Call. 2010b. Two-year-old children copy more reliably and more often than nonhuman great apes in multiple observational learning tasks. *Primates* 51: 337–351.

Tennie, C., D. Hedwig, J. Call, and M. Tomasello. 2008. An experimental study of nettle feeding in captive gorillas. *American Journal of Primatology* 70: 584–593.

Tennie, C., K. Jensen, and J. Call. 2016. The nature of prosociality in chimpanzees. *Nature Communications* 7: 13915.

Terrace, H. S. 1979. *Nim*. New York: Knopf.

Thorndike, E. L. 1911. *Animal Intelligence*. New York: Macmillan.

Thornton, A., and K. McAuliffe. 2006. Teaching in wild meerkats. *Science* 313: 227–229.

Thornton, A., and N. J. Raihani. 2008. The evolution of teaching. *Animal Behaviour* 75: 1823–1836.

Tomasello, M. 1994. The question of chimpanzee culture. In R. W. Wrangham, W. C. McGrew, F. B. M. de Waal, and P. G. Heltne, eds., *Chimpanzee Cultures*, 301–317. Cambridge, MA: Harvard University Press.

Tomasello, M. 1999a. *The Cultural Origins of Human Cognition*. Cambridge, MA: Harvard University Press.

Tomasello, M. 1999b. The human adaptation for culture. *Annual Review of Anthropology* 28: 509–529.

Tomasello, M., and J. Call. 1997. *Primate Cognition*. New York: Oxford University Press.

Tomasello, M., S. Savage-Rumbaugh, and A. C. Kruger. 1993. Imitative learning of actions on objects by children, chimpanzees, and enculturated chimpanzees. *Child Development* 64: 1688–1705.

van de Waal, E., C. Borgeaud, and A. Whiten. 2013. Potent social learning and conformity shape a wild primate's foraging decisions. *Science* 340: 483–485.

van de Waal, E., N. Renevey, C. M. Favre, and R. Bshary. 2010. Selective attention to philopatric models causes directed social learning in wild vervet monkeys. *Proceedings of the Royal Society B: Biological Sciences* 277: 2105–2111.

van Leeuwen, E. J. C., K. A. Cronin, S. Schütte, J. Call, and D. B. M. Haun. 2013. Chimpanzees *(Pan troglodytes)* flexibly adjust their behaviour in order to maximize payoffs, not to conform to majorities. *PLoS ONE* 8: e80945.

van Leeuwen, E. J. C., and D. B. M. Haun. 2013. Conformity in nonhuman primates: Fad or fact? *Evolution and Human Behavior* 34: 1–7.

van Leeuwen, E. J. C., R. L. Kendal, C. Tennie, and D. B. M. Haun. 2015. Conformity and its look-a-likes. *Animal Behaviour* 110: e1–e4.

van Schaik, C. P., M. Ancrenaz, B. Gwendolyn, B. Galdikas, C. D. Knott, I. Singeton, A. Suzuki, S. S. Utami, and M. Merrill. 2003. Orangutan cultures and the evolution of material culture. *Science* 299: 102–105.

van Schaik, C. P., and J. M. Burkart. 2011. Social learning and evolution: The cultural intelligence hypothesis. *Philosophical Transactions of the Royal Society B: Biological Sciences* 366: 1008–1016.

van Schaik, C. P., K. Isler, and J. M. Burkart. 2012. Explaining brain size variation: From social to cultural brain. *Trends in Cognitive Sciences* 16: 277–284.

van Schaik, C. P., and G. R. Pradhan. 2003. A model for tool-use traditions in primates: Implications for the coevolution of culture and cognition. *Journal of Human Evolution* 44: 645–664.

Visalberghi, E. 1987. Acquisition of nut-cracking behaviour by 2 capuchin monkeys *(Cebus apella)*. *Folia Primatologica* 49: 168–181.

Vonk, J., S. F. Brosnan, J. B. Silk, J. Henrich, A. S. Richardson, S. P. Lambeth, S. J. Schapiro, and D. J. Povinelli. 2008. Chimpanzees do not take advantage of very low cost opportunities to deliver food to unrelated group members. *Animal Behaviour* 75: 1757–1770.

Wakano, J. Y., K. Aoki, and M. W. Feldman. 2004. Evolution of social learning: A mathematical analysis. *Theoretical Population Biology* 66: 249–258.

Walker, R. S., M. V. Flinn, and K. R. Hill. 2010. Evolutionary history of partible paternity in lowland South America. *Proceedings of the National Academy of Sciences* 107: 19195–19200.

Warneken, F., F. Chen, and M. Tomasello. 2006. Cooperative activities in young children and chimpanzees. *Child Development* 77: 640–663.

Warneken, F., and M. Tomasello. 2006. Altruistic helping in human infants and young chimpanzees. *Science* 311: 1301–1303.

Watts, J., S. J. Greenhill, Q. D. Atkinson, T. E. Currie, J. Bulbulia, and R. D. Gray. 2015. Broad supernatural punishment but not moralizing high gods precede the evolution of political complexity in Austronesia. *Proceedings of the Royal Society B: Biological Sciences* 282: 20142556.

Whiten, A. 1998. Imitation of the sequential structure of actions by chimpanzees *(Pan troglodytes)*. *Journal of Comparative Psychology* 112: 270–281.

Whiten, A. 2011. The scope of culture in chimpanzees, humans and ancestral apes. *Philosophical Transactions of the Royal Society B: Biological Sciences* 366: 997–1007.

Whiten, A. 2012. *Culture Evolves*. Oxford: Oxford University Press.

Whiten, A., D. M. Custance, J.-C. Gomez, P. Teixidor, and K. A. Bard. 1996. Imitative learning of artificial fruit processing in children *(Homo sapiens)* and chimpanzees *(Pan troglodytes)*. *Journal of Comparative Psychology* 110: 3–14.

Whiten, A., J. Goodall, W. C. McGrew, T. Nishida, V. Reynolds, Y. Sugiyama, C. E. G. Tutin, R. W. Wrangham, and C. Boesch. 1999. Cultures in chimpanzees. *Nature* 399: 682–685.

Whiten, A., J. Goodall, W. C. McGrew, T. Nishida, V. Reynolds, Y. Sugiyama, C. E. G. Tutin, R. W. Wrangham, and C. Boesch. 2001. Charting cultural variation in chimpanzees. *Behaviour* 138: 1481–1516.

Whiten, A., V. Horner, and F. B. M. de Waal. 2005. Conformity to cultural norms of tool use in chimpanzees. *Nature* 437: 737–740.

Whiten, A., N. McGuigan, S. Marshall-Pescini, and L. M. Hopper. 2009. Emulation, imitation, over-imitation and the scope of culture for child and chimpanzee. *Philosophical Transactions of the Royal Society B: Biological Sciences* 364: 2417–2428.

Whiten, A., and C. P. van Schaik. 2007. The evolution of animal "cultures'" and social intelligence. *Philosophical Transactions of the Royal Society B: Biological Sciences* 362: 603–620.

Wiessner, P. 2002. Hunting, healing, and hxaro exchange: A long-term perspective on !Kung (Ju/'hoansi) large-game hunting. *Evolution and Human Behavior* 23: 407–436.

Wilson, M. L., C. Boesch, B. Fruth, T. Furuichi, I. C. Gilby, C. Hashimoto, C. L. Hobaiter, G. Hohmann, N. Itoh, K. Koops, J. N. Lloyd, T. Matsuzawa, J. C. Mitani, D. C. Mjungu, D. Morgan, M. N. Muller, R. Mundry, M. Nakamura, J. Pruetz, A. E. Pusey, J. Riedel, C. Sanz, A. M. Schel, N. Simmons, M. Waller, D. P. Watts, F. White, R. M. Wittig, K. Zuberbuhler, and R. W. Wrangham. 2014. Lethal aggression in Pan is better explained by adaptive strategies than human impacts. *Nature* 513: 414–417.

Wobber, V., and B. Hare. 2011. Psychological health of orphan bonobos and chimpanzees in African sanctuaries. *PLoS ONE* 6: e17147.

Wood, L. A., R. L. Kendal, and E. G. Flynn. 2013. Whom do children copy? Model-based biases in social learning. *Developmental Review* 33: 341–356.

Wrangham, R., and R. Carmody. 2010. Human adaptation to the control of fire. *Evolutionary Anthropology* 19: 187–199.

Yamakoshi, G. 1998. Dietary responses to fruit scarcity of wild chimpanzees at Bossou, Guinea: Possible implications for ecological importance of tool use. *American Journal of Physical Anthropology* 106: 283–295.

Yamakoshi, G., and M. Myowa-Yamakoshi. 2004. New observations of ant-dipping techniques in wild chimpanzees at Bossou, Guinea. *Primates* 45: 25–32.

Yamamoto, S., T. Humle, and M. Tanaka. 2013. Basis for cumulative cultural evolution in chimpanzees: Social learning of a more efficient tool-use technique. *PLoS ONE* 8: e55768.

Zentall, T. R. 2006. Imitation: Definitions, evidence, and mechanisms. *Animal Cognition* 9: 335–353.

19

Chimpanzee Cognition and the Roots of the Human Mind

ALEXANDRA G. ROSATI

The origins of the human mind have been a puzzle ever since Darwin (1871, 1872). Despite striking continuities in the behavior of humans and nonhumans, our species also exhibits a suite of abilities that diverge from the rest of the animal kingdom: we create and utilize complex technology, pass cultural knowledge from generation to generation, and cooperate across numerous and diverse contexts. Why do humans exhibit these abilities, but other animals (mostly) do not? This is a fundamental question in biology, psychology, and philosophy. This puzzle involves two main parts. The first is concerned with identifying the psychological capacities that are unique to humans. This phylogenetic question can be addressed through careful comparisons of humans and other animals to pinpoint the cognitive traits that are likely derived in our species. The second is concerned with the function of these capacities, and the context in which they arose. This evolutionary question examines why, from an ultimate perspective, we evolved these specialized capacities in the first place.

Solving these puzzles poses a special challenge because it is only possible to directly measure the cognition of living animals. The bodies of extinct species leave traces in the fossil record, and even some behavioral traits exhibit well-understood relationships with physical traits—such as relationships between dentition and dietary ecology, or mating system and sexual size dimorphism. These relationships provide important benchmarks when biologists infer the behavior of extinct species. Unfortunately, cognition does not fossilize, and neither do the brains that generate cognitive abilities. Even those features of neuroanatomy that do leave some trace in the fossil record—such as brain size or particular anatomical landmarks—are often related to the kinds of complex cognitive capacities potentially unique to humans in a coarse fashion. As such, identifying derived human cognitive traits requires reconstructing the mind of the last common ancestor of chimpanzees *(Pan troglodytes),* bonobos *(Pan paniscus),* and humans *(Homo sapiens).* This reconstruction then can be used to infer what cognitive characteristics have changed in the human lineage.

The best current model for the last common ancestor's psychology is the psychology of the living great apes. This review therefore aims to identify features of human cognition that differ from that of other apes, with a particular focus on our closest living relatives, chimpanzees and bonobos. Most research to date has been specifically devoted to understanding the capacities of chimpanzees, and when comparisons of multiple ape species do exist, there are often broad similarities in cognitive abilities across the extant great apes. Consequently, much of this chapter is focused on chimpanzee cognition. However, there are important exceptions to this generalization. For example, recent research comparing chimpanzee and bonobo cognition, with sufficient power to detect variation in their capacities, has revealed important variation in their skills in some domains (Hare et al. 2012; Hare and Yamamoto 2015). These differences clearly impact inferences about the last common ancestor's psychology, given that humans are equally related to both *Pan* species, who diverged from each other less than one million years ago (Won and Hey 2005; Prüfer et al. 2012). As such, differences in apes' abilities will warrant closer scrutiny through the chapter.

The first goal of the present review is to evaluate hypotheses for unique features of human cognition using empirical evidence on the psychological capacities of the living apes. In fact, there are several diverse hypotheses concerning potentially human-unique skills (Tomasello and Call 1997; Hermer-Vazquez et al. 1999; Pinker and Jackendoff 2005; Tomasello et al. 2005; Herrmann et al. 2007; Suddendorf and Corballis 2007; Penn et al. 2008; Hill et al. 2009; Hare 2011; Shettleworth 2012; Sterelny 2012), but here I will examine a subset of three hypotheses:

1. **The social cognition hypothesis.** Humans have unique abilities to infer others' (unobservable) mental states from their superficial behaviors, as well as the motivation to share such mental states with others.
2. **The mental time travel hypothesis.** Humans are unique in our ability to engage in episodic memory and prospection, projecting a representation of the self into the past or future to solve problems.
3. **The executive function hypothesis.** Humans have specializations in self-control and decision-making capacities, which allow us to engage in flexible, goal-directed behavior across diverse contexts.

I propose that there is currently evidence for both deep similarities and profound differences in human and ape cognition across all of these domains, in contrast to approaches that focus on identifying a single core difference between humans and other animals.

The second goal of this review is to assess the evolutionary history of these unique cognitive capacities, illuminating the factors that engendered their emergence. To address this issue, I first show that the seemingly diverse cognitive abilities covered in the first part—social cognition, episodic mental time travel, and executive function—are in fact linked through shared psychological and neurobiological substrates. I then argue that one path to understanding the emergence of these skills is to examine what aspects of human behavior they actually support. In fact, many natural behaviors of deep and abiding interest to psychologists and biologists—such as hunting and foraging—require the seamless integration of skills that cut across typical cognitive distinctions. Consequently, I evaluate differences in the behavior of great apes and traditional human societies, with an eye toward the particular psychological capacities that may be necessary for individuals to engage in these behaviors. As human behavioral ecology is uniquely dependent on exploiting difficult-to-acquire food resources that are brought back to a central location and shared with others (following Kaplan et al. 2000, 2012; Marlowe 2005), I propose that successfully utilizing these types of resources requires multiple cognitive capacities, including prospection and memory to locate food in a large, complex home range; executive functions to avoid the temptation to eat that food on the go; and social cognition to engage in the cooperative acts that characterize human food sharing (see also Rosati 2017a). As such, I argue that human cognitive uniqueness is not defined by a single core difference. Rather, the behavioral problems posed by our species' behavioral niche require a suite of cognitive traits across psychological domains.

The First Puzzle: Identifying Unique Features of Human Cognition

How is human cognition similar to or different from that of our closest living relatives? This section will examine three hypotheses concerning human-unique cognition: our social cognitive abilities to think about others' behaviors and metal states; our episodic memory and prospection, linked capacities that allow individuals to project themselves backward or forward in time to simulate events; and finally, our executive function abilities, which allow us to flexibly update and control our responses to a changing world. It is important to note that understanding psychological processes requires controlled experiments that can rule out alternative explanations for observed patterns of results. As any given behavior can be psychologically implemented by many different possible mechanisms, observational research is inherently limited in terms of possible inferences about the cognitive abilities underlying that behavior (see also Tomasello and Call 2008). For example, studies of social cognition rule out the possibility that animals are utilizing the direct behavioral cues other actors provide, rather than actually inferring their (unobservable) mental states; studies of planning rule out that animals are acting to meet a need they are currently experiencing, rather than a future need; and studies of executive function rule out that animals learn simple rules without being able to flexibly adjust in new contexts. Such controlled experiments are pragmatically (and ethically) challenging to conduct in wild populations, and therefore quite rare (Zuberbühler 2014). Research aimed at understanding cognition in nonhuman primates is therefore conducted primarily in populations living in environments where such controlled experimentation is possible, such as in zoos or sanctuaries (see Figure 19.1).

Hypothesis 1: Social Cognition

One influential claim about cognitive differences between humans and other animals concerns the unique ways in which we think about other individ-

FIGURE 19.1. Chimpanzees and the evolutionary origins of human cognition. Chimpanzees living in sanctuaries and zoos, such as these individuals living at Tchimpounga Chimpanzee Sanctuary in Republic of Congo, can be tested in controlled experiments that rule out different possible explanations for behavioral responses.

uals. Humans possess a complex belief-desire psychology that represents not only others' observable actions, but also others' internal and therefore unobservable psychological states. For example, imagine that your friend suggested several times that the two of you eat at a specific restaurant. An inference based on that person's behavior alone would allow you to predict that they will eat at that restaurant again. However, an inference about that person's mental state would illuminate that they *like* the restaurant and desire to eat a particular type of food—an inference that would allow you to make better predictions about your friend's actions in the future across others contexts—such as what kind of food they will order at a different restaurant, or what they will prepare at their own home. Many other animals clearly engage in complex social interactions, but what cognitive capacities do they use to do so? Several important theoretical views have argued that nonhumans can pick up on complex contingencies in others' behaviors, but do not represent subjective mental states (Cheney and Seyfarth 1990; Povinelli and Eddy 1996b; Tomasello and Call 1997; Heyes 1998; Tomasello 1999). That is, animals can learn about others' concrete actions ("she eats at that restaurant a lot") but do not infer the underlying mental state mediating those actions ("she likes French food"). However, increasing evidence indicates that apes also possess at least some theory of mind capacities.

Understanding Intentions, Perceptions, and Knowledge
What evidence supports the claim that apes can reason about mental states? First, they seem to conceive of others' actions in terms of their underlying goals, intentions, and desires. For example, chimpanzees and orangutans are sensitive to whether a human's behavior is purposeful versus accidental, even when different actions have similar qualities (Call and Tomasello 1998; Tomasello and Carpenter 2005; see also Warneken and Tomasello 2006). Similarly, when watching a human trying but failing to accomplish some task, chimpanzees tend to copy the (uncompleted) goal, rather than the observable (but failed) action sequence (Tomasello et al. 1987) (see also Horner and Whiten 2005; Buttelmann et al. 2007). That is, chimpanzees seem to conceive of these actions in terms of their underlying purpose, not just their visible components. Chimpanzees also modulate their responses to a human's behavior based on that person's intentions (Call et al. 2004). For example, when a human presents food but repeatedly does not give it, chimpanzees are more likely to stick around and beg for the food when the human is trying but failing to hand over the food, compared to when he is teasing the chimpanzee. Indeed, all four species of great apes can use contextual information to distinguish between the goals underlying completely identical actions (Buttelman et al. 2012). Finally, apes predict that others will act in line with their desires. If a person initially responds more positively to one type of food, apes later predict that the person will choose to eat from a container with that food, rather than an alternative with different contents (Buttelman et al. 2009).

Second, there is now strong convergent evidence that apes can reason about what others can see. Many primate species follow others' gaze, for example, by looking in a particular direction when they see another do so. Chimpanzees and other apes, in contrast, exhibit particularly flexible gaze-following behaviors that suggest that they are not just exhibiting a reflexive response, but rather that they understand that others look because they see something interesting (Rosati and Hare 2009). For example, all four great ape species will follow others' gaze—and even reorient to look around barriers or check back with the actor to verify the correct location of the target (Povinelli and Eddy 1996a; Tomasello et al. 1999, 2001, 2007; Okamoto et al. 2002; Braeuer et al. 2005; Herrmann et al. 2007, 2010; Kano and Call 2014). However, there seems to be some important variation in how different species of apes conceive of others' visual perspective. For example, chimpanzees and bonobos seem to have a richer understanding of others' line of sight when they have to infer what others are looking at in complex physical situations, whereas gaze-following responses in gorillas and orangutans are more reflexive (Okamoto-Barth et al. 2007). Similarly, chimpanzees are more successful than bonobos at using their knowledge about others' past experiences to infer what others are really looking at (MacLean and Hare 2012). Yet the strongest evidence to date that apes reason about what others can see comes from studies examining how chimpanzees behave when competing for food either with other conspecifics (Hare et al. 2000; Braeuer et al. 2007) or with humans (Hare et al. 2006; Melis et al. 2006a). For example, if a barrier prevents a competitor from seeing specific locations in space, chimpanzees will specifically pursue food hidden in that "safe" location. There is even some evidence that chimpanzees can infer others' auditory perceptions, indicating that this skill may extend across modalities (Melis et al. 2006a) (but see Braeuer et al. 2008).

Humans do not merely infer what others can and cannot see—we also can use our knowledge about others' perspective to assess what others do or do not know about the world. In fact, there is good evidence that at least chimpanzees do so as well. For example, chimpanzees realize that other conspecifics do not know about the presence of food that was baited out of their sight (Hare et al. 2001). Chimpanzees can also make sophisticated inferences about what choices others will make based on their knowledge. In one setup, two chimpanzees played a competitive back-and-forth game involving sequential choices for food locations in an array of three locations (Kaminski et al. 2008; see also Schmelz et al. 2011). Both chimpanzees might see one of those locations baited, whereas only the subject chimpanzee would see another location baited (while the competitor's view was blocked). If the subject then got to choose first, they were equally likely to choose either of the baited spots. Yet if the subject knew the competitor had already made one choice—but did not know what location the competitor had actually chosen—the subject

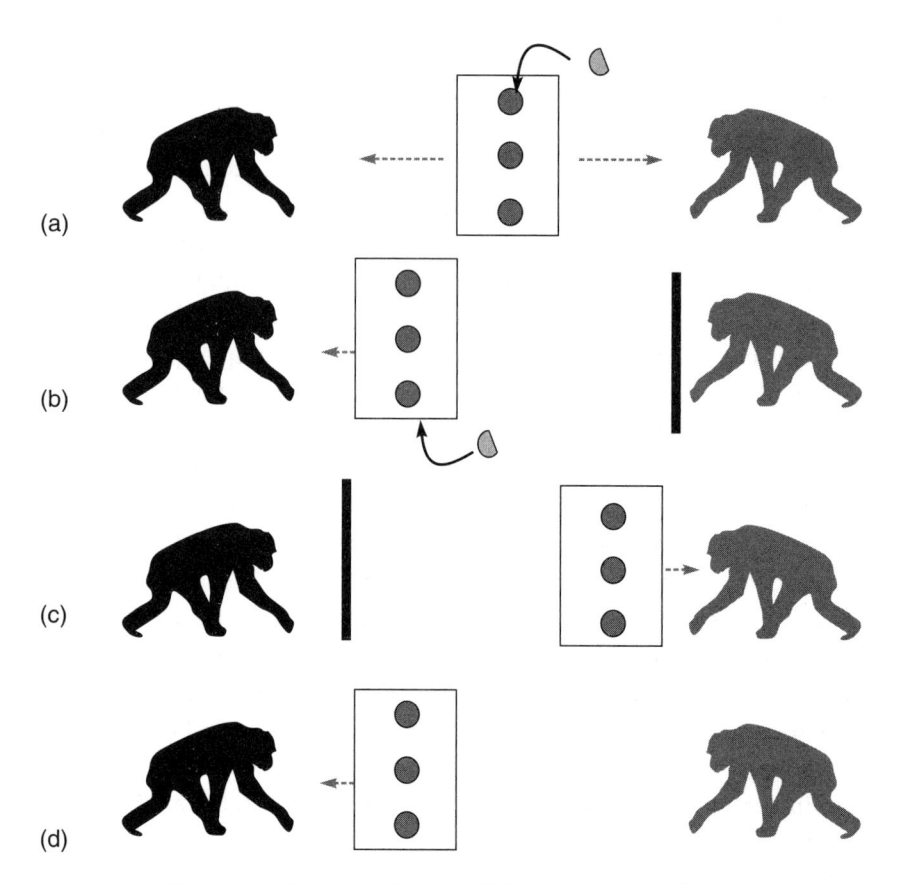

FIGURE 19.2. Understanding mental states: Chimpanzees infer how seeing leads to knowing. Setup for competitive back-and-forth game, adapted from Kaminski et al. (2008). (a) Both the chimpanzee subject (on left) and the competitor (right) watched as one container was baited with food. (b) Only the subject saw that a second location was also baited; the competitor's view was blocked. (c) In some trials, the competitor made the first choice, but their choice could not be seen by the subject. (d) If the subject chose second, they inferred that the competitor must have chosen the location that both had seen baited, and therefore chose the alternative location only they had seen baited.

could nonetheless deduce that the competitor must have already chosen the piece both had seen hidden (see Figure 19.2).

Despite these successes, there is currently limited evidence that other primates understand that other individuals can hold beliefs about the world that can be false (Rosati et al. 2010; Martin and Santos 2016). False-belief comprehension is sometimes treated as the benchmark test for human-like theory of mind, and studies using competition paradigms have generally failed to show that apes attribute false beliefs (Call and Tomasello 1999; Krachun et al. 2007; Kaminski et al. 2008). For example, in the back-and-forth game described above, chimpanzees do not infer that others may have a false belief about the location of food if they observe it being placed in one location, but then it is subsequently moved (Kaminski et al. 2008). However, some recent work using eye-tracking methods suggests that apes may be at least implicitly sensitive to others' false beliefs (Krupenye et al. 2016). Here, apes watched short movies in which an actor viewed something hidden in one location, but then it moved while the actor was absent. The question concerned where apes thought the actor would search when they returned. In fact, chimpanzees, bonobos, and orangutans made anticipatory looks to locations that the actor (incorrectly) thought was the hiding place. Given that human children also succeed at such implicit looking tasks earlier in development than in explicit response tasks (Baillargeon et al. 2010), it may be that anticipatory looking measures reduce executive function demands, as discussed in subsequent sections.

Overall, these results indicate that apes do not merely reason about the superficial qualities of other individuals' behavior. Rather, they can make surprisingly complex inferences about the goals, desires, and perceptions that motivate that behavior, although they may be relatively limited in their ability to attribute false beliefs to others (for reviews, see Tomasello et al. 2003; Call and Tomasello 2008; Rosati et al. 2010; Hare 2011; Martin and Santos 2016). However, there may be some important differences in how the four great apes reason about others' mental states. For example, chimpanzees and bonobos appear to have more sophisticated gaze-following capacities than other apes (Okamoto-Barth et al. 2007), but some of the best evidence that apes reason about others' visual perspective and knowledge comes from chimpanzees alone (Hare et al. 2000, 2001, 2006). While there is some evidence that other species such as rhesus macaques can infer perceptions and knowledge in competitive contexts (Flombaum and Santos 2005; Santos et al. 2006; Marticorena et al. 2011), there is currently little data addressing whether these capacities are more widely shared among primates.

Sharing Mental States

The particular tasks used to assess social cognitive capacities in apes highlight another important difference with humans: most successful demonstrations of theory of mind have focused on the cognitive skills animals use in competitive contexts to outwit conspecifics or humans. In fact, some theoretical views suggest that primates might show more sophisticated cognitive abilities in such contexts, given the importance of competition with others over food or other resources in many primate species' daily lives (Hare 2001; Hare and Tomasello 2004; Lyons and Santos 2006). In contrast, humans regularly engage in complex forms of cooperation in which individuals work together to reach collective goals. Consequently, recent proposals have begun to focus on differences between the social-cognitive abilities humans and apes utilize when multiple individuals are collaborating (Tomasello et al. 2012).

One proposal suggests that human-specific forms of cooperation require human-unique cognitive capacities. In particular, humans may be unique in our ability to represent joint activities as underpinned by shared goals or motivations that both individuals know are shared (Tomasello et al. 2005; Tomasello and Carpenter 2007). What constitutes such shared mental states? Take the example of gaze following. As noted previously, both humans and chimpanzees can expertly follow gaze ("I see you looking at something"). However, humans also exhibit joint attention—where both individuals knowingly share attention with each other toward the target ("You and I both know we are looking at something together"). While joint attention emerges within the first year of life in human infants, there is little evidence that chimpanzees engage in such triadic interactions (for review, see Carpenter and Call 2013). Indeed, this early difference in social cognitive development may result in later divergence in the trajectories of human and ape gaze-following skills into adulthood (Rosati et al. 2014).

The lack of "togetherness" in the way that other animals conceive of joint activities may have particularly important repercussions for cooperative behaviors. As is the case with joint attention in gaze following, young human children form joint intentions when engaging in cooperative actions. In particular, humans assume that the joint task needs to be completed together—and even act to help a partner achieve a successful outcome after their own individual needs have been met (Warneken et al. 2006; Greenberg et al. 2010; Hamann et al. 2011). That is, humans seem to act as though success at a joint task is not only about me getting my own reward, but both of us being successful together. Chimpanzees also engage in a variety of complex cooperative behaviors (Melis et al. 2006b, 2006c)—but they seem to conceive of their joint actions in terms of individual, but parallel, goals. Indeed, chimpanzees prefer to solve collaborative problems individually if possible (Bullinger et al. 2011; Rekers et al. 2011). To date, there is little evidence that chimpanzees or bonobos form joint intentions when engaging in cooperative activities (Herrmann et al. 2007, 2010; Wobber et al. 2013) (but see Pika and Zuberbuhler 2008; MacLean and Hare 2013 for work with different criteria for shared intentionality). Importantly, these differences in human and ape social cognition may have quite far-reaching consequences for a wide variety of behaviors. For example, shared intentionality may support many forms of cultural learning (see Rolian and Carvalho, this volume; Henrich and Tennie, this volume) as well as human-specific forms of communication (Slocombe and Scott Philips, this volume). Thus, this cognitive difference may constrain the forms of social interaction that chimpanzees can engage in compared to humans (Tomasello et al. 2005; Herrmann et al. 2007; Wobber et al. 2013).

Hypothesis 2: Mental Time Travel

Another important proposal for human-unique cognition concerns our ability to recall the past and imagine the future (Suddendorf and Corballis 1997, 2007; Suddendorf and Bussey 2003). As is the case for theories of human social cognition, this hypothesis concerns whether animals share a specific cognitive mechanism with humans—not whether other animals generally lack memory or never act in accordance with future events. This hypothesis particularly focuses on whether animals possess episodic memory and prospection. In contrast to other forms of memory, such as semantic memory for facts about the world, episodic memory allows individuals to mentally reexperience life events—with the feeling of having been there personally (Tulving 1983). For example, imagine the difference between knowing the date of your birth, and knowing what happened on your most recent birthday—while you may know many details about the day you were born, you (probably!) only recall a set of facts that others have reported to you. In contrast, you likely can recreate your own personal experiences on your most recent birthday. This ability to mentally simulate past events is also involved in anticipating what is likely to happen in future (Buckner and Carroll 2007; Schacter et al. 2007; Suddendorf and Corballis 2007), and episodic memory and future planning are sometimes together referred to as *mental time travel.* Thus, the claim is that animals are mentally "stuck in time" (Roberts 2002): they can track intervals, remember semantic information, and tailor their behavior to likely future occurrences—but they cannot represent particular episodes occurring in the past, nor simulate new events that will occur in the future.

Episodic Memory

Theoretical claims about human-unique memory skills go back to Wolfgang Koehler (1927): in a major synthesis of his work with captive chimpanzees on Tenerife Island, he noted that their sense of time was "limited in past and future." The issue of whether other animals can engage in mental time travel has been contentious ever since. Some theories about human episodic memory explicitly define this ability in terms of the phenomenal quality of these memories, in which individuals have conscious awareness of being engaged in the act of recollection (Tulving 1983, 1985; Tulving and Markowitsch 1998). This sort of definition obviously makes it quite difficult to test such abilities in nonlinguistic organisms. Consequently, comparative psychologists have developed behavioral criteria that animals must meet for the underlying cognitive abilities to be considered episodic memory. To meet this formal definition of episodic-like memory, animals must demonstrate knowledge of three pieces of information—what happened, when it happened, and where it happened—all structured into a single, coherent representation (Clayton et al. 2003; Hampton and Schwartz 2004). But it is also important to note that both episodic memory and spatial memory more generally depend on the hippocampus (Burgess et al. 2002), and hippocampal neuroanatomy is highly conserved across mammals (Manns and Eichenbaum 2006; see also Murray et al. 2017). Indeed, recalling the spatial context of in which items were previously encountered is an important component of human episodic memory (Davachi 2006). This suggests that evidence for shared mechanisms supporting spatial memory and navigation in humans and other animals can also illuminate the evolution of episodic memory.

To date, some of the strongest evidence for episodic-like memory in nonhumans comes from studies of caching behavior in corvids (Clayton and Dickinson 1999; Emery and Clayton 2001; Clayton et al. 2003; see Martin-Ordas and Call 2013 for a review). What about apes? There is strong evidence that apes do use fairly sophisticated cognitive skills to encode spatial locations. For example, chimpanzees can recall the locations of multiple items that were hidden in a large space, and use a "cognitive map" of the optimal route between these locations when retrieving food from these locations (Menzel 1973; Menzel et al. 2002). Moreover, these memories persist over long periods. In one study, a language-trained chimpanzee could communicate the type and location of food they had seen hidden in the enclosure up to sixteen hours earlier (Menzel 1999), and chimpanzees preferentially searched at locations where they had previously received food even three months previously (Mendes and Call 2014). This suggests that the apes' representations of the spatial locations are quite temporally durable. However, there are some potential differences between ape spatial memory capacities. For example, while juvenile chimpanzees exhibit more accurate spatial memory than do infant chimpanzees, bonobos show no such developmental improvement over the same age range (Rosati and Hare 2012). This indicates that humans and chimpanzees, but not bonobos, may share similar patterns of ontogenetic improvement in spatial memory during early childhood (Newcombe and Huttenlocher 2006).

Is there evidence that apes can also recall specific events in the past by binding items with their context? One approach to answering this question has been to expose apes to unique events and then test their memory. For example, apes might be exposed to an unusual event—such as an unfamiliar person playing a guitar—and then later be asked to identify that person. In fact, gorillas are able to successfully select cards associated with that person and item after a delay (Schwartz et al. 2002, 2004). Another approach is to examine whether apes can recall unique episodes over very long timescales. For example, in one study chimpanzees and orangutans were presented with the unique room setup used in a tool-finding task the apes had completed three years previously. These situational cues were enough to trigger the apes' recall of the location of specific tools hidden in the room (Martin-Ordas et al. 2013). Both of these approaches show that apes have sophisticated, temporally durable memories for unique events they experienced in their past. However, it is unclear whether they use episodic-like memory in these contexts, as they do not require bound representations of what, when, and where. For example, recalling an unfamiliar person who played a guitar required an association of "who" with "what" information, but it is not necessary to think about when specifically this event happened. Similarly, apes had to recall a unique event that had happened in the past to solve the tool-finding task, but not necessarily use information about *how long ago* the event happened.

The best evidence to date for apes binding what-when-where information comes from a study involving choices between foods that decayed over different timescales (Martin-Ordas et al. 2010). Chimpanzees, bonobos, and orangutans saw a container in one location baited with a preferred but perishable food (frozen juice), whereas another was baited with a less-preferred but nonperishable food (a grape). Apes were then given the opportunities to choose one of these containers before seeing their current contents (see Figure 19.3). When given the opportunity to choose after five minutes, apes selected the container with the frozen juice—the option that was generally more preferred. However, after one hour—a duration during which the frozen juice would have melted away—they preferentially selected the container with the grape. This shift in responses indicates that apes recalled what food was baited where, as well as how long ago these baiting events occurred. Yet it is important to note that these abilities may be somewhat fragile, unlike the more automatic encoding and recall of episodes that humans exhibit. For example, in a more complex situation in which food artificially decayed over

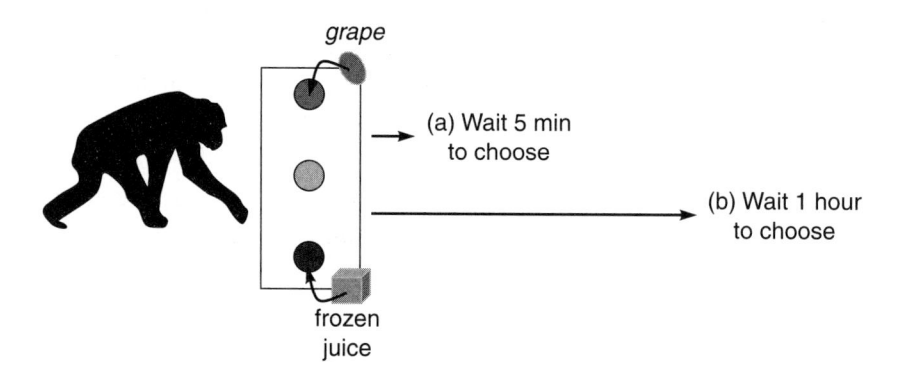

FIGURE 19.3. Episodic memory: Apes can remember where and how long ago different items were hidden. Setup for memory task, adapted from Martin-Ordas et al. (2010). Chimpanzees, bonobos, and orangutans chose between a location baited with a preferred but perishable food item (frozen juice, which melted and became inaccessible over time) and less preferred but temporally stable food item (a grape). Apes had to choose before observing the current contents of the containers. (a) If apes could make their choice after only a five-minute delay, they preferred the location baited with the perishable juice. (b) If apes could not choose until one hour after the baiting, they preferred the location with the grape.

different timescales, chimpanzees failed to form such integrated memories (Dekleva et al. 2011). Thus, apes show durable long-term memories, robust recall of spatial information, and can encode at least some memories that integrate what-when-where information—but they do not always do so.

Future Planning

As is the case for episodic memory, comparative psychologists have developed fairly stringent behavioral criteria for what qualifies as evidence for future prospection in nonhumans. Animals may show many future-oriented behaviors that are not planning; this test concerns whether animals can act to satisfy their future rather than immediate needs. To understand this difference, imagine you are rushing between two meetings. If you are hungry when you leave the first meeting, you buy a snack—but if you are running late, you might not have a chance to eat it until after the second meeting. Contrast this with an alternative scenario where you feel fine after the first meeting, but know you tend to feel hungry in the middle of the afternoon— so you buy a snack in anticipation of how you will feel after the second meeting. In both cases, you completed the same actions (buying a snack and eating it later)—but in the latter case you planned for a future need, whereas in the former you acted due to your current state. Similarly, animals may seem to act for the future but actually be thinking only about their current motivational state. For example, chimpanzees have been observed to transport a tool some distance until they find suitable nuts to crack (Boesch and Boesch 1984), but this may be because their current hunger state is not yet satisfied. Thus, the most rigorous criteria for future prospection requires animals to anticipate a future need that they are not currently experiencing, termed the Bischof-Köhler criteria (Suddendorf and Corballis 1997; Mulcaly and Call 2006; Raby et al. 2007).

Studies of future planning in corvids have focused on food caching contexts, in which birds store food for retrieval at a later time (e.g., Raby et al. 2007). As apes do not store food, studies of ape future planning have focused on tool-selection tasks in which apes must select and save a tool in anticipation of a future opportunity to use it. In one influential study, both bonobos and orangutans were presented with an out-of-reach apparatus where they needed a tool to acquire food rewards (Mulcahy and Call 2006). In this situation, apes anticipated that they would have access to this apparatus, selected the correct tool from a set of possibilities in advance, and retained this tool for long periods. Indeed, some apes were even able to select the correct tool and hang on to it overnight, carrying it into their sleeping room and then back to the testing room in the morning. This shows that apes can select tools not only on the basis of their current need, but also in anticipation of the future. Indeed, both species showed similar performance, suggesting that this capacity might be widely shared among great apes.

However, some have argued that this situation does not meet the strictest definition of the Bischof-Köhler criteria (Suddendorf 2006). Even though apes might have known they could not yet use the tool, they may nonetheless have experienced an internal desire for the food rewards throughout the duration of the test (at least while they were awake!). That is, it is unclear whether the apes were actually planning for a need they were not currently experiencing. However, some evidence suggests that apes can overcome immediate needs when faced with similar tool-selection choices (Osvath and Osvath 2008). For example, chimpanzees and orangutans were able to select between a tool (a straw) that would allow future access to juice, when also presented with an alternative of taking a preferred fruit that provided immediate benefit. Furthermore, if given a second choice between another tool and another fruit, apes then selected the fruit—indicating that the tool was merely an instrument for satisfying future desires, not motivationally equivalent to food. Indeed, even though apes do not normally store food in their natural behaviors, some evidence suggests that chimpanzees will spontaneously save raw food for several minutes, in order to place it in a device that appeared to transform it into a more desirable cooked item. In contrast, chimpanzees without access to this device in a control condition consumed all the raw food in their possession (Warneken and Rosati 2015).

Together with evidence of anticipatory tool selection, these studies provide clear evidence that apes can plan for their future states. However, there is still debate concerning whether these findings satisfy the more conservative Bischof-Köhler criteria, requiring that animals plan for future states that they are not currently experiencing (see Suddendorf et al. 2008; Osvath 2010 for this discussion). That is, while apes show a clear capacity to plan for future desires related to food or hunger, it is unclear whether they can plan for activities that are not strongly related to such central biological needs. Indeed, the challenge of devising valid tests for nonlinguistic organisms means that most studies thus far have focused on mental time travel specifically in food-acquisition contexts.

Hypothesis 3: Executive Control

Social cognition and mental time travel are examples of specific cognitive domains where human cognition may differ from that of other apes. That is, these hypotheses suggest that humans may have special representational capacities: perhaps our species alone can think about certain types of mental states in others, or form coherent memories of what happened when and where. The final hypothesis discussed in this chapter is that humans exhibit exceptional executive function. In contrast to those discussed in previous sections, this hypothesis stems not from comparisons of cognition in humans and animals, but rather primarily from theories about human brain evolution. In particular, humans exhibit a suite of anatomical changes in the size and structure of our frontal cortex—a region of the brain that supports executive function and decision making (Goldman-Rakic 1996; Rilling and Insel 1999; Miller and Cohen 2001; Semendeferi et al. 2002; Schoenemann et al. 2005). However, this hypothesis is receiving increasing attention from comparative psychologists as well (Rumbaugh et al. 1996; Barkley 2001; Siegal and Varley 2008). Executive functions allow individuals to flexibly regulate and control their behavior and goals, overriding responses that would otherwise be carried out automatically. That is, executive processes encompass a diverse set of cognitive skills allowing individuals to consider relevant information, inhibit currently inappropriate responses, and shift to new responses when the rules of the game change (Rosati 2017b). As such, executive functions encompass the cognitive skills that enable behavioral flexibility, and are an important regulator of many other behaviors.

Working Memory

One important component of human executive functioning is working memory (Kane and Engle 2002; Baddeley 2003). In contrast to the episodic memory skills discussed previously, working memory concerns the ability to consider and manipulate multiple pieces of information at once. That is, working memory is a system that processes information over the short term—and because attention is limited, this system inherently involves dealing with interference from multiple, possibly conflicting sources of information. In humans, working memory is thought to consist of storage systems for visual and auditory (or linguistic) information, as well as a central control process that organizes different information (Baddeley and Hitch

1974; Baddeley 2000). Working memory can therefore constrain other aspects of executive control if individuals cannot access the necessary information to make an appropriate response. For example, humans might be able to make more flexible, context-appropriate decisions if they are able to consider greater amounts of information at any given point.

One approach to measuring working memory in apes has been to examine their ability to remember multiple items over short time spans. For example, in one set of studies chimpanzees briefly saw up to nine Arabic numerals that were located randomly on a touch-screen computer (Biro and Matsuzawa 1999; Kawai and Matsuzawa 2000). In order to receive a treat, the chimpanzees had to touch the numbers in their correct order. The trick was that the remaining numbers were masked after the chimpanzee first touched the number "1"—so they had to recall the location of the entire sequence from memory. In fact, chimpanzees were quite successful at completing this ordering task, sometimes able to recall all the remaining numerals after they were masked. One chimpanzee even outperformed adult humans when the numbers were presented for very short time periods (Inoue and Matsuzawa 2007). However, the chimpanzees may have benefited from their extensive experience, as people who receive additional training perform more like those chimpanzees (Cook and Wilson 2010). There may also be important individual differences in chimpanzees' working memory abilities, as other chimpanzees recalled the location of only one numeral after the initial masking—despite similar experience with Arabic numerals (Beran et al. 2004). Together, this evidence suggests that apes and humans may have similar limits on working memory capacities, at least for visual information.

Response Control

Another important component of executive function is controlling responses—inhibiting actions when they are undesirable, and selecting the response that is appropriate given one's current goals (Mostofsky and Simmonds 2008). While working memory is therefore critical for processing information in order to determine how to proceed in a given situation, response control concerns how this course of action is actually implemented. In fact, a plethora of studies have shown that many other animals can have quite difficult times inhibiting prepotent responses. For example, one task used to study response control is the A-not-B test, derived from a set of observations by Piaget (1954). Here individuals initially experience that a reward is repeatedly hidden in one location (A), but on the critical test trials it is placed in a different location (B). In fact, even if they have seen the reward being placed in the second location, human infants and monkeys tend nonetheless to search at the old location (Diamond 1990). That is, they fail to inhibit their prepotent response to the spot that previously provided a reward. In contrast to the performance of these groups, apes and older children show near-ceiling levels of success on the A-not-B task (Barth and Call 2006). Moreover, two large comparative studies examining a wild variety of ape, monkey, and lemur species have indicated that apes are generally the most successful at inhibiting responses on the A-not-B task and related measures of response inhibition (Amici et al. 2008; MacLean et al. 2014). Thus, there is a clear difference in the level of control that apes have over their responses compared to primates more generally.

In terms of addressing human uniqueness, however, the question is whether humans nonetheless outpace nonhuman apes. In fact, in some contexts apes and human children exhibit similar levels of response control. In addition to their similar performance on the A-not-B task, apes and young children exhibit similarities in other reaching tasks. For example, apes and children both have some difficulty when they must inhibit direct reaching for a reward, but rather detour around the food to reach for it from another direction (Vlamings et al. 2010). However, when faced with more complex situations requiring the flexible inhibition of responses, children soon outpace apes. For example, when three-year-olds, six-year-olds, and chimpanzees initially learn to obtain a reward by raking it out of a tube or box, they can quickly acquire this skill. However, when they are then subsequently faced with a slightly different situation—where there is now a "trap" or hole in the bottom of the tube such that the reward will fall out of reach if they rake it out in the same way—six-year-olds are much more adept at shifting their response to acquire the reward with a new action than either three-year-olds or chimpanzees (Herrmann et al. 2015). Similarly, when the outcome of an event is ambiguous, such that a reward may come out of two possible locations, children rapidly outpace apes in preparing for both possible outcomes, rather than just one (Redshaw and Suddendorf 2016).

Even when human children and apes show similar performance, they may differ in the extent to which they can successfully inhibit their responses. For example, in one task individuals were faced with a reward trapped in a vertical maze (Voelter and Call 2014). Because the maze had various dead ends, apes and children had to plan their moves up to two steps in advance and inhibit prepotent actions in order to get the reward. Although apes performed similarly to children in terms of rate of success, apes had more difficulties than the children in inhibiting motor responses that led to dead ends. Another task tested how apes responded to a different vertical maze task in which individual had to move internal platforms so that a food reward would drop through to an accessible location at the bottom. Here, both bonobos and orangutans could succeed in situations where they could access the food by acting on the barrier that the food was initially resting on, but failed if they had to plan out an initial series of moves before touching to barrier with the food on it (Tecwyn 2013). That is, it was difficult for the apes to inhibit their prepotent response to reach toward the baited platform, even if doing so made it impossible to access the food subsequently. Together, these results indicate that, even though children and apes show similar performance on more simple reaching tasks, children may outpace chimpanzees in inhibiting motor responses in more complex contexts.

However, many of these tasks have only been implemented in a few species, making it difficult to make broader comparative claims. One response control test that has been widely studied across species is known as the reverse contingency task. Here an individual is faced with a choice between different value rewards—for example, one candy or five candies—but the trick is that they receive the item they *do not* choose. Thus, it is necessary to select the smaller reward in order to actually receive the larger reward (see Figure 19.4). This problem is challenging for chimpanzees and other primates, who generally cannot inhibit their tendency to approach or reach for the more valuable reward. In fact, primates are only able to overcome this prepotent response when faced with symbolic or nonconcrete representations of the rewards (Boysen and Berntson 1995; Boysen et al. 1996, 1999; Vlamings et al. 2006; Uher and Call 2008; Shifferman 2009). In contrast, children also improve when faced with symbolic rewards, but even many three-year-old children make the correct response when faced with real candy—and four-year-olds are even more successful (Carlson et al. 2005). That is, young children can easily solve an inhibitory control problem that is challenging (or impossible) for other primates.

Unfortunately, there have been few studies of chimpanzee performance in more typical cognitive neuroscience tasks used to assess response control in adult humans, such as the flanker task or Stroop task that examine how

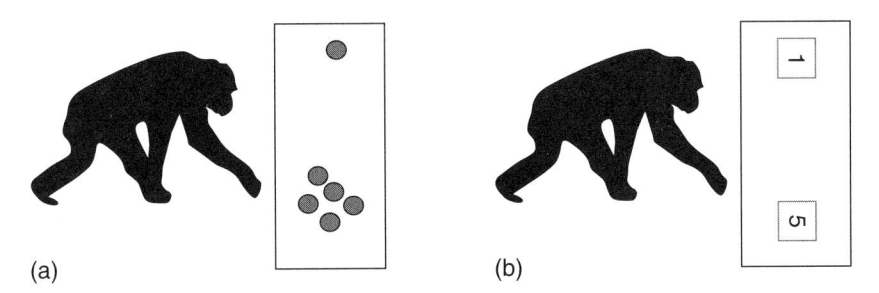

FIGURE 19.4. Response control: Symbolic representations improve primates' inhibition of prepotent responses. Setup for reverse contingency task, adapted from Boysen et al. (1996). Apes chose between a larger and smaller amount, but were given the option they did *not select*. (a) When apes saw different amounts of food, they had difficulty inhibiting choices for the larger amount, even though this resulted in their receiving the smaller amount. (b) When apes chose between symbolic representations of the food amounts (Arabic numerals), they were more successful at inhibiting. A variety of other species show similar improvements when faced with cues to the reward (such as color) rather than the concrete rewards.

individuals can control their responses under situations where there is conflict between different possible responses (Botvinick et al. 2004). For example, in the classic version of the Stroop task, individuals are asked to report the font color of a word they see, and the trick is that sometimes the word itself is a color term—that is, a person might see the word "brown" written in red font, so there is a conflict between correctly reporting the color written out (brown) and incorrectly reporting the actual color of the word itself (red). However, the little evidence that exists also suggests important differences between humans and apes. For example, a language-trained chimpanzee was tested on an adapted Stroop task in which she learned to associate geometric symbols with specific colors, and then had to report the color of stimuli she saw by choosing the associated lexigram (Beran et al. 2007). In fact, she made close to 50 percent errors when she faced conflicting stimuli—which is striking, as adult humans tend to make few actual errors in the Stroop task, but rather show slowed responses on trials involving response conflict (MacLeod and MacDonald 2000).

Strategic Control

Response control problems consist of decision-making situations in which individuals face conflict between different potential responses, as in the

Stroop task. In this case, individuals face a conflict between producing different possible motor actions, such as pushing the button to indicate red, versus pushing a different button to indicate brown. The "correct" response is not under debate—the conflict concerns whether individuals can actually produce this response. Most studies of animal cognition often measure response control in tasks tapping into this sort of control over motor actions, such as which direction to reach or which container to point at in order to successfully acquire food. But this type of conflict between competing motor responses is not the only sort of conflict humans face when making decisions. Other situations involve a choice between different cognitive strategies—that is, the problem is not whether individuals can produce the (correct) motor response, but how they assess what the "correct" thing to do is in the first place. Many kinds of real-world decisions require that individuals decide which *strategy* to pursue, prioritizing different aspects of the available information (Venkatraman et al. 2009a, 2009b). Executive control systems play a critical role in this kind of decision-making by updating the strategies or rules guiding behavior when new information comes to light (Mansouri et al. 2009). The ability to change and adapt responses in a changing world is therefore a key skill underlying behavioral flexibility.

One important example of a problem involving strategic control is intertemporal choices between rewards with different timings: individuals can pursue immediate gratification, or choose to wait for larger payoffs (Stevens and Stephens 2008). While humans often favor waiting for future gains—we can wait weeks or even months for larger rewards in some contexts (Frederick et al. 2002)—many other species seem quite impulsive and are willing to wait only a few seconds (Rachlin 2000). However, recent work suggests that chimpanzees and other apes can wait longer durations for future payoffs. For example, chimpanzees can sometimes wait over ten minutes in a delay of gratification task where rewards are accumulating over time (Beran 2002; Evans and Beran 2007). In fact, apes outwait monkeys and lemurs in accumulation situations (Stevens et al. 2011; Evans et al. 2012; Parrish et al. 2014), when they make a series of repeated choices between immediate and delayed options (Stevens et al. 2005; Rosati et al. 2007; Amici et al. 2008; Addessi et al. 2011; Stevens and Muhlhoff 2012), and in exchange situations where smaller rewards can later be traded for larger ones (Ramseyer et al. 2006; Dufour et al. 2007; Pelé et al. 2010, 2011).

Furthermore, although humans can outwait other species when faced with decisions about abstract rewards involving money (see Santos and Rosati 2015 for a review), chimpanzees and humans exhibit similar preferences when faced with matched choices involving delayed food rewards (Rosati et al. 2007).

These results indicate that chimpanzees can exert strategic control over their temporal choices, and sometimes even do so in ways comparable to humans. But can chimpanzees update and shift their strategies when faced with a changing world? A simple example of this is reversal learning, where individuals initially learn one rule (for example, that the correct response is to choose the blue item), but then the rule switches (now they should choose the red item). This situation therefore examines whether animals persevere in choosing the previously correct option, or update their representation of the rules of the game. Chimpanzees (and apes in general) outperform many other primate species in terms of picking up on the new contingencies (Rumbaugh and Pate 1984; Rumbaugh et al. 1996; Deaner et al. 2006). This suggests that apes exhibit a much more flexible ability to learn and apply new rules relative to other nonhuman primates—similar to their improved ability to control their motor responses.

It is also important to note that this type of reversal problem is markedly less complex than common tasks used to measure set shifting in humans. For example, one problem commonly used to assess human executive functioning is the Wisconsin card sorting task (Grant and Berg 1948; Milner 1968). Here individuals must sort modified playing cards, and the correct response changes once the player demonstrates initial learning of the rule. However, in contrast to primate reversal learning tasks, the images on these cards actually vary on multiple dimensions (such as number, color, and shape), and thus detecting the rule and updating responses is significantly more complicated. Grade-school children can already solve this task with the same level of accuracy as adults (Chelune and Baer 1986), but there has been no demonstration of how chimpanzees perform in a comparable task (but see Moore et al. 2005 for a simplified version developed for rhesus macaques). That is, many of the tasks developed to test these skills in nonhumans are quite simplified compared to problems that humans must solve. Consequently, future research comparing humans and apes on matched tasks will be critical to address the evolution of flexibility.

The Second Puzzle: Evolving the Human Mind

Current comparative evidence indicates that humans may have specialized cognitive abilities across several domains, including social cognition, mental time travel, and executive functioning. Although chimpanzees and other nonhuman apes share many homologous capacities with humans, they also exhibit important divergences that point to derived features in our lineage (see Table 19.1 for a summary). These comparisons are critical to address the phylogenetic problem of pinpointing which human cognitive traits are derived. But the second question about human cognitive uniqueness concerns why this suite of novel traits arose. In the following sections, I first argue that these diverse cognitive abilities—social cognition, prospection and memory, and executive control—are intrinsically connected, with overlapping psychological and neurobiological substrates. Then I argue that understanding the evolution of human cognition requires a careful examination of what aspects of human behavior they actually support. In fact, complex behaviors often recruit diverse cognitive skills, and many behavioral differences between great apes and humans cut across traditional cognitive distinctions.

Interrelationships between Abilities

The previous sections examined evidence for similarities and divergences between humans and chimpanzees in particular cognitive "traits" spanning the range of theories concerning human uniqueness (Shettleworth 2012). While social cognition is a classic example of a domain-specific skill that processes a particular type of information, executive functioning is a set of domain-general skills that influence a wide range of behaviors. Most of the experimental paradigms described in previous sections (necessarily) aim to tease apart discrete skills, while controlling for other possible influences on animals' performance. Yet increasing evidence from psychology and neuroscience suggests that these skills are in fact linked. It is therefore important to understand which cognitive abilities share underlying mechanisms, and which covary across individuals or species, in order to understand how these abilities evolved.

In fact, even abilities that seem quite distinct can sometimes recruit shared underlying mechanisms. For example, skills such as theory of mind

TABLE 19.1. Summary of empirical evidence for ape cognitive abilities.

Domain	Skill	Chimpanzees and Other Apes
Social cognition	Goal and intention understanding	Reason about behavior in terms of underlying goals, not just superficial actions
	Gaze following and perspective taking	Understand that others see things, and can track their line of sight
	Knowledge and beliefs	Reason about how seeing leads to knowing, but only implicit understanding of false beliefs
	Joint attention	Little evidence for human-like triadic gaze interactions
	Joint intentions and collaboration	Prefer to work alone, and seem to show parallel (not joint) goals with partner
Mental time travel	Spatial memory	Rich representations of location of items in their spatial context
	Long-term memory	Some evidence that apes can recall unique events for several weeks or even years
	Episodic memory	Can bind item with spatial and temporal context, but may be a fragile representation
	Save artifacts for the future	Can select and retain a useful tool for a future action over hours or overnight
	Plan for future needs	Some evidence that apes abstain from immediately eating food in order to obtain something more desirable in the future
Executive control	Working memory	Can hold in mind and manipulate complex information over short timescales
	Simple response control	Flexible inhibition of undesirable reaching actions compared to other primates
	Complex response control	Some control over undesirable actions, but less robust than humans
	Temporal discounting	Greater capacity to delay gratification than other primates, but less robust than humans
	Updating rules and set shifting	Greater capacity to update rules than other primates, but little research on more complex human tasks

and spatial memory might be considered distinct domains—one encompassing how we think about the social world, and the other encompassing how we think about the physical world (Tomasello and Call 1997; Herrmann et al. 2007). However, there is increasing evidence that social cognition and episodic memory or planning may actually share some core cognitive systems. For example, there is broad overlap in the brain regions supporting social cognition, episodic memory and prospection, and even spatial navigation and memory (Buckner and Carroll 2007). In fact, this proposal suggests that a key constituent process underlying these cognitive abilities is simulation: theory of mind involves simulating the content of another individual's mind; mental time travel involves simulating past or future events; and navigation involves simulating a path through a complex environment. That is, these diverse abilities may depend on a core computational problem of simulating events that are not directly witnessed in the moment.

Executive function systems, by their very nature, are concerned with regulating and controlling other cognitive processes, and there is clear evidence that executive control influences both social cognition and mental time travel. In terms of social cognition, developmental psychologists have long recognized that executive function may constrain theory of mind, preventing individuals from displaying the full extent of their abilities in complex tasks (Hughes 1998; Perner and Lang 1999; Carlson et al. 2002; Baillargeon et al. 2010). For example, correctly indicating that another individual has a particular false belief about the world with a clear behavioral response (such as a reach or a point) requires the ability to hold two conflicting versions of the world in mind at once, and then inhibit responses concerning the (true) state of the world. As such, one possibility is that apes' failures at explicit false belief tasks may be partially due to the high executive control demands posed by those situations (see Krachun et al. 2007; Kaminski et al. 2008). This may also explain why apes do exhibit sensitivity to false beliefs in anticipatory looking tasks (Krupenye et al. 2016). Executive function demands in such situations are reduced, given that individuals do not need to inhibit a direct response toward the true location of an object.

The relationship between executive function and mental time travel is perhaps even stronger, as many studies of prospection in nonhumans involve heavy inhibitory control demands. For example, planning for the future inherently requires that an individual exhibit inhibitory control, subsuming present needs for future gains (Osvath and Osvath 2008). While the previous sections explored whether apes can construct mental simulations of future events (Suddendorf and Corballis 2007), successful future planning could not occur without the ability to suppress immediate gratification in favor of delayed rewards. Similarly, increasing cognitive and neurobiological evidence indicates that successful recall of episodic memory engages multiple cognitive control processes in order to make an appropriate response (Baddeley

2000; Dobbins et al. 2002; Wagner et al. 2005). Accordingly, apes may be more successful at some episodic memory tasks because they require less executive control resources.

Defining Cognitive Adaptations

A major challenge for understanding the evolution of uniquely human cognitive abilities concerns identifying what constitutes a cognitive adaptation in the first place. Given that even seemingly disparate cognitive abilities share underlying psychological and neurobiological substrates, defining a coherent functional unit in the mind is a tricky task. Beyond these intrinsic relationships between psychological processes, many complex natural behaviors require the seamless integration of multiple abilities. Experimental cognitive tasks are explicitly designed to disentangle the contributions of a particular capacity, but real-world behavioral problems often require the flexible deployment of multiple skills in tandem. In fact, some of the behaviors of the greatest interest to psychologists, biologists, and anthropologists span these cognitive domains. Consequently, understanding why human-unique cognitive abilities evolved requires understanding the role of those abilities in contexts in which they are actually used (Rosati 2017a).

Consider the example of chimpanzee hunting, a behavior that has long been a focus of both theoretical ideas and experimental research on the origins of human behavior (Boesch and Boesch 1989; Stanford 1999; Muller and Mitani 2005; Gilby and Wrangham 2007; Tomasello et al. 2012). What cognitive capacities do chimpanzees use when hunting? Hunting clearly has a strong social component: often only one chimpanzee captures the monkey, but other individuals also tend to get at least a little meat (Mitani and Watts 2001; Gilby 2006). These hunting behaviors have therefore inspired psychological insights into the social cognitive capacities that support cooperation. However, this sort of complex behavior likely recruits cognitive skills from other domains as well. For example, successful hunting clearly requires components of executive function, including working memory to track the different visual paths of monkeys and chimpanzees during the chase, or response control to stop oneself from lunging directly at a monkey when doing so might result in catastrophic falls. Chimpanzees' abilities to plan and anticipate the future location of monkeys as they are being chased could also impact their hunting success in the wild.

From an evolutionary perspective, it is therefore critical to consider cognitive abilities in terms of the real-world behaviors. By integrating experimental comparisons of cognitive skills with information about different species' natural history, it is therefore possible to address *why* differences in cognitive traits may arise. In particular, the comparative method—one of the most powerful tools in evolutionary biology—can provide insight into the ultimate function of cognitive capacities. The comparative method illuminates the historical process of natural selection by examining the traits of different species in relationship to their socioecological context (Clutton-Brock and Harvey 1979; Mayr 1982; Harvey and Purvis 1991). Although such species comparisons are used widely for understanding the evolution of morphological characters, only more recently have they been applied to the problem of cognitive evolution (Balda and Kamil 1989; Clayton and Krebs 1994; Bond et al. 2003; Sherry 2006; Amici et al. 2008; MacLean et al. 2012). In terms of understanding human evolution specifically, this approach can assess whether the unique behavioral problems humans face in their natural world leave a recognizable psychological signature in our minds—much like the types of morphological signatures that can be identified in the dentition or other physical characteristics.

Human Cognitive Specializations

Human behavioral ecology differs from that of other great apes in several important ways (Marlowe 2005; Hill et al. 2009). First, human hunter-gatherers tend to consume higher-quality resources that are more difficult to acquire and may involve additional processing. For example, humans eat more meat, and consume more foods that involve extractive techniques such as nuts or honey than do chimpanzees (Cordain et al. 2000; Kaplan et al. 2000; Marlowe et al. 2014). Indeed, humans are characterized as inhabiting a cognitive ecological niche in terms of their foraging behavior. Humans also uniquely depend on starchy foods such as roots that must be cooked to become palatable and energetically valuable (Wrangham et al. 1999; Marlowe and Berbesque 2009; Wrangham 2009). In contrast, while chimpanzees possess many of the basic cognitive capacities needed to cook, such as the ability to delay gratification and understand the causal transformation that occurs during the cooking process (Warneken and Rosati 2015), they do not engage in these kinds of cooking behaviors in the wild—possibly because of the social risk of theft associated with engaging in cooking.

Second, human hunter-gatherers have larger home ranges than other apes (Marlowe 2005). While chimpanzees and other apes are more reliant on fruits and leaves, humans' reliance on distant, patchily distributed resources means that they must travel farther to find them. Moreover, while apes tend to consume food on the go, humans exhibit a unique pattern of central-place foraging, in which foods are brought back to and shared at a central camp. Indeed, humans engage in several complex cooperative behaviors involving the redistribution of food. Importantly, a diet that focuses on high-quality, hard-to-acquire resources is a diet that carries great inherent risk (Hill and Hawkes 1983; Hawkes et al. 2001): foragers exploit a food supply that is superabundant on some days, but absent on others. Accordingly, although humans are relatively risk-averse when faced with decisions about money, some evidence suggests we are more risk-prone when making decisions about food (Rosati and Hare 2016). Food sharing may represent an additional mechanism for humans to reduce the inherent variability that results from exploiting a high-quality diet (Kaplan et al. 2012).

Successfully utilizing these complex resources—locating food, processing that food, and finally distributing that food within a social group—requires a diverse array of cognitive abilities. For example, hunting and food processing are complex behaviors that are often learned socially. Some activities, such as hunting, may also occur in social groups where members cooperate with others to be successful, and therefore involve social cognitive skills supporting collaboration (Gurven 2004). Utilizing distantly located food resources across humans' large ranges requires identifying the location of food and navigating between locations. Thus, memory and planning systems allow individuals to successfully recall the locations of food and plan a route between them and camp. Finally, all of these behaviors require that humans exert executive control. For example, central-place foraging patterns mean that individuals must successfully avoid the immediate temptation of consuming all the food once they have access to it, but rather carry major portions of food as they navigate back to camp in order to share it. Similarly, many behaviors that are necessary to process food—such as cooking—require that individuals inhibit immediate consumption. Critically, all of these skills must be used in order to successfully exploit human-specific resources. Thus, human cognitive uniqueness may be best conceptualized as a suite of capacities cutting across psychological domains (Rosati 2017a).

Comparative studies further support the hypothesis that these cognitive abilities can exhibit evolutionary coherence in their emergence—at least when an animal's socioecological niche requires the integration of these skills. In fact, many differences between human ecology and that of other apes also exist (albeit in less dramatic form) between chimpanzees and bonobos. For example, chimpanzees depend more on spatially dispersed food resources with greater seasonal variability, face higher levels of social competition, and engage in more costly hunting and extractive foraging behaviors than bonobos (Kano 1992; Wrangham and Peterson 1996; Hare et al. 2012). Accordingly, chimpanzees exhibit more accurate spatial memory (Rosati and Hare 2012), higher levels of temporal self-control (Rosati et al. 2007; Rosati and Hare 2013), and more consistently demonstrate robust theory of mind skills in competitive contexts (see Rosati et al. 2010 for a review). Some of these same skills also seem to emerge in tandem in other taxa. For example, corvids exhibit robust skills in both social cognition and episodic memory (Emery and Clayton 2004; Clayton et al. 2007)—and their natural caching behaviors likely require both sets of skills to sucessfully retrieve cached foods without theft by compeitiors. Unfortunately, to date there has been little research systematically investigating capacities across all these domains in other species of interest, such as elephants, cetaceans, or hyenas—all of which exhibit complex social cognitive abilities (Connor 2007; Holekamp et al. 2007; Plotnik et al. 2011). However, comparative studies of apes and corvids provide tantalizing clues that these abilities may emerge together phylogenetically. Together, this provides convergent evidence for hypotheses about why human and ape cognition differs, as well as how different cognitive skills are mechanistically linked across evolutionary history.

Summary

In contrast to theories focused on identifying a single fundamental difference between the cognition of humans and apes, I have proposed that there are important divergences across a suite of psychological abilities including social cognition, episodic memory and planning, and executive control. At first glance, this range of differences presents a challenging problem for understanding human cognitive evolution: if humans and other apes differ psychologically in so many diverse ways, how is it possible to identify the ultimate evolutionary roots of these differences? I have argued that it is important to consider cognitive capacities not in isolation, but in terms of their interrelationships as well as their contributions to evolutionarily relevant behaviors. In fact, even diverse skills like social cognition, memory, and executive function have clear links in the mind and brain—and often must be used in tandem for the implementation of complex human behaviors such as hunting, foraging, and food sharing. As such, these suites of cognitive traits may emerge together in phylogeny. Comparisons of cognition in humans and chimpanzees—when integrated with information about these species' natural socioecology and behavior—can therefore begin to identify the evolutionary origins of human cognitive adaptations. Consequently, understanding the evolution of human cognitive uniqueness must involving relating those cognitive traits to the behavioral problems that humans uniquely must solve in the natural world.

Acknowledgments

I thank Felix Warneken and Alyssa Arre for comments on an earlier version of the manuscript.

References

Addessi, E., F. Paglieri, and V. Focaroli. 2011. The ecological rationality of delay tolerance: Insights from capuchin monkeys. *Cognition* 119: 142–147.

Amici, F., F. Aureli, and J. Call. 2008. Fission-fusion dynamics, behavioral flexibility and inhibitory control in primates. *Current Biology* 18: 1415–1419.

Baddeley, A. 2000. The episodic buffer: A new component of working memory. *Trends in Cognitive Sciences* 11: 417–423.

Baddeley, A. D. 2003. Working memory: Looking back and looking forward. *Nature Reviews Neuroscience* 4: 829–838.

Baddeley, A. D., and G. Hitch. 1974. Working memory. In G A. Bower, ed., *The Psychology of Learning and Motivation*, 48–79. New York: Academic Press.

Baillargeon, R., R. Scott, and Z. He. 2010. False-belief understanding in infants. *Trends in Cognitive Sciences* 14: 110–118.

Balda, R. P., and A. C. Kamil. 1989. A comparative study of cache recovery by three corvid species. *Animal Behaviour* 38: 486–495.

Barkley, R. A. 2001. The executive functions and self-regulation: An evolutionary neuropsychological perspective. *Neuropsychology Review* 11: 1–30.

Barth, J., and J. Call. 2006. Tracking the displacement of objects: A series of tasks with great apes and young children. *Journal of Experimental Psychology: Animal Behavior Processes* 32: 239–252.

Beran, M. J. 2002. Maintenance of self-imposed delay of gratification by four chimpanzees *(Pan troglodytes)* and an orangutan *(Pongo pygmaeus)*. *Journal of General Psychology* 129: 49–66.

Beran, M. J., J. L. Pater, D. A. Washburn, and D. M. Rumbaugh. 2004. Sequential responding and planning in chimpanzees *(Pan troglodytes)* and rhesus macaques *(Macaca mulatta)*. *Journal of Experimental Psychology: Animal Behavior Processes* 2004: 203–212.

Beran, M. J., D. A. Washburn, and D. M. Rumbaugh. 2007. A stroop-like effect in color-naming of color-word lexigrams by a chimpanzee *(Pan troglodytes)*. *Journal of General Psychology* 134: 217–228.

Biro, D., and T. Matsuzawa. 1999. Numerical ordering in a chimpanzee *(Pan troglodytes)*: Planning, executing, and monitoring. *Journal of Comparative Psychlogy* 113: 178–185.

Boesch, C., and H. Boesch. 1984. Mental map in wild chimpanzees: An analysis of hammer transports for nut cracking. *Primates* 25: 160–170.

Boesch, C., and H. Boesch. 1989. Hunting behavior of wild chimpanzes in the Taï National Park. *American Journal of Physical Anthroplogy* 78: 547–573.

Bond, A. B., A. C. Kamil, and R. P. Balda. 2003. Social complexity and transitive inference in corvids. *Animal Behaviour* 65: 479–487.

Boysen, S. T., and G. G. Berntson. 1995. Responses to quantity: Perceptual versus cognitive mechanisms in chimpanzees *(Pan troglodytes)*. *Journal of Experimental Psychology: Animal Behavior Processes* 21: 82–86.

Boysen, S. T., G. G. Berntson, M. B. Hannan, and J. T. Cacioppo. 1996. Quantity-based interference and symbolic representations in chimpanzees *(Pan troglodytes)*. *Journal of Experimental Psychology: Animal Behavior Processes* 22: 76–86.

Boysen, S. T., K. L. Mukobi, and G. G. Berntson. 1999. Overcoming response bias using symbolic representations of number by chimpanzees *(Pan troglodytes)*. *Animal Learning & Behavior* 27: 229–235.

Braeuer, J., J. Call., and M. Tomasello. 2005. All great ape species follow gaze to distant locations and around barriers. *Journal of Comparative Psychology* 119: 145–154.

Braeuer, J., J. Call, and M. Tomasello. 2007. Chimpanzees really know what others can see in a competitive situation. *Animal Cognition* 10: 439–448.

Braeuer, J., J. Call, and M. Tomasello. 2008. Chimpanzees do not take into account what others can hear in a competitive situation. *Animal Cognition* 11: 175–178.

Buckner, R. L., and D. C. Carroll. 2007. Self-projection and the brain. *Trends in Cognitive Sciences* 11: 49–57.

Bullinger, A. F., A. P. Melis, and M. Tomasello. 2011. Chimpanzees *(Pan troglodytes)* prefer individual over collaborative strategies toward goals. *Animal Behaviour* 82: 1135–1141.

Burgess, N., E. A. Maguire, and J. O'Keefe. 2002. The human hippocampus and spatial and episodic memory. *Neuron* 35: 625–641.

Buttelmann, D., J. Call, and M. Tomasello. 2009. Do great apes use emotional expressions to infer desires? *Developmental Science* 12: 688–698.

Buttelmann, D., M. Carpenter, J. Call, and M. Tomasello. 2007. Enculturated chimpanzees imitate rationally. *Developmental Science* 10: F31–F38.

Buttelmann, D., S. Schuette, M. Carpenter, J. Call, and M. Tomasello. 2012. Great apes infer others' goals based on context. *Animal Cognition* 15: 1037–1053.

Call, J., B. Hare, M. Carpenter, and M. Tomasello. 2004. "Unwilling" versus "unable": Chimpanzees' understanding of human intentional action. *Developmental Science* 7: 488–498.

Call, J., and M. Tomasello. 1998. Distinguishing intentional actions in orangutans *(Pongo pygmaeus)*, chimpanzees *(Pan troglodytes)*, and human children *(Homo sapiens)*. *Journal of Comparative Psychology* 112: 192–206.

Call, J., and M. Tomasello. 1999. A nonverbal false belief task: The performance of children and great apes. *Child Development* 70: 381–395.

Call, J., and M. Tomasello. 2008. Does the chimpanzee have a theory of mind? 30 years later. *Trends in Cognitive Sciences* 12: 187–192.

Carlson, S. M., A. C. Davis, and J. G. Leach. 2005. Less is more: Executive function and symbolic representation in preschool children. *Psychological Science* 16: 609–616.

Carlson, S. M., M. B. Moses, and C. Breton. 2002. How specific is the relation between executive function and theory of mind? Contributions of inhibitory control and working memory. *Infant and Child Development* 11: 73–92.

Carpenter, M., and J. Call. 2013. How joint is the joint attention of apes and human infants? In J. Metcalf and H. S. Terrace, ed., *Agency and Joint Attention*, 49–61. New York: Oxford University Press.

Chelune, G. J., and R. A. Baer. 1986. Developmental norms for the Wisconsin card sorting task. *Journal of Clinical and Experimntal Neurophysiology* 8: 219–228.

Cheney, D. L., and R. M. Seyfarth. 1990. *How Monkeys See the World*. Chicago: Chicago University Press.

Clayton, N. S., T. J. Bussey, and A. Dickinson. 2003. Can animals recall the past and plan for the future? *Nature Neuroscience* 4: 685–691.

Clayton, N. S., J. M. Dally, and N. J. Emery. 2007. Social cognition by food-caching corvids: The western scrub-jay as a natural psychologist. *Philosophical Transactions of the Royal Society B: Biological Sciences* 362: 507–522.

Clayton, N. S., and A. Dickinson. 1999. Motivational control of caching behaviour in the scrub jay, *Aphelocoma coerulescens*. *Animal Behaviour* 57: 435–444.

Clayton, N. S., and J. R. Krebs. 1994. Memory for spatial and object-specific cues in food-storing and non-storing birds. *Journal of Comparative Physiology A* 174: 371–379.

Clutton-Brock, T. H., and P. H. Harvey. 1979. Comparison and adaptation. *Proceedings of the Royal Society B* 205: 547–565.

Connor, R. C. 2007. Dolphin social intelligence: Complex alliance relationships in bottlenose dolphins and a consideration of selective environments for extreme brain size evolution in mammals. *Proceedings of the Royal Society B: Biological Sciences* 362: 587–602.

Cook, P., and M. Wilson. 2010. Do young chimpanzees have extraordinary working memory? *Psychonomic Bulletin and Review* 17: 599–600.

Cordain, L., J. B. Miller, S. B. Eaton, N. Mann, S. A. H. Holt, and J. D. Speth. 2000. Plant-animal subsistence ratios and macronutrient energy estimations in worldwide hunter-gatherer diets. *American Society for Clinical Nutrition* 71: 682–692.

Darwin, C. 1871. *The Descent of Man, and Selection in Relation to Sex*. London: Murray.

Darwin, C. 1872. *The Expression of the Emotions in Man and Animals*. New York: Oxford University Press.

Davachi, L. 2006. Item, context, and relational episodic encoding in humans. *Current Opinion in Neurobiology* 16: 693–700.

Deaner, R., C. van Schaik, and V. Johnson. 2006. Do some taxa have better domain-general cognition than others? A meta-analysis of nonhuman primate studies. *Evolutionary Psychology* 4: 149–196.

Dekleva, M., V. Dufour, H. de Vries, B. M. Spruijt, and E. H. M. Sterck. 2011. Chimpanzees *(Pan troglodytes)* fail a what-where-when task but find rewards by using a location-based association strategy. *PLoS ONE* 6: e16593.

Diamond, A. 1990. Developmental time course in human infants and infant monkeys, and the neural bases of, inhibitory control in reaching. *Annals of the New York Academy of Sciences* 608: 637–676.

Dobbins, I. G., H. Foley, D. L. Schacter, and A. D. Wagner. 2002. Executive control during episodic retrieval: Multiple prefrontal processes subserve source memory. *Neuron* 35: 989–996.

Dufour, V., M. Pelé, E. H. M. Sterck, and B. Thierry. 2007. Chimpanzee *(Pan troglodytes)* anticipation of food return: Coping with waiting time in an exchange task. *Journal of Comparative Psychology* 121: 145–155.

Emery, N. J., and N. S. Clayton. 2001. Effects of experience and social context on prospective caching strategies by scrub jays. *Nature* 414: 443–446.

Emery, N. J., and N. S. Clayton. 2004. The mentality of crows: Convergent evolution of intelligence in corvids and apes. *Science* 306: 1903–1907.

Evans, T. A., and M. J. Beran. 2007. Chimpanzees use self-distraction to cope with impulsivity. *Biology Letters* 3: 599–602.

Evans, T. A., M. J. Beran, F. Paglieri, and E. Addessi. 2012. Delaying gratification for food and tokens in capuchin monkeys *(Cebus apella)* and chimpanzees *(Pan troglodytes)*: When quantity is salient, symbolic stimuli do not improve performance. *Animal Cognition* 15: 539–548.

Flombaum, J. I., and S. Santos. 2005. Rhesus monkeys attribute perceptions to others. *Current Biology* 15: 447–452.

Frederick, S., G. Loewenstein, and T. O'Donoghue. 2002. Time discounting and time preference: A critical review. *Journal of Economic Literature* 2: 351–401.
Gilby, I. C. 2006. Meat sharing among the Gombe chimpanzees: Harassment and reciprocal exchange. *Animal Behaviour* 71: 953–963.
Gilby, I. C., and R. W. Wrangham. 2007. Risk-prone hunting by chimpanzees *(Pan troglodytes schweinfurthii)* increases during periods of high diet quality. *Behavioral Ecology and Sociobiology* 61: 1771–1779.
Goldman-Rakic, P. S. 1996. The prefontal landcape: Implications of functional architecture for understanding human mentation and the central executive. *Philosophical Transactions of the Royal Society B: Biological Sciences* 351: 1445–1453.
Grant, D. A., and E. A. Berg. 1948. A behavioral analysis of degree of reinforcement and ease of shifting to new responses in a Weigl-type card sorting problem. *Journal of Experimental Psychology* 34: 404–411.
Greenberg, J. R., K. Hamann, F. Warneken, and M. Tomasello. 2010. Chimpanzee helping in collaborative and noncollaborative contexts. *Animal Behaviour* 80: 873–880.
Gurven, M. 2004. To give and to give not: The behavioral ecology of human food transfers. *Behavioral and Brain Sciences* 27: 543–583.
Hamann, K., F. Warneken, J. R. Greenberg, and M. Tomasello. 2011. Collaboration encourages equal sharing in children but not chimpanzees. *Nature* 476: 328–331.
Hampton, R. R., and B. L. Schwartz. 2004. Episodic memory in nonhumans: What, and where, is when? *Current Opinion in Neurobiology* 14: 192–197.
Hare, B. 2001. Can competitive paradigms increase the validity of experiments on primate scial cognition? *Animal Cognition* 4: 269–280.
Hare, B. 2011. From hominoid to hominid mind: What changed and why? *Annual Review of Anthropology* 40: 293–309.
Hare, B., J. Call, B. Agnetta, and M. Tomasello. 2000. Chimpanzees know what conspecifics do and do not see. *Animal Behaviour* 59: 771–785.
Hare, B., J. Call, and M. Tomasello. 2001. Do chimpanzees know what conspecifics know? *Animal Behaviour* 61: 139–151.
Hare, B., J. Call, and M. Tomasello. 2006. Chimpanzees deceive a human competitor by hiding. *Cognition* 101: 495–514.
Hare, B., and M. Tomasello. 2004. Chimpanzees are more skillful in competitive than cooperative cognitive tasks. *Animal Behaviour* 68: 571–581.
Hare, B., V. Wobber, and R. Wrangam. 2012. The self-domestication hypothesis: Evolution of bonobo psychology is due to selection against aggression. *Animal Behaviour* 83: 573–585.
Hare, B., and S. Yamamoto. 2015. Moving bonobos off the scientifically endangered list. *Behaviour* 152: 247–258.
Harvey, P. H., and A. Purvis. 1991. Comparative methods for explaining adaptations. *Nature* 351: 619–624.
Hawkes, K., J. F. O'Connell, and N. G. Blurton Jones. 2001. Hadza meat sharing. *Evolution and Human Behavior* 22: 113–142.

Hermer-Vazquez, L., E. Spelke, and A. S. Katsnelson. 1999. Sources of flexibility in human cognition: Dual-task studies of space and language. *Cognitive Psychology* 39: 3–36.

Herrmann, E., J. Call, M. V. Hernadez-Lloreda, B. Hare, and M. Tomasello. 2007. Humans have evolved specialized skills of social cognition: The cultural intelligence hypothesis. *Science* 317: 1360–1366.

Herrmann, E., B. Hare, J. Call, and M. Tomasello. 2010. Differences in the cognitive skills of bonobos and chimpanzees. *PLoS ONE* 5: e12438.

Herrmann, E., A. Misch, V. Hernadez-Lloreda, and M. Tomasello. 2015. Uniquely human self-control begins at school age. *Developmental Science* 18: 979–993.

Heyes, C. M. 1998. Theory of mind in nonhuman primates. *Behavioral and Brain Sciences* 21: 101–148.

Hill, K., and K. Hawkes. 1983. Neotropical hunting among the Ache of Eastern Paraguay. In *Adaptive Responses of Native Amazonians,* ed. R. Hames and W. Vickers, 139–188. New York: Academic Press.

Hill, K., M. Barton, and M. Hurtado. 2009. The emergence of human uniqueness: Characters underlying behavioral modernity. *Evolutionary Anthropology* 18: 187–200.

Holekamp, K. E., S. T. Sakai, and B. L. Lundrigan. 2007. Social intelligence in the spotted hyena *(Crocuta crocuta). Proceedings of the Royal Society B: Biological Sciences* 362: 523–538.

Horner, V., and A. Whiten. 2005. Causal knowledge and imitation/emulation switching in chimpanzees *(Pan troglodytes)* and children *(Homo sapiens). Animal Cognition* 8: 164–181.

Hughes, C. 1998. Executive function in preschoolers: Links with theory of mind and verbal ability. *British Journal of Developmental Psychology* 16: 233–253.

Inoue, S., and T. Matsuzawa. 2007. Working memory of numerals in chimpanzees. *Current Biology* 17: R2004–R1005.

Kaminski, J., J. Call, and M. Tomasello. 2008. Chimpanzees know what others know, but not what they believe. *Cognition* 109: 224–234.

Kane, M. J., and R. W. Engle. 2002. The role of prefrontal cortex in working-memory capacity, executive attention, and general fluid intelligence: An individual-differences perspective. *Psychonomic Bulletin & Review* 9: 637–671.

Kano, F., and J. Call. 2014. Cross-species variation in gaze following and conspecific preference among great apes, human infants and adults. *Animal Behaviour* 91: 137–150.

Kano, T. 1992. *The Last Ape: Pygmy Chimpanzee Behavior and Ecology.* Stanford, CA: Stanford University Press.

Kaplan, H., K. Hill, J. Lancaster, and A. M. Hurtado. 2000. A theory of human life history evolution: Diet, intelligence, and longevity. *Evolutionary Anthropology* 9: 156–185.

Kaplan, H., E. Schiniter, V. L. Smith, and B. J. Wilson. 2012. Risk and the evolution of human exchange. *Proceedings of the Royal Society B: Biological Sciences* 279: 2930–2935.

Kawai, N., and T. Matsuzawa. 2000. Numerical memory span in a chimpanzee. *Nature* 403: 39–40.

Koehler, W. 1927. *The Mentality of Apes*. London: Routledge & Kegan Paul.

Krachun, C., M. Carpenter, J. Call, and M. Tomasello. 2007. A competitive nonverbal false belief task for children and apes. *Developmental Science* 12: 521–535.

Krupenye, C., F. Kano, S. Hirata, J. Call, and M. Tomasello. 2016. Great apes anticipate that other individuals will act according to false beliefs. *Science* 354: 110–114.

Lyons, D. E., and L. R. Santos. 2006. Ecology, domain specificity, and the origins of theory of mind: Is competition the catalyst? *Philosophy Compass* 1: 481–492.

MacLean, E. L., and B. Hare. 2012. Bonobos and chimpanzees infer the target of another's attention. *Animal Behavior* 83: 345–353.

MacLean, E. L., and B. Hare. 2013. Spontaneous triadic engagement in bonobos *(Pan paniscus)* and chimpanzees *(Pan troglodytes)*. *Journal of Comparative Psychology* 127: 245–255.

MacLean, E. L., B. Hare, C. L. Nunn, E. Addessi, F. Amici, R. C. Anderson, F. Aureli, J. M. Baker, A. E. Bania, A. M. Barnard, N. J. Boogert, E. M. Brannon, E. E. Bray, J. Bray, L. J. N. Brent, J. M. Burkart, J. Call, J. F. Cantlon, L. G. Cheke, N. S. Clayton, M. M. Delgado, J. R. DiVincenti, K. Fujita, E. Herrmann, C. Hiramatsu, L. J. Jacobs, K. E. Jordan, J. R. Laude, K. L. Leimgruber, E. J. E. Messer, A. C. de A. Moura, L. Ostojic, A. Picard, M. L. Platt, J. M. Plotnik, F. Range, S. M. Reader, R. B. Reddy, A. R. Sandel, L. R. Santos, K. Schumann, A. M. Seed, K. R. Sewall, R. C. Shaw, K. E. Slocombe, Y. Su, A. Takimoto, J. Tan, R. Tao, C. P. van Schaik, Z. Viranyi, E. Visalberghi, J. C. Wade, A. Watanabe, J. Widness, J. K. Young, T. R. Zentall, and Y. Zhao. 2014. The evolution of self-control. *Proceedings of the National Academy of Sciences* 111: E2140–E2148.

MacLeod, C. M., and P. A. MacDonald. 2000. Interdimensional interference in the Stroop effect: Uncovering the cognitive and neural anatomy of attention. *Trends in Cognitive Sciences* 4: 383–391.

Manns, J. R., and H. Eichenbaum. 2006. Evolution of declarative memory. *Hippocampus* 16: 795–808.

Mansouri, F. A., K. Tanaka, and M. J. Buckley. 2009. Conflict-induced behavioural adjustment: A clue to the executive functions of the prefrontal cortex. *Nature Reviews Neuroscience* 10: 141–152.

Marlowe, F. W. 2005. Hunter-gatherers and human evolution. *Evolutionary Anthropology* 14: 54–67.

Marlowe, F. W., and J. C. Berbesque. 2009. Tubers as fallback foods and their impact on Hadza hunter-gatherers. *American Journal of Physical Anthropology* 140: 751–758.

Marlowe, F. W., J. C. Berbesque, B. Wood, A. Crittenden, C. Porter, and A. Mabulla. 2014. Honey, Hazda, hunter-gatherers, and human evolution. *Journal of Human Evolution* 71: 119–128.

Marticorena, D. C., A. M. Ruiz, C. Mukerji, A. Goddu, and L. R. Santos. 2011. Monkeys represent others' knowledge but not their beliefs. *Developmental Science* 14: 1406–1416.

Martin, A., and L. R. Santos. 2016. What cognitive representations support primate theory of mind? *Trends in Cognitive Sciences* 20: 375–382.

Martin-Ordas, G., D. Berntsen, and J. Call. 2013. Memory for distant past events in chimpanzees and orangutans. *Current Biology* 23: 1438–1441.

Martin-Ordas, G., and J. Call. 2013. Episodic memory: A comparative approach. *Frontiers in Behavioral Neuroscience* 7: 63.

Martin-Ordas, G., D. Haun, F. Colmenares, and J. Call. 2010. Keeping track of time: Evidence for episodic-like memory in great apes. *Animal Cognition* 13: 331–340.

Mayr, E. 1982. *The Growth of Biological Thought*. Cambridge, MA: Harvard University Press.

Melis, A., J. Call, and M. Tomasello. 2006a. Chimpanzees *(Pan troglodytes)* conceal visual and auditory information from others. *Animal Behaviour* 120: 154–162.

Melis, A. P., B. Hare, and M. Tomasello. 2006b. Chimpanzees recruit the best collaborator. *Science* 311: 1297–1300.

Melis, A. P., B. Hare, and M. Tomasello. 2006c. Engineering cooperation in chimpanzees: Tolerance constraints on cooperation. *Animal Behaviour* 72: 275–286.

Mendes, N., and J. Call. 2014. Chimpanzees form long-term memories of food locations after limited exposure. *American Journal of Primatology* 76: 485–495.

Menzel, C. R. 1999. Unprompted recall and reporting of hidden objects by a chimpanzee *(Pan troglodytes)* after extended delays. *Journal of Comparative Psychology* 113: 426–434.

Menzel, C. R., E. S. Savage-Rumbaugh, and E. W. Menzel. 2002. Bonobo *(Pan paniscus)* spatial memory and communication in a 20-hectare forest. *International Journal of Primatology* 23: 601–619.

Menzel, E. W. 1973. Chimpanzee spatial memory organization. *Science* 182: 943–945.

Miller, E. K., and J. D. Cohen. 2001. An integrative theory of prefrontal cortex function. *Annual Review of Neuroscience* 24: 167–202.

Milner, B. 1968. Effects of different brain lesions on card sorting: The role of the frontal lobes. *Archives of Neurology* 9: 90–100.

Mitani, J. C., and D. P. Watts. 2001. Why do chimpanzees hunt and share meat? *Animal Behaviour* 61: 915–924.

Moore, T. L., R. J. Killiany, J. G. Herndon, D. L. Rosene, and M. B. Moss. 2005. A nonhuman primate test of abstraction and set shifting: An automated adaptation of the Wisconsin Card Sorting Test. *Journal of Neuroscience Methods* 146: 165–173.

Mostofsky, S. H., and D. J. Simmonds. 2008. Response inhibition and response selection: Two sides of the same coin. *Journal of Cognitive Neuroscience* 20: 751–761.

Mulcahy, N., and J. Call. 2006. Apes save tools for future use. *Science* 312: 1038–1040.

Muller, M. N., and J. C. Mitani. 2005. Conflict and cooperation in wild chimpanzees. *Advances in the Study of Behavior* 35: 275–331.

Murray, E. A., S. P. Wise, and K. S. Graham. 2017. *The Evolution of Memory Systems: Ancestors, Anatomy, and Adaptations*. Oxford: Oxford University Press.

Newcombe, N. S., and J. Huttenlocher. 2006. Development of spatial cognition. In D. Kuhn and R. S. Siegler, eds., *Handbook of Child Psychology*, 6th ed., 734–776. Hoboken, NJ: John Wiley.

Okamoto, S., M. Tomonaga, K. Ishii, N. Kawai, M. Tanaka, and T. Matsuzawa. 2002. An infant chimpanzee *(Pan troglodytes)* follows human gaze. *Animal Cognition* 5: 107–114.

Okamoto-Barth, S., J. Call, and M. Tomasello. 2007. Great apes' understanding of other individuals' line of sight. *Psychological Science* 18: 462–468.

Osvath, M. 2010. Great ape foresight is looking great. *Animal Cognition* 13: 777–781.

Osvath, M., and H. Osvath. 2008. Chimpanzees *(Pan troglodytes)* and orangutans *(Pongo abelii)* forethought: Self-control and pre-experience in the face of future tool use. *Animal Cognition* 11: 661–674.

Parrish, A. E., B. M. Perdue, E. E. Stromberg, A. E. Bania, T. A. Evans, and M. J. Beran. 2014. Delay of gratification by orangutans *(Pongo pygmaeus)* in the accumulation task. *Journal of Comparative Psychology* 128: 209–214.

Pelé, M., V. Dufour, J. Micheletta, and B. Thierry. 2010. Long-tailed macaques display unexpected waiting abilities in exchange tasks. *Animal Cognition* 13: 263–271.

Pelé, M., J. Micheletta, P. Uhlrich, B. Thierry, and V. Dufour. 2011. Delay maintenance in Tonkean macaques *(Macaca tonkeana)* and brown capuchin monkeys *(Cebus apella)*. *International Journal of Primatology* 32: 149–166.

Penn, D. C., K. J. Holyoak, and D. J. Povinelli. 2008. Darwin's mistake: Explaining the discontinuity between human and nonhuman minds. *Behavioral and Brain Sciences* 31: 109–178.

Perner, J., and B. Lang. 1999. Development of theory of mind and executive control. *Trends in Cognitive Sciences* 3: 337–344.

Piaget, J. 1954. *The Construction of Reality in the Child*. New York: Basic Books.

Pika, S., and K. Zuberbuhler. 2008. Social games between bonobos and humans: Evidence for shared intentionality? *American Journal of Primatology* 70: 207–210.

Pinker, S., and R. Jackendoff. 2005. The faculty of language: What's special about it? *Cognition* 95: 201–236.

Plotnik, J. M., R. Lair, W. Suphackoksahakun, and F. B. M. de Waal. 2011. Elephants know when they need a helping trunk in a cooperative task. *Proceedings of the National Academy of Sciences* 108: 5116–5121.

Povinelli, D. J., and T. J. Eddy. 1996a. Chimpanzees: Joint visual attention. *Psychological Science* 7: 129–135.

Povinelli, D. J., and T. J. Eddy. 1996b. What young chimpanzees know about seeing. *Monographs of the Society for Research in Child Development* 61: 1–152.

Prüfer, K., et al. 2012. The bonobo genome compared to the chimpanzee and human genomes. *Nature* 486: 527–531.

Raby, C. R., D. M. Alexis, A. Dickinson, and N. S. Clayton. 2007. Planning for the future by western scrub jays. *Nature* 445: 919–921.

Rachlin, H. 2000. *The Science of Self-Control*. Cambridge, MA: Harvard University Press.

Ramseyer, A., M. Pelé, V. Dufour, C. Chauvin, and B. Thierry. 2006. Accepting loss: The temporal limits of reciprocity in brown capuchin monkeys. *Proceedings of the Royal Society B: Biological Sciences* 273: 179–184.

Redshaw, J., and T. Suddendorf. 2016. Children's and apes' preparatory responses to two mutually exclusive possibilities. *Current Biology* 26: 1758–1762.

Rekers, Y., D. B. M. Haun, and M. Tomasello. 2011. Children, but not chimpanzees, prefer to collaborate. *Current Biology* 21: 1756–1758.

Rilling, J. K., and T. R. Insel. 1999. The primate neocortex in comparative perspective using magnetic resonance imaging. *Journal of Human Evolution* 37: 191–223.

Roberts, W. A. 2002. Are animals stuck in time? *Psychological Bulletin* 128: 473–489.

Rosati, A. G. 2017a. Foraging cognition: Reviving the ecological intelligence hypothesis. *Trends in Cognitve Sciences* 21: 691–702.

Rosati, A. G. 2017b. The evolution of primate executive function: From response control to strategic decision-making. In J. Kaas and L. Krubitzer, eds., *Evolution of Nervous Systems,* 2nd ed., 423–437. New York: Elsevier.

Rosati, A. G., and B. Hare. 2009. Looking past the model species: Diversity in gaze-following skills across primates. *Current Opinion in Neurobiology* 19: 45–51.

Rosati, A. G., and B. Hare. 2012. Chimpanzees and bonobos exhibit divergent spatial memory development. *Developmental Science* 15: 840–853.

Rosati, A. G., and B. Hare. 2013. Chimpanzees and bonobos exhibit emotional respones to decision outcomes. *PLoS ONE* 8: e63058.

Rosati, A. G., and B. Hare. 2016. Reward type modulates human risk preferences. *Evolution and Human Behavior* 37: 159–168.

Rosati, A. G., L. R. Santos, and B. Hare. 2010. Primate social cognition: Thirty years after Premack and Woodruff. In M. L. Platt and A. A. Ghazanfar, eds., *Primate Neuroethology.* Oxford: Oxford University Press.

Rosati, A. G., J. R. Stevens, B. Hare, and M. D. Hauser. 2007. The evolutionary origins of human patience: Temporal preferences in chimpanzees, bonobos, and human adults. *Current Biology* 17: 1663–1668.

Rosati, A. G., V. Wobber, K. Hughes, and L. R. Santos. 2014. Comparative developmental psychology: How is human cognitive development unique? *Evolutionary Psychology* 12: 448–473.

Rumbaugh, D. M., and J. Pate. 1984. The evolution of primate cognition: A comparative primate perspective. In H. L. Roitblat, H. S. Terrace, and T. G. Bever, eds., *Animal Cognition,* 569–587. Hillsdale, NJ: Lawrence Erlbaum Associates.

Rumbaugh, D. M., E. S. Savage-Rumbaugh, and D. A. Washburn. 1996. Toward a new outlook on primate learning and behavior: Complex learning and emergent processes in comparative perspective. *Japanese Psychological Research* 38: 113–125.

Santos, L. R., A. G. Nissen, and J. A. Ferrugia. 2006. Rhesus monkeys, *Macaca mulatta,* know what others can and cannot hear. *Animal Behavior* 71: 1175–1181.

Santos, L. R., and A. G. Rosati. 2015. The evolutionary roots of human decision making. *Annual Review of Psychology* 66: 3221–3347.

Schacter, D. L., D. R. Addis, and R. L. Buckner. 2007. Remembering the past to imagine the future: The prospective brain. *Nature Reviews Neuroscience* 8: 657–661.

Schmelz, M., J. Call, and M. Tomasello. 2011. Chimpanzees know that others make inferences. *Proceedings of the National Academy of Sciences* 108: 3077–3079.

Schoenemann, P. T., M. J. Sheehan, and L. D. Glotzer. 2005. Prefrontal white matter volume is disproportionately larger in humans than other primates. *Nature Neuroscience* 8: 242–252.

Schwartz, B. L., M. R. Colon, I. C. Sanchez, I. A. Rodriguez, and S. Evans. 2002. Single-trial learning of "what" and "who" information in a gorilla *(Gorilla gorilla gorilla)*: Implications for episodic memory. *Animal Cognition* 5: 85–90.

Schwartz, B. L., C. A. Meissner, M. Hoffman, S. Evans, and L. D. Frazier. 2004. Event memory and misinformation effects in a gorilla *(Gorilla gorilla gorilla)*. *Animal Cognition* 7: 93–100.

Semendeferi, K., A. Lu, N. Schenker, and H. Damasio. 2002. Humans and great apes share a large frontal cortex. *Nature Neuroscience* 5: 272–276.

Shettleworth, S. J. 2012. Modularity, comparative cognition and human uniqueness. *Philosophical Transactions of the Royal Society B: Biological Sciences* 367: 2794–2802.

Shifferman, E. M. 2009. Its own reward: Lessons to be drawn from the revsered-reward contingency paradigm. *Animal Cognition* 12: 547–558.

Siegal, M., and R. Varley. 2008. If we could talk to the animals. *Behavioral and Brain Sciences* 31: 146–147.

Stanford, C. B. 1999. *The Hunting Apes: Meat Eating and the Origins of Human Behavior*. Princeton NJ: Princeton University Press.

Sterelny, K. 2012. *The Evolved Apprentice*. Cambridge, MA: MIT Press.

Stevens, J. R., E. V. Hallinan, and M. D. Hauser. 2005. The ecology and evolution of patience in two New World monkeys. *Biology Letters* 1: 223–226.

Stevens, J. R., and N. Muhlhoff. 2012. Intertemporal choice in lemurs. *Behavioural Processes* 89: 121–127.

Stevens, J. R., A. G. Rosati, S. R. Heilbronner, and N. Muelhoff. 2011. Waiting for grapes: Expectancy and delayed gratification in bonobos. *International Journal of Comparative Psychology* 24: 99–111.

Stevens, J. R., and D. W. Stephens. 2008. Patience. *Current Biology* 18: R11–R12.

Suddendorf, T. 2006. Foresight and the evolution of the human mind. *Science* 312: 1006–1007.

Suddendorf, T., and J. Bussey. 2003. Mental time travel in animals? *Trends in Cognitive Sciences* 9: 391–396.

Suddendorf, T., and M. C. Corballis. 1997. Mental time travel and the evolution of the human mind. *Genetic, Social, and Pychology Monographs* 123: 133–168.

Suddendorf, T., and M. C. Corballis. 2007. The evolution of foresight: What is mental time travel, and is it unique to humans? *Behavioral and Brain Sciences* 30: 299–351.

Suddendorf, T., M. C. Corballis, and E. Collier-Baker. 2008. How great is ape foresight? *Animal Cognition* 12: 751–754.

Tecwyn, E. C. 2013. A novel test of planning ability: Great apes can plan step-by-step but not in advance of action. *Behavioral Processes* 100: 174–184.

Tomasello, M. 1999. *The Cultural Origins of Human Cognition*. Cambridge, MA: Harvard University Press.

Tomasello, M., and J. Call. 1997. *Primate Cognition*. Oxford: Oxford University Press.

Tomasello, M., and J. Call. 2008. Assessing the validity of ape-human comparisons: A reply to Boesch (2007). *Journal of Comparative Psychology* 122: 449–452.

Tomasello, M., J. Call, and B. Hare. 2003. Chimpanzees understand psychological states: The question is which ones and to what extent. *Trends in Cognitive Sciences* 7: 153–156.

Tomasello, M., and M. Carpenter. 2005. The emergence of social cognition in three young chimpanzees. *Monographs of the Society for Research in Child Development* 70.

Tomasello, M., and M. Carpenter. 2007. Shared intentionality. *Developmental Science* 10: 121–125.

Tomasello, M., M. Carpenter, J. Call, T. Behne, and H. Moll. 2005. Understanding and sharing intentions: The origins of cultural cognition. *Behavioral and Brain Sciences* 28: 675–735.

Tomasello, M., M. Davis-Dasilva, L. Camak, and K. Bard. 1987. Obervational learning of tool-use by young chimpanzees. *Human Evolution* 2: 175–183.

Tomasello, M., B. Hare, and B. Agnetta. 1999. Chimpanzees, *Pan troglodytes,* follow gaze direction geometrically. *Animal Behaviour* 58: 769–777.

Tomasello, M., B. Hare, and T. Fogleman. 2001. The ontogeny of gaze following in chimpanzee, *Pan troglodytes,* and rhesus macaques, *Macaca mulatta. Animal Behaviour* 61: 335–343.

Tomasello, M., B. Hare, H. Lehmann, and J. Call. 2007. Reliance on head versus eyes in the gaze following of great apes and human infants: The cooperative eye hypothesis. *Journal of Human Evolution* 52: 314–320.

Tomasello, M., A. P. Melis, C. Tennie, E. Wyman, and E. Herrmann. 2012. Two key steps in the evolution of human cooperation: The interdependence hypothesis. *Current Anthropology* 53: 673–692.

Tulving, E. 1983. *Elements of Episodic Memory*. New York: Oxford University Press.

Tulving, E. 1985. Memory and consciousness. *Canadian Journal of Psychology* 26: 1–12.

Tulving, E., and J. Markowitsch. 1998. Episodic and declarative memory: Role of the hippocampus. *Hippocampus* 8: 198–204.

Uher, J., and J. Call. 2008. How the great apes *(Pan troglodytes, Pongo pygmaeus, Pan paniscus, Gorilla gorilla)* perform on the reversed reward contingency task, II: Transfer to new quantities, long-term retention, and the impact of quantity ratios. *Journal of Comparative Psychology* 122: 204–212.

Venkatraman, V., J. W. Payne, J. R. Bettman, M. F. Luce, and S. A. Huettel. 2009a. Separate neural mechanisms underlie choices and strategic preferences in risky decision making. *Neuron* 62: 593–602.

Venkatraman, V., A. G. Rosati, A. A. Taren, and S. A. Huettel. 2009b. Resolving response, decision, and strategic control: Evidence for a functional topography in dorsomedial prefrontal cortex. *Journal of Neuroscience* 29: 13158–13164.

Vlamings, P. H. M., B. Hare, and J. Call. 2010. Reaching around barriers: The performance of the great apes and 3–5 year old children. *Animal Cognition* 13: 273–285.

Vlamings, P. H. M., J. Uher, and J. Call. 2006. How the great apes (*Pan troglodytes, Pongo pygmaeus, Pan paniscus,* and *Gorilla gorilla*) perform on the reversed contingency task: The effects of food quantity and food visibility. *Journal of Experimental Psychology: Animal Behavior Processes* 32: 60–70.

Voelter, C. J., and J. Call. 2014. Younger apes and human children plan their moves in a maze task. *Cognition* 130, 186–203.

Wagner, A. D., B. J. Shannon, I. Kahn, and R. L. Buckner. 2005. Parietal lobe contributions to episodic memory retrieval. *Trends in Cognitive Sciences* 9: 445–453.

Warneken, F., F. Chen, and M. Tomasello. 2006. Cooperative activities in young children and chimpanzees. *Child Development* 77: 640–663.

Warneken, F., and A. G. Rosati. 2015. Cognitive capacities for cooking in chimpanzees. *Proceedings of the Royal Society B: Biological Sciences* 282: 20150229.

Warneken, F., and M. Tomasello. 2006. Altruistic helping in human infants and young chimpanzees. *Science* 311: 1301–1303.

Wobber, V., E. Herrmann, B. Hare, R. Wrangham, and M. Tomasello. 2013. Differences in the early cognitive development of children and great apes. *Developmental Psychobiology.* 56: 547–573.

Won, Y. J., and J. Hey. 2005. Divergence population genetics of chimpanzees. *Molecular Biology and Evolution* 22: 297–307.

Wrangham, R. W. 2009. *Catching Fire: How Cooking Made Us Human.* New York: Basic Books.

Wrangham, R. W., J. H. Jones, G. Laden, D. Pilbeam, and N. Conklin-Brittain. 1999. The raw and the stolen: Cooking and the ecology of human origins. *Current Anthropology* 40: 567–594.

Wrangham, R. W., and D. Peterson. 1996. *Demonic Males: Apes and the Origins of Human Violence.* New York: Houghton Mifflin.

Zuberbühler, K. 2014. Experimental field studies with non-human primates. *Current Opinion in Neurobiology* 28: 150–156.

20

Ancestral Precursors, Social Control, and Social Selection in the Evolution of Morals

CHRISTOPHER BOEHM

Humans blush with shame. We judgmentally apply standards of right and wrong to others in our groups, and we gossip about reputations. In addition, we mobilize collectively to cope with deviant behavior through applications of moralistic aggression. On this basis it is easy to conclude that we are distinctively moral, and that chimpanzees lack a conscience. However, the preponderance of scientific opinion holds that while morality is human (e.g., Krebs 2011; Boehm 2012c; Tomasello 2016), as proxies for our last common ancestor (LCA) chimpanzees at least exhibit some important evolutionary "building blocks," ones that were useful to moral evolution (Flack and de Waal 2000; see also von Rohr et al. 2011).

In this sense, morality can be seen as a bastion of difference between ourselves and chimpanzees and also bonobos, the apes with which we share that ancestor. Things about being moral that seem to be uniquely ours include having a shame-oriented, self-judgmental conscience that is emotionally potent in effecting individual self-control (Bandura 1991). Furthermore, as groups we are able to engage in collective moralistic aggression against wrongdoers (Durkheim 1933; Trivers 1971), and this makes our community life distinctive in the natural world. It also makes our inner psychological life distinctive. After seriously breaking a rule, we feel badly about ourselves, because we have identified with the rules of our groups (Simon 1990; Gintis 2003). In contrast, chimpanzees merely worry about affronting a punitive higher-ranking ape, who is imposing what amount to rules on others through dominance behavior (Boehm 2014).

For humans, defining the emotions involved in morality is difficult; in fact, when psychiatrist Gerhart Piers and anthropologist Milton Singer (1971) wrote their book *Shame and Guilt,* they discovered that both cross-culturally and on psychiatrists' couches, people referred to these emotions in contexts of *both* private, internalized psychological malaise and distress that came from public discovery. In short, on the ground, the distinctions we normally rely upon become seriously blurred.

Many conscience functions can be localized in the prefrontal cortex (Damasio 2002; see also Anderson et al. 1999), where brain damage can cause social and moral abnormalities. The paralimbic system also seems important as a locus of emotional connections (Kiehl 2008), and a small percentage of humans, classified as strongly psychopathic, are born with anomalies in this brain area. These "natural deviants" seem to be born dominators, who lack both feelings of right and wrong, and an ability to identify with the feelings of others (Hare 1993). Remorseless psychopaths are so morally different from the rest of us that when they become serial killers we shudder at their very names.

Of Normal People, Apes, and Psychopaths

If we compare chimpanzees with psychopaths, there are not only some marked similarities, but significant differences. As with psychopaths, chimpanzees seem to lack an internalized, self-judgmental sense of right and wrong, and feelings of remorse. During six years of frequently visiting the Gombe Stream research site in Tanzania, I was watching constantly for any sign of moral feelings, but as a cultural anthropologist I came away convinced that chimpanzees are not troubled by their own hurtful past actions, at least not in a way that suggests conscience functions. Experimental studies with captive chimpanzees suggest the same thing[1] (e.g., Engelmann et al. 2012).

My focus in the field was on pacifying interventions, while in brief laboratory visits I witnessed communication using computer keyboards with arbitrary symbols and saw no sense of shame. I also read accounts of young chimpanzees being raised by moralistic humans, sometimes from infancy (e.g., M. Temerlin 1975; J. Temerlin 1980; see also Fouts 1997). None of this suggested that chimpanzees were moral by my Darwinian, conscience-based definition.

A similar status is imputed to human psychopaths. Robert Hare (1993) coped brilliantly with the psychopath's sometimes remarkable capacity to dissemble and lie, by creating a penetrating and lengthy interview protocol that reliably identifies this elusive type of deviant. This was an important boon for parole boards, and also for the understanding of psychopathy and morals. Some humans appear to have strong psychopathic symptoms (Kiehl et al. 2006), and these rare born "immoralists" do not internalize the rules of their groups as normal people do. Emotionally they do not feel remorse or shame, though they can manipulate such feelings in others, and even though often they can dissemble to make others think they are morally normal. In their social interactions, compared to the rest of us, strong psychopaths are deceitful, dominating, self-centered, and heartless.

How do these "people without morality" compare with chimpanzees (see Lilienfeld et al. 1999; Brüne et al. 2006)? The fit is only partial. With respect to feeling the pain or needs of others, chimpanzees are like normal humans. Both species show a capacity for empathy and taking on another's perspective from an early age (Warneken and Tomasello 2006), and both exhibit prosocial helping behavior (reviewed in Melis and Warneken 2016).

In this one important respect, chimpanzees appear to be strikingly different from psychopaths. As an example of empathetic chimpanzee perspective-taking (see Whiten and Byrne 1988), there are field reports of badly wounded chimpanzees that cannot reach all of their wounds to lick them, and are serviced by others who appear to understand their plight (Boesch and Boesch-Ackerman 2000). Goodall (1986) provides other examples of empathy-based altruism at Gombe, and we may assume that such behavior is species-specific. However, with humans a major difference is that we preach moralistically in favor of altruism (Boehm 2008).

There is no sign that chimpanzees have a conscience with moral feelings that involve internalizing group rules (e.g., Gintis 2003), or a shameful, self-judgmental sense of right and wrong. As a result, chimpanzees understand very well the rules imposed on them by a dominant other, but when rule-breaking becomes attractive, and they believe there are no rule enforcers present, there is no sign that they are reining themselves in (Boehm 2012c).

Assuming that some kind of empathetic perspective taking that leads to altruistic helping was ancestral, such behavior certainly would have been strongly involved with moral evolution in providing useful social capabilities. However, in my opinion the core of being moral involves three things. One is a conscience based on rule internalization. Another is the capacity to act as moralistic consensus groups in the sanctioning of wrongdoers. Last is a symbolic capacity to make explicit the shared rules involved and, through gossiping, to evaluate individuals in that light and determine their reputations.

The Evolution of Conscience

Human life is made of moral dilemmas. An evolutionary conscience guides people in their social situations, enabling them to use self-control to behave properly, be well thought of by their fellows (Darwin 1871), and avoid punishment. According to biologist Richard D. Alexander (1987), the evolutionary conscience can be taken, functionally, as a kind of personal behavior calculator, which tells us not only what will bring a good reputation, but which corners we may safely cut without becoming too morally tarnished and without being punished.

Intrinsic to conscience functions is an evolved sense of shame, which works for us in conjunction with the hardwired capacity for social blushing that fascinated Darwin (1865, 1871). This naturalist was intrigued to the point that he wrote to colonial administrators and missionaries all around the world, to see how indigenous people behaved in shameful situations. Their body language, facial expressions, and blushing behaviors were similar, so he concluded that heredity was at work.

As a result of having a shameful conscience, we are morally judgmental. We make "right or wrong" judgments about others, and also about ourselves, based on shame and a moralistic sense of remorse. It is mainly because chimpanzees and psychopaths lack this key reaction that I believe they lack a sense of morality. If we look at the conscience from a social perspective, humans are evolved to docilely internalize the rules and values of their groups (Simon 1990), and these rules then reside in our conscience to orient our self-judgment. When morally normal people break these rules, their personal sense of right and wrong can make them feel culpable on both a present and past basis, while anticipation of future culpability is another basic conscience function.

Morally based feelings of public social discomfort are closely associated with facial reddening as a hardwired response, so it makes sense to use "shame" as a cross-cultural cover term for the various and heavily overlapping vernacular and academic meanings of both shame and guilt. A primacy for such *externally* oriented moral discomfort is also seen linguistically: Casimir and Schnegg (2002) found that terms for "shame" appeared worldwide in indigenous lexicons, while "guilt" has a narrower distribution. In many cases, the shame terms included metaphoric reference to facial darkening, so this physiological-social connection is widely appreciated.

Chimpanzees do not blush socially, and like human psychopaths these African great apes appear not to be subject to feelings of shame, guilt, or remorse. Of course, a pair of chimpanzees who have quarreled may reconcile afterward (de Waal 1982), but there is no sign that this species is expressing mutual regret, let alone remorse informed by a sense of morals. Rather, they are reestablishing a comfortable feeling of social equilibrium in the face of social tension they both find disturbing (de Waal 1996).

Even when chimpanzees like Lucy are raised humanly in our homes, they appear to feel neither shame nor remorse. They certainly may *look* "guilty" when they are breaking rules unaware of being observed, but their concern is with being detected by superiors, which means they are merely sneaking. They do understand the rules of their human masters, but they do not seem to internalize them in ways that make them subject to negative self-judgment. This also defines strong psychopaths.

In the wild, chimpanzees definitely seem to understand "rules" inferable from predictable patterns of punishment from their superiors. However, whenever such punishment does not pose a direct threat, the ape will go ahead with the action, showing no signs of the ambivalence that humans are likely to feel because of shame or anticipated remorse. For instance, in the wild I have seen a low-ranking adult male meet with an estrous female when the alpha male was feeding in a tree above them, and then the pair would separately move to a ravine where the alpha could not see them to copulate. This also takes place in captivity (de Waal 1996). What is evident is a *fearful* awareness of the potential punisher. Normal humans who break rules also watch out to avoid discovery and punishment, and they do so *partly* out of fear. But a psychopath, like a chimpanzee, will do so entirely to avoid punishment.

As a social calculator, a conscience helps most of us to advance our fitness by keeping us out of serious trouble socially. A psychopathic immoralist lacks the emotional side of this guidance system, which generally serves to damp antisocial behavior in the group. He is motivated by an unusually predominant degree of egocentric self-interest that, in a way similar to a chimpanzee, simply makes him look to self-gratification and punishment avoidance.

The social assessments of normal humans are complicated by the fact that we have internalized the rules we are conforming to, or are breaking. For that reason, moral emotions enter the picture strongly: rule breaking brings on special feelings of psychological malaise that are either internalized or connected with public exposure, or very often both (Piers and Singer 1971).

Like empathy, fear of punishment very likely was present in a hierarchical ancestral species, because dominance and submission were present and salient (Boehm 1999). These evolutionary building blocks technically are called "preadaptations" (see Mayr 2002), and in terms of selection mechanics they helped to make the evolution of human morality more likely. What is lacking, in living chimpanzees, in psychopaths, and by proxy in our chimpanzee-like ancestor, is a conscience that internalizes the rules of the group, contributes to personal self-control, and involves us strongly with moral emotions (Boehm 2012b).

In summary, by nature humans are both rule followers and rule breakers, and our evolutionary conscience is the "organ" that enables us to weigh our social opportunities in complex ways and resolve our moral dilemmas. We understand the social benefits of moral conformity and gaining a favorable reputation, and we try to avoid the group punishment that can descend hard on serious moral deviants. At the same time, we figure out how to cut corners, for often enough our groups are prone to wink at certain degrees of transgression even as the same type of behavior is punished severely when it seriously disrupts group life (e.g., adultery). Having a conscience that makes fine distinctions of this kind is important to our personal welfare, and, ultimately, to our fitness.

The Question of Group Outrage

If we look for just a single behavior of the LCA that helped us (preadaptively) to evolve our capacity to be *morally self-judgmental,* it would be the individual capacity for sensing rules and fearing punishment by higher-ranking individuals. However, there were also collective aspects of ancestral social life that were relevant.

We now turn from self-control and self-judgment to the human capacity to judge and punish others when people are acting as a moral community (see Durkheim 1933). Trivers (1971) noted that humans in small groups are capable of moralistic aggression, which figures in the collective side of moral evolution (see also Boehm 2000). As we look to chimpanzees to see if there may be ancestral precursors, we are interested now in the "social control" side of morality. In defining and explaining moral life, these collective behaviors are just as important as the individual reactions discussed above (Boehm 2012c).

When individuals internalize the rules of their groups, they not only apply these rules to themselves, they also join with others, through gossip, to arrive at joint opinions about individuals whose behavior raises eyebrows. Group opinion can lead to group action, and the result is a powerful moral community that can aggressively (and dominantly) act against any deviant whose actions arouse the entire group. When the group coalesces as a political coalition, and is in a position to act consensually, the disruptive free-riding behavior of cheaters and bullies can be curbed.

Coalitionary behavior is found in many primates, including especially baboons (Pandit and van Schaik 2003) and *Pan* (Wrangham 1999). In chimpanzees, male coalitions cooperatively patrol community borders, and some of them gang-attack lone enemies from neighboring groups (Muller and Mitani 2005; see also Wilson and Wrangham 2003). This behavior may be useful to group members because, with sufficient imbalance of power (Wrangham 1999), neighboring groups can be weakened and their territory appropriated, along with females in some cases (Williams et al. 2004; Mitani et al. 2010; Wilson and Glowacki, this volume).

Collective Intentions

Male chimpanzees are noted for competing for power in small, group-internal coalitions, and for going on these often lethal border patrols (Muller and Mitani 2005). However, less fully analyzed are their sometimes lethal coalitionary attacks on members of the same group (Boehm 2017a). These attacks by sizable coalitions are based on shared intentionality, which of course becomes highly developed in large-brained humans (Tomasello 2014, 2016). I believe these internal gang attacks by chimpanzees to be suggestive of group social control in humans, and similar behaviors can be conservatively suggested for the LCA, in part because bonobos also exhibit this behavior (Boehm 2012a).

However, Tomasello (2016) denies a capacity for shared intentions in chimpanzees. My view (Boehm 2012c) differs sharply as a result of having watched wild chimpanzees on patrol. For instance, in analyzing field videotapes I noted that when a Gombe patrol stopped to reconnoiter and then moved forward, not just the alpha male was controlling these movements (see Boehm 1991). Rather, the cautious leading moves were made alternately by several high-ranking males, who shared intentions in that the group had a past history of ambushing strangers and was proceeding quietly.

Tomasello has used captive chimpanzees, often in comparison with human children, to explore a number of possibilities relevant to moral evolution. He has shown that both children and chimpanzees show tendencies to assist adults who are frustrated in performing a task that requires a helper (Warneken and Tomasello 2006; reviewed in Melis and Warneken 2016), and this is important in comparing altruism in the two species. However, in my view conscience and social control are central to moral behavior, while I see altruism more as an important behavior that is supported by moral communities, and is closely tied to the conscience in terms of empathy (Boehm 2008, 2012b).

Directly relevant to social control is the fact that Tomasello's experimental work has shown that chimpanzees do not act as third-party rule enforcers (see Riedl et al. 2012), and this is in accord with my field observations in the wild. Where we diverge is in his belief, on the basis of dyadic captive experiments, that chimpanzees are not capable of collective intentionality (Tomasello 2016). Even if one is limited just to larger captive groups, de Waal's (1996) description of an alpha male's bullying behavior being curbed by a coalition of females who are all joining together to issue threatening vocalizations provides a vivid description of shared intentions in action, and suggestive instances of similar female behavior have been recorded at one wild field site (Newton-Fisher 2006). In the wild, patrols attack strangers with a discernible division of labor and a shared goal, and in a species this intelligent an assumption of shared intentions is both logical and consistent with behavior cues that are readily observed. Descriptions of intracommunity gang attacks suggest that similar political and psychological dynamics are operative (Boehm 2012b).

I agree strongly with Tomasello (2014) that norm enforcement is central to morality, and that collective intentions are important for understanding humans. However, I think that in the abovementioned contexts, and perhaps especially in their coalitionary attacks within their communities, the *Pan* species are showing strong signs of collective intentionality. I also believe that this was an important preadaptation for moral evolution, that humans do not totally transcend chimpanzees in this respect, and that wild groups are the first place to look.

Intracommunity Gang Attacks by *Pan*

Chimpanzees appear to form intracommunity attack coalitions far more often than bonobos, even considering the substantially greater research time devoted to them. Here we examine the limited but important published case histories and the patterns they reflect, to evaluate this instance of collective intentional behavior as a possible major preadaptation for social control in humans, and also to see if there may even be some moral component in such political cooperation.

Field studies of wild chimpanzees began in 1960, but it required ten years of improving field methods before it was established that the Gombe community's males went on patrol and killed strangers they were able to ambush (see Goodall 1986). With many more field sites today, we now know that attacks on strangers are reported as a pattern in the better-studied groups (Wilson et al. 2014).

It was thirty years after research began at Gombe that Jane Goodall published the first account of a *within*-community attack by a sizable coalition (Goodall 1992). Goblin, the long-standing alpha male, got into a fight with his political rival Wilkie over an estrous female, and was grievously wounded by Wilkie's bite to his groin. Goblin retreated to a day nest, and his wounds began to suppurate, even though Gombe field assistants gave him bananas laced with antibiotics. Thus, this situation was not entirely "natural," and it became even less so when a valiant veterinarian climbed into the nest and drained the infected wounds. Goblin, who otherwise would have died, began to recover.

After hundreds of hours of field observation, I knew Goblin well because my daily follows were oriented to watching high-ranking males in hopes of observing conflict interventions. This long-term alpha male was quite small, and at the same time extremely competitive with other males. His dominance displays probably had to be particularly exuberant and aggressive, because his group included males much larger and stronger than himself (see Goodall 1986).

A few days after he began to recover, Goblin returned to the group, but Wilkie had established an alpha relationship over the group, and on his rival's return Wilkie didn't even have to defend his position. A sizable coalition formed to mount a prolonged gang attack on Goblin (Goodall 1992), doing so almost as fiercely as they would have attacked a stranger when on patrol. A newly wounded Goblin escaped and survived for several months living in dangerous peripheral areas. Then, pant-grunting submissively, he rejoined the group as a low-ranking member of the male hierarchy.

Also at Gombe, years later an unusually large, aggressive, and socially dominant male named Frodo was similarly attacked by a coalition of his peers after losing his alpha position due to illness. He was becoming a contender again when he was gang attacked, and, like Goblin, Frodo went into exile for a time (Fallow 2003). Sheldon succeeded Frodo as alpha, and after he no longer controlled the position he was attacked by three other males and briefly pinned to the ground, again in a manner reminiscent of attacks on strangers. After a period of tension, Sheldon removed himself from the group (Wilson et al. 2004). Recently (at this writing), alpha male Ferdinand received similar treatment from a sibling rival, another sibling, and Fifi, their mother, who were acting as an aggressive coalition with the rest of the group backing them up (Mjungu and Collins 2016).

If we compare the four attacks at Gombe, the one against Goblin strongly resembled the fierce attacks against strangers by patrols having great numerical superiority, while the attack against Frodo also involved a sizable coalition inflicting wounds. Numerically, the attack against Sheldon was closer to the routine competition of high-ranking males through dyadic coalitions, but his being pinned to the ground strongly suggests an attack on a stranger. The vicious, nearly fatal attack on Ferdinand also involved a small coalition, but with the group backing it.

Both chimpanzees and bonobos have the potential to make gang attacks on members of their own groups. Table 20.1 shows the currently recorded

TABLE 20.1. Intracommunity attacks for *Pan*.

Attackee	Attacking Coalition	Outcome	Comments
Pan troglodytes schweinfurthii—Gombe (Kasakela), Tanzania			
Goblin (wounded alpha attempting comeback)	Six-male coalition	Escaped to temporary exile	Recently dethroned alpha attacked, wounded, exiled. Group was backing strong rival Wilkie (Goodall et al. 1992).
Frodo (ex-alpha)	All the group's males	Escaped to temporary exile	Former alpha attacked, wounded, exiled. Victim normally was unusually aggressive, but was weakened by illness; as alpha didn't groom others (Fallow 2003).
Sheldon (ex-alpha)	Three males	Escaped to temporary exile	Partially absentee alpha dethroned, wounded, and exiled. Attack was similar to intercommunity attack with holding down (Wilson et al. 2004).
Ferdinand (alpha)	Sizable male group backing three attackers up in tree	Escaped badly wounded, had to avoid the same group	Alpha dethroned, severely wounded, exiled. Victim was unusually aggressive as alpha; inflicted wounds on subordinates (Mjungu and Collins 2016).
Pan troglodytes schweinfurthii—Gombe (Mitumba), Tanzania			
Vincent (ex-alpha)	Two males (only two in small group)	Dead	In an unusually small community, alpha Vincent injured in fall, loses rank, goes into exile, rejoins group, and is lethally attacked by one male joined by another (Mjungu 2010; Wilson et al. 2005).
Pan troglodytes schweinfurthii—Mahale, Tanzania			
Ntologi (ex-alpha)	Sizable male group	Escaped to temporary exile	Former alpha attacked, wounded, exiled. Aggressive male (Nishida et al. 1995).
Ntologi (alpha)	Sizable male group	Dead: observed dying	Attack not observed, but former and current alpha wounded to death. Aggressive male (Nishida 1996).
Jilba (young adult, middle rank)	Eight male attackers	Escaped to temporary exile for three-plus months	Politically potent younger male attacked, wounded, exiled (Nishida et al. 1995).
Pimu (alpha)	Five attackers, while two passively support alpha		Alpha attacked until dead. Attack was similar to intercommunity attack with pinning, severe biting. Began with dyadic fight with two interveners, then loser solicited help versus Pimu. During attack by four males other than the loser, two males vocalized in support of Pimu (Kaburu et al. 2013).

Species / Individual	Attackers	Outcome	Notes
Pan troglodytes schweinfurthii—Kyambura Gorge, Uganda			
Hatari (adult male, second rank)	Four adult males attack, Hatari wounded and leaves group	Escaped to exile	Small, isolated group (Kyambura Chimpanzee Community 2011).
Pan troglodytes schweinfurthii—Ngogo, Uganda			
Grapeli (young adult male, middle rank)	Three main attackers and at least four others	Dead	Socially peripheral but politically active male attacked, wounded, and exiled. Victim had failed to groom with top males (Watts 2004).
Pan troglodytes schweinfurthii—Budongo (Sonso), Uganda			
Zesta (young adult male, low rank)	At least three attackers	Dead	Possible redirection of aggression at group level. Victim was not politically competitive; there were unusually few females cycling and male sexual competition was high; apparently not connected with alpha competition (Fawcett and Muhumuza 2000).
Pan troglodytes verus—Fongoli, Senegal			
Foudoko (ex-alpha)	Five males attack, partially eat body	Dead	Foudoko was alpha thirteen years ago, lost strong coalition partner, exiled for many years, rejoined group but chased by younger adult males and finally attacked and killed and partly cannibalized (Pruetz 2017).
Pan paniscus—Lomako, Congo			
Volker (high-ranking male; son of alpha female)	Sizable male/female group with fifteen attackers, with one ally abstaining	Disappeared; suspected death	High-ranking male attacked, probably wounded fatally. Volker was vying for alpha male position with backing of his mother the alpha female; he approached a high-ranking female with infant, who attacked him, and group joined in. Protracted savage attack with immobilization; reminiscent of chimpanzee attacks on strangers. Fearful alpha mother did not support him (Hohmann and Fruth 2011).

published attacks, with one attack for bonobos and thirteen for chimpanzees, almost all in eastern Central Africa. There, potentially lethal gang attacks usually are directed at active or former alpha males; only the attack at Sonso was directed at a young male. In that case, there was an unusual shortage of estrous females in the group, so the case can be made that this simply involved the redirection of aggression in a situation of unusual mating competition. Striking is the general resemblance of these attacks to patrolling chimpanzees' gang attacks against strangers, in that they are prolonged, and submissive victims sometimes are held down by some members of the attacking team while others inflict bites.

Intercommunity attacks benefit the aggressors because they reduce their competitors from other groups, leading to shared advantages in feeding and mating. By the same token, intracommunity attacks would appear to weaken a group's male fighting force (Watts 2004). Thus, from an adaptive perspective, intracommunity attacks may be difficult to account for. However, two other hypotheses can be entertained, the first involving piggybacking. It seems possible that the genetic benefits accruing from patrolling attacks may be "carrying" the rarer maladaptive intracommunity attacks, so that the latter are being genetically "subsidized" on a pleiotropic basis (e.g., Boehm 1981). Another hypothesis would be that when a number of group members gang up against a higher-ranking individual to eliminate him or reduce his rank, in effect they will be sharing the reproductive gains.

These intracommunity gang attacks are genetically consequential. Six of the chimpanzee attacks were observed to be fatal, while the wounded bonobo victim was never seen again. In seven other chimpanzee cases, the target was obliged to spend a few months away from the group in exile, doing so riskily in peripheral areas, so there always were fitness costs.

Just as the intercommunity attacks may be a basis for genetic piggybacking, cognitively and culturally they would seem to provide a repeated opportunity for chimpanzees to learn gang attack coalitionary skills, while mobbing predators would have provided rather similar learning experiences. We may assume that the trigger that unleashes these attacks on members of the same community is that hostilities rise within the group, and they are basically directed at a single target who lacks strongly active coalition partners. The same balance-of-power calculations are likely in effect as when patrols attack singleton strangers who have no allies nearby.

Some of these fourteen attacks include individuals showing some support for the victim, and this can be likened to the usual small-coalitional rivalry for rank, because both alphas and their rivals rely on allies. However, in some cases the entire foraging party makes the attack unanimously. For instance, at Gombe, the group attack on Ferdinand was savage and the motivation was so high that after he managed to slip away, with very serious wounds, he desperately avoided his group and acted as though he were being hunted. Likewise, when Goblin was attacked and escaped into exile, there was no report of support for him; only by behaving very submissively was he able to reintegrate into the group, months later. At Mahale, something similar took place with ex-alpha Ntologi; he went into exile but then, after returning to the group, as reestablished alpha he received a fatal attack.

Chimpanzees are behaviorally quite flexible, and in the case of these attacks it would appear that, in effect, they were combining within-group male rivalry behavior, which usually involves just small coalitions, with the routine pattern of making large-scale, vicious attacks on strangers. It is of interest that xenophobic tendencies were not involved, yet some of the attacks against insiders were as violent as attacks on strangers.

What were the likely motives of the attackers? In part these sizable and mostly male coalitions could have been supporting an incumbent alpha male against a former alpha who wanted the position back, or supporting the incumbent against a new rival. However, we must ask why normally such coalitions are merely dyadic (e.g., Goodall 1986; de Waal 1982), and what triggers the mobilization of much larger coalitions. Additionally, we may ask whether, competitive political dynamics aside, there was something "personal" about these males that made them intensely and widely disliked compared to other chimpanzees. There are hints about such behaviors in the field reports, and this will be explored further in the Discussion.

Once the intracommunity gang attack against Goblin was documented, similar attacks started to be reported from other field sites rather frequently. First came attacks described at Mahale, also in Tanzania and a hundred miles from Gombe. In Table 20.1, aside from Ntologi, one was against a young male who was politically ambitious, bullied females more than usual, and also failed to show submissive deference to his superiors (Nishida et al. 1995). He fled the attack and stayed away from the group for several months, similar to Goblin, after which he was able to return. This attack was led by the alpha male with seven in his coalition, and in human terms the result was long-term group rejection from the community—as an extreme form of ostracism. Quite recently, at Mahale the savage killing of the reigning alpha male by a sizable male coalition was observed and photographically documented by tourists (Kaburu et al. 2013). In this case, the details are extremely reminiscent of intercommunity attacks, in which patrol members immobilize, bite, and stomp their foreign male victim, and usually wound him to the point that he soon dies (e.g., Watts et al. 2006).

Earlier, as an unusually aggressive alpha male, Ntologi, was attacked twice by sizable and mostly adult male coalitions, and after the second attack he was observed dying (Nishida 1996). Uehara et al. (1994) suggest that these Mahale attacks may be due to the victims having formed fewer social bonds with other males than with group members not attacked. The attack described for Ngogo in Uganda also fits with this hypothesis, as does the Gombe attack against Frodo, who seldom groomed other males.

At the Ngogo field site in Uganda's Kibale forest, a young adult male was killed in a situation in which the study group was so large that there were functional subcommunities within it. Watts (2004) points out that this young adult male was low-ranking but rising rapidly in the hierarchy, that he lived in a subgroup of the large main group, and that he associated little with the high-ranking males in the main group and had few grooming partners there. After a large group had fed on figs and was dispersing, a scream in the young male's vicinity drew the main group to run in that direction, and then he was attacked seriously by at least six adult males, some more intensively and some less so. He was wounded but did not die.

Another attack against a low-ranking male took place in the Sonso study group in the Budongo forest. It was unlikely that this male was reacted to as a threat to higher-ranking males, as was possible at Gombe, Kibale, and Ngogo. Distinguishing this attack was a situation in which an unusually low number of estrous females were available to the rivalrous males, and it was surmised (see Fawcett and Muhumuza 2000) that it was simply intensified male sexual competition that led to the attack. A similar situation prevailed at Fongoli (Pruetz et al. 2017).

These attacks on community members appear to be both connected with male rivalry within the group, and similar to patrols attacking strangers. They also may bear some resemblance to the actions of hunter-gatherer groups when they gang up to ostracize or kill disliked or threatening male deviants in their midst. However, as when we focused on chimpanzees as individuals, specifically moral emotions surface nowhere in these case studies of viciously hostile collective attacks.

There is only a single report of bonobos making a gang attack, one suspected to have been fatal. The attack was made against the high-ranking son of the alpha female when he approached a high-ranking female and she immediately attacked him; obviously, some previous history is lacking in the field report. The entire foraging party of more than two dozen bonobos joined in the attack, while the male's mother (the alpha female) fearfully did not try to intervene. The attack suggested ferocity on a level with chimpanzee patrolling attacks, or the more severe chimpanzee intracommunity attacks treated above, and tactics appeared to be similar.

Bonobos have not been recorded to engage in gang attacks against strangers, though some significant vocalizing hostilities may result when two groups from different communities meet (Kano et al. 1994). Thus, with bonobos, it is more difficult to hypothesize that intracommunity attacks are a behavioral spinoff from intercommunity attack behavior.

Ancestral Assessment

*Intra*community attacks by sizable coalitions are very likely to be ancestral, for they are found not only in today's chimpanzees and humans (Boehm 2012a), but also in bonobos (Hohmann and Fruth 2011). If so, they could have helped to set the stage for the evolution of moralistic group social control in humans. However, in the chimpanzee behaviors we have reviewed, I emphasize that there has been no discernible trace of moral sentiment in the form of individual shame or remorse, nor of contagiously shared *moralistic* outrage. This is my judgment, even though these attacking groups of apes surely exhibit a high level of arousal as they try to tear their victim to pieces and sometimes manage to do so, as with the pervasively wounded dead alpha males at Mahale and Fongoli.

Morality seems to have intensified and partially transformed this kind of group behavior, for, as will be seen, a human group consumed with moral outrage can become a still more efficient killing machine (e.g., Boehm 2014). Unlike chimpanzees, our species clearly understands death and how to kill, and deliberately inflicts death as an ultimate form of social distancing. And as will be seen, when human foraging groups practice capital punishment, in the majority of cases this is aimed at eliminating a male whose socially aggressive behavior has become unacceptable, and who has no one to strongly defend him.

The targeting pattern in wild chimpanzee communities may be similar, for we have seen that the five observed attacks at Gombe (including Mitumba), four at Mahale, and one each at Fongoli and Kyambura Gorge were aimed at politically prominent males, while elsewhere two other attacks were aimed at lower-ranking adult males whose bonding behavior fell short even as in social competition they were politically aggressive. None were aimed at females.

Coalitions of Females

Female chimpanzees form sizable coalitions when they have time to bond. This has been routine in sizable captive chimpanzee groups, where they predictably unite to prevent male bullying of females, and often can actually control who will become alpha (de Waal 1982, 1996). Collective threats by females, expressed through concerted *waa* vocalizations, are effective in suppressing male bullying of females. To date, de Waal (personal communication) has not recorded actual coalitional attacks, but he has indicated that the potential is present and that the alpha male anticipates this and breaks off the disliked behavior.

In the wild, females tend to be socially isolated, but at one field site where food was relatively plentiful, the females apparently had more time for socializing, and similar but less sizable agonistic coalitions formed similarly (Newton-Fisher 2006), as a response to males redirecting their aggression at females. To date, as in captivity, there is no instance of one of these wild female coalitions' threats turning into an actual attack that seriously wounded or killed a male. However, the one suspected lethal attack shown by bonobos in Table 20.1 involved a very large coalition of both males and females. Given the political importance of females and female coalitions in bonobo groups (Kano et al. 1994), this is not unexpected.

Preadaptive Interpretations

To summarize, similarly to egalitarian humans when they crack down on males who are trying to exert too much power, the great majority of attacks by *Pan troglodytes* have been directed at higher-ranking males, and often at ex-alphas or incumbent alphas; in bonobos, the single such attack was directed against the alpha female's son. These attacks were recorded mainly at the two longest-running field sites in Africa, in the easternmost part of the chimpanzee range, but the attack at Fongoli includes *P. troglodytes verus* as well as *P. troglodytes schweinfurthii* in the pattern.

Looking forward, in these attacks by apes there are definite parallels with the human use of ostracism and capital punishment in collective acts of moral outrage against overly aggressive males. In any egalitarian society, these take place occasionally (Boehm 1999). This type of attack can be confidently included in the LCA's likely behavioral repertoire, but in the absence of moral feelings. Thus a reasonable assumption is that moralistic group sanctioning of certain socially aggressive individuals had an ancestral precursor, and it can be hypothesized that in earlier human foraging societies this preadaptation became useful not only to the evolution of morality, but more specifically to the evolution of the morally based, cooperative egalitarianism that typified the life of most behaviorally modern humans in the Late Pleistocene (Boehm 1999) and possibly much earlier (Knauft 1991).

LCA and Humans: Sanctioners' Costs and Benefits

Moral feelings definitely appear to be missing in individual chimpanzees as they cope with their superiors on the basis of fear of punishment. The same is true in their coalitionary attacks on fellow group members. However, in these collective actions some basic elements of human social control do seem to be present, including cognitively shared intentions to gang up and attack (e.g., Boehm 1991). Accompanying these intentions are shared hostile emotions, evident in the high arousal that accompanies gang attacks of apes.

Wrangham's (1999) imbalance of power hypothesis applies nicely to the chimpanzee group attack pattern, be it within or between communities. It applies equally to chimpanzees and to the one intracommunity attack by bonobos, and it also applies to human foragers if, rarely, they actively unite to physically attack a deviant. However, the common pattern is to delegate the execution duty to a close male relative of the target who kills the unsuspecting target by ambush.

The numerical preponderance of the ape enforcers is great, and wounding of these coalitional attackers will be rare because (in apes) the overwhelmed, submissive targets are so unlikely to fight back. In hunter-gatherers, the males carry lethal weapons and their kin are prone to avenge them; it is for this reason that the group as a whole will delegate the executioner's role to a single close kinsman of the target—who will not be attacked.

The considerable literature on costs and benefits of punishing deviants is based largely on experimental economics research (e.g., Fehr and Gächter 2002; see also Gintis 2000; Boyd et al. 2003; Egas and Riedl 2008). Human enforcers' overall costs are low, especially if one takes into account the rarity of the behavior (Boyd et al. 2010), but they are not negligible, since time and energy must be added to possible risks. Compensation comes through denying an aggressively deviant free-rider his fitness advantages; in effect, the enforcers are sharing the proceeds.

There remains a possible secondary free-rider problem (Panchanathan and Boyd 2004), but it may well be that outside of the laboratory, foragers' defections from gang attacks are merely situational, because defectors are so likely to be kinsmen of the deviant (Boehm 2012b). Furthermore, for chimpanzees this question may be all but irrelevant if their intracommunity attack behavior is a piggybacking evolutionary by-product of the profitable cooperative attacks made against strangers or mobbing predators.

The human situation would appear to be more complicated. There are two types of attack that result in capital punishment. The preponderant pattern is for the group to appoint an executioner who can use hunting weapons to ambush the deviant without too much personal risk. However, there is also the possibility that, as with *Pan,* the group as a whole actively participates in the execution. The great majority of executions involve individuals delegated by the group to ambush the deviant, and the risks are low.

There is also another factor that works against secondary effects. If a small human band cannot reach a consensus about killing a serious deviant, usually no action will be taken. Thus, while a defector avoids personal risks, his decision also makes the group unable to act, so risks are equally avoided for other group members (see also Boehm 2012b).

The Problem of Coercive Social Predators

Half a century after Williams's (1966) anti-group selection insights, the genetic riddle of human altruism continues to be a major challenge, one that has resulted in multiple explanatory paradigms (see Nowak 2006; West et al. 2007), which, I believe, can still be augmented. Deceptive free-riding has been a long-term focus (Price et al. 2002; Cosmides and Tooby 2004), and because altruism is so closely connected to cooperation (Sober and Wilson 1998; Nesse 2007), these dedicated anti-altruistic "cheaters" have become negative icons in our evolutionary analyses.

Here I shall argue that *coercive* free-riders loom still larger in this important evolutionary conversation. Unlike Williams's cheaters, who are "designed" to hoodwink altruists, a coercive free-rider depends not on guile but on competitive power (Boehm and Flack 2010) to advance fitness, and the targets are not restricted to altruists, though their generosity makes them prime targets.

In any species that has a social dominance hierarchy, such coercive free-riding will be paying off on a routine basis: the aggressively powerful will be outcompeting those they can dominate. Essentially, any individual less strong, less politically adept, less selfishly motivated, more generous, or in any other way less selfishly competitive is a likely loser, while the prize will be an advantage in feeding, a mating opportunity, or simply a better position in space. All go to the "alphas," at the expense of lower-ranking individuals in general, and of altruists in particular, since their innate generosity poses a special handicap in competition (see also Muller and Mitani 2005).

Dominance interactions develop because individuals condition one another, usually by bluffing, but sometimes by fighting, in ways that routinize an orderly hierarchical access to resources (e.g., Schjelderup-Ebbe 1935; Lorenz 1966). In hierarchical primates, because high rank correlates significantly with reproductive success (Ellis 1995), dominance behavior has had a profound impact on the genomes of the three African great apes in our *Pan-Homo* clade.

As intimidators, coercive free-riders are likely to be more effective competitors than deceptive free-riders, for when they use dominance to win contests, their competitive strategy is more generalized in seeking targets: the victims include not only generous, gullible altruists, but anyone else who is less prepared to compete. Furthermore, within the group an identified cheating free-rider can be avoided socially, whereas to avoid a coercive free-rider, victims must pay costs by sacrificing their established group membership and leaving if they can (Vehrencamp 1983). In the case of a male chimpanzee, such an exit is impossible.

Evolution of Cooperation

Very likely, the ape ancestral to humans was capable of significant cooperation as it formed small and sometimes large coalitions to express aggression within and between its groups (e.g., Wrangham and Peterson 1996), and presumably mobbed predators. It also would have shared the small game it hunted, at least in limited ways (e.g., Muller and Mitani 2005). In contrast, humans today have arrived at a distinctive, morally based potential for altruistic group cooperation (Campbell 1975), which sometimes can parallel that of the eusocial insects (Wilson 1975).

A long-standing challenge for evolutionary scholars has been to identify the mechanisms that brought this about. At the level of phenotype, punishment can help cooperation to stay in place (Boyd and Richerson 1992; Boehm 2000; Krasnow et al. 2015), which means that in theory, immediate social sanctioning of predatory free-riders could account for cooperation through fear of punishment alone.

However, humans' remarkable degree of cooperation could have been biologically distinctive in its recent modes of evolution. It is proposed here that one major responsible agency has been a special type of "social selection" (e.g., West-Eberhard 1979), namely, "sanctioning selection" that takes place through active, moralistic group punishment (Boehm 1997; see also Boehm 2012c, 2014). I shall hypothesize that this variant of social selection increased our cooperative potential, and that as it did so self-domestication (see Hare and Wrangham, this volume) was taking place.

This sanctioning selection hypothesis looks beyond subsistence cooperation to highly collectivized social control; the aim was to contain obvious social predators who threatened others or undermined cooperation. Ultimately it is suggested that this socially actuated genetic selection process could have been potent in helping to decisively shape the human behavioral genome, and the fact that punishment can take a social predator's life (Boehm 2014) suggests that these social selection effects could have been quite robust.

Once fully moralized social control arrived, potent group sanctioning could have helped to make possible the rapid genetic evolution of altruistic behavior traits associated with cooperation. The sanctioning of social predators who took free rides could have made this possible, leading to the evolution of a conscience as a uniquely complex and sophisticated agency of self-control (Boehm 2012c). This socially motivated type of selection was focused by human cognitive abilities and a capacity for collective intentionality, and, because decisions by both enforcers and deviants were involved, there might have been "runaway" effects (Fisher 1930; Flinn and Alexander 2007; Nesse 2007).

Social Selection: The Details

Ancestral coalitional basics have carried forward, but humans use symbols so efficiently that they can explicitly agree that certain types of social predation are a group concern, and effectively outlaw them. Sometimes, in effect, groups can even form a practical consensus about the kind of basic social order they wish to live in, and actively promote it. For instance, many hunter-gatherers are determined to remain egalitarian (Boehm 1999), with less powerful individuals uniting against any would-be dominators to protect their personal autonomy.

The more aggressive individuals are being culled at the level of phenotype, so this political leveling works against the cultural growth of social hierarchy expected in a species like ours. With respect to the genotype, because aggressively competitive behavior tendencies are also being reduced or modified there is a "parallel effect," insofar as the deliberate impact on the phenotype is so similar to the unintended effect on the genotype. The result is less competition in everyday life, and also reproductive leveling (Boehm 1997; Bowles 2006) and autodomestication (see Hare et al. 2012; Hare and Wrangham, this volume).

The original version of social selection theory was Darwin's sexual selection (West-Eberhard 1979). Females chose mating partners, while males evolved to fulfill female choice criteria. Next came selection-by-reputation (Alexander 1987), with male and female choices of worthy, altruistic partners (e.g., in marriage) playing a significant selection role in systems of indirect reciprocity. Alexander's selection scenario goes a long way toward explaining the evolution of altruistic traits that feed into human cooperation.

Here my intention is to advance sanctioning selection theory as a less appreciated social selection mechanism. This involves individuals in groups aggressively making decisions about individuals they consider to be deviant, and the responses of such deviants to their judgmental groups (Boehm 2014). Something that sexual selection, selection-by-reputation, and sanctioning selection have in common is that individual and collective social choices, including decisions to sanction, are making an immediate and potent impact on natural selection process. As a drastic *collective* reaction, capital punishment could have modified our genome (Boehm 1986, 1997; see also Otterbein 1988), and the three tables on hunter-gatherer deviance and punishment will suggest that the main targeted behaviors were ones involving male intimidation of others.

Capital punishment obviously is final, while shaming and ostracism allow deviants to adjust, reform, and conform if they can properly regulate their own behavior. The conscience is their social calculator. Primitive forms of "self-regulation" (see Nicholson 2012) are evident in apes when they submit to a superior out of fear. However, in humans, self-monitoring and self-control rise to the level of having a sophisticated evolutionary conscience.

When consistent social sanctioning stays in place over thousands of generations, its effects on gene pools may be considered in terms of a special, moralistically based type of selection that culls certain behavioral genotypes, and ancestrally it would appear that it was in the area of political coalitionary behavior that the preadaptations were striking.

Reversing Dominance Hierarchies

Noncomplex foragers can police powerful, coercive free-riders only by acting as groups, for in being egalitarian they lack strong or authoritative leaders who might do this for them. When would-be dominators are held down decisively by their peers, I have called the result a "reverse dominance hierarchy" (Boehm 1993, 1999; Gintis et al. 2015).[2]

Human societies in which hierarchies are definitively reversed include not only mobile hunter-gatherers and sedentary complex foragers like the moderately hierarchical California Indians, but also most of the tribal agriculturalists who followed them and never centralized, as well as some of the more ephemeral "big-man" societies found among New Guinea horticulturalists (e.g., Roscoe 2002) and a few "democratic" chiefdoms like the Iroquois Nation. To qualify, this reversal of power must be so definitive that alpha types dare not express their dominance tendencies at all strongly (see Boehm 1999).

The predictable result includes rule by coalition based on consensus seeking (Boehm 1996), rather than decisive or autocratic individual leadership. Yet these groups can deal effectively with would-be dominators and other free-riders, because the entire band can act as a socially solid coalition. In humans, we have seen that relatively complicated collective intentions constitute a potent evolutionary force (Boehm 1991).

It is very likely that our chimpanzee-like ape ancestors ganged up on disliked individuals to temporarily or permanently eliminate them from the group. By building on these preadaptations, humans could have reached a point, at least 45,000 years ago, that their punitive coalitions were creating a special, social selection force. This not only aided in the evolution of our uniquely human conscience, it also improved our capacity for cooperation by keeping bullies and deceivers like thieves under control and by suppressing the genes that motivated these free-riders.

As we were moving into behavioral and cultural modernity, these selection patterns could have been responsible for some major changes in our behavioral genome (e.g., Klein 2009), while cultural factors were involved as well (McBrearty and Brooks 2000). This surely was an important phase of our evolution, and social selection by means of group sanctioning could have been an important factor.

Quantitative Assessments for Human Foragers

By use of ethnographic analogy, it will be possible to build prehistoric social selection scenarios based on the behavior of today's hunter-gatherers. The purpose will be to explain the probably rather rapid later evolution of both social control and morally based subsistence cooperation in our small, egalitarian bands. In turn, this will illuminate the previous discussion of similarities and differences between ourselves and chimpanzees.

Today's moralistic hunter-gatherer egalitarianism surely reaches back at least 45,000 years, and likely through the entire period of behavioral modernity (see Shea 2011), and possibly earlier (Boehm 2012c). This remarkable political-structural outcome was possible not only because of weapons (Gintis et al. 2015; see also Bingham and Souza 2009), but because a symbolic and moral species developed the capacity for well-strategized group social control as it routinized the expression of an ancestrally based dislike of being dominated (Boehm 1999).

Our large brains led to socially sophisticated, shared group intentions, which efficiently adjusted punishments to the seriousness of the crimes (Boehm 2000). Group punishment was a culturally shaped expression of our political potential, and at the same time it presumably helped to shape that potential, because on average the individuals singled out for punishment were more prone to antisocially aggressive competition. Of course, with humans, contemporary social control by hunter-gatherers reaches beyond the suppression of predatory competitive free-riding by coercive or deceptive free-riders; there are also sex crimes, violations of supernaturally based rules, and other noncompetitive transgressions that can stir an entire band to act as a policing unit.

In the following three tables, a preliminary, nonrandomized sample of sixty-five nomadic "Late Pleistocene-appropriate" foraging societies is used to yield information about these categories of social deviance, which are recognized by band members today and, by ethnographic analogy (see Ames 2004), were likely to have been salient 45,000 years ago, or possibly earlier.

Ideally, we should have full ethnographic coverage of all hunter-gatherer behavior so that any universal behaviors could be confidently projected backward in time. In fact, the ethnographic coverage in single field reports is notoriously uneven, and for certain regions there are only a few such reports. However, in spite of a presently limited, nonrandomized sample size of only sixty-five hunter-gatherer societies,[3] certain frequently occurring, well-reported behaviors can be projected back into our recent evolutionary past on a conservative basis.

The database is now large enough to at least identify the stronger central tendencies for contemporary foragers; however, the measure I use for analogizing will be statistically rather crude. If a behavior is reported at least once in all six of the world regions in which foragers are found, then by ethnographic analogy it can be projected backward in time as being *widespread*, as long as the people involved were behaviorally modern.

The assumption is that if the behavior in question is represented in all world regions, conceivably it could have been universal or at least heavily represented in Late Pleistocene prehistoric populations. With respect to the regional distributions we are concentrating on, intimidating behaviors, including murder, sorcery, and beating another person are reported ethnographically as deviant in all six of the world regions populated by foragers; condemnation for theft, lying, failure to share, incest, adultery, and taboo violation also is reported worldwide. The fact that Brown (1991) reports comparable but more abstract moral proscriptions to exist today in *all* human societies provides strong supporting evidence.

Ethnographic Analysis

Dominant Free-Riding

Table 20.2 shows that punishment aside, three basic types of competitive, antisocial free-riding deviance are recognized and condemned by foragers worldwide: intimidation, deception, and shirking when sharing is expected.

As a major type of intimidation, sorcery often involves a shamanistic healer's turning a role that is culturally defined as being altruistic into an agency for personal gain. Malicious shamans thought to kill people supernaturally are feared, and if unrestrained they can use this fear to despotically take free rides. Were such behavior not dealt with, in extreme cases the intimidators' personal fitness advantages could be substantial. Fortunately for their potential victims, these coercive free-riders are easily recognized and, if necessary, targeted for capital punishment.

A second basis for coercive deviance worldwide is murder, for serial killers are feared and submitted to until they are executed. Basically, it is repeat killers who will intimidate their groups and dominantly outcompete others until they are killed. In contrast, one-time situational killings are frequent (Knauft 1991), but normally they are not punished by the group even though this may disrupt its social life or cause it to split because of a revenge killing.

Knauft (1991) found that with his "simple forager" type, the main cause of single homicides was male competition over females, and, of course, eliminating a rival and gaining a desirable partner helps one's fitness. At first blush, this type of coercive free-riding appears to have major payoffs even though it usually goes unpunished by the group. However, such calculations are complicated by the fact that a retaliatory killing is likely (Boehm 2011), and sometimes there will even be a chain of such killings that eliminates additional family members (see Kelly 2000). This is fraternal interest group revenge (Boehm 2011), not deliberate punishment by the entire band, but the specter of an angry retaliatory killing does function deterrently, as a form of *indirect* social control.

TABLE 20.2. Type of male and female deviance as mentioned in ethnographies for sixty-five forager societies

Types of Deviance	Number of Regions Reporting	World Regions				
		AFR (6)	ARC (27)	ASIA (6)	AUS (12)	NA/SA (14)
Intimidation						
Murder	5	49	111	9	27	18
Sorcery	5	6	64	5	48	16
Beating someone	5	16	13	1	12	10
Bullying	4	4	11		4	1
Power Abuse	1				7	
Trying to grab power or position	1				4	
Leader's trying to boss others around	1				3	
Intimidation Total	5	75	210	15	116	46
Deception						
Thievery	5	27	56	6	24	25
Lying	4	2	13	3		3
Adultery	5	43	31	11	47	23
Cheating	3	2	5			1
Cheating an individual	1	2				
Cheating the group	3	17	5			2
Deception Total	5	93	110	20	71	54
Failure to share/cooperate	5	19	9	1	8	3
Failure to Share Total	5	19	9	1	8	3
Sexual Deviance						
Incest	5	31	34	5	38	11
Man dishonors a woman	4	3	15		3	1
Sexual Deviance Total	5	34	49	5	41	12

It is when a man kills more than once, or if he competitively makes people fearful as a sorcerer and is thought to seriously threaten their lives, that the group has a problem to solve and the entire group must conspire to eliminate him physically. For efficiency, often band members will ask the deviant's close kinsman to do the job, so that lethal retaliation and the further strife it brings will be out of the question (Boehm 2011). Such retaliation has been documented among the Aranda (Strehlow 1970), where a group of young men who did the execution were then killed by their victim's relatives (see also

Bowles et al. 2012). Because such executions are rather rare, it seems likely that the custom of the group's delegating a close male relative of the deviant to kill him has been invented, repeatedly in different groups, to cope with such anticipated complications.

Unless punished, this coercive free-riding by social predators can degrade not only the political autonomy of band members, but their nutrition—for instance, if predatory behavior disrupts equalized meat distributions. However, people can protect themselves by using social sanctioning to reform or eliminate these intimidators, and this policing enhances their own fitness even as they degrade the deviants' fitness. The same is true of recidivist thieves.

Noncoercive Free-Riding

Thieves cover their tracks by feigning innocence, and cheaters use active deception to hoodwink their victims, while shirkers brazenly fail to share without any deception. A main form of cooperation in foraging bands entails a roughly equalized distribution of meat from large-game carcasses (Stanford 1998; see also Winterhalder 2001; Kelly 2013), and one way to cheat other band members deceptively is to immediately hide the carcass (which in effect belongs to the entire band) and consume it secretly.

Normally, this "public" meat (including large-package marine life) is handed over for community-wide distribution, and each family then owns the share it receives. This private property can then be stolen by thieves, who consume it surreptitiously. Often such private meat is shared between families, and expectations of reciprocation develop. If these are not met by shirkers, their refusal to cooperate can lead not only to conflict, but to a bad general reputation or even (rarely) to noncapital punishment by the entire group.

Capital Punishment

Table 20.3 shows the types of behavior that are punished definitively by killing the deviant. Although overall the reported numbers are rather limited, capital punishment of serious intimidators is reported at least once in all five world regions, which by the definition used here qualifies as being "widespread."

TABLE 20.3. Capital punishment of male and female free-riders, by type of crime

Types of Deviance	Number of Regions Reporting	World Regions				
		AFR (6)	ARC (27)	ASIA (6)	AUS (12)	NA/SA (14)
Intimidation						
Murder	3	2	3	0	3	0
Sorcery	4	0	7	1	2	5
Power Abuse	1	0	2	0	0	0
Bullying	1	0	1	0	0	0
Intimidation Total	5	2	13	1	5	5
Deception						
Thievery	1	0	3	0	0	0
Cheating	1	1	0	0	0	0
Adultery	1	0	0	0	2	0
Deception Total	3	1	3	0	2	0
Other Deviance						
Incest	2	1	0	0	4	0
Taboo broken	3	1	1	0	12	0
Malicious Gossip	1	0	2	0	0	0
Betray Group	1	0	3	0	0	0
Mentally Ill (dangerous)	1	0	1	0	0	0
Unspecified	2	2	2	0	0	0
Other Deviance Total	3	4	9	0	16	0

With respect to deceivers, with this size-limited sample of societies capital punishment is reported in only four regions against thieves and in just two against cheaters, while shirking is not reported as being punished capitally. Although these numbers may be suggestive, a substantially larger sample would be needed to reduce sampling error and more reliably evaluate these lesser trends. The same is true of other behaviors reported in just one or a few regions.

By ethnographic analogy, we may plausibly suggest that 45,000 years ago the most widespread use of capital punishment would have been to prevent fear-based social predation, and to keep intimidating male upstarts from gaining a political hold on the band as a whole (Boehm 1999). The consequences would have been not only social but genetic, and even a pattern of relatively rare capital punishment could have been impactful on human gene pools; being executed part way through the life cycle not only prevents further reproduction; it also cuts off support of offspring and kin by a male who often is in the prime of life.

Lesser Sanctions against Free Riders

Table 20.4 shows the deviant targets for ostracism and shaming combined. Although being ostracized or shamed can be quite costly to deviants, usually the band is willing to relent if the antisocial behavior is reduced or stopped, so personal reform is encouraged.

If we compare these figures for lesser sanctioning actions with those for capital punishment, at the present sample size the numbers strongly suggest that these lesser types of social control are used much less widely than capital punishment even though some ethnographic reporting biases may exist because capital punishment is so noteworthy. Again, a substantially larger ethnographic sample would be needed to see exactly how widespread these less frequently reported sanctions are likely to be, but the total absence of reports for shaming for the six mostly Andaman foragers in the Asian region raises questions that cannot be answered with this limited database.

Social-distancing sanctions that deprive deviants of useful social contacts can be temporary, so moderate degrees of ostracism often permit a social predator to reform. Shaming is still more immediate, and it too can be accompanied by loss of reputation, but again personal reform can repair some of the damage.

Social Selection by Reputation

Selection scenarios for chimpanzees' social behavior are multilevel and relatively straightforward (e.g., Muller and Mitani 2005; Watts et al. 2006). In contrast, for humans, strongly interactive gene-culture evolution (Durham 1991; see also Richerson and Boyd 2005) greatly complicates the selection scenarios used to explain why humans have become so cooperative. For instance, with respect to reputations, these are constantly reassessed by gossiping (Boehm 2017b; e.g., Wiessner 2005, 2014).

Selection-by-reputation is a mode of social selection proposed by Alexander (1987) in the context of building a theory to explain altruism. The idea

TABLE 20.4. Ostracism and shaming of male and female free-riders

		World Regions				
Types of Deviance	Number of Regions Reporting	AFR (6)	ARC (27)	ASIA (6)	AUS (12)	NA/SA (14)
Intimidation						
Murder	3	0	4	1	0	3
Sorcery	1	0	3	0	0	0
Power Abuse	2	0	2	0	1	0
Bullying	3	3	6	0	1	0
Intimidation Total	5	3	15	1	2	3
Deception						
Thievery	3	1	4	0	0	3
Lying	1	0	2	0	0	0
Deception Total	3	1	6	0	0	3
Failure to share						
Shirking	3	2	0	0	1	1
Failure to Share Total	3	2	0	0	1	1
Sexual Deviance						
Incest	4	4	1	0	1	2
Adultery	2	2	1	0	0	0
Other Sexual Deviance	4	1	2	0	2	1
Sexual Deviance Total	4	7	4	0	3	3
Other Deviance						
Taboo broken	2	1	0	0	0	1
Unspecified	5	6	23	2	13	1
Other Deviance Total	5	7	23	2	13	2

is that in bands people altruistically help families in need even though payback is highly contingent; in their own time of need, these altruists will be helped by other altruists in whatever band they are in, again without expectations of payback in kind. This cooperative system involves altruism not only because the help is not reciprocated exactly, but because demographically dynamic hunter-gatherer bands include unrelated families (Hill et al. 2011).

At the core of Alexander's theory of "selection-by-reputation" (Wilson and Dugatkin 1997; see also Barclay 2006) is the hypothesis that while a group's altruists pay obvious costs for being generous, the costs are more than compensated because individuals with generous reputations are favored in marriage, and in forming economic partnerships. Thus, these fitness gains can more than repay generosity's costs. Indirect reciprocity has been documented for forager-horticulturalists by Gurven (2004), and Alexander's (1987) seminal selection-by-reputation theory has been further elaborated in terms of costly signaling (Zahavi 1995). However, ancestral precursors, if they exist, are far less obvious than in the case of selection through gang attacks.

Sanctioning Selection and Autodomestication

The second notable type of social selection in our recent evolutionary past is sanctioning by groups. The most dramatic instance of this has to be capital punishment (Boehm 2014), but ostracism at least deprives the deviant of social contacts valuable to cooperation. Both have ancestral precedents, and ostracism results in a bad social reputation. As a much more recent development (Boehm 2012b), shaming is more transitory than ostracism; but it too leads to a bad reputation. Furthermore, in the absence of personal reform it can also lead to ostracism or worse.

Ancestrally, group sanctioning of coercively free-riding deviants could have resulted in reproductive leveling, which is favorable to genetic group selection that favors cooperation within groups (Boehm 1997). Group sanctioning also would have favored self-domestication (Wrangham and Peterson 1996; see also Wilkins et al. 2014), in that exceptionally aggressive individuals were being targeted and punished, with drastic reproductive consequences.

Of course, humans are far from being "fully domesticated," in the sense of possessing *markedly* diminished aggressive impulses compared to what might be estimated for the LCA. In fact, if our general capacity for cooperation and our capacity for genocidal warfare are both taken into account, it may be more realistic simply to speak of some major differences, including a significant reduction of certain aspects of aggression due to sanctioning selection.

Wrangham and Glowacki (2012) have extensively compared chimpanzee intergroup aggression with that of humans. It is likely that there has been considerable reduction in innately well-prepared intimidation displays, replete with the long black hair and dramatic piloerection for which *Pan* is well known. When humans reach for and brandish their hunting weapons this is similar, but also different (see Gintis et al. 2015) in that we have the cognitive capacity to understand that our fashioned weapons are lethal, and even to conceive of genocide against another group as an explicit aim.

We too are capable of contagious lethal gang attacks, yet as hunter-gatherers most of our group executions have been relatively cool and deliberate. As a prime example, we have the practice of the band's delegating a close relative of the victim to efficiently carry out an ambush execution, but more generally, when social decisions are made in the interest of group cohesion and cooperation, the foragers' consensual decision process tends to be deliberate with a careful weighing of alternatives. The often-cited, chaotically inefficient Ju/'hoansi execution described by Lee (1979) would appear to be an exception.

We surely cooperate better than the LCA. However, one aspect of this cooperation is that we are more prone to fight in large groups, and that even in the face of power imbalances (see Wrangham and Glowacki 2012) we may bravely attack; thus we enter pitched battles in which the outcomes are uncertain. We also mount genocidal attacks, fully aware of the resources at stake. Our use of symbols and our large brains help to make all of this possible, and morality figures in the underlying ethnocentrism that allows us often to prey on neighbors even as we outlaw social predation within our groups.

I have suggested that the LCA created primitive "sanctioning selection" forces in the form of the intracommunity coalitionary attacks we have examined; the effects can be likened to capital punishment and ostracism, respectively. As with today's chimpanzees, it is likely that such attacks were fairly rare and directed uniformly or chiefly against males, while predominantly the targets would have been unusually politically aggressive, dominance-seeking individuals. As an ancestral preadaptation, this could at least have assisted whatever process of self-domestication may have taken place in the chimpanzee lineage. (In *Pan troglodytes,* it is likely that this development has not gone nearly as far as with moralistic humans, or bonobos.)

By proxy, this chimpanzee type of group sanctioning, also seen more rarely in bonobos, could define a critical ancestral preadaptation that went on to serve as an important building block as moralistic social control was evolving. Natural selection could have made use of it as humans eventually evolved to include in their behavioral repertoire the lethal *moral* sanctioning of overly aggressive deviants, which would have intensified the self-domestication process.

Discussion

Morality involves individuals controlling themselves, but also groups controlling individuals. In the LCA, fear-based self-control was one important preadaptation, and group attacks on individuals within the community, emphasized here, were another.

Based on fear, today's chimpanzees are able to routinely read predictable "rules" into the behavior of their intimidating superiors and stay out of harm's way. Similarly, conscience functions of humans are at least in part based on simple fear of punishment, and in both cases fear leads to self-control. In the case of humans, however, it is the distinctively *moral* aspects of the evolutionary conscience that catch our attention because they are so distinctive, and it is a sense of shame combined with the inhibitory internalization of group rules that enables most of us to preemptively control ourselves and avoid group punishment.

What about groups actively controlling individuals? When captive female chimpanzees gang up to prevent female victimization by males, Flack and de Waal (2000) refer to "community concern" as the motivator for their gang threats. In the wild we may assume that similar concerns are motivating attacks like those reported in Table 20.1. The capacity of a local *Pan troglodytes* group to organize itself for a coalitionary attack against a disliked member is another important precursor for human morality, and more specifically for the evolution of aggressive social control by groups.

Thus it would appear that the moralistic social sanctioning described by Durkheim had its phylogenetic roots in this collective behavior of the LCA. The hypothesis is that the self-inhibitory human conscience evolved because group social control was exacting serious penalties from those who did not control their antisocial tendencies. Targeted in particular were aggressive social predators, and prominently this included capital punishment of bullies like sorcerers or recidivist killers.

The analysis suggests that the LCA was gang attacking and killing certain types of males, namely alphas and former alphas, but that at the same time the great majority of high-ranking males (including alphas) were not being attacked. It bears mentioning that some of the males killed were ones whose social bonding was below average, or whose intimidation tactics were unusually harsh, while a more general hypothesis would be that the targets were individuals whose behavior *unusually* limited the political autonomy and reproductive fitness of other group members. However, there are only some major hints in these directions in the field reports Table 20.1 was based on, because often these reports lacked any detailed description of what led up to the gang attack. What is definite is that the targets were of high rank and therefore effective as intimidators, and that the gang attacks would have reduced whatever distinctive genetic traits these particular targets shared.

If human evolution generally seems to have accelerated over the past several hundred thousand years, the two morally connected types of social selection we have discussed (sanctioning and reputational) could have been major contributors as we became increasingly moral and increasingly different from the LCA. The fact that both sanctioning selection and reputational selection involve both decision makers and recipients of decisions suggests that, as with sexual selection, the traits being favored might have become strongly selected.

Sophisticated collective intentions have figured prominently in the analysis, and they may have helped to make humans so distinctive as a species. In fact, in a broad sense our uniquely human morals might be compared to an improbable peacock's tail, as it flourishes in mating season. However, the individual intentions of peacocks are simply to mate, and their resplendent plumage can be considered to be a simple evolutionary by-product, the result of random variation undergoing natural selection. However, with humans in egalitarian moral communities the variation is different because it is shaped by sophisticated intentions of threatened group members, who also look to improve social life and cooperation.

When intentions to improve egalitarian social life are implemented by eliminating a sorcerer or serial killer who is acting the bully, the evolutionary by-products run *in parallel*. By this is meant that when prehistoric foragers sanctioned bullies or thieves, inadvertently they were also reducing bullying and thieving tendencies at the genome level—which also had the effect, long-term, of improving egalitarian social life. Because of the inadvertence this is nothing like genetic engineering; nonetheless, the selection process would appear to be influenced by sophisticated intentionality. Could this be suggestive of some lower-level teleology?

Campbell's (1975) *"blind variation-and-selective-retention"* characterization of the genetic selection process is perfectly parsimonious, universally accepted by evolutionists, and serves science well. However, if an evolutionary by-product is going in exactly the same direction as the intentions of the actors who cause it, this is at least suggestive of some "directionality"—even though these actors do not understand genetics. Indeed, because a significant impact on the genome was going in very much the same direction as the intentional group sanctions, some special modeling consideration may be needed.

Very recent and obviously teleological exceptions to Campbell's blind model are genetic engineering and "eugenics," but otherwise in human evolution his blind paradigm is generally assumed to have held, and scientifically this assumption is consistent with trying to reduce such a complex process in a useful explanatory framework. However, we may need to think about how standard evolutionary models can do justice to the parallel effects described here, especially when groups are killing individuals because of disliked behavior patterns that are heritable, and they are doing so consistently over time.

Two quite different types of parallel effects have been introduced. The first parallel effect involved entire groups making consensual decisions to punish certain coercive free-riders, and the preadaptation reaches back for seven million years. This provided a well-focused and long-standing *negative* selection force—a collectively activated force that would have both reduced such predators' gains and at the same time reduced predatory competitive tendencies in the human genome. This effect could have made a major contribution to our species' autodomestication, and because the selection effects involved prematurely terminating a social predator's life history, it could have acted rapidly.

The second parallel effect involved intentions to partner with altruists. With respect to Alexander's reputation-based theory, the ancestral baseline shows no apparent precedent for the evolution of individually based selection-by-reputation, which stems from moralistic group opinions about peers. These decisions favor altruists as partners, so altruistic genes profit. This social agency of selection could have become significant whenever people began to deal in moral reputations.

Both of these parallel effects contributed to today's behavioral repertoire. The result is an often feisty but unusually cooperative moralistic species that has become distinctive in the animal kingdom. However, this position of uniqueness does not amount to an unexplainable saltation in our natural history, for the LCA's often lethal group sanctioning of aggressive males, based on collective intentions, provided a key preadaptation that led to our becoming a moral species.

Endnotes

1. Engelmann et al. (2012) showed that when given the opportunity to engage in either cooperative or selfish behavior, chimpanzees behaved the same whether they were being watched by conspecifics or not.
2. Evolutionary psychologists Erdal and Whiten (1994) have criticized the use of this "reverse dominance hierarchy" terminology, possibly because they are focusing mainly on the motives involved, which are aptly called "counter-dominant." However, I am considering a hierarchy's political structure and political dynamics, and the broader sociological point is that when the suppression of dominance behavior becomes definitive and routinized, the flow of power actually reverses because the united subordinates are so firmly in control (see Boehm and Flack 2011).
3. The database is a work in progress. Its purpose is to identify current widespread or universal hunter-gatherer behavior patterns, and project them as reliably as possible into the behaviorally modern past. There are currently sixty-five foraging societies included, which have been coded in detail for social and political information. The criteria for selection have included maximizing world regional distribution (see Burton and White 1991), selection of societies with unusually high-quality or multiple ethnographies, all of which are coded, and, where possible, maximizing geographic distribution within regions. Most of the societies selected are of the nomadic/egalitarian type that predominated in the Late Pleistocene, but they include not only Knauft's (1991) "simple foragers" with their immediate-return economies, but also highly egalitarian or moderately hierarchical nomadic or sedentary foragers that engage with food storage. Basically I have used continents as cultural units, and of course Australia and the Arctic are two vast regions of hunter-gatherers with scores of societies reported. Africa and Asia have only six eligible societies each, with little opportunity to maximize geographic distribution within the area. North America has a dozen societies but

South America has only two that are fully eligible, and in terms of Galton's Problem they shared a relatively recent migration route so the two regions were merged. This set of five regions provides the basis for analogical reconstruction, the rule being that any behavior found in all five of them was likely to have been widespread prehistorically.

Acknowledgments

Gratitude is expressed to the Templeton Foundation and to the Goodall Research Center at USC for funding the hunter-gatherer database project. I thank Martin Muller, Frans de Waal, Michael L. Wilson, and Richard Wrangham for comments on or other assistance with the manuscript.

References

Alexander, R. D. 1987. *The Biology of Moral Systems*. New York: Aldine de Gruyter.
Ames, K. M. 2004. Review: Supposing hunter-gatherer variability. *American Antiquity* 69: 364–374.
Anderson, R. D. 1987. *The Biology of Moral Systems*. New York: Aldine de Gruyter.
Anderson, S. W., A. Bechara, H. Damasio, D. Tranel, and A. R. Damasio. 1999. Impairment of social and moral behavior related to early damage in human prefrontal cortex. *Nature Neuroscience* 2: 1032–1037.
Bandura, A. 1991. Social cognitive theory of self-regulation. *Organizational Behavior and Human Decision Processes* 50: 138–187.
Barclay, P. 2006. Reputational benefits for altruistic punishment. *Evolution and Human Behavior* 27: 325–344.
Bingham, P., and J. Souza. 2009. *Death from a Distance and the Birth of a Humane Universe: Human Evolution, Behavior, History, and Your Future*. Charleston, SC: Book-Surge.
Boehm, C. 1981. Parasitic selection and group selection: A study of conflict interference in rhesus and Japanese macaque monkeys. In A. B. Chiarelli and R. S. Corruccini, eds., *Primate Behavior and Sociobiology: Proceedings of the International Congress of Primatology*. Heidelberg: Springer.
Boehm, C. 1986. Capital punishment in tribal Montenegro: Implications for law, biology, and theory of social control. *Ethology and Sociobiology* 7: 305–320.
Boehm, C. 1991. Lower-level teleology in biological evolution: Decision behavior and reproductive success in two species. *Cultural Dynamics* 4: 115–134.
Boehm, C. 1993. Egalitarian behavior and reverse dominance hierarchy. *Current Anthropology* 34: 227–254.

Boehm, C. 1996. Emergency decisions, cultural selection mechanics, and group selection. *Current Anthropology* 37: 763–793.

Boehm, C. 1997. Impact of the human egalitarian syndrome on Darwinian selection mechanics. *American Naturalist* 150: 100–121.

Boehm, C. 1999. *Hierarchy in the Forest: The Evolution of Egalitarian Behavior.* Cambridge, MA: Harvard University Press.

Boehm, C. 2000. Conflict and the evolution of social control. *Journal of Consciousness Studies* 7: 79–101.

Boehm, C. 2008. Purposive social selection and the evolution of human altruism. *Cross-Cultural Research* 42: 319–352.

Boehm, C. 2011. Retaliatory violence in human prehistory. *British Journal of Criminology* 51: 518–534.

Boehm, C. 2012a. Ancestral hierarchy and conflict. *Science* 336: 844–847.

Boehm, C. 2012b. Costs and benefits in hunter-gatherer punishment. *Behavioral and Brain Sciences* 35: 19–20.

Boehm, C. 2012c. *Moral Origins: The Evolution of Altruism, Virtue, and Shame.* New York: Basic Books.

Boehm, C. 2014. The moral consequences of social selection. *Behaviour* 51: 167–183.

Boehm, C. 2017a. Collective intentionality: A basic and early component of moral evolution. *Philosophical Psychology* 30: 222–246.

Boehm, C. 2017b. The evolutionary impact of hunter-gatherer gossiping behaviour. In F. Giardini and R. Wittek, eds., *The Oxford Handbook of Gossip and Reputation.* Oxford: Oxford University Press.

Boehm, C., and J. Flack. 2010. The emergence of simple and complex power structures through social niche construction. In A. Guinote, ed., *The Social Psychology of Power.* New York: Guilford.

Boesch, C., and H. Boesch-Achermann. 2000. *The Chimpanzees of the Taï Forest: Behavioural Ecology and Evolution.* Oxford: Oxford University Press.

Bowles, S. 2006. Group competition, reproductive leveling, and the evolution of human altruism. *Science* 314: 1569–1572.

Bowles, S., R. Boyd, and S. Mathew. 2012. The punishment that sustains cooperation is often coordinated and costly. *Behavioral and Brain Sciences* 35: 20–21.

Boyd, R., S. Gintis, and S. Bowles. 2010. Coordinated punishment of defectors sustains cooperation and can proliferate when rare. *Science* 328: 617–620.

Boyd, R., H. Gintis, S. Bowles, and P. J. Richerson. 2003. The evolution of altruistic punishment. *Proceedings of the National Academy of Sciences* 100: 3531–3535.

Boyd, R., and P. J. Richerson. 1992. Punishment allows the evolution of cooperation (or anything else) in sizable groups. *Ethology and Sociobiology* 13: 171–195.

Brown, D. 1991. *Human Universals.* New York: McGraw-Hill.

Brüne, M., U. Brüne-Cohrsa, W. C. McGrew, and S. Preuschoft. 2006. Psychopathology in great apes: Concepts, treatment options and possible homologies to human psychiatric disorders. *Neuroscience & Biobehavioral Reviews* 8: 1246–1259.

Burton, M. L., and D. R. White. 1991. Regional comparisons, replications, and historical network analysis. *Cross-Cultural Research* 25: 55–78.

Campbell, D. T. 1975. On the conflicts between biological and social evolution and between psychology and moral tradition. *American Psychologist* 30: 1103–1136.

Casimir, M. J., and M. Schnegg. 2002. Shame across cultures: The evolution, ontogeny, and function of a "moral emotion." In H. Keller, Y. H. Poortinga, and A. Scholmerich, eds., *Between Culture and Biology: Perspectives on Ontogenetic Development*. Cambridge: Cambridge University Press.

Cosmides, L., and J. Tooby. 2004. Social exchange: The evolutionary design of a neurocognitive system. In M. S. Gazzaniga, ed., *The Cognitive Neurosciences*. Cambridge, MA: MIT Press.

Damasio, A. R. 2002. The neural basis of social behavior: Ethical implications. Paper presented at the conference Neuroethics: Mapping the Field, San Francisco, California, May 13–14.

Darwin, C. 1865. *The Expression of the Emotions in Man and Animals*. Chicago: University of Chicago Press.

Darwin, C. 1871. *The Descent of Man, and Selection in Relation to Sex*. Princeton, NJ: Princeton University Press.

de Waal, F. B. M. 1982. *Chimpanzee Politics: Power and Sex Among Apes*. New York: Harper and Row.

de Waal, F. B. M. 1996. *Good Natured: The Origins of Right and Wrong in Humans and Other Animals*. Cambridge, MA: Harvard University Press.

Durham, W. H. 1991. *Coevolution: Genes, Culture, and Human Diversity*. Stanford, CA: Stanford University Press.

Durkheim, É. 1933. *The Division of Labor in Society*. Translated by George Simpson. New York: Macmillan.

Egas, M., and A. Riedl. 2008. The economics of altruistic punishment and the maintenance of cooperation. *Proceedings of the Royal Society B: Biological Sciences* 275: 871–978.

Ellis, L. 1995. Dominance and reproductive success among nonhuman animals: A cross-species comparison. *Ethology and Sociobiology* 16: 257–333.

Engelmann, J. M., E. Herrmann, and M. Tomasello. 2012. Five-year olds, but not chimpanzees, attempt to manage their reputations. *PLoS ONE* 7: 48433.

Erdal, D., A. Whiten, C. Boehm, and B. Knauft. 1994. On human egalitarianism: An evolutionary product of Machiavellian status escalation? *Current Anthropology* 35: 175–183.

Fallow, A. 2003. Frodo: The alpha male. National Geographic Online, http://ngm.nationalgeographic.com/ngm/0304/feature4/online_extra2.html.

Fawcett, K., and G. Muhumuza. 2000. Death of a wild chimpanzee community member: Possible outcome of intense sexual competition. *American Journal of Primatology* 51: 243–247.

Fehr, E., and S. Gächter. 2002. Altruistic punishment in humans. *Nature* 415: 137–140.

Fisher, R. A. 1930. *The Genetical Theory of Natural Selection*. New York: Dover.

Flack, J. C., and F. B. M. de Waal. 2000. "Any animal whatever": Darwinian building blocks of morality in monkeys and apes. *Journal of Consciousness Studies* 7: 1–29.

Flinn, M. V., and R. D. Alexander. 2007. Runaway social selection in human evolution. In S. Gangestad and J. Simpson, eds. *The Evolution of Mind: Fundamental Questions and Controversies*. New York: Guilford.

Fouts, R., with S. T. Mills. 1997. *Next of Kin: My Conversations with Chimpanzees*. New York: Avon.

Gintis, H. 2000. Strong reciprocity and human sociality. *Journal of Theoretical Biology* 206: 169–179.

Gintis, H. 2003. The hitchhiker's guide to altruism: Gene-culture coevolution and the internalization of norms. *Journal of Theoretical Biology* 220: 407–418.

Gintis, H., C. van Schaik, and C. Boehm. 2015. The evolutionary roots of human sociopolitical systems. *Current Anthropology* 56: 327–353.

Goodall, J. 1986. *The Chimpanzees of Gombe: Patterns of Behavior*. Cambridge, MA: Harvard University Press.

Goodall, J. 1992. Unusual violence in the overthrow of an alpha male chimpanzee at Gombe. In T. Nishida, W. C. McGrew, P. Marler, M. Pickford, and F. B. M. de Waal, eds., *Topics in Primatology*, vol. 1: *Human Origins*. Tokyo: University of Tokyo Press.

Gurven, M. 2004. Reciprocal altruism and food sharing decisions among Hiwi and Ache hunter-gatherers. *Behavioral Ecology and Sociobiology* 56: 366–380.

Hare, B., V. Wobber, and R. Wranghman. 2012. The self-domestication hypothesis: Evolution of bonobo psychology is due to selection against aggression. *Animal Behaviour* 83: 1–13.

Hare, R. D. 1993. *Without Conscience: The Disturbing World of the Psychopaths among Us*. New York: Guilford.

Hill, K. R., R. Walker, M. Božičević, J. Eder, T. Headland, B. Hewlett, A. M. Hurtado, F. Marlowe, P. Wiessner, and B. Wood. 2011. Coresidence patterns in hunter-gatherer societies show unique human social structure. *Science* 331; 1286–1289.

Hohmann, G., and B. Fruth. 2011. Is blood thicker than water? In M. M. Robbins and C. Boesch, eds., *Among African Apes: Stories and Photos from the Field*. Berkeley: University of California Press.

Kaburu, S. S. K., S. Inoue, and N. E. Newton-Fisher. 2013. Death of the alpha: Within-community lethal violence among chimpanzees of the Mahale Mountains National Park. *American Journal of Primatology* 75: 789–797.

Kano, T., G. Idani, and C. Hashimoto. 1994. The present situations of bonobos at Wamba, Zaire. *Primate Research* 10: 191–214.

Keller, L. 1997. Indiscriminate altruism: Unduly nice parents and siblings. *Trends in Ecology and Evolution* 12: 99–103.

Kelly, R. C. 2000. *Warless Societies and the Evolution of War*. Ann Arbor: University of Michigan Press.

Kelly, R. L. 2013. *The Lifeways of Hunter-Gatherers: The Foraging Spectrum*. Cambridge: Cambridge University Press.

Kiehl, K. A. 2008. Without morals: The cognitive neuroscience of criminal psychopaths. In W. Sinnott-Armstrong, ed., *Moral Psychology*, vol. 1: *The Evolution of Morality: Adaptations and Innateness*. Cambridge, MA: MIT Press.

Kiehl, K. A., A. T. Bates, K. R. Laurens, R. D. Hare, and P. F. Liddle. 2006. Brain potentials implicate temporal lobe abnormalities in criminal psychopaths. *Journal of Abnormal Psychology* 115: 443–453.

Klein, R. G. 2009. *The Human Career: Human Biological and Cultural Origins.* Chicago: University of Chicago Press.

Knauft, B. M. 1991. Violence and sociality in human evolution. *Current Anthropology* 32: 391–428.

Krasnow, M. M., A. W. Delton, L. Cosmides, and J. Tooby. 2015. Group cooperation without group selection: Modest punishment can recruit much cooperation. *PLoS ONE* 10: 1–17.

Krebs, D. L. 2011 *The Origins of Morality.* Oxford: Oxford University Press.

Kyambura Gorge Chimpanzee Community. 2011. Facebook, https://www.facebook.com/permalink.php?story_fbid=127295474009286&id=119478481457652.

Lee, R. B. 1979. *The !Kung San: Men, Women, and Work in a Foraging Society.* Cambridge: Cambridge University Press.

Lilienfeld, S. O., J. Gershon, M. Duke, L. Marino, and F. B. M. de Waal. 1999. A preliminary investigation of the construct of psychopathic personality (psychopathy) in chimpanzees *(Pan troglodytes)*. *Journal of Comparative Psychology* 113: 365–375.

Lorenz, K. 1966. *On Aggression.* New York: Harcourt, Brace & World.

Mayr, E. 2002. *What Evolution Is.* New York: Basic Books.

McBrearty, S., and A. Brooks. 2000. The revolution that wasn't: A new interpretation of the origin of modern human behavior. *Journal of Human Evolution* 39: 453–563.

Melis, A. P., and F. Warneken. 2016. The psychology of cooperation: Insights from chimpanzees and children. *Evolutionary Anthropology* 25: 297–305.

Mitani, J. C., D. P. Watts, and S. J. Amsler. 2010. Lethal intergroup aggression leads to territorial expansion in wild chimpanzees. *Current Biology* 20: R507–R508.

Mjungu, D. C. 2010. Dynamics of intergroup competition in two neighboring chimpanzee communities. Ph.D. dissertation. University of Minnesota.

Mjungu, D., and A. Collins. 2016. *Gombe Gets a New Alpha: The Fall of Ferdinand.* Toronto: Jane Goodall Institute of Canada.

Muller, M., and J. C. Mitani. 2005. Conflict and cooperation in wild chimpanzees. *Advances in the Study of Behavior* 35: 275–331.

Nesse, R. M. 2007. Runaway social selection for displays of partner value and altruism. *Biological Theory* 2: 143–155.

Newton-Fisher, N. E. 2006. Female coalitions against male aggression in wild chimpanzees of the Budongo forest. *International Journal of Primatology* 27: 1589–1599.

Nicholson, N. 2012. The evolved self, self-regulation, and the co-evolution of leadership. *Biological Theory* 6: 399–412.

Nishida, T. 1996. The death of Ntologi, the unparalleled leader of M group. *Pan Africa News* 3: 4.

Nishida, T., K. Hosaka, M. Nakamura, and M. Hamai. 1995. A within-group gang attack on a young adult male chimpanzee: Ostracism of an ill-mannered member? *Primates* 36: 207–211.

Nowak, M. A. 2006. Five rules for the evolution of cooperation. *Science* 314: 1560–1563.

Otterbein, K. F. 1988. Capital punishment: A selection mechanism. Comment on Robert K. Dentan, on Semai homicide. *Current Anthropology* 29: 633–636.

Panchanathan, K., and R. Boyd. 2004. Indirect reciprocity can stabilize cooperation without the second-order free rider problem. *Nature* 432: 499–502.

Pandit, S. A., and C. P. van Schaik. 2003. A model for leveling coalitions among primate males: Toward a theory of egalitarianism. *Behavioral Ecology and Sociobiology* 55: 161–168.

Piers, G., and M. B. Singer. 1971. *Shame and Guilt: A Psychoanalytic and a Cultural Study.* New York: W. W. Norton.

Price, M. E., L. Cosmides, and J. Tooby. 2002. Punitive sentiment as an anti-free rider psychological device. *Evolution and Human Behavior* 23: 203–231.

Pruetz, J. D., K. B. Ohr, E. Cleaveland, S. Lindshield, J. Marshak, and E. G. Wessling. 2017. Intragroup lethal aggression in West African chimpanzees *(Pan troglodytes verus):* Inferred killing of a former alpha male at Fongoli, Senegal. *International Journal of Primatology,* http://www.eva.mpg.de/documents/Springer/Pruetz_Intragroup_IntlJPrim_2017_2391861.pdf.

Richerson, P. J., and R. Boyd. 2005. *Not by Genes Alone: How Culture Transformed Human Evolution.* Chicago: University of Chicago Press.

Riedl, K., K. Jensen, J. Call, and M. Tomasello. 2012. No third-party punishment in chimpanzees. *Proceedings of the National Academy of Sciences* 109: 4824–4829.

Roscoe, P. B. 2002. The hunters and gatherers of New Guinea. *Current Anthropology* 43: 153–162.

Schjelderup-Ebbe, T. 1935. Social behavior of birds. In C. Murchison, ed., *A Handbook of Social Psychology.* Worcester, MA: Clark University Press.

Shea, J. J. 2011. *Homo sapiens* is as *Homo sapiens* was: Behavioral variability versus "behavioral modernity" in paleolithic archaeology. *Current Anthropology* 52: 1–35.

Simon, H. A. 1990. A mechanism for social selection and successful altruism. *Science* 250: 1665–1668.

Sober, E., and D. S. Wilson. 1998. *Unto Others: The Evolution and Psychology of Unselfish Behavior.* Cambridge, MA: Harvard University Press.

Stanford, C. B. 1998. *The Hunting Apes: Meat Eating and the Origins of Human Behavior.* Princeton, NJ: Princeton University Press.

Strehlow, T. G. H. 1970. Geography and the totemic landscape in central Australia: A functional study. In R. M. Berndt, ed., *Australian Aboriginal Anthropology,* 93–140. Perth: University of Western Australia Press.

Temerlin, J. W. 1980. The self-concept of a home-reared chimpanzee. In M. A. Roy, ed., *Species Identity and Attachment: A Phylogenetic Evaluation.* New York: Garland.

Temerlin, M. K. 1975. *Lucy: Growing Up Human: A Chimpanzee Daughter in a Psychotherapist's Family.* Palo Alto, CA: Science and Behavior Books.

Tomasello, M. 2014. *A Natural History of Human Thinking.* Cambridge, MA: Harvard University Press.

Tomasello, M. 2016. *A Natural History of Human Morality*. Cambridge, MA: Harvard University Press.

Trivers, R. L. 1971. The evolution of reciprocal altruism. *Quarterly Review of Biology* 46: 35–57.

Uehara, S., et al. 1994. The fate of defeated alpha male chimpanzees in relation to their social networks. *Primates* 35: 49–55.

Vehrencamp, S. L. 1983. A model for the evolution of despotic versus egalitarian societies. *Animal Behavior* 31: 667–682.

von Rohr, C. R., J. M. Burkart, and C. P. van Schaik. 2011. Evolutionary precursors of social norms in chimpanzees: A new approach. *Biological Philosophy* 26: 1–30.

Warneken, F., and M. Tomasello. 2006. Altruistic helping in human infants and young chimpanzees. *Science* 311: 1301–1303.

Watts, D. P. 2004. Intracommunity coalitionary killing of an adult male chimpanzee at Ngogo, Kibale National Park, Uganda. *International Journal of Primatology* 25: 507–521.

Watts, D. P., M. Muller, S. J. Amsler, G. Mbabazi, and J. C. Mitani. 2006. Lethal intergroup aggression by chimpanzees in Kibale National Park, Uganda. *American Journal of Primatology* 68: 161–180.

West-Eberhard, M. J. 1979. Sexual selection, social competition, and evolution. *Proceedings of the American Philosophical Society* 123: 222–234.

West, S. A., A. S. Griffin, and A. Gardner. 2007. Social semantics: Altruism, cooperation, mutualism, strong reciprocity and group selection. *Journal of Evolutionary Biology* 20: 415–432.

Whiten, A., and R. W. Byrne. 1988. Tactical deception in primates. *Behavioural and Brain Sciences* 11: 233–244.

Wiessner, P. 2005. Norm enforcement among the Ju/'hoansi Bushmen: A case of strong reciprocity? *Human Nature* 16: 115–145.

Wiessner, P. 2014. Embers of society: Firelight talk among the Ju/'hoansi Bushmen. *Proceedings of the National Academy of Sciences* 111: 14027–14035.

Wilkins, A. S., R. W. Wrangham, and T. Fitch. 2014. The "domestication syndrome" in mammals: A unified explanation based on neural crest cell behavior and genetics. *Genetics* 197: 795–808.

Williams, G. C. 1966. *Adaptation and Natural Selection: A Critique of Some Current Evolutionary Thought*. Princeton, NJ: Princeton University Press.

Williams, J., G. Oehlert, J. Carlis, and A. Pusey. 2004. Why do male chimpanzees defend a group range? *Animal Behaviour* 68: 523–532.

Wilson, D. S., and L. A. Dugatkin. 1997. Group selection and assortative interactions. *American Naturalist* 149: 336–351.

Wilson, E. O. 1975. *Sociobiology: The New Synthesis*. Cambridge, MA: Harvard University Press.

Wilson, M. L., C. Boesch, B. Fruth, T. Furuichi, I. C. Gilby, C. Hashimoto, C. Hobaiter, G. Hohmann, N. Itoh, K. Koops, J. Lloyd, T. Matsuzawa, J. C. Mitani, D. C. Mjungu,

D. Morgan, R. Mundry, M. N. Muller, M. Nakamura, J. D. Pruetz, A. E. Pusey, J. Riedel, C. Sanz, A. M. Schel, N. Simmons, M. Waller, D. P. Watts, F. J. White, R. M. Wittig, Z. Zuberbühler, and R. W. Wrangham. 2014. Lethal aggression in *Pan* is better explained by adaptive strategies than human impacts. *Nature* 513: 414–417.

Wilson, M. L., D. A. Collins, W. R. Wallauer, and S. Kamenya. 2005. *Gombe Stream Research Centre 2005 Annual Report*. Kigoma, Tanzania: Jane Goodall Institute.

Wilson, M. L., S. Kamenya, D. A. Collins, and W. R. Wallauer. 2004. *Gombe Stream Research Centre Annual Report 2004*. Kigoma, Tanzania: Jane Goodall Institute.

Wilson, M. L., and R. W. Wrangham. 2003. Intergroup relations in chimpanzees. *Annual Review of Anthropology* 32: 363–392.

Winterhalder, B. 2001. Intragroup resource transfers: Comparative evidence, models, and implications for human evolution. In C. B. Stanford and H. T. Bunn, eds., *Meat-Eating and Human Evolution*. Oxford: Oxford University Press.

Wrangham, R. W. 1999. Evolution of coalitionary killing. *Yearbook of Physical Anthropology* 42: 1–30.

Wrangham R. W., and L. Glowacki. 2012. Intergroup aggression in chimpanzees and war in nomadic hunter-gatherers: Evaluating the chimpanzee model. *Human Nature* 23: 5–29.

Wrangham, R. W., and D. Peterson. 1996. *Demonic Males: Apes and the Origins of Human Violence*. New York: Houghton Mifflin.

Wrangham, R. W., M. L. Wilson, and M. N. Muller. 2006. Comparative rates of violence in chimpanzees and humans. *Primates* 47: 14–26.

Zahavi, A. 1995. Altruism as a handicap: The limitations of kin selection and reciprocity. *Journal of Avian Biology* 26: 1–3.

21

Communication and Language

KATIE E. SLOCOMBE *and* THOM SCOTT-PHILLIPS

Human language is an incredibly powerful communication system reliant on complex cognitive processes. Tracing the evolutionary origins of this remarkable ability is the focus of considerable interdisciplinary research effort. Within this field, comparative research offers important insights into when and how the building blocks for language may have evolved in our evolutionary past. For instance, by comparing the communication systems and the related cognitive processes of humans and chimpanzees, we may be able to identify shared elements that were likely present in our last common ancestor (LCA), and uniquely human elements that likely evolved after the divergence of the human lineage. Such comparisons may, in turn, allow us to estimate the communicative and cognitive skills of our LCA, and to reconstruct the key changes that occurred after the divergence of the two species. In this chapter, we focus on three such comparisons: neurological aspects, the cognitive foundations of communication, and, first, the extent to which several key structural properties of language are shared with chimpanzees.

Structural Features of Language

A considerable amount of research effort has been dedicated to trying to establish which structural features of language are unique to human communication, and which are shared with our close living relatives. Here we focus on three key features in particular: (1) language as an open and generative system, where new signals can be generated and existing signals can be combined to create new meanings; (2) human signals being referential and conveying specific meaning (semantics); and (3) language being hierarchically structured and governed by syntactic rules.

An Open, Generative Communication System

Two central features of human language result in it being an open and generative system. The first of these is that new signals can be generated to convey new meanings. This requires not only the ability to physically generate new signals, which in the vocal domain relies on the ability to generate novel vocalizations and to vocally imitate, but also the ability to assign meaning to a new signal and for other language users to learn and share the meaning of the new signal. In this regard, it has long been claimed by researchers examining great ape gestural signals in captivity that new gestures can be invented, added to the repertoire of individuals, and spread through groups, and that this represents continuity between the communication systems of humans and other great apes (Arbib et al. 2008; Tomasello 2008). Both idiosyncratic and group-specific gestures have been reported in chimpanzees and bonobos (McGrew and Tutin 1978; Nishida 1980; Tomasello et al. 1997; Pika et al. 2005). For instance, "high-arm" grooming (see Figure 21.1) is reported in some wild chimpanzee communities (McGrew and Tutin 1978; Wrangham et al. 2016), but is absent from others (Whiten et al. 1999), indicating that individuals have the capacity learn them from others in their social group and to generate new signals. Tomasello (2008) suggests that novel gestures develop through an interactive social process called ontogenetic ritualization. In this process, an individual performs an action (for instance, a slap to initiate play), and with repeated interaction, the recipient comes to anticipate the action from the initial movements (raising of the arm) and thus over time, the signaler only produces the initial, ineffective movement (arm raise), and this gesture then elicits an appropriate response from the receiver.

Although great apes seem to have the capacity to generate new gestural signals, the frequency with which this occurs has been challenged recently by researchers conducting longer-term studies across more sites, including

FIGURE 21.1. Two individuals from the Kanyawara community, Uganda, performing high-arm grooming, where two individuals mutually groom while each raising one arm. (Photo by Katie Slocombe.)

wild populations (Genty et al. 2009; Hobaiter and Byrne 2011a; Graham et al. 2016). Byrne and colleagues argue that great apes have a large innately specified repertoire, with some gestures in the repertoire used only rarely; these rare gestures may have been misclassified by early short-term studies as group-specific or idiosyncratic gestures. One detailed study of wild chimpanzee gestures found that individual repertoire size was correlated heavily with observation time, and no idiosyncratic gestures were found (Hobaiter and Byrne 2011a). The same study also found little support for the idea that gestures arise from ontogenetic ritualization. The authors hypothesized that a gesture ritualized from an originally effective action should retain some aspects of the physical form of that action, and they tested this idea with two common gestures. First, they looked for physical similarities, in terms of hand orientation and finger position, between the begging-reach gesture and a taking action that it may reasonably be expected to have originated from. Second, they examined the part of the hand used to perform a position gesture (used to indicate a desired position during grooming) and a position action that achieves the same ends through physical manipulation of the partner. In both cases, they found significant differences in the physical form of the gestures and the actions from which they may have ritualized, leading them to argue that these gestures were unlikely to be a product of ontogenetic ritualization. There is growing support for the idea that early evidence of an open generative gestural system may have been an artifact of methodology, and that the chimpanzees are better characterized as possessing a large yet fixed gestural repertoire (Hobaiter and Byrne 2011a; Roberts et al. 2012; Graham et al. 2016). Having said that, the matter is not definitively settled, and debate remains ongoing (e.g., Halina et al. 2013).

While the degree to which the chimpanzee gestural repertoire is open and generative is currently under debate, it is widely accepted that chimpanzees lack the ability to generate novel vocalizations; instead, they operate with a fixed, relatively small, genetically constrained vocal repertoire. Recent evidence suggests there may be some flexibility to alter the structure of existing calls in their repertoire, even when these calls function referentially: chimpanzees have been shown to modify the structure of their meaningful food calls to match those of other group members, and that this change in call structure occurred independently of their preference for the specific food type that elicited the call (Watson et al. 2015a, 2015b). However, in terms of novel vocalization generation, attempts to teach Viki the chimpanzee to speak as part of one of the first ape language projects failed (Hayes 1951), and although chimpanzees have been documented to generate novel raspberry sounds or extended grunts to beg for food from caregivers in captivity (Hopkins et al. 2007), these sounds are not thought to engage the larynx, which plays such a pivotal role in human speech production. Indeed, it is humans, not chimpanzees, that are the outliers in terms of vocal plasticity in the primate family (see also section "Summary and Conclusions").

The second feature of language that makes it an open and generative system is the capacity to convey an infinite number of meanings through finite means. In other words, an infinite number of sentences that each convey different meaning can be constructed from a finite vocabulary of words (Hurford 2011). While chimpanzees may be able to generate new gestural signals, there is no evidence that they combine their natural gestures in meaningful ways to extend the number of messages they can convey (Hobaiter and Byrne 2011b). In both gorillas and chimpanzees, sequences of gestures are commonly repetitions of the same gesture, such as production of several branch shakes in a sequence by a male in a copulation context (Liebal et al. 2004a; Genty and Byrne 2010), and where different gestures are produced in a sequence the meaning changes little or not at all (Genty and Byrne 2010). Rapid-fire sequences of gestures are most commonly used by younger chimpanzees, and they seem to be a strategy used to maximize the chances of producing an effective gesture to elicit the desired response from the recipient, rather than any attempt to convey different meanings (Hobaiter and Byrne 2011b). In the vocal domain, although some basic call combinations that convey a different meaning to listeners compared to the constituent calls have been reported in monkey species (Arnold and Zuberbühler 2006), similar findings have not been forthcoming in chimpanzees. Chimpanzees do commonly produce call sequences (Crockford and Boesch 2005); however, whether these combinations convey different meanings to their constituent parts has not yet been established. Surprisingly, chimpanzee vocalizations are less studied and understood than the vocal systems of many monkey species (Slocombe et al. 2011), and our lack of understanding of what single call types mean to listeners currently hampers our ability to tackle the bigger question of what call combinations may mean. It is also possible that new meanings can be generated by the combination of different types of signals (Partan and Marler 2005), and indeed recent papers document wild and captive chimpanzees flexibly combining vocalizations, gestures, and facial expressions into multimodal signals (Tagllialatela et al. 2015; Wilke et al. 2017). Wilke and colleagues compared recipient responses to one common multimodal combination and its component parts when given in isolation, and found one individual produced component and multimodal signals that elicited different responses. Although little can be concluded from this preliminary observation, it does suggest that multimodal combinations of signals may be a fruitful avenue for future research.

When chimpanzees have been taught to use an artificial language system to communicate with, there have only been a few anecdotal reports of novel combinations produced to convey new meanings. For instance, Washoe is reported to have combined the signs for "water" and "bird" the first time she saw a swan (Lieberman 1984); however, without a systematic record of all the sign sequences Washoe had received from her caregivers, it is difficult to establish whether such combinations are truly novel. Indeed, declarative utterances are rare in language-trained apes (Lyn et al. 2011), and thus there may be limited motivation to convey new meanings (through generation of new signals or combinations of existing ones), once basic desires and requests can be successfully communicated.

In summary, although chimpanzees have traditionally been thought to have an open generative gestural system, more recent evidence has challenged this view, and it currently seems likely that wild chimpanzees operate with fixed vocal and gestural repertoires. This is an important difference from human language, and critically, chimpanzees may lack the motivation to convey new meanings. A further general issue is that while signal combinations have been observed in some primate species, as summarized above, similar findings have also been described in bacterial communication (Scott-Phillips et al. 2014). As we discuss further below, this raises important questions about what comparisons between communication *systems* can actually tell us, and indeed about what "meaning" itself is (Scott-Phillips 2014, 2015a).

A Referential Communication System

Most human words have semantic meaning, and we can refer to specific events and objects in the world with both words and pointing gestures. These abilities are crucial for language, and the extent to which they are present in our primate cousins has been the focus of considerable research effort.

Although chimpanzees in the wild do not point (except for one reported incident in a wild bonobo: Veà and Sabater-Pi 1998), chimpanzees in captivity learn to point for human caregivers in order to request food. These points, which can be extensions of the whole hand or the index finger, have a referential function, and are produced intentionally (Leavens et al. 1996, 2005). Unlike the points of young children, however, the motivation behind the chimpanzee pointing gestures seems to be imperative rather than declarative or truly informative (Liszkowski et al. 2004; Bullinger et al. 2011). While chimpanzees will point to inform an experimenter of the location of a hidden tool that can be used to provide the chimpanzee with food, they fail to point to inform the experimenter of the location of a tool that can only be used to provide the experimenter with food (Bullinger et al. 2011). Another indication that the cognitive mechanisms that underlie pointing production in chimpanzees is likely different from those in humans comes from the remarkable failure of chimpanzees to spontaneously comprehend human informative pointing gestures (Miklósi and Soproni 2006). For instance, if presented with a basic object choice task involving two containers, one of which contains hidden food, chimpanzees will usually choose randomly, ignoring the pointing gesture of the human toward the correct container (Miklósi and Soproni 2006).

Although chimpanzees readily learn to point for human caregivers to refer to food items or tools that they wish to gain access to, there has only been one report of a gesture used with conspecifics that may be referential. Pika and Mitani (2006) suggest that the loud, conspicuous scratching gesture that chimpanzees produce when grooming may be used by signalers in the Ngogo community to refer to the body part that they wish to be groomed. They found that recipients were more likely to groom in the location indicated by the scratch; however, this could be a product of stimulus enhancement (e.g., visual attention is drawn to that location by the conspicuous movement, thus making it more likely that when grooming recommences it will be in the area the individual is attending to). In contrast to the human-directed pointing gestures that show markers of goal-directed, intentional usage, there was an absence of any data suggesting that chimpanzees produced scratches with the goal of requesting those body parts to be groomed. Persistence in signaling in the face of a recipient who did not groom the indicated location would be one way of empirically demonstrating this.

In the vocal domain, there is considerable evidence that primate vocalizations can function referentially. Primate calls given in alarm, feeding, and social contexts have all been shown to provide listening individuals with information about specific events or objects in the world. The seminal study in this regard focused on predator-specific alarm calls of vervet monkeys: this species produces acoustically distinct calls for pythons, eagles, and leopards, and playback experiments show that when recipients hear the alarm call of a group member, they react in the same adaptive manner to the call as they do when they detect the predator directly (Seyfarth et al. 1980). Although several chimpanzee calls have been shown to be context-specific (Crockford and Boesch 2003; Slocombe and Zuberbühler 2005a, 2007), only a single call type, the rough grunt, has been empirically shown to meet the criteria for functional reference. This graded call type is highly context-specific, with 93 percent of rough grunts produced in a feeding context (Schel et al. 2013a), and within this context, acoustically distinct subtypes of this call are produced reliably in regard to foods of different value (Slocombe and Zuberbühler 2006; see Figure 21.2). A playback experiment showed that a listening chimpanzee could distinguish between rough grunts produced for a high- and a low-value food, and this information informed the listener's own search for food (Slocombe and Zuberbühler 2005b).

It is possible that nonhuman primates produce context-specific calls as part of an automatic, inflexible response to specific trigger stimuli (e.g., leopard), and that listeners, using general learning mechanisms, have simply learned that certain calls reliably predict the occurrence of certain environmental events (e.g., leopard) and so react appropriately (Tomasello 2008). Listeners may therefore be extracting information from calls that producers never intended to provide (Cheney and Seyfarth 2005). It is vital, therefore, that researchers focus on elucidating the cognitive mechanisms underlying the production of these calls (see below). In chimpanzees, recent studies have attempted to examine the motivation and mechanisms underlying rough grunt production in chimpanzees (Slocombe et al. 2010, Schel et al. 2013a). Observational data from the wild has indicated that chimpanzees were more likely to produce rough grunts in the presence of individuals with whom they enjoyed strong affiliative relationships (Slocombe et al. 2010; Fedurek and Slocombe 2013; see Figure 21.3).

In order to rigorously test whether rough grunt production was selective and directed at specific individuals, rather than an automatic reaction to

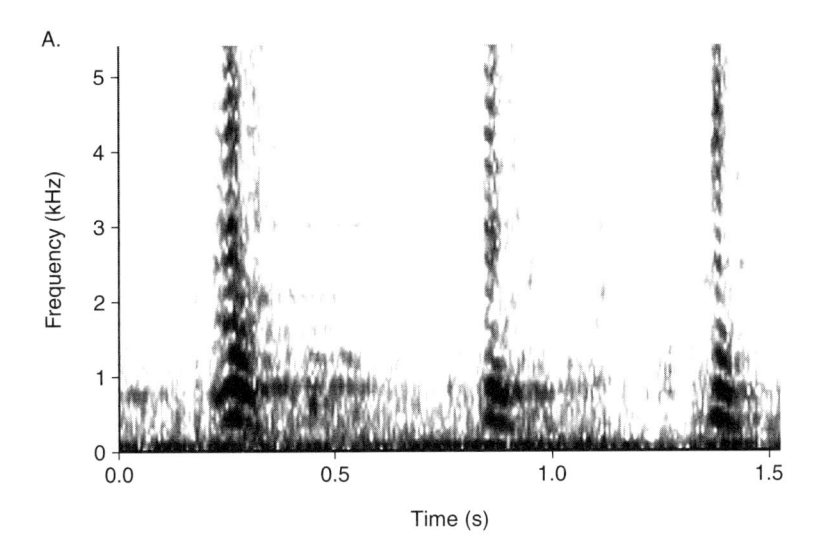

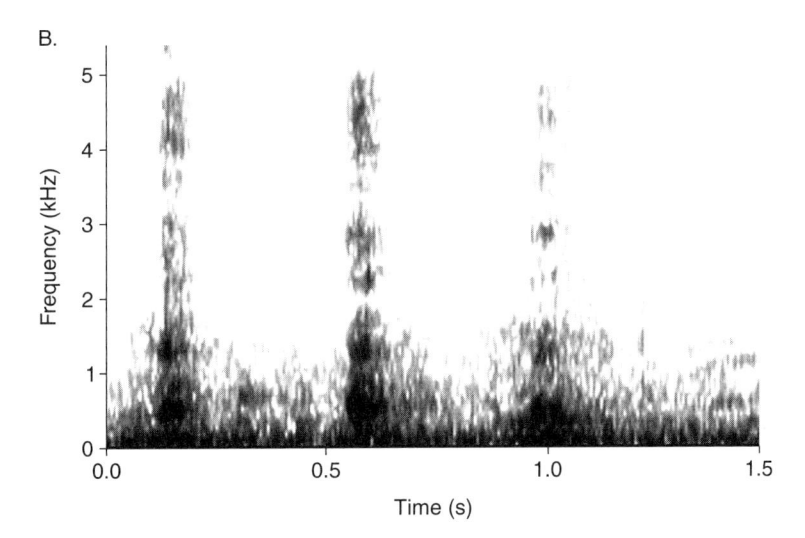

FIGURE 21.2. Time-frequency spectrograms of rough grunt call bouts produced by captive chimpanzee Louis to (a) a high-value food (bread) and (b) a low-value food (apples). Grunts given to the high-value food have more energy (depicted by the darkness of the image) at higher frequencies and a much clearer harmonic structure compared to the lower-pitched, noisy grunts produced in response to low-value food. Figure taken from Slocombe and Zuberbühler (2005b).

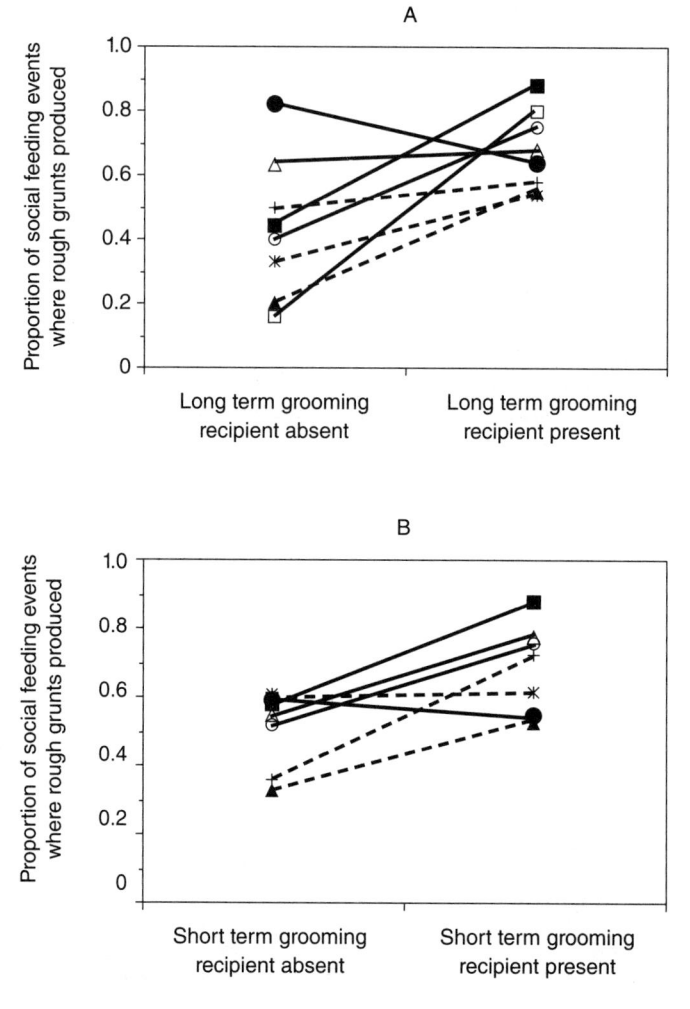

FIGURE 21.3. Line graphs illustrating the proportion of social feeding events where the focal chimpanzee produced rough grunts as a function of the presence of an important social partner, determined in terms of (a) long-term grooming recipients and (b) short-term grooming recipients. Solid lines denote adult males ($n = 4$) and dotted lines denote subadult males ($n = 3$). Adapted from Slocombe et al. (2010).

food discovery, a field experiment was conducted (Schel et al. 2013a). Lone male chimpanzees who were feeding silently in fruit trees were presented with a playback of a group member's pant-hoot call that simulated the arrival of that individual into the vicinity of the feeding tree. Neither the food alone, nor the simulated arrival of an audience, was sufficient to elicit rough grunts;

instead, the decision of males to produce a rough grunt bout was mediated by who they thought had arrived and their relationship with that individual. Males were more likely to call for individuals who were higher ranking than themselves and with whom they enjoyed strong friendships. This indicates that chimpanzees do have control over the production of these functionally referential calls, and they use them in tactical ways. Future research still needs to ascertain, however, whether chimpanzees are producing rough grunts to selectively inform ignorant individuals about a food source. This would be evidence of true referential communication directly analogous to human language, where we routinely inform each other about events in the world, while taking into consideration the mental states of our recipients.

Studies with artificial languages have shown that chimpanzees are capable of both understanding and using arbitrary symbols as referents for specific objects and events (Liebal et al. 2013). Although many ape language studies have produced apes with repertoires of referential signs or symbols, how the apes learned the meaning of these signs varied across studies. While many projects explicitly taught associations between symbols and referents (e.g., Gardner and Gardner 1969; Terrace 1979), Savage-Rumbaugh and colleagues made the remarkable discovery that Kanzi, exposed to a language-rich environment from birth, spontaneously started to use and understand the arbitrary lexigram signals that other apes in the facility, including his mother, were being explicitly trained to use. Kanzi developed a formidable vocabulary of over four hundred arbitrary symbols that he readily uses in an intentional and referential manner (Savage-Rumbaugh and Lewin 1996).

In summary, it seems that chimpanzees are capable of referential communication, since in a human environment they can use and understand artificial referential symbols, and they can produce pointing gestures in captivity. The extent to which they use this capacity to communicate referentially in their natural environment is less clear. While natural rough grunt vocalizations function referentially and are selectively directed at specific recipients, whether they take the mental state of the receiver into account, as humans do when communicating referentially, is still an empirical question.

As in the previous section, this summary comes with a caveat. The similarities and differences between functional referential communication and human referential signals are much debated. Although the surface behaviors may look similar, there are theoretical reasons why the cognitive mechanisms underlying the production and reception of these calls may in fact be different from those that underlie language. If so, then it is not clear that linguistic terms such as "meaning" can be applied in the same way to human and to nonhuman communication (Rendall et al. 2009; Wheeler and Fischer 2012; Scott-Phillips 2014, 2015a, 2015b). These debates are ongoing, and their resolution will determine what can and cannot be inferred from comparisons between the communication of humans and other species (see also "Cognitive Foundations of Language," below).

A Rule-Governed Syntactic System

Human language is characterized by its hierarchical grammatical structures, which allow an infinite number of utterances to be constructed from a finite set of signals. Relatively little progress has been made in identifying even simple rule-based structures in natural chimpanzee communication that could indicate that the precursors of human syntactical abilities were likely present in the LCA. While various rule-governed combinations of calls have recently been identified in putty-nosed monkeys and Campbell monkeys (Arnold and Zuberbühler 2008; Ouattara et al. 2009), in the vocal domain, similar findings have not been forthcoming for great apes. Chimpanzees do combine different calls into sequences, and some sequences do seem to have basic rules governing the order of the calls (e.g., in a sequence a pant-hoot will normally precede a pant-grunt; Crockford and Boesch 2005); however, how this affects the meaning of the call sequence is yet to be investigated. Equally multimodal sequences of signals (facial expression + gesture; gesture + vocalization) have yet to be systematically examined, but may be a promising avenue for future research (Liebal et al. 2013; Wilke et al. 2017). As reviewed above, although great ape gestures are sometimes produced in sequences, there appears to be no hint of a rule-governed order or syntactic structure to these sequences (Genty and Byrne 2010; Hobaiter and Byrne 2011b).

A different approach to understanding syntactic competencies in nonhumans is to focus on the comprehension rather than the production of rule-governed sequences. Monkeys have been shown to be sensitive to sequential order in both natural call sequences and artificial sound sequences (Hauser et al. 2002; Zuberbühler 2002; Fitch and Hauser 2004); however, they do not seem able to process more complex hierarchically dependent structures embedded in artificial sound sequences (Fitch and Hauser 2004). Similar studies have yet to be completed with chimpanzees, so we do not currently know the extent to which they can process sequential or hierarchical rules within sequences.

Production of grammatical utterances in language-trained apes is relatively rare or inflexible. Some apes learned set structures that they could slot new symbols or signs into to create new utterances (e.g., Lana frequently used the structure Please + (person) + (action); Rumbaugh and Gill 1977); however, more complex grammatical structures were rarely detected in the symbol sequences produced by the apes (e.g., Terrace et al. 1979). In contrast, in terms of comprehension, Kanzi, the language-competent bonobo, showed performance comparable to a two-and-a-half-year-old child in blind tests of understanding novel request sentences (Savage-Rumbaugh et al. 1993). Tomasello (1994) suggests that in order to perform so well across a variety of different types of requests, using different grammatical structures, Kanzi must at a minimum understand object labels, action words, and word order.

In summary, although important findings concerning the production and comprehension of rule-based sequences in monkeys have appeared in recent years, comparable progress has not been made with chimpanzees. When introduced to an artificial language, although spontaneous production of grammatically structured utterances is rare, one bonobo at least shows quite sophisticated understanding of spoken English and certain grammatical rules.

Neurological Bases of Communication

Evidence from brain injuries, anatomical studies, direct brain stimulation, and more recently a range of neuroimaging studies, has helped to localize language production functions in the human brain and identified Broca's area as critical to the production of language (reviewed in Démonet et al. 2005). Damage to this area of the cortex results in severe disruption to language production (Benson and Geschwind 1985), and most humans show significant lateralization of language production functions in the left hemisphere of the brain (Pujol et al. 1999). Unsurprisingly, our understanding of signal production in chimpanzees is severely limited in comparison to the considerable research effort that has been dedicated to understanding the neurological bases of human language; however, some progress is now being made. Until recently we had to rely on extrapolation of results obtained from monkeys using invasive techniques to infer the brain mechanisms likely involved in the production and processing of communicative signals in chimpanzees. This body of work, largely conducted on vocalizations in squirrel monkeys and rhesus macaques, indicated that production of vocalizations was largely mediated by noncortical structures in the limbic system (Jürgens 2002), in stark contrast to the cortical control of language in humans. Groundbreaking research by Jared Taglialatela and colleagues has challenged these ideas, and for the first time has revealed the brain structures involved in intentional signal production in chimpanzees. Taglialatela and colleagues built on promising anatomical evidence of a lateralized and enlarged inferior frontal gyrus (IFG; homologue of Broca's area, which is critical for language production in humans) (Cantalupo and Hopkins 2001) and used PET scans to provide functional data on the areas involved in signal production. They found that the IFG was active during the production of vocal and gestural signals directed at a human with desirable food (Taglialatela et al. 2008). In an attempt to partition the contribution of the gestural and vocal signals to the IFG activation, in a follow-up study they tested two chimpanzees who produced gestures only, and two chimpanzees who produced vocalizations and gestures (Taglialatela et al. 2011). Significant IFG activation was observed only in the subjects producing vocal and gestural signals, indicating that multimodal signaling or vocal signaling is driving this activation. Interestingly, the vocal signals produced by these chimpanzees in this context were "novel" attention-getting sounds (e.g., raspberries and extended grunts), not species-typical vocalizations, so the neural basis of these calls is still unknown. These studies have highlighted important parallels in the brain areas recruited in chimpanzees and humans for the production of communicative signals, but far more waits to be discovered in this domain.

Chimpanzees and humans communicate with a wide range of facial expressions, and anatomical studies have revealed greater similarity in the fine neuronal architecture underlying cortical control of orofacial muscles in chimpanzees, gorilla, and humans compared to Old World monkeys (Sherwood et al. 2004). The greater density of certain types of neurons indicates that great apes may have higher degrees of voluntary, dexterous control over facial movements during communication than other members of the primate order. Systematic development of Facial Action Coding Systems (FACS) for humans (Ekman et al. 2002) and chimpanzees (Vick et al. 2007) has allowed detailed cross-species comparisons of the muscle movements involved in different facial expressions and identification of morphological similarities. The human smile, for instance, uses many of the same muscle movements as the chimpanzee bared-teeth expression (Ekman et al. 2002; Vick et al. 2007), yet they are not identical. FACS analysis indicates that the human smile may represent a blend of the bared teeth and play face expressions shown by most primates, including chimpanzees (Liebal et al. 2013).

In summary, anatomical and functional neurological data indicate that signal productions in humans and chimpanzees share several key neurological bases; however, there are still far more unanswered questions than answered ones in this area. With the advance of functional imaging technology, we may hope that future research will be able to shed more light onto the similarities and differences of the brain structures underlying signal production in humans and chimpanzees.

Cognitive Foundations of Language

As a discipline, linguistics tends to emphasize the importance of structural aspects of language, particularly phonetics, phonology, morphology, syntax, and semantics. It is thus natural and unsurprising that a great deal of comparative research focuses on the presence or absence of these features in other species, particularly primates (see sections above). However, human communication is not simply the sum of these structural parts. What we say is not the same as what we mean. Languages allow human communication to be very expressive and precise, but what makes human communication possible in the first place is the cognitive ability to engage in a type of communication that involves much more than encoding and decoding. Human communication is not just intentional, it is overtly so: we make manifest our intentions to inform one another (Origgi and Sperber 2000; Scott-Phillips 2014). Correspondingly, it is a mistake for comparative research to focus on structural aspects of language alone. At least as much attention should be given to the sociocognitive ability to engage in overtly intentional communication (also called "ostensive communication"). This section therefore reviews the extent to which chimpanzees possess various sociocognitive skills relevant to this ability.

Intentional Signal Production

Humans produce language intentionally; that is, they produce signals voluntarily with the goal of changing the behavior and mental states of their intended receivers. As mentioned in the section on referential communication above, behavior can often appear similar on the surface, but if the psychological mechanisms underlying the production of the behaviors are different, the behaviors are not really comparable. Thus it is crucial that we try to understand the mechanisms and intentions underlying communicative behavior in chimpanzees, to identify similarities and differences in the psychological processes that underpin the production of chimpanzee and human signals. This is, however, no easy task: identifying the mental processes of any nonlinguistic being is a considerable challenge, and there is not even cross-disciplinary consensus on how intentionality itself should be defined. In this regard, Dennett (1987) offers a constructive framework with which to describe different levels of intentionality, and these can be usefully applied to communicative behavior. It has been suggested that many animal signals are the product of zero-order intentional processes, such as emotional processes or automatic responses to specific stimuli. For instance, the rough grunts produced by a feeding chimpanzee may simply be an automatic excitement response to food. First-order intentionality requires voluntary control over signal production that is directed toward specific recipients to achieve a goal, which is usually changing the behavior of another individual. For instance, rough grunts may be selectively produced to invite certain individuals to share a food source with the caller. Second-order intentionality requires the signaler to have the goal not only of changing the behavior, but also of influencing the mental state of the receiver. For instance, rough grunts may be produced selectively for ignorant individuals to inform them about the presence or quality of the food source, thus changing their knowledge state.

There is convergent evidence that chimpanzees produce a range of signals with first-order intentionality. Comparative researchers have used criteria established by developmental psychologists examining gesture use in preverbal infants (Bates et al. 1979) to distinguish between zero- and first-order intentional signal production in chimpanzees (Leavens and Hopkins 1998; Liebal et al. 2004a; 2004b; Leavens et al. 2005; Hopkins et al. 2007; Hobaiter and Byrne 2011a). The criteria indicating first-order intentionality are shown in Table 21.1; they have been used primarily to examine the mechanisms underlying gesture production in great apes.

It is important to note that all of these criteria, if considered in isolation, can in principle be explained by zero-order intentional processes. For instance, the presence of an audience is known to affect arousal rates in humans (Zajonc 1965), so calling in the presence rather than absence of an audience could simply be a product of increased arousal. Equally, gaze alternation between an object and a recipient may simply reflect competing foci of attention for the caller (Liebal et al. 2013). It is therefore crucial that convergent evidence from a number of these criteria be forthcoming before we conclude that signal production is likely a product of first-order intentional processes (Townsend et al. 2016). There is, unfortunately, great variation in the number and type of criteria used by different studies before

TABLE 21.1. Criteria used for identifying intentional production of communicative signals, as outlined in the study of great ape gestures.

Criteria	Explanation
Social use	The signal is directed at a recipient. This can be assessed at various levels: 1. Presence/absence audience effect: the signal is only produced in the presence of a recipient. 2. Composition of audience: the signal is only produced in the presence of certain recipients (e.g., kin, dominants, friends). 3. Behavior of audience: signal production is contingent on the behavior of the recipient.
Sensitivity to attentional state of recipient*	Visual signals are only produced in the field of view of recipients. If signaler does not have a recipient's visual attention, tactile or auditory signals should be produced. This can also be considered a level 3 audience effect.
Manipulation of attentional state of recipient*	Before a visual signal is produced, attention-getting behaviors are directed toward a recipient who is not visually attending to the signaler.
Audience checking and gaze alternation	Signaler monitors the audience and visually orients toward the recipient before producing a signal. If a third entity is involved, gaze alternation may occur between recipients and this entity.
Persistence or elaboration	Goal-directed signaling shown by repetition of the same signal (persistence), or production of different signals (elaboration) until the desired goal is met.

* Indicates applicable only to visual signals and therefore not relevant for vocal production. Taken from Schel et al. (2013b).

intentional signal production is assumed (Liebal et al. 2013). Nevertheless, across multiple studies, chimpanzee gestures have been shown to meet all of these criteria. For instance, in conspecific interactions, chimpanzees show sensitivity to the attentional state of others and will move into the recipient's visual field before producing a visual gesture (Liebal et al. 2004b), and individuals will persist through repetition of a gesture until the recipient responds in a manner that meets their goal (Liebal et al. 2004a). In experimental situations where chimpanzees can beg for food from a human experimenter, their gestures and attention-getting vocalizations (e.g., raspberries and extended grunts) are accompanied with gaze alternation, and, in the face of an unresponsive human, individuals will elaborate, through the use of different signals, to try and achieve their goal of obtaining the desirable food (Leavens et al. 2004, 2005).

In contrast to chimpanzee gestures that are commonly branded as intentional, until very recently chimpanzee vocalizations have been assumed to be the product of zero-order intentional processes (Tomasello 2008). This assumption was based on extrapolating findings in monkeys, in the absence of data directly addressing this issue in chimpanzees. Schel and colleagues recently provided direct data on this issue, successfully challenging this assumption by applying the applicable criteria commonly used in gesture studies (see Table 21.1) to the production of alarm calls. In a field experiment, Schel et al. (2013b) presented chimpanzees with a moving model python and collected detailed data on their subsequent behavior. One call type, the soft *huu*, at least immediately after the caller discovered the snake, did not seem to be directed toward conspecifics, as its production was not mediated by the presence or absence of an audience, and it was rarely associated with visual monitoring of recipients. Two other call types, the alarm *huu* and the *waa* bark, did, however, meet several key intentionality criteria and seemed to be the product of first-order intentional processes. Callers selectively increased the rate of these calls when friends who had not been present when the snake was discovered arrived in the party, and alarm *huu* and *waa* bark bouts were commonly preceded by visual monitoring of the audience. Seventy-five percent of individuals exhibited gaze alternation between the snake and a recipient while calling, a level comparable to the percentage of chimpanzees that engage in this behavior when gesturing for food from a human experimenter (Leavens and Hopkins 1998). Finally, Schel et al. attempted to identify stopping rules for calling in an attempt to extrapolate the goal of the callers.

They found that stopping calling was not associated with any change in the caller's own position or risk relative to the snake. Instead, all other individuals in the caller's party were significantly more likely to be safe from the snake when calling stopped compared to the rest of the trial, indicating that the goal of alarm call production may be to warn others of danger. This study indicates that species' typical vocalizations can be produced intentionally in chimpanzees, and thus it seems that both gestures and vocalizations can be the product of first-order intentional processes.

Although we arguably have better evidence for first-order intentionality in chimpanzee signal production than in any other nonhuman species, humans are still routinely operating on levels above this. A necessary prerequisite for communicating with the goal of altering the mental state of another is an awareness or understanding of others' mental states. Considerable controversy surrounds the extent to which chimpanzees understand the minds of others, or have a "theory of mind" (Rosati, this volume), but several lines of research suggest that they do understand that others have intentions and knowledge states, and there is mixed evidence as to whether they understand beliefs (see review by Call and Tomasello 2008, but more recently Krupenye et al. 2016). Crockford and colleagues set out to test whether these abilities are used in chimpanzee vocal communication with a field experiment in which they presented a static snake model to chimpanzees. Crockford et al. (2012) distinguished between receivers who were ignorant of the snake, had partial knowledge (assumed to have heard alarm calls but not seen the snake), or had total knowledge (had seen the snake), and tested whether production of alarm calls was mediated by the knowledge state of the receivers. They found that calls were more common when receivers were ignorant, but in these cases the callers had also just discovered the snake and so elevated calling could have simply reflected an individualistic expression of fear toward the snake. This would fit with the finding of Schel et al. (2013b) that in immediate response to snake discovery, soft *huus* were associated with visual fixation on the snake and were produced regardless of whether an audience was present or not. In a further analysis, Crockford and colleagues excluded the discovery events and focused on a small subset of their data ($n = 32$ cases), where a receiver was always approaching the snake, and they found that calls were more likely when receivers were only partially knowledgeable compared to fully knowledgeable. The authors claim that chimpanzee alarm call production is therefore mediated by an understanding of receiver knowledge states. While this is a promising result, lower-level explanations have not been satisfactorily excluded. In particular, callers may find the behavior of the individual approaching the snake alarming in itself if it is inappropriate (e.g., they are approaching too fast or without sufficient caution). To exclude this possibility, the same pattern of behavior would need to be shown in the eighty remaining cases where no individual was approaching. In addition, given that Schel et al. (2013b) found that the arrival of specific individuals and their friendship level with the potential caller affected calling rates, it is also possible that the identity of the approaching receiver mediated the call production, not their knowledge state. A more controlled experimental approach to this question may be better able to control for confounding variables and alternative explanations, but for now it seems premature to conclude that chimpanzee vocal production is influenced by the knowledge state of receivers.

In the discussion of intentional communication so far, we have drawn no distinctions between different types of intention that may be present in a communicating individual. However, contemporary pragmatic theory makes a key theoretical distinction between communicative and informative intentions in human communication (Sperber and Wilson 1995). An informative intention is an intention that the audience recognize *what* the signaler is trying to communicate (for example, that she wants more coffee). The communicative intention, on the other hand, is an intention that the audience recognize *that* the signaler is trying to communicate in the first place—in other words, that the signaler actually has an informative intention. This is where the overtness mentioned above is relevant. Human communication does not just involve intentional informing of others; it involves making those intentions overt to the audience. Signals that do this are called *ostensive* signals. Humans communicate ostensively in many ways, such as with eye gaze, pointing—and language. These behaviors reveal both that the signaler is trying to communicate, and what they wish to communicate. In other words, ostensive communication is the foundation on which linguistic communication is built (Sperber and Wilson 1995; Tomasello 2008; Scott-Philips 2014). An important comparative question, then, is whether any other species communicates ostensively, that is, by using communicative and informative intentions (Tomasello 2008).

One type of behavior that provides clear evidence of an understanding of communicative and informative intentions, and the distinction between them, is *hidden authorship*. Here, one individual provides a stimulus for another, but hides the fact that it is for them. For instance, a dinner party guest might move her empty wine glass to a conspicuous location while her host's back is turned, so that when he turns around, he will see it and refill her glass, while remaining unaware of the fact that she had moved the wine glass precisely so that he would do this. Here, the guest intends that the hosts should know that the glass is empty, but, at the same time, the guest also intends that the host does *not* know that the guest has this intention. Expressed in full: the guest intends* that the host does not know that the guest intends that the host knows that the glass is empty! The intention marked * is not a communicative intention per se, but it does have the same relationship to an informative intention that a communicative intention normally has (see Scott-Phillips 2014 for extended discussion). In this way, the competent use of hidden authorship is one behavior that is good evidence of an understanding of what a communicative intention should consist of, its relationship to informative intentions, and hence of the cognitive basis of human ostensive communication.

Children are able to hide authorship in appropriate circumstances (Grosse et al. 2013). In an experiment, children were told an adult was looking for a hidden toy, but that they did not want help to find it. Children as young as three years old were able to make it easier for the adult to find the toy (the content of the stimulus), at the same time concealing their actions and their intent to reveal the location of the toy to the adult (i.e., hiding their informative intention). This shows that children from an early age are aware of the distinction between informative and communicative intentions. The extent to which chimpanzees can recognize these two levels of intention in humans or even have markers of communicative intent in their natural communication system has yet to be empirically investigated. Due to the advanced theory of mind skills required for this type of intentional communication (communicative intent means that I intend that you understand that I intend to communicate with you), it is widely assumed that this is likely to be a uniquely human aspect of language, but empirical tests of this assumption are needed (Scott-Phillips 2015b).

In summary, chimpanzees produce gestural and vocal signals that meet established behavioral markers of first-order intentionality, although there is currently no convincing evidence that signal production is mediated by mental state understanding. This may represent a solid foundation from which higher orders of intentional communication that characterize language may have arisen. It is clear that the development of a more sophisticated theory of mind was likely a crucial cognitive development that ultimately allowed linguistic and other forms of ostensive communication to emerge.

Joint Attention

Due to the inherent ambiguity of language, listeners rely on shared common ground they have created with the speaker in order to understand the meaning of utterances and pointing gestures (Tomasello 2008). Establishment of shared common ground relies heavily on our capacity for joint attention and our strong motivation to share attention with others about our surroundings. Joint attention skills emerge early in infancy, with reliable gaze following, gaze alternation, and sharing emotional reactions about objects appearing at around nine to twelve months, before pointing emerges at about twelve to fifteen months (Carpenter et al. 1998). The ability to follow, share. and manipulate the attention of others seems to represent a crucial foundation for human communication and the emergence of language: infants with deficits in joint attention, such as individuals with autism spectrum disorder, often go on to have difficulties with language acquisition and communication (Colombi et al. 2009). The extent to which joint attention is a special feature of human cognition, or is shared with chimpanzees, is an issue that has attracted considerable research effort.

Chimpanzees certainly seem to possess some of the skills necessary for joint attention: chimpanzees are able to follow the gaze of a human experimenter to external objects, even when the target objects are located outside of their visual field or behind barriers (Povinelli and Eddy 1996; Tomasello et al. 1999), and they also follow the gaze of conspecifics (Tomasello et al. 1998). Chimpanzees are capable of directing the attention of a human experimenter and will produce vocal and gestural signals together with gaze alternation to request desirable food (Leavens et al. 2005). Chimpanzees are also capable of social referencing—that is, gaining information about an object from another's reaction or interaction with it—and they have been shown to do this with human experimenters (Russell et al. 1997), and in one case with their natural mother (Tomonaga et al. 2004). Despite having the ability to follow and direct the attention of others and to coordinate their attention triadically between objects and others, chimpanzees have not been observed to engage these skills either during play with human experimenters (Tomasello et al. 2005) or when exploring a novel object with their mother (Tomonaga et al. 2004). It seems possible that although chimpanzees may be capable of joint attention, in contrast to human infants, they may lack the motivation to engage these skills to share attention in a declarative way (Liszkowski et al. 2004). In line with this hypothesis, the motivation behind pointing gestures in captive chimpanzees seems to be imperative rather than informative or declarative (Bullinger et al. 2011), and although language-competent apes such as Kanzi have been recorded to produce declarative utterances, these are exceedingly rare occurrences (Lyn et al. 2011). It is, however, necessary to issue one significant warning about the type of evidence on which our current understanding of joint attention in chimpanzees is based: only one study to date has given chimpanzees the opportunity to demonstrate joint attention when interacting with members of their own species; all the other studies required chimpanzees to interpret the signals of and share attention with another species (humans), making the results incomparable to studies focused on human infants interacting with other humans. Thus, before strong conclusions about the abilty and motivation of chimpanzees to engage in joint attention are reached, more studies examining chimpanzee-chimpanzee interactions are necessary.

Conclusions

Table 21.2 summarizes many of the comparisons discussed in this chapter. It shows that while chimpanzees demonstrate some communicative and cognitive abilities that may indicate some building blocks for language were present in the last common ancestor, it is also apparent that a great many changes must have occurred in terms of our cognition and communication abilities since divergence to result in the advanced linguistic abilities that characterize modern humans. How, why, and when these changes occurred is a matter of ongoing debate (Hurford 2007; Tomasello 2008; Fitch 2010; Scott-Phillips 2014). It is not possible in the scope of this chapter to give an overview of the many theories and scenarios that have been proposed regarding the social and ecological factors that may have made these cognitive changes adaptive, or of when language may have emerged, but what can and should

TABLE 21.2. Summary table indicating the extent to which key aspects of language and cognition are shared between chimpanzees and humans, or seem unique to humans.

Aspect of Language or Relevant Cognition	Shared Features	Distinctive in Humans, Based on Current Data
Structural		
Open and generative system: new signals generated	Some capacity to invent new gestures, but this may have been overestimated.	Extensive vocal learning ability.
Open and generative system: existing signals combined to create new meanings	Few anecdotal reports of novel sign combinations in language-trained apes, but not mirrored in natural communication.	Highly motivated to convey new meanings with powerful system that allows infinite meanings to be created from finite means.
Reference	Naturally occurring calls that function referentially and are directed at specific individuals. Language-trained apes can learn to use and understand arbitrary referential signs.	Extensive repertoire of referential words used to change mental state of receiver.
Pointing	Production of imperative points in captivity.	Production of declarative points and understanding of informative pointing.
Syntactic structure	Ape language studies show certain grammatical rules can be understood, but limited structure in production.	Hierarchically structured system with complex grammatical rules.
Neurological		
Brain structures involved in communication	Lateralized and enlarged Broca's area, which is involved in production of intentional signals.	Cortical control of all language production (spoken or signed).
Cognitive		
Intentional signal production	First-order intentionality: natural gestures and calls produced voluntarily to change the behavior of others.	Signals produced to change mental state of the receiver. Ostensive communication (expression and comprehension of communicative and informative intent).
Joint attention	Gaze following, directing attention of others to items requested.	Sharing attention about objects of interest.

be stressed is that research in this area must be and increasingly is a multi-disciplinary endeavor (see, for instance, the proceedings of recent EVOLANG conferences, which are greatly interdisciplinary: Scott-Phillips et al. 2012; Cartmill et al. 2014).

One potential source of data that is not much represented in Table 21.2, or indeed in the language evolution literature in general, is archaeology. Languages do not fossilize, nor do the cognitive abilities that make it possible. Cognitive archaeology may in the future be able to infer when particular brain regions relevant to language appear in the fossil record, but few conclusions of this sort can be made at present (see Falk 2014 for an example of such arguments). The fossil record could inform us about the evolution of speech, but speech is not the same as language. A hypothetical chimpanzee-like ancestor, with a vocal canal quite unlike humans, and equipped for the right cognitive abilities, could still communicate linguistically; they would just sound very different. Languages make use of whatever modal-specific features are available, whether that modality is gesture (i.e., sign languages), speech using human vocal tracts, or, hypothetically, speech using a different vocal tract. As such, fossil data about what type of vocal tract different ancestral primate species might have possessed tells us little about the presence or absence of language in those species.

In final summary, while chimpanzee communication does share some structural characteristics with human language, such as referential gestures and vocalizations and the potential capacity to generate novel signals, overall, chimpanzees are not more sophisticated in these structural aspects than more distantly related monkey species. Chimpanzee communication has received less research effort than monkey communication, so it is possible that with more research, more of a linguistic hierarchy will be discovered, but based on our current knowledge, such a hierarchy seems absent (Pinker 1994). In contrast, if one considers neurological evidence or the cognitive foundations of language, particularly in terms of theory of mind and intentional communication, chimpanzees show much greater similarity to humans than do monkey species. Investigating the cognitive processes underlying communication is a real challenge, and some scholars will argue that it is impossible to ever truly know the intentions, motivations, and psychological processes harbored in the mind of another species, and that scientific studies should therefore not aim to address such issues. In contrast, while acknowledging that it is a challenge, we advocate that researchers focus their attention on tackling this issue, rather than being satisfied with identifying behaviors that share surface similarities with aspects of human language. It seems that it is in the cognitive domain that the greatest similarities and informative differences may be revealed, which will greatly enhance our understanding of language evolution.

We would like to conclude with a methodological comment. In comparison to other aspects of chimpanzee behavior, we still know relatively little about their natural communication system, and our understanding of this may have been hampered by researchers focusing exclusively on single modalities (e.g., vocalizations, gestures, or facial expressions). Chimpanzee communication, like that of humans, is inherently multimodal, and by taking a holistic, integrated multimodal approach to this communication system we may challenge existing assumptions and reveal new complexities (Slocombe et al. 2011; Liebal et al. 2013). Looking ahead, studying different types of chimpanzee signals together across a range of contexts is likely to enrich our understanding of their capabilities and the potential complexity of their communication system. It seems critical, however, that researchers try where possible to design studies where some insight into the cognitive processes driving communication is generated. Improving our understanding of chimpanzee communication and cognition is fundamental to identifying aspects of language that are unique to humans and that likely arose since the divergence of humans from the LCA with chimpanzees.

Summary

Language is one of humanity's defining characteristics. Comparisons with chimpanzee communication can offer crucial insights into how our LCA may have communicated, and the extent to which cognitive abilities that underpin language may have been present in this ancestral primate. Understanding the aspects of communication and associated cognitive processes that are shared between chimpanzees and humans, and those that are uniquely human, can help us to identify the key changes that must have occurred in the human lineage since the divergence of the two species. Various approaches have been used to assess the similarities and differences between human and chimpanzee communication, from studying their natural communicative abilities to trying to teach them to master a human communication system. In this chapter we examined several different aspects of communication and language from this comparative perspective. We focused in particular on structural, neurological, and cognitive factors. We emphasized the importance of the last of these, and the need for more comparative research on the cognitive foundations of human communication.

References

Arbib, M. A., K. Liebal, and S. Pika. 2008. Primate vocalization, gesture, and the evolution of human language. *Current Anthropology* 59: 1053–1076.

Arnold, K., and K. Zuberbühler. 2006. Language evolution: Semantic combinations in primate calls. *Nature* 441: 303.

Arnold, K., and K. Zuberbühler. 2008. Meaningful call combinations in a non-human primate. *Current Biology* 18: R202–R203.

Bates, E., L. Benigni, I. Bretherton, L. Camaioni, and V. Volterra. 1979. *The Emergence of Symbols: Cognition and Communication in Infancy*. New York: Academic Press.

Benson, D. F., and N. Geschwind. 1985. Aphasia and related disorders: A clinical perspective. In M. Mesulam, ed., *Principles of Behavioural Neurology*, 193–238. Philadelphia: F. A. Davis.

Bullinger, A. F., F. Zimmermann, J. Kaminski, and M. Tomasello, M. 2011. Different social motives in the gestural communication of chimpanzees and human children. *Developmental Science* 14: 58–68.

Call, J., and M. Tomasello. 2008. Does the chimpanzee have a theory of mind? 30 years later. *Trends in Cognitive Sciences* 12: 187–192.

Cantalupo, C., and W. D. Hopkins. 2001. Asymmetric Broca's area in great apes. *Nature* 414: 505.

Carpenter, M., K. Nagell, M. Tomasello, G. Butterworth, and C. Moore. 1998. Social cognition, joint attention, and communicative competence from 9 to 15 months of age. *Monographs of the Society for Research in Child Development* 63(4).

Cartmill, E. A., S. Roberts, H. Lyn, and H. Cornish. 2014. *The Evolution of Language. Proceedings of the 10th International Conference on the Evolution of Language (EVOLANGX)*. Singapore: World Scientific.

Cheney, D. L., and R. M. Seyfarth. 2005. Constraints and preadaptations in the earliest stages of language evolution. *Linguistic Review* 22: 135–159.

Colombi, C., K. Liebal, M. Tomasello, G. Young, F. Warneken, and S. J. Rogers. 2009. Examining correlates of cooperation in autism: Imitation, joint attention, and understanding intentions. *Autism* 13: 143–163.

Crockford, C., and C. Boesch. 2003. Context-specific calls in wild chimpanzees, *Pan troglodytes verus*: Analysis of barks. *Animal Behaviour* 66: 115–125.

Crockford, C., and C. Boesch. 2005. Call combinations in wild chimpanzees. *Behaviour* 142: 397–421.

Crockford, C., R. M. Wittig, R. Mundry, and K. Zuberbühler. 2012. Wild chimpanzees inform ignorant group members of danger. *Current Biology* 22: 142–146.

Démonet, J. F., G. Thierry, and D. Cardebat. 2005. Renewal of the neurophysiology of language: Functional neuroimaging. *Physiological Reviews* 85: 49–96.

Dennett, D. C. 1987. *The Intentional Stance*. Cambridge, MA: MIT Press.

Ekman, P., W. V. Friesen, and J. C. Hager, J. C. 2002. *The Facial Action Coding System*, 2nd ed. Salt Lake City: Research Nexus.

Enard, W., M. Przeworski, S. E. Fisher, et al. 2002. Molecular evolution of FOXP2, a gene involved in speech and language. *Nature* 418: 869–872.

Falk, D. 2014. Interpreting sulci on hominin endocasts: Old hypotheses and new findings. *Frontiers in Human Neuroscience* 8.

Fedurek, P., and K. E. Slocombe. 2013. The social function of food-associated calls in male chimpanzees. *American Journal of Primatology* 75: 726–739.

Fitch, W. T. 2010. *The Evolution of Language*. Cambridge: Cambridge University Press.

Fitch, W. T., and M. D. Hauser. 2004. Computational constraints on syntactic processing in nonhuman primates. *Science* 303: 377–380.

Gardner, R. A., and B. T. Gardner. 1969. Teaching sign language to a chimpanzee. *Science* 165: 664–672.

Genty, E., T. Breuer, C. Hobaiter, and R. W. Byrne. 2009. Gestural communication of the gorilla *(Gorilla gorilla):* Repertoire, intentionality and possible origins. *Animal Cognition* 12: 527–546.

Genty, E., and R. W. Byrne. 2010. Why do gorillas make sequences of gestures? *Animal Cognition* 13: 287–301.

Graham, K. E., T. Furuichi, and R. W. Byrne. 2016. The gestural repertoire of the wild bonobo *(Pan paniscus):* A mutually understood communication system. *Animal Cognition* 20: 171–177.

Grosse, G., T. C. Scott-Phillips, and M. Tomasello. 2013. Three-year-olds hide their communicative intentions in appropriate contexts. *Developmental Psychology* 49: 2095–2101.

Halina, M., F. Rossano, and M. Tomasello. 2013. The ontogenetic ritualization of bonobo gestures. *Animal Cognition* 16: 155–163.

Hauser, M. D., S. Dehaene, G. Dehaene-Lambertz, and A. L. Patalano. 2002. Spontaneous number discrimination of multi-format auditory stimuli in cotton-top tamarins (*Saguinus oedipus*). *Cognition* 86: B23–B32.

Hayes, C. 1951. *The Ape in Our House*. Oxford: Harper.

Hobaiter, C., and R. W. Byrne. 2011a. The gestural repertoire of the wild chimpanzee. *Animal Cognition* 14: 745–767.

Hobaiter, C., and R. W. Byrne. 2011b. Serial gesturing by wild chimpanzees: Its nature and function for communication. *Animal Cognition* 14: 827–838.

Hopkins, W. D., J. P. Taglialatela, and D. A. Leavens. 2007. Chimpanzees differentially produce novel vocalizations to capture the attention of a human. *Animal Behaviour* 73: 281–286.

Hurford, J. R. 2007. *Origins of Meaning*. Oxford: Oxford University Press.

Hurford, J. R. 2011. *The Origins of Grammar: Language in the Light of Evolution*, vol. 2. Oxford: Oxford University Press.

Jürgens, U. 2002. Neural pathways underlying vocal control. *Neuroscience and Biobehavioral Reviews* 26: 235–258.

Krupenye, C., F. Kano, S. Hirata, J. Call, and M. Tomasello. 2016. Great apes anticipate that other individuals will act according to false beliefs. *Science* 354: 110–114.

Leavens, D. A., and W. D. Hopkins. 1998. Intentional communication by chimpanzees: A cross-sectional study of the use of referential gestures. *Developmental Psychology* 34: 813–822.

Leavens, D. A., W. D. Hopkins, and K. A. Bard. 1996. Indexical and referential pointing in chimpanzees *(Pan troglodytes)*. *Journal of Comparative Psychology* 110: 346–353.

Leavens, D. A., A. B. Hostetter, M. J. Wesley, and W. D. Hopkins. 2004. Tactical use of unimodal and bimodal communication by chimpanzees, *Pan troglodytes*. *Animal Behaviour* 67: 467–476.

Leavens, D. A., J. L. Russell, and W. D. Hopkins. 2005. Intentionality as measured in the persistence and elaboration of communication by chimpanzees *(Pan troglodytes)*. *Child Development* 76: 291–306.

Liebal, K., J. Call, and M. Tomasello. 2004a. Use of gesture sequences in chimpanzees. *American Journal of Primatology* 64: 377–396.

Liebal, K., S. Pika, J. Call, and M. Tomasello. 2004b. To move or not to move: How apes adjust to the attentional state of others. *Interaction Studies* 5: 199–219.

Liebal. K., B. Waller, A. Burrows, and K. E. Slocombe. 2013. *Primate Communication: A Multimodal Approach*. Cambridge: Cambridge University Press.

Lieberman, P. 1984. *The Biology and Evolution of Language,* Cambridge, MA: Harvard University Press.

Liszkowski, U., M. Carpenter, A. Henning, T. Striano, and M. Tomasello. 2004. Twelve-month-olds point to share attention and interest. *Developmental Science* 7: 297–307.

Lyn, H., P. M. Greenfield, S. Sayage-Rumbaugh, K. Gillespie-Lynch, and W. D. Hopkins. 2011. Nonhuman primates do declare! A comparison of declarative symbol and gesture use in two children, two bonobos, and a chimpanzee. *Language and Communication* 31: 63–74.

McGrew, W. C., and C. E. G. Tutin. 1978. Evidence for a social custom in wild chimpanzees? *Man* 13: 234–251.

Miklósi, Á., and K. Soproni. 2006. A comparative analysis of animals' understanding of the human pointing gesture. *Animal Cognition* 9: 81–93.

Nishida, T. 1980. The leaf-clipping display: A newly discovered expressive gesture in wild chimpanzees. *Journal of Human Evolution* 9: 117–128.

Origgi, G., and D. Sperber. 2000. Evolution, communication and the proper function of language. In P. Carruthers and A Chamberlain, eds., *Evolution and the Human Mind: Language, Modularity and Social Cognition*, 140–169. Cambridge: Cambridge University Press.

Ouattara, K., A. Lemasson, and K. Zuberbühler. 2009. Campbell's monkeys concatenate vocalizations into context-specific call sequences. *Proceedings of the National Academy of Sciences* 106: 22026–22031.

Partan, S. R., and P. Marler. 2005. Issues in the classification of multimodal communication signals. *American Naturalist* 166: 231–245.

Pika, S., K. Liebal, and M. Tomasello. 2005. Gestural communication in subadult bonobos *(Pan paniscus):* Repertoire and use. *American Journal of Primatology* 65: 39–61.

Pika, S., and J. Mitani. 2006. Referential gestural communication in wild chimpanzees *(Pan troglodytes)*. *Current Biology* 16: R191–R192.

Pinker, S. 1994. *The Language Instinct: The New Science of Language and Mind*. New York: HarperCollins.

Povinelli D. J., and T. J. Eddy. 1996. Chimpanzees: Joint visual attention. *Psychological Science.* 7: 129–135.

Pujol, J., J. Deus, J. M. Losilla, and A. Capdevila. 1999. Cerebral lateralization of language in normal left-handed people studied by functional MRI. *Neurology* 52: 1038–1043.

Rendall, D., M. J. Owren, and M. J. Ryan. 2009. What do animal signals mean? *Animal Behaviour* 78: 233–240.

Roberts, A. I., S. J. Vick, and H. M. Buchanan-Smith. 2012. Usage and comprehension of manual gestures in wild chimpanzees. *Animal Behaviour* 84: 459–470.

Rumbaugh, D. M., and T. V. Gill. 1977. Lana's acquisition of language skills. In D. M. Rumbaugh, ed., *Language Learning by Chimpanzee: The Lana Project*, 165–192. New York: Academic Press.

Russell, C. L., K. A. Bard, and L. B. Adamson. 1997. Social referencing by young chimpanzees *(Pan troglodytes)*. *Journal of Comparative Psychology* 111: 185–193.

Savage-Rumbaugh, E. S., and R. Lewin. 1996. *Kanzi: The Ape at the Brink of the Human Mind*. New York: Wiley.

Savage-Rumbaugh, E. S., J. Murphy, R. A. Sevcik, K. E. Brakke, S. L. Williams, D. M. Rumbaugh, and E. Bates. 1993. Language comprehension in ape and child. *Monographs of the Society for Research in Child Development* 58(3–4).

Schel, A., M., Z. Machanda, S. W. Townsend, K. Zuberbühler, and K. E. Slocombe. 2013a. Chimpanzee food calls are directed at specific individuals, *Animal Behaviour* 86: 955–965.

Schel, A., M., S. W. Townsend, Z. Machanda, K. Zuberbühler, and K. E. Slocombe. 2013b. Chimpanzee alarm call production meets key criteria for intentionality. *PLoS ONE* 8: e76674

Scott-Phillips, T. C. 2014. *Speaking Our Minds: Human Communication and the Evolutionary Origins of Language*. London: Palgrave MacMillan.

Scott-Phillips, T. C. 2015a. Meaning in animal and human communication. *Animal Cognition* 18: 801–805.
Scott-Phillips, T. C. 2015b. Non-human primate communication, pragmatics, and the origins of language. *Current Anthropology* 56: 56–80.
Scott-Phillips, T. C., S. Diggle, J. Gurney, A. Ivens, and R. Popat. 2014. Combinatorial communication in bacteria: Implications for the origins of linguistic generativity. *PLoS ONE* 9: e95929.
Scott-Phillips, T. C., M. Tamariz, E. A. Cartmill, and J. R. Hurford. 2012. *The Evolution of Language: Proceedings of the 9th International Conference on the Evolution of Language (EVOLANG9)*. Singapore: World Scientific.
Seyfarth, R. M., D. L. Cheney, and P. Marler. 1980. Vervet monkey alarm calls: Semantic communication in a free-ranging primate. *Animal Behaviour* 28: 1070–1094.
Sherwood, C. C., R. L. Holloway, J. M. Erwin, et al. 2004. Cortical orofacial motor representation in Old World monkeys, great apes, and humans, I: Quantitative analysis of cytoarchitecture. *Brain, Behavior and Evolution* 63: 61–81.
Slocombe, K. E., T. Kaller, L. Turman, S. W. Townsend, S. Papworth, P. Squibbs, and K. Zuberbühler. 2010. Production of food-associated calls in wild male chimpanzees is dependent on the composition of the audience. *Behavioral Ecology and Sociobiology* 64: 1959–1966.
Slocombe, K. E., B. M. Waller, and K. Liebal. 2011. The language void: The need for multimodality in primate communication research. *Animal Behaviour* 81: 919–924.
Slocombe, K. E., and Z. Zuberbühler. 2005a. Agonistic screams in wild chimpanzees (*Pan troglodytes schweinfurthii*) vary as a function of social role. *Journal of Comparative Psychology* 119: 67–77.
Slocombe, K. E., and K. Zuberbühler. 2005b. Functionally referential communication in a chimpanzee. *Current Biology* 15: 1779–1784.
Slocombe, K. E., and K. Zuberbühler. 2006. Food-associated calls in chimpanzees: Responses to food types or food preferences? *Animal Behaviour* 72: 989–999.
Slocombe, K. E., and K. Zuberbühler. 2007. Chimpanzees modify recruitment screams as a function of audience composition. *Proceedings of the National Academy of Sciences* 104: 17228–17233.
Sperber, D. 2000. Metarepresentations in an evolutionary perspective. In D. Sperber, ed., *Metarepresentations: An Interdisciplinary Perspective*, 117–137. Oxford: Oxford University Press.
Sperber, D., and G. Origgi. 2010. A pragmatic perspective on the evolution of language. In R. K. Larson, V. Déprez, and H. Yamakido, eds., *The Evolution of Human Language: Biolinguistic Perspectives*, 124–131. Cambridge: Cambridge University Press.
Sperber, D., and D. Wilson. 1995. *Relevance: Communication and Cognition*, 2nd ed. Oxford: Blackwell.
Taglialatela, J. P., J. L. Russell, S. M. Pope, T. Morton, S. Bogart, L. A. Reamer, L. A., S. J. Schapiro, and W. D. Hopkins. 2015. Multimodal communication in chimpanzees. *American Journal of Primatology* 77: 1143–1148.

Taglialatela, J. P., J. L. Russell, J. A. Schaeffer, and W. D. Hopkins. 2008. Communicative signaling activates "Broca's" homolog in chimpanzees. *Current Biology* 18: 343–348.

Taglialatela, J. P., J. L. Russell, J. A. Schaeffer, and W. D. Hopkins. 2011. Chimpanzee vocal signaling points to a multimodal origin of human language. *PLoS ONE* 6: e18852.

Terrace, H. S. 1979. *Nim.* New York: Knopf.

Terrace, H. S., L. A. Petitto, R. J. Sanders, and T. G. Bever. 1979. Can an ape create a sentence? *Science* 206: 891–902.

Tomasello, M. 1994. Can an ape understand a sentence? A review of language comprehension in ape and child by E. S. Savage-Rumbaugh et al. *Language & Communication* 14: 377–390.

Tomasello, M. 2008. *Origins of Human Communication.* Cambridge, MA: MIT Press.

Tomasello, M., J. Call, and B. Hare. 1998. Five primate species follow the visual gaze of conspecifics. *Animal Behaviour* 55: 1063–1069.

Tomasello, M., J. Call, J. Warren, T. Frost, M. Carpenter, and K. Nagell. 1997. The ontogeny of chimpanzee gestural signals. In S. Wilcox, B. King, and L. Steels, eds., *Evolution of Communication,* 224–259. Amsterdam: John Benjamins.

Tomasello, M., M. Carpenter, and R. P. Hobson. 2005. The emergence of social cognition in three young chimpanzees. *Monographs of the Society for Research in Child Development* 70(1).

Tomasello, M., B. Hare, and B. Agnetta. 1999. Chimpanzees, *Pan troglodytes,* follow gaze direction geometrically. *Animal Behaviour.* 58: 769–777.

Tomonaga, M., M. Tanaka, T. Matsuzawa, M. Myowa-Yamakoshi, D. Kosugi, Y. Mizuno, and K. A. Bard. 2004. Development of social cognition in infant chimpanzees *(Pan troglodytes):* Face recognition, smiling, gaze, and the lack of triadic interactions. *Japanese Psychological Research* 46: 227–235.

Townsend, S. W., S. E. Koski, R. W. Byrne, K. E. Slocombe, B. Bickel, M. Boeckle, et al. 2016. Exorcising Grice's ghost: An empirical approach to studying intentional communication in animals. *Biological Reviews,* doi: 10.1111/brv.12289

Veà, J. J., and J. Sabater-Pi. 1998. Spontaneous pointing behaviour in the wild pygmy chimpanzee *(Pan paniscus). Folia Primatologica* 69: 289–290.

Vick, S. J., B. M. Waller, L. A. Parr, M. C. Smith Pasqualini, and K. A. Bard. 2007. A cross-species comparison of facial morphology and movement in humans and chimpanzees using the Facial Action Coding System (FACS). *Journal of Nonverbal Behavior* 31: 1–20.

Watson, S. K., S. W. Townsend, A. M. Schel, C. Wilke, E. K. Wallace, L. Cheng, V. West, and K. E. Slocombe. 2015a. Reply to Fischer et al. *Current Biology* 25: R1030–R1031.

Watson, S. K., S. W. Townsend, A. M. Schel, C. Wilke, E. K. Wallace, L. Cheng, V. West, and K. E. Slocombe. 2015b. Vocal learning in the functionally referential food grunts of chimpanzees. *Current Biology* 25: 495–499.

Wheeler, B. C., and J. Fischer. 2012. Functionally referential signals: A promising paradigm whose time has passed. *Evolutionary Anthropology* 21: 195–205.

Whiten, A., J. Goodall, W. C. McGrew, T. Nishida, V. Reynolds, Y. Sugiyama, C. E. G. Tutin, R. W. Wrangham, and C. Boesch. 1999. Cultures in chimpanzees. *Nature* 399: 682–685.

Wilke, C., E. Kavanagh, E. Donnellan, B. M. Waller, Z. P. Machanda, and K. E. Slocombe. 2017. Production of and responses to unimodal and multimodal signals in wild chimpanzees, *Pan troglodytes schweinfurthii*. *Animal Behaviour* 123: 305–316.

Wrangham, R. W., K. Koops, Z. P. Machanda, S. Worthington, A. B. Bernard, N. F. Brazeau, R. Donovan, J. Rosen, C. Wilke, E. Otali, and M. N. Muller. 2016. Distribution of a chimpanzee social custom is explained by matrilineal relationship rather than conformity. *Current Biology* 26: 3033–3037.

Zajonc, R. B. 1965. Social facilitation. *Science* 149: 269–274.

Zuberbühler, K. 2002. A syntactic rule in forest monkey communication. *Animal Behaviour* 63: 293–299.

Contributors

Joyce Benenson
Department of Psychology, Emmanuel College

Christopher Boehm
Department of Biological Sciences, University of Southern California

Ann E. Caldwell
Anschutz Health and Wellness Center, University of Colorado, Denver

Rachel N. Carmody
Department of Human Evolutionary Biology, Harvard University

Susana Carvalho
Institute of Cognitive and Evolutionary Anthropology, University of Oxford

Bernard Chapais
Department of Anthropology, University of Montreal

Peter T. Ellison
Department of Human Evolutionary Biology, Harvard University

Melissa Emery Thompson
Department of Anthropology, University of New Mexico

Ian C. Gilby
School of Human Evolution and Social Change, and Institute of Human Origins, Arizona State University

Luke Glowacki
Institute for Advanced Study in Toulouse
Department of Anthropology, Pennsylvania State University

Cristina M. Gomes
School of Environment, Science and Arts, Florida International University

Michael D. Gurven
Department of Anthropology, University of California, Santa Barbara

Marian I. Hamilton
Department of Anthropology, University of New Mexico

Brian Hare
Department of Evolutionary Anthropology, and Center for Cognitive Neuroscience, Duke University

Joseph Henrich
Department of Human Evolutionary Biology, Harvard University

Paul L. Hooper
The Santa Fe Institute

Adrian V. Jaeggi
Department of Anthropology, Emory University

Hillard S. Kaplan
Department of Anthropology, University of New Mexico

Jane B. Lancaster
Department of Anthropology, University of New Mexico

Daniel E. Lieberman
Department of Human Evolutionary Biology, Harvard University

Martin N. Muller
Department of Anthropology, University of New Mexico

Sherry V. Nelson
Department of Anthropology, University of New Mexico

David R. Pilbeam
Department of Human Evolutionary Biology, Harvard University

Herman Pontzer
Hunter College, City University of New York
New York Consortium for Evolutionary Primatology

Campbell Rolian
Department of Comparative Biology and Experimental Medicine, University of Calgary

Alexandra G. Rosati
Department of Psychology, University of Michigan

Thom Scott-Phillips
Department of Anthropology, Durham University
Department of Cognitive Science, Central European University, Budapest

Katie E. Slocombe
Department of Psychology, University of York

Claudio Tennie
School of Psychology, University of Birmingham

Michael L. Wilson
Department of Anthropology and Department of Ecology, Evolution, and Behavior, University of Minnesota

Brian M. Wood
Department of Anthropology, University of California, Los Angeles

Richard W. Wrangham
Department of Human Evolutionary Biology, Harvard University

Index

Aché: aggression, 389–391; arboreality, 268; child care, 556; cooperation, 522–523; divorce, 385; fertility, 220, 559; foraging returns, 404–405, 558; life expectancy, 185–186; male-male competition, 389–391; marriage, 385; mortality, 197; sexual dimorphism, 396–397; spousal abuse, 580, 583; strength, 397, 404–405
Acheulean tools, 492, 628
aerobic capacity, 264, 275, 404–405
Afropithecus, 50, 57–59
age at first birth, 49, 183, 232–234, 405
aggression: chimpanzee/bonobo differences, 14, 148; conditioning, 579, 585; domestication and selection against, 149–152, 160–161; female-female, 391–392; male-male, 389–391; moralistic, 752, 761–762; sex differences in, 510, 528; sexual dimorphism and, 404–410; within-group, 389–392, 754–761. *See also* coalitionary killing; intergroup aggression; male-female aggression
aging, 196–199, 201–202, 205, 209

Agta: child care, 556; divorce, 385; hunting, 556; mortality, 270
Aka: aggression, 391, 409; child care, 556; divorce, 385; hunting, 349–350, 356, 556; mortality, 270; paternal care, 556; prey, 347, 356; scavenging, 315, 355; sharing, 353, 367; spousal abuse, 580, 583
alarm calls, 798, 808–810
Alexander, Richard, 749
alliances: between groups, 437–438, 440–441, 448–452, 498; bonobo, 10, 148, 159, 518–519, 530–533; in the LCA, 10. *See also* female alliances; male alliances
allometry, 38–45, 73, 79, 400
ambush hunting, 355, 361, 363–364
ambush raiding, 479, 495
amenorrhea, 229–232, 322, 575–576
androgens, 153, 155–156, 160, 198, 408
Ankarapithecus, 64–66
Anoiapithecus, 63
ant dipping, 655, 665–668, 672–673
anteater, 109, 259
antifeedants, 6, 287–288, 325
aquatic plants, 295, 297–299, 301–304

827

Aranda, 772
arboreality, 266–270
Ardipithecus: bipedalism, 92–94, 272–273, 275; carpals, 107–109; clavicle, 102; cranial base, 80–83; cranium, 77–80; dating, 36, 70, 87; dentition, 73–76; face, 83–86; feeding niche, 299–302; femur, 94–96; foot, 98–100; forelimb, 101–102; hand, 624–625; humerus, 104–105; pelvis, 92–94; radius, 107; sexual dimorphism, 12, 398, 401; ulna, 105–107; vertebrae, 88, 92
attention (biases), 656–657
attention (joint), 712–713, 812–813
Australopithecus: bipedalism, 273–274, 276–277; cranial base, 80–83; cranium, 77–80; dentition, 73–76; diet, 288; feeding niche, 299–302; hand, 625–626; isotope data, 295–298; ranging, 486; scapula, 102–104; sexual dimorphism, 398–403, 442; stone tool use, 626, 632; territoriality, 486–487; vertebrae, 89
autism, 155, 812

baboon: bipedalism, 275–276; birth rate, 49; body fat, 396; as chimpanzee prey, 346–347; as referential model, 6–7, 16; sexual coercion, 588; social systems, 394, 437, 442, 483–485, 493–494; tool use, 611
Badjau, 320
balance of power. *See* imbalance of power
baldness, 406–407
bands, 384, 437, 468, 475–479, 486, 493–494, 512–514
Battle of the Somme, 464–465
beards, 406–407
Belyaev, Dmitry, 151
bipedalism: anatomical adaptations, 96, 100, 260–263; distribution, 259; early hominin, 92–94, 272–274; energetics of, 263–266; hypotheses for, 9, 271, 275–277; weapon use and, 401–402
birth. *See* parturition
birth rate: baboon, 49; body size and, 5–6; human, 5–6, 217–219; primates, 5–6, 217–218
blushing, 746, 749–750

body size: birth rate and, 5–6; driver of chimpanzee/gorilla differences, 13–14, 37–45; fertility and, 391; Miocene ape, 57; sexual dimorphism in, 395–397, 399, 484
Bofi, 315, 355
bonobo: body fat, 396; cranial base, 80–83; as derived, 14–15, 82–83; differences with chimpanzees, 146, 515–516, 518–519, 530–532; female-female alliances, 10, 159, 518–519; gang attacks, 761; infanticide, 241; juvenilized traits in, 157; maternal support, 159, 515, 553; mortality, 191; muscle morphology, 114; paedomorphism, 82, 86, 114, 147–152; prosociality, 152–155; reconciliation, 532; scramble competition, 159–161; sex differences, 530–533; sexual coercion, 586; sexual dimorphism, 397; sexual swelling, 237; similarities with chimpanzees, 144–145, 147–148; tool use, 608–609; vertebrae, 88–90
"bonobo-like" hypothesis, 146–147
"bottom-up" inference, 22–23, 34–37
brain size, 80, 186, 369–370, 443, 561, 680–681, 704
breastfeeding, 229–230
Broca's area, 803–804
bushbaby, 348, 358, 605

Calment, Jeanne, 181–182
canine dimorphism, 73–75, 272, 397–402
canine honing complex, 73–75, 401
capital punishment, 520, 763–764, 768, 771–775, 778
capitate, 98, 108–109
captivity: coalitions in, 753, 762, 779; constraints of, 178; invention in, 671–673, 674, 677, 792, 795; mortality in, 191–192, 196, 560; pointing in, 797, 801, 813; reproduction in, 220, 231; tool use in, 603, 609, 612, 619, 629, 631
capuchin: teaching, 660; tool use, 611–612, 620–622, 629–631, 663, 672
carpals, 107–109
cassava, 357
causality, 619–620, 622
CG (chorionic gonadotropin), 226, 228

INDEX

character selection, 11–12, 31–32
chewing, 76, 85, 288, 312, 315–317, 325–326, 399, 401
"chimpanzee-like" hypothesis, 146–147
Chororapithecus, 35, 38, 60, 76
cladistics, 31
Classification and Human Evolution (Washburn), 26
clavicle, 102
climbing, 94, 96, 101–106, 266–270, 273
club fighting, 390–391
coalitionary attacks. *See* coalitionary killing
coalitionary killing: in chimpanzees, 471–472; effects of grouping patterns, 481–482; evolution of, 480–488; in foragers, 479–480; in gorillas, 466; sexual dimorphism, 410; within groups, 754–761
coalitions: in bonobos, 148, 515, 530–531, 553; in chimpanzees, 148, 526–528; female, 528–529, 553, 762; male, 483–485, 526–528, 752; meat sharing, 351, 366, 527. *See also* alliances
coercive mate guarding. *See* indirect coercion
cognition: ape, 727; bonobo vs. chimpanzee, 145, 156–158; executive function, 719–725; foraging, 160; interrelationships between abilities, 726–729; language and, 805–813; mental time travel, 713–718; social, 154–155, 161–162, 706–713; temperament and, 154–155; tool use and, 618–622; unique human, 730–732
collective intentionality, 752–754, 769, 780
communication. *See* gestural communication; language; vocal communication
composite tools, 490, 603–604, 606, 620, 628
concealed ovulation, 238–240
conception, 221–224, 228, 236, 238, 386, 392
conceptual modeling, 15–16
conformity, 652–654, 657–658, 674–675
Congo River, 144–145, 151, 159
conscience, 746–751, 767–769
conservatism, 653–654
consortship, 386–389, 554, 578
convergence. *See* homoplasy
cooking: cognition for, 730–732; dietary diversity and, 321–322; effects on lipids, 318; effects on protein, 318–319; effects on starch, 317–318; energy availability and, 318–319; food preferences and, 322; weaning and, 203
cooperative breeding, 180, 243, 339, 560
cooperative foraging, 208, 492, 521–523
cortisol, 153, 226, 575–576
corvids, 714, 717, 732
C-peptide, 223, 230–232
CpG mutations, 52–53
cranial base, 40, 80–83, 86
cranial trauma, 574, 581
cranium, 14, 39, 77–80, 84
CRH (corticotropin-releasing hormone), 225–226
C_3/C_4 plants, 288–289
cuckoldry. *See* extrapair paternity
culture: chimpanzee, 661–669; group size and, 677; LCA, 680–681; phylogenetic signal, 664–665; reinvention, 671–673
cumulative culture, 602, 669–682

dating: the LCA, 45–54; regular meat consumption, 314
day range, 263–266, 276–277, 360, 368
"deep structure," 5, 427
delayed gratification, 723–725, 728, 730
Denisovans, 326
dental maturation, 50
dental microwear, 67–68, 287–288, 303
dentition, 73–76, 399, 482, 625
Descent of Man (Darwin), 3, 259, 371, 611
desertion, 240, 550, 558
developmental window, 676–677
DHEAS, 198
diarrhea, 194
diet: *Australopithecus*, 288; canines and, 399; chimpanzee, 242, 286–287, 509, 550; diversity, 321–322; enamel thickness and, 67–68, 76, 276, 287–288; gorilla, 13–14; human, 208, 242–243, 268, 313–328; intergroup aggression and, 483; niche breadth, 298–303; proportion meat, 314, 356–357; ranging and, 487; sexual dimorphism and, 552; stable isotopes and, 289–303; tool use and, 610

digestion, 312, 315–319, 325–326
digging stick, 349, 371, 489
digit ratios, 110–111, 616
dinosaurs, 259
direct coercion: chimpanzee, 576–578; defined, 572. *See also* rape
disease, 192–194, 196–197, 206, 236
dispersal: bisexual, 450–451; sex-biased, 431–432, 484; stress and, 234
divorce, 385–386, 512, 559
dolphins, 554
domestication syndrome, 149–152, 156, 158
Dryopithecus, 35, 63–64, 87, 95–96, 104
dual inheritance. *See* gene-culture coevolution

ecological validity, 603, 650
Efe: arboreality, 268; day range, 360; fertility, 220; hunting, 350, 357, 361, 363, 365; prey, 347; rape, 584–585; scavenging, 355; sharing, 353, 367
Ekembo, 50, 57–59, 68–69, 88–96, 98, 103–104, 108, 110
"Embodied Capital" model, 202–204, 207–208
empathy, 748–749, 751, 753
enamel thickness: diet and, 67–68, 76, 276, 287–288; lack of phylogenetic signal, 67
enculturation, 650
endometrium, 222
endurance, 264, 266, 274–278
epididymis, 410–411
Equatorius, 57, 59, 63, 96, 104, 108
Eskimo, 48, 589–591
estradiol. *See* estrogen
estrogen, 198, 221–222, 226, 230, 237
estrus, 238, 392, 554
execution. *See* capital punishment
extractive foraging, 604–606, 609–611, 620–622, 625, 629–634, 730, 732
extrapair paternity, 241–242, 558, 583–584, 588, 592
eye tracking, 711

face: ape vs. early hominin, 83–86; sexual dimorphism, 402–403, 588
Facial Action Coding System (FACS), 804–805
facial expression, 804–805
fallback foods, 76, 159, 288
falling, 270
false beliefs, 143–144, 710–711, 728
fat (body): menarche and, 233; sexual dimorphism in, 395–396
fat (dietary), 205, 313–316, 327–328, 356
fecundity: defined, intimate partner violence and, 219–220, 581–582; regulation of, 222–224
female alliances, 10, 148, 159, 518–519, 530–533
female choice, 383, 386, 408–409, 551, 553
femur, 40–41, 58, 68, 71, 87, 94–96, 272, 400
fermentation, 326–328
fertility: body size and, 391; defined, 219–220; sex differences, 559–560
fetal loss, 224, 227–228, 236
fire: earliest evidence for, 322–323, 370; use in honey gathering, 319–320. *See also* cooking
fission-fusion grouping, 384, 466–467, 481–482, 487–488, 515–516, 676
flexor pollicis longus, 616–617, 623–625
food caching, 717
food processing. *See* cooking; fermentation; pounding
food sharing: chimpanzees and bonobos, 153–154, 157, 351–353, 365–366; cognition and, 705, 731–733; foragers, 351–353, 363, 366–367, 556–558; hunting and, 242, 351–353, 356; language and, 359; mortality reduction and, 205–209; pooled energy budgets, 243; sex differences in, 523, 531; strategic, 351–353
foot, 97–100
foraging returns: by age, 361–363, 405; by sex, 556
foramen magnum, 41, 80–81, 260, 272–273
forced copulation, 577–578. *See also* rape
fox, silver, 151–152, 154
fractionation, 289–291
free riding, 764–765, 773–775
frontal cortex, 719
FSH, 222, 226, 234

INDEX

galago. *See* bushbaby
Gambians, 229–230
gang attacks. *See* coalitionary killing
gaze following, 143–144, 708–709, 711–712, 812
Gee, Henry, 7
gene-culture coevolution, 646–647
generalist, 289, 299–300, 302, 304
generation intervals, 48–51
geometric morphometrics, 36, 41, 77, 98, 102
Gesher Benot Ya'aqov, 323
gestation, 183, 226–228
gestural communication: flexibility, 792–796; intentionality, 806–808; neurological basis, 803–805; semantics, 796–797
GG rubbing, 531
gibbon: as conservative, 8, 34; extrapair paternity, 583; monogamy, 445–446, 476; referential model, 23; sexual dimorphism, 395; sexual swelling, 237; tool use, 610–611; vertebrae, 89
glenoid fossa, 260, 273–274
gluconeogenesis, 316
Goodall, Jane, 3, 754
gorilla: body mass, 38; coalitionary killing, 466; cranial base, 80–83; cranium, 77–80; culture, 677; dentition, 72–76; as derived, 13–14; diet, 13, 287; distribution, 159; ecology, 13; foot, 98–100; gaze following, 709; hand, 617–618, 624–625; infanticide, 515–516; knuckle-walking, 274; as large chimpanzees, 8, 27, 37–45; male-male relationships, 533–535; ovarian cycle, 222; reconciliation, 534; scapula, 102–103; sexual coercion, 586–588; sexual dimorphism, 400–403; sexual swelling, 237; social learning, 649; social system, 13–14, 476, 516–518; speciation time, 53; territoriality, 516–517; tool use, 609–610, 619; vertebrae, 89, 92
gossip, 525, 746, 749, 752, 775–777
Gould, Stephen J., 161
grades, evolutionary, 24–27
grandfathers, 434, 556
"Grandmother" hypothesis, 201–203

grandmothers, 148, 178, 201, 243, 432, 451, 556, 679
gregariousness, 515, 518, 521, 523, 527, 529, 532, 535, 554–555
Griphopithecus, 62–63, 104
grooming: alliance formation and, 447–448, 534, 553, 591; high-arm, 792–793; meat sharing and, 351–352, 366; paternity recognition and, 433; sex difference in, 467, 516; tool use and, 604, 609, 632–633
/Gui-//Gana: day range, 360; home range, 360; hunting, 349–350, 356–357, 362, 365; prey, 348, 356; scavenging, 355; sharing, 367
guilt, 747, 750

Hadza: aggression, 390–391; arboreality, 268–270; cooking, 323; day range, 360; fertility, 220; honey gathering, 319–320; hunting, 349–350, 356–357, 361–365; life expectancy, 185–186; marriage, 385; prey, 347; rape, 584–585; scavenging, 315, 355; sharing, 353, 367, 558; spousal abuse, 580, 583; strength, 405–406
hairlessness, 274
hand axe, 320, 404, 680
hand grips: chimpanzee, 614–616; human, 616–617; LCA, 617–618; power, 613; precision, 613
Hare, Robert, 748
Hawkes, Kristen, 201
heart disease, 194, 196
Hennigian revolution, 26
herding, 443, 577–578, 587
heterochrony, 82–83, 149, 151, 156–158, 161
"heterogeneity" hypothesis, 197–198
hidden authorship, 811
Hispanopithecus, 63–64, 68–69, 76, 91, 96, 105
Hiwi: aggression, 390; divorce, 385; life expectancy, 186; mortality, 197
home range, 241, 263–265, 287, 360, 445, 467, 485–486, 731
homicide, 195, 206, 390, 524, 771
Homo erectus: hunting, 314; life span, 186; niche space, 297–302; sexual dimorphism, 400, 402, 442; shoulder, 314; territoriality, 487

Homo habilis: hands, 626; life span, 186; tool use, 626–628
homology: African apes and, 9–12, 37–45, 69; sexual coercion and, 585–588
homoplasy: African apes and, 101, 112, 271; Miocene apes and, 66–69; problem for phylogenetic reconstruction, 29–33, 113; tool use and, 629
homosexual behavior, 530–531
honey, 268, 270, 317, 319–321, 367, 605, 662, 673, 730
human nature, 4–5
humerus, 104–105
hunting: ambush, 363–364; big game, 355–356; cognition, 729–730; cooperation in, 349–351, 364–365, 444; day range effects, 360; early hominin, 367–371; female, 556; language and, 359; male bias in, 348–349, 404; prey, 341–348; returns by age, 353–354, 361–363, 405–406; seasonality, 361; sexual division of labor and, 550, 556; success rates, 315; technology, 358–359, 488–491; territoriality and, 483; time spent, 357–358
Hutterites, 220
Hylobatidae. *See* gibbon

ILS (incomplete lineage sorting), 29, 48
imbalance of power hypothesis, 475, 480, 752, 758, 763
implantation, 222, 225
impulsivity, 723–725
incest avoidance, 432–433, 449, 451, 577
indirect coercion: bonobo, 586; chimpanzee, 577–579; defined, 572; gorilla, 587; humans, 579–584
infanticide: chimpanzee, 148, 195, 471, 553; gorilla, 14, 516–517; pair bonds and, 446, 558; promiscuity and, 240–241, 383, 393, 553; sexual coercion and, 579
infant mortality, 183, 185, 188–189, 192, 228, 517
infidelity: divorce and, 385–386; male, 558, 591–592; violence and, 390, 582–583
inhibition, 156–157, 720–723
insulin, 223, 230–232

intentionality (shared). *See* collective intentionality
intentionality (signal production), 806–812
interbirth interval: body size and, 5–6; chimpanzee, 220, 224, 473, 475; determinants of, 229–231, 473, 475; forager, 206, 220; gorilla, 49
intergroup aggression: benefits of, 473–475; in chimpanzees, 467–475; cooperation within groups and, 518–519, 581; effects of grouping patterns, 481–482; in foragers, 475–480; in hominins, 480–488
intimate partner violence, 580–583, 591–592
Inuit, 661
invention, 443, 465–466, 669, 671–673, 674, 677, 792, 795
ischium, 94, 273, 275–276
Ituri forest, 268, 289

jealousy, 390, 521, 583
Ju/'hoansi: aggression, 389–391; amenorrhea, 231; birth interval, 220, 224; birth weight, 226; fertility, 220; hunting, 349–351, 356, 358, 362, 365; life expectancy, 186; marriage, 385; mortality, 197; prey, 347–348, 354, 356; scavenging, 315, 355; sharing, 353, 367; spousal abuse, 580, 583; wife exchange, 591

Kamoyapithecus, 56
Kenyapithecus, 35, 59, 62–63
Khoratpithecus, 65–66, 69
kin recognition, 428–436, 446–447, 455, 494
kinship: human configuration, 428–429; matrilineal, 432–433; patrilineal, 433–434; sharing and, 366–367; social learning and, 655
Kipling, Rudyard, 5
knuckle-walking: body mass and, 43; humerus and, 104–105; in the LCA, 86–88, 108–109, 111, 271–272, 274; radius and, 107; scapula and, 103–104

labor (parturition), 227
language: cognitive foundations, 805–813; hunting and, 359; intergroup relations

and, 494–496; neurological basis, 803–805; semantics, 796–802; structural features of, 791–805; syntax, 802–803
larynx, 406–409, 795
LCA (Last Common Ancestor, chimpanzee/human): dating of, 45–54; history of ideas about, 24–28; knuckle-walking in, 86–88, 108–109, 111, 271–272, 274
leaf-swallowing, 671–672
Leakey, Louis, 3–4, 16
lemur: cognition, 160, 721, 724; female dominance, 4, 573
LH, 222, 234
life expectancy: chimpanzee, 188–191; forager, 186; modern increase, 205; sex differences in, 560
life history theory, 219
life span: extrinsic mortality and, 197–198; maximal, 181–186; postreproductive, 199–205; prehistoric, 187–188; reproductive, 232–236
limbic system, 804
lithic technology. See stone tools
Little Mama, 181–182
locomotion. See bipedalism; climbing; knuckle-walking
Lufengpithecus, 50, 65–66, 69
lumbar column. See vertebrae

macaque: body fat, 396; cognition, 711, 725; kin recognition, 432–433; tool use, 611
majoritarian bias, 657–658
male alliances: in bonobos, 518–519, 530–533; in chimpanzees, 553, 560; in gorillas, 533–534; intergroup aggression and, 581; patriarchy and, 511; wife exchange and, 589
male-female aggression: in bonobos, 586; in chimpanzees, 573–579; in gorillas, 586–588; in humans, 579–585, 591–592
Mardu, 580, 583
Marks, Jonathan, 4–5
marmosets, 160
marriage, 384–386, 394, 452–454. See also monogamy; pair-bonding; polygyny
mate guarding, 433, 554, 558. See also indirect coercion

mate preferences: effects of pair-bonding, 392–393; female, 393, 577–578, 585; male, 392–393; strength of, 585–586, 588
mating effort, 392, 405, 549–552, 558–559
matriarchy, 520
matriliny, 431–433
Mauna Loa, 291
Mbuti: honey consumption, 321; hunting, 340, 349–350, 356–357, 361–365; prey, 347, 356; ranging, 360; rape, 584; sharing, 353
Mead, Margaret, 4
meat eating: archaeological evidence for, 313–314; dentition for, 315–316; genetic adaptations for, 205; proportion of diet, 314, 356–357; sexual division of labor, 561. See also hunting
meat sharing, 351–353, 365–367, 371–372, 444, 527, 554
melanocytes, 158
memory: episodic, 713–717; spatial, 160, 715, 727–728; working, 719–720
menarche, 183, 232–234, 236, 316
menopause, 178, 198–205, 234–236, 243, 393, 559
menstrual cycle. See ovarian cycle
microbiome, 321, 327–328
migration, 452, 467, 560, 664
milk, 229–230, 242–243
Miocene, 56
Miocene apes: African, 56–62; ages, 55; Asian, 64–66; enamel thickness, 67–68; European, 62–64; hands, 623–624; life history, 57; paleoecology, 293–297, 302–303; phylogeny, 58–69
molecular clock, 46–54, 113
monogamy: evolution of, 430, 438–446; forager, 384–385, 394, 398
morality, 746–754, 761, 763, 778–779
Morotopithecus, 57–59, 69, 90–91
mortality: causes, 192–196; extrinsic, 205–207
mortality rates: in bonobos, 191; in chimpanzees, 188–191; doubling time, 197; in humans, 184–188; intergroup aggression and, 478–479; in paleodemography, 187; plasticity in, 191–192

"mosaic hypothesis," 146–147
multi-level societies, 430, 437–438, 481, 492 494
multimodal communication, 796, 816
muscle: conservatism in *Pan*, 114, 260; locomotor, 264, 266, 275; sexual dimorphism in, 397, 404–410
mutation rate: estimating, 51–53; generation intervals and, 50
myosin, 312

Nacholapithecus, 35, 57–59, 63, 69, 90–91, 103, 108, 110
Nakalipithecus, 35, 60–61
Neanderthals, 52, 323, 326, 402, 484, 490–491
neural crest, 158
niche breadth, 298–303
Nsungwepithecus, 56
nubility, 392
nut cracking, 605–607, 609, 616, 620, 632, 655, 659, 668–669, 672

oblique transmission, 654–657
Oldowan tools, 614–615, 626–628
olecranon process, 105–106
oligomenorrhea, 223
opportunistic mating, 386
orangutan: cognition, 708–709, 711, 715–718, 722; generation interval, 49; genital swelling, 237; hand, 617; model for the LCA, 27; ovarian cycle, 222; relationship to *Sivapithecus*, 64–69; sexual swelling, 237; tool use, 610, 621–622, 628, 663; vertebrae, 89
Oreopithecus: anatomy, 35, 63–64, 69, 87, 91, 105–106, 108, 110; feeding niche, 293–294, 297, 302–303
orphans, 162–163, 553
Orrorin: bipedalism, 71, 272; dates, 35, 70; dentition, 73–76; femur, 94–96; forelimb, 101–102; hand, 623; humerus, 104–105; paleoenvironment, 624
os peroneum, 99–100
ostensive communication, 805, 810
ostracism, 653, 760, 763, 768, 775–778
Otavipithecus, 76

Ouranopithecus, 35, 38, 50, 61, 63–64, 68, 76, 86
ovarian cycle: length, 222; regulation, 221–222; sexual behavior across, 238–240
overimitation, 674–675
oxytocin, 154–155

paedomorphism: bonobo, 14–15, 86, 114, 149–152, 156–158; human, 161
pair-bonding: concealed ovulation and, 240; consortships and, 387–388; defined, 383–384; evolution, 394, 438–446, 561; infanticide and, 446, 558; kin recognition with, 435–436; mate preferences and, 392–393; sexual division of labor and, 392, 550–551, 559, 561; sexual jealousy and, 583–584; stability, 385, 559
paleodemography, 187–188
parasites, 194, 314, 671
parental investment, 203, 548–553. See also paternal care
parsimony, 12, 29, 33, 37, 72, 75, 89, 111–112, 608, 617, 629
parturition, 225–227, 233
paternal care, 385, 394, 438–440, 550–552, 556
paternity recognition, 388–389, 433–434
"Patriarch" hypothesis, 202–203
patriarchy, 510–511, 520
patriliny, 431, 433–434; evolution of, 446–448
patrilocality, 450, 512, 517–518
patrol (border), 263, 467, 470, 475, 495, 527, 536, 752–755, 758
patrol (hunting), 360
pedigree method, 51–53
pelvis: bipedalism and, 92–94, 260–263, 273–274; parturition and, 227, 233
penis, 411
peramorphism, 158
periovulatory period, 237–240, 575
phylogenetic methods: problems with, 11–12, 28–34; reconstructing the LCA, 9–12, 28–34
phytoliths, 288, 303
Pierolapithecus, 63, 69, 87, 91, 94, 106, 108
Piers, Gerhart, 747
placenta, 225–226

INDEX

planning, 605, 620, 706, 717–718, 727–728
plantar arch, 260, 264, 273
Pliobates, 63, 69
pointing, 796–797, 801, 810, 812–814
poison, 355, 359, 390, 491, 589
policing, 574–575, 770, 773
polygyny, 385, 394–395, 398, 413, 437, 440, 442–446, 551, 559, 588
pooled energy budgets, 243–244
possessive mating, 386
pounding, 324–325
predation: mortality risk, 191–193, 195, 206; social, 767, 774, 778. *See also* hunting
prefrontal cortex, 747
pregnancy: loss (*see* fetal loss); maintenance of, 225–226; reduction in intimate partner violence during, 582; sexual swelling during, 237
prestige: hunting reputation and, 371; social learning bias, 654–657, 661, 670; warfare and, 480, 492
Proconsul. See *Ekembo*
progesterone, 221–223, 226, 230
prognathism, 82, 84–86
prolactin, 226, 229–231
promiscuity: anti-infanticide strategy, 240–241; chimpanzee and bonobo, 147, 237, 240–241, 383, 386, 389, 392, 476; kin recognition and, 435–436; paternity certainty and, 241–242, 553; sperm competition and, 410–412
prosimians, 225, 612
prosociality: chimpanzee vs. bonobo, 151–155; cultural evolution and, 677–678; social cognition and, 154–155
provisioning, 201, 203, 242–243, 275, 394, 438–440, 482, 549, 553, 556
psychopathy, 747–751
punishment, 577–578, 749–752, 763–764, 766–768, 771–773
python, 319, 798, 808–809

Qesem Cave, 323

"rabbit starvation," 315
radius, 107, 109, 259
rape, 584–585
ratchet effect, 670
rate test, 45
raw foodists, 322
reconciliation, 529, 532, 534
red colobus monkey, 340–341, 346–350, 355–358, 360–361, 363, 365, 369
referential communication, 792, 794, 796–802, 815
referential modeling, 15–16, 28, 45, 111
reinvention, 671–673
reproductive rate. *See* birth rate
reproductive skew, 483–485, 488, 497, 560
reputation, 392, 405, 525, 746, 749, 751, 768, 775–777
revenge, 498, 589, 771
reverse dominance hierarchy, 768–769
ritualization, 792, 794
rough grunt, 798–801
Rudapithecus, 63–64, 68–69, 104–105, 108
Rukwapithecus, 56
running: adaptations for, 264–266, 274–275; fecundity and, 223

Sahelanthropus: bipedalism, 272–273; cranial base, 80–83; cranium, 77–80; dates, 70–71; dentition, 73–76, 401; face, 83–86; paleoenvironment, 624
Samburupithecus, 38, 60–61, 76, 86
"savanna" chimpanzees, 287, 292, 605
"savanna hypothesis," 275
scaling (ontogenetic), 38–45
scaphoid, 107–108, 260
scapula, 43–44, 59, 102–104, 274
scavenging, 276, 313, 315, 317, 319, 327, 354–355, 368–369, 439, 561
scramble competition, 159–161, 410, 475
self-control, 720–723
self-domestication: bonobo, 15, 151–154, 158, 160, 162–163; human, 161–162, 766–768, 777–779
semantics, 796–802
semenogelin, 411–412
senescence, 196–199
sequestration, 577–578
serial killers, 771–772, 780

sex differences: aggression, 404, 510, 524; bonobo, 530–533; competition, 524–535; cooperation, 519–535; fat, 395–396; food sharing, 523; gregariousness, 515, 518, 521, 523, 527, 529, 532, 535, 554–555; social networks, 521; strength, 396–397, 405; tree climbing, 268; vocal pitch, 406–409.
See also sexual dimorphism
sex ratio, 385, 395, 398, 514–515
sex segregation, 519–520
sexual coercion: bonobo, 586; chimpanzee, 575–579; defined, 572; gorilla, 586–587; as homology, 585–588; human, 579–585, 589–592. *See also* direct coercion; indirect coercion; rape
sexual conflict, 572
sexual dimorphism: *Ardipithecus*, 12; *Australopithecus*, 399–400; body size, 395–397, 399; canines, 397–399; diet, 552; evolution of, 404–410; face size, 402–403; fat, 395–396; strength, 397; vocal pitch, 406–409
sexual division of labor, 178, 404–406, 408, 412–413, 477, 513, 520, 555–559, 561, 592–593
sexual swelling, 147, 222, 234, 237–238, 471, 553–554, 579
shamans, 771
shame, 747–750, 768
Shame and Guilt (Piers and Singer), 747
sharing. *See* food sharing; wife exchange
Singer, Milton, 747
Sivapithecus: anatomy, 50, 64–66, 68–69, 86, 94, 101, 104–105; feeding niche, 293–295; relationship to orangutan, 64–69
sleep, 178
Small, Meredith, 8
social carnivores, 483
social cognition, 706–713
social learning: chimpanzee, 648–652; conformity, 652–654; group size, 675–677; mechanisms for, 652–658; teaching, 658–661; transmission biases, 654–658; transmission fidelity, 649, 651, 674–675
social networks: cultural evolution and, 675–677; multilevel societies, 476

sociosexual behavior, 14, 153, 158
speciation times, 53–54
sperm competition, 241, 410–412
spine. *See* vertebrae
spousal abuse. *See* intimate partner violence
stable breeding bond. *See* pair-bonding
stable isotopes: carbon, 288–289, 292–297; fractionation, 289–291; oxygen, 288–289
status competition: bonobos, chimpanzees, 527–529, 560; humans, 524–526; social learning and, 656
sterility, 236
stone tools: *Australopithecus*, 626; bonobo, 609; capuchin, 611–612; chimpanzee, 605–608, 622; earliest, 323–324, 368–369; hand grips and, 614–618; *Homo*, 626–629; long-tailed macaques, 611
stress, 576
Stroop task, 722–724
supraorbital torus, 80, 83–84, 401
Sussman, Robert, 8
syntax, 802–803

tamarins, 160
tapeworm, 314
taurine, 314
teaching: chimpanzee, 658–661, 675, 678; defined, 658–659; human, 659; meerkat, 659; prosociality and, 677
telomeres, 198–199
termite fishing, 655, 665–668, 672
territoriality, 10, 430, 483, 485–487, 514–516
testosterone. *See* androgens
thenar muscles, 616–617
theory of mind, 161, 453, 708–713, 726–729, 809–812, 815
thermoregulation, 274
throwing, 314, 404
Toba, 230–232
tolerated theft, 366
tool use: bonobo, 608–609; capuchin, 611, 620–622, 629–631; chimpanzee, 604–608; cognition for, 618–622; defined, 603; environmental effects on, 630–632; gibbon, 610; gorilla, 609–610; hand grips, 613–618; hunting, 358–359; in the LCA,

628–633; old world monkeys, 611–612; orangutan, 610, 621–622; stone, 323–324, 368–369; "terrestriality effect," 631–632. *See also* stone tools
"top-down" inference, 22–23, 28–34
total fertility rate (TFR), 183, 220
trade, 357, 495–496, 498, 552
trichotomy, 25–27, 46
trochlear notch, 105–106
Tsimané: divorce, 386; food sharing, 556–558; production, 204; sexual division of labor, 555–558; spousal abuse, 592
tuber. *See* underground storage organs (USOs)
Twain, Mark, 114
2D4D ratio, 155

Ugandapithecus, 58
ulna, 63, 105–107, 259
underground storage organs (USOs), 295, 297, 325, 477, 489

vaginal fatty acids, 237
vas deferens, 411
vasopressin, 155
vegetarians, 316, 322, 326
vertebrae, 58, 60, 63, 88–93
vervet monkeys, 656, 798
violence. *See* aggression
vocal communication: flexibility, 794–796; intentionality, 808–812; neurological basis, 804; semantics, 796–797; testosterone and, 408
vocal pitch, 406–409

wading, 275–276
walking speed, 263–265
warfare: benefits, 492; evolution of, 465–466; forager, 477–480. *See also* intergroup aggression
weapons, 315, 359, 368, 371–372, 390, 402, 404, 443–444, 488–491, 769
WEIRD societies, 659
wetlands, 301–303
wife beating. *See* intimate partner violence
wife exchange, 589–591
Williams, George, 201
Wisconsin card sorting task, 725
Wonderwerk Cave, 322
World Health Organization, 580–583

Yerkes, Robert, 144